GEOMICROBIOLOGY

Third Edition, Revised and Expanded

Henry Lutz Ehrlich

Rensselaer Polytechnic Institute
Troy, New York

 MARCEL DEKKER, INC.   NEW YORK · BASEL · HONG KONG

Library of Congress Cataloging-in-Publication Data

Ehrlich, Henry Lutz
Geomicrobiology / Henry Lutz Ehrlich.—3rd ed., rev. and expanded.
   p. cm.
   Includes bibliographical references and index.
   ISBN 0-8247-9541-5 (alk. paper)
   1. Geomicrobiology.   I. Title.
QR103.E37   1995
550'.1'576—dc20   95-40328
                                                                           CIP

The publisher offers discounts on this book when ordered in bulk quantities. For more information, write to Special Sales/Professional Marketing
at the address below.

This book is printed on acid-free paper.

Marcel Dekker, Inc.
270 Madison Avenue, New York, New York 10016

Current printing (last digit):
10 9 8 7 6 5 4 3 2

PRINTED IN THE UNITED STATES OF AMERICA

GEOMICROBIOLOGY

To my former and present students,
from whom I have learned
as much as I hope they have learned from me.

Preface to the Third Edition

The need for a third edition of *Geomicrobiology* has arisen because of some important advances in the field since the second edition. Of particular significance are advances in the areas of subsurface microbiology as it relates to groundwater, carbonate deposition, rock weathering, methylmercury formation, oxidation and reduction of iron and manganese, chromate reduction, oxidation and reduction of molybdenum, reduction of vanadate (V) and uranium (VI), oxidation and reduction of sulfur compounds, reduction of selenate and selenite, methanogenesis, microbial attack of coal, and degradation of hydrocarbons. These advances have been integrated into the treatment of these subjects. The chapter dealing with the biochemistry and physiology of geomicrobial processes has been updated to convey the basis for our current understanding of how and why microbes are involved in these processes.

Because this book is meant to serve as a reference as well as a textbook, very little material from the second edition has been eliminated. By retaining this information, an overview of the growth of the field of geomicrobiology since its inception is retained. It enables newcomers to learn what has been accomplished in the field and to gain an introduction to the literature. The literature citations on the different subjects are not exhaustive, but include the most important ones, making it possible to locate other works by cross referencing. As in the previous editions, a glossary is included to aid in definition of unfamiliar scientific terms.

In preparing this edition, I have retained some of the line drawings prepared by Stephen Chiang for the first edition that were also included in the second edition. Some other illustrations from the second edition have been replaced, and a few entirely new illustrations have been included. I am indebted to a number of persons and publishers for making available original photographs or allowing reproduction of previously published material. They are acknowledged in the legends of the individual illustrations.

I wish to thank Marcel Dekker, Inc., for their continued belief in the importance of this book by encouraging the preparation of this third edition. I want to express special thanks to Bradley Benedict, Assistant Production Editor, and the editorial staff for their assistance in preparing this edition.

Responsibility for the presentation and interpretation of the subject matter in this edition rests entirely with me.

Henry Lutz Ehrlich

Preface to the Second Edition

As in the first edition of this book, geomicrobiology is presented as a field distinct from microbial ecology and microbial biogeochemistry. The stress remains on examination of specific geomicrobial processes, microorganisms responsible for them, and the pertinence of these processes to geology.

Most chapters from the earlier edition have been extensively revised and updated. As far as possible, new discoveries related to geomicrobiology reported by various investigators since the writing of the first edition have been integrated into the new edition. Two new chapters have been added, one on the geomicrobiology of nitrogen and the other on the geomicrobiology of chromium. The second chapter of the first edition has been divided into two to allow for a more concise development of the two topics: Earth as microbial habitat and the origin of microbial life on Earth.

In the new edition, Chapters 2–6 are intended to provide the background needed for understanding the succeeding chapters, which deal with specific aspects of geomicrobiology. An understanding of microbial physiology and biochemistry is very important for a full appreciation of how specific microbes can act as geomicrobial agents. For this reason, Chapter 6 was extensively revised from its antecedent, Chapter 5 in the first edition.

Like its predecessor, the present edition is meant to serve not only as a text, but also as a general introduction and guide to the geomicrobial literature for microbiologists, ecologists, geologists, environmental engineers, mining engineers, and others interested in the subject. The literature citations are not intended to be exhaustive, but cross-referencing, especially in cited review articles, should lead the reader to many other pertinent references not mentioned in this book.

Some of the revisions in this edition, especially those relating to bioenergetics, were significantly influenced by a number of stimulating informal discussions with my colleague and research collaborator John C. Salerno.

In preparing this edition, I have retained some of the line drawings by Stephen Chiang. I have, however, replaced many of the other illustrations, and added some new ones that I prepared on a Macintosh Plus computer with Cricket Draw and Cricket Graph applications. I wish to thank the Voorhees Computer Center of Rensselaer Polytechnic Institute for allowing me to use the Laser Printer Facility and George Clarkson for making the necessary arrangements. Once again, I am indebted to a number of persons and publishers for making available original photographs or allowing reproduction of previously published material. They are acknowledged in the legends of the individual illustrations.

I wish to thank Marcel Dekker, Inc., for deeming the subject matter of this book of sufficient continued importance to publish this second edition. Special thanks go to Judith DeCamp, Production Editor, and the editorial staff for their help in bringing this edition to fruition.

Responsibility for the presentation and interpretation of the subject matter in this edition rests entirely with me.

Henry Lutz Ehrlich

Preface to the First Edition

This book deals with geomicrobiology as distinct from microbial ecology and microbial biogeochemistry. Although these fields overlap to some degree, each emphasizes different topics (see Chapter 1). A reader of this book should not, therefore, expect to find extensive discussions of ecosystems, food chains, nutritional cycles, mass transfer, or man-made pollution problems as such, because these topics are not at the heart of geomicrobiology. Geomicrobiology is the study of the role that microbes play or have played in specific geological processes.

This book arose out of a strong need I felt in teaching a course in geomicrobiology. As of this writing, no single text is available that deals with the group of topics presented in this book. Previously, students in my geomicrobiology course needed to be referred to the many primary publications on the various topics. These publications are very numerous and are scattered among a plethora of journals and books that are often not readily available. Some are written in languages other than English. This book is an attempt to glean the basic geomicrobial principles from this literature and to illustrate these principles with many different examples.

Some readers of this book will have a stronger background in Earth and marine science than in microbial physiology, while others will have a stronger background in microbial physiology than in Earth and marine sciences. To enable all these readers to place the geomicrobial discussions in the later chapters in proper context, the introductory Chapters 2–5 were written. They are not meant to be definitive treatises on their subjects, and as a result any one of them will appear elementary to a person already knowledgeable in its field. However, I have found the material in these chapters to be essential in teaching my students.

As for the rest of the book, Chapter 6 summarizes the methods used in geomicrobiology, and Chapters 7–17 examine specific geomicrobial activities in relation to geologically important classes of substance or elements. A single basic theme pervades these last 11 chapters: biooxidation and bioreduction and/or bioprecipitation and biosolution. This may seem an unnecessary reiteration of a common set of principles, but closer examination will show that the manifestations of these principles in different geomicrobial phenomena differ so strikingly as to require separate examination. In discussing geomicrobial processes, I have tended to emphasize the physiological more than the geological aspects. This is in part because the former is my own area of greater expertise, but also, and more importantly, because I feel that the physiological and biochemical nature of geomicrobial processes has to be understood to fully appreciate why some microbes are capable of these activities.

In citing microorganisms in the text, the names employed by the investigators whose work is described are used. In the case of bacteria, these names may have changed subsequently. The currently used names of the bacteria may be found by referring to *Bergey's Manual of Determinative Bacteriology* (8th edition, edited by R. E. Buchanan and N. E. Gibbons, 1974, Williams and Wilkins, Baltimore) and to the *Index Bergeyana* (R. E. Buchanan, J. G. Holt, and E. F. Lessel, 1966, Williams and Wilkins, Baltimore). In some instances, however, it may be impossible to find a bacterial organism listed in the *Manual* or the *Index* because the organism was never sufficiently described to achieve taxonomic status. The current names of renamed bacteria may also be found in the index of organisms at the end of this book.

It is hoped that this book will serve not only as a text but also as an introduction and guide to the geomicrobiological literature for microbiologists, ecologists, geologists, environmental engineers, and others interested in the subject.

The preparation of this book was greatly aided by discussion with, and review of the manuscript by, Galen E. Jones, R. Schweisfurth, William C. Ghiorse, Edward J. Arcuri, Paul A. LaRock, and many students in my geomicrobiology course. Responsibility for the presentation and interpretation of the subject matter as found in this book rests, however, entirely with me. I am indebted to a number of persons and publishers for making available original photographs or allowing reproduction of previously published material for illustration. They are acknowledged in the legends of the individual illustrations furnished by them. I wish to thank Stephen Chiang for his preparation of the finished line drawings from the crude sketches I furnished him. I also wish to thank the editorial staff of Marcel Dekker, Inc., for their help in readying my manuscript for publication.

Henry Lutz Ehrlich

Contents

1

Introduction

GEOMICROBIOLOGY examines the role that microbes have played in the past and are currently playing in a number of fundamental geological processes: for example, in the weathering of rocks, in soil and sediment formation and transformation, in the genesis and degradation of minerals, and in the genesis and degradation of fossil fuels. GEOMICROBIOLOGY should not be equated with microbial ecology or microbial biogeochemistry. **Microbial ecology** is the study of interrelationships between different microorganisms; among microorganisms, plants, and animals; and between microorganisms and their environment. **Microbial biogeochemistry** is the study of microbially influenced chemical reactions, enzymatically catalyzed or not, and their kinetics. These reactions are often studied in the context of mineral cycles, with emphasis on environmental mass transfer and energy flow. These three subjects do overlap to some degree, as shown in Figure 1.1.

The origin of the word "geomicrobiology" is obscure. It obviously derived from the term "geological microbiology." Beerstecher (1954) defined geomicrobiology as the study of the relationship between the history of the earth and the microbial life upon it." Kuznetsov et al. (1963) defined it as "the study of microbial processes currently taking place in the modern sediments of various bodies of water, in ground waters circulating

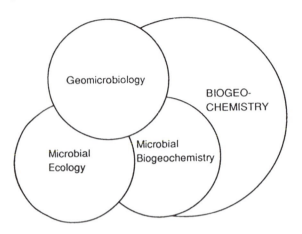

Figure 1.1 Interrelationships among geomicrobiology, microbial ecology, microbial biogeochemistry, and biogeochemistry.

through sedimentary and igneous rocks, and in weathered earth crust . . . [and also] the physiology of specific microorganisms taking part in presently occurring geochemical processes.'' Neither author traced the history of the word, but they pointed to the important roles that scientists such as Winogradsky, Waksman, and ZoBell have played in the development of the field.

Geomicrobiology is not a new field of study. Although certain early investigators in soil and aquatic microbiology may not have thought of themselves as geomicrobiologists, they have, nevertheless, had an influence on the subject. One of the first contributors to geomicrobiology was Ehrenberg (1838), who in the second quarter of the nineteenth century discovered an association of *Gallionella ferruginea* with ochreous deposits of bog iron. He believed that the organism, which he thought to be an infusorian but which we now recognize as a stalked bacterium, was important in the formation of such deposits. Another important early contributor to geomicrobiology was Winogradsky, who discovered that *Beggiatoa* could oxidize H_2S to elemental sulfur (1887), and that *Leptothrix ochracea* promoted oxidation $FeCO_3$ to ferric oxide (1888). He believed that both organisms gained energy from these processes. Still other important early contributors to geomicrobiology were Harder (1919), a researcher trained as a geologist and microbiologist, who studied the significance of microbial iron oxidation and precipitation in relation to the formation of sedimentary iron deposits, and Stutzer (1912), Vernadsky (1908–1922), and others, whose studies led to recognition of the significance of microbial oxidation of H_2S and elemental sulfur in the formation of sedimentary sulfur deposits [see Ivanov (1967), Lapo (1987), and Bailes (1990) for a discussion of early Russian geomicrobiology and its practitioners]. Our understanding of the role of bacteria in sulfur deposition in nature received a further boost from the discovery of bacterial sulfate reduction by Beijerinck (1895) and van Delden (1903). Starting with the Russian investigator Nadson (1903; see Nadson, 1928) at the end of the nineteenth century, and continuing with such investigators as Bavendamm (1932), the important role of microbes in $CaCO_3$ precipitation began to be noted. Microbial participation in manganese oxidation and precipitation in nature was first indicated by Beijerinck (1913), Soehngen (1914), Lieske (1919), and Thiel (1925). This was later related to sedimentary ore formation by Zappfe (1931). The microbial role in methane formation became apparent through the observation and studies of Béchamp (1868), Tappeiner (1882), Popoff (1875), Hoppe-Seyler (1886), Omeliansky (1906), and Soehngen (1906). The role of bacteria in rock weathering was first suggested by Muentz (1890) and Merrill (1895). Later, involvement of acid-producing microorganisms such as nitrifiers, and of crustose lichens and fungi was suggested (see Waksman, 1932). Thus by the beginning of the twentieth century, many of the important areas of geomicrobiology had begun to receive serious attention from microbiologists. In general, it may be said that most of the geomicrobiologically important discoveries of the nineteenth century were made through physiological studies in the laboratory which revealed the capacity of specific organisms for geomicrobiologically important transformations, causing later workers to study the extent of the microbial activities in the field.

Geomicrobiology in the United States can be said to have begun with the work of E. C. Harder (1919) on iron-depositing bacteria. Other early American investigators of geomicrobial phenomena include J. Lipman, S. A. Waksman, R. L. Starkey, and H. O. Halvorson, all prominent in soil microbiology, and G. A. Thiel, C. Zappfe, and C. E. ZoBell, all prominent in aquatic microbiology.

Very fundamental discoveries in geomicrobiology continue to be made, some basic ones having been made comparatively recently. For instance, the concept of environmental limits of pH and E_h for microbes in their natural habitats was first introduced by Baas Becking et al. in 1960 (see Chapter 6). The discovery of a specific acidophilic, iron-oxidizing bacterium and its identification as the primary cause of acid coal-mine drainage came only in 1950, the result of studies by Colmer, Temple, and Hinkle. The subsequent demonstration of the presence of these same organisms in acid mine drainage from a copper sulfide ore body in Utah (Bingham Canyon open-pit mine) and the experimental demonstration that these organisms can promote the leaching of metals from various metal sulfide ores (Bryner et al., 1954) led to the first industrial application of geomicrobially active organisms (Zimmerley et al., 1958) (Chapter 18). The first attempts at visual detection of Precambrian prokaryotic fossils in sedimentary rocks were made by Tyler and Barghoorn in 1954, by Schopf et al. in 1965, and by Barghoorn and Schopf in 1965 (see Chapter 3). Paleontological discoveries resulting from these studies were to have a profound influence on current theories about the evolution of life on Earth (Schopf, 1983). The discovery of geomicrobially active microorganisms around submarine hydrothermal vents (Jannasch and Mottle, 1985; Tunnicliffe, 1992) and the demonstration of a significant viable microflora with a potential for geomicrobially important activity in the deep subsurface (Ghiorse and Wilson, 1988; Sinclair and Ghiorse, 1989) are opening up previously unsuspected new areas for geomicrobiological study.

As this book will show, many areas of geomicrobiology remain to be fully explored or developed further.

REFERENCES

Baas Becking, L. G. M, I. R. Kaplan, and D. Moore. 1960. Limits of the environment in terms of pH and oxidation and reduction potentials. J. Geol. 68:243–284.

Bailes, K. E. 1990. Science and Russian Culture in an Age of Revolution. V. I. Vernadsky and His Scientific School, 1863–1945. Indiana University Press, Bloomington.

Barghoorn, E. S., and J. W. Schopf. 1965. Microorganisms from the late Precambrian of Central Australia. Science 150:337–339.

Barker, H. A. 1956. Bacterial Fermentations. CIBA Lectures in Microbial Biochemistry. Wiley, New York.

Bavendamm, W. 1932. Die mikrobiologische Kalkfaellung in der tropischen See. Arch. Mikrobiol. 3:205–276.

Béchamp, E. 1868. Ann. Chim. Phys. 13:103 (as cited by Barker, 1956).

Beerstecher, E. 1954. Petroleum Microbiology. Elsevier, New York.

Beijerinck, M. W. 1895. Ueber *Spirillum desulfuricans* als Ursache der Sulfatreduktion. Zentralbl. Bakteriol. Parasitkd. Infektionskr. Hyg. Abt. I Orig. 1:1–9; 49–59; 104–114.

Beijerinck, M. W. 1913. Oxydation des Mangancarbonates durch Bakterien und Schimmelpilzen. Folia Microbiol. (Delft) 2:123–134.

Bryner, L. C., J. V. Beck, D. B. Davis, and D. G. Wilson. 1954. Microorganisms in leaching sulfide minerals. Ind. Eng. Chem. 46:2587–2592.

Colmer, A. R., K. L. Temple, and H. E. Hinkle. 1950. An iron-oxidizing bacterium from the acid drainage of some bituminous coal mines. J. Bacteriol. 59:317–328.

Ehrenberg, C. G. 1838. Die Infusionsthierchen als vollkommene Organismen. L. Voss. Leipzig, East Germany.

Ghiorse, W. C., and J. T. Wilson. 1988. Microbial ecology of the terrestrial subsurface. Adv. Appl. Microbiol. 33:107–172.

Harder, E. C. 1919. Iron depositing bacteria and their geologic relations. U.S. Geol. Surv. Prof. Pap. 113. 89 pp.

Hoppe-Seyler, F. 1886. Z. Physiol. Chem. 10:201; 401 (as cited by Barker, 1956).

Ivanov, M. V. 1967. The development of geological microbiology in the U.S.S.R. Mikrobiologiya 31:795–799.

Jannasch, H. W., and M. J. Mottl. 1985. Geomicrobiology of deep-sea hydrothermal vents. Science 229:717–725.

Kuznetsov, S. I., M. V. Ivanov, and N. N. Lyalikova. 1963. Introduction to Geological Microbiology. English transl. McGraw-Hill, New York.

Lapo, A. V. 1987. Traces of Bygone Biospheres. Mir Publishers, Moscow.

Lieske, R. 1919. Zur Ernaehrungsphysiologie der Eisenbakterien. Zentralbl. Bakteriol. Parasitenkd. Infektionskr. Hyg. Abt. II 49: 413–425.

Merrill, G. P. 1895. Geol. Soc. Am. Bull. 6:321–332 (as cited by Waksman, 1932).

Muentz, A. 1890. Sur la decomposition des roches et la formation de la terre arable. C. R. Acad. Sci. (hebd. séances) (Paris) 110:1370–1372.

Nadson, G. A. 1903. Microorganisms as geologic agents. I. Tr. Komisii Isslect. Min. Vodg. Slavyanska, St. Petersburg.

Nadson, G. A. 1928. Beitrag zur Kenntnis der bakeriogenen Kalkablagerung. Arch. Hydrobiol. 19:154–164.

Omeliansky, 1906. Zentralbl. Bakteriol. Parasitenkd. Infektionskr. Hyg. Abt. II 15:673 (as cited by Barker, 1956).

Popoff, L. 1875. Arch. Ges. Physiol. 10:142 (as cited by Barker, 1956).

Schopf, J. W., ed. 1983. Earth's Earliest Biosphere. Its Origin and Evolution. Princeton University Press, Princeton, NJ. 543 pp.

Schopf, J. W., E. S. Baarghoorn, M. D. Maser, and R. O. Gordon. 1965. Electron microscopy of fossil bactreria two billion years old. Science 149:1365–1367.

Sinclair, J. L., and W. C. Ghiorse. 1989. Distribution of aerobic bacteria, protozoa, algae, and fungi in deep subsurface sediments. Geomicrobiol. J. 7:15–31.

Soehngen, N. L. 1906. Het outstaan en verdwijnen van waterstof en meth-

aan ouder den invloed van het organische leven. Thesis. Technical
   University, Delft (Vis, Jr., Delft, publisher).
Soehngen, N. L. 1914. Umwandlung von Manganverbindungen unter dem
   Einfluss mikrobiologischer Prozesse. Zentralbl. Bakteriol. Para-
   sitenk. Infektionskr. Hyg. Abt. II 40:545–554.
Stutzer, O. 1912. Origin of sulfur deposits. Econ. Geol. 7:733–743.
Tappeiner, W. 1882. Ber. Dtsch. Chem. Ges. 15:999 (as cited by Barker,
   1956).
Thiel, G. A. 1925. Manganese precipitated by microorganisms. Econ.
   Geol. 20:301–310.
Tunnicliffe, V. 1992. Hydrothermal-vent communities of the deep sea.
   Am. Sci. 80:336–349.
Tyler, S. A., and E. S. Barghoorn. 1954. Occurrence of structurally pre-
   served plants in Precambrian rocks of the Canadian Shield. Science
   119:606–608.
Van Delden, A. 1903. Beitrag zur Kenntnis der Sulfatreduktion durch
   Bakterien. Zentralbl. Bakteriol. Parasitenkd. Infektionskr. Hyg.
   Abt. II 11:81–94.
Vernadsky, V. I. 1908–1922. An attempt at descriptive mineralogy. Iz-
   brannye Trudy, Vol. 2. Izdatel'stvo. Akad. Nauk SSSR, Moscow,
   1955.
Waksman, S. A. 1932. Principles of Soil Microbiology, 2nd. ed. rev. Wil-
   liams & Wilkins, Baltimore, MD.
Winogradsky, S. 1887. Ueber Schwefelbakterien. Bot. Ztg. 45:489–600;
   606–616.
Winogradsky, S. 1888. Ueber Eisenbakterien. Bot. Ztg. 46:261–276.
Zappfe, C. 1931. Deposition of manganese. Econ. Geol. 26:799–832.
Zimmerley, S. R., D. G. Wilson, and J. D. Prater. 1958. Cyclic leaching
   process employing iron oxidizing bacteria. U.S. patent US
   2,829,964.

2

The Earth as a Microbial Habitat

2.1 GEOLOGICALLY IMPORTANT FEATURES

The interior of the planet Earth consists of three major concentric shell-like structures (Fig. 2.1), the innermost one being the **core**. It is surrounded by the **mantle**, which, in turn, is surrounded by the outermost shell, the **crust**. The crust is surrounded by a gaseous envelope, the **atmosphere**. The core, which is estimated to have a radius of about 3,450 km, is believed to consist of an Fe-Ni alloy with an admixture of small amounts of the siderophile elements cobalt, rhenium, and osmium and perhaps some sulfur and phosphorus (Mercy, 1972; Anderson, 1992). The inner portion of the core, which has an estimated radius of 1,250 km, is solid, having a density of 13 g cm^{-3} and being subjected to a pressure of 3.7×10^{12} dyn cm^{-2}. The outer portion of the core has a thickness of 2,200 km and is molten, owing to the high temperature but lower pressure than at the central core ($1.3-3.2 \times 10^{12}$ dyn cm^{-2}). The density of this portion is $9.7-12.5$ g cm^{-3}. The mantle, which has a thickness of about 2,870 km, has a very different composition from the core and is separated from it by the Wickert-Gutenberg discontinuity (Mercy, 1972). The mantle rock is dominated by the elements O, Mg, and Si with lesser amounts of Fe, Al, Ca, and Na. Its consistency, although not molten, is thought to be

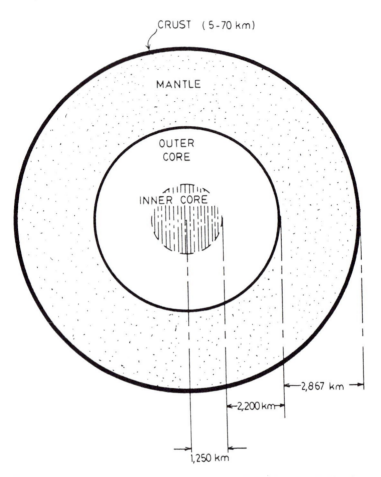

Figure 2.1 Diagrammatic cross section of the Earth. Radii of core and mantle drawn to scale.

plastic, especially in the uppermost portion called the **asthenosphere**. Upper mantle rock penetrates the crust on rare occasions and may be recognized as an outcropping, as in the case of some ultramafic rock on the bottom of the western Indian Ocean (Bonatti and Hamlyn, 1978). The crust is separated from the mantle by the Mohorovicic discontinuity. Its thickness varies from as little as 5 km under ocean basins to as great as 70 km under continental mountain ranges. Under an average continental shield area, its thickness is 33 km (Mercy, 1972). The rock of the crust is dominated by O, Si, Al, Fe, Mg, Na, and K. These elements make up

98.6% of the weight of the crust (Williams, 1962) and occur predominantly in the rocks and sediments. The bedrock under the ocean is generally basaltic, whereas that of the continents is granitic to an average crustal depth of 25 km. Below this depth it is basaltic to the Mohorovicic discontinuity. Sediment covers most of the bedrock under the oceans. It ranges in thickness from 0 to 4 km. Sedimentary rock and sediment (soil in a non-aquatic context) cover the bedrock of the continents; their thickness may exceed that of marine sediments. The continents make up about 64% of the crustal volume, the oceanic crust makes up 21%, and the shelf and subcontinental crust make up the remaining 15%.

Although until the 1960s it was usually viewed as a coherent structure which rests on the mantle, the Earth's crust is now seen to consist of a series of moving and interacting **plates** of varying sizes and shapes. Some plates support the continents and parts of the ocean floor; others support only parts of the ocean floor. Figure 2.2 shows the outlines of major crustal plates of the ocean basins and adjacent continents. The present estimate of the number of major plates involved is 16 (National Geographic Society, 1995). The plates float on the asthenosphere of the mantle. The crust plus asthenosphere is sometimes considered to be the lithosphere of the Earth. Convection resulting from thermal gradients in the plastic rock of the asthenosphere is believed to be the cause of move-

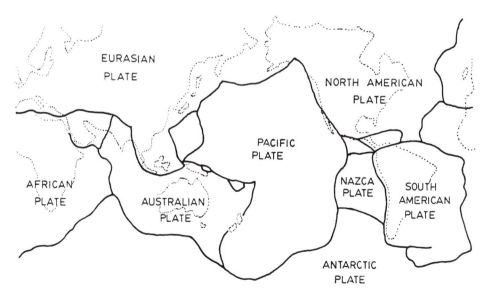

Figure 2.2 Major crustal plates.

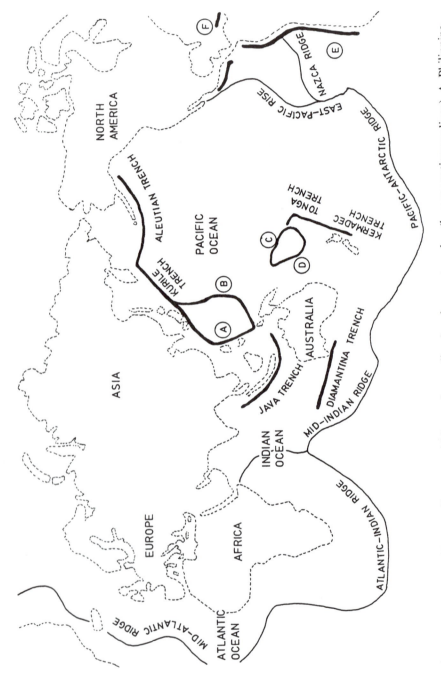

Figure 2.3 Major midocean rift systems (thin continuous lines) and ocean trenches (heavy continuous lines). A, Philippine Trench; B, Marianas Trench; C, Vityaz Trench; D, New Hebrides Trench; E, Peru–Chile Trench; F, Puerto Rico Trench. The East Pacific Ridge is also known as the East Pacific Rise.

ment of the crustal plates. This movement may manifest itself in some locations in collision of plates, in other locations in a sliding past each other along transform faults, and in still other locations in a sliding over each other. The last process is called **subduction** (crustal convergence). It may result when a denser oceanic plate slides below a lighter continental plate, or when adjacent oceanic plates of nearly equal density interact. Either interaction may lead to formation of a trench-volcanic island arc system. In the case of oceanic–continental plate collisons, the resulting arc system may eventually accrete to the continental margin due to the movement of the subducting oceanic plate in the direction of the continental plate. The island arc system results from a sedimentary wedge formed by the overriding plate (Van Andel, 1992; Gurnis, 1992).

Oceanic plates grow along **oceanic ridges** (crustal divergence), such as the Mid-Atlantic Ridge, the East Pacific Rise, and others (Fig. 2.3). The oldest portions of growing oceanic plates are destroyed through subduction with formation of deep-sea trenches, such as the Marianas, Kurile and Philippine Trenches in the Pacific Ocean, and the Puerto Rico Trench in the Atlantic Ocean. Growth of the oceanic plates at the midocean ridges is the result of submarine volcanic eruptions of **magma** (molten rock from the deep crust or upper mantle), which is added to opposing plate margins along a midocean ridge causing adjacent parts of the plates to be pushed away from the ridge in opposite directions (Fig. 2.4). The oldest portions of oceanic plates are consumed by subduction more or less in proportion to the formation of new oceanic plate at the midocean ridges, thereby maintaining a fairly constant plate size.

Volcanism occurs not only at midocean ridges but also in the regions of subduction where the sinking crustal rock undergoes melting. The molten rock may then erupt through fissures in the crust and contribute to mountain building at the continental margins (**orogeny**). It is plate collision and volcanic activity associated with subduction at continental plate margins that accounts mainly for the existence of coastal mountain ranges. The origin of the Rocky Mountains and Andes on the North and South American Continents, respectively, is associated with subduction activity, whereas the Himalayas are the result of collision of the plate holding the Indian subcontinent with that holding the Asian continent.

Volcanic activity may also occur away from crustal plate margins, at so-called **hot spots**. In the Pacific Ocean, one such hot spot is represented by the island of Hawaii with its active volcanoes. The remainder of the Hawaiian island chain had its origin at the same hot spot where the island of Hawaii is presently located. Crustal movement of the Pacific Ocean plate westward caused the remaining islands to be moved away from the hot spot so that they are no longer volcanically active.

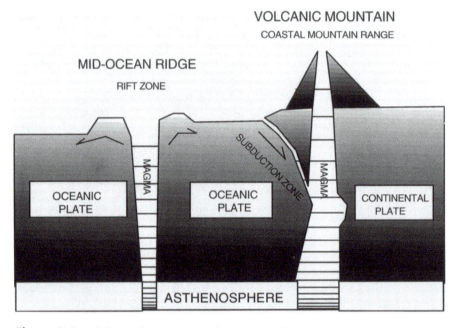

Figure 2.4   Schematic representation of sea floor spreading and plate subduction. New oceanic crust is formed at the rift zone of the midocean ridge. Old oceanic crust is consumed in the subduction zone near a continental margin or island arc.

The continents as they exist today are thought to have derived from a single continental mass, **Pangaea**, which broke apart due to crustal movement less than 200 million years ago, first into **Laurasia** (which included present-day North America, Europe, and most of Asia) and **Gondwana** (which included present-day Africa, South America, Australia, Antarctica, and the Indian subcontinent). These separated subsequently into the continents we know today, except for the Indian subcontinent, which did not join the Asian continent until some time after this breakup (Fig. 2.5) (Dietz and Holden, 1970; Fooden, 1972; Matthews, 1973; Palmer, 1974; Hoffman, 1991; Smith, 1992). The continents that evolved became modified by accretion of small landmasses through collision with plates bearing them. The separation of Pangaea into various continents is attributed to the movement of the crustal plates. Pangaea itself is thought to have originated 250–260 million years ago from an aggregation of crustal plates bearing continental landmasses including Baltica (consisting of Russia west of the Ural Mountains, Scandinavia, Poland, and northern Germany),

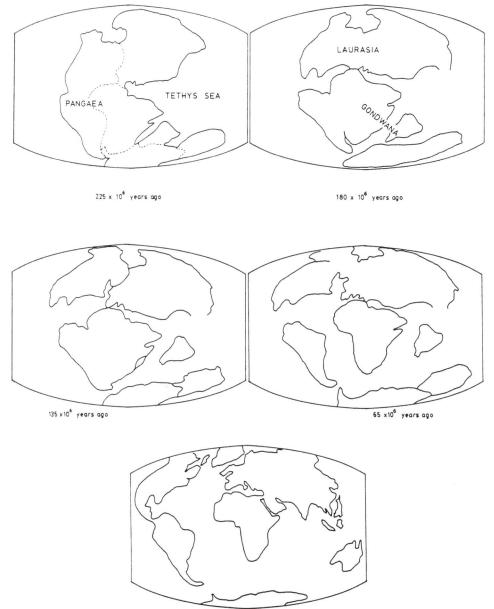

Figure 2.5 Continental drift. Simplified representation of the breakup of Pangaea to present time. (Reproduced from R. S. Dietz, and J. C. Holden, *J Geophys. Res.* 75:4939–4956, 1970.

China, Gondwana, Kazakhstania (consisting of present-day Kazakhstan), Laurentia (consisting of most of North America, Greenland, Scotland, and the Chukotski Peninsula of eastern Russia), and Siberia (Bambach et al., 1980). Mobile continental plates are believed to have existed as long ago as 3.5 billion years (Kroener and Layer, 1992).

The evidence for the origin and movement of the present-day continents rests on paleomagnetic and seismic studies of the Earth's crust; on comparative sediment analyses of deep-sea cores obtained from drillings by the *Glomar Challenger*, an ocean-going research vessel; and on paleoclimatic considerations (Bambach, 1980; Nierenberg, 1978; Vine, 1970). Although the separation of the present-day continents had probably no significant effect on the evolution of prokaryotes (they had pretty much evolved to their present complexity by this time), it did have a profound effect on the evolution of metaphytes and metazoans (McKenna, 1972; Raven and Axelrod, 1972). Flowering plants, birds and mammals, for example, had yet to establish themselves.

2.2   THE BIOSPHERE

The **biosphere**, that portion of the planet which supports life, is restricted to the uppermost part of the crust and to a degree, to the lowermost part of the atmosphere. It includes the the land surface, i.e., the exposed sediment or soil and rock, sometimes called the **lithosphere**, and the **hydrosphere**, that portion of the crust covered by water. Life in the exposed crust has been detected to a maximal depth of around 4,000 m according to Pokrovskiy (cited by Kuznetsov et al., 1963, p. 26). In confirmation, a more recent controlled study revealed the presence of thermophilic, anaerobic, fermenting bacteria in groundwater recovered at 3,500 m from a borehole drilled into granitic rock in the Siljan Ring in central Sweden (Szewzyk et al., 1994). The bacteria were capable of growth in a temperature range of 45–75°C (65°C optimum) in the laboratory. Most life exists at the surface of the exposed crust. Microbes have now also been detected in very significant numbers in some sedimentary rock strata at depths of several hundred meters (Ghiorse and Wilson, 1988). An important factor limiting life to the upper portion of the crust is undoubtedly temperature, which increases with depth in the crust. At a depth of 1,000 m, the temperature may range from − 20 to + 100°C, depending on location (Kuznetsov et al., 1963, p. 25). The heat in the crust is the result of its diffusion from the interior of the Earth, where it is generated primarily as a by-product of natural radioactivity and pressure. Other factors restricting life to the uppermost portion of the crust are lack of porosity of most of the rock and lack of sufficient moisture and nutrients.

Unlike the lithosphere, the hydrosphere is inhabited by life at all depths, some as great as 11,000 m, the depth of the Marianas Trench. Some special **ecosystems** have evolved in some parts of the hydrosphere whose primary energy source is geothermal rather than radiant energy from the sun (Jannasch, 1983). These ecosystems occur around hydrothermal vents at midocean rift zones where heat from magma chambers diffusing upward into the overlying basalt causes seawater that has penetrated deep into the basalt to react with it (see Chapter 15, Fig. 15.16 for a diagrammatic representation of this process). Hydrogen sulfide results from this interaction as well as the solution of some metals, such as iron and manganese, and in some cases other metals such as copper and zinc. The altered seawater (now a **hydrothermal solution**) charged with these solutes is eventually forced up through cracks and fissures in the basalt to enter the overlying ocean through hydrothermal vents. Autotrophic bacteria living free around the vents or in symbiotic association with some metazoa use the hydrogen sulfide as an energy source for converting carbon dioxide into organic matter. Some of this organic matter is used as food by heterotrophic microorganisms and metazoa in the immediate environment around the vents (Jannasch, 1983; Tunnicliffe, 1992). The hydrogen-sulfide-oxidizing bacteria are the **primary producers** in these ecosystems, taking the place of photosynthesizers such as cyanobacteria, algae, or plants, the usual primary producers on Earth. Photosynthesizers cannot operate in the location of hydrothermal vent communities because of the perpetual darkness that prevails at these sites (see also Chapter 18).

Not all submarine communities featuring chemosynthetic, hydrogen sulfide oxidizers as the primary producers are based on hydrothermal discharge. On the Florida Escarpment in the Gulf of Mexico, ventlike biological communities have been found at abyssal depths around hydrogen sulfide seeps whose discharge is at ambient temperature. The sulfide in this instance may originate from an adjacent carbonate platform containing fluids with 250% solids and temperatures up to 115°C (Paul et al., 1984).

In some other instances, such as at the Oregon Subduction Zone or at some sites on the Florida Escarpment, methane of undetermined origin expelled from the pore fluid of the sediments, rather than hydrogen sulfide, is the basis for primary production on the seafloor. Metazoa share in the carbon fixed by free-living or symbiotic methane-oxidizing bacteria (Kulm et al., 1986; Childress et al., 1986; Cavanaugh et al., 1987) (see also Chapter 20).

Finally, the biosphere includes the lower portion of the atmosphere. Living microbes have been recovered from it at heights as great as 48–77 km above the Earth's surface (Imshenetsky et al., 1978; Lysenko, 1979).

Whether the atmosphere constitutes a true microbial habitat is very debatable. Although it harbors viable vegetative cells and spores, it is generally not capable of sustaining growth and multiplication of these organisms because of lack of sufficient moisture and nutrients and, especially at higher elevations, because of lethal radiation. At high humidity in the physiological temperature range, however, some bacteria may propagate to a limited extent (Dimmick et al, 1979; Straat et al., 1977). The residence time of microbes in air may also be limited, owing to eventual fallout. In the case of microbes associated with solid particles suspended in still air, the fallout rate may range from 10^{-3} cm sec^{-1} for 0.5-μm particle sizes to 2 cm sec^{-1} for 10-μm particle sizes (Brock, 1974, p. 541). Yet even if not a true habitat, the atmosphere is, nevertheless, important to microbes. It is a vehicle for spreading microbes from one site to another; it is a source of oxygen for strict and facultative aerobes; it is a source of nitrogen for nitrogen-fixing microbes; and its ozone layer screens out most of the harmful ultraviolet radiation from the sun.

Although the biosphere is restricted to the very uppermost crust and the atmosphere, the core of the Earth does exert an influence on some forms of life. The core, with its solid center and molten outer portion acts like a dynamo in generating the magnetic field surrounding the Earth (Strahler, 1976, p. 36; Gubbins and Bloxham, 1987). Magnetotactic bacteria, because they form magnetite crystals or iron sulfide particles in their cells, which behave like compasses, can utilize the Earth's magnetic field for purposes of orientation in seeking their preferred habitat, which is a partially reduced environment. They are able to align themselves with respect to the Earth's magnetic field (Blakemore, 1982; DeLong et al., 1993).

2.3 SUMMARY

The surface of the Earth includes the lithosphere, hydrosphere, and atmosphere, all of which are habitable by microbes to a greater or lesser extent and constitute the biosphere of the Earth.

The structure of the Earth can be separated into the core, the mantle, and the crust. Of these, only the uppermost portion of the crust is habitable by living organisms. The crust is not a continuous solid layer over the mantle, but consists of a number of crustal plates afloat on the mantle, or more specifically on the asthenosphere of the mantle. Some of these plates lie entirely under the oceans. Others carry parts of a continent or of a continent and an ocean. Oceanic plates are growing along midocean spreading centers, while old portions are being destroyed by subduction under or collision with continental plates. The crustal plates are in constant, albeit slow, motion that accounts for continental drift.

REFERENCES

Anderson, D. L. 1992. The earth's interior. *In*: G. C. Brown, C. J. Hawksworth, and R. C. L. Wilson, eds. Understanding the Earth. Cambridge University Press, Cambridge, U.K., pp. 44–66.

Bambach, R. K., C. R. Scotese, and A. F. Ziegler. 1980. Before Pangaea: The geography of the Paleozoic world. Am. Sci. 68:26–38.

Blakemore, R. P. 1982. Magnetotactic bacteria. Annu. Rev. Microbiol. 36:217–238.

Bonatti, E., and P. R. Hamlyn. 1978. Mantle uplifted block in the western Indian Ocean. Science 201:249–251.

Brock, T. D. 1974. Biology of Microorganisms. 2nd ed. Prentice Hall, Englewood Cliffs, NJ.

Cavanaugh, C. M., P. R. Levering, J. S. Maki, R. Mitchell, and M. E. Lidstrom. 1987. Symbiosis of methylotrophic bacteria and deep-sea mussels. Nature (Lond.) 325:346–348.

Childress, J. J., C. R. Fischer, J. M. Brooks, M. C. Kennecutt II, R. Bidigare, and A. E. Anderson. 1986. A methanotrophic marine molluscan (Bivalvia, Mytilidae) symbiosis: mussels fueled by gas. Science 233:1306–1308.

DeLong, E. F., R. B. Frankel, and D. A. Bazylinski. 1993. Multiple evolutionary origin of magnetotaxis. Science 259:803–806.

Dietz, R. C., and J. C. Holden. 1970. Reconstruction on Pangaea: Breakup and dispersion of continents, Permian to present. J. Geophys. Res. 75:4939–4956.

Dimmick, R. L., H. Wolochow, and M. A. Chatigny. 1979. Evidence that bacteria can form new cells in air-borne particles. Appl. Environ. Microbiol. 37:924–927.

Fooden, J. 1972. Breakup of Pangaea and isolation of relict mammals in Australia, South America, and Madagascar. Science 175:894–898.

Ghiorse, W. C., and J. T. Wilson. 1988. Microbial ecology of the terrestrial subsurface. Adv. Appl. Microbiol. 33:107–171.

Gubbins, D., and J. Bloxham. 1987. Morphology of the geomagnetic field and implications for the geodynamo. Nature (Lond.) 325:509–511.

Gurnis, M. 1992. Rapid continental subsidence following the initiation and evolution of subduction. Science 255:1556–1558.

Hoffman, P. F. 1991. Did the breakout of Laurentia turn Gondwanaland inside-out? Science 252:1409–1412.

Imshenetsky, A. A., S. V. Lysenko, and G. A. Kazatov. 1978. Upper boundary of the biosphere. Appl. Environ. Microbiol. 35:1–5.

Jannasch, H. W. 1983. Microbial processes at deep sea hydrothermal vents. pp. 677–710. *In*: P. A. Rona et al., eds. Hydrothermal Processes at Sea Floor Spreading Centers. Plenum Press, New York.

Kroener, A., and P. W. Layer. 1992. Crust formation and plate motion in the Early Archean. Science 256:1405–1411.

Kulm, L. D., E. Suess, J. C. Moore, B. Carson, B. T. Lewis, S. D. Ritger, D. C. Kadko, T. M. Thornburg, R. W. Embley, W. D. Rugh, G. J. Maasoth, M. G. Langseth, G. R. Cochrane, and R. L. Scamman. 1986. Oregon subduction zone: venting, fauna, and carbonates. Science 231:561–566.

Kuznetsov, S. I., M. V. Ivanov, and N. N. Lyalikova. 1963. Introduction to Geological Microbiology. English translation. McGraw-Hill, New York.

Lysenko, S. V. 1979. Microorganisms in the upper atmospheric layers. Mikrobiologiya 48:1066–1074 (Engl. transl. pp. 871–877).

Matthews, S. W. 1973. This changing Earth. Nat. Geogr. 143:1–37.

McKenna, M. C. 1972. Possible biological consequences of plate tectonics. BioScience 22:519–525.

Mercy, E. 1972. Mantle geochemistry. *In*: R.W. Fairbridge, ed. The Encyclopedia of Geochemistry and Environmental Sciences, Encyclopedia of Earth Science Series, Vol. IVA. Van Nostrand Reinhold, New York, pp. 677–683.

National Geographic Society. 1995. The Earth's fractured surface (map supplement). National Geographic 187(4).

Nierenberg, W. A. 1978. The deep sea drilling project after ten years. Am. Sci. 66:20–29.

Palmer, A. R. 1974. Search for the Cambrian world. Am. Sci. 62:216–224.

Paul, C. K., B. Hecker, R. Commeau, R. P. Freeman-Lynde, C. Neumann, W. P. Corso, S. Golubic, J. E. Hook, E. Sikes, and J. Curray. 1984. Biological communities at the Florida Escarpment resemble hydrothermal vent taxa. Science 226:965–967.

Raven, P. H., and D. I. Axelrod. 1972. Plate tectonics and Australian Paleogeography. Science 176:1379–1386.

Smith, A. G. 1992. Plate tectonics and continental drift. *In*: G. Brown, C. Hawkesworth, C. Wilson, eds. Understanding the Earth. Cambridge University Press, Cambridge, U.K.

Straat, P. A., H. Woodrow, R. L. Dimmick, and M. A. Chatigny. 1977. Evidence for incorporation of thymidine into deoxyribonucleic acid in air-borne bacterial cells. Appl. Environ. Microbiol. 34:292–296.

Strahler, N. 1976. Principles of Physical Geology. Harper & Row, New York.

Szewzyk, U., R. Szewzyk, and T-A. Stenstroem. 1994. Thermophilic, anaerobic bacteria isolated from a deep borehole in granite in Sweden. Proc. Natl. Acad. Sci. USA 91:1810–1813.

Tunnicliffe, V. 1992. Hydrothermal-vent communities. Am. Sci. 80: 336–349.

Van Andel, T. H. 1992. Seafloor spreading and plate tectonics. *In*: G. Brown, C. Hawkesworth, and C. Wilson, eds. Understanding the Earth. Cambridge University Press, Cambridge, U.K., pp. 167–186.

Vine, F. J. 1970. Spreading of the ocean floor: New evidence. Science 154:1405–1415.

Williams, J. 1962. Oceanography. Little, Brown, Boston.

3

The Origin of Life and Its Early History

3.1 THE BEGINNINGS

The Earth is thought to be about 4.6×10^9 years old (4.6 eons). One accepted view holds that it derived from an accretion disc that resulted from gravitational collapse of interstellar matter. A major portion of the matter condensed to form the Sun. Other components in it subsequently accreted to form planetesimals of various sizes. These in turn accreted to form our Earth and the other inner (rocky) planets of our solar system, namely Mercury, Venus, and Mars. As accretion of the Earth proceeded, its internal temperature could have risen sufficiently to result ultimately in separation of silicates and iron, leading to a differentiation into mantle and core. Alternatively, and more likely, a primordial rocky core could have been displaced by a liquid iron shell that surrounded it. Displacement of the rocky core would have been made possible if it fragmentated as a result of nonhydrostatic pressures, and as a consequence, the iron core became surrounded by a hot, well-mixed mantle of rock material in a **catastrophic process**. Whichever process actually took place, much heat must have been released during this formational process, resulting in outgassing from the mantle to form a primordial atmosphere and hydrosphere. All this is thought to have occurred in the span of about 10^8 years. As the planet cooled, segregation of mantle components is thought to have occurred and a thin crust to have formed by 4.0–3.8 eons ago. Accretion by meteorite bombardment is believed to have become insignificant by this time. The first crustal plates may have formed between 3.8 and 2.7 eons ago, initiating tectonic activity in the crust which continues to this day. Protocontinents may have emerged at this time to be subsequently followed by the development of true continents. (For details about these early steps in the formation of the Earth, see Stevenson, 1983; Ernst, 1983; and Taylor, 1992).

Insofar as the beginning of life on Earth is concerned, two views have been expressed. One is that of **panspermia** according to which pre-formed life arrived on this planet in the form of a spore(s) from another world. As unrealistic as this view may seem, recent laboratory experiments have suggested that panspermia is not an impossibility (Weber and Greenberg, 1985). These studies employed spores of *Bacillus subtilis* enveloped in a mantle of 0.5 μm thickness or greater, derived from equal parts of H_2O, CH_4, NH_3, and CO (presumed interstellar condition). The mantle shielded them from ultraviolet (UV) irradiation (100–200-μm wavelength) in ultrahigh vacuum ($<1 \times 10^6$ torr) at 10°K but not from far-UV rays (200–300 μm). From experimentally determined survival rates, it was calculated that if spores were enveloped in a mantle of 0.9 μm thickness having a refractive index of 0.5, which would protect them from short- and long-wavelength UV, they could survive in sufficient numbers over a period of 4.5–45 Myr in outer space to allow them to travel from one solar system to another. Spores could have entered outer space in high-speed ejecta as a result of a collision between a life-bearing planet and a meteorite or comet (Weber and Greenberg, 1985). A more widely held view is, however, that life began *de novo* on Earth, although certain critical organic molecules, trapped in comets, could have arrived from outer space when some of these comets bombarded Earth in the early stages of its formation (see report by Cohen, 1995).

For life to have originated *de novo* on Earth, the existence of a primordial, reducing atmosphere was of primary importance. Constituents may have included H_2O, H_2, CO_2, CO, CH_4, N_2, and NH_3 (see Table 4.3 in Chang et al., 1983). The exact composition of the atmosphere will have changed with time due to photochemical reactions and reactions driven by electric discharge (lightning), due to interaction of some of the gases with mineral constituents at high temperature, and due to escape of the lightest gases (e.g., hydrogen) into space (Chang et al., 1983; Schopf et al., 1983).

Two opposing views are currently held on how life may have arisen *de novo* on Earth. The older view is that life arose in a dilute "organic soup'' (broth) which covered the surface of the planet. The biologically important organic molecules in this soup are thought to have been formed in abiotic chemical reactions among some of the atmospheric gases with heat, electric discharge, and/or light energy as the driving force (see discussion by Chang et al., 1983). Clays could have played an important role as catalysts and templates in the assembly of some polymeric molecules (Cairns-Smith and Hartman, 1986). If, as has been recently theorized, the surface of the early Earth was frozen because the sun was less luminous than it was to become later, bolide impacts could have caused episodic melting, during which time the abiotic reactions took place (Bada et al., 1994). Special polymeric molecules finally arose with the ability to self-reproduce (beginning of true life) at the expense of building blocks which continued to be abiotically synthesized and added to the organic soup. Cellular life arose subsequently as a result of the ''encapsulation'' in a lipid membrane of the self-reproducing entities in the organic soup that were vital to the continuation of life. The feasibility of enzyme-catalyzed RNA synthesis by a polymerase protein, a reaction of key importance in the evolution of early life, in an artificially formed lipid vesicle was recently demonstrated. Substrate adenosine diphosphate penetrated the vesicles readily from the exterior solution to be acted upon by the polymerase (Chakrabarti et al., 1994). The primitive cells that arose in this way are thought to have been heterotrophs which continued to use abiotically formed organic building blocks in the organic soup until the evolution of the first autotrophs which acquired the ability to form their own organic building blocks from inorganic constituents, using chemical or radiant energy as the driving force of these reactions.

The newer, more elegant view, which is also more rational thermodynamically, is that life may have started as a form of surface chemistry (Waechtershaeuser, 1988). In this view, building blocks were synthesized and polymerized starting with inorganic constituents (carbon dioxide, phosphate, ammonia) on the surface of minerals with a positive (anodic) surface charge (e.g., iron pyrite). According to Waechtershaeuser (1988), polymerizations of surface-bound molecules are thermodynamically favorable, whereas polymerizations of the same molecules in the organic soup scenario are thermodynamically unfavorable. In the latter case, the water in which the monomers and polymers were dissolved favors hydrolytic cleavage of polymers. In the surface chemistry model, some of the surface reactions were autocatalytic and, because of their ability to replicate the reaction product, constituted the first two-dimensional life forms (''surface metabolists'' in Waechtershaeuser terminology). With the emergence of isoprenoid lipid synthesis, surface metabolists eventually became covered by lipid-membranes (half-cells) and ultimately were complete enclosed by them. This membrane encapsulation thus produced the first cells. They featured a membrane-enclosed cytosol in which initially the vital chemistry still occurred on the surface of a mineral grain. However, with the passage of time and the development of some critical molecules, the vital chemistry became progressively independent of the mineral surface and took on a distinct existence in the aqueous phase of the cytosol. In the original cells, the mineral grain may have consisted of pyrite (FeS_2), which may have been the product of early energy- and reducing-power generation according to Waechtershaeuser (1988) (see below). Its surface had anodic properties. Development of a genetic apparatus consisting of deoxyribonucleic acid and ribonucleic acid that ultimately enabled firm control of metabolic behavior and its perpetuation began as a part of surface metabolism in the pre-cellular stage, but at that time did not exert direct control in metabolists. Further evolution of the genetic apparatus in the cellular stage led to the incorporation of structural information and its decyphering by transcription and translation, required for the synthesis of specific catalytic proteins (enzymes). After metabolism became independent of the mineral grain, these enzymes exerted needed control over cellular metabolism.

In contrast to the organic soup theory, the surface metabolism theory proposes that the first life in the attached and the detached state was **autotrophic**, i.e., it depended on CO_2, CO, NH_3, H_2S, H_2, and H_2O. Driving energy and reducing power for autotrophic metabolism may have come from an interaction of ferrous iron with hydrogen sulfide (Waechtershaeuser, 1988):

$$Fe^{2+} + 2H_2S \Rightarrow FeS_2 + 2H^+ + 2e \ (\Delta G°, \ -2.62 \text{ kcal or } -11 \text{ kJ})$$

$$(3.1)$$

Both Fe(II) and H_2S should have been plentiful on the primitive Earth. The formation of pyrite and H_2 from a reaction between FeS and H_2S under fastidiously anaerobic conditions was experimentally demonstrated by Drobner et al. (1990):

$$FeS + H_2S \Rightarrow FeS_2 + H_2 \ (\Delta G°, \ -6.14 \text{ kcal or } -25.7 \text{ kJ}) \quad (3.2)$$

Current thinking based on some fossil finds is that first life, i.e., the first living entities, whatever their from, could have originated as early as 3.8 billion years ago if not earlier (e.g., 4 billion years ago) (Schopf et al., 1983). One fossilized form of "well developed" cellular life in the form of a **stromatolite** (in this instance a fossilized mat of filamentous organisms) has been found in chert of the Warrawoona Group in the Pilbara Block of Western Australia, whose age has been determined to be ~3.5 billion years (Fig. 3.1)(Lowe, 1980; Walter et al., 1980). Slightly younger microfossils (3.3 to 3.5 billion years old) having a recognizable cell type resembling cyanobacteria have been found in the Early Archean Apex Basalt and Towers Formation of the Warrawoona Group (Fig. 3.2) (Schopf and Packer, 1987). Some other stromatolites of approximately similar age have been reported from limestone in the Fort Victoria greenstone belt of the Rhodesian Archean Craton within Zimbabwe (Orpen and Wilson, 1981). Finding fossils of once-living organisms of such an age leads to the conclusion that single-celled life and noncellular life (in the form of self-reproducing molecules, whether in an organic soup or attached to minerals with anodic surface charge) must have preceded the emergence of stromatolites by a span of 500 million years.

Figure 3.1 Fossil remnant of ancient life: domical stromatolite ($\times 0.35$) from a stratum of the 3.5-billion-year-old Warrawoona Group in the North Pole Dome region of northwestern Australia. A stromatolite is formed from fossilization of a mat of filamentous microorganisms such as cyanobacteria. (Photograph reproduced from the fronticepiece of The Earth's Earliest Biosphere: Its Origin and Evolution, J. W. Schopf, ed., Princeton University Press, 1983, with publisher's permission.)

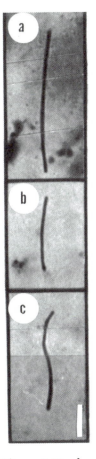

Figure 3.2(a, b, c)   Rod-shaped, threadlike juvenile forms of apparently nonseptate bacteria in petrographic thin section of stromatolitic black chert from the 3.5-billion-year-old Warrawoona Group (Pilbara Subgroup) of Western Australia. Scale mark in panel c is 5 μm and also applies to panels a and b. (Reproduced from photo 9-4 C, D, and E in The Earth's Earliest Biosphere: Its Origin and Evolution, J. W. Schopf, ed., Princeton University Press, 1983, with publisher's permission.)

3.2   EVOLUTION OF LIFE THROUGH THE PRECAMBRIAN: BIOLOGICAL AND BIOCHEMICAL BENCHMARKS

Organic Soup Scenario   The Precambrian (4.5–0.57 eons b.p.) may be divided into Archean (4.5–2.5 eons b.p.) and Proterozoic (2.5–0.57 eons b.p.). The Archean may be subdivided into the Hadean (4.5–3.9 eons b.p.), Early Archean (3.9–2.9), and Late Archean (2.9–2.5 eons b.p.). The Proterozoic may be subdivided into Early Proterozoic (2.5–1.6 eons b.p.), Middle Proterozoic (1.6–0.9 eons b.p.) and Late Proterozoic (0.9–0.57 eons b.p.) (Fig. 3.3) (Harrison and Peterman, 1980). In the organic soup scenario, the precursor molecules such as amino acids, purine and pyrimidine bases, sugars, etc. from which life must have originated, first appeared in the middle to late Hadean, according to this time scale. These monomers later polymerized into heteropolymers such as proteins and nucleic acids. The energy driving these monomer syntheses and polymerizations was heat, sunlight, and electric discharges.

If we accept the ability to self-reproduce as the basic definition of life, then the appearance of the first proteins and/or nucleic acids with this ability marked the beginning of life. Clays may have played a role in the production of self-reproducing molecules, especially proteins (Cairns-Smith and Hartman, 1986). If proteins were the first living molecules, they would have developed independently of abiotically produced nucleic acids at this stage. Recent findings of self-splicing ribonucleic acids (RNA) in cells suggests, however, that RNAs may have been the first self-reproducing molecules and that some of them, in turn, evolved into templates for the synthesis of catalytic proteins important to life (Cech, 1986; Doudna and Szostak, 1989; Joyce, 1991).

To cope more effectively with the environment and to become more efficient in the synthesis of monomers and polymers, lipid vesicles (primitive cells) evolved which enclosed the self-reproducing molecules. If proteins were the first self-reproducing molecules, they and ribonucleic acids formed a replication system in these cells in which the RNA replaced clays as templates in protein synthesis and DNA subsequently became the repository for structural information of proteins. Many of these proteins were catalysts needed for the synthesis of monomers and the synthesis of polymers from them. If RNAs were the first living molecules, their enclosure in primitive cells would have led to the synthesis of proteins, many of them with catalytic functions, for which some of the RNA served as templates (Schopf, et al., 1983; Miller and Orgel, 1974; Oparin, 1938; Haldane, 1929; Joyce, 1991). DNA evolved subsequently and became a more stable repository for template and, therefore, protein structure.

YEARS BEFORE PRESENT

EON: PRECAMBRIAN

ERA: ARCHEAN $5\text{-}2.5 \times 10^9$

PERIOD: Hadean $4.5\text{ -}3.9 \times 10^9$

Early Archean $3.9\text{-}2.9 \times 10^9$

Late Archean $2.9\text{-}2.5 \times 10^9$

PROTEROZOIC $2.5\text{-}0.57 \times 10^9$

Early Proterozoic $2.5\text{-}1.6 \times 10^9$

Middle Proterozoic $1.6\text{-}0.9 \times 10^9$

Late Proterozoic $0.9\text{-}0.57 \times 10^9$

PHANEROZOIC

PALEOZOIC $570\text{-}225 \times 10^6$

Cambrian $570\text{-}500 \times 10^6$
Ordovician $500\text{-}430 \times 10^6$
Silurian $430\text{-}395 \times 10^6$
Devonian $395\text{-}345 \times 10^6$
Carboniferous $345\text{-}280 \times 10^6$
Permian $280\text{-}225 \times 10^6$

MESOZOIC $225\text{-}65 \times 10^6$

Triassic $225\text{-}190 \times 10^6$
Jurassic $190\text{-}136 \times 10^6$
Cretaceous $136\text{- }65 \times 10^6$

CENOZOIC 65×10^6 - present

Tertiary $65\text{-}1 \times 10^6$
Quaternary 1×10^6 - present

Figure 3.3 Geological time scale.

As the primitive cells evolved, they soon must have developed the traits we associate with modern **prokaryotic** cells, as suggested by micropaleontological evidence. This implies that they possessed a plasma membrane surrounded by a cell envelope and enclosing an interior that featured a large deoxyribonucleic acid strand (repository of genetic information), nucleoprotein granules (ribosomes), and other proteins and smaller polymers and monomers. In the organic soup scenario, these first cells must have been anaerobic **heterotrophs**, cells that could live without free oxygen, because the Earth was surrounded at this time by a reducing atmosphere. They depended on externally available organic molecules (at this time abiotically synthesized monomers). As Archean time progressed, the slow rate of abiotic synthesis, the gradual disappearance of conditions needed for abiotic synthesis, and the continual increase in cell populations, which up to then depended on the limited supply of abiotically produced monomers, required an evolutionary step that would make them independent of abiotic synthesis. This step was the emergence of **autotrophy**, a process in which an organism can obtain needed energy either from the oxidation of an inorganic compound (**chemosynthetic autotrophy** or **chemolithotrophy**) or from the transformation of radiant energy from sunlight into chemical energy (**photosynthetic autotrophy** or **photolithotrophy**) to convert CO_2 into a form of reduced (organic) carbon. Such autotrophs could not only satisfy completely their own needs for reduced carbon monomers from inorganic matter, but they could also feed the already existing heterotrophs by excreting reduced carbon monomers in excess of their needs, and with organic matter (reduced carbon) from the remains of any cell after its death. This would eliminate dependence on the abiotic supply of reduced carbon, which was being exhausted.

Whether these first autotrophs were chemosynthetic or photosynthetic is currently a matter of debate. One school of thought favors chemosynthetic autotrophs in the form of **methanogens**, which formed methane according to the following reaction:

$$4H_2 + CO_2 \Rightarrow CH_4 + 2H_2O \tag{3.3}$$

Such bacteria exist today. They are strict anaerobes. Since a number of the extant methanogens live at relatively high temperature (60 to $+90°C$), not unlike conditions which must have prevailed on Earth at this time (Schopf et al., 1983), and since the large majority of them can use hydrogen as an energy source, a gas thought to have been sufficiently abundant in the primordial atmosphere, they could have been the first autotrophs to evolve. Analyses by the techniques of molecular biology have shown that methanogens have a very ancient origin and that they should be grouped in a special kingdom of the **Archaebacteria** (Fox et al.,

1977, 1980; Woese and Fox, 1977). If they were the first autotrophs, they were, however, soon displaced by photosynthetic autotrophs as **primary producers**.

The other school of thought favors photosynthetic eubacterial prokaryotes as representative of the first autotrophs with equal justification. As mentioned in Section 3.1, the microfossil record suggests that **cyanobacteria** (formerly called blue-green algae) are very ancient organisms, remains of them having been found in sedimentary rock of the Middle Archean, about 3.5 billion years ago (Schopf et al., 1983; Schopf and Packer, 1987). Since modern cyanobacteria are aerobes, which usually release oxygen when photosynthesizing, and since the primordial atmosphere at this time is thought to have been reducing, they must have been preceded by anaerobic, photosynthetic bacteria like the modern purple and green bacteria which photosynthesize without producing oxygen. These latter bacteria transduce radiant energy from sunlight into chemical energy which enables them to reduce CO_2 to organic carbon with H_2 or H_2S. Elemental sulfur or sulfate are produced if H_2S is the reductant, e.g.,

$$2H_2S + CO_2 \Rightarrow (CH_2O) + H_2O + 2S \qquad (3.4)$$

Such organisms could have coped in the reducing environment existing in the Archean. The emergence of anaerobic photosynthesis required the appearance of chlorophyll, the light harvesting and energy-transducing pigment, in the process (Chapter 6).

Further evolution of the photosynthetic machinery through the development of new types of chlorophyll must have led to the emergence of the prototypes of cyanobacteria which could substitute H_2O for H_2 or H_2S as a reducing agent of CO_2 leading to the introduction of oxygen into the atmosphere:

$$H_2O + CO_2 \Rightarrow (CH_2O) + O_2 \qquad (3.5)$$

Indeed, a few cyanobacteria are now known which have the capacity to carry out **oxygenic** photosynthesis in air and **anoxygenic** photosynthesis in the absence of oxygen when H_2S is present. In the latter instance they photosynthesize like purple or green bacteria, using the H_2S rather than water to reduce fixed CO_2 (Cohen et al., 1975). Assuming that the interpretation by Schopf and Packer (1987) that the microfossils they found in 3.3 to 3.5 billion year-old chert from the Warrawoona Group represent cyanobacteria with a capacity for oxygenic photosynthesis is correct, photosynthetic oxygen evolution could have first occurred more than 3.5 billion years ago. Buick (1992) has inferred from stromatolites in the Tumbiana Formation of the late Archean located in Western Australia that oxygenic photosynthesis occurred about 2.7 billion years ago.

As the cyanobacteria achieved dominance in the Proterozoic as carbon-fixing and oxygen-evolving microorganisms, some filamentous forms among them tended to aggregate in the form of mats which trapped siliceous and carbonaceous sediment that in many instances contributed to their ultimate preservation by silicification in the geologic record. Environmental conditions at this time seemed to favor the mat-forming growth habit. Continental emergence, development of shallow seas, climatic and atmospheric changes resulting from oxygenic photosynthesis probably exerted selective pressure favoring this growth habit of cyanobacteria (Knoll and Awramik, 1983). Most microfossil finds representing this period have been stromatolites, possibly because they are among the more easily recognized microfossils. Since unicellular or inconspicuous multicellular microfossils would be much harder to find and identify, one should not draw the conclusion that mat-forming cyanobacteria were necessarily the only common form of life at this time.

Initially the oxygen that was evolved in oxygenic photosysnthesis probably reacted with oxidizable inorganic matter such as iron [Fe(II)], forming iron oxides such as magnetite (Fe_3O_4) and hematite (Fe_2O_3) (see Chapter 14). But eventually, about 2.3 billion years ago (Schopf, 1978), free oxygen began to accumulate in the atmosphere, gradually changing it from a reducing to an oxidizing one. As the atmospheric oxygen concentration increased, some organisms began to evolve biochemical catalysts and molecules capable of electron transfer that included **cytochromes**, which are proteins containing iron porphyrins that could have arisen from mutation of the magnesium porphyrins of the chlorophylls of photosynthetic prokarykotes. The cytochromes were incorporated into electron transport systems in the plasma membrane, thereby enabling these cells to dispose of excess reducing power (electrons) to molecular oxygen instead of partially reduced organic compounds (see Chapter 6). The cytochrome system made possible a reaction sequence of discrete steps (**respiratory chain**), some of which became coupled to energy transduction. In such a process, some chemical energy is trapped in special chemical (anhydride) bonds for subsequent utilization in energy-requiring reactions (syntheses, polymerizations). Since biochemical one-step reduction of oxygen leads to the formation of very toxic superoxide radicals (O_2^-) (Fridovich, 1977), the oxygen-utilizing organisms also had to evolve a special protective system against them. Such a system included **superoxide dismutase** and **catalase**, metalloenzymes that together catalyze the reduction of superoxide to water and oxygen. Superoxide dismutase catalyzes

$$2O_2^- + 2H^+ \Rightarrow H_2O_2 + O_2 \qquad (3.6)$$

and catalase catalyzes

$$2H_2O_2 \Rightarrow 2H_2O + O_2 \tag{3.7}$$

Peroxidase may replace catalase as the enzyme for disposing of the toxic hydrogen peroxide,

$$RH_2 + H_2O_2 \Rightarrow 2H_2O + R \tag{3.8}$$

where RH_2 represents an oxidizable organic molecule.

Schopf et al. (1983) suggest that the first oxygen utilizing prokaryotes were **amphiaerobic**; i.e., they retained the ability to live anaerobically even though they had acquired the ability to metabolize aerobically. (Present-day amphiaerobes are called **facultative** organisms.) From them, at a time before 2 billion years ago, the obligate aerobes evolved according to Schopf et al. (1983). Towe (1990) has proposed that aerobes could have first appeared in the Early Archean. In any case, they became well established only 1.5 to 1.7 billion years ago. However, the evolutionary sequence was probably more complex, since present-day facultative organisms include some which use nitrate, ferric iron, manganese oxides, and some other oxidized inorganic species as terminal electron acceptors that could not have occurred in significant quantities before the atmosphere became oxidizing. Amphiaerobes that evolved in the late Archean to Early Proterozoic are best defined as organisms which had the capacities to respire aerobically in presence of oxygen or to ferment in its absence. Two exceptions are the methanogens that appeared much earlier, which, though strict anaerobes, respired anaerobically using CO_2 as terminal electron acceptor, and sulfate reducers, obligate anaerobes that respire on SO_4^{2-}.

Some new strict anaerobes probably evolved subsequent to the appearance of oxygen in the atmosphere. These were organisms with a capacity to respire anaerobically. The reason for this evolutionary development was that the atmospheric oxygen must have led to extensive accumulation of oxidized inorganic species such as sulfate and those mentioned above, which could serve as terminal electron acceptors in a respiratory process. Eubacterial sulfate reducers have been viewed as having made their first appearance in the Late Archean (Ripley and Nicol, 1981). They have been important ever since their first appearance for the reductive segment of the sulfur cycle in the mesophilic temperature range (~ 15–$40°C$) (Schidlowski et al., 1983). At ambient temperature and pressure, bacterial sulfate reduction is the only process whereby sulfate can be reduced to hydrogen sulfide. The likely existence of a group of extremely thermophilic, archaebacterial sulfate reducers in the Early Archean was recently predicted as a result of the isolation of living cultures of such organisms from marine hydrothermal systems in Italy. These isolates grew under laboratory conditions in a temperature range of 64° to 92°C with an optimum near 83°C. They were found to reduce sulfate, thiosulfate and sulfite with H_2, formate, formamide, lactate, and pyruvate (Stetter et al., 1987). According to their discoverers, these types of bacteria may have inhabited Early Archaean hydrothermal systems containing significant amounts of sulfate of magmatic origin.

The accumulation of oxygen in the atmosphere must also have led to the buildup of an ozone (O_3) shield which screened out UV components of sunlight. This would have stopped any abiotic synthesis dependent on UV radiation, and at the same time would have allowed the emergence of life forms onto the land surfaces of the planet, which were directly exposed to sunlight. This emergence would have been impossible before because of the lethality of UV radiation (Schopf et al., 1983).

With the appearance of oxygen-producing cyanobacteria and aerobic heterotrophs, the stage was set for cellular compartmentalization of such vital processes as photosynthesis and respiration. New types of photosynthetic cells evolved in which photosynthesis was carried on in special organelles, the **chloroplasts**, and new types of respiring cells evolved in which respiration was carried on in other special organelles, the **mitochondria**. It is very likely that these organelles arose by **endosymbiosis**, a process in which primitive cells incapable of photosynthesis or respiration were invaded by cyanobacteria (e.g., **Prochloron**, which contains both chlorophyll a and b like most chloroplasts, whereas other cyanobacteria contain only chlorophyll a) and/or by aerobically respiring bacteria which established a permanent symbiosis with their host (see the discussion by Margulis, 1970; Dodson, 1979). Eventually, the relationship of some of the host cells with their endosymbionts became one of absolute interdependence, the endosymbionts having lost their capacity for an independent existence. The result was the appearance of **eukaryotic** cells about 1 billion years ago or a little earlier (Schopf et al., 1983). They could even have appeared as early as 2.1 billion years ago (Han and Runnegar, 1992), which would change the date for the initial buildup of oxygen in the atmosphere. The organization of these cells contrasts with the less-compartmentalized **prokaryotic** cells (eubacteria including cyanobacteria, and archaebacteria).

Supporting evidence for a probable endosymbiotic origin of chloroplasts and mitochondria is the discovery in them of genetic substance—deoxyribonucleic acid—and of cell particles called **ribosomes**, each of a type that is otherwise found only in prokaryotes. Furthermore, in the case of chloroplasts, molecular biological analytical comparison of the highly conserved 16S ribonucleic acid fraction of some cyanobacterial ribosomes and chloroplast ribosomes has revealed close relatedness (Giovannoni et al., 1988). The emergence of eukaryotic cells was evidently needed for the evolution of more complex forms of life, such as the protozoa, fungi, plants, and animals.

The Surface Metabolism Scenario   If surface metabolism as described by Waechtershaeuser (1988) preceded cellular metabolism, the origin-of-life scenario based on the organic soup theory must be extensively revised. Surface metabolism, as described by Waechtershaesuer (1988), is an autocatalytic process that does not involve enzymes or templates. Any dissolved organic molecules synthesized abiotically with heat, electrical, or radiant energy could not have played any role in the surface processes. All organic molecules of the surface metabolists were the result of interactions between inorganic species on positively charged mineral surfaces. Waechtershaeuser (1988) favors pyrite (FeS_2) surfaces. The metabolism was **autotrophic** in that the starting molecules were CO_2, CO, NH_3 (or N_2 after emergence of nitrogen fixation), H_2S, H_2, and H_2O. Detachment of metabolists from the positively charged mineral surface took place after they became enveloped in a lipid membrane, also formed by them on the positively charged mineral surface. Each membrane enclosed a cytosol in which enzymes and templates for replicating them appeared gradually with the development of a genetic apparatus. But until the advent of enzymes and templates in these membrane vesicles, surface metabolism was still central to continued life. As before, it took place on a mineral grain with a positively charged surface, but now located in the cytosol. The grains were still very likely pyrite, now formed in the energy metabolism involving ferrous iron and hydrogen sulfide (see section 3.1, equation 3.1). After the appearance of enzymes in the cytosol, cells gradually dispensed with surface metabolism on the intracellular mineral particle, probably by shifting to energy-yielding reactions which did not form the particles (see below).

The cytosol and the appearance of enzymes in it made possible for the first time the utilization of accumulated, surface-detached reaction products (salvaging action according to Waechtershaeuser, 1988) that were not needed in biosynthesis. Surface metabolists had been completely unable to use these products. Since the first cells were autotrophs, this represented a form of **autotrophic catabolism**. The evolution of a membrane-bound respiratory system, which followed, was a consequence of the development of the cell membrane. According to Waechtershaeuser (1988), the respiratory chain liberated organisms which developed it from using the reaction of ferrous iron with H_2S as a source of energy and reducing power by enabling them to use reactions such as one in which H_2

reduced elemental S° or sulfate to hydrogen sulfide, instead. Autotrophic catabolism is also seen as the basis for the emergence of substrate-level phosphorylation (see Chapter 6) as an alternate means of conserving energy. Heterotrophy is believed by Waechtershaeuser (1988) to have evolved from autotrophic catabolism in some types of cells which developed a transport mechanism for importing dissolved organic molecules from the surrounding environment. Waechtershaeuser believes that this occurred at a late stage in the evolution of early cells. The first prokaryotic anoxygenic photosynthesizers probably appeared some time before the first heterotrophs appeared. Thus, in the surface metabolism scenario, autotrophy preceded heterotrophy, and anaerobic chemosynthetic autotrophy preceded anoxygenic, photosynthetic autotrophy.

3.3   THE EVIDENCE

The scenarios for the origin and early evolution of life outlined in the previous section is at best based on educated guesses. The scenario for the evolution of cellular life was constructed in part on the basis of observations in the geologic record and in part from comparisons between fossilized microorganisms and their present-day counterparts. The geologic record has contributed evidence in the form of microfossils and geochemical data. Morphological similarities between microfossils and certain present-day organisms as well as some geochemical data permit inferences to be drawn concerning likely physiological and biochemical activities of fossilized organisms when they were alive. Molecular biological analysis of present-day organisms has permitted the construction of evolutionary trees which reflect biochemical evolution, using as a basis the conservation of certain genetic information over geologic periods of time (Woese, 1987; Olsen et al., 1994).

The identification of Precambrian microfossils is difficult because structures which appear to be fossilized microorganisms may actually be modern contaminants or abiotically formed structures resembling microfossils in appearance. A true microfossil should meet the following criteria (Schopf and Walter, 1983): (a) the sedimentary rock in which it was found must be of a scientifically established age; (b) the purported fossil must be indigenous to the rock sample and not a modern contaminant; (c) the purported fossil must have been formed at the same time as the enclosing rock, i.e., **syngenetically**; and (d) the purported fossil had a truly biogenic in origin. On the basis of these criteria, some previously identified microfossils have now been rejected as bona fide and reclassified as **dubiomicrofossils** (uncertainty about biogenicity), **nonfossils** (may be present-day microbes which invaded a particular rock in situ or contaminants introduced while sampling rock; or may be abiotic structures resembling a fossil). Walter and Schopf (1983) and Hofmann and Schopf (1983) have reclassified Precambrian microfossil finds before 1983 in the above categories based on the criteria of Schopf and Walter (1983).

Microfossils arose as a result of entrapment in siliceous sediment. This was followed by impregnation of cell structures like those of cyanobacteria with dissolved silica from rock diagenesis. Subsequent dewatering under conditions of moderately elevated temperature and pressure resulted in precipitation of silica in the cells. This process has been reproduced in the laboratory on a time scale obviously compressed over many orders of magnitude. (Oehler and Schopf, 1971). Fossilization of bacteria may also have involved concentration of certain metallic ion species in their cell envelope and subsequent crystallization of a specific sulfide, phosphate, oxide, carbonate, silicate or other mineral from the accumulated metal (Beveridge et al., 1983; Ferris et al., 1986). Observations by Ferris et al. (1986) suggest that microfossils formed as a result of mineral precipitation were probably best preserved if they had been previously embedded by in a fibrous silica matrix (**premineralization**). Growth of crystals of metal containing minerals would otherwise have caused rupture of the cells. Microfossils are visualized by thin-sectioning of sedimentary rock containing them and examining such sections, after suitable treatment, by light and/or electron microscopy.

Bona fide Archean microfossils are represented by Warrawoona specimens from the North Pole Dome region of the Pilbara block in Western Australia (~3.5 billion years old) (Fig.3.1), in which have been found examples of "filamentous bacteria" (see Schopf and Walter, 1983) and microfossiliferous stromatolites (Walter, 1983). Unicell-like spheroids which are currently classed as dubiofossils have also been seen in these specimens (see Schopf and Walter, 1983). Spheroid structures of an age similar to the Warrawoona specimens, classed as dubiofossils, have been found in the Onverwacht Group of the Swaziland System, eastern Transvaal, South Africa. Filamentous fossil bacteria, ~2.7 billion years old, have been identified in specimens of the Fortescue Group, Tumbiana Formation in Western Australia (see Schopf and Walter, 1983).

Bona fide Proterozoic microfossils have been identified in greater numbers than Archean microfossils. They include filamentous organisms named *Gunflintia minuta* Barghoorn in Pokegama Quartzite (1.8–2.1 billion years old), coccoid, septate filamentous, and tubular unbranched, and budding bacteria-like microfossils in the Gunflint formation of north-

western Ontario, Canada (1.8–2.1 billion years old), coccoid, septate filamentous, and tubular, unbranched microfossils in the Tyler Formation of
Gogebic County, northern Michigan, U.S.A. (1.6–2.5 billion years old),
spheroidal, planktonic organic walled microfossils in the Krivoirog Series
of the Ukranian Shield, Ukraine, and a number of other examples from
other parts of the world (Fig. 3.4). Many of these microfossils were associated with stromatolites, which abounded in the Proterozoic (1.7–0.57 bil-
lions years ago) (Krylov and Semikhatov, 1976). These stromatolites were
formed chiefly by cyanobacteria. Their abundance in the Proterozoic is
attributed to environmental conditions that favored formation of cyanobacterial mats and in which organisms capable of grazing on such mats
were absent (Knoll and Awramik, 1983). With the emergence of grazers
in the Phanerozoic, stromatolites became rare and have remained so
to the present day. Modern stromatolites are confined to very special
locations.

The finding of microfossils in Precambrian sedimentary rock formations is evidence of the presence of life at that time. Because of the resemblance of some of these microfossils to present-day cyanobacteria, inferences can be drawn concerning the physiology and biochemistry of these
microfossils. It can be inferred, for instance, that photosynthesis, possibly
oxygenic photosynthesis, occurred at the time corresponding to the geological age of these particular fossils. On the other hand, if independent
geochemical data indicate that a reducing atmosphere prevailed at the
time, the fossils may represent anoxygenic precursors of oxygenic cyanobacteria, or they may represent the time of emergence of oxygenic cyanobacteria.

No microfossils of the very earliest life forms have been found, nor
are they likely ever to be found on Earth because of the weathering or
diagenetic processes to which sedimentary rock on our planet has been
subjected from the start and to which it is continuing to be subjected. This
weathering was originally a physicochemical process involving water and
various reactive substances in the planet's atmosphere. With the emergence of prokaryotic life, microbes also became important agents of
weathering. Their present-day weathering activities will be discussed in
Chapters 4 and 9.

Geochemical studies of Precambrian rocks can also tell us something
about early life. For instance, measurements of stable isotope ratios of
major elements important to life, namely C, H, N, and S in appropriate
inorganic or organic samples can give an indication if a biological agent
was involved in their formation or transformation (Schidlowski et al.,
1983). Such interpretation rests on observations that some present-day

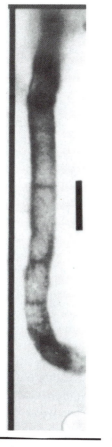

Figure 3.4 *Gunflintia*? sp.: a septate filament with unusually elongate cells, preserved in dark-brown organic matter from Duck Creek Dolomite, Mount Stuart area, Western Australia: ~2.02 billion years old. Scale mark on right is 10 μm. (Reproduced from photo 14-3 J in The Earth's Earliest Biosphere: Its Origin and Evolution, J. W. Schopf, ed., Princeton University Press, 1983, with publisher's permission.)

microrganisms can discriminate among stable isotopes by metabolizing the lighter species faster than the heavier species. For instance, they attack ^{12}C more readily than ^{13}C, hydrogen more readily than deuterium, ^{14}N than ^{15}N, or ^{32}S than ^{34}S. Abiotic reactions do not discriminate among stable isotopes in this way. As a result, products of such fractionation reactions will show an enrichment in respect to the lighter isotope whereas residual substrates will show an enrichment in the heavier isotope in initial stages of a reaction in a closed system or in an open system with low rate of substrate consumption (Chapter 6). Carbon isotope study of many sediment samples of Archean age indicate that life played a dominant role in the carbon cycle as far back as 3.5 billion years ago (Schidlowski et al., 1983). Sulfur isotopic data for Early Archean sediments indicate the likely activity of photosynthetic sulfur bacteria [barites ($BaSO_4$) of this time were only slightly enriched in ^{34}S compared to sulfides from the same sequence, Schidlowski et al., 1983]. They also suggest that sulfate respiration by prokaryotes may not have occurred to a significant extent before 2.7 billion years ago. Some more recent geochemical evidence indicates that some sulfate reduction occurred as long ago as 3.4 billion years ago (Ohmoto et al., 1993).

Organic geochemistry provides another approach to seeking clues to early life on Earth. Organic matter trapped in sediment subjected to abiological transformation due to heat and pressure may be transformed into products which are stable in situ over geologic time. Organic matter which underwent this kind of transformation is likely to have been in a form which was not rapidly degraded by biological means as are carbohydrates, nucleic acids and nucleotides, and most proteins. Nevertheless, amino acids, fatty acids, porphyrins, n-alkanes and isoprenoid hydrocarbons have been identified in sediments of Archean age (Kvenvolden cited by Schopf, 1977; Hodgson and Whiteley, 1980). If the source compound of any stable organic product identified in an ancient sample of sedimentary rock is known, the latter can be used as an indicator or "biological marker" of the source compound. If the source compound such as porphyrin is a key compound in a particular physiologic process, it indicates that photosynthesis or respiration or both were occurring when the source material was trapped in sediment. **Kerogen** is an example of a stabilized substance formed from ancient organic matter. Finding it in an ancient sedimentary rock indicates the existence of life contemporaneous with the age of that rock.

Studies in molecular biology have revealed that the proportion and sequence of certain monomers in some bioheteropolymers such as ribosomal RNA's are highly conserved in different organisms, i.e., they have not become detectably modified over very long times. Such conserved sequences can be used to study the degree of relatedness among different organisms and also the relative times of divergence in the evolution of groups of organisms (e.g., see Fox et al., 1980; Woese, 1987; Olsen et al., 1994). These studies have led to the conclusion that Archaebacteria and

Eubacteria diverged early in Archean times from a common prokaryotic ancestor and have evolved ever since along independent parallel lines. They also indicate that gram-positive bacteria had most likely a photosynthetic ancestry (Woese et al., 1985). The fact that certain physiological processes such as protein synthesis, energy conservation by chemiosmosis, and some biodegradative as well as biosynthetic pathways are held in common by these two groups (kingdoms) but differ in some details, suggests that these pathways may have existed in the common ancestor but became modified during divergent evolution. Convergent evolution cannot, however, be ruled out at present in all cases.

Combining several lines of paleontological evidence can lend strong support for a model of an ancient biological process and/or microbe responsible for it. For instance, Summons and Powell (1986) found in a certain Canadian petroleum deposit of Silurian age (ca. 400×10^6 ago) the presence of (1) characteristic biological markers indicating an ancient presence of aromatic carotenoids from green sulfur bacteria (Chlorobiaceae) and (2) enrichment of ^{13}C in these markers relative to the saturated oils. Relating these findings to the paleoenvironmental setting of the oil deposit, the investigators deduced that microbial communities that included Chlorobiaceae must have existed in the ancient restricted sea in which the source material from which the oil was derived was emplaced.

3.4 SUMMARY

The Earth is about 4.6 eons old. Primitive life probably arose *de novo*, first appearing 0.5 to 1 eon after formation of the planet. Initially it was surrounded by a reducing atmosphere until oxygen-generating, photosynthetic microorganisms that became the progenitors of cyanobacteria caused a net accumulation of free oxygen, which began about 2.3 eons ago. Thus, the earliest forms of cellular life were anaerobic prokaryotes. Aerobic prokaryotic forms did not evolve until free oxygen began to accumulate in the atmosphere, and eukaryotic forms did not appear until the atmosphere had become fully oxidizing, by about 1 eon ago. The evolutionary sequence of prokaryotes in terms of physiological types according to the organic soup scenario started with fermenting heterotrophs and progressed with the development of anaerobic photo- and chemoautotrophs, some anaerobic respirers, oxygenic photoautotrophs, aerobically respiring heterotrophs and autotrophs and other anaerobic respirers. Eukaryotes evolved by endosymbiosis involving anaerobic heterotrophic prokaryotes as host cells and oxygenically photosynthetic and aerobically

<u>EVENT</u>	<u>YEARS BEFORE PRESENT</u>
A. *Organic Soup Scenario* :	
Origin of the Earth	4.6×10^9
First self-reproducing molecules	$4.3\text{-}4.0 \times 10^9$ (?)
First primitive heterotrophic cells	$4.0\text{-}3.8 \times 10^9$ (?)
First autotrophs (methanogens and/or anoxygenic photosynthesizers)	$\sim 3.8 \times 10^9$ (?)
Warrawoona stromatolite	$\sim 3.5 \times 10^9$
First oxygenic photosynthesizers	$3.5\text{-}3.0 \times 10^9$
First anaerobic respirers (sulfate-reducers)	$\sim 2.7 \times 10^9$
Fully oxidizing atmosphere	$\sim 2.1 \times 10^9$
First aerobic respirers	$\sim 2.0 \times 10^9$
First eukaryotic cells	$1.4\text{-}0.85 \times 10^9$

Figure 3.5 Milestones in Precambrian evolution of life: (A) organic soup scenario; (B) surface metabolism scenario. Dates followed by (?) represent guesses without any palentological backup. Their sequence is based on the evolutionary description of Waechtershaeuser (1988).

respiring prokaryotes as intracellular symbionts. According to the surface metabolism theory, the sequence of emergence of prokaryotic physiological types was autotrophic surface metabolists, semi-cellular surface metabolists, membrane-bound, detached chemosynthetically autotrophic cells, anoxygenic photoautotrophic cells, anaerobic fermenting and respiring heterotrophs, oxygenic photosynthetic autotrophs, and aerobically respiring heterotrophs, followed by the emergence of eukaryotes. Major steps in Precambrian evolution according to the organic soup theory and the surface metabolism theory are summarized in Fig. 3.5.

Evidence supporting the currently held view of how cellular life evolved on Earth derives from the interpretation of the microfossil record

B. *Surface Metabolism Scenario:* [a]

Origin of the Earth	4.6×10^9
First surface metabolists (chemo-autotrophic)	$\sim 4.3 \times 10^9$ (?)
First primitive cells (chemo-autotrophic)	$\sim 4.1 \times 10^9$ (?)
First anaerobic S & SO_4 respirers (autotrophic)	$\sim 4.0 \times 10^9$ (?)
First photo-autotrophs	$\sim 3.8 \times 10^9$ (?)
First heterotrophs	$\sim 3.5 \times 10^9$ (?)
Warrawoona stromatolite	$\sim 3.5 \times 10^9$ (?)
First oxygenic photosynthesizers	$\sim 3.5\text{-}3.0 \times 10^9$
Fully oxidizing atmosphere	$\sim 2.1 \times 10^9$
First aerobic respirers	$\sim 2.0 \times 10^9$
First eukaryotic cells	$1.4\text{-}0.85 \times 10^9$

Figure 3.5 Continued.

in geologically dated sedimentary rock, inorganic and organic geochemical studies of Precambrian rocks, and molecular biological analysis of highly conserved polymers in living cells such as some nucleic acids and proteins.

REFERENCES

Bada, J. L., C. Bingham, and S. L. Miller. 1994. Impact melting of frozen oceans on the early Earth: Implications for the origin of life. Proc. Natl. Acad. Sci. USA 91:1248–1250.

Beveridge, T. J., J. D. Meloche, W. S. Fyfe, and R. G. E. Murray. 1983. Diagenesis of metals chemically complexed to bacteria: Laboratory formulation of metal phosphates, sulfides, and organic condensates in artificial sediments. Appl. Environ. Microbiol. 45:1094–1108.

Buick, R. 1992. The antiquity of oxygenic photosynthesis: Evidence from stromatolites in sulfate-deficient Archean lakes. Science 255:74–77.

Cairns-Smith, A. G., and H. Hartman, eds. 1986. Clay Minerals and the Origin of Life. Cambridge University Press, Cambridge, 193 pp.

Cech, T. R. 1986. A model for the RNA-catalyzed replication of RNA. Proc. Natl. Acad. Sci. USA 83:4360–4363.

Chakrabarti, A. C., R. R. Breaker, G. F. Joyce, and D. W. Deamer. 1994. Production of RNA by a polymerase protein encapsulated within phospholipid vesicles. J. Molec. Evol. 39:555–559.

Chang, S., D. DesMarais, R. Mack, S. L. Miller, and G. E. Strathearn. 1983. Prebiotic organic syntheses and the origin of life. *In*: J. W. Schopf, ed. The Earth's Earliest Biosphere. Its Origin and Evolution. Princeton University Press, Princeton, NJ, pp. 53–92.

Cohen, Y., E. Padan, and M. Shilo. 1975. Facultative anoxygenic photosynthesis in the cyanobacterium *Oscillatoria limnetica*. J. Bacteriol. 123:855–861.

Cohen, J. 1995. Getting all turned around over the origins of life on Earth. Science 267:1265–1266.

Dodson, E. O. 1979. Crossing the procaryote-eucaryote border: Endosymbiosis or continuous development. Can. J. Microbiol. 25: 651–674.

Doudna, J. A., and J. W. Szostak. 1989. RNA-catalyzed synthesis of complementary-strand RNA. Nature (Lond.) 339:519–522.

Drobner, E., H. Huber, G. Waechtershaeuser, D. Rose, and K. O. Stetter. 1990. Pyrite formation linked with hydrogen evolution under anaerobic conditions. Nature (Lond.) 346:742–744.

Ernst, W. G. 1983. The early Earth and the Archean rock record. pp. 41–52. *In*: J. W. Schopf, ed. Earth's Earliest Biosphere. Its Origin and Evolution. Princeton University Press, Princeton, NJ.

Ferris, F. G., T. J. Beveridge, and W. S. Fyfe. 1986. Iron-silica crystallite nucleation by bacteria in a geothermal sediment. Nature (Lond.) 320: 609–611.

Fox, G. E., L. J. Magrum, W. E. Balch, R. S. Wolfe, and C. R. Woese. 1977. Classification of methanogenic bacteria by 16S ribosomal RNA characterization. Proc. Natl. Acad. Sci. USA 74:4537–4541.

Fox, G. E., E. Stackebrandt, R. B. Hespell, J. Gibson, J. Maniloff, T. A. Dyer, R. S. Wolfe, W. E. Balch, R. S. Tanner, L. J. Magrum, L. B. Zablen, R. Blakemore, R. Gupta, L. Bonen, B. J. Lewis, D. A. Stahl, K. R. Luehrsen, K. N. Chen, and C. R. Woese. 1980. The phylogeny of prokaryotes. Science 209:457–463.

Fridovich, I. 1977. Oxygen is toxic! BioScience 27:462–466.

Giovannoni, S. J., S. Turner, G. J. Olsen, S. Barns, D. J. Lane, and N. R. Pace. 1988. Evolutionary relationships among cyanobacteria and green chloroplasts. J. Bacteriol. 170:3584–3592.

Haldane, J. B. S. 1929. The origin of life. Rationalist Annu. 1929:148–169. (Reprinted in John Maynard Smite, ed. On Being the Right Size and Other Essays. Oxford University Press, Oxford, U.K., pp. 101–112.)

Han, T-M., and B. Runnegar. 1992. Megascopic eukaryotic algae from the 2.1-billion-year-old Negaunee iron-formation, Michigan. Science 257:232–235.

Harrison, J. E., and Z. E. Peterman. 1980. North American Commission on Stratigraphic Nomenclature Note 52—A preliminary proposal for a chronometric scale for the Precambrian of the United States and Mexico. Geol. Soc. Am. Bull. Part 1 91:377–380.

Hodgson, G. W., and C. G. Whiteley. 1980. The universe of porphyrins. pp. 35–44. *In*: P. A. Trudinger, M. R. Walter, and B. J. Ralph, eds. Biogeochemistry of Ancient and Modern Environments. Australian Academy of Science, Canberra, and Springer-Verlag, Berlin.

Hofman, H. J., and J. W. Schopf. 1983. Early Proterozoic microfossils. *In*: J. W. Schopf, ed. Earth's Earliest Biosphere. Its Origin and Evolution. Princeton University Press, Princeton, NJ, pp. 321–360.

Joyce, G. F. 1991. The rise and fall of the RNA world. New Biol. 3: 399–407.

Knoll, A. H., and S. M. Awramik. 1983. Ancient microbial ecosystems. pp. 287–315. *In*: W. E. Krumbein, ed. Microbial Geochemistry. Blackwell Scientific Publications, Oxford.

Krylov, I. N., and M. A. Semikhatov, 1983 Appendix II. Table of time ranges of the principal groups of Precambrian stromatolites. *In*: M. R. Walter, ed. Stromatolites. Elsevier Scientific Publishing Company, Amsterdam, pp. 693–694.

Lowe, D. R. 1980. Stromatolites 3,400 myr old from the Archaean of Western Australia. Nature (Lond.) 284:441–443.

Margulis, L. 1970. Origin of Eukaryotic Cells. Yale University Press, New Haven, CT.

Miller, S. L., and L. E. Orgel. 1974. The Origins of Life on the Earth. Prentice-Hall, Englewood Cliffs, NJ.

Oehler, J. H., and J. W. Schopf. 1971. Artificial microfossils: Experimental studies of permineralization of blue-green algae in silica. Science 174:1229–1231.

Ohmoto,, H., T. Kakegawa, and D. R. Lowe. 1993. 3.4 billion-year-old biogenic pyrites from Barberton, South Africa: sulfur isotope evidence. Science 262:555–557.

Olsen, G. J., C. R. Woese, and R. Overbeek. 1994. The winds of evolutionary change: Breathing new life into microbiology. J. Bacteriol. 176: 1–6.

Oparin, A. I. 1938. The Origin of Life. Macmillan, New York. (Reprinted by Dover Publications, Inc., New York, 1953.)

Orpen, J. L., and J. F. Wilson. 1981. Stromatolites at ~3,500 Myr and a greenstone-granite unconformity in the Zimbabwean Archaen. Nature (Lond.) 291:218–220.

Ripley, E. M., and D. L. Nicol. 1981. Sulfur isotopic studies of Archean slate and graywacke from northern Minnesota: Evidence for the existence of sulfate reducing bacteria. Geochim. Cosmochim. Acta 45: 839–846.

Schidlowski, M., J. M. Hayes, and I. R. Kaplan. 1983. Isotopic inferences of ancient biochemistries: Carbon, sulfur, hydrogen, and nitrogen. *In*: J. W. Schopf, ed. Earth's Earliest Biosphere. Its Origin and Evolution. Princeton University Press, Princeton, NJ, pp. 149–186.

Schopf, J. W. 1977. Evidences of Archean life. *In*: C. Ponnamperuma, ed. Chemical Evolution of the Early Precambrian. Academic Press, New York, pp. 101–105.

Schopf, J. W. 1978. The evolution of the earliest cells. Am. Sci. 238: 110–139.

Schopf, J. W., and B. M. Packer. 1987. Early Archean (3.3-billion-year-old) mirofossils from Warrawoona Group, Australia. Science 237: 70–73.

Schopf. J. W., and M. R. Walter. 1983. Archaean microfossils: New evidence of ancient microbes. *In*: J. W. Schopf, ed. Earth's Earliest Biosphere. Its Origin and Evolution. Princeton University Press, Princeton, NJ, pp. 214–239.

Schopf, J. W., J. M. Hayes, and M. R. Walter. 1983 Evolution of Earth's earliest ecosystems: recent progress and unsolved problems. *In*: J.W. Schopf, ed. Earth's Earliest Biosphere. Its Origin and Evolution. Princeton University Press, Princeton, NJ, pp. 360–384.

Stetter, K. O., G. Lauerer, M. Thomm, and A. Neuner. 1987. Isolation of extremely thermophilic sulfate reducers: Evidence for a novel branch of Archaebacteria. Science 236:822–824.

Stevenson, D. J. 1983. The nature of the Earth prior to the oldest known rock record: The Hadean Earth. *In*: J. W. Schopf, ed. Earth's Earliest Biosphere. Its Origin and Evolution. Princeton University Press, Princeton, NJ, pp. 32–40.

Summons, R. E., and T. G. Powell. 1986. Chlorobiaceae in Paleozoic seas revealed by biological markers, isotopes, and geology. Nature (Lond.) 319:763–765.

Taylor, S. R. 1992. The origin of the Earth. *In*: G. Brown, C. Hawkesworth, and C. Wilson, eds. Understanding the Earth. Cambridge University Press, Cabridge, U.K., pp. 25–43.

Towe, K. M. 1990. Aerobic respiration in the Archean. Nature (Lond.) 348:54–56.

Waechtershaeuser, G. 1988. Before enzymes and templates: Theory of surface metabolism. Microbiol. Rev. 52:452–484.

Walter, M. R. 1983. Archean stromatolites: Evidence of the Earth's earliest benthos. *In*: J. W. Schopf, ed. The Earth's Earliest Biosphere. Its Origin and Evolution. Princeton University Press, Princeton, NJ, pp. 187–213.

Walter, M. R., R. Buick, and J. S. R. Dunlop. 1980. Stromatolites 3,400–3,500 Myr old from the North Pole area, Western Australia. Nature 284:443–445.

Weber, P., and J. M. Greenberg. 1985. Can spores survive in interstellar space? Nature (Lond.) 316:403–497.

Woese, C. R. 1987. Bacterial evolution. Microbiol. Rev. 51:221–271.

Woese, C. R., and G. E. Fox. 1977. Phylogenetic structure of the prokaryotic domain: The primary kingdoms. Proc. Natl. Acad. Sci. USA 74:5088–5090.

Woese, C. R., B. A. Debrunner-Vossbrinck, H. Oyaizu, E. Stackebrandt, and W. Ludwig. 1985. Gram-positive bacteria: Possible photosynthetic ancestry. Science 229:762–765.

4

The Lithosphere as a Microbial Habitat

4.1 ROCK AND MINERALS

To understand how the lithosphere supports the existence of microbes on and in it and how microbes exert influence on the formation or transformation of its rocks and minerals, we must familiarize ourselves with some general chemical and physical features of rocks and minerals.

Geologically speaking, the term **rock** refers to massive, solid, inorganic matter consisting usually of two or more intergrown minerals. Rock may be **igneous** in origin; that is, it may arise by cooling of **magma** (molten rock material) from the interior of the Earth (crust and/or asthenosphere). The cooling may be a slow or a fast process. In slow cooling, different minerals begin to crystallize at different times, owing to their different melting points. In the process of crystallization, they intergrow and thereby evolve into rock with visually distinguishable crystals, as seen in granite, for example (Fig. 4.1). In fast cooling, rapid crystallization occurs, and rock forms which contains only tiny crystals, not visible with the naked eye. Basalt is an example of rock formed in this way.

Rock may also be **sedimentary** in origin; that is, it may arise through the accumulation and compaction of sediment that consists mainly of mineral matter derived from other rock. In other instances, it may arise as a

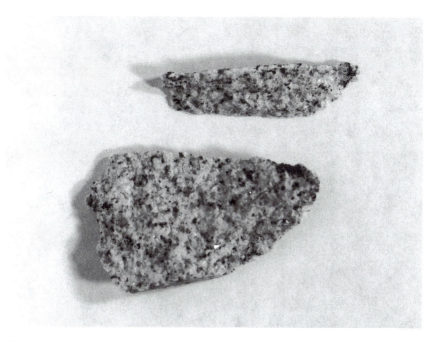

Figure 4.1 Pieces of granite showing phenocrysts, i.e., visible crystals of mineral in a fine crystalline ground mass. An igneous rock. Fragment in back is 5 cm long.

result of cementation of accumulated inorganic sediment by carbonate, silicate, aluminum oxide, ferric oxide, or combinations thereof. The cementing substance may result from microbial activity. These transformations of loose sediment into sedimentary rock are called **lithification**. Examples of sedimentary rock are limestone, sandstone, and shale. Sedimentary rock facies often exhibit a layered structure in vertical section, reflecting changes in composition as the sediment accumulated. Analysis of the different layers may tell something of the environmental conditions during which they accumulated.

Finally, rock may be of **metamorphic** origin; that is, it may be produced through alteration of igneous or sedimentary rock by the action of heat and pressure. Examples of metamorphic rock are marble, derived from limestone; slate, derived from shale; quartzite, derived from sandstone; and gneiss, derived from granitic rock.

Geochemically speaking, **minerals** are usually defined as inorganic compounds, usually crystalline but sometimes amorphous, of specific

Table 4.1   Minerals Classified as to Mode of Formation

Primary minerals
  Feldspars
  Pyroxenes and amphiboles
  Olivines
  Micas
  Silica
Secondary minerals
  Clay minerals
    Kaolinites
    Montmorillonites
    Illites
  Hydrated iron and aluminum oxides
  Carbonates

Source: Lawton (1955), p. 54ff.

chemical composition and structure. Sometimes the term "mineral" is also applied to certain organic compounds in nature, such as asphalt and coal. Inorganic minerals may be very simple in composition, as in the case of sulfur ($S°$) or quartz (SiO_2), or very complex, as in the case of the igneous mineral biotite [$K(Mg,Fe,Mn)_3AlSi_3O_{10}(OH)_2$]. Minerals that result from crystallization during the cooling of magma are **primary** or **igneous minerals**. Minerals that result from chemical alteration (weathering or diagenesis) of primary minerals are known as **secondary minerals**. Microbes play a role in this transformation of primary to secondary minerals (Chapter 9). Examples of primary and secondary mineral groups are listed in Table 4.1. Minerals may also result from precipitation from solution, in which case they are called **authigenic minerals**. Microbes may also play a role in their formation (e.g., ferromanganese concretions; see Chapter 15).

4.2   MINERAL SOIL

Origin of Mineral Soil   The mineral constituents of mineral soil ultimately derived from rock that underwent weathering. **Weathering**, that which leads to soil formation, is a process whereby rock is eroded and/or broken down into ever smaller particles and finally into constituent minerals. Some or all minerals may become chemically altered. Some forms of weathering involve physical processes. For example, freezing and thawing of water in cracks and crevices may cause expansion and hence exert pressure in rock fissures. Sand carried by wind may cause sand blasting of rock. Alternate heating by the sun's rays and cooling at night may also cause expansion and contraction of rock leading to widening of cracks and crevices. Water-borne abrasives or rock collisions may cause rock to break. Seismic activity may cause crumbling of rock. Evaporation of hard water in cracks and fissures of rock and resultant formation of crystals from the solutes may cause rock to break because the crystals occupy a larger volume than the original water solution from which they formed, thereby widening the cracks and fissures through the pressure they exert. Mere alternate wetting and drying may also cause such breakup.

Rock weathering processes may also be chemical, where the reagents are of nonbiological origin. Examples are the solvent action of water, CO_2 of volcanic origin, mineral acids such as H_2SO_3, HNO_2, and HNO_3 formed from gases of nonbiological origin such as SO_2, NO, and NO_2, respectively. Chemical weathering may also be caused by redox reagents of nonbiological origin, such as H_2S of volcanic origin or nitrate of atmospheric origin.

Finally, rock weathering may be the result of biological activity. Roots of plants penetrating cracks and fissures in rock may force it apart. Rock surfaces and the interior of porous rock are frequently inhabited by a flora of algae, fungi, lichens and bacteria. The microorganisms on the surface of rocks may exist in biofilms, especially in a moist or wet environment. In biofilms consisting of mixed microbial populations, the different organisms may arrange themselves in distinct zones where conditions are most favorable for their existence (Costerton et al., 1994). Some microorganisms, so-called boring organisms, may form cavities in limestone rock that they occupy (Golubic et al., 1975). In other cases, opportunistic microorganisms invade preformed cavities in rock (chasmolithic organisms) (Friedman, 1982). Invertebrates, snails in particular, may feed on boring organisms (Golubic and Schneider, 1979; Shachak et al., 1987) or chasmoliths by grinding away the superficial rock to expose them and then feeding on them. The rock debris which the snails generate then becomes part of a soil (Shachak et al., 1987; Jones and Shachak, 1990).

Microbes dissolve rock minerals through production of reactive metabolic products such as NH_3; HNO_3; H_2SO_4; CO_2 (forming H_2CO_3 in water); and oxalic, citric, and gluconic acids and others. Organic compounds formed by microorganisms such as lichens have been shown to cause distinct weathering in studies using scanning electron microscopy (Jones et al., 1981). Waksman and Starkey as long ago as 1931 cited the following reactions as examples of how microbes can affect weathering of minerals:

$$2KAlSi_3O_8 + 2H_2O + CO_2 \longrightarrow H_4Al_2Si_2O_9 + K_2CO_3 + 4SiO_2$$
orthoclase   kaolinite

(4.1)

$$12MgFeSiO_4 + 26H_2O + 3O_2$$
olivine

$$\longrightarrow 4H_4Mg_3Si_2O_9 + 4SiO_2 + 6Fe_2O_3{\cdot}3H_2O \quad (4.2)$$
serpentine

The first reaction is promoted by CO_2 production in the metabolism of heterotrophic microorganisms, and the second reaction is promoted by O_2 production in photosynthesis, as by cyanobacteria, algae, and lichens. Recent investigations have extended these observations (e.g., Hiebert and Bennett, 1992; Welch and Ullman, 1993) and examined possible mechanisms whereby organic acids such as those formed and excreted by microbes promote weathering of primary minerals such as feldspars and and secondary minerals such as clays (Browne and Driscoll, 1992; Lucas et al., 1993; Brady and Carroll, 1994; Oelkers et al., 1994), but general consensus on the mechanisms has not yet been reached. Some current models favor protonation as a means of displacement of cationic components from the crystal lattice and subsequent cleavage of Si-O and Al-O bonds (e.g., Berner and Holdren, 1977; Chou and Wollast, 1984). Others favor complexation, for instance, of Al and/or Si in aluminosilicates, as a primary mechanism of dissolution (Wieland and Stumm, 1992; Welch and Vandevivere, 1995).

Mineral soil may derive from aquatic sediment or **alluvium** left behind after the water which carried it from its place of origin to its final site of deposition has receded. Mineral soil can also form in place as a result of progressive weathering of parent rock and subsequent differentiation of the weathering products. Soils originating by either mechanism undergo **eluviation** (removal of some products by washing out) and/or **alluviation** (addition of some products by water transport). Any soil once formed, undergoes further gradual transformation due to biological activity which it supports (Buol et al., 1980).

Some Structural Features of Mineral Soil   Mineral soil will vary in composition, depending on the source of the parent material, the extent of weathering, the amount of organic matter introduced or generated in the soil, and the amount of moisture it holds. Its texture is affected by the particle sizes of its inorganic constituents (stones, >2 mm; sand grains, 0.05–2 mm; silt, 0.002–0.05 mm, clay particles, <0.002 mm), which determine its porosity and thus its permeability to water and gases.

Many, but not all, soils tend to be more or less obviously stratified. As many as three or four major strata or **horizons** may be recognizable in agricultural and forest soil **profiles**. A soil profile is a vertical section through soil (Fig. 4.2). The strata are called O, A, B, and C horizons. The O horizon represents the litter zone, consisting of much undecomposed and partially decomposed organic matter. It may be absent from a soil profile. The A and B horizons represent the true soil, and the C horizon represents the parent material from which the soil was formed. It may be bedrock or an earlier soil. The A and B horizons are often further subdivided, although these divisions are somewhat arbitrary. The A horizon is the biologically most active zone, containing most of the root systems of plants growing on it and the microbes and other life forms that inhabit soil. As is to be expected, the carbon content is also highest in this horizon. The biological activity in the A horizon may cause solubilization of organic and inorganic matter, some or all of which, especially the inorganic matter is carried by soil water into the B horizon. The A horizon is, therefore, known at times as the **leached layer**, and the B horizon is at times known as the **enriched** layer. Both biological and abiological factors play a role in soil profile formation.

Effects of Plants and Animals on Soil Evolution   Biological agents such as plants assist soil evolution by contributing organic matter through excretions from their root systems and as dead organic matter. The plant excretions may react directly with some soil mineral constituents, or they may be modified together with the dead plant matter by microbial activity forming products which react with soil mineral constituents. During their lifetime, plants remove some minerals from soil and contribute to water movement through the soil by water absorption via their roots and transpiration from their leaves. Their root system may also help to prevent destruction of the soil through wind and water erosion by anchoring it.

Burrowing invertebrates, from small mites to large earthworms, help to break up soil, keep it porous, and redistribute organic matter. Some of these invertebrates have a habitat which is restricted to specific regions of a soil profile.

Effects of Microbes on Soil Evolution   Microbes contribute to soil evolution by mineralizing some or all of any added organic matter during the decay process. Some of the metabolic products of this decay, such as organic and inorganic acids, CO_2, or NH_3, interact slowly with soil minerals and cause their alteration or solution, an important step in soil profile formation (Berthelin, 1977; Welch and Ullman, 1993). For instance, the mineral chlorite has been reported to be bacterially altered in this manner through loss of Fe and Mg and an increase in Si, and the mineral vermiculite has been reported to be bacterially altered through solubilization of

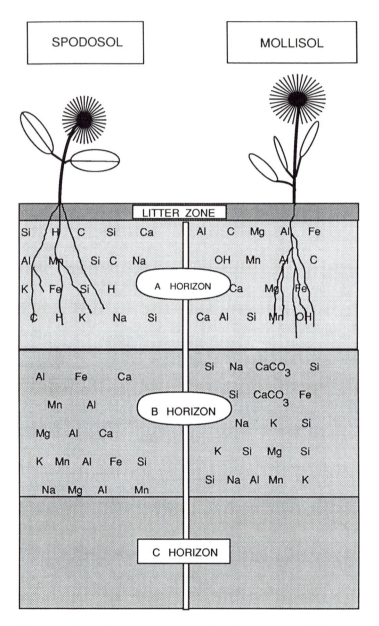

Figure 4.2 Schematic representation of the major soil horizons of spodosol and mollisol. The litter zone is also called the O horizon. The A and B horizons may be further subdivided on the basis of soil chemistry.

Si, Al, Fe, and Mg, thereby forming montmorillonite (Berthelin and Boymond, 1978). Certain microbes may interact directly (i.e., enzymatically) with certain inorganic soil constituents by oxidizing or reducing them (see Chapters 11 and 13–17), resulting in their solution or precipitation (Berthelin, 1977). Microbes may also play an important role in **humus** formation. Humus is an important constituent of soil, consisting of humic and fulvic acids as well as amino acids, lignin, amino sugars, and other compounds of biological origin (Campbell and Lees, 1967, p. 209). These constituents represent, in part, components of soil organic matter that are only slowly decomposed, and in part, products of microbial attack of the metabolizable constituents of soil organic matter. Humus gives proper texture to soil and plays a significant role in regulating the availability of those mineral elements which are important in plant nutrition and in detoxifying those which are harmful to plants by complexing them. Humus also contributes to the water-holding capacity of soil.

Effects of Water in Soil Evolution   Water from rain or melting snow may mobilize and transport some soluble soil components and cause precipitation of others. This can contribute to horizon development as the water permeates the soil. Precipitates, especially inorganic ones, may promote soil clumping. In addition, water may affect the distribution of soil gases by displacing the rather insoluble ones, such as nitrogen and oxygen, and by absorbing the more soluble ones, such as CO_2, NH_3, and H_2S.

Water Distribution in Mineral Soil   Only about 50% of the volume of mineral soil is solid matter. The other 50% is pore space occupied by water and gases such as CO_2, N_2, and O_2. As might be expected, owing to the biological activity in soil and slow gas exchange with the external atmosphere, the CO_2 concentration in the gas space in soil usually exceeds that in air, whereas the O_2 concentration is less than that in air. According to Lebedev (see Kuznetsov et al, 1963, pp. 39–41), soil water may be distributed in distinct zones among soil particles (Fig. 4.3). Surrounding a soil particle is **hygroscopic water**, a thin film 3×10^{-2} μm in thickness when surrounding a 25-mm-diameter particle. This water never freezes and never moves as a liquid. It is adsorbed by soil particles from water vapor in the soil atmosphere. In a water-saturated atmosphere, **pellicular water** surrounds the hygroscopic water. Pellicular water may move from soil particle to soil particle by intermolecular attraction but not by gravity (Fig. 4.3B). It may contain dissolved salts which may depress its freezing point to $-1.5°C$. **Gravitational water** surrounds pellicular water in Lebedev's model when moisture in excess of what the soil atmosphere can

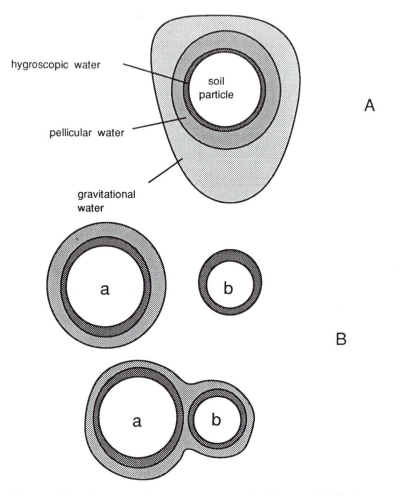

Figure 4.3 Diagrammatic representation of soil water distribution according to Lebedev. For explanation of (A) and (B) see text. (Adapted from Kuznetsov et al., 1963.)

hold is present. It moves by gravity and responds to hydrostatic pressure, unlike hygroscopic and pellicular water. So far, it is unclear which of these forms of water is available to microorganism. A reasonable guess is that gravitational water can be used by them and probably pellicular water but not hygroscopic water. The water need of microorganisms is usually studied in terms of moisture content, water activity, or water potential without regard to the form of soil water (Dommergues and Mangenot, 1970).

Water activity of a soil sample is a measure of the degree of water saturation of its vapor phase expressed in terms of relative humidity (as a fractional number; pure water has a water activity of 1.0). Except for extreme halophiles, bacteria have a higher water activity requirement (above 0.85) than many fungi (above 0.60) (Brock et al., 1984).

Water potential of soil is a measure of the difference in free energy between a soil sample and pure water at the same temperature. The more negative the water-potential value, the lower the water availability. A zero potential is equivalent to pure water. Either adsorption to surfaces (matric effect) or the presence of solutes (osmotic effect) can lower the water potential. Osmotic water potential can be calculated from the formula of Lang (1967):

$$\text{Water potential (J kg}^{-1}) = 1.322 \times \text{freezing-point depression}$$
(4.3)

where 100 J kg^{-1} is equal to 1 bar. Matric water potential requirements can be determined by the method of Harris et al. (1970). In this method, NaCl or glycerol solutions of desired water potentials, solidified with agar, are used to equilibrate matric material on which microbial growth is to occur. (For further discussion of water potential see Brock et al., 1984; Brown, 1976).

The water potential requirement for two strains of the acidophilic iron oxidizer *Thiobacillus ferrooxidans* has been determined by Brock (1975). Using NaCl as osmotic agent, strain 57–5 exhibited a minimum water potential requirement at −18 to −32 bars, whereas strain 59–1 exhibited it at −18 to −20 bars. Using glycerol as osmotic agent, strain 57–5 exhibited minimum water potential at −8.8 bars, whereas strain 59–1 exhibited it at −6 bars (Table 4.2). In the same study, it was shown that significant amounts of CO_2 were assimilated by *T. ferrooxidans* on coal refuse material with water potentials between −8 and −29 bars whereas none was assimilated when the water potential of the coal refuse was > -90 bars.

Nutrient Availability in Mineral Soil   Organic or inorganic nutrients required by soil microbes are distributed between the soil solution and the surface of mineral particles. Partitioning effects will determine their relative concentration in these two phases. Their presence on the surface of soil particles may be the result of adsorption or ion exchange. Non-ionizable molecules will tend to be adsorbed while ionizable ones will bind as

Table 4.2   Effect of Osmotic Water Potential (Glycerol) on Growth of
T. ferrooxidans[a]

Glycerol (g liter^{-1})	Total water potential (bars)	Strain 57-5	Strain 59-1
184	-61	$-$	$-$
147	-49	$-$	$-$
92	-32	$-$	$-$
74	-26	$-$	$-$
55	-20	$-$	$-$
37	-15	$-$	$-$
18.4	-8.8	$+$	$-$
9.2	-6	$+$	$+$
3.7	-4.2	$+$	$+$
0	-3	$+$	$+$

[a] All experiments were done with replicate tubes which showed the same results. The incubation period was 2 weeks. Iron concentration in the medium, 10 g of $FeSO4 \cdot 7H_2O$ per liter. $+$, Visible iron oxidation and microscopically visible growth; $-$, no iron oxidation and microscopically visible growth.
Source: Brock (1975); reproduced by permission.

a result of surface charges of opposite sign and may involve ion exchange. Binding by ion exchange is pH dependent.

Clay particles are especially important in ionic binding of organic or inorganic cationic solutes (those having a positive charge). They exhibit mostly negative charges except at their edges, where positive charges may appear. Their capacity for ion-exchange depends on their crystal structure. The partitioning of solutes between soil solution and mineral surfaces often results in their greater concentration on mineral surfaces than in the soil solution, and, as a result, the mineral surfaces may be the preferred habitat of soil microbes that require these solutes in more concentrated form. On the other hand, ionically bound solutes, on clay particles for example, may be less available to soil microbes because the microbes may not be able to dislodge them from the clay surface. In that instance soil solution may be the preferred habitat for microbes that have a major requirement for such solutes. Ionic binding may be beneficial if a solute subject to such binding is toxic (see Chapter 9).

Some Major Soil Types   Distinctive soil types may be identified by and correlated with climatic conditions and with the vegetation they support (Bunting, 1967; Buol et al., 1980). Climatic conditions determine the kind of vegetation that may develop. Thus, in the high northern latitudes, **tun-**

dra soil prevails, which in that cold climate is often frozen and therefore supports only limited plant and microbial development. It has a poorly developed profile. It may be slightly acid to slightly alkaline. Examples of tundra soil are Arctic Brown Soil and Bog Soil. In the cool (i.e., temperate), humid zones at midlatitudes, **spodosols** (Figs. 4.2, 4.4) prevail, which support extensive forests, particularly of the coniferous type. Spodosols tend to be acidic, having a strongly leached, grayish A horizon depleted in colloids, iron, aluminum, and a brown B horizon enriched in the iron, aluminum, and colloids leached from the A horizon. In regions of moderate rainfall in temperate climates at midlatitudes, **mollisols** (Figs. 4.2, 4.5) prevail. These are soils that support grasslands (i.e., they are prairie soils). They exhibit rich black topsoil and show lime accumulation in the B horizon because they have a neutral to alkaline pH. **Oxisols** are found at low latitudes in tropical, humid climates. They are poorly zonated jungle soils. Owing to the hot, humid climatic conditions under which they exist, these soils are intensely active microbiologically and require constant replenishment of organic matter by the vegetation and from animal excretions and remains to stay fertile. The neutral-to-alkaline pH conditions of jungle soils promote leaching of silicate and precipitation of iron and aluminum. When jungle soils are denuded of their arboreal vegetation, as in slash-and-burn agriculture, they quickly lose their fertility as a result of the intense microbial activity which rapidly destroys soil organic matter. Since little organic matter is returned to the soil in its agricultural exploitation, conditions favor **laterization**, a process in which iron and aluminum oxides, silica, and carbonates are precipitated which cement the soil particles together and greatly reduce its porosity and water-holding capacity and make it generally unfavorable for plant growth.

Aridisols and **entisols** are desert soils that occur mostly in hot, arid climates at low latitudes. Aridisols have an ochreous surface soil and may show one or more subsurface horizons as follows: argillic horizon (a layer with silica and clay minerals dominating), cambic horizon (an altered, light-colored layer, low in organic matter, with carbonates usually present), natric horizon (dominant presence of sodium in exchangeable cation fraction), salic horizon (enriched in water-soluble salts), calcic horizon (secondarily enriched with $CaCO_3$), gypsic horizon (secondarily enriched with $CaSO_4 \cdot 2H_2O$), and duripan horizon (primarily cemented by silica and secondarily by iron oxides and carbonates) (see Fuller, 1974; Buol et al., 1980). Entisols are poorly developed, immature desert soils without subsurface development. They may arise from recent alluvial deposits or from rock erosion (Fuller, 1974; Buol et al., 1980).

Desert soils are not infertile. It is primarily the lack of sufficient moisture that prevents the development of lush vegetation. However, in-

Figure 4.4   Soil profile: spodosol (podsol). (Courtesy U.S.D.A. Soil Conservation Service.)

Figure 4.5   Soil profile: mollisol (chernozem). (Courtesy U.S.D.A. Soil Conservation Service.)

sufficient nitrogen as a major nutrient, and zinc, iron, and sometimes copper, molybdenum, or manganese as minor nutrients may also limit plant growth. Desert soils support a specially adapted macroflora and fauna that can cope with the stressful conditions in such an environment. They also harbor a characteristic microflora of bacteria, fungi, algae, and lichens. Actinomycetes, algae and lichens may sometimes be dominant. Cyanobacteria seem to be more important in nitrogen fixation in desert soils than are bacteria. Desert soils can sometimes be converted to productive agricultural soils by irrigation. Such watering often results in extensive solubilization of salts from the subhorizons where they have accumulated during soil-forming episodes. As a consequence the salt level in the available water in the growth zone of the soil may increase to a level at which it becomes inhibitory to plant growth. The drainage water from such irrigated soil will also develop increasing salt concentrations and present a disposal and reuse problem.

Types of Microbes and Their Distribution in Mineral Soil Microorganisms found in mineral soil include bacteria, fungi, protozoans, algae, and viruses associated with these groups. A great variety of different bacteria may be encountered. Morphological types include gram-positive rods and cocci, gram-negative rods and spirals, sheathed bacteria, stalked bacteria, mycelial bacteria (**actinomycetes**), budding bacteria, and others. Physiological types include cellulolytic, pectinolytic, saccharolytic, proteolytic, ammonifying, nitrifying, denitrifying, nitrogen-fixing, sulfate-reducing, iron-oxidizing and -reducing, manganese–oxidizing and -reducing, and other types. Morphologically the dominant forms of culturable bacteria seem to be gram-positive cocci, probably representing the coccoid phase of *Arthrobacter* or possibly microaerophilic cocci related to *Mycococcus* (Casida, 1965). At one time, nonsporeforming rods were held to be the dominant form. Sporeforming rods are not very prevalent despite the fact that they are readily encountered when culturing soil in the laboratory. Numerical dominance of a given type does not necessarily speak for its biochemical importance in soil. Thus, nitrifying and nitrogen-fixing bacteria are less numerous than some others but of vital importance to the nitrogen cycle in soil. A given soil under a given set of conditions will harbor an optimum number of individuals of each resident bacterial group. These numbers will change with modification in prevailing physical and chemical conditions. Total viable counts in soils generally range from 10^5 per gram in poor soil to 10^8 per gram in garden soil.

The bacteria in soil are primarily responsible for mineralization of organic matter, for fixation of nitrogen, for nitrification, for denitrification, for mineral mobilization and immobilization, and for other processes. Some bacterial types, especially **copiotrophs** (microorganisms requiring a nutrient-rich environment), often reside in microcolonies or in biofilms on soil particles because it is there where they find nutritional and other conditions for their existence optimal (see Water Distribution in Mineral Soil). Conditions of nutrient supply, oxygen supply, moisture availability, pH, and E_h may vary widely from particle to particle, owing in part to the activity of different bacteria or other micro- or macroorganisms. Thus, soil contains many different microenvironments. The colonization of soil particles by bacteria may cause some particles to adhere to one another (Martin and Waksman, 1940, 1941), which means that bacteria can affect soil texture.

Fungi reside mainly in the upper A horizon of soil because they are principally strict aerobes and find their richest food supply there. In numbers, they represent a much smaller fraction of the total microbial population than do the bacteria. Their mycelial growth habit causes them to grow over soil particles and penetrate the pore space of soil. They may also cause clumping of soil particles. The soil fungi include members of all the major groups: Phycomycetes, Ascomycetes, Basidiomycetes, and Deuteromycetes, and also the slime molds. The last named are usually classified separately from fungi and protozoans, although they have attributes of both. Total numbers of fungi in soil, expressed as propagules (spores, hyphae, hyphal fragments), may range from 10^4 to 10^6 per gram of soil. Fungi are important in soil because their ability to attack cellulose and lignin is greater than that of the bacterial flora. Some fungi are also predaceous and help control the protozoan (Alexander, 1977, p. 67) and nematode population (Pramer, 1964) in soil.

Protozoa are also found in soil. Like fungi, they are less numerous than the bacteria, ranging typically from 7×10^3 to 4×10^5 per gram of soil (Alexander 1977). They are represented by flagellates (Mastigophora), amoebae (Sarcodina), and ciliates (Ciliata). Although both saprozoic and holozoic types occur, it is the latter that are of ecological importance in soil. Being predators, the holozoic forms help to keep the bacteria and, to a much lesser extent, other protozoa, fungi, and algae in check (Paul and Clark, 1989). The protozoa inhabit mainly the upper portion of soil where their food source (prey) is most abundant. The types and numbers of protozoa in a given soil depend on soil type and soil condition.

Although cyanobacteria and algae are associated mostly with aquatic environments, they do occur in significant numbers in the uppermost portion of A horizons in soils (Alexander, 1977) where sufficient light penetrates through translucent minerals and pore space. They are usually the least numerous of the various microbial groups, ranging from 10^2 to 10^4 per gram just below the soil surface. Besides cyanobacteria, the most important algal groups represented in soil are green algae (chlorophytes), and diatoms (chrysophytes). Xanthophytes may also be present. Since cyanobacteria and algae are photosynthetic organisms, they grow mostly on or just below the soil surface, where light can penetrate. However, cyanobacteria and algae are also found below the light zone, where they may grow heterotrophically. Some cyanobacteria, green algae and diatoms have been shown to be capable of heterotrophic growth in the dark. Some of the cyanobacteria in soil are capable of nitrogen fixation and may in some cases be more important in enriching the soil in fixed nitrogen than other bacteria. The growth of algae in soil is dependent on adequate moisture and CO_2 supply. The latter is rarely limiting. The pH will influence which types of algae will predominate. While cyanobacteria prefer neutral to alkaline soil, green algae will also grow in acid soil. When grow-

Table 4.3   Bacterial Densities at Different Depths of Different Soil Types on
July 7, 1915

Soil depth (in.)	Bacterial densities (bacteria per gram of air-dried soil $\times 10^6$)			
	Garden soil	Orchard soil	Meadow soil	Forest soil
1	7.76	6.23	6.34	1.00
4	6.22	3.70	5.20	0.34
8	2.81	1.01	3.80	0.27
12	0.80	0.82	1.11	0.060
20	0.31	0.075	0.10	0.040
30	0.30	0.052	0.70	0.023

Source: Waksman (1916).

ing photosynthetically, cyanobacteria and algae are primary producers, generating at least some of the reduced carbon on which heterotrophs depend. Because of their role as primary producers, cyanobacteria and algae are the pioneers in the formation of new soils, for instance in volcanic areas.

Bacterial distribution in mineral soils is illustrated in Table 4.3. Characteristically, the largest number of organisms occur in the upper A horizon and the smallest in the B horizon. Aerobic bacteria are generally more numerous than anaerobic bacteria, actinomycetes, fungi, or algae. In enumerations as shown in Table 4.3, anaerobes decrease with depth to about the same degree as aerobes. This seems contradictory but may reflect the fact that most of the anaerobes that were enumerated by the methodology then in use were facultative.

Determination of microbial distribution in soil, when done by culturing as in Table 4.3, never yields an absolute estimate because no universal medium exists on which all living soil microbes can grow. A somewhat better estimate can be obtained by fluorescence microscopy of soil preparations treated with special fluorescent reagents (see Chapter 7), but even this method does not provide an absolute estimate because, for one, it may not distinguish living and dead cells.

4.3   ORGANIC SOILS

In some special locations, **organic soils** or **histosols** are found. They form from rapid accumulation and slow decomposition of organic matter, especially plant matter, as a result of displacement of air by water, which prevents rapid and extensive microbial decomposition of the organic matter. These soils are thus of sedimentary origin and are never the result of rock weathering. These soils consist of 20% or more organic matter (Lawton, 1955; Buol et al., 1989). Their formation is associated with the evolution of swamps, tidal marshes, bogs, and even shallow lakes. An organic soil such as peat may have an ash content of 2–50% and contain cellulose, hemicellulose, lignin and derivatives, heterogeneous complexes, fats, waxes, resins, and water-soluble substances such as polysaccharides, sugars, amino acids, and humins (Lawton, 1955). The pH of organic soils may range from 3 to 8.5. Examples of such soils are peats and "mucks." They accumulate to depths ranging from less than a meter to more than 8 m (Lawton, 1955) and are not stratified like mineral soils. They are rare. Some are agriculturally very productive.

4.4 SUMMARY

The lithosphere of the Earth consists of rock which may be igneous, metamorphic, or sedimentary. Rock is composed of intergrown minerals. The rock surface, and in the case of porous rock, its interior may be habitats for microbes. Rock may be broken down by weathering, which may ultimately lead to formation of mineral soil. Some of the rock minerals become chemically altered in the process. Weathering may be biological, especially microbiological, as well as chemical and physical.

Progress of mineral soil development is recognizable in the soil profile. A vertical section through mineral soil may reveal more or less well-developed horizons. Typical horizons of Spodosols and Mollisols include the litter zone (O horizon), a leached layer (A horizon), an enriched layer (B horizon), and the parent material (C horizon). The appearance of these horizons varies with soil type. Climate is one of several important determinants of soil type. The horizons are the result of intense biological activity in the litter zone and A horizon. Much of the organic matter in the litter zone is microbially solubilized and at least partly degraded. Soluble components are washed into the A horizon or transported there by some invertebrates, where they may be further metabolized and where they contribute directly or indirectly to transformation of some of the mineral matter. Soluble products, especially inorganic ones, formed in the A horizon may be washed into the B horizon. The more refractory organic matter in the soil accumulates as humus, which contributes to soil texture, water-holding capacity, and general fertility. Mineral soil may be 50% solid matter and 50% pore space. The pore space is occupied by gases, such as N_2,

CO_2, and O_2, and by water. Water also surrounds soil particles to varying degrees. Microbes, including bacteria, fungi, protozoans, and algae, may inhabit the soil pores or live on the soil particles. They are most numerous in the upper layers of soil.

Not all soils can be classified as mineral soils. A few are organic and have a different origin. They arise from the slow decomposition of organic matter, mainly plant residues, which accumulates by sedimentation, as in swamps, marshes, or shallow lakes. They are not stratified and usually have a low mineral content.

Soil represents a most important microbial habitat in the Earth's lithosphere.

REFERENCES

Alexander, M. 1977. Introduction to Soil Microbiology. 2nd ed. Wiley, New York.

Berner, R. A., and G. R. Holdren, Jr. 1977. Mechanism of feldspar weathering: Some observational evidence. Geology 5:369–372.

Berthelin, J. 1977. Quelques aspects des mécanismes de transformation des minéraux des sols par les micro-organismes hétérotrophes. Sci. Sol, Bull. A.F.E.S., no. 1, pp. 13–24.

Berthelin, J., and D. Boymond. 1978. Some aspects of the role of heterotrophic microorganisms in the degradation of minerals in waterlogged soils. *In*: W. E. Krumbein, ed. Environmental Biogeochemistry and Geomicrobiology. Ann Arbor Science Publishers, Ann Arbor, MI, pp. 659–673.

Brady, P. V., and S. A. Carroll. 1994. Direct effects of CO_2 and temperature on silicate weathering: possible implications for climate control. Geochim. Cosmochim. Acta 58:1853–1856.

Brock, T. D. 1975. Effect of water potential on growth and iron oxidation by *Thiobacillus ferrooxidans*. Appl. Microbiol. 29:495–501.

Brock, T. D., D. W. Smith, and M. T. Madigan. 1984. Biology of Microorganisms. 4th ed. Prentice-Hall, Englewood Cliffs, NJ, 847 pp.

Brown, A. D. 1976. Microbial water stress. Bacteriol. Rev. 40:803–846.

Browne, B. A., and C. T. Driscoll. 1992. Soluble aluminum silicates: Stoichiometry, stability, and implications for environmental geochemistry. Science 256:1667–1670.

Bunting, B. T. 1967. Geography of Soil. Hutchinson University Library, London.

Buol, S. W., F. D. Hole, and R. J. McCracken. 1980. Soil Genesis and Classification. 2nd ed. Iowa State University Press, Ames, 404 pp.

Campbell N. E. R., and H. Lees. 1967. The nitrogen cycle. *In*: A. D. McLaren and G. H. Petersen, eds. Soil Biochemistry. Vol. 1. Marcel Dekker, New York, pp. 194–215.

Casida, E. L. 1965. Abundant microorganism in soil. Appl. Microbiol. 13: 327–334.

Chou, L., and R. Wollast. 1984. Study of the weathering of albite at room temperature and pressure with a fluidized bed reactor. Geochim. Cosmochim. Acta 48:2205–2218.

Costerton, J. W., Z. Lewandowski, D. deBeer, D. Caldwell, D. Korber, and G. James. 1994. Biofilms, the customized microniche. J. Bacteriol. 176:2137–2142.

Dommergues, Y., and F. Mangenot. 1970. Ecologie Microbienne du Sol. Masson, Paris.

Friedman, E. I. 1982. Endolithic microorganisms in the Antarctic cold desert. Science 215:1045–1053.

Fuller, W. H. 1974. Desert soils. *In*: G. W. Brown, ed. Desert Biology, Vol. II. Academic Press, New York, pp. 31–101.

Golubic, S., and J. Schneider. 1979. Carbonate dissolution. *In*: P. A. Trudinger and D. J. Swaine, eds. Biogeochemical Cycling of Mineral Forming Elements. Elsevier Scientific Publishing Co., Amsterdam, pp. 107–129.

Golubic, S., D. D. Perkins, and K. J. Lukas. 1975. Boring microorganisms and microborings in carbonate substrates. *In*: R. W. Frey, ed. The Study of Trace Fossils. Springer-Verlag, New York, pp. 229–259.

Harris, R. F., W. R. Gardner, A. A. Adebayo, and L. E. Sommers. 1970. Agar dish isopiestic equilibration method for controlling the water potential of solid substrates. Appl. Microbiol. 19:536–537.

Hiebert, F. K., and P. C. Bennett. 1992. Microbial control of silicate weathering in organic-rich groundwater. Science 258:278–281.

Jones, C. G., and M. Shachak. 1990. Fertilization of the desert soil by rock-eating snails. Nature (Lond.) 346:839–841.

Jones, D., M. J. Wilson, and W. J. McHardy. 1981. Lichen weathering of rock-forming minerals: Application of scanning electron microscopy and microprobe analysis. J. Microsc. 124:95–104.

Kuznetsov, S. I., M. V. Ivanov, and N. N. Lyalikova. 1963. Introduction to Geological Microbiology. English translation. McGraw-Hill, New York.

Lang, A. R. G. 1967. Osmotic coefficients and water potentials of chloride solutions from 0–40°C. Aust. J. Chem. 20:2017–2023.

Lawton, K. (1955). Chemical composition of soils. *In*: F. E. Bear, ed. Chemistry of the Soil. Van Nostrand Reinhold, New York, pp. 53–84.

Lucas, Y., F. J. Luizao, A. Chauvel, J. Rouiller, and D. Nahon. 1993. The relation between biological activity of the rain forest and mineral composition of soils. Science 260:521–523.

Martin, J. P., and S. A. Waksman. 1940. Influence of microorganisms on soil aggregation and erosion. Soil Sci. 50;29–47.

Martin, J. P., and S. A. Waksman. 1941. Influence of microorganisms on soil aggregation and erosion. II. Soil Sci. 52:381–394.

Oelkers, E. H., J. Schott, and J.-L. Devidal. 1994. The effect of aluminum, pH, and chemical affinity on the rates of aluminosilicate dissolution reactions. Geochim. Cosmochim. Acta 58:2011–2024.

Paul, E. A., and F. E. Clark. 1989. Soil Microbiology and Biochemistry. Academic Press, San Diego.

Pramer, D. 1964. Nematode trapping fungi. Science 144:382–388.

Shachak, M., C. G. Jones, and Y. Granot. 1987. Herbivory in rocks and the weathering of a desert. Science 236:1098–1099.

Waksman, S. A. 1916. Bacterial numbers in soils at different depths and in different seasons of the year. Soil Sci. 1:363–380.

Waksman, S. A., and R. L. Starkey. 1931. The Soil and the Microbe. Wiley, New York.

Welch, S. A., and W. J. Ullman. 1993. The effect of organic acids on plagioclase dissolution rates and stoichiometry. Geochim. Cosmochim. Acta 57:2725–2736.

Welch, S. A., and P. Vandevivere. 1995. Effect of microbial and other naturally occurring polymers in mineral dissolution. Geomicrobiol. J. 12:227–238.

Wieland, E., and W. Stumm. 1992. Dissolution kinetics of kaolinite in acidic aqueous solutions at 25°C. Geochim. Cosmochim. Acta 56: 3339–3355.

5

The Hydrosphere

5.1 THE OCEANS

Physical Attributes   The oceans are a habitat for various forms of life, ranging from the largest anywhere on Earth to the smallest. The **fauna** includes various vertebrates (mammals, birds, reptiles, fish) as well as a wide range of invertebrates. The **flora** includes the algae—from the macroscopic kelps to the small unicellular forms. The **plankton** includes the floating biota. The **phytoplankton** includes free-floating algae such as diatoms and dinoflagellates, and the **zooplankton** includes the free-floating microscopic invertebrates and protozoans. The **bacterioplankton** consist of free, unattached bacterial forms that include some cyanobacteria.

The oceans cover about 70% of the Earth's surface, occupying an area of 3.6×10^8 km^2 and a volume of 1.37×10^9 km^3, which amounts to 1.41×10^{21} kg of water. (The total mass of the Earth is estimated to be 5.98×10^{24} kg.) Because of the unequal distribution of the continents between the northern and southern hemispheres of the present Earth, only 60.7% of the northern hemisphere is covered by oceans, whereas 80.9% of the southern hemisphere is covered by them. The world's major oceans include the Atlantic Ocean (16.2% of the Earth's surface), the Pacific

Ocean (32.4% of the Earth's surface), the Indian Ocean (14.4% of the Earth's surface), and the Arctic Ocean (2.8% of the Earth's surface). The average depth of all oceans is 3,795 m. The average depth of the Atlantic Ocean is 3,296 m, that of the Pacific Ocean 4,282 m, that of the Indian Ocean is 3,693 m, and that of the Arctic Ocean is 1,205 m. The greatest depth in the oceans occurs in **ocean trenches**. For instance, in the Pacific Ocean, the water depth at the Marianas Trench is 10,500 m. In the Atlantic Ocean, the Puerto Rico Trench has a water depth of 8,650 m, and in the Indian Ocean, the Java Trench water depth is 7,450 m. Shallow ocean depths are encountered in the marginal seas along the coasts of the continents. These seas are usually less than 2,000 m and frequently less than 1,000 m deep. Figure 5.1 shows the oceans of the world.

An ocean includes a basin which is surrounded by a continental margin with several structural features. Projecting from each continental shore is the **continental shelf**, encompassing about 7.5% of the ocean area. It slopes gently downward toward the ocean basin at an average angle of 7 min to a water depth of about 130 m. Its average width is about 65 km, but may range from 0 to 1,290 km, the greatest distance being represented by the shelf projecting from the coast of Siberia into the North Polar Sea. The waters over the continental shelf are a biologically important part of the ocean. They are the site of high biological productivity. This is easily explained by the contribution of nutrients in general runoff from the adja-

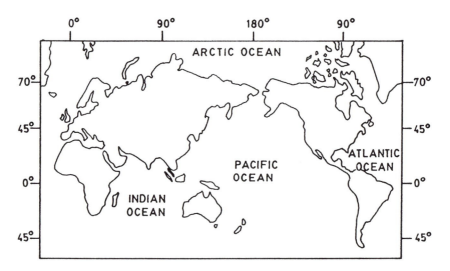

Figure 5.1   Oceans of the world.

cent land and in particular from rivers emptying into the waters over the shelf area.

At the edge of the continental shelf, the ocean floor drops sharply at an average angle of 4° (range 1–10°) to abyssal depths of about 3,000 m. This is the region of the **continental slope** and constitutes about 12% of the ocean area. In some places the slope is cut by deep canyons. They often occur at the mouths of large rivers (e.g., Hudson River, Amazon River). Many canyons were cut over geologic time by **turbidity currents**. Such currents consist of strong water movements that carry a high sediment load picked up as a river flows oceanward. As the river meets the sea, the sediment is dropped, and as it settles, it abrades the slope. Marine canyons may also be cut as a result of slumping of an unstable sediment deposit on a portion of a continental slope and the consequent abrading of the slope. On occasion, the continental slope may be interrupted by a terraced region, as in the case of Blake Plateau off the southern Atlantic coast of the United States. This particular shelf is about 302 km wide and drops gradually from a depth of 732 m to one of 1,100 m over this distance. It was gouged out of the continental slope by the northward-flowing Gulf Stream.

At the foot of the continental slope lies the **continental rise**, consisting of accumulations of sediment carried downslope by turbidity currents. Such deposits may be a few kilometers thick, forming fanlike structures in some places and wedges in others. An idealized profile of a continental margin is shown in Figure 5.2.

The **ocean basin** takes up 80% of the ocean area. Its floor, far from being a flat expanse, as some once believed, often exhibits a rugged topography. Mountain ranges cut by fracture zones and rift valleys stretch over thousands of kilometers as the midocean ridge system where new ocean floor is created (see Chapter 2). Elsewhere, somewhat more isolated submarine mountains, some active and others dormant volcanoes, dot the ocean floor. Some of the seamounts have flattened tops and have been given the special name of **guyots**. Some of flattened tops of seamounts, especially in the Pacific Ocean, reach surface waters at depths of 50 to 100 m where the temperature is about 21°C. In these instances, the flattened tops may furnish the substratum for colonization by corals (coelenterates) and coralline algae, resulting in the formation of atolls and reefs.

Covering the ocean floor almost everywhere are **sediments**. They range in thickness from 0 to 4 km, with an average thickness of 300 m. Their rate of accumulation varies, being slowest in midocean (less than 1 cm per 10^3 years) and fastest on continental shelves (10 cm per 10^3 years). These rates may be even greater in some inland seas and gulfs (e.g., 1 cm per 10–15 years in the Gulf of California and 1 cm per 50 years

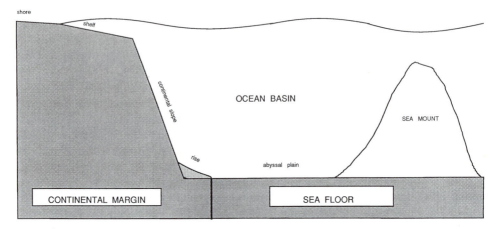

Figure 5.2   Schematic representation of a profile of an ocean basin.

in the Black Sea). In some regions in the deep ocean, the sediments consist mainly of deposits of siliceous and/or calcareous remains of marine organisms. The siliceous remains are derived from the frustules of diatoms (algae) and the support skeleton and spines of radiolarians (protozoans). The calcareous remains are derived from the tests of foraminifera (protozoans), carbonate platelets from the walls of coccolithophores (algae), and shells from pteropods (mollusks). **Diatomaceous oozes** predominate in colder waters (e.g., in the North Pacific between 40° and 70° north latitude and 140° west to 145° east longitude, according to Horn et al., 1972), whereas **radiolarian oozes** predominate in warmer waters (e.g., in the North Pacific between 5° and 20° north latitude and 90° and 180° west longitude, according to Horn et al., 1972). **Calcareous oozes** are found mainly in warmer waters on ocean bottoms no deeper than 4,550–5,000 m (e.g., in the North Pacific between 0° and 10° north latitude and 80° and 180° west longitude, according to Horn et al., 1972). At greater depths, the CO_2 concentration in the water is large enough to cause dissolution of carbonate unless the structures are enclosed in a protective membrane.

Other vast areas of the ocean floor are covered by clays (**red clays** or **brown mud**), which are probably of terrigenous origin and washed into the sea by rivers and general runoff from continents and islands and carried into the ocean basins by ocean currents, mudflows, and turbidity currents. At high latitudes in both hemispheres, particularly on and near continental shelves, ice-rafted sediments are found. They were dropped into the ocean by melting icebergs that had previously separated from glacier fronts that had picked up terrigenous debris during glacial progression. Except for ice-rafted detritus, only the fine portion of terrigenous debris (clays and fine silts) is carried out to sea. The clay particles are defined as less than 0.004 mm in diameter, and the silt particles are defined in a size range from 0.004 to 0.1 mm in diameter. Figure 5.3 shows the appearance of some Pacific Ocean sediments under the microscope.

The Ocean in Motion A significant portion of the ocean is in motion at all times. The causes of this motion are (a) wind stress on surface waters,

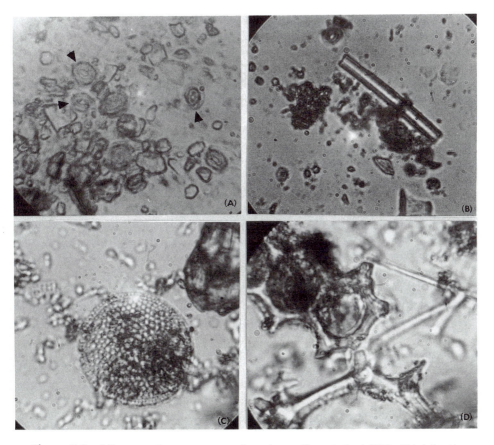

Figure 5.3 Microscopic appearance of marine sediments ($\times 1,750$). (A) Atlantic sediment showing coccoliths ($CaCO_3$) (arrows) and clay particles. (B) Atlantic sediment showing diatom frustules (SiO_2) and other debris. (C) Pacific sediment showing a centric diatom frustule (SiO_2) and other debris. (D) Pacific sediment showing fragments of radiolarian tests (SiO_2).

(b) Coriolis force arising from the rotation of the Earth, (c) density variations of seawater resulting from temperature and salinity changes, and (d)
tidal movement due to gravitational influences on the water by the sun
and moon. Surface currents (Fig. 5.4) are prominent in regions of prevailing winds, such as the trade winds, which blow from east to west about
20° north and south latitude; the westerlies, which blow from west to east
between 40° and 60° north and south latitude; and the easterly polar winds,
which blow in a westerly direction south of the Arctic Ocean. The effect
of these winds, together with the deflecting influence of the continents
and the Coriolis force, is to set up surface circulations in the form of
gyrals between the north and south poles in each major ocean. They are
the North Subpolar Gyral (small), North Subtropical Gyral (large), North
Tropical Gyral (small), South Tropical Gyral (small), South Subtropical
Gyral (large); and the Antarctic Current, which circulates around Antarctica from west to east (Fig. 5.4A). Thus, the Gulf Stream, together with
the Canary Current and the North Equatorial Current, is part of the North
Subtropical Gyral of the Atlantic Ocean (Fig. 5.4B). The flow rates of the
waters in these gyrals and their parts vary. The flow rate of the water in
the Gulf Stream is the fastest for any surface current, 250 cm sec^{-1}. Other
currents have flow rates that fall mostly in the range of 25–65 cm sec^{-1}.

Meanders in the Gulf Stream in the Atlantic and the Kuroshio Current in the Pacific Ocean may give rise to so-called **rings**, a small, closed
current system that may measure as much as 300 km in diameter and may
have a depth as great as 2 km. Such rings may move 5–10 km per day. The
chemical, physical, and biological characteristics of the water enclosed in
the ring may be significantly different from the surrounding water. A slow
exchange of solutes and biota as well as heat transfer may take place
across the boundary of a ring. Rings thus constitute means of nutrient
transport from ocean currents. The rings may ultimately rejoin the current
that spawned them (Gross, 1982; Richardson, 1993; Ring Group, 1981).
More recently **anticyclones** have been reported to arise from the Gulf
Stream in addition to the rings, and **meddies** north of the Strait of Gibralta
(Richardson, 1993). Whereas rings have a cold-water core surrounded by
a warm-water layer and a counterclockwise rotation, anticyclones have
a warm-water core surrounded by colder water and a clockwise rotation.
Meddies have a core that is more saline than the surrounding ocean water
and a clockwise rotation. Collectively these formations are known as
ocean eddies.

Deep water is also in motion. Its movement appears to be caused
by slow diffusion due to density differences of water masses through broad
zones in the oceans. Some deep currents that have been measured in the
Atlantic Ocean have a velocity between 1 and 2 cm sec^{-1} (Dietrich and

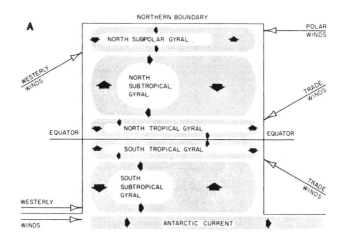

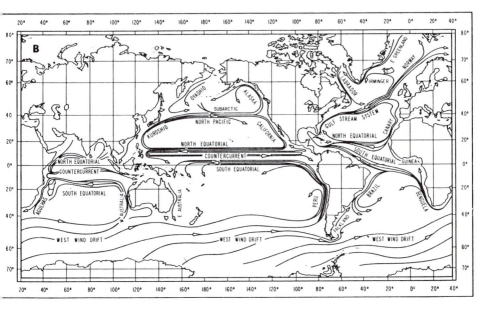

Figure 5.4 Oceanic surface currents. (A) Schematic representation of the prevailing winds and their effects on the surface currents of an imaginary rectangular ocean. (B) Average surface currents of the world's oceans. (From Williams, 1962, by permission.)

Kalle, 1965, pp. 399, 407; Gross, 1982). Occasional short bursts in velocity may occur. The bottom current movement is influenced by bottom topography.

In regions where the movement of a water mass diverges into different directions, deep water rises into the zone of **divergence**, resulting in **upwelling**. The same upwelling of deep water may also result when winds blow large surface water masses away from coastal regions (Smith, 1968). Deep water is relatively rich in mineral nutrients, including nitrate and phosphate, and thus upwelling is of great ecological consequence because it replenishes biologically depleted nutrients in the surface waters. Regions of upwelling are, therefore, very fertile. An important region of upwelling in the eastern Pacific Ocean occurs off the coast of Peru. A disturbance in the surface water circulation in the southern Pacific can result in failure of upwelling in this region (El Nino) and can spell disaster for the fisheries of the area.

In regions where two water masses of different density meet in a **convergence**, the heavier water will sink to a level where it meets water of its own density. This phenomenon is important in the oxygenation of deep waters. Important convergences in the oceans occur at high latitude in both hemispheres.

Chemical and Physical Properties of Seawater   Seawater is saline. Some important chemical components of seawater, listed in decreasing order of concentration, are presented in Table 5.1 (see also Marine Chemistry, 1971). Of these components, chloride (55.2%), sodium, (30.4%), sulfate (7.7%), magnesium (3.7%), calcium (1.16%), potassium (1.1%), bromide (0.1%), strontium (0.04%), and borate (0.07%) account for 99.5% of the total salts in solution. Because these components generally occur in constant proportions relative to each other in true ocean waters, it has been possible to estimate salt concentration in seawater samples by merely measuring the chloride concentration. The chloride concentration in grams per kilogram (chlorinity, Cl) is related to the total salt concentration in grams per kilogram (salinity, S) by the following empirical relationship[*]:

$$S\ (\%_o) = 0.030 + 1.8050Cl\ (\%_o) \tag{5.1}$$

The salinity so determined is an estimate of the total amount of solid material in a unit mass of seawater in which all carbonate has been converted to oxide and all bromide and iodide has been replaced by chloride, and in which all organic matter has been completely oxidized. For refer-

---

[*] The symbol $\%_o$ represents parts per thousand or grams per kilogram.

Table 5.1 Some Constituents of Seawater (μg liter^{-1})

Major constituents		Minor constituents		Minor constituents	
Cl	1.9×10^7	Si	3×10^3	Cu	3
Na	1.1×10^7	N	6.7×10^2	Fe	3
Mg	1.3×10^6	Li	1.7×10^2	U	3
S (SO$_4$)	9.0×10^5	Rb	1.2×10^2	As	2.6
Ca	4.1×10^5	P	90	Mn	2
K	3.9×10^5	I	60	Al	1
Br	6.7×10^4	Ba	20	Co	0.4
C (CO$_3$, HCO$_3$)	2.8×10^4	Mo	10	Se	9×10^{-2}
B	4.5×10^3	Zn	10	Pb	3×10^{-2}
		Ni	7	Ra	1×10^{-7}

Source: Values taken from *Marine Chemistry* (1971).

ence purposes, the salinity of standard seawater has been taken as 34.3‰. The actual salinity of different parts of the world oceans can vary from less than 34‰ to almost 36‰ (Dietrich and Kalle, 1965; p. 156). Table 5.2 lists the salinities of some different marine waters as well as of some saline lakes. It must be pointed out that while the Great Salt Lake in Utah (U.S.A.) has a salt composition that is qualitatively similar to that of oceans, some other inland hypersaline water bodies, such as the Dead Sea at the mouth of the Jordan River, have a different salt composition. In the Dead Sea the dominant cations are in descending order Mg (approximately 44 g L^{-1}), Na (40 g L^{-1}), Ca (17 g L^{-1}), and K (7.5 g L^{-1}) and the dominant anions are Cl (225 g L^{-1}) and Br (5.5 g L^{-1}) (Nissenbaum, 1979).

Table 5.2 Salinities (‰) of Some Marine Waters and Salt Lakes

Gulf of Bothnia	2–6[a]
Baltic Sea	6–17[a]
Black Sea	16–18[a]
Mediterranean Sea	37–39[b]
Red Sea	40–41[b]
Dead Sea	320[c]
Great Salt Lake	320[c]
Ocean bottoms	34.66–34.92[d]

[a] Smith, 1974; [b] Bowden, 1975; [c] ZoBell, 1946; [d] Defant, 1961.

Although some portion of the salts in seawater are contributed in the runoff from the continents and the weathering of minerals in the surficial sediments, a most important contribution to seawater solutes is made by hydrothermal discharges from vents at the midocean spreading centers. These discharges are the consequence of seawater penetration deep into the porous basalt (up to 1 to 3 km depths) where the seawater then reacts with constituents of the basalt in various ways. The reactions include the reduction of seawater-sulfate to hydrogen sulfide by ferrous iron in the basalt, the removal of magnesium from seawater as a magnesium hydroxide and the incorporation of seawater calcium into minerals such as plagioclase to form new aluminosilicate minerals, such as clinozoisite or Ca-rich amphibole, accompanied by the generation of acidity. In the case of Ca incorporation into plagioclase this can be illustrated by the reaction

$$3CaAl_2Si_2O_8 + Ca^{2+} + 2H_2O \Rightarrow 2Ca_2Al_3Si_3O_{12}(OH) + 2H^+$$

$$(5.2)$$

The resultant acidity is the cause of leaching of some other components from the basalt such as hydrogen sulfide from pyrrhotite and base metals from other basalt constituents. All these reactions are possible because of high hydrostatic pressure exerted on the solution in the basalt at these depths, and because of high temperature ($\sim$350°C) from heat diffusing from underlying magma chambers into the reaction zone in the basalt. The resultant chemically altered seawater is forced upward by hydrostatic pressure through porous channels and fissures in the basalt and is ultimately discharged as hydrothermal solution from vents at the spreading centers into the overlying seawater and mixed with it (Bischoff and Rosenbauer, 1983; Edmond et al, 1982; Seyfried and Janecky, 1985; Shanks et al, 1981) (see also Chapters 2 and 18). These reactions contribute significantly to the stability of seawater composition.

Seawater contains a pH buffering system. It consists of bicarbonate and carbonate ions, which constitute 0.35% of the solute components in seawater, and borates and silicates. Together, they keep the pH of seawater in the range of 7.5–8.5. Surface seawater pH tends to fall into a narrow range of 8.0–8.5. At depth, seawater pH may approach 7.5. The variation in pH of seawater with depth may be related to oxygen utilization by marine organisms to a major extent and to carbonate dissolution to a lesser extent (Park, 1968). Figure 5.5 illustrates pH variation with depth at one particular station in the Pacific Ocean.

The salts dissolved in seawater impart a special osmotic property to it. The **osmotic pressure** of seawater is of the order of magnitude of the internal pressure of bacterial cells or the of cell sap of eukaryotic cells.

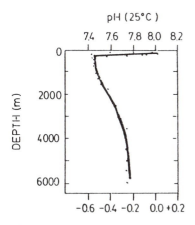

Figure 5.5 Vertical profile of pH at station 54°46′N, 138°36′W in the northeast Pacific Ocean. (Adapted from P. K. Park, Seawater hydrogen-ion concentration: vertical distribution, Science 162:357–368, Copyright © 1968 by the American Association for the Advancement of Science, by permission.)

At a salinity of 35‰ and a temperature of 0°C, seawater has an osmotic pressure of 23.37 bars (23.07 atm), whereas at the same salinity but at 20°, it has an osmotic pressure of 25.01 bars (24.69 atm). Clearly, then, the osmotic pressure of seawater is not deleterious to living cells.

With increasing depth in a water column, **hydrostatic pressure** becomes a significant factor. On average it increases about 1.013 bars (1 atm) for every 10 m of depth. Related to the weight of overlying water at a given depth, hydrostatic pressure ranges from 0 to more than 1013 bars (1,000 atm). Thus, the highest pressures are experienced in the deep ocean trenches. Living organisms in the sea are strongly affected by high hydrostatic pressure. Among the fauna, some are adapted to live only in surface waters, others at intermediate depths, and still others at abyssal depths. Generally, none are known that can live over the whole depth range of the open ocean. Although microorganisms such as bacteria appear to be more adaptable, facultative and obligate barophilic bacteria are known (see later section in this chapter).

Salinity and temperature affect the **density** of seawater. At 0°C, seawater with a salinity of 30–37‰ has a corresponding density range of 1.024–1.030 g cc^{-1}. A variation in seawater density due to variation in salinity is one cause of water movement in the ocean, because denser water will sink below lighter water, or conversely, lighter water will rise above denser water (upwelling). The following processes may cause changes in salinity and, therefore, density: (a) dilution of seawater by runoff or less saline waters; (b) dilution by rain or snow; (c) concentration through surface evaporation by the sun's heat; (d) freezing, which excludes salt from ice and thus leaves any residual water more saline; or (e) thawing of ice, which dilutes already existent saline water.

As already stated, variation of salinity of seawater is not the sole cause of its variation in density. The other important cause of density variation of seawater is temperature. Unlike fresh water, whose density is greatest at 4°C (Fig. 5.6B), seawater with a salinity of 24.7‰ or greater has maximum density at its freezing point (Fig. 5.6A). A body of fresh water thus freezes from its surface downward because fresh water at its freezing point is lighter than at temperatures up to 4°C. Ocean water in the Arctic and Antarctic seas also freezes from the sea surface downward, in this case because ice, which excludes salt as it forms from seawater, is lighter than the seawater and will thus float on it.

The **temperature** of seawater ranges from about $-2°$ (the freezing point at 36‰ salinity) to $+30°C$, in contrast to the temperature of air over the ocean, which ranges from -65 to $+65°C$. The narrower temperature range for seawater can be related to (a) its high heat capacity, (b) its latent heat of evaporation, and (c) the heat transfer from lower to higher latitudes by surface currents on both hemispheres. The major source of heat in the ocean is solar radiation. More than half of the surface waters of the ocean are at 15–30°C. Only 27% of the surface waters are below 10°C. From about 50°N latitude to 50°S latitude, the ocean is thermally stratified. In this range of latitudes, the water temperature below about 1,000 m is below 4°C (**deep water**). At depths from about 300 to 1,000 m, the temperature drops rapidly with increasing depth. The zone of this rapid temperature change is called the **thermocline**. The thickness and position of the thermocline varies with geographic location and season of the year. Above the thermocline lies the warm surface water, the **mixed layer**, which is continually agitated by wind and water currents and thus exhibits relatively little temperature change with increasing depth.

At latitudes higher than 50°N and 50°S, seawater is not thermally stratified. The waters around Antarctica, being cold ($-1.9°C$) and hypersaline (34.62‰) due to ice formation, are hyperdense and thus sink below warmer, less dense water to the north and flow northward along the bottom of the ocean basin. This is an example of convergence. Similarly, Atlantic waters from the subarctic region having a temperature in the range 2.8–3.3°C and a salinity in the range 34.9–34.96‰ sink and flow southward at near-bottom to bottom levels of the ocean. Because the Arctic Ocean bottom is separated from the other oceans by barriers such as the shallow Bering Strait in the case of the Pacific Ocean, and a shallow ridge in the case of the Atlantic Ocean, it does not influence the water

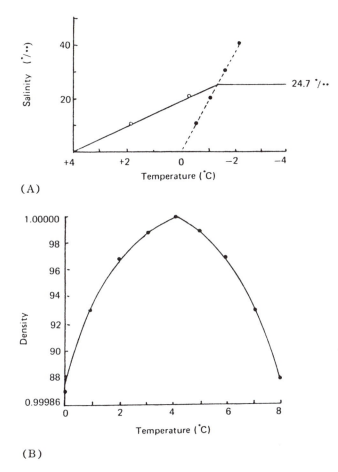

Figure 5.6 Density relationships in seawater and freshwater. (A) Relationship of seawater salinity to freezing point. Symbols: open circles, temperature of maximum density and a given salinity; closed circles, freezing-point temperature at a given salinity. Note that above a salinity of 24.7‰ seawater freezes at its maximum density since its temperature at maximum density cannot be lower than its freezing point. (B) Relationship of freshwater density (in g cc^{-1}) to temperature. Data points for chemically pure water are shown. Note that in the case of freshwater, its density at its freezing point is lower than its density at 4°C.

masses of the Atlantic and Pacific oceans directly. Other convergences occur in the world's oceans in both hemispheres because of interaction of waters of different densities. In these instances, the heavier waters sink to lesser depths because they have lower densities than the heavier waters at high latitudes.

The water convergences alluded to above help to explain why generally ocean water is oxygenated at all depths (Fig. 5.7). Of all ocean waters, only some coastal or near-coast waters (e.g., estuarine waters, Cariaco Trench) may, as a result of intensive biological activity, be devoid of oxygen at depth. At some sites, intensive biological activity is sometimes the direct result of human pollution. Surface waters of the open ocean tend to be saturated with oxygen because of oxygenation by the atmosphere and, equally important, by the photosynthetic activity of the phytoplankton. Oxygenation by phytoplankton can occur to depths of about 100 m (200 m in exceptional cases), where light penetration is 1% of surface illumination. Seawater at a salinity of 34.352‰ is saturated at 5.86 ml or 8.40 mg of oxygen per liter at 760 mm Hg and 15°C. The higher the salinity and the higher the temperature, the lower is the oxygen solubility in seawater.

Starting at the top of a water column, the oxygen concentration in seawater will at first decrease with depth, owing mainly to oxygen consumption by the respiration of living organisms (Fig. 5.7). Since many life forms in the oceans tend to be concentrated in the upper waters, oxygen concentration will fall to a minimum at about 600–900 m of depth, where respiration (oxygen consumption) by zooplankton and other animal forms as well as bacterioplankton but not photosynthesis (oxygen production) by phytoplankton will occur. Below this depth, because of rapidly

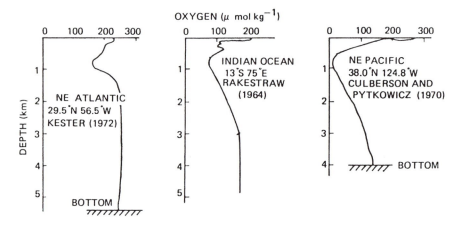

Figure 5.7 Vertical distribution of oxygen in the ocean. Profiles from three ocean basins. [From Kester, 1975 with permission from Chemical Oceanography (J. P. Riley and G. Skirrow, Eds.), Vol. 1, 2nd ed., 1975. Copyright by Academic Press Inc. (London) Ltd.]

decreasing biological activity, the oxygen concentration may at first increase once more and then slowly decrease again toward the bottom. Bottom water may, however, still be half-saturated with oxygen relative to surface water. This oxygen is not supplied by in situ photosynthesis, which cannot occur in the absence of light at these depths, nor is it the result of significant oxygen diffusion from the atmosphere to these depths. As previously indicated, these oxygenated waters derive from the Antarctic and Subarctic convergences. The oxygen-carrying waters from the Antarctic convergences flow northward along the bottom and at intermediate depths of the ocean basins whereas the waters from the subarctic convergence in the Atlantic flow southward at more intermediate depths. The oxygen content of these waters is only slowly depleted because of the low numbers of oxygen-consuming organisms in these deep regions of the oceans and the low rates of oxygen consumption in the upper sediments.

Photosynthetic activity of phytoplankton is dependent on penetration of sunlight into the water column since phytoplankton derives its energy almost exclusively from sunlight. It has been shown that light absorption by pure water in the visible range between 400 and 700 μm increases greatly toward the red end of the spectrum. It has also been shown that light which penetrates transparent water has been 60% absorbed at a depth of 1 m. The same light has been 80% absorbed at 10 m and 99% absorbed at 140 m. In less-transparent coastal water, 95% of the light may have been absorbed at 10 m. Although the photosythetic process of phytoplankters can use light over the whole visible spectrum, action spectra show peaks in the red and blue end of the spectrum, where chlorophylls absorb optimally. Accessory pigments, such as carotenoids, absorb light at intermediate visible wavelengths. Clearly, light penetration limits the depth at which phytoplankton can grow. This depth is about 80–100 m on average (200 m maximally), and often much less in less-transparent waters. Two exceptions were recently noted. One was observed off the northern border of San Salvador Island in the Bahamas where crustose coralline algae (Rhodophyta) were observed from a submersible growing attached to rock at a depth of 268 m where the light intensity was only about 0.0005% of that at the surface (Littler et al., 1985). The other exception was noted in the Black Sea where the photosynthetic sulfur bacterium *Chlorobium phaeobacteroides* was found to grow in a chemocline at an 80-m depth where light transmission from surface irradiance has been calculated to be 0.0005%, as at the station at San Salvador Island (Overmann et al., 1992).

The water layer to the depth below which photosynthesis cannot take place constitutes the **euphotic zone**. Zooplankton and bacteria may abound to somewhat lower depths (about 750 m) than phytoplankton, being scavengers and able to feed on dying and dead phytoplankters and their remains in the process of settling.

Microbial Distribution in the Water Column and Sediments Microbial distribution in the open oceans is not uniform geographically throughout the water column (Fig. 5.8). Factors affecting this distribution are energy, carbon, nitrogen, and phosphorus limitations, temperature, hydrostatic pressure, and salinity. Accessory growth factors, such as vitamins, may also be limiting to those microbes that cannot synthesize them themselves. Phytoplankton distribution is limited to the euphotic zone of the water column primarily by available sunlight, the energy source for these organisms. However, phytoplankton distribution in the euphotic zone may also

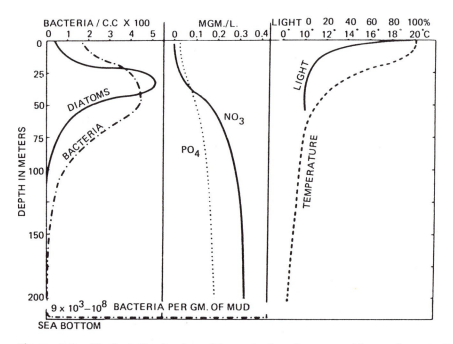

Figure 5.8 Vertical distribution of bacteria (number per cubic centimeter of water), diatoms (number per liter of water), PO_4, NO_3 (milligrams per liter), light, temperature in the sea based upon average results at several different stations off the coast of southern California. (Reproduced from ZoBell, C. E., Bacteria of the Marine World, Scientific Monthly 55:320–330 Copyright © 1942 by the American Association for the Advancement of Science; by permission.)

be limited by temperature and to some extent by salinity, and by available dissolved nitrogen (nitrate) and phosphorus (phosphate). Nonphotosynthetic microorganisms are limited to certain zones in the water column of the oceans by nutrient availability, and in addition, by temperature, salinity, and hydrostatic pressure. The nonphotosynthetic microorganisms include predators (zooplankton), scavengers (zooplankton, fungi), and decomposers (bacteria and fungi and, to a possibly small extent, zooplankton). Zooplankters therefore dominate the euphotic zone, where they can feed optimally on phytoplankton, zooplankton and bacteria. Bacteria and fungi are also very prevalent here, because they find sufficient sources of nutrients produced by phytoplankton and zooplankton. Very high bacterial populations occur at the air-water interface of the ocean as a result of concentration of organic carbon in the surface film, where it is as much as 1,000-fold greater than in the water column below. The bacterial population in this film is known as **bacterioneuston** (Sieburth, 1976; Sieburth et al., 1976; Wangersky, 1976).

Marine sediments contain significant numbers of bacteria, fungi, and other benthic microorganisms, as well as higher forms of life. Viable bacteria have been recovered from 350 cm below the sediment surface (Rittenberg, 1940). Their numbers usually decrease with increasing depth in the sediment column. Fungi seem commonly to be restricted to sediments at shallow water depths, whereas bacteria, protozoans, and metazoans are found associated with sediments of shallow as well as abyssal depths. The chief function of the microbes is to aid in scavenging or decomposing organic matter that has settled undecomposed or partially decomposed from the overlying regions of biological productivity. Most of the organic matter from the euphotic zone settles to the bottom in the form of fecal pellets from metazoans. It should be pointed out that not all settled organic matter in deep-sea sediments is utilizable by microbes, for reasons that are not yet clearly understood. This unutilizable organic matter constitutes a significant part of **sedimentary humus**.

The metabolic activity of free-living bacteria of deep-sea sediments has been shown to be at least 50 times lower than that of microorganisms in shallow waters or on sediments at shallow depths (Jannasch et al, 1971; Jannasch and Wirsen, 1973; Wirsen and Jannasch, 1974). Environmental factors contributing to this slow rate of bacterial metabolism seem to be the low temperature ($<5°C$) and, especially, hydrostatic pressure. Turner (1973) observed that pieces of wood left on sediment at a station in the Atlantic Ocean at a depth of 1,830 m for 104 days was rapidly attacked by two species of wood borers (mollusks). This observation has led to the suggestion that the primary attackers of organic matter in the deep sea, including sediment, are metazoans, and that bacteria and other microbes harbored in the digestive tract of these metazoans decompose this organic matter only after ingestion by the metazoans (see, for instance, Jannasch, 1979). The bacteria appear essential in the digestion of cellulose to enable these metazoans to assimilate it.

Schwartz and Colwell (1976) were the first to report that the bacterial flora of the intestines of amphipods (crustaceans), which they collected in the Pacific Ocean at a depth of 7,050 m, was able to grow and metabolize nearly as rapidly at 780 atm and 3°C as at 1 atm and 3°C in laboratory experiments. Their study suggested that these types of gut bacteria behave very differently from free-living bacteria from the same depths. These findings have been extended by other observations on amphipod microflora. Deming and Colwell (1981) reported finding barophilic and barotolerant bacteria in the intestinal flora of amphipods living at 5,200–5,900 m. Yayanos and co-workers (1979) isolated a barophile from a decomposing amphipod that grew optimally at ~500 bars and 2–4°C and poorly at atmospheric pressure in this temperature range. The same workers (Yayanos et al., 1981) isolated an obligately barophilic bacterium from an amphipod recovered from 10,746 m in the Marianas Trench.

The generally low rate of metabolism of the biological community (benthos) on deep-sea sediments is also reflected by respiratory measurements carried out at 1,850 m. The measurements revealed a rate of oxygen consumption that was orders of magnitude less than in sediments at shallow shelf depths (Smith and Teal, 1973).

Phytoplankton, zooplankton, bacteria, and fungi are not found in very significant numbers at intermediate depths, chiefly because of a lack of an adequate nutrient supply, including a suitable source of energy, or because of the low temperature. Kriss (1970), having examined a north-south transect of the Atlantic Ocean, found an uneven distribution of bacteria at intermediate depths, which he attributed to the different origins and characteristics of particular water masses. He interpreted his findings on the basis of available metabolizable nutrients, claiming that higher concentrations occur in water masses of equatorial-tropical origin, owing to **autochthonous** (of native, e.g., planktonic, origin) and **allochthonous** (from runoff from continents and islands) contributions, than in water masses of Arctic and Antarctic origin.

Bacterial and fungal growth and reproduction in ocean water also occurs on surfaces of some living organisms and on the surface of suspended organic and inorganic particles (epiphytes) because at these sites essential nutrients may be very concentrated (Sieburth, 1975, 1976; Hermansson and Marshall, 1985). They may form a biofilm on these surfaces. Detritus, even if by itself not a nutrient, usually has adsorption capacity, which helps to concentrate nutrients on its surface and thus make for a preferred microbial habitat. The beneficial effect which that buildup of nutrients by adsorption to particulate surfaces has on microbial growth is great because the concentration of these nutrients in solution in seawater is very low (0.35–0.7 mg liter^{-1}; Menzel and Rhyter, 1970). ZoBell (1946) long ago showed a significant increase in the bacterial population in natural seawater during 24 h of storage in an Erlenmeyer flask. He attributed this to the adsorption of essential nutrients in the seawater to the walls of the flask where the bacteria actually grew.

Effects of Temperature and Pressure on Microbial Distribution Temperature and pressure may have profound influence on where a given nonphotosynthetic microbe may live in the ocean. Some will grow only in the temperature range 15–45°C (**mesophiles**); others will grow only in the range 0°C or slightly below to 20°C with an optimum at 15°C or below (**psychrophiles**); and still others will grow in the range 0–30° or even higher (e.g., 37°C; Ehrlich, 1983) with an optimum near 25°C (**psychrotrophs**) (Morita, 1975). Thus, mesophiles would be expected to grow only in waters of the mixed zone and near active hydrothermal vents, whereas psychrophiles would grow only below the thermocline and in the polar seas. Psychrotrophs would be expected to grow above and below the thermocline and in the polar seas, although they might do better above the thermocline. Mesophiles may be recovered from cold waters and deep sediments, where they are able to survive but cannot grow (i.e., they are **psychrotolerant**).

Many bacteria that normally grow at atmospheric pressure are not inhibited by hydrostatic pressures up to about 197 bars (200 atm), but are retarded at 296 bars (300 atm) and will not grow above 395 bars (400 atm). Many bacteria isolated from waters at 494 bars (500 atm) and 592 bars (600 atm) were found to grow better at these pressures under laboratory conditions than at atmospheric pressure. Such organisms are called **barophiles**. Some organisms described in 1957 in a pioneering study by ZoBell and Morita, which had been recovered from extreme depths (10,000 m), were suspected of having been obligate barophiles. Since that time, an obligately barophilic bacterium has been isolated from an amphipod taken at 10,476 m in the Marianas Trench and studied (Yayanos et al., 1981). It exhibited an optimal growth rate (generation time of 25 h) at 2°C and 690 bars (699 atm) of hydrostatic pressure. As already mentioned, Yayanos et al. (1979) also isolated a facultatively barophilic spirillum from 5,700 m depth which grows fastest at about 494 bars (500 atm) and 2–4°C, with a generation time of 4–13 h.

In prokaryotes, the most pressure-sensitive biochemical process is protein synthesis. It determines the degree of barotolerance and limits growth under pressure. Other processes, even nucleic acid synthesis, in the same cell are less pressure sensitive (Pope and Berger, 1973). The most pressure sensitive step in protein synthesis is translation, and the site of action is the 30S ribosomal subunit (Smith et al., 1975). (For a more complete discussion of ecological implications of temperature and pressure in the marine environment, see Marquis, 1982; Morita, 1967, 1980; Jannasch and Wirsen, 1977).

Marine microorganisms, especially bacteria, vary in their salinity requirements. Those which can grow only in a narrow range of salinities are called **stenohaline**, and those which can grow in a wide range of salinities are called **euryhaline**. Both types are found in the open ocean. Their salt requirement is not explained on the basis of osmotic pressure but by a specific requirement for one or more of the ions Na^+, K^+, Mg^{2+}, and Ca^{2+}. These ions may affect cell permeability or specific enzyme activities or both (MacLeod, 1965). They may also affect cell integrity.

Dominant Phytoplankters and Zooplankters   Diatoms, dinoflagellates, coccolithophores, and other flagellates are the dominant phytoplankters of the sea (Fig. 5.9) (Sieburth, 1979). The first two are the chief source of food for the herbivorous organisms in the seas. Diatoms are also important agents in the control of Si and Al concentrations in seawater (Mackenzie et al., 1978). Kelps and other sessile algae are mostly restricted to the shelf areas of the seas since they cannot grow at depths below about 30 m. A few kelps can float in the open ocean (e.g., Sargasso weed). The dominant members of zooplankton include not only protozoa but also invertebrates such as coelenterates, pteropods, and crustaceans, some of which are not found free-floating as adults but have planktonic larval stages. Among protozoa of the zooplankton, dominant forms include foraminifers and radiolarians. Their place in the ecology of the oceans is chiefly as predators on bacteria and some other members of the plankton population. Some of these forms also are the food for higher predatory animals. The phytoplankters are the principal **primary producers** (i.e., the chief synthesizers of organic carbon by photosynthesis) in the euphotic zone of the oceans. (For further discussion see Sieburth, 1979.)

At special sites at abyssal depths around hydrothermal vents or seeps from which H_2S or methane are discharged, primary production is the result of chemoautotrophic bacteria that obtain energy from such sources as hydrogen sulfide (Paul et al., 1984; Jannasch and Taylor, 1984; Stein, 1984; Tunnicliffe, 1992) or methane (Jannasch and Mottl., 1985; Kulm et al., 1986). This primary production is the basis for the existence of biologi-

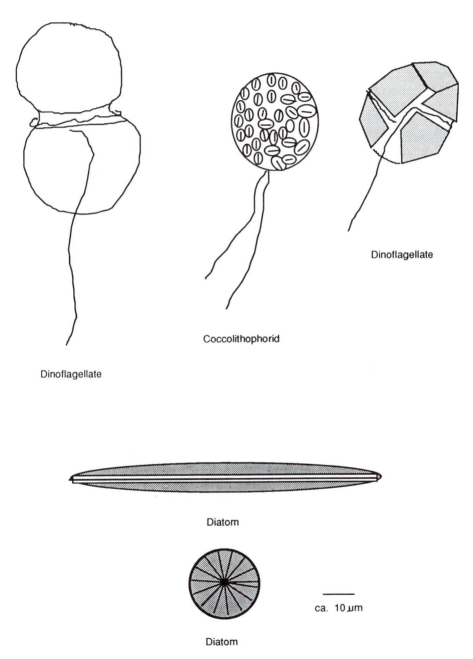

Dinoflagellate

Coccolithophorid

Dinoflagellate

Diatom

Diatom

ca. 10 μm

Figure 5.9 Sketches of important marine phytoplankters (adapted from Wood, 1965, and from Sieburth, 1979).

cal communities including metazoa and even vetebrates that are spacially restricted to the site of activity of the primary producers and their energy sources.

Plankters of Geomicrobial Interest   The phytoplankters of special geomicrobial interest include the diatoms, coccolithophores, and silicoflagellates. The zooplankters of special geomicrobial interest include the foraminifers and radiolarians. It is these organisms which precipitate much of the $CaCO_3$ and SiO_2 in the open sea that upon their death settle out and become incorporated into the sediments (see Chapters 8 and 9).

Bacterial Flora   The bacterial flora of the seas consists primarily of aerobic or facultative, gram-negative bacteria. Members of the genera *Shewanella, Pseudomonas, Vibrio,* and *Oceanospirillum* are most commonly found, although many other gram-negative genera are also found, albeit in lesser numbers and in special niches. Gram-positive bacteria such as *Arthrobacter* or sporeforming rods are relatively less common and encountered frequently in regions directly influenced by runoff from land. Most of the bacteria in the sea are aerobic or facultative. Strict anaerobes, such as the sulfate reducers are encountered mainly where organic matter accumulates in significant quantities such as in salt marshes, estuaries, and some nearshore waters. Although heterotrophic and mixotrophic bacteria are the most numerous, autotrophic bacteria are encountered in significant numbers in certain niches, such as around deep-sea hydrothermal vents and as facultative or strict symbionts of some invertebrates, residing, for instance, in trophosomes in the coelomic cavity of vestimentiferan worms and on the gills of certain mollusks. Nonsymbiotic heterotrophic, mixotrophic, or autotrophic bacteria in the marine environment may be free-living or attached to inert particles suspended in the water column (e.g., inorganic matter such as clay particles and fecal pellets) (for bacterial attachment to fecal pellets see Turner and Ferrante, 1979). They can also be detected in the upper sediment column in all parts of the ocean. Marine bacteria are usually defined as types which will not grow in media prepared in fresh water; they need one or more of the salt constituents of sea water. Bacteria which do not require seawater for growth can, however, be readily isolated from seawater and marine sediment samples far from any shore. Many of these organisms may grow very readily in media prepared in seawater. They may represent terrestrial forms. The active bacteria in the marine environment are important as decomposers and in special marine niches as primary producers. Some may also play an important role in mineral formation and diagenesis (e.g., in ferromanganese deposits)

(For further discussion see Ehrlich, 1975; Jannasch, 1984; Kulm et al., 1986; Sieburth, 1979.)

5.2 FRESHWATER LAKES

Structural Features of Lake Basins   Fresh waters are also important habitats for certain life forms. Among these habitats are lakes. They are part of the **lentic** environments, the standing waters, which include lakes, ponds, and swamps. Lakes represent 0.009% of the total water in the world (van der Leeden et al., 1990). They have arisen in various ways. Some resulted from glacial action—an advancing glacier gouged out a basin which, when the glacier retreated, filled with water from the melting ice, and was later kept filled by runoff from the surrounding watershed. Other basins resulted from landslides which obstructed valleys and blocked the outflow from their watershed, or from crustal up-and-down movement (dip-slip faulting) which formed dammed basins for the collection of runoff water. Still others resulted from solution of underlying rock, especially limestone, which led to the formation of basins in which water collected. Lakes have also been formed by the collection of water from glacier melts in craters of extinct volcanoes and by the obstruction of river flow or changes in river channels (Welch, 1952; Strahler and Strahler, 1974).

Lakes vary greatly in size. The combined Great Lakes in the United States cover an area of 328,000 km^2, an unusually great expanse. More commonly, lakes cover areas of 26–520 km^2, but many are smaller. Most lakes are less than 30 m deep. However, the deepest lake in the world, Lake Baikal in southern Siberia (U.S.S.R.), has a depth of 1,700 m. The average depth of the Great Lakes is 700 m and that of Lake Tahoe on the California-Nevada border (U.S.A.) is 487 m. The elevation of lakes ranges from below sea level (e.g., the Dead Sea at the mouth of the Jordan River) to elevations as high as 3,600 m (Lake Titicaca in the Andes on the border between Bolivia and Peru).

Some Physical and Chemical Features of Lakes   Some of the water of lakes may be in motion, at least intermittently. Most prevalent are horizontal currents, which result from wind action and the deflecting action of shorelines. Vertical currents are rare in lakes of average or small size. They may result from thermal, morphological or hydrostatic influences. Thermal influence can result in changes in water density such that heavier (denser) water sinks below lighter water. Morphological influence can result from rugged bottom topography, which may deflect horizontal water flow downward or upward. Hydrostatic influences can result from springs at the lake bottom which force water upward into the lake. Besides horizontal and vertical currents, return currents may occur as a result of water being forced against a shore by wind and piling up. Depending on the type of lake and the season of the year, only a portion of the total water mass of a lake, or all of it, may be circulated by the wind. (For a further discussion, see Welch, 1952; Strahler and Strahler, 1974).

The waters of lakes vary in composition from a very low salt content (e.g., Lake Baikal) to a very high salt content (e.g., Dead Sea between Israel and Jordan; Lake Natron in Africa), and from low organic content to high organic content. Salt accumulation in lakes is the result of input of runoff from the watershed, including stream flow, slow solution of sediment components and rock minerals in the lake bed, and evaporation.

The waters of lakes may or may not be thermally stratified, depending upon various factors: geographic location, the season of the year and lake depth and size. Stratification, when it occurs, may or may not be permanent. When the waters are not stratified, wind action can cause complete mixing or **turnover**. When the waters are thermally stratified into a warmer layer at the top (**epilimnion**) and a cooler layer below (**hypolimnion**), complete mixing is not possible because of a density difference between the two layers. Lakes may be classified according to whether and when they turn over (Reid, 1961). The categories can be defined as follows. **Amictic** lakes are bodies of water that never turn over, being permanently covered by ice. Such lakes are found in Antarctica and at high altitudes. **Cold monomictic** lakes are bodies of water that contain waters never exceeding 4°C, which turn over once during the summer, being thermally stratified the rest of the year. **Dimictic** lakes turn over twice each year, in spring and fall. They are thermally stratified at other times. These are typically found in temperate climates and at higher altitudes in subtropical regions. **Warm monomictic** lakes have water that is never colder than 4°C. They turn over once a year in winter and are thermally stratified the rest of the year. **Oligomictic** lakes contain water that is significantly warmer than 4°C and turn over irregularly. Such lakes are found mostly in tropical zones. **Polymictic** lakes have water just over 4°C and turn over continually. Such lakes occur at high altitudes in equatorial regions. **Meromictic** lakes are deep, narrow lakes whose bottom waters never mix with the waters above. The bottom waters usually have a relatively high concentration of dissolved salts, which makes them dense and separates them from the overlying waters by a **chemocline**. The upper waters in temperate climates may be thermally stratified in summer and winter and may undergo turnover in spring and fall.

A dimictic lake in a temperate zone during spring thaw accumulates water near 0°C, which, because of its lower density, floats on the remaining denser water which is near 4°C. As the season progresses, the colder surface water is slowly warmed by the sun to near 4°C. At this point, all water has a more-or-less uniform temperature and thus uniform density. This allows the water to be completely mixed or turned over by wind agitation if the lake is not excessively deep like a meromictic lake. As the surface water undergoes further warming by the sun, segregation of water masses recurs as warmer, lighter water comes to lie over colder, denser water. A **thermocline** is established between the two water masses, separating them into epilimnion and hypolimnion. The temperature of the water in the epilimnion may be greater by 10°C and vary little with depth (perhaps 1°C m^{-1}). The water in the thermocline, on the other hand, will show a rapid drop in temperature with depth. This drop may be as drastic as 5.5°C 0.3 m^{-1} but is more usually 2.4°C 0.3 m^{-1}. The thickness of the thermocline varies with position in the lake and between different lakes, an average value being around 1 m. The water in the hypolimnion will have a temperature well below that in the epilimnion and show a small drop in temperature with depth, usually less than 1°C m^{-1}. The water in the epilimnion but not in the hypolimnion is subject to wind agitation and is thus fairly well mixed at all times. It is the greater density of the water of the hypolimnion that prevents its mixing by the wind. Continual warming by the sun and mixing by the wind produces horizontal currents and, in larger lakes, return currents over the thermocline, resulting in some exchange with water of the thermocline. This water exchange progressively increases the volume of the epilimnion and causes a progressive drop in the position of the thermocline. At fall turnover, the thermocline will have touched bottom in the lake and disappeared, the water now having uniform warm temperature. With the approach of winter, the lake water will cool. Once the surface water has cooled below 4°C, a thermocline will be reestablished. Ice may form on the epilimnion if the water temperature reaches the freezing point. The winter thermocline will usually remain near the lower surface of any ice cover on the lake. Figure 5.10 shows in idealized form the seasonal cycle of thermal stratification of a dimictic lake.

The thermocline of a lake acts as a barrier between the epilimnion and the hypolimnion. It prevents easy exchange of salts, dissolved organic matter, and gases because the two water masses which it separates do not readily mix owing to their difference in density. The oxygen content in the epilimnion is usually around the saturation level. At times of intense photosynthetic activity of phytoplankton, oxygen supersaturation may be achieved. The source of oxygen in the epilimnion is photosynthesis and

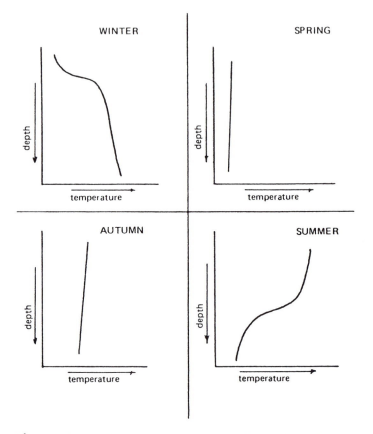

Figure 5.10   Schematic representation of thermal stratification in a dimictic lake at different seasons in a year.

aeration, especially during wind agitation. The oxygen concentration in fresh water at saturation at 0°C is 14.62 ppm or 10.23 ml liter^{-1}; at 15°C it is 10.5 ppm or 7.10 ml liter^{-1}; and at 20°C it is 9.2 ppm or 6.5 ml liter^{-1}. Optimal conditions of light and oxygen concentration together with adequate nutrient supply permit phytoplankton, zooplankton, bacteria, fungi and fish life to attain their greatest numbers in the waters of the epilimnion. The phytoplankters are the primary producers on which the remaining life forms depend directly or indirectly for food.

Oxygen that was introduced into lake waters by aeration during spring turnover will be gradually depleted from the hypolimnion of a fertile (eutrophic) lake after thermal stratification because of biological activity, especially on and in the sediment. Thus, the hypolimnion may be anoxic during a shorter or longer period of time before fall turnover. Only anaerobic or facultative organisms will carry on active life processes under these anoxic conditions. Such organisms include bacteria and protozoans as well as certain nematodes, annelids, immature stages of some insects, mollusks, and some fishes (Welch, 1952; Strahler and Strahler, 1974). Figure 5.11 illustrates the oxygen distribution measured in a dimictic lake during summer stratification.

Lake Bottoms   The nature of lake bottoms is highly variable, depending on the location and history of the lake. The basins of many smaller lakes are flat expanses of sediment overlying bedrock. On the other hand, the basins of larger lakes (e.g., the Great Lakes) have a more rugged topography in many places. The bottom of lakes may be dominated by sand and grit, by clay, by a brown mud rich in humus, by diatom oozes, by ochreous mud rich in limonitic iron oxide, or by calcareous deposits. The organic components of any sediment may derive from dead or dying plankters that have sunk to the bottom, or from plant or animal remains. Some inorganic and some organic components may have been introduced into the lake and its sediment by the wind. Much silt and clay is washed into

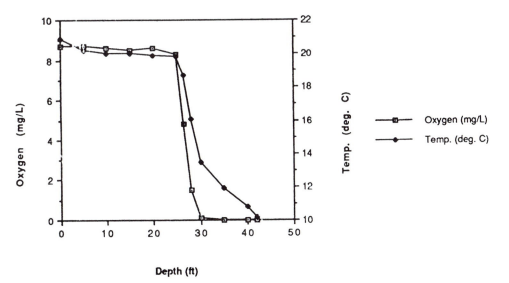

Figure 5.11   Oxygen profile in Tomhannock Reservoir (near Troy, NY) on September 5, 1967. (From LaRock, 1969, with permission.)

lakes by runoff. Some is also contributed by wind erosion of the shoreline. The sediments are a major habitat of microbes.

Lake Fertility   Lakes may be classified in terms of their fertility or their nutritional status (i.e., their ability to support a flora and fauna). **Oligotrophic** lakes have an impoverished nutrient supply in which phosphorus or nitrogen are in short supply and in which the oxygen concentration is high at all depths. **Eutrophic** lakes, on the other hand, are fertile lakes in which phosphorus and nitrogen are in more plentiful supply than in oligotrophic lakes. **Mesotrophic** lakes are intermediate between oligotrophic and eutrophic lakes. **Dystrophic** lakes are defined as having an oversupply of organic matter which cannot be completely decomposed because of an insufficiency of oxygen and alternative terminal electron acceptors such as ferric iron, nitrate or sulfate, and sometimes assimilable nitrogen or phosphorus. The waters of such lakes is turbid and often acid. The origin of the dystrophic condition may be the encroachment of the shoreline by plants, including reeds, shrubs, and trees.

Lake Evolution   Lakes have an evolutionary history. Once fully matured, they age progressively. Their basins slowly fill with sediment, partly contributed by the surrounding land through runoff and erosion and partly by the biological activity in the lake. The size of the contribution that each process makes depends on the fertility of the lake. Changes in climate may also contribute to lake evolution (e.g., through lessened rainfall, which can cause a drop in water level, or through increase in temperature, which can cause more rapid evaporation). These effects usually make themselves felt slowly. Ultimately, a lake may change into a swamp.

Microbial Populations in Lakes   The microbial population in eutrophic lakes tends to be orders of magnitude greater than in the seas. Numbers of culturable bacteria may range from 10^2 to 10^5 per milliliter of lake water and be in the order of 10^6 per gram of lake sediment. The size of the bacterial population may be affected by runoff, which contributes soil bacteria. The bacterial population of lakes consists predominantly of gram-negative rods (Wood, 1965, p. 36), although gram-positive sporeforming rods and actinomycetes can be readily isolated, especially from sediments. Few, if any, of the bacterial types found in lakes are exclusively limnetic organisms. The main activity of the bacteria, other than the cyanobacteria, is that of decomposers. The cyanobacteria along with the algae serve as the primary producers. Fungi and protozoans are also found. Important functions of the former are as scavengers and decomposers and of the latter as predators.

Cyanobacteria and algae are abundant in eutrophic lakes. The algae include green forms as well as diatoms and pyrrhophytes. **Cyanobacterial** and **algal blooms** may occur at certain times when one species suddenly multiplies explosively and becomes the dominant algal form temporarily, often forming a carpetlike layer or mat on the water surface. After having reached a population peak, most of the cells in the algal bloom die off and are attacked by scavengers and decomposers. Especially favorable growth conditions appear to be the stimulus for such blooms.

5.3 RIVERS

Rivers, which account for only 0.0001% of the total water in the world (van der Leeden et al., 1990), are part of the **lotic** environment in which water moves in channels on the land surface. Such flowing water may start as a brook, then widen into a stream and ultimately into a river. The source of the water is surface runoff and ground water reaching the surface through springs or, more important, through general seepage. A riverbed is shaped and reshaped by the flowing water that scours the bottom and walls, especially with the help of suspended particles from clays to small stones. Young rivers may feature rapids and steep valley slopes. Mature rivers lack rapids and feature more uniform stream flow, owing to a smoothly graded riverbottom and an ever-widening riverbed. Old rivers may develop meanders in their wide, flat floodplains. The flow of the water is caused by gravity because the head of a river always lies above its mouth. Average flow rates of rivers range from 0 to 9 m sec^{-1}. However, the flow of water in a river cross-section is not uniform. Some portions in such a section flow much faster than others. This can be attributed to frictional effects related to the riverbed topography as well as to density differences of different parts of the water mass. Density variations may arise from temperature differences or from solute concentration differences between parts of a river. Portions of river water may exhibit strong turbulence engendered in part by river topography. Water velocity, turbulence, and terrain determine the size of particles a river may sweep along (see, e.g., Strahler and Strahler, 1974; Stanley, 1985).

Most river water ultimately is discharged into an ocean, but exceptions exist. The Jordan River, whose headwaters originate in the mountains of Syria and Lebanon, empties into a lake called the Dead Sea, which has no connection with any ocean. The Dead Sea does not overflow because it loses its water by evaporation, which accounts, in part, for its high salt accumulation. Its waters are nearly saturated with salts, which makes life impossible except for specially adapted organisms. When river water is discharged into an ocean, an estuary is frequently formed where the less dense water will flow over the denser saline water from the sea with incomplete mixing. Tidal effects of the sea may alter the water level of the discharging river, sometimes for a considerable distance upstream. Estuaries form special habitats for microbes and higher forms of life, which must cope with periodic changes in salinity, water temperature, nutrient and oxygen availability, and so forth, engendered by tidal movement.

Because of constant water movement, the water temperature of rivers tends to be rather uniform (i.e., rivers generally are not thermally stratified when examined in cross section). Only where a tributary at a different water temperature enters a stream may there be local temperature stratification. Different segments of a river may, however, differ in temperature. The pH of river water can range from very acid (pH 3), for instance in streams receiving acid mine drainage, which is the result of microbial activity, to alkaline (pH 8.6) (Welch, 1952; p. 413). Unless heavily polluted by human activities, rivers are generally well aerated. It has been thought that in unpolluted rivers most organic and inorganic nutrients supporting microbial as well as higher forms of life are largely introduced by runoff (allochthonous). More recently, it was suggested that a significant portion of fixed carbon in such streams and rivers may be contributed by autotrophs, mainly algae, growing in quiet waters (autochthonous) (Minshall, 1978). Pollution may cause organic overloading, which because of excessive oxygen demand will result in anoxic conditions with the consequent elimination of many micro- as well as macroorganisms.

Planktonic organisms tend to be found in greater numbers in the more stagnant or slower-flowing waters of a river than in fast-flowing portions. The plankters include algae such as diatoms, cyanobacteria, green algae, protozoa, and rotifers. The proportions depend on the condition of a particular river and its sections. Sessile plants or algae tend to develop to significant extents only in sluggish streams or in the backwaters of otherwise rapidly flowing streams. Bacteria are represented in significant numbers where physical and chemical conditions favor them. Rheinheimer (1980) indicated total bacterial numbers in the River Elbe in Germany to range from 4.7 to 6.9 $\times$ 10^9 L^{-1} and bacterial biomass to range from 0.55 to 0.71 mg L^{-1} in an unspecified year. As in lakes, no unique flora occurs in unpolluted rivers.

5.4 GROUNDWATERS

Water that collects below the land surface, in soil, sediment, and permeable rock strata is called **groundwater**. It represents 0.61% of all water in the world (van der Leeden et al., 1990). Groundwater derives mainly from **surface water** whose origin is meteoritic precipitation such as rain and melted snow. Surface water includes the water of rivers, lakes and the like (Fig. 5.12). A minor amount of groundwater derives from **connate water, water of dehydration**, or **juvenile water**. Connate water, often of marine origin and therefore saline, is water which became trapped in rock strata in the geologic past by up- or downwarping or faulting and, as a consequence, became isolated as a stagnant reservoir. Its salt composition frequently became highly altered from that of the original water from which it derived as a result of long-term interaction with the enclosing rock. Connate waters are often associated with oil formations. Waters of dehydration are derived from waters of crystallization, which are part of the structure of certain crystalline minerals. They are released as a result of the action of heat and pressure in the lithosphere. Juvenile waters are associated with magmatism, which causes them to escape from the interior of the Earth. They had never before reached the Earth's surface.

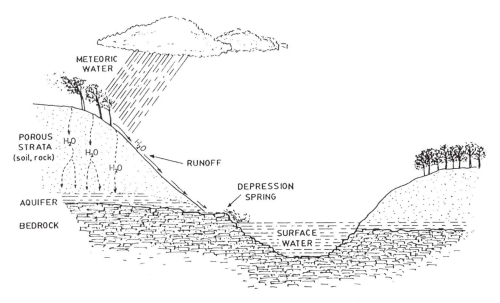

Figure 5.12 Interrelation of meteoric, surface, and ground waters.

Surface water slowly infiltrates permeable ground as long as the ground is not already saturated. It passes through a zone of aeration or unsaturation called the **vadose zone** to the zone of saturation or **aquifer**, which lies over an impermeable rock stratum (Strahler and Strahler, 1974). The vadose zone may include the soil, an intermediate zone, and the capillary fringe and can range in thickness from from a few centimeters to one hundred meters or more. The resident water in the vadose zone is pore water (pellicular water; see Chapter 4) which is under less than atmospheric pressure and held there by capillarity. In soil this water supports the soil micro- and macroflora and fauna and plant growth. The water in the aquifer but not that in the vadose zone can be extracted through human intervention (Hackett, 1972). The rate of infiltration of permeable strata depends not only on the surface water supply but also on the porosity of the permeable strata. Similarly, the water-holding capacity of the aquifer depends on the porosity of its matrix rock. The cause of water movement below ground level is not only gravity but also intermolecular attraction between water molecules, capillary action, and hydrostatic head (see Chapter 4).

At a given location, two or more aquifers may occur, one over the other separated by one or more impervious strata. An example of such multiple aquifers is to be found in the Upper Atlantic Coastal Plain Province in South Carolina, U.S.A. (Sargent and Fliermans, 1989). In such a sequence of aquifers, the uppermost may be directly rechargeable with surface water from above and is then called an **unconfined aquifer**, or it may be a **perched aquifer** if it is separated from a larger aquifer just below by a thin lens of material with low or no porosity, called an **aquiclude**. The lens material may be clay or shale. The underlying aquifers are rechargeable only where the overlying confining stratum is absent. They are called **confined aquifers**. The matrix of aquifers may consist of one of the following materials: sand and gravel, limestone and other soluble rock, basalt and other volcanic rocks, sandstone and conglomerate, crystalline and metamorphic rocks, and other porous but poorly permeable materials (Hackett, 1972).

Groundwater may escape to the surface or into the atmosphere through **springs** or by evaporation with or without the mediation of plants (**transpiration**). Some water will be accumulated by the vegetation itself. Depending on the relative rates of water infiltration and water loss, the level of the **water table** in the ground may rise, fall, or remain constant. Groundwater that reaches the surface through springs may do so under the influence of gravity, which may create sufficient head to force the water to the surface through a channel, as in an **artesian spring**. Groundwater may also reach the surface as a result of an intersection of the water table with the land surface, as in a **depression spring**. Finally, groundwater may reach the surface in springs under the influence of thermal energy applied to reservoirs deep underground. Such **hot springs** in their most spectacular form are **geysers** from which hot water spurts forth intermittently. Some hot springs emit not only water derived from infiltration of surface water but, in addition, from juvenile water.

As it infiltrates permeable soil and rock strata, surface water will undergo changes in the composition of dissolved and suspended organic and inorganic matter. These changes are the result of adsorption and ion exchange by surfaces of soil and rock particles and the biochemical action of microbes, including bacteria, fungi, and protozoa, which exist mainly in biofilms on the surface of many of the rock particles and metabolize the adsorbed organic and (to a limited extent) inorganic matter (Cullimore, 1992; Costerton et al., 1994). Polluted water infiltrating the ground may thus become thoroughly purified, provided that it moves through a sufficient depth and does not encounter major cracks and fissures which, because of reduced surface area, would exert only limited "filtering" action. Under some circumstance, groundwater may also become highly mineralized during infiltration or after reaching the water table. Such mineralized water when reaching the land surface may leave extensive deposits of calcium carbonate, iron oxide, and other material as it evaporates.

Systematic studies of groundwater microbes are now being undertaken. As many as 10^6 bacteria per gram have been recovered from the vadose zone and some shallow-watertable aquifers. They included gram-positive and gram-negative types, the former in apparent greater numbers (Wilson et al., 1983). Evidence of fungal spores and yeast cells was also seen in one instance (Ghiorse and Balkwill, 1983), but eukaryotic microbes were generally thought to be absent from subsurface samples associated with groundwater. Sinclair and Ghiorse (1987) showed that the number of protozoans, mostly flagellates and amebae, decreased sharply to 28 g^{-1} (dry wt) with increasing depth in the vadose zone at the bottom of a clay loam subsoil material taken from a site in Lula, Oklahoma, and were absent in the saturated zone except in a gravelly, loamy sand matrix at a depth of 7.5 m, which also contained significant numbers of bacteria.

Special deep-drilling methods that minimize the possibility of microbial contamination were recently developed (see Fredrickson et al., 1993; Russell et al., 1992). They are allowing the study of the microbiology of deep vadose zones and aquifers. Frederickson et al. (1991) reported as many as 10^4 and 10^6 colony-forming heterotrophic bacteria per gram of Middendorf (ca. 366–416 m deep) and Cape Fear sediments (ca. 457–470 m deep) of Cretaceous age, respectively, which were obtained by drilling into the Atlantic Coastal Plain. The isolates from the two sediments were physiologically distinct. Contrary to what they expected from their study of a shallow aquifer at Lula, Oklahoma (see above), Sinclair and Ghiorse (1989) found significant numbers of fungi and protozoa in samples of deep aquifer material, the numbers being highest where the prokaryotic population was high. The bacteria in these deep subsurface samples were represented by diverse physiological groups that included autotrophs, such as sulfur oxidizers, nitrifiers, and methanogens (Frederickson et al., 1989; R.E. Jones, 1989), as well as heterotrophs, such as aerobic and anaerobic mineralizers. The latter included denitrifiers, sulfate reducers, and methanogens (Balkwill, 1989; Hicks and Frederickson, 1989; Phelps et al., 1989; Madsen and Bollag, 1989; Francis et al., 1989; Jones et al., 1989).

Much remains to be learned about the microbial populations of pristine aquifers and their associated vadose zones and the response of this population to environmental stress (pollution).

5.5 SUMMARY

The hydrosphere is mainly marine. It occupies more than 70% of the Earth's surface. The world's oceans reside in basins surrounded by the margins of continental land masses which project into the sea by way of the continental shelf, the continental slope, and the continental rise, and bottoming out at the ocean floor. The ocean floor is traversed by mountain ranges cut by fracture zones and rift valleys—the midocean spreading centers—where new ocean floor is being formed. The ocean floor is also cut by deep trenches, the zones of subduction, where the margin of an oceanic crustal plate slips beneath continental crustal plate. Parts of the ocean floor also feature isolated mountains, which are live or extinct volcanoes. They may project above sea level as islands. The average world ocean depth is 3,975 m, the greatest depth is about 11,000 m in the Marianas Trench.

Most of the ocean floor is covered by sediment of 300 m average thickness, accumulating at rates of less than 1 cm to greater than 10 cm per 10^3 years. Ocean sediments may consist of sand, silt, and/or clays of terrigenous origin and of oozes of biogenic origin, such as diatomaceous radiolarian or calcareous oozes.

Different parts of an ocean are in motion at all times, driven by wind stress, the Earth's rotation, density variations, and gravitational effects exerted by the sun and the moon. Surface, subsurface, and bottom currents have been found in various geographical locations.

Where water masses diverge, upwelling of deeper waters, which replenish nutrients for plankton in the surface waters, occurs. Where the water masses converge, surface water sinks and carries oxygen to deeper levels of the ocean, ensuring some degree of oxygenation at all levels.

Seawater is saline (average salinity about 35‰) owing to the presence of chloride, sodium, sulfate, magnesium, calcium, potassium, and some other ions. Variations in total salt concentrations affects the density of seawater as does variation in water temperature. The ocean is thermally stratified between 50° north and 50° south latitude into a mixed layer (about 300 m deep), with water at more or less uniform temperature between 15° and 30°C depending on latitude; a thermocline (about 700 m deep) in which the temperature drops to about 4°C with depth; and the deep water (from the thermocline to the bottom), where the temperature is uniform between less than 0° and 4°C. Hydrostatic pressure in the water column increases by about 1.013 bars (1 atm) for every 10 m of increase in depth. Light penetrates to an average depth of about 100 m, which restricts phytoplankton to shallow depths. Zooplankton and bacterioplankton can exist at all depths but are found in greatest numbers at the seawater-air interface, near where the phytoplankton abounds, and on the ocean sediment. Intermediate depths are at most sparsely inhabited because of limited nutrient supply.

Marine phytoplankton is constituted of algal forms, mainly diatoms, dinoflagellates, and coccolithophores, while marine zooplankton is constituted mainly of flagellated and amoeboid protozoans as well as some small invertebrates. Bacterioplankton is composed of bacteria, chiefly heterotrophs. The phytoplankton organisms are the primary producers, the zooplankton organisms are the predators and scavengers, and the heterotrophic bacteria are the decomposers. The metabolic rate of microorganisms decreases markedly with depth, probably as a result of the effects of high hydrostatic pressure and low temperature. Different life forms in the ocean show different tolerances to salinity, temperature, and pressure.

Fresh water is found on land in lakes and streams above ground and in aquifers below ground. Lakes are standing bodies of water, usually of low salinity which may be thermally stratified into epilimnion, thermocline, and hypolimnion. The degrees of stratification may vary with the season of the year. Water below the thermocline (i.e., in the hypolimnion) may develop anoxia because the thermocline is an effective barrier to diffusion of oxygen into it. Only after the disappearance of the thermocline do these waters become reoxygenated again due to total mixing by wind agitation. Lakes vary in their nutrient quality. Phosphorus is usually the most limiting element to lake life. Phytoplankton, zooplankton, and bacterioplankton are important life forms in lakes. Phytoplankton is restricted to the epilimnion, whereas zooplankton and bacterioplankton together with fungi are found in the entire water column and in the sediment.

Rivers constitute moving fresh waters, which are generally not thermally stratified. Abundant life forms, such as phytoplankton, zooplankton, and bacterioplankton, are concentrated mainly in the quieter portions of the streams, especially when unpolluted.

Groundwaters are derived from surface waters that seep into the ground and accumulate above impervious rock strata as aquifers. Water from an aquifer may come to the surface again by way of springs or seepage. In passing through the ground, water is purified. Microorganisms as well as organic and inorganic chemicals are removed by adsorption to rock and soil particles. Organic matter may be mineralized by the microbial decomposers. Sediment samples from shallow as well as deep aquifers have revealed the presence of a significant microbial population with very diverse physiological potentials.

REFERENCES

Balkwill, D. L. 1989. Numbers, diversity, and morphological characteristics of aerobic, chemoheterotrophic bacteria in deep subsurface sediments from a site in South Carolina. Geomicrobiol. J. 7:33–52.

Bischoff, J. L., and R. J. Rosenbauer. 1983. A note on the chemistry of seawater in the range of 350°-500°C. Geochim. Cosmochim. Acta 47:139–144.

Bowden, K. F. 1975. Oceanic and estuarine mixing processes. *In*: J. P. Riley and G. Skirrow, eds. Chemical Oceanography, Vol. 1, 2nd edition, Academic Press, London.

Costerton, J. W., Z. Lewandowski, D. DeBeer, D. Caldwell, D. Korber, and G. James. 1994. Biofilms, the customized microniche. J. Bacteriol. 176:2137–2142.

Cullimore, D. R. 1992. Practical Groundwater Microbiology. Lewis Publishers, Chelsea, MI.

Defant, A. 1961. Physical Oceanography, Vol. 1, Pergamon Press, Oxford.

Deming, J. W., and R. R. Colwell. 1981. Barophilic bacteria associated with deep-sea animals. Bioscience 31:507–511.

Dietrich, G., and K. Kalle. 1965. Allgemeine Meereskunde. Eine Einfuehrung in die Ozeanographie. Gebrueder Borntraeger, Berlin Nikolassee.

Edmond, J. M., K. L. Von Damm, R. E. McDuff, and C. I. Measures. 1982. Chemistry of hot springs on the East Pacific Rise and their effluent dispersal. Nature 297:187–191.

Ehrlich, H. L. 1975. The formation of ores in the sedimentary environment of the deep sea with microbial participation: The case for ferromanganese concretions. Soil Sci. 119:36–41.

Ehrlich, H. L. 1983. Manganese oxidizing bacteria from a hydrothermally active area on the Galapagos Rift. Ecol. Bull. (Stockholm) 35: 357–366.

Francis, A. J., J. M. Slater, and C. J. Dodge. 1989. Denitrification in deep subsurface sediments. Geomicrobiol. J. 7:103–116.

Frederickson, J. K., T. R. Garland, R. J. Hicks, J. M. Thomas, S. W. Li, and K. M. McFadden. 1989. Lithotropohic and heterotrophic bacteria in deep subsurface sediments and the relationship to sediment properties. Geomicrobiol. J. 7:53–66.

Fredrickson, J. K., D. L. Balkwill, J. M. Zachara, S-M. W. Li, F. J. Brockman, and M. A. Simmons. 1991. Physiological diversity and distributions of heterotrophic bacteria in deep Cretaceous sediments in the Atlantic Coastal Plain. Appl. Environ. Microbiol. 57:402–411.

Fredrickson, J. K., F. J. Brockman, B. N. Bjornstad, P. E. Long, S. W. Li, J. P. McKinley, J. V. Wright, J. L. Conca, T. L. Kieft, and D. L. Balkwill. 1993. Microbial characteristics of pristine and contaminated deep vadose sediments from an arid region. Geomicrobiol. J. 11:95–107.

Ghiorse, W. C., and D. L. Balkwill. 1983. Enumeration and morphological characterization of bacteria indigenous to subsurface environments. Dev. Indust. Microbiol. 24:213–224.

Gross, M. G. 1982. Oceanography: A View of the Earth. 3rd ed. Prentice-Hall, Englewood Cliffs, NJ.

Hackett, O. M. 1972. Groundwater. pp. 470–478. *In*: R.W. Fairbridge, ed. The Encyclopedia of Geochemistry and Environmental Sciences. Encylopedia to the Earth Sciences series Vol; IVA. Van Nostrand Reinhold, New York.

Hermansson, M., and K. C. Marshall. 1985. Utilization of surface localed substrate by non-adhesive marine bacteria. Microb. Ecol. 11: 91–105.

Hicks, R. J., and J. K. Frederickson. 1989. Aerobic metabolic potentials of microbial populations indigenous to deep subsurface environments. Geomicrobiol. J. 7:67–77.

Horn, D. R., B. M. Horn, and M. N. Delach. 1972. Sedimentary Provinces, North Pacific (Map). Lamont-Doherty Observatory of Columbia University, Palisades, NY.

Jannasch, H. W. 1979. Microbial turnover of organic matter in the deep sea. BioScience 29:228–232

Jannasch, H. W. 1984. Microbial processes at deep sea hydrothermal vents. pp. 677–709. *In*: P. A. Rona, K. Bostrom, L. Laubier, and K. L. Smith, Jr., eds. Hydrothermal Processes at Seafloor Spreading Centers. Plenum Press, New York.

Jannasch, H. W., and M. J. Mottl. 1985 Geomicrobiology of deep-sea hydrothermal vents. Science 229:717–725.

Jannasch, H. W., and C. D. Taylor. 1984. Deep-sea microbiology. Annu. Rev. Microbiol. 38:487–514.

Jannasch, H. W., and C. O. Wirsen. 1973. Deep-sea microorganisms: in situ response to nutrient enrichment. Science 180:641–643.

Jannasch, H. W., and C. O. Wirsen. 1977. Microbial life in the deep sea. Sci. Am. 236:2–12.

Jannasch, H. W., K. Eimhjellen, C. O. Wirsen, and A. Farmanfarmaian. 1971. Matter in the deep sea. Science 171:672–675.

Jones, R. E., R. E. Beeman, and J. M. Suflita. 1989. Anaerobic metabolic processes in deep terrestrial subsurface. Geomicrobiol. J. 7:117–130.

Kester, D. R. (1975). Dissolved gases other than CO_2, pp. 497–547. *In*: J. P. Riley and G. Skirrow, eds. Chemical Oceanography, Vol. 1, 2nd ed. Academic Press, New York.

Kulm, L. D., E. Suess, J. C. Moore, B. Carson, B. T. Lewis, S. D. Ritger, D. C. Kado, T. M. Thornburg, R. W. Embley, W. D. Rugh, G. J. Massoth, M. G. Langseth, G. R. Cochrane, and R. L. Scamman. 1986. Oregon subduction zone: Venting, fauna, and carbonates. Science 231:561–566.

Kriss, A. E. 1970. Ecological-geographic patterns in the distribution of heterotrophic bacteria in the Atlantic Ocean. Mikrobiologiya 39: 362–371 (Engl. transl. pp. 313–320).

LaRock, P. A. 1969. The bacterial oxidation of manganese in a fresh water lake. Ph.D. thesis. Rensselaer Polytechnic Institute, Troy, NY.

Littler, M. M., D. S. Littler, S. M. Blair, J. N. Norris. 1985. Deepest known plant life discovered on an uncharted seamount. Science 227: 57–59.

Mackenzie, F. T., M. Stoffym, and R. Wollast. 1978. Aluminum in seawater: Control by biological activity. Science 199:680–682.

MacLeod, R. A. 1965. The question of the existence of specific marine bacteria. Bacteriol. Rev. 9–23.

Madsen, E. L., and J. M. Bollag. Aerobic and anaerobic microbial activity in deep subsurface sediments from the Savannah River Plant. Geomicrobiol. J. 7:93–101.

Marine Chemistry. 1971. A report to the Marine Chemistry Panel of the Committee of Oceanography. National Academy of Sciences, Washington, DC.

Marquis, R. E. 1982. Microbial barobiology. BioScience 32:267–271.

Menzel, D. W., and J. H. Rhyter. 1970. Distribution and cycling of organic matter in the ocean, pp. 31–54. *In*: D. W. Wood, ed. Organic Matter in Natural Waters. Institute of Marine Science. Occasional Publication No. 1. University of Alaska, Fairbanks.

Minshall, G. W. 1978. Autotrophy in stream ecosystems. BioScience 28: 767–771.

Morita, R. Y. 1967. Effects of hydrostatic pressure on marine bacteria. Oceanogr. Mar. Biol. Annu. Rev. 5:187–203.

Morita, R. Y. 1975. Psychropilic bacteria. Bacteriol. Rev. 39:144–167.

Morita, R. Y. 1980. Microbial life in the deep sea. Can. J. Microbiol. 26: 1375–1385.

Nissenbaum, A. 1979. Life in the Dead Sea—Fables, allegories, and scientific research. BioScience 29:153–157.

Overmann, J., H. Cypionka, and N. Pfennig. 1992. An extremely low-light-adapted phototrophic sulfur bacterium from the Black Sea. Limnol. Oceanogr. 37:150–155.

Park, P. K. 1968. Seawater hydrogen-ion concentration: Vertical distribution. Science 162:357–358.

Paull, C. K., B. Hecker, R. Commeau, R. P. Freeman-Lynde, C. Neumann, W. P. Corso, S. Golubic, J. E. Hook, E. Sikes, and J. Curray. 1984. Biological communities at the Florida Escarpment resemble hydrotherma vent taxa. Science 226:965–967.

Phelps, T. J., E. G. Raione, D. C. White, and C. B. Fliermans. 1989. Microbial activities in deep subsurface environments. Geomicrobiol. J. 7:79–91.

Pope, D. H., and L. R. Berger. 1973. Inhibition of metabolism by hydrostatic pressure:what limits microbial growth? Arch. Mikrobiol. 93: 367–370.

Reid, G. K. 1961. Ecology of Inland Waters and Estuaries. Reinhold Publishing Corp., New York.

Rheinheimer, G. 1980. Aquatic Microbiology. 2nd ed. Wiley, Chichester.

Richardson, P. L. 1993. Tracking ocean eddies. Am. Sci. 81:261–271.

Ring Group. 1981. Gulf-Stream cold-core rings: Their physics, chemistry, and biology. Science 212:1091–1100.

Rittenberg, S. C. 1940. Bacteriological analysis of some long cores of marine sediments. J. Mar. Res. 3:191–201.

Russell, B. F., T. J. Phelps, W. T. Griffin, and K. A. Sargent. 1992. Procedures for sampling deep subsurface microbial communities in unconsolidated sediments. Ground Water Monitor Rev. 12:96–104.

Sargent, K. A., and C. B. Fliermans. 1989. Geology and hydrology of the deep subsurface microbiology sampling sites at the Savannah River Plant, South Carolina. Geomicrobiol. J. 7:3–13.

Schwartz, J. R., and R. R. Colwell. 1976. Microbial activities under deep-ocean conditions. Dev. Indust. Microbiol. 17:299–310.

Seyfried, W. E., Jr., and D. R. Janecky. 1985. Heavy metal and sulfur transport during subcritical and supercritical hydrothermal alteration of basalt: Influence of fluid pressure and basalt composition and crystallinity. Geochim. Cosmochim. Acta 49:2545–2560.

Shanks, W. C., III, J. L. Bischoff, and R. J. Rosenbauer. 1981. Seawater sulfate reduction and sulfur isotope fractionation in basaltic systems: Interaction of seawater with fayalite and magnetite at 200–350°C. Geochim. Cosmochim. Acta 45:1977–1995.

Sieburth, J. McN. 1975. Microbial Seascapes. A Pictorial Essay on Marine Microorganisms and Their Environments. University Park Press, Baltimore, MD.

Sieburth, J. McN. 1976. Bacterial substrates and productivity in marine ecosystems. Annu. Rev. Ecol. Syst. 7:259–285.

Sieburth, J. McN. 1979. Sea Microbes. Oxford University Press, New York, 491 pp.

Sieburth, J. McN. P.-J. Willis, K. M. Johnson, C. M. Burney, D. M. Lavoie, K. R. Hinga, D. A. Caron, F. W. Franck III, P. W. Johnson, and P. G. Davis. 1976. Dissolved organic matter and heterotrophic microneuston in the surface microlayers of the North Atlantic. Science 194:1415–1418.

Sinclair, J. L., and W. C. Ghiorse. 1987. Distribution of protozoa in subsurface sediments of a pristine groundwater study site in Oklahoma. Appl. Environ. Microbiol. 53:1157–1163.

Sinclair, J. L., and W. C. Ghiorse. 1989. Distribution of aerobic bacteria, protozoa, algae, and fungi in deep subsurface sediments. Geomicrobiol. J. 7:15–31.

Smith, F. G. W., ed. 1974. Handbook of Marine Science, Vol. 1. CRC Press, Cleveland, OH, p. 617.

Smith, K. L., and J. M. Teal. 1973. Deep-sea benthic community respiration: an in situ study at 1850 meters. Science 179:282–283.

Smith, R. L. 1968. Upwelling. Oceanogr. Marine Biol. Annu. Rev. 6:11–46.

Smith, W., D. Pope, and J. V. Landau. 1975. Role of bacterial ribosome subunits in barotolerance. J. Bacteriol. 124:582–584.

Stanley, S. M. 1985. Earth and Life Through Time. W. H. Freeman, New York.

Stein, J. L. 1984. Subtidal gastropods consume sulfur-oxidizing bacteria: evidence from coastal hydrothermal vents. Science 223:696–698.

Strahler, A. N., and A. H. Strahler. 1974. Introduction to Environmental Science. Hamilton Publishing Company, Santa Barbara, CA.

Tunnicliffe, V. 1992. Hydrothermal-vent communities of the deep sea. Am. Sci. 80:336–349.

Turner, J. T., and J. G. Ferrante. 1979. Zooplankton fecal pellets in aquatic ecosystems. BioScience 29:670–677.

Turner, R. D. 1973. Wood-boring bivalves, opportunistic species in the deep-sea. Science 180:1377–1379.

van der Leeden, F., F. L. Troise, and D. K. Todd. 1990. The Water Encyclopedia. 2nd ed. Lewis Publishers, Chelsea, MI.

Wangersky, P. J. 1976. The surface film as a physical environment. Annu. Rev. Ecol. Syst. 7:161–176.

Welch, P. H. 1952. Limnology. 2nd edition. McGraw-Hill, New York.

Williams, J. 1962. Oceanography. Little, Brown, Boston.

Wilson, J. T., J. F. McNabb, D. L. Balkwill, and W. C. Ghiorse. 1983. Enumeration and characterization of bacteria indigenous to a shallow water-table aquifer. Ground Water 21:134–142.

Wirsen, C. O., and H. W. Jannasch. 1974. Microbial transformation of some ^{14}C-labeled substrates in coastal water and sediment. Microb. Ecol. 1:25–37.

Wood, E. J. F. 1965. Marine Microbial Ecology. Reinhold Publishing Corp., New York.

Yayanos, A. A., A. S. Dietz, and R. van Boxtel. 1979. Isolation of a deep-sea barophilic bacterium and some of its growth characteristics. Science 205:808–810.

Yayanos, A. A., A. S. Dietz, and R. van Boxtel. 1981. Obligately barophilic bacterium from the Mariana Trench. Proc. Natl. Acad. Sci. USA 78:5212–5215.

ZoBell, C. E. 1942. Bacteria in the marine world. Sci. Mon. 55:320–330.

ZoBell, C. E. 1946. Marine Microbiololgy. Chronica Botanica, Waltham, MA.

Zobell, C. E., and R. Y. Morita. 1957. Barophilic bacteria in some deep sea samples. J. Bacteriol. 73:563–568.

6

Geomicrobial Processes: A Physiological and Biochemical Overview

6.1 TYPES OF GEOMICROBIAL AGENTS

Various microorganisms, both **prokaryotes** and **eukaryotes**, play an active role in certain geological processes, a fact that seems not always sufficiently appreciated by some microbiologists and geologists. Prokaryotic microorganisms include all types of bacteria, and eukaryotic microorganisms include algae, fungi, protozoa, and slime molds. Geomicrobially active prokaryotes include members of the **archaebacteria** (now also known as **archaeobacteria** and **archaea**) and **eubacteria**. Both archaebacteria and eubacteria are prokaryotes because they lack a true nucleus but have their genetic information encoded in a large, circular polymeric molecule of dexoyribonucleic acid (DNA). This structure is often called the **bacterial chromosome**, which, unlike the chromosomes of eukaryotic cells, is not surrounded by a nuclear membrane. The molecular size of a prokaryotic chromosome is measured in the order of 10^9 daltons. Some genetic information in prokaryotes may also be located in extrachromosomal, circularized DNA, or **plasmids**, which have a molecular size ranging around 10^7 daltons.

Archaebacteria and eubacteria are also both classified as prokaryotes because they lack **mitochondria**, membranous organelles that carry out respiration in eukaryotic cells, and **chloroplasts**, membranous organelles that carry out photosynthesis in eukaryotic cells. Respiratory activity in prokaryotes is carried out in the plasma membrane. Photosynthetic activity in prokaryotes may be carried out by internal membranes derived from the plasma membrane (purple bacteria, cyanobacteria) or special internal membranes (green bacteria).

Archaebacteria are distinguishable from eubacteria only on a submicroscopic basis. When compared to eubacteria at that level, they exhibit distinct differences in structure and composition of their cell wall and plasma membrane, and in structure and composition of their ribosomes, which are the sites of protein synthesis. They also differ in key enzymes of nucleic acid and protein synthesis when compared with eubacteria (Atlas, 1988; Brock and Madigan, 1988).

Eukaryotic cells differ from prokaryotic cells in possessing a **true nucleus**, which is a double-membrane-bound organelle in which the chromosomes and the nucleolus (center for ribosomal RNA synthesis) are located. They also feature mitochondria, chloroplasts, and vacuoles. The structure and mechanism of operation of their **flagella**, organelles of locomotion, if present, also differ from prokaryotic flagella. In eukaryotic cells, some key metabolic processes are highly compartmentalized unlike those in prokaryotes.

Examples of geomicrobially important archaebacteria include **methanogens** (methane-forming bacteria), **extreme halophiles, thermophiles**, and **thermoacidophiles**. Examples of geomicrobially important eubacteria include some aerobic and anaerobic hydrogen metabolizing bacteria, iron-oxidizing and -reducing bacteria, manganese-oxidizing and -reducing bacteria, nitrifying and denitrifying bacteria, sulfate-reducing bacteria, sulfur-oxidizing and -reducing bacteria, anaerobic photosynthetic sulfur bacteria, oxygen-producing cyanobacteria, and many others. Examples of geomicrobially important eukaryotes include fungi, which can attack silicate, carbonate, and phosphate minerals, among others, and which are important in initiating degradation of somewhat recalcitrant natural polymers such as lignin, cellulose and chitin, as in the O and A horizon of soil, or on and in surface sediments; algae, which contribute oxygen to the atmosphere and which can promote calcium carbonate precipitation or dissolution or which can form silica frustules; and protozoa which lay down siliceous, calcium carbonate, strontium sulfate or manganous oxide tests, or which may accumulate preformed iron oxide on their cell surface.

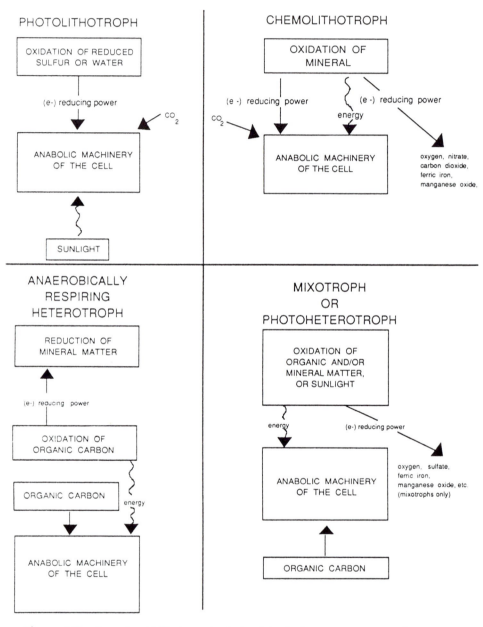

Figure 6.1 Geomicrobially important physiological groups among prokaryotes.

6.2 GEOMICROBIALLY IMPORTANT PHYSIOLOGICAL GROUPS OF PROKARYOTES

Prokaryotes can be divided into various physiological groups such as chemolithotrophs, photolithotrophs, mixotrophs, photoheterotrophs, and heterotrophs (Fig. 6.1). Each of these groups includes some geomicrobially important organisms. **Chemolithotrophs** (chemosynthetic autotrophs) include both eubacteria and archaebacteria. They are microorganisms that derive energy for doing metabolic work from the oxidation of inorganic compounds, and that assimilate carbon as CO_2, HCO_3^-, or CO_3^{2-} (see, e.g., Wood, 1988). **Photolithotrophs** (photosynthetic autotrophs) include a variety of eubacteria but no known archaebacteria. They are microorganisms that derive energy for doing metabolic work by converting radiant energy from the sun into chemical energy, and that assimilate carbon as CO_2, HCO_3^-, or CO_3^{2-} (photosynthesis). Some of these microbes are **anoxygenic** (do not produce oxygen from photosynthesis), while others are **oxygenic** (produce oxygen from photosynthesis). **Mixotrophs** include eubacteria and archaebacteria. They may derive energy simultaneously from oxidation of reduced carbon compounds and oxidizable inorganic compounds, or they may derive their carbon simultaneously from organic carbon and CO_2, or they may derive their energy totally from the oxidation of an inorganic compound but their carbon from organic compounds. **Photoheterotrophs** include mostly eubacteria, but also a few archaebacteria (extreme halophiles). They derive their energy from sunlight but derive their carbon by assimilating organic compounds. **Heterotrophs** include both eubacteria and archaebacteria. They derive their energy from the oxidation of organic compounds and most or all of their carbon from the assimilation of organic compounds. They may respire aerobically or anaerobically, or they may ferment (see later discussion of catabolic reactions).

6.3 ROLE OF MICROBES IN INORGANIC CONVERSIONS IN THE LITHOSPHERE AND HYDROSPHERE

A number of microbes in the biosphere can be considered to be **geological agents**. They may serve as agents of concentration, dispersion, or fractionation of matter. As **agents of concentration**, they cause localized accumulation of inorganic matter by (1) intracellular deposition, as do purple sulfur bacteria and other sulfur bacteria that form sulfur granules inside their cells during oxidation of hydrogen sulfide, (2) by passive accumulation involving adsorption or ion exchange. This may be subsequently followed by formation of salts of inorganic cations (alkaline earths, metal ions) in or on specific cell surface structures which then serve as nuclei around which a corresponding mineral forms (e.g., iron accumulation in the sheath of *Leptothrix* spp.) (Beveridge et al., 1983; Beveridge and Doyle, 1989; Doyle, 1989; Ferris, 1989; Geesey and Jang, 1989), or (3) by promoting precipitation of insoluble compounds external to the cell, as in most bacterial $CaCO_3$ precipitation. The net effect of these localized precipitations is the manifold local increase in concentration of an inorganic substance.

As **agents of dispersion**, microbes promote dissolution of insoluble mineral matter as, for example, in the dissolution of $CaCO_3$ by respiratory CO_2, or in the biochemical reduction of insoluble ferric oxide or manganese dioxide to soluble compounds.

As **agents of fractionation**, microbes may act selectively on a mixture of inorganic compounds by promoting selective chemical change of one or a few compounds of the mixture, causing a selective concentration or dispersion; for example in the oxidation of arsenopyrite by *Thiobacillus ferrooxidans* (Ehrlich, 1964), in which some oxidized soluble iron and arsenate may precipitate as ferric arsenate, or in the preferential reduction of Mn(IV) over Fe(III) in ferromanganese nodules (Ehrlich et al., 1973). Microbes may also cause fractionation by attacking a light stable isotope of an element in a compound in preference to a heavier isotope of the same element, as in the reduction of $^{32}SO_4^{2-}$ in preference to $^{34}SO_4^{2-}$, the assimilation of $^{12}CO_2$ in preference to $^{13}CO_2$, or the oxidation of ^{12}C in preference to ^{13}C in organic matter by *Desulfovibrio desulfuricans* under conditions of slow growth (see discussion by Doetsch and Cook, 1973). Other isotopes that may be fractionated by microbes include hydrogen and deuterium (Estep and Hoering, 1980), ^{14}N and ^{15}N (Wada and Hattori, 1978), and ^{16}O and ^{18}O (Duplessy et al. 1981). In the laboratory, the magnitude of these effects is relatively large and may involve significant changes in isotopic ratios in relatively short time. In some natural settings, microbial isotope fractionation has also been found to be of readily detectable magnitude. Studies so far lead to the impression that only a few mostly unrelated microorganisms have the capacity fractionate stable isotope mixtures.

6.4 TYPES OF MICROBIAL ACTIVITIES INFLUENCING GEOLOGICAL PROCESSES

A number of geological processes at the Earth's surface are under influence of microbes. **Lithification** is a type of geologic process in which microbes may produce the cementing substance which binds inorganic sedimentary particles together to form sedimentary rock. The microbially produced cementing substance may be calcium carbonate, iron or aluminum oxide, or silicate. Some types of **mineral formation** may be the result of microbial activity. Iron sulfides such as pyrite, iron oxides such as ochre, manganese oxides such as vernadite, psilomelane, and others, calcium carbonates such as calcite and aragonite, and silica may be generated **authigenically** by microbes (for a survey see Lowenstam, 1981). In some instances, microbes may be responsible for mineral **diagenesis**, in which microbes may cause alteration of rock structure and transformations of primary into secondary minerals, as in the conversion of orthoclase to kaolinite (see Chapter 4). **Rock weathering** may be promoted by microbes through production of metabolic products which attack the rock and cause solubilization or diagenesis of some mineral constituents of the rock, or through direct enzymatic attack of certain oxidizable or reducible rock minerals, thereby causing their solubilization or diagenesis. Microbes may contribute to **sediment accumulation** in the form of calcium carbonate tests like those from coccolithophores or foraminifera, silica frustules from diatoms, or silica test from radiolaria or actinopods in pelagic or limnetic oozes. The aging of lakes may also be influenced by microbes through their rock weathering activity and/or their generation of organic debris from incomplete decomposition of organic matter (see Chapter 5).

Geological processes that are not influenced by microbes include **magmatic activity** or **volcanism**; **rock metamorphism**, resulting from heat and pressure; **tectonic activity**, related to crustal formation and transformation, and the allied process of **orogeny** or **mountain building**. **Wind** and **water erosion** should also be included, although these activities may be facilitated by prior or concurrent microbial weathering activity. Even though these geological processes are not influenced by microbes, microbes may be influenced by them because these processes may create new environments that may be more or less favorable for microbial growth and activities than before these happenings.

6.5 MICROBES AS CATALYSTS OF GEOCHEMICAL PROCESSES

Most of the influence that microbes exert on geologic processes is physiological; i.e., they may act as **catalysts** in some geochemical processes, or they may act as a producers or consumers of certain geochemically active substances and thereby influence the rate of geochemical reactions in which such substances are reactants or products. In either case, they act through their **metabolism**. This metabolism has two aspects. One, **catabo-**

lism, is that portion that provides the cell with needed energy through **energy conservation**, and it may also yield to the cell some compounds that can serve as building blocks for polymers. A key reaction in energy conservation is the oxidation of a suitable nutrient or **metabolite** (a compound metabolically derived from a nutrient). The other, **anabolism**, is that portion of metabolism that deals with **assimilation** (synthesis, polymerization) and leads to the formation of organic polymers such as nucleic acids, proteins, polysaccharides, lipids, and others, and of "inorganic polymers" such as the polysilicate in diatom frustules and radiolarian tests or the polyphosphate granules which are formed by some bacteria and yeasts as energy storage compounds within their cells. Anabolism, by contributing to an increase in cellular mass and duplication of vital molecules, makes growth and reproduction possible. Catabolism and anabolism are linked to each other in that catabolism provides the energy and some of the building blocks that make anabolism, which is an energy-requiring process, possible. Both catabolism and anabolism may play a geomicrobial role. Catabolism is involved, for instance, in large-scale iron, manganese, and sulfur oxidation. Anabolism is involved, for instance, in the formation of organic compounds from which fossil fuels (peat, coal, petroleum) are generated. It is also the process by which polysilicate of diatom frustules and radiolarian tests is formed.

Catabolism may take the form of aerobic respiration, anaerobic respiration, or fermentation. Catabolism may thus be carried on in the presence or absence of oxygen in air. The role of oxygen is that of **terminal electron acceptor**. Indeed microorganisms can be grouped as **aerobes** (oxygen-requiring organisms), **anaerobes** (oxygen-shunning organisms), **microaerophilic organisms** (requiring low concentrations of oxygen), or **facultative organisms**. The facultative microbes can adapt their catabolism to operate in the presence or absence of oxygen in air; i.e., they use oxygen as terminal electron acceptor when available. When oxygen is not available, they use a reducible inorganic (e.g., nitrate, sulfate, ferric iron) or organic (e.g., fumarate) compound as substitute.

Catabolic Reactions: Aerobic Respiration   In **aerobic respiration**, reducing power in the form of hydrogen atoms or electrons is removed in the oxidation of organic or inorganic compounds and conveyed by an **electron transport system** to oxygen to form water. In the transfer of reducing power via the electron transport system, some of the energy that is liberated is conserved in special phosphate anhydride bonds of adenosine 5'-triphosphate (ATP) by a chemiosmotic process (see below). These bonds (Fig. 6.2), upon hydrolysis, yield 7.3 kcal of free energy per mole at pH 7.0 and 25°C (Lehninger, 1975), as opposed to ordinary phosphate ester bonds, which release only about 2 kcal mol^{-1} of free energy under these

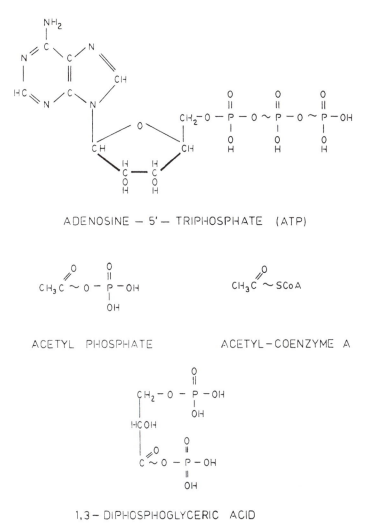

ADENOSINE — 5' — TRIPHOSPHATE (ATP)

ACETYL PHOSPHATE   ACETYL — COENZYME A

1,3 — DIPHOSPHOGLYCERIC ACID

Figure 6.2 Examples of compounds containing one or more high-energy phosphate bonds (~).

conditions. The energy in high-energy bonds is used by cells for driving energy-consuming reactions, such as syntheses or polymerizations, forward.

Typical components of the electron transport system include nicotinamide adenine dinucleotide (NAD), flavoprotein (FP), iron-sulfur protein

(Fe-S), quinone (CoQ), cytochromes (cyt Fe) and cytochrome oxidase (cyt oxid). They are arranged in complexes in the plasma membrane of some bacteria [e.g., *Paracoccus denitrificans* (Payne et al., 1987; Onishi et al., 1987); marine bacterial strain SSW_{22} (Graham, 1987)] (Fig. 6.3) and in the inner mitochondrial membrane in eukaryotes. The types of electron carriers and enzymes and their arrangement into complexes, if any, differs in different kinds of bacteria. Indeed, in the same bacterium, the kinds of carriers may vary quantitatively and/or qualitatively, depending on growth conditions. Whatever the make-up of the assemblage of electron carriers, they interact in a specific sequence such as the one shown in Fig. 6.4. Hydrogen or electrons enter the electron transport system where the E_h of the half-reaction by which they are removed from a substrate is near and below the E_h of the appropriate component of the system. For example, the electrons from the oxidation of H_2 or pyruvate may enter the transport system at the level of complex I via NAD^+ as carrier and complex III via and flavin carrier, whereas electrons from the oxidation of ferrous iron enter the transport system at the level of complex IV. Table 6.1 lists the E_h values for some geomicrobially important enzyme catalyzed oxidations, the level at which their hydrogens or electrons are probably fed into the electron transport system upon their oxidation, and also the maximum number of high-energy phosphate bonds (ATP) that may be generated in the transfer of hydrogen or electron pairs to oxygen.

It is important to recognize that in prokaryotic cells the electron transport system is located in the plasma membrane (Fig. 6.5A), whereas in eukaryotic cells it is located internally in special organelles called mitochondria (Fig. 6.5B). As a result bacteria endowed with appropriate oxidoreductases (enzymes which transfer hydrogen atoms or electrons) in their cell envelope are able to oxidize or reduce insoluble substrates which cannot be taken into the cell, such as elemental sulfur, iron sulfide, iron oxide, and manganese oxide. Because the essential enzymes are located in the periplasmic space and/or plasma membrane of prokaryotes or even the outer membrane in the case of at least one gram-negative bacterium (Myers and Myers, 1992), they are able to make direct contact with the substrate. Eukaryotic cells, on the other hand, are unable to carry out such reactions because their electron transport system is located on the inner membrane of mitochondria, which reside in the cytoplasm of these cells. The mitochondrial electron transport system being removed from the cell surface thus lacks direct access to insoluble substrates (Ehrlich, 1978).

Catabolic Reactions: Anaerobic Respiration In aerobic respiration, oxygen is always the terminal electron acceptor, but in anaerobic respiration

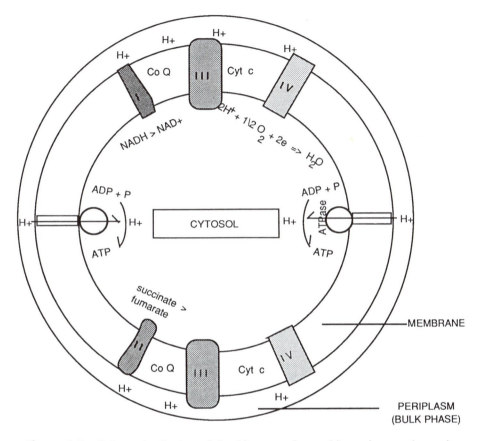

Figure 6.3 Schematic display of the bioenergetic machinery in a prokaryotic cell envelope. Structures labeled I, II, III, and IV represent specific electron transport complexes involved in the orderly transfer of electrons from substrate donor to oxygen in some prokaryotes. Complex I, reactive with NADH, includes a flavoprotein and Fe-S protein; complex II, reactive with succinate, includes succinic dehydrogenase (another flavoprotein) and Fe-S protein; complex III includes cytochromes b and c_1 and an Fe-S protein; complex IV includes cytochrome oxidase (e.g., cytochrome a + a_3). Coenzyme Q (CoQ) and cytochrome c (cyt c) shuttle electrons between respective complexes. Proton translocation from the cytosol to the periplasm involves complex I or II, CoQ + complex III, and often complex IV. Oxygen reduction to water occurs on the inner surface of the plasma membrane. ATPase is the site of ATP synthesis.

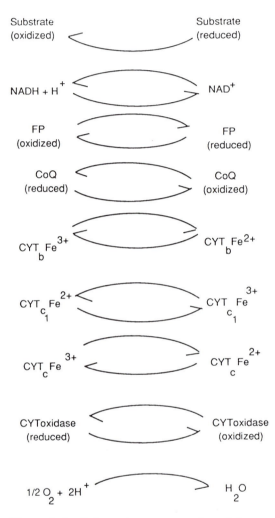

Figure 6.4   Schematic representation of the sequence of interactions of components of an electron transport system in a bacterial membrane by which reducing power is transferred from a substrate to oxygen.

other reducible compounds such as nitrate, sulfate, sulfur, carbon dioxide, ferric oxide and other ferric compounds, and manganese oxides, or an organic compound such as fumarate serve as terminal electron acceptors instead. The process is normally associated with some archaebacteria and some eubacteria including some cyanobacteria. Anaerobically respiring

Table 6.1 Microbially Catalyzed Oxidations of Geological Significance: Some Characteristics of Their Interaction with the Electron Transport System

Reaction	E'_h at pH 7.0 (volts)	Entrance level into electron transport system	ATP/2e or 2H
$Fe^{2+} \rightarrow Fe^{3+} + e$	$+0.77$	Complex IV	1
$S^0 + 4H_2O \rightarrow SO_4^{2-} + 8H^+ + 6e$	-0.20	Complex III or IV	2 or 1[a]
$H_2S \rightarrow S^0 + 2H^+ + 2e$	-0.27	Complex I or III	3 or 2
$H_2 \rightarrow 2H^+ + 2e$	-0.42	Complex I or III	3 or 2
$Mn^{2+} + 2H_2O \rightarrow MnO_2 + 4H^+ + 2e$	$+0.46$	Complex IV (?)	1

[a] Add 0.5 mol of ATP per mol of SO_3^{2-} oxidized to SO_4^{2-} if substrate-level phosphorylation is part of the oxidation process.

bacteria may be facultative (e.g., nitrate reducers, some iron reducers), or obligate anaerobes (e.g., sulfate-reducing bacteria, methanogens, acetogens, some other iron reducers). In two cases, anaerobic respirers may reduce O_2 and concurrently an alternate inorganic electron acceptor (certain nitrate- and MnO_2-reducers; see Chapters 11 and 15, respectively). In most cases of anaerobic respiration by facultative organisms, oxygen competes with the other possible electron acceptors and thus must be absent or present at lower concentration than in normal air. Anaerobic respiration usually employs some of the hydrogen and electron carriers of aerobic respiration but usually substitutes a suitable terminal reductase for cytochrome oxidase to convey electrons to the terminal electron acceptor which replaces oxygen. The best characterized of these systems are those in which sulfate and nitrate are reduced. Like aerobic respiration, anaerobic respiration takes place in the bacterial plasma membrane.

Catabolic Reactions: Fermentation Fermentation is a catabolic process in which energy is conserved in a disproportionation process in which part of the energy-yielding substrate consumed is oxidized by reducing the remainder of the substrate consumed. No externally supplied, terminal electron acceptor is involved in this redox process. Glucose fermentation to lactic acid is a typical example (Fig. 6.6). Pairs of hydrogen atoms are removed in an oxidation step from an intermediate metabolic product, glyceraldehyde 3-phosphate, resulting in the formation of 1,3-diphosphoglyceric acid. The hydrogen pairs that are removed are transferred to

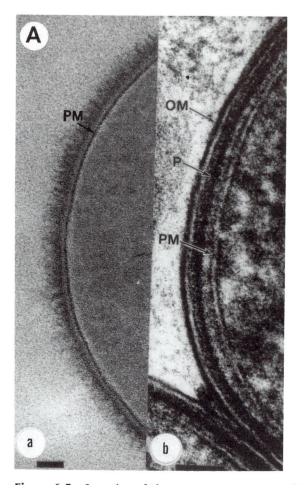

Figure 6.5 Location of electron transport system (ETS) in prokaryotic and eukaryotic cells. (A) Thin sections of the gram-positive cell wall of *Bacillus subtilis* (a) and the gram-negative cell wall of *Escherichia coli* (b). Both sections were prepared by freeze substitution. OM, outer membrane; PM, plasma membrane; P, periplasmic gel containing the peptidoglycan located between the outer and plasma membranes. In both types of cells, the ETS is located in the plasma membrane. The bars in (a) and (b) equal 25 nm. (From Beveridge, 1989, by permission.) (B) Cross section of a dormant conidium (spore) of *Aspergillus fumigatus*, a fungus and eukaryote ($\times 64,000$). ETS is located in the mitochrondria. Mi, mitochondrion; PM, plasma membrane; M, thin layer of electron-dense material; PS, polysaccharide storage material; II, membrane-bound storage body. (Courtesy of W. C. Ghiorse.)

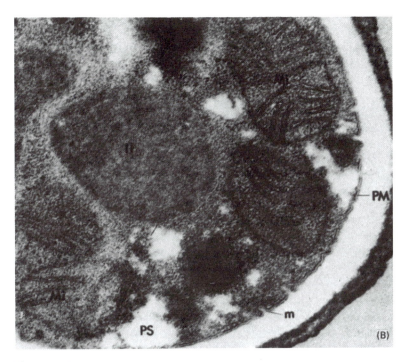

Figure 6.5 Continued.

pyruvate, reducing it thereby to lactic acid. The source of the pyruvate is the enzymatic transformation of 1,3-diphosphoglycerate. Energy in fermentation is conserved by substrate-level phosphorylation (see next section); a membrane-bound electron transport system is never involved. Fermentation always occurs in the cytoplasm of a cell. It is a process of which a number of bacteria, both facultative and anaerobic, are capable, but it is relatively rare among eukaryotic microbes. Certain fungi, such as the yeast *Saccharomyces cerevisiae*, are exceptions.

How Energy Is Generated by Aerobic and Anaerobic Respirers and by Fermenters During Catabolism In aerobic and anaerobic respiration, most useful energy is trapped in high-energy phosphate bonds conserved in ATP as a result of **oxidative phosphorylation**. The reaction leading to the formation of high-energy bond may be summarized by*

$$ADP + P_i \rightarrow ATP. \tag{6.1}$$

---

* AMP, adenosine 5′-monophosphate; ADP, adenosine 5′-diphosphate; ATP, adenosine 5′-triphosphate; P_i, inorganic orthophosphate.

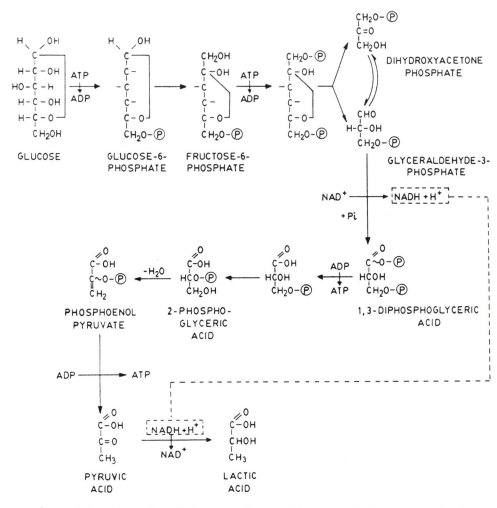

Figure 6.6 Conversion of glucose to lactic acid by glycolysis, an example of fermentation. Note that in this reaction sequence, biochemically useful energy in the form of ATP is generated exclusively by substrate-level phosphorylation.

Reaction (6.1) is energy-consuming and is made possible by charge separation between the inside and outside of the plasma membrane. This charge separation results in most respirers from the passage of electrons down the electron transport chain to oxygen which is coupled with the concurrent pumping of protons from the inside of the cell to the outside (periplasm or its equivalent) across the plasma membrane in most prokaryotes or from the interior (matrix) to the outside of the inner membrane of the mitochondrion in eukaryotes. In actively respiring prokaryotic cells or mitochondria of eukaryotic cells, the inside is alkaline relative to the outside and therefore more electronegative. The plasma membrane and the inner mitochondrial membrane is impermeable to protons except at the sites where the enzyme complex ATPase is located. ATPase permits the reentry of protons into the prokaryotic cell or mitochondrion through a proton channel (Fig. 6.3). The enzyme complex couples this proton reentry with ATP synthesis [Reaction (6.1)]. Proton reentry via ATPase is facilitated in aerobes by the consumption of protons in the reduction of O_2 to water by cytochrome oxidase on the inside of the plasma or inner mitochondrial membrane. In anaerobically respiring bacteria, protons may be consumed in the reduction of the electron acceptor which replaces oxygen and is catalyzed by an enzyme other than cytochrome oxidase. The energy which drives Reaction (6.1) comes from the proton gradient and from the membrane potential:

$$PMF = \Delta\psi - [2.3 \ RT \ (\Delta pH)/F], \tag{6.2}$$

where PMF is the proton motive force, $\Delta\psi$ is the transmembrane potential, ΔpH is the pH gradient across the membrane, and R is the gas constant, T is absolute temperature, and F is the Faraday constant. The overall process by which energy is generated in aerobic or anaerobic respiration is called **chemiosmosis** (see Hinkle and McCarty, 1978 for a further discussion of this process). A maximum of 3 molecules of ATP may be formed per electron pair transferred from donor to terminal acceptor in aerobic respiration, and a probable maximum of 2 in anaerobic respiration.

In methanogens, the electron transport system is as yet incompletely characterized. It is generally agreed, however, that they use a chemiosmotic mechanism for the production of ATP. Many methanogens achieve charge separation in the form of a pH gradient and transmembrane potential by generating protons by the oxidation of H_2 in a periplasmic-space-equivalent outside their plasma membrane and by conducting electrons to the cell interior for use in CO_2 assimilation. This model is supported by evidence developed by Blaut and Gottschalk (1984), Butsch and Bachofen (1984), Mountford (1978), and Sprott et al. (1985). Since the interior of the cells of the methanogens has a pH near neutrality, H_2 oxidation in the periplasmic-space equivalent generates a pH gradient in actively metabolizing cells. This pH gradient appears to be utilized in the generation of ATP by ATPase complexes projecting from the inside surface of the membrane. The electrons removed in the oxidation of hydrogen are conveyed to the cell interior, probably by a membrane-bound hydrogenase, and used in the reduction of CO_2 to methane (see Chapter 20). The charge separation mechanism of methanogens for generation of ATP by oxidative phosphorylation is probably illustrative of the earliest chemiosmotic mechanisms from which the more elaborate systems utilizing a variety of different membrane-bound electron carriers and enzymes found in modern aerobic and anaerobic respirers evolved. Some methanogens appear to use Na^+ rather than H^+ to achieve charge separation (see Chapter 20).

In fermentation, useful energy is formed by **substrate-level phosphorylation**, a process in which a high-energy bond, which traps some of the total free energy released during oxidation, is formed on the substrate which is being oxidized. An example is the oxidation of glyceraldehyde 3-phosphate to 1,3-diphosphoglycerate in glucose fermentation, illustrated in Fig. 6.6. Substrate-level phosphorylation may also occur during aerobic and anaerobic respiration, but it contributes only a small portion of the total energy conserved in high-energy bonds by cells. Clearly, aerobic and anaerobic respiration are much more efficient energy yielding processes than fermentation for a cell. It takes much less substrate to satisfy a fixed energy requirement of a cell if a substrate is oxidized by aerobic or anaerobic respiration than by fermentation. If the energy-yielding substrate is organic, this greater efficiency may also be due to the fact that respirers may oxidize the substrate completely to CO_2 and H_2O whereas fermenters cannot.

Whereas many of the microbes that oxidize inorganic substrates to obtain energy are aerobes, a few are not. All autotrophically growing methanogens oxidize hydrogen gas (H_2) anaerobically by transferring electrons from H_2 to CO_2 to form methane (CH_4), generating ATP in the process by oxidative phosphorylation. **Acetogens** carry out a similar reduction of CO_2 by hydrogen but form acetate instead of methane (Eden and Fuchs, 1983)

$$4H_2 + 2CO_2 \Rightarrow CH_3COOH + 2H_2O \ (\ \Delta G°, \ -25 \text{ kcal or } 104.8 \text{ kJ})$$

$$(6.3)$$

Some oxidizers of sulfur compounds can transfer electrons from a reduced sulfur substrate such as thiosulfate or elemental sulfur to nitrate in the absence of oxygen. In the presence of oxygen, these sulfur oxidizing organisms transfer the electrons from the reduced sulfur compound to oxygen. The maximum ATP yield in methane formation from H_2 reduction of CO_2 and in the oxidation of reduced sulfur by nitrate, two examples of anaerobic respiration, has not yet been clearly established.

How Chemolithotrophic Bacteria (Chemosynthetic Autotrophs) Generate Reducing Power for Assimilating CO_2 and Converting It to Organic Car-

bon Unlike heterotrophs, most chemolithotrophs when reducing CO_2 with NADPH have a special problem in generating NADPH. These chemolithotrophs which posses an electron transport chain containing Fe-S proteins, quinones, and cytochromes in their plasma membrane, whether they are aerobic or anaerobic respirers, depend on **reverse electron transport**. In this process, electrons travel against a redox gradient through the expenditure of energy in the form of ATP because the source of the electrons which is also the source of energy, usually has a midpoint potential (E_h) associated with its redox reaction which is significantly higher than that for the NADP/NADPH couple. Methanogens which do not use NADPH for reducing fixed CO_2 to organic carbon but instead employ unique hydrogen carriers such as factor F_{420} (8-OH-5-deazaflavin) and carbon dioxide reducing (CDR) factor, appear not to consume ATP in CO_2 reduction by hydrogen. Acetogens are another exception. They employ ferredoxin and other Fe-S proteins, whose reduction does not require the expenditure of energy (Pezacka and Wood, 1984).

How Photosynthetic Microbes Generate Energy and Reducing Power Anoxygenic bacteria, such as the purple and green sulfur bacteria, purple non-sulfur bacteria, and some cyanobacteria generate their ATP by transducing light energy of appropriate wavelengths into chemical energy, which they conserve in high-energy phosphate bonds by **cyclic or noncyclic photophosphorylation** (Gottschalk, 1985). They work on a chemiosmotic principle analogous to respiration. In cyclic photophosphorylation, as in purple sulfur bacteria, electrons pass from a reduced, low-potential Fe-S protein (E_h − 530 mV) along an electron transport pathway which includes membrane-bound quinones and cytochromes, ultimately ending up on bacteriochlorophyll. High energy phosphate bonds are generated in this phase of electron passage and conserved in ATP. The passage of the electrons is coupled to proton pumping and a resultant proton gradient, as in respiration except that in this case the proton pumping is in a direction opposite of that in respiration, i.e., from the outside of the membrane-barrier to the inside. The proton motive force thus generated causes ATPase in the photosynthetic membrane to generate ATP from ADP and P_i. For the electrons to return to the low potential Fe-S protein from the high-potential bacteriochlorophyll in cyclic photophosphorylation, they have to be energized by light absorption at an appropriate wavelength (Fig. 6.7). It is light, therefore, which drives the cyclic movement of the electrons.

In green sulfur bacteria, photophosphorylation is noncyclic as well as cyclic (Fig. 6.7). The cyclic mechanism is similar to that in the purple sulfur bacteria. In the noncyclic photophosphorylation process, ATP is

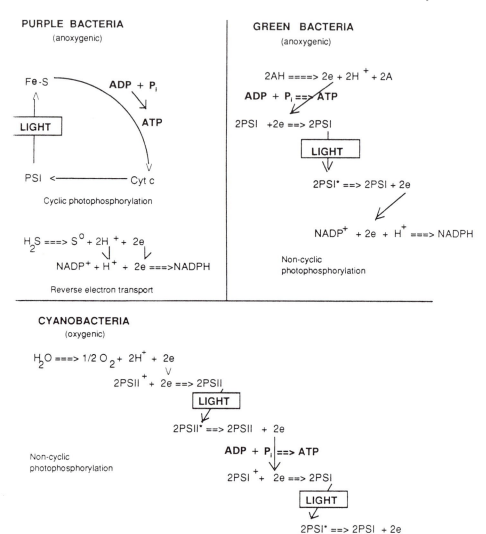

PURPLE BACTERIA
(anoxygenic)

Fe-S   **ADP** + **P**$_i$

LIGHT

ATP

PSI <—————— Cyt c

Cyclic photophosphorylation

$H_2S ===> S^0 + 2H^+ + 2e$

$NADP^+ + H^+ + 2e ===>NADPH$

Reverse electron transport

GREEN BACTERIA
(anoxygenic)

$2AH ====> 2e + 2H^+ + 2A$

ADP + **P**$_i$ ==> **ATP**

$2PSI + 2e ==> 2PSI$

LIGHT

$2PSI^* ==> 2PSI + 2e$

$NADP^+ + 2e + H^+ ===> NADPH$

Non-cyclic
photophosphorylation

CYANOBACTERIA
(oxygenic)

$H_2O ===> 1/2 O_2 + 2H^+ + 2e$

$2PSII^+ + 2e ==> 2PSII$

LIGHT

$2PSII^* ==> 2PSII + 2e$

Non-cyclic
photophosphorylation

ADP + **P**$_i$ ==> **ATP**

$2PSI^+ + 2e ==> 2PSI$

LIGHT

$2PSI^* ==> 2PSI + 2e$

$NADP^+ + H^+ + 2e ==> NADPH$

Figure 6.7 Diagrammatic representation of the mechanisms of photophosphorylation and generation of reducing power (NADPH) in purple and green photosynthetic bacteria and in cyanobacteria. PSI, photosystem I; PSII, photosystem II. (Adapted from Stanier et al., 1986.)

synthesized in a reaction sequence in which an external electron donor like H_2S, $S°$, or $S_2O_3{}^{2-}$, for instance, reduces chlorobium chlorophyll. The electrons that have reduced the chlorophyll are then used to reduce $NADP^+$ to NADPH. This requires input of light energy because the midpoint potential for chlorophyll reduction is much higher ($\sim +400$ mV) than that for NADP reduction (~ -350 mV). As in most known chemolithotrophs, the NADPH is needed for CO_2 assimilation. The ATP-synthesizing mechanisms in both cyclic and noncyclic photophosphorylation of green sulfur bacteria involve chemiosmosis. The extent to which the cyclic and noncyclic mechanisms operate depend on the demands for ATP and NADH by the cell.

Cyanobacteria also use a noncyclic photophosphorylation process for generating ATP, but they use a more complex pathway than the green sulfur bacteria (Fig. 6.7). Their photosynthetic machinery, which is normally oxygenic because it uses H_2O as a source of reducing power unlike that of either the purple or green sulfur bacteria, involves two major components, photosystems I and II. These are linked to each other by a series of interacting electron carriers which transfer electrons from photosystem II to photosystem I and promote proton pumping, which permits chemiosmotic ATP synthesis. Photosystem II generates electrons by the photolysis of water with transduced light energy. Photosystem I sends the electrons received from photosystem II to $NADP^+$ to form NADPH with another boost from transduced light energy. As in green sulfur bacteria, NADPH is required for reducing fixed CO_2. Cyanobacteria also perform cyclic photophosphorylation involving only photosystem I.

The purple sulfur bacteria also need to generate reducing power in the form of NADPH for assimilating CO_2 and converting it to organic carbon (i.e., reduced carbon). Purple bacteria accomplish this by **reverse electron transport** that is not *directly* dependent on light energy input (Fig. 6.7). They transfer electrons from a high potential membrane carrier to low- potential membrane carriers and finally to NADP by the consumption of ATP. The energy consumption is needed because the electrons in this instance travel against a redox gradient. For further discussion of these photosynthetic processes the reader is referred to Stanier et al. (1986).

Anabolism: How Energy Trapped in High-Energy Bonds Is Used to Drive Energy-Consuming Reactions   As an example of how aerobic chemolithotrophs couple ATP formation to CO_2 assimilation and reduction to organic carbon, the process in *Thiobacillus ferrooxidans*, an acidophilic iron oxidizer, will be considered. This organism oxidizes ferrous to ferric iron at acid pH:

$$2Fe^{2+} \rightarrow 2Fe^{3+} + 2e^- \tag{6.4}$$

Some of the reducing power (e$^-$) generated in this way is transferred to oxygen:

$$\tfrac{1}{2}O_2 + 2H^+ + 2e^- \rightarrow H_2O \tag{6.5}$$

with the simultaneous chemiosmotic production of ATP:

$$ADP + P_i \rightarrow ATP. \tag{6.6}$$

At maximum efficiency, one ATP is formed for every electron pair (2e$^-$) transferred to oxygen. The remaining reducing power from the oxidation of the ferrous iron is used to reduce pyridine nucleotide (NAD$^+$, NADP$^+$). Since, however, the electrons in this case have to travel against a redox gradient from +800 mV (high-potential Q-cycle intermediate in *T. ferrooxidans* at pH 2; Ingledew, 1982; Ehrlich et al., 1991) to −305 mV (E$_m$ at pH 6.5 for NADP$^+$/NADPH), energy stored in high energy phosphate bonds of ATP has to be consumed:

$$NAD^+ + 2H^+ + 2e^- + 2ATP \rightarrow NADH + H^+ + 2ADP + P_i \tag{6.7}$$

$$NADH + H^+ + NADP^+ \rightarrow NAD^+ + NADPH + H^+ \tag{6.8}$$

The NADPH + H$^+$, together with some ATP, are used in the assimilation of CO_2 and its reduction to organic carbon:

$$\text{Ribulose 5-phosphate} + ATP \rightarrow \text{ribulose 1,5-diphosphate} + ADP \tag{6.9}$$

$$\text{Ribulose 1,5-diphosphate} + CO_2 \rightarrow 2(\text{3-phosphoglycerate}) \tag{6.10}$$

$$\begin{aligned}
&2(\text{3-phosphoglycerate}) + 2NADPH + 2H^+ \\
&+ 2ATP \rightarrow 2(\text{glyceraldehyde 3-phosphate}) + 2NADP^+ \\
&+ 2ADP + 2P_i
\end{aligned} \tag{6.11}$$

From glyceraldehyde 3-phosphate the various organic constituents of the cell are then manufactured, including the building blocks for polymers such as proteins, nucleic acids, lipids, polysaccharides, and so on, and subsequently combined into corresponding polymers with the expenditure of additional ATPs because polymerizations are energy requiring reactions. Also, some ribulose 5-phosphate is regenerated to permit continued CO_2 fixation [Reactions (6.10–6.11)]. Although generally, chemolithotrophs can grow in the complete absence of organic matter under laboratory conditions, many, if not all of these organisms can assimilate some types of organic compounds, such as amino acids and vitamins. Some chemolithotrophs are able to use organic carbon as a sole energy source under some conditions (**facultative chemolithotrophs**), others can not (**obligate chemolithotrophs**).

Anaerobic chemolithotrophs, such as methanogens, acetogens, and some sulfate reducers use a different mechanism for assimilating CO_2. They form acetate (CH_3COO^-) from two molecules of CO_2. This involves the reduction of one CO_2 stepwise to a methyl carbon and of the other to carbon monoxide (CO) and then forming a carboxyl carbon and combining the methyl and carboxyl carbons to form acetate in an ATP-consuming process. They then carboxylate the acetate with another molecule of CO_2 in another ATP-consuming process to form pyruvate (Fig. 6.8). This product is a key precursor for the formation of all other monomeric building blocks from which the various polymers are formed by ATP-consuming processes.

Anoxygenic photolithotrophs assimilate CO_2 by one of several different mechanisms. Purple sulfur bacteria usually fix CO_2 and reduce it to organic carbon by the same set of reactions (6.8–6.10) as aerobic chemolithotrophs such as *T. ferrooxidans*. They obtain the needed reduced NADPH for this process from reverse electron transport with H_2, H_2S, $S°$, or $S_2O_3^{2-}$ as electron donors (see above).

The filamentous green photolithotroph *Chloroflexus aurantiacus* fixes CO_2 by a 3-hydroxypropionate cycle (Ivanovsky et al., 1993; Strauss and Fuchs, 1993; Eisenreich et al., 1993) in which glyoxalate is the fixation product with ATP consumption. In the proposed cycle, CO_2 is incorporated into acetate to form malate, which in turn is transformed into hydroxypropionate, and from which prionate is formed. The propionate is transformed to succinate via methyl malonate by further CO_2 fixation. The succinate is transformed into oaxalacetate, which is then cleaved into acetate and glyoxalate, completing the cycle (Fig. 6.9).

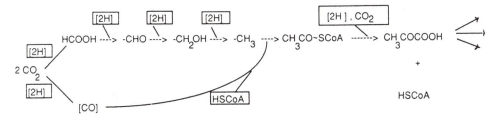

Figure 6.8 Pathway for carbon assimilation in methanogens [the activated acetate ($CH_3CO\sim SCoA$) pathway]. Pyruvate ($CH_3COCOOH$) is a key intermediate for forming various building blocks for the cell, including sugars, amino acids, fatty acids, and so on.

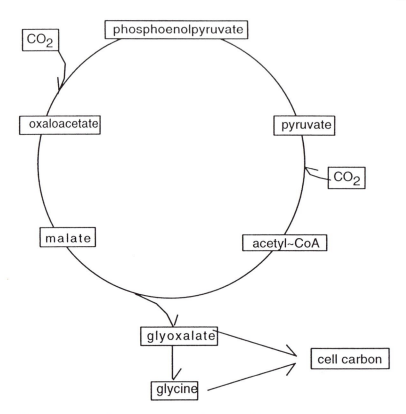

Figure 6.9 Pathway of autotrophic CO_2 fixation by *Chloroflexus aurantiacus* as proposed by Ivanovsky (1993). (Adapted from Ivanovsky et al., 1993.)

Green sulfur bacteria of the genus *Chlorobium*, on the other hand, fix CO_2 and reduce it by a reverse tricarboxylic acid cycle (Fig. 6.10). In this process, CO_2 is combined with pyruvate in an ATP-consuming process to form oxalacetate, which is then converted via malate, fumarate, and succinate to 2-ketoglutarate, the last step requiring consumption of ATP. The 2-ketoglutarate is a key precursor in amino acid synthesis as well as being a precursor for citrate synthesis. Formation of citrate involves fixation of another molecule of CO_2. The citrate is cleaved to oxalacetate and acetate. The acetate serves as precursor in the synthesis of pyruvate by CO_2 fixation with ATP consumption, thus completing the tricarboxylic acid cycle. The pyruvate is a key intermediate for the synthe-

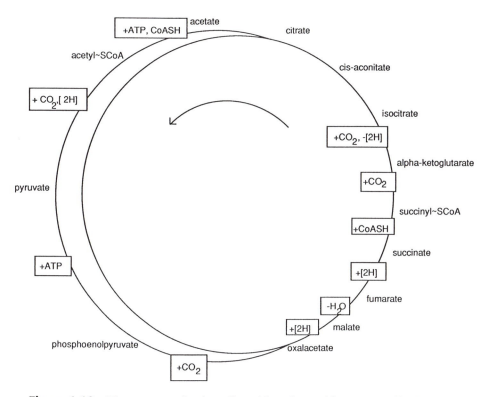

Figure 6.10 The reverse tricarboxylic acid cycle used by green sulfur bacteria for carbon assimilation. (Modified from Stanier et al., 1986; scheme originally presented by M. C. W. Evans, B. B. Buchanan, and D. I. Arnon, A new ferredoxin-dependent carbon reduction cycle in a photosynthetic bacterium, *Proc Natl. Acad. Sci. USA* 55:928–934, 1966.)

sis of other biochemical building blocks. Reduced NAD and NADP needed in the operation of this cycle is generated by a noncyclic photoreduction mechanism (see above).

Although once thought unique to *Chlorobium*, the reverse tricarboxylic acid cycle has since been found to operate as the mechanism of CO_2 assimilation in some chemosynthetic autotrophs, e.g., *Aquifex pyrophilus*, a chemolithotrophic, thermophilic, H_2-oxidizing eubacterium, and *Thermoproteus neutrophilus*, a chemolithotrophic, thermophilic, H_2-oxidizing, and $S°$-reducing archaebacterium (Beh et al., 1993).

Oxygenic photolithotrophs fix CO_2 and reduce it to organic carbon by a reaction sequence similar to Reactions (6.9–6.11). They obtain the reduced NADP and the ATP for this process via noncyclic photophosphorylation (see above).

Carbon Assimilation by Mixotrophs, Photoheterotrophs, and Heterotrophs
Because they fashion some monomeric building blocks by catabolism and acquire others preformed among their nutrients, heterotrophs use most of the ATP that they generate catabolically for polymerization reactions as in the formation of proteins, polysaccharides, nucleotides and nucleic acids, lipids, and others. Mixotrophs and photoheterotrophs perform anabolic reactions which are similar to those performed by chemolithotrophs, photolithotrophs, and/or heterotrophs.

Nothing is known so far about how algae or protozoa form inorganic polymers such as polysilicate. Such a process is expected to involve the consumption of ATP or its equivalent.

6.6   MICROBIAL MINERALIZATION OF ORGANIC MATTER

Microbes play a major role in the transformation of organic matter in the upper lithosphere (soils and sediments) and in the hydrosphere (oceans and bodies of fresh water). Since biological availability of carbon as well as of other nutritionally vital inorganic elements in the biosphere is limited, it is essential that these elements be recycled for the continuation of life. This recycling requires complete degradation of dead organic matter into inorganic matter, whether the dead organic matter is in the form of remains of dead organisms or metabolic wastes. This process is called **mineralization**. It may be aerobic or anaerobic.

In **aerobic mineralization**, organic matter is completely degraded (oxidized) to CO_2 and H_2O, and if N, S, and P are present in the original organic molecule, to NO_3^-, SO_4^{2-}, PO_3^{4-}, and so on. The process by which this mineralization comes about is aerobic respiration, i.e., oxygen serves as terminal electron acceptor in the oxidations. In many instances of aerobic mineralization, a single organism may be responsible for the complete degradation of a compound, although in other instances consortia of two or more organisms may collaborate in the process, especially in the degradation of polymers.

In **anaerobic mineralization**, products of complete degradation of organic matter are CH_4 and/or CO_2, H_2, NH_3, H_2S, PO_4^{3-}, and so on. The process by which this degradation is accomplished is usually anaerobic respiration in which NO_3^-, Fe(III), Mn(IV), SO_4^{2-}, and CO_2 among others serve as terminal electron acceptor. However, in some special instances, certain methanogens may generate CH_4 through disproportionation (fermentation) from acetate, methanol, or methylamine. In these instances, externally supplied CO_2 is not used as terminal electron acceptor. Anaerobic mineralization frequently involves a succession of different microorganisms, although in some instances a single organism is responsible for the entire process. In a succession of microorganisms, one group may provide hydrolytic enzymes to depolymerize proteins, carbohydrates, nucleic acids, lipids, and so forth. The same group or a different one may then convert the products of hydrolysis to organic acids, alcohols, and some CO_2 and H_2. In the presence of sufficient NO_3^-, Mn(IV), Fe(III), or SO_4^{2-}, denitrifying, manganese-, iron-, or sulfate-reducing bacteria, respectively, may then oxidize these products to CO_2. In the absence of sufficient NO_3^-, Mn(IV), Fe(III) or SO_4^{2-}, another group of microorganisms may convert the organic acids and alcohols to acetate, CO_2, and H_2 by fermentation, and methanogens may then convert the acetate, and the CO_2, and H_2 to CH_4. Depending on availability of the corresponding terminal electron acceptor, nitrate, iron(III), and Mn(IV) respiration, and methanogenesis are important processes of anaerobic mineralization in soils and sediments where sulfate is limiting, whereas sulfate respiration will be important in estuarine and anoxic, coastal marine environments, where sulfate occurs in plentiful amounts.

In some environments, accumulations of nonliving organic matter are noted. This may be because the particular organic matter is difficult to degrade, as for instance, lignin in wood. In other cases the accumulation may be the result of limited availability of O_2, NO_3^-, Mn(III) or (IV), Fe(III), SO_4^{2-}, or CO_2 as terminal electron acceptors. In still other instances, incomplete degradation of organic matter may be the result of enzymatic deficiencies of the organisms in the environment in which the organic matter is accumulating. In soil, incomplete oxidation of some organic matter results in the formation of an important soil constituent, **soil humus**, a mixture of polymeric substances derived from the partial decomposition of plant, animal and microbial remains and from microbial syntheses. It is usually recognizable as a brownish-black organic complex, only portions of which are soluble in water. A larger fraction is soluble in alkali. Soil humus includes aromatic molecules, often in polymerized form, whose origin is mostly lignin, bound and free amino acids, uronic acid polymers, free and polymerized purines and pyrimidines, and other forms of bound phosphorus.

Organic matter accumulating in marine sediments has been called **marine humus** because of a similarity to soil humus in its C/N ratio (Waksman, 1933) and because of its relative resistance to aerobic microbial decomposition (Waksman and Hotchkiss, 1937; Anderson, 1940). However, more detailed chemical analysis of marine humus indicates differences from soil humus, which is not surprising in view of differences in origin (Jackson, 1975; Moore, 1969, p. 271). Whereas soil humus is formed mainly from plant remains, typical marine humus in sediments far from land derives mainly from phytoplankton remains and fecal residues. Marine humus from three northern Pacific sediments contains from 0.14 to 0.34% organic matter, including 20 to 1,145 ppm of alkali-soluble humic acids, 40 to 55% of benzene soluble bitumen, and 50 to 180 ppm of amino acids. The remainder is kerogen, a material insoluble in aqueous and nonpolar organic solvents (Palacas et al., 1966). The organic matter of deep-sea sediments contains a fraction which, although refractory to microbial attack in situ, is readily attacked by microbes when brought to the surface (see, e.g., Ehrlich et al., 1972). Presumably high hydrostatic pressure (>300 atm) and low temperature (<4°) prevent rapid microbial in situ decomposition (Jannasch and Wirsen, 1973; Wirsen and Jannasch, 1975). Metabolizable organic matter in shallow-water sediments will undergo more complete decomposition provided that it does not accumulate too rapidly.

6.7 PRODUCTION OF MICROBIAL PRODUCTS THAT CAN CAUSE GEOMICROBIAL TRANSFORMATION

Many heterotrophic bacteria, whether aerobic, facultative, or anaerobic, form significant quantities of **organic acids** among the products from their catabolism in addition to CO_2. At least some of the CO_2 will hydrolyze to form carbonic acid in aqueous solution. Some chemolithotrophs and photolithotrophs form significant amounts of sulfuric or nitric acids, depending on the substrate which they use as their source of energy and/or reducing power. These acids may react chemically with certain minerals resulting in their partial or complete dissolution or alteration. Other heterotrophs when growing at the expense of nitrogenous carbon and energy sources such as protein or peptides, generate ammonia which forms NH_4OH in aqueous solution. This base can solubilize some silicates.

Various prokaryotes and some eukaryotes form ligands which can complex inorganic ions. Some, like siderophores, are very specific as to the ion they complex. When the ligand complex is stabler than the source mineral for the ion complex, the ligand can withdraw the ion from the source mineral causing its diagenesis or dissolution.

6.8 PHYSICAL PARAMETERS THAT INFLUENCE GEOMICROBIAL ACTIVITY

Temperature is an important parameter which influences geomicrobial activity. In fact, it influences biological activity in general. This is because biochemical reaction rates, as do all chemical reaction rates, increase with a rise in temperature, except that with enzyme-catalyzed reactions, this positive response is confined to a relatively narrow temperature range because of the heat-stability limit of enzyme proteins. Proteins denature, i.e., they become structurally randomized above the maximum temperature tolerated by them. If they are enzymes, this means that they lose their catalytic activity. Denaturation of some enzymes at temperatures slightly above the temperature maximum can be prevented by a moderate increase in hydrostatic pressure (Haight and Morita, 1962).

The lipid phase of cell membranes also responds to temperature. It is more fluid at higher than at lower temperatures. A certain degree of membrane fluidity is essential for proper cell functioning. Cells can control membrane fluidity by the degree of saturation of the fatty acids in the membrane lipids. The more saturated the fatty acids of a given chain length, the higher the temperature required for a desirable degree of fluidity, and conversely, the more unsaturated these fatty acids, the lower the temperature required to maintain a similar degree of fluidity.

At present, life is known to exist in a temperature range from slightly below 0° to as high as +130°C. No organism exists, however, which spans this entire range. That is because proteins and some other structural components require somewhat different composition and structure for stability and activity for different temperature intervals within the overall temperature range in which life exists. Intact organisms reflect the heat stability range of their enzymes and critical cell structures such as cell membranes in terms of the temperature range in which they grow. Key molecules in organisms with different temperature requirements evidently have different heat labilities (Brock, 1967; Tansey and Brock, 1972). **Psychrophiles** grow in a range from slightly below 0 to about 20°C with an optimum at 15°C or lower (Morita, 1975). **Psychrotrophs** grow over a wider temperature range than do psychrophiles (e.g., 0–30°C), with an optimum near 25°C. **Mesophiles** are microbes that grow in the range of 10–45°C, with an optimum range for some of about 25–30°C and for others about 37–40°C. **Thermophiles** are microbes that live in a temperature range from 42–130°C, but the range for any given thermophile is considerably narrower. The temperature optimum for any one organism depends on its identity and usually corresponds to the predominant temperature of its normal habitat.

Extreme thermophiles, those growing optimally above 60°C seem to be mostly archaebacteria. Generally, thermophilic photosynthetic prokaryotes cannot grow at temperatures higher than 73°C, whereas thermophilic eukaryotic algae cannot grow at temperatures higher than 56°C (Brock, 1967, 1974, 1978). Thermophilic fungi generally exhibit temperature maxima around 60°C and thermophilic protozoa around 50°C. Only nonphotosynthetic, thermophilic bacteria have temperature maxima that may be as high as 130°C. For growth at temperatures at or above the boiling point of water, elevated pressure is needed to keep the water liquid. Liquid water is a requirement for life.

The parameters of pH and E_h also exert important influences on geomicrobial activity, as they do on biological activity in general. Each enzyme has its characteristic pH optimum, and E_h optimum in the case of redox enzymes, at which it catalyzes most efficiently. That is not to say that in the cell or, in the case of extracellular enzymes outside the cell, an enzyme necessarily operates at its optimal pH and E_h. The interior of living cells tends to have a pH around neutrality and an E_h that is higher than that of its external environment. Enzymes with higher or lower pH optima will operate at less than optimal efficiency. This helps a cell to integrate individual enzyme reactions in a sequence. Changes in external pH within the physiological range of a microorganism does not affect its internal pH because of the plasma membrane barrier and its ability to control internal pH. Extreme changes will, however, affect it adversely.

Environmental pH and E_h control the range of distribution of microorganisms (see, however, Ehrlich, 1993, for environmental significance of E_h). Figure 6.11 summarizes the pH and E_h ranges in which certain microbial groups are found to grow. An important feature shown in the diagram is the prevalence of iron-oxidizing bacteria and, to some extent, of thiobacteria in environments of relatively reduced potential and elevated pH.

As mentioned in Chapter 5, hydrostatic pressures in excess of 405 bars (400 atm) at a fixed, physiologically permissive temperature below the boiling point of water generally prevent growth of **nonbarophilic** microbes. Pressures between 203 and 405 bars (200 and 400 atm) at such a temperature tend to interfere reversibly with cell division of bacteria (ZoBell and Oppenheimer, 1950). **Barophilic** organisms can grow at pressures above 405 bars (400 atm) at physiologically permissive temperatures. **Facultative barophiles** grow progressively slower with increasing pressure under such conditions, whereas **obligate barophiles** grow best at or near the pressure and temperature of their native environment and grow progressively more slowly with decreasing pressure and usually not at all at atmospheric pressure at the same temperature (Yayanos et al., 1982). The growth inhibiting

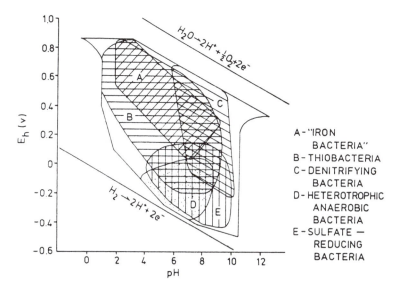

Figure 6.11 Environmental limits of E_h and pH for some bacteria. A, "iron bacteria"; B, thiobacteria; C, denitrifying bacteria; D, facultative and anaerobic heterotrophic bacteria and methanogens; E, sulfate-reducing bacteria. (Adapted from Baas Becking et al., *J. Geol.* 68:243–284. Copyright 1960, by permission of the University of Chicago Press.)

effect of hydrostatic pressure is attributable to its effect on protein synthesis (Schwarz and Landau, 1972a; Schwarz and Landau, 1972b; Pope et al., 1975; Smith et al., 1975). Many other biochemical reactions are much less pressure sensitive (Pope and Berger, 1973) (see also Chapter 5).

6.9 SUMMARY

Microbes may be geological active in lithification, mineral formation, mineral diagenesis, and sedimentation, but not in volcanism, tectonic activity, orogeny, or wind and water erosion. They may act as agents of concentration, dispersion, or fractionation of mineral matter. Their influence may be direct, through action of their enzymes, or indirect, through chemical action of their metabolic products, through passive concentration of insoluble substances on their cell surface, and through alteration of pH and E_h conditions in their environment. Their metabolic influence may involve anabolism or catabolism under aerobic or anaerobic conditions. Respiratory activity of prokaryotes may cause oxidation or reduction of certain inorganic compounds, resulting in their precipitation or solubilization. Some chemolithotrophic and mixotrophic bacteria can obtain useful energy from the oxidation of some inorganic substances, such as H_2, Fe(II), Mn(II), H_2S, $S°$, and others. Photolithotrophic bacteria can use H_2S as a source of reducing power in the assimilation of CO_2 and, in the process, deposit sulfur. Anaerobically respiring organisms, which use certain oxidized substances as terminal electron acceptors, are important in the mineralization of organic matter in environments devoid of atmospheric oxygen. Mineralization of organic matter by microbes under aerobic conditions results in the formation of CO_2, H_2O, NO_3^-, SO_4^{2-}, PO_4^{2-}, and so on, while under anaerobic conditions it leads to the formation of CH_4, CO_2, NH_3, H_2S, and so on. Some organic matter is refractory to mineralization under anaerobic conditions and is microbially converted to humus. All microbial activities are greatly influenced by pH and E_h conditions in the environment.

REFERENCES

Anderson, D. Q. 1940. Distribution of organic matter in marine sediments and its availability to further decomposition. J. Mar. Res. 2:225–235.

Atlas, R. M. 1988. Microbiology. Fundamentals and Applications. 2nd ed. Macmillan, New York.

Baas Becking, L. G. M., I. R. Kaplan, and D. Moore. 1960. Limits of the natural environment in terms of pH and oxidation-reduction potentials. J. Geol. 68:243–284.

Beh, M., G. Strauss, R. Huber, K-O. Stetter, and G. Fuchs. 1993. Enzymes of the reductive citric acid cycle in the autotropic eubacterium *Aquifex pyrophilus* and in the archaebacterium *Thermoproteus neutrophilus*. Arch. Microbiol. 160:306–311.

Beveridge, T. J. 1989. Metal ions and bacteria. *In*: T. J. Beveridge and R. J. Doyle, eds. Metal Ions and Bacteria. Wiley, New York, pp. 1–29.

Beveridge, T. J., and R. J. Doyle, eds. 1989. Metal Ions and Bacteria. Wiley, New York.

Beveridge, T. J., J. D. Meloche, W. S. Fyfe, and R. G. E. Murray. 1983. Diagenesis of metals chemically complexed to bacteria: laboratory formation of metal phosphates, sulfides and organic condensates in artificial sediments. Appl. Environ. Microbiol. 45:1094–1108.

Blaut, M., and G. Gottschalk. 1984. Protonmotive force-driven synthesis of ATP during methane formation from molecular hydrogen and formaldehyde or carbon dioxide in *Methanosarcina barkeri*. FEMS Microbiol. Lett. 24:103–107.

Brock, T. D. 1967. Life at high temperatures. Science 158:1012–1019.

Brock, T.D. 1974. Biology of Microorganisms. 2nd ed. Prentice-Hall, Englewood Cliffs, NJ.

Brock, T.D. 1978. Thermophilic Microoganisms and Life at High Temperatures. Springer-Verlag, New York.

Brock, T. D., and M. T. Madigan. 1988. Biology of Microorganisms. 5th ed. Prentice-Hall, Englewood Cliffs, NJ.

Butsch, B. M., and R. Bachofen. 1984. The membrane potential in whole cells of *Methanobacterium thermoautotrophicum*. Arch. Microbiol. 138:293–298.

Doetsch, R. N. and T. M. Cook. 1973. Introduction to Bacteria and Their Ecobiology. University Park Press, Baltimore, MD.

Doyle, R. J. 1989. How cell walls of gram-positive bacteria interact with metal ions. *In*: T. J. Beveridge and R. J. Doyle, eds. Metal Ions and Bacteria. Wiley, New York, pp. 275–293.

Duplessy, J-C., P-L. Blanc, and A. W. H. Be. 1981. Oxygen-18 enrichment of planktonic foraminifera due to gametogenic calcification below the euphotic zone. Science 213:1247–1250.

Eden, G., and G. Fuchs. 1983. Autotrophic CO_2 fixation in *Acetobacterium woodii*. II. Demonstration of enzymes inolved. Arch. Microbiol. 135:68–73.

Ehrlich, H.L. 1964. Bacterial oxidation of arsenopyrite and enargite. Econ. Geol. 59:1306–1312.

Ehrlich, H.L. 1978. Inorganic energy sources for chemolithotrophic and mixotrophic bacteria. Geomicrobiol. J. 1:65–83.

Ehrlich, H. L., 1993. Bacterial mineralization of organic carbon under anaerobic conditions. *In*: J-M. Bollag and G. Stotzky, eds. Soil Biochemistry. Vol. 8. Marcel Dekker, New York, pp. 219–247.

Ehrlich, H. L., W. C. Ghiorse, and G. L. Johnson II. et al. 1972. Distribution of microbes in manganese nodules from the Atlantic and Pacific Oceans. Dev. Indust. Microbiol. 13:57–65.

Ehrlich, H. L., S. H. Yang, and J. D. Mainwaring, Jr. 1973. Bacteriology of manganese nodules. VI. Fate of copper, nickel, cobalt and iron during bacterial and chemical reduction of the manganese (IV). Z. Allg. Mikrobiol. 13:39–48.

Ehrlich, H. L., W. J. Ingledew, and J. C. Salerno. 1991. Iron- and manganese-oxidizing bacteria. *In*: J. M. Shiveley and L. L. Barton, eds. Variations in Autotrophic Life. Academic Press, London, pp. 147–170.

Eisenreich, W., G. Strauss, U. Werz, G. Fuchs, and A. Bacher. 1993. Retrobiosynthetic analysis of carbon fixation in the phototrophic eubacterium *Chloroflexus auriantiacus*. Eur. J. Biochem. 215: 619–632.

Estep, M. F. and T. C. Hoering. 1980. Biogeochemistry of the stable hydrogen isotopes. Geochim. Cosmochim. Acta 44:11997–1206.

Ferris, F. G. 1989. Metallic ion interactions with the outer membrane of gram-negative bacteria. *In*: T. J. Beveridge and R. J. Doyle, eds. Metal Ions and Bacteria. Wiley, New York, pp. 295–323.

Geesey, G. G., and L. Jang. 1989. Interactions between metal ions and capsular polymer. *In*: T. J. Beveridge and R. J. Doyle, eds. Metal Ions and Bacteria. Wiley, New York, 325–357.

Gottschalk, G. Bacterial Metabolism, 2nd ed. Springer-Verlag, New York.

Graham, L. A. 1987. Biochemistry and electron transport of a manganese-oxidizing bacterium. Ph.D. dissertation. Rensselaer Polytechnic Institute, Troy, NY.

Haight, R. D., and R. Y. Morita. 1962. Interaction between the parameters of hydrostatic pressure and temperature on aspartase of *Escherichia coli*. J. Bacteriol. 83:112–120.

Hinkle, P. C., and R. E. McCarty. 1978. How cells make ATP. Sci. Am. 238:104–123.

Ingledew, W. J. 1982. *Thiobacillus ferrooxidans*. The bioenergetics of an acidophilic chemolithotroph. Biochim. Biophys. Acta 683:89–117.

Ivanovsky, R. N., E. N. Krasilnikova, and Y. I. Fal. 1993. A pathway of the autotrophic CO_2 fixation in *Chloroflexus auriantiacus*. Arch. Microbiol. 159:257–264.

Jackson, T. A. 1975. Humic matter in natural waters and sediments. Soil Sci. 119:56–64.

Jannasch, H. W., and C. O. Wirsen. 1973. Deep-sea microorganisms: in situ response to nutrient enrichment. Science 180:641–643.

Lehninger, A. L. 1975. Biochemistry. 2nd ed. Worth Publishers, New York.

Lowenstam, H. A. 1981. Minerals formed by organisms. Science 211: 1126–1131.

Moore, L. R. 1969. Geomicrobiology and geomicrobial attack on sedimented organic matter, pp. 264–303. *In*: G. Eglinton and M. T. J. Murphy, eds. Organic Geochemistry: Methods and Results. Springer-Verlag, New York.

Morita, R. Y. 1975. Psychrophilic bacteria. Bacteriol. Rev. 39:144–167.

Mountford, D. O. 1978. Evidence for ATP synthesis driven by a proton gradient in *Methanobacterium barkeri.* Biochem. Biophys. Res. Commun. 85:1346–1351.

Myers, C. R., and J. M. Myers. 1992. Localization of cytochromes to the outer membrane of anaerobically grown *Shewanella putrefaciens* MR-1. J. Bacteriol. 174:34290–3438.

Onishi, T. S. W. Meinhardt, T. Yagi and T. Oshima. 1987. Comparative studies on the NADH-Q oxidoreductase segment of the bacterial respiratory chain, pp. 237–248. *In*: C. H. Kim, H. Tedeschi, J. J. Diwan, and J. C. Salerno, eds. Advances in Membrane Biochemistry and Bioenergetics. Plenum Press, New York.

Palacas, J. G., V. E. Swanson, and G. W. Moore. 1966. Organic geochemistry of three North Pacific deep sea sediment samples. U.S. Geol. Surv. Prof. Pap. 550C, pp. C102-C107.

Payne, W. E., X. Yang, and B. L. Trumpower. 1987. Biochemical and genetic approaches to elucidating the mechanism of respiration and energy transduction in *Paracoccus denitrificans*, pp. 273–284. *In*: C. H. Kim, H. Tedeschi, J. J. Diwan and J. C. Salerno, eds. Advances in Membrane Biochemistry and Bioenergetics. Plenum Press, New York.

Pezacka, E., and H. G. Wood. 1984. The synthesis of acetyl-CoA by *Clostridium thermoaceticum* from carbon dioxide, hydrogen, coenzyme A and methyltetrahydrofolate. Arch. Microbiol. 137:63–69.

Pope, D. H., and L. R. Berger. 1973. Inhibition of metabolism by hydrostatic pressure: What limits microbial growth? Arch. Mikrobiol. 93: 367–370.

Pope, D. H., W. P. Smith, R. W. Swartz, and J. V. Landau. 1975. Role of bacterial ribosomes in barotolerance. J. Bacteriol. 121:664–669.

Schwarz, J. R., and J. V. Landau. 1972a. Hydrostatic pressure effects on *Escherichia coli*: site of inhibition of protein synthesis. J. Bacteriol. 109:945–948.

Schwarz, J. R., and J. V. Landau. 1972b. Inhibition of cell-free protein synthesis by hydrostatic pressure. J. Bacteriol. 112:1222–1227.

Smith, W., D. Pope, and J. V. Landau. 1975. Role of bacterial ribosome subunits in barotolerance. J. Bacteriol. 124:582–584.

Sprott, G. D., S. E. Bird, and I. J. McDonald. 1985. Proton motive force as a function of the pH at which *Methanobacterium bryantii* is grown. Can. J. Microbiol. 31:1031–1034.

Stanier, R. Y., J. L. Ingraham, M. L. Wheelis, P. R. Painter. 1986. The Microbial World, fifth edition. Prentice-Hall, Englewood Cliffs, NJ.

Strauss, G., and G. Fuchs. 1993. Enzymes of a novel autotrophic CO_2 fixation parthway in the phototrophic bacterium *Chloroflexus auran-*

tiacus, the 3-hydroxypropionate cycle. Eur. J. Biochem. 215: 633–643.

Tansey, M. R., and T. D. Brock. 1972. The upper temperature limit for eukaryotic organisms. Proc. Natl. Acad. Sci. USA 69:2426–2428.

Wada, E., and A. Hattori. 1978. Nitrogen isotope effects in the assimilation of inorganic nitrogen compounds by marine diatoms. Geomicrobiol. J. 1:85–101.

Waksman, S. A. 1933. On the distribution of organic matter in the sea and chemical nature and origin of marine humus. Soil Sci. 36:125–147.

Waksman, S. A., and M. Hotchkiss. 1937. On the oxidation of organic matter in marine sediments by bacteria. J. Mar. Sci. 36:101–118.

Wirsen, C. O., and H. W. Jannasch. 1975. Activity of marine psychrophilic bacteria at elevated hydrostatic pressure and low temperature. Mar. Biol. 31:201–208.

Wood, P. 1988. Chemolithotrophy, pp. 183–230. *In*: C. Anthony, ed. Bacterial Energy Transduction. Academic Press, London.

Yayanos, A. A., A. S. Dietz, and R. Van Boxtel. 1982. Dependence of reproduction rate on pressure as a hallmark of deep-sea bacteria. Appl. Environ. Microbiol. 44:1356–1361.

ZoBell, C. E., and C. H. Oppenheimer. 1950. Some effects of hydrostatic pressure on the multiplication and morphology of marine bacteria. J. Bacteriol. 60:771–781.

7

Methods in Geomicrobiology

Geomicrobial phenomena can be studied in the field (**in situ**) and in the laboratory and/or in microcosms (**in vitro**). Field study of a given geomicrobial phenomenon should ideally involve recognition and enumeration of the geomicrobially active agent(s), in situ measurement of the growth rate of the agent(s), chemical and physical identification of the substrate(s), i.e., reactant(s) attacked, and the product(s) formed in the geomicrobial process, measurement of the rate of reaction of the process, and assessment of the impact of different environmental factors on the process. In practice, however, it may happen that a suspected geomicrobial process is no longer operating at a given site but took place in the geological past. In that instance the role of microorganisms in the process has to be reconstructed from microscopic observations (e.g., searching for microfossils associated with the starting compound and especially the product(s) of the process), and from geochemical observations (e.g., searching for "fingerprint" compounds in sedimentary rock samples that indicate the past existence of an organism or a group of organisms which could have been the geomicrobial agent responsible, and, if applicable, searching for evidence of isotopic fractionation of a key element relevant to the geomicrobial process under consideration).

In situ observation of an ongoing geomicrobial process should include a study of the setting. In a terrestrial environment, the nature of rocks, soil, or sediment, whichever are involved, and their constituent minerals ought to be considered together with prevailing temperature, pH oxidation-reduction potential (E_h), sunlight intensity, seasonal cycles, and the source and availability of moisture, oxygen or other potential terminal electron acceptors, and nutrients. In an aqueous environment, water depth, availability of oxygen or other terminal electron acceptors, turbidity, light penetration, thermal stratification, pH, E_h, chemical composition of the water, nature of the sediment if part of the habitat, and nutrient source and availability should be examined.

For laboratory study of a geomicrobial process in a **microcosm**, a large sample of the soil, sediment, or rock on or in which the process is occurring is collected and placed in a suitable vessel (e.g., a flow-through chamber, a glass column, a battery jar, or other kind of suitable vessel). Filter-sterilized water from the site at which the sample was collected, or a synthetic nutrient solution of a composition that approximates qualitatively and quantitatively the nutrient supply available at the sampling site, is added intermittently or continuously. The added nutrient solution should displace an equivalent volume of spent solution from the culture vessel. The setup may be placed in the same environment from which the sample was taken, or it may be incubated at the temperature with the illumination and access to air and humidity to which the sample was exposed at the sampling site. Measurement of the concentration of nutrients and products in the influent and effluent critical to the process under study will give a measure of the process rate. Solid products that are not recoverable in the effluent can be identified and measured in representative samples taken from the microcosm. Continuous or intermittent measurement of temperature, pH, E_h, and oxygen availability in the microcosm will give information about any changes in these parameters as a result of microbial activity in the microcosm.

The microcosm will probably contain a **mixed population** of bacteria, not all of which are likely to play a role in the geomicrobial process of interest. Manipulation of the microcosm through qualitative or quantitative change in nutrient supply, adjustment of pH or temperature, or a combination of these factors may cause selective increase of the organisms directly responsible for the geomicrobial process of interest and intensify the process.

In vitro laboratory study of a geomicrobial process may be done by isolating the responsible microorganism(s) in **pure culture** from a representative sample from the geomicrobially active site. The process originally observed in the field is recreated in batch culture and/or continuous culture with the isolate(s). Characterization of the process mechanism will involve qualitative and quantitative measurements of the biogeochemical transformation(s). It may include enzymatic study, where appropriate, as well as an assessment of environmental effects on the in vitro process. In vitro laboratory study may be important in lending support to field interpretations of geomicrobial processes that are occurring at present or have ocurred in the past.

7.1 DETECTION AND ISOLATION OF GEOMICROBIALLY ACTIVE ORGANISMS

A geomicrobial process may be the result of a single microbial species or an association of two or more. An association of microbial species is often called a **consortium**. The basis for the association may be **synergism** in which no one organism is capable of carrying out the complete process but in which each member of the consortium carries out part of the process in a sequential interaction. It is also possible that not all members of an association of microbes contribute directly to the overall geomicrobial process but rather carry out reactions which create environmental conditions relating, for instance, to pH or E_h, which facilitate the geomicrobial process under consideration.

Even if a geomicrobial process is the result of a single organism, that organism rarely occurs as a pure culture in the field. It will usually be accompanied by other organisms, which may play no direct role in the geomicrobial process under study, although they may compete with the geomicrobially active agent for living space and nutrients and may even produce metabolites that stimulate or inhibit the geomicrobial agent to a degree. Three types of microorganisms may be found associated with a geomicrobial sample taken in the field: (a) **indigenous organisms**, whose normal habitat is being examined and which include the geomicrobially active organism(s); (b) **adventitious organisms**, which were introduced by chance into the habitat by natural circumstances and which may or may not grow in the new environment but do survive in it; and (c) **contaminants**, which were introduced in manipulating the environment during in situ geomicrobial study or in sampling. Distinctions among these groups are frequently difficult to make experimentally. A criterion for identifying indigenous organisms may be their frequency of occurrence in a given habitat and in similar habitats at different sites. A criterion for identifying adventitious organisms may be their inability to grow successfully in the habitat under study and their lower frequency of occurrence than in their normal habitat. Neither of these criteria are absolute, however. Identification of a contaminant may simply be based on knowledge about the organisms concerned that would make unlikely its natural existence in the habitat under study.

In Situ Observation of Geomicrobial Agents   To detect geomicrobially active microorganisms in situ, a visual approach by direct observation or by light microscopy, or transmission or scanning electron-microscopy is possible. Direct visual observation can be made only in rare instances. Such observation is possible when the microbes occur so massively as to be easily observed as, for instance, in the case of algal and bacterial mats in hot springs (e.g., Yellowstone Park; see Brock, 1978) or in the case of lichen growth on rocks. In most instances, observations of microbes in their natural habitat require special examination of a field sample under laboratory conditions. In soil, sediment or rock samples, such observations may be made by the buried slide method or the capillary technique of Perfil'ev (Perfil'ev and Gabe, 1969). The first method involves insertion of a clean microscope slide in soil or sediment and leaving it undisturbed for a number of days. It is then withdrawn, washed, and suitably stained. Examination with a light microscope will then reveal microorganisms (especially bacteria and fungi which had attached to the glass surface during burial (Fig. 7.1).

The capillary technique involves insertion of one or more glass capillaries with optically flat sides into soil (pedoscope) or sediment (peloscope). Each capillary takes up soil or sediment solution and very small soil or sediment particles from the surround which become the culture medium for microbes which may also have entered the capillary lumen. The capillaries can be periodically withdrawn and their content examined under a light microscope. They may also be perfused with special nutrient solutions. The capillaries permit observation of trapped microbes in a living or nonliving state (Fig. 7.2). A drawback to the capillary technique is that it will reveal mostly microbes living in the pore solution and may miss microbes which grow principally on the surface of soil or sediment particles. Using this technique, Perfil'ev and Gabe (1965) discovered several previously unknown bacteria in soil and sediment, including *Metallogenium*, *Kuznetsova*, and *Caulococcus*.

Microbes may be observed directly on the surface of soil or sediment particles by fluorescence microscopy in conjunction with staining with fluorescent dyes, or in conjunction with fluorescent antibody reaction if specific microbes are being sought (Fig. 7.3) (Bohlool and Brock, 1974; Casida, 1962, 1965, 1971; Eren and Pramer, 1966; Huber et al., 1985; Kemper and Pratt, 1994; Muyzer et al., 1987; Schmidt and Bankole, 1965). Such observations can also be made by transmission electron microscopy,

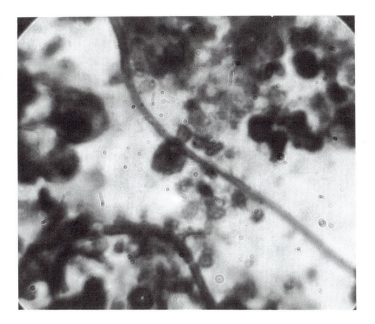

Figure 7.1 Demonstration of microbes in soil by the buried slide method (×5,240). This slide was stained with methylene blue and shows evidence of rod-shaped bacteria and a hypha.

especially in identifying fossilized microbes (Ghiorse and Balkwill, 1983; Jannasch and Mottl, 1985; Jannasch and Wirsen, 1981), and scanning electron microscopy (e.g., Fig. 15.14) (Jannasch and Wirsen, 1981; LaRock and Ehrlich, 1975; Sieburth, 1975).

Recent advances in the techniques of molecular biology have led to the development of powerful methods for identifying microorganisms. These methods can also be adpated to locate and enumerate specific microorganisms in environmental samples, even those that, although they are viable, have not been cultured in the laboratory (Ward et al., 1990). They are also useful in identifying geomicrobially active isolates obtained in pure culture in the laboratory. Depending on the method employed, DNA/DNA hybridization, DNA/RNA hybridization, and DNA fingerprinting may be involved (Saylor and Layton, 1990; Amann et al., 1995). DNA fingerprinting involves extraction of the DNA from a pure culture of the unkown organism, digesting the DNA with specific restriction enzymes, subjecting the digest to gel electrophoresis, and comparing the

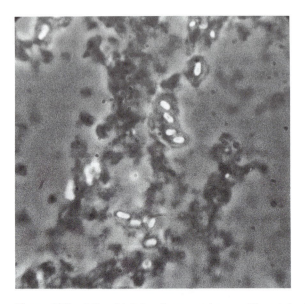

Figure 7.2   Microbial development in a capillary tube inserted into lake sediment contained in a beaker and incubated at ambient temperature ($\times$ 5,270). The oval, refractile structures are bacterial spores.

resultant pattern of different nucleotide fragments with patterns obtained in the same way from DNA extracted from pure cultures of known organisms.

With the recognition of the existence of one or more highly conserved nucleotide sequences in 16S ribosomal RNA in bacteria whose composition from different kinds of bacteria is different, it became possible to synthesize oligodeoxynucleotide probes containing the structural information of such a highly conserved sequence in complementary form. These probes can be applied directly to intact cells that have been treated to make them permeable to the probes and allow hybridization between the probes and the corresponding nucleotide sequence in the ribosomal 16S RNA. By labeling the probe with a fluorescent dye or with a radioactive isotope, cells that have reacted with the probe can be readily located (Amann et al., 1990; 1992; 1995; Braun-Howland et al., 1992, 1993; DeLong et al., 1989; Giovannoni et al., 1988; Tsien et al., 1990; Ward et al., 1990). The 16S ribosomal RNA probes can be group-, genus-, and/or species-specific, but other kinds of probes can be made that are kingdom-,

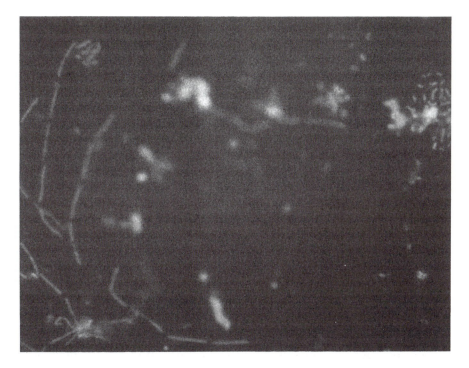

Figure 7.3 Bacteria growing on the surface of 304L stainless steel immersed in tap water (Troy, NY) for 3 days, stained with fluorescein isothiocyanate. (Courtesy of Daniel H. Pope.)

genus-, or species-specific, the last two being the most useful in most geomicrobial applications.

Application of any of these techniques in taxonomic microbial identification presupposes that DNA, RNA, or 16S ribosomal probes of known bacteria are available or that the critical nucleotide sequences are known. Application for purposes of locating and/or enumerating requires probes that come from a similar population of organisms but does not require that the organisms have been taxonomically identified.

If geomicrobial agent(s) cannot be identified in situ by visual inspection, microscopy, or molecular probing, isolation by culturing and subsequent characterization from representative samples in the laboratory may be necessary (for technical approach, see below). Isolation of specific geomicrobial agents will be necessary in any case if laboratory reconstruction of a specific geomicrobial process is to be undertaken.

Sampling   Whether for in situ or laboratory study, sampling at the on-going site of a geomicrobial process is usually required. On land, collection of rock samples may require chipping with a rock hammer or chisel under as aseptic conditions as possible. Asepsis in this instance means thorough washing and alcohol flaming of critical surfaces of sampling gear. Steriliza-tion of the rock surface may be necessary if the rock interior is to be sampled. This can be accomplished by using a disinfectant such as a car-bol-gentian violet-methanol spray (Bien and Schwartz, 1965) or by flaming for 1 min with a bunsen flame (Weirich and Schweisfurth, 1985). If a sampling device cannot be sterilized, the sample it gathers should be sub-sampled to obtain an aliquot which is least likely to have been contami-nated. Rock chips should be collected in sterile plastic containers. If the collecting is done manually, hands should be covered by sterile, surgical gloves. Weirich and Schweisfurth (1985) devised a special method for obtaining an undisturbed core of rock with a hollow drill under sterile conditions and with cooling by sterile tap water. The extracted rock core was aseptically cut into sections with a flamed chisel, each of which was then aseptically crushed in a flamed mortar mill in sterile dispersing solu-tion, and microbiologically tested.

Soil or sediment samples from shallow depths on land surfaces may be collected manually with an auger or other coring device, observing aseptic conditions. Cores should be subdivided aseptically for sampling at different depths. If the cores cannot be obtained with a sterilized sampling device, they should be subsampled so as to obtain a least contaminated sample.

Uncontaminated sediment samples representative of a depth range as great as 3,000–4,000 m can now be obtained with minimal contamina-tion by a special drilling method (Phelps and Russell, 1990). This method involves use of modified wireline coring tools with cores collected in Lexan or PVC lined barrels. The drill rig, rods, and tools are steam-cleaned. The drilling fluid system includes a recirculation tank, the drilling fluid being chlorinated water. Tracers added to the drilling fluid, including potassimum bromide, the dye rhodamine T, fluorescent beads (~2 μm diam.), and perfluorocarbons aided in determining to what extent, if any, cores had become contaminated by drilling fluid. The assessment is made by measuring the extent of their presence, if any, in the core samples. The extent of bacterial contamination can be determined by quantitative enumeration of bacteria such as coliforms that were not expected as part of the normal flora of the sediment samples run on drilling fluids and core samples (Beeman and Suflita, 1989).

To obtain aquatic samples, special gear is often required. Water sam-ples at any given depth below the surface, including deep-water samples, may be obtained with a Van Dorn sampler. It consists of a piece of large diameter plastic tube, open at both ends and mounted in such away on a cable or rope that it can be lowered vertically in a water column. While the device is being lowered, water will pass through it. When a desired depth has been reached, the plastic tube is closed at both ends with rubber closures, which are activated by a spring mechanism that can be released by a messenger (brass weight) that is allowed to slide down on the cable or rope. Below-surface water samples can also be collected with a Niskin sampler, consisting of a collapsible, sterile plastic bag with a tubelike opening, mounted between two hinged metal plates. The sampler, with the bag in a collapsed state between the metal plates, is lowered on a cable or rope to a desired depth, where a spring mechanism of the sampler is activated by a messenger that causes the metal plates to open, expanding the bag, which now draws water into it.

Aquatic sediment samples may be obtained with dredging or coring devices. Lake sediment can be collected with an Ekman dredge (Fig. 7.4) or a Peterson dredge if surface sediment is desired. A corer needs to be used if different levels of the sediment column are to be examined. Ocean sediment may be collected by dragging a bucket dredge over a desired

Figure 7.4   Ekman dredge. The brass messenger on the rope is 5 cm long.

area on the ocean floor if the sediment surface only is to be sampled. Such a sample will however, consist of combined, mixed surface sediment encompassing the total surface area sampled. To obtain samples representing different sediment depths at a given location, a gravity corer (Fig. 7.5), a piston corer, or a box corer (Fig. 7.6) has to be used. Such devices are rammed into the sediment. Box corers of sufficient cross section provide least disturbed cores. All cores need to be subsampled to obtain representative, minimally contaminated samples representative of the sediment at a give depth. Large mineral fragments or concretions on the sediment surface may be collected with a chain dredge or similar device dragged over the sediment surface in a desired area (Fig. 7.7).

If samples cannot be examined immediately after collection, they

Figure 7.5 Gravity corer. This is simply a hollow pipe containing a removable plastic liner and having a cutting edge at the lower end with a "core catcher" to retain the sediment core when the device is pulled out of the sediment. A heavy lead weight at the top helps to ram the corer into the sediment when allowed to freefall just above the sediment surface.

Figure 7.6 Box corer. After the frame hits bottom, the coring device is forced into the sediment mechanically. (Courtesy of Mark Sand.)

If samples cannot be examined immediately after collection, they should be stored so as to minimize microbial multiplication or death. Cooling the sample is usually the best way to preserve it temporarily in its native state, but the degree of cooling may be critical. Freezing may be destructive to the microbes. On the other hand, icing may not prevent the growth of psychrophiles or psychrotrophs. The duration of storage before examination should not be longer than absolutely necessary.

Isolation and Characterization of Active Agents in Samples   To study the geomicrobial agent(s) active in a process of interest, culture enrichment and pure culture isolations may be necessary. Not all cultures obtained by these procedures may be geomicrobially active. Each isolate must be tested for its ability to perform the particular geomicrobial activity under investigation. Since some geomicrobial processes are the result of the activity of a microbial consortium, no one of the pure culture isolates may exhibit the desired activity but may have to be tested with others of the isolates in different combinations. Examples of geomicrobial cooperation between microorganisms in microbial manganese oxidation which have been described include (a) the bacterium *Metallogenium symbioticum* in association with the fungus *Coniothyrium carpaticum* (Zavarzin, 1961; Dubinina, 1970; but see Schweisfurth, 1969), (b) the bacteria *Corynebacte-*

Figure 7.7   Seafloor samplers. (A) Chain dredge. (B) Dredge for collecting manganese nodules from the ocean floor. A conical bag of nylon netting is attached to a pyramidal frame.

rium sp. and *Chromobacterium* sp. (Bromfield and Skerman, 1950; Bromfield 1956), and (c) two strains of *Pseudomonas* (Zavarzin, 1962).

Enrichment of and isolation from a mixed culture require selective conditions. If a microbial agent with a specific geomicrobial attribute is sought, the selective culture medium should have ingredients incorporated which favor the geomicrobial activity performed by the organism(s). Apart from special nutrients, special pH and E_h, and temperature conditions may also have to be chosen to favor selective growth of the geomicrobial agent(s).

Isolation and characterization of pure cultures from enrichments should follow standard bacteriological technique and will not be discussed further (see, for instance, Gerhardt et al., 1981, 1993; Skerman, 1961 for details).

7.2 IN SITU STUDY OF PAST GEOMICROBIAL ACTIVITY

Microbial geochemical activity that occurred in the geologic past may under certain circumstances be identified indirectly in terms of isotope fractionation. Certain prokaryotic and eukaryotic microbes have been shown to distinguish between stable isotopes of elements such as S, C, O, N, and H. These microbes prefer to metabolize substrates containing the lighter isotope of these elements (e.g., ^{32}S, in preference to ^{34}S, ^{12}C in preference to ^{13}C, ^{16}O in preference to ^{18}O, ^{14}N in preference to ^{15}N, and H in preference to D), especially under conditions of limited growth (see Jones and Starkey, 1957; Emiliani et al., 1978; Wellman et al., 1968; Estep and Hoering, 1980). Thus, products of metabolism will be enriched in a lighter isotope when compared to the starting compound or to some reference standard that has not been subject to isotope fractionation. In practice, isotope fractionation is measured by determining isotopic ratios (e.g., $^{34}S/^{32}S$, $^{13}C/^{12}C$, $^{18}O/^{16}O$, $^{15}N/^{14}N$, and D/H) by mass spectrometry and then calculating the amount of isotope enrichment from the relationship

$$\delta \text{ isotope (\textperthousand)}$$
$$= \frac{\text{isotope ratio of sample } - \text{ isotope ratio of standard}}{\text{isotope ratio of standard}} \times 1,000 \quad (7.1)$$

If the enrichment value (δ) is negative, it means that the sample tested was enriched in the lighter isotope relative to a reference standard, and if the value is positive, the sample tested was enriched in the heavier isotope relative to a reference standard. Thus, to determine, for instance, if a certain metal sulfide mineral deposit is of biogenic origin, various parts of the deposit are sampled and ($\delta^{34}S$) values of the sulfide determined. If the values are generally negative (although the magnitude of the $\delta^{34}S$ values may vary from sample to sample and fall in the range of -5 to -50 ‰), the deposit can be viewed as of biogenic origin because a chemical explanation for such ^{32}S enrichment under natural conditions is not likely. If the $\delta^{34}S$ values are positive and fall in a narrow range, the deposit is viewed as being of abiogenic origin.

7.3   IN SITU STUDY OF ONGOING GEOMICROBIAL ACTIVITY

Ongoing geomicrobial activity may be measurable in situ. Such activity may be followed by use of radioisotopes. For instance, bacterial sulfate reducing activity may be determined by adding a small, measured quantity of $Na_2{}^{35}SO_4$ to a water, soil, or sediment sample of known sulfate content. After incubating in a closed vessel under in situ conditions (e.g., a water sample may be held in a closed reaction vessel in the water column at the depth from which it was taken), the sample is analyzed for loss of $^{35}SO_4{}^{2-}$ and buildup of $^{35}S^{2-}$ by separating these two entities and measuring their quantity in terms of of their radioactivity. A direct application of this method is that of Ivanov (1968). It allows estimation of the rate of sulfate reduction in the sample without having any knowledge of the number of physiologically active organisms. A modified method is that of Sand et al. (1975). It allows an estimation of sulfate reducing activity in terms of the number of physiologically active bacteria in the sample as distinct from an estimation of the sum of physiologically active and inactive bacteria. The assay for the estimation of active bacteria can be set up either to measure the percentage of sulfate reduced in a fixed amount of time, which is proportional to the logarithm of active cell concentration, or to measure the length of time required to remove a fixed amount of sulfate, which is related to the concentration of physiologically active sulfate-reducing bacteria in the sample. Ivanov's method can be adapted to measure the formation of sulfur and sulfates from sulfide by adding $^{35}S^{2-}$ to a sample and, after incubation in situ, separating ^{35}S and $^{35}SO_4{}^{2-}$ and measuring their quantity in terms of their radioactivity.

Microbial action on manganese (Mn^{2+} fixation by biomass; Mn^{2+} oxidation) in situ is another example of a geomicrobial process which can be followed by the use of a radioisotope, $^{54}Mn^{2+}$ in this case (LaRock, 1969; Emerson et al., 1982; Burdige and Kepkay, 1983). One approach is to measure manganese oxidation in terms of decrease in dissolved $^{54}Mn(II)$. It assumes that the oxidized manganese is insoluble. Decreases in dissolved ^{54}Mn(II) are measured on acidified samples of reaction mixture from which the oxidized manganese has been removed by centrifugation. Acidification of samples insures resolubilization of any adsorbed Mn(II). The differences in radioactivity counts between a 0-time sample and a sample taken at a subsequent time are a measure of the amount of Mn(II) oxidized by a combination of biological and chemical means over this time interval. Manganese oxidation due to biological activity alone can be estimated by subtracting determinations of chemical oxidation from the total Mn oxidation for a corresponding time interval. An estimate of the amount of chemical oxidation can be obtained from a reaction mixture in which biological activity is inhibited by autoclaving the sample, by adding one or more chemical inhibitors to the reaction mixture, or by excluding air (if enzymatic manganese-oxidation is oxygen-dependent). This experimental approach has been used to assess the manganese oxidizing activity potential in lake sediment (LaRock, 1969). It makes no assumptions about binding of unoxidized or oxidized manganese to the bacterial cells.

Another approach is to measure manganese oxidation in terms of product formation. Manganese binding by metabolically active bacteria in a water sample may be followed in terms of ^{54}Mn accumulation by the cells. The bound manganese in this case represents oxidized manganese. To this end, a sample of the bacterial suspension is filtered through a 0.2-μm membrane and the radioactivity deposited on the membrane (assumed to be bound to the cells) determined in a suitable counter. The results from this experiment are then compared to those of a parallel experiment in which the cells were inhibited by a suitable reagent such as formaldehyde or a mixture of sodium azide, penicillin G, and tetracycline-HCl. The difference in binding between these two experiments represents binding by actively metabolizing cells. It includes Mn^{2+} which is bound as Mn(II) and that which was oxidized by the cells (Emerson et al, 1982). Manganese binding by bacteria in sediment can be followed in a modified form of this method in a special reaction vessel called a peeper by Burdige and Kepkay (1983). $^{54}Mn^{2+}$ adsorbed by the cells deposited on the filter membrane is displaced by washing with $CuSO_4$ solution. The radioactivity recovered in the wash is then counted. Residual ^{54}Mn associated with the cells (assumed to be oxidized manganese) is dissolved by washing with hydroxylamine-HCl solution followed by $CuSO_4$ solution and the radioactivity in the resultant solution determined.

Biological manganese binding in lake sediments may be studied by a method that requires controls in which biological inhibitors are used to account for abiological manganese binding by sediment constituents in the overall manganese budget (Burdige and Kepkay, 1983).

One advantage in using radioisotopes in quantitative assessment of a specific geomicrobial transformation in nature is that their detection is extremely sensitive so that only minute amounts of radiolabeled substrate need to be added that do not significantly change the naturally occurring concentration of the substrate. Another advantage is that in cases where the rate of transformation is very slow even though the natural substrate concentration is high, **spiking** the reaction with radiolabeled substrate allows analysis after relatively brief reaction time because of the sensitivity of radioisotope detection.

The use of radioisotopes is, however, not essential for quantitative assessment in all instances of biogeochemical transformations in a natural environment. Other analytical methods with sufficient sensitivity may be applicable (see, for instance, Jones et al., 1983, 1984; Hornor, 1984; Kieft and Caldwell, 1984; Tuovila and LaRock, 1987).

7.4 LABORATORY RECONSTRUCTION OF GEOMICROBIAL PROCESSES IN NATURE

It is often important to reconstruct a naturally occurring geomicrobial process in the laboratory in order to investigate the mechanism whereby the process operates. Laboratory reconstruction can permit optimization of a process through the application of more favorable conditions than in nature, such as, for instance, use of a pure culture or a purified consortium to eliminate interference by competing microorganisms, optimization of substrate availability, temperature, pH, E_h, oxygen and carbon dioxide supply, and so forth.

The activity of organisms growing on solid substrates, such as soils, sediments rocks, and ore, may be investigated in batch culture, in airlift columns, in percolation columns, or in a chemostat. A **batch culture** represents a closed system in which an experiment is started with a finite amount of substrate that is continually depleted during growth of the organism. Cell population and metabolic products build up, and changes in pH and E_h are likely to occur. Conditions within the culture are thus continually changing and become progressively less favorable. Batch experiments may be least representative of a natural process, which usually occurs in an open system with continual or intermittent replenishment of substrate and removal of at least some of the metabolic wastes. A culture in an **air-lift column** (Fig. 7.8) is a partially open system, in that the microbes grow and carry on their biogeochemical activity on a mineral charge of the column. They are fed with a continually recirculated nutrient solution from which nutrients are depleted and which removes metabolic

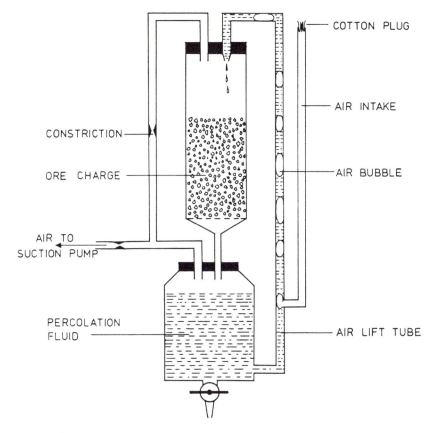

Figure 7.8 Air-lift column for ore leaching.

products from the solid substrate charge in the column. **Percolation columns** (Fig. 7.9) are even more open systems than air-lift columns. In them, microbes grow on the solid substrate charge of the column, but they are fed with nutrient solution that is not recirculated. Thus, fresh nutrient is added continually or at intervals, and wastes are removed at the same time in the effluent without recirculation, while pH, E_h, and temperature are held constant or nearly so. Steady-state conditions such as exist in a **chemostat** idealize the open culture system. They also do not imitate nature because conditions are too constant. In open systems in nature, some fluctuation in various environmental parameters occurs over time.

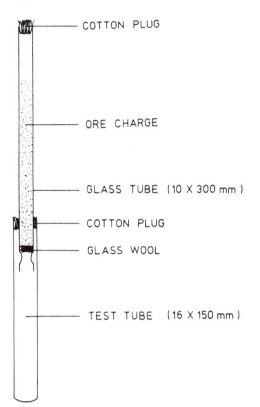

COTTON PLUG

ORE CHARGE

GLASS TUBE (10 X 300 mm)

COTTON PLUG

GLASS WOOL

TEST TUBE   (16 X 150 mm)

Figure 7.9   Percolation column for ore leaching.

The chemostat is a liquid culture system of constant volume in which nutrient supply and actively growing cell population as well as metabolic wastes can be held constant by introducing fresh nutrient solution at such a rate that the outflow removes cells and wastes at a rate that does not change the population in and composition of the medium in the culture vessel (steady-state conditions). If the culture phase is completely liquid, this can be expressed mathematically as

$$\frac{dx}{dt} = \mu x - Dx \tag{7.2}$$

where dx/dt is the rate of cell population change in the chemostat, μ the instantaneous growth rate constant, D the dilution rate, and x the cell concentration or cell number in the chemostat. The dilution rate is defined as the flow rate of the influent feed or the effluent waste divided by the liquid volume of the culture in the chemostat. Under steady-state conditions, $dx/dt = 0$ and therefore $\mu x = Dx$ (i.e., instantaneous growth rate equals dilution rate). Under conditions where $D > \mu$, the cell population in the chemostat will decrease and may ultimately be washed out. Conversely, if $\mu > D$, the cell population in the chemostat will increase until a new steady state is reached, which is determined by the growth-limiting concentration of an essential substrate.

The steady state in the chemostat can also be expressed in terms of the rate of change of growth-limiting substrate concentration (ds/dt) on the principle that the rate of change in substrate concentration is dependent on the rate of substrate addition to the chemostat, the rate of washout from the chemostat and the rate of substrate consumption by the growing organism:

$$\frac{ds}{dt} = D(S_{inflow} - S_{outflow}) - \mu(S_{inflow} - S_{outflow}) \tag{7.3}$$

where D is the dilution rate, S_{inflow} the substrate concentration entering the chemostat, $S_{outflow}$ the concentration of unconsumed substrate, and μ the instantaneous growth-rate constant. At steady state, $ds/dt = 0$. The substrate consumed, i.e., ($S_{inflow} - S_{outlow}$) is related to the cell mass produced (x) according to the relationship

$$x = y(S_{inflow} - S_{outflow}) \tag{7.4}$$

where y is the growth-yield constant (mass of cells produced per mass of substrate consumed). These relationships require modification if a solid substrate is included in the chemostat (see, for instance, section 7.5).

The chemostat can be used, for example, to determine limiting substrate concentrations for growth of bacteria under simulated natural conditions. Thus, the limiting concentrations of lactate, glycerol, and glucose required for growth at different relative growth rates (D/μ_m) of *Archromobacter aquamarinus* (strain 208) and *Spirillum lunatum* (strain 102) in seawater have been determined by this method (Table 7.1) (Jannasch, 1967). The chemostat principle can also be applied to a study of growth rates of microbes in their natural environment by laboratory simulation (Jannasch, 1969) or directly in their natural habitat. For instance, the algal population in an algal mat of a hot spring in Yellowstone Park, Montana, was found to be relatively constant, implying that the algal growth rate equaled its washout rate from the spring pool. When a portion of the algal mat was darkened by blocking access of sunlight, thereby stopping photosynthesis and thus algal growth and multiplication in that part of the mat, the algal cells were washed out from it at a constant rate after a short lag (Fig. 7.10). The washout rate under these conditions equaled the growth rate in the illuminated part of the mat. This follows from Equation (7.2) when $dx/dt = 0$ (Brock and Brock, 1968). Similarly, the population of the sulfur-oxidizing thermophile *Sulfolobus acidocaldarius* has been found to be in

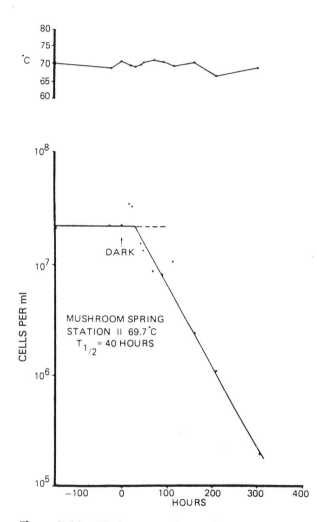

Figure 7.10   Washout rate of part of an algal mat at a station at Mushroom Spring in Yellowstone Park, Montana, after shading it experimentally. (From Brock and Brock, 1968, by permission.)

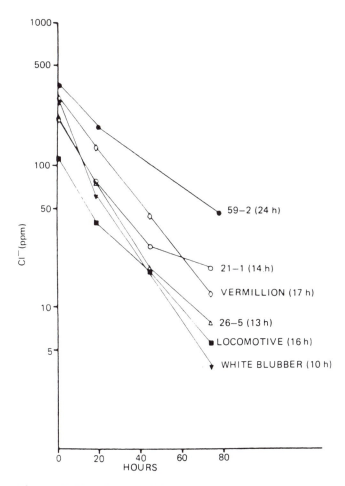

Figure 7.11 Chloride dilution in several small springs in Yellowstone Park, Montana. Estimated half-times for chloride dilution are given in parentheses. At site 21-1, chloride concentration had reached the natural background level by the final sampling time, and the dilution rate was estimated from the data from the first three sampling times. (From Mosser et al., 1974, by permission.)

steady state in certain hot springs in Yellowstone Park, implying, as in the case of the algae, that the growth rate of the organism equals its washout rate from the spring. In this instance, the washout rate was measured by following the water turnover rate in terms of dilution rate of a small, measured amount of NaCl added to the spring pool (Fig. 7.11).

Table 7.1 Threshold Concentrations of Three Growth-Limiting Substrates (mg liter^{-1}) in Seawater at Several Relative Growth Rates of Six Strains of Marine Bacteria and the Corresponding Maximum Growth Rates (μ_m) (hr^{-1})

Strain	D/μ_m	Lactate	Glycerol	Glucose
208	0.5	0.5	1.0	0.5
	0.1	0.5	1.0	0.5
	0.005	1.0	5.0	1.0
μ_m:		0.15	0.20	0.34
102	0.3	0.5	No growth	0.5
	0.1	1.0		5.0
	0.05	1.0		10.0
μ_m:		0.45		0.25

Source: Excerpted from Table 2 in Jannasch (1967), with permission.

The dilution rate of NaCl was then translated into the growth rate of *S. acidocaldarius* in the spring (Mosser et al., 1974).

7.5 QUANTITATIVE STUDY OF GROWTH ON SURFACES

The chemostat (or its principle of operation) is not applicable to all culture situations. In the study of geomicrobial phenomena, the central microbial activity often occurs on the surface of inorganic or organic solids. Indeed, the solid may be the growth-limiting substrate upon which the organism acts. Under these conditions, the microbial population as it colonizes the surface will increase geometrically once it has settled on it, approximating the relationship

$$\log N = \log N_0 + \frac{\Delta t}{g} \log 2 \qquad (7.5)$$

where N is the final cell concentration per unit area, N_0 the initial cell concentration per unit area, Δt the time required for N_0 to multiply to N cells, and g the average doubling time (generation time). Once all available space on the surface has been occupied, however, the cell population on it will remain constant (provided that the surface area does not decrease significantly due to solid substrate consumption or dissolution). The cell population in the liquid in contact with the solid, will then show an arithmetic increase in cell numbers according to the relationship:

$$N_{\text{final,liquid}} = N_{\text{initial,liquid}} + z N_{\text{solid}} \qquad (7.6)$$

where z equals the number of cell doublings on the solid and N_{solid} equals N cells on the solid when all attachment sites are occupied. Equation (7.6) states that for every cell doubling on the surface of the cell-saturated solid, one daughter cell will be displaced into the liquid medium for lack of space on the solid. This model assumes that the liquid medium cannot support growth of the organism. As long as there is no significant change in surface area of the solid phase, the model applies, for instance, to the growth of autotrophic thiobacteria using water-insoluble, elemental sulfur or appropriate metal sulfides as their energy source in a mineral salts solution that satisfies all nutrient requirements except the energy source (Chapters 17 and 18).

To introduce the element of time into equation 7.6, the following relationship should be considered:

$$g = \frac{\Delta t}{z} \tag{7.7}$$

where g represents doubling time, and Δt the time interval between $N_{initial,liquid}$ and $N_{final,liquid}$ determinations. If Eqs. (7.6) and (7.7) are combined, we get

$$N_{final,liquid} = N_{initial,liquid} + \left(\frac{\Delta t}{g}\right)(N_{solid}) \tag{7.8}$$

When solving for g, we get

$$g = \frac{(\Delta t)(N_{solid})}{(N_{final,liquid} - N_{initial,liquid})} \tag{7.9}$$

Espejo and Romero (1987) have taken a different mathematical approach to bacterial growth on a solid surface. They developed the following relationship to describe growth before surface saturation is reached:

$$\frac{dN_a}{dt} = \mu N_a \frac{N_s - N_a}{N_s} \tag{7.10}$$

where N_a is the number of attached bacteria, N_s is the limit value of bacteria that can attach to the surface under consideration, and μ is the specific growth rate. After surface saturation by the bacteria, they propose the following relationship:

$$\frac{dN_f}{dt} = \mu N_s \tag{7.11}$$

where N_f represents the number of free bacteria in the liquid phase with which the surface is in contact. From Eq. (7.10), they derive the relationship

$$\mu = \frac{\Delta N_f}{\Delta t * N_s} \tag{7.12}$$

for the specific growth rate of the culture after surface saturation. These relationships were tested by them in a real experiment growing *Thiobacillus ferrooxidans* on elemental sulfur in which the sulfur was the only available energy source for the bacterium. They observed a logarithmic population increase before surface saturation and a linear increase thereafter. They concluded, however, that the value for N_s was not constant in their experiment but changed gradually after linear growth had begun.

A still different mathematical model was presented by Konishi et al. (1994), in which change in particle size of the solid substrate with time was taken into account as well as cell adsorption to the solid substrate.

7.6 TEST FOR DISTINGUISHING BETWEEN ENZYMATIC AND NONENZYMATIC GEOMICROBIAL ACTIVITY

To determine if a geomicrobial transformation is an enzymatic or a nonenzymatic process, attempts to reproduce the phenomenon with cell-free extract should be tried. If active catalysis is observed, identification of the enzyme system involved should be undertaken by classical techniques. Spent culture medium, from which all cells have been removed, e.g., by filtration, or inactivated by heating, should also be tested for activity. If equal levels of activity are observed with untreated and treated spent medium, operation of a nonenzymatic process may be inferred. If the level of activity in treated cell-free spent medium is lower than in untreated cell-free spent medium, extracellular enzyme activity may be present in the untreated, cell-free spent medium. This can be checked by testing activity in unheated cell-free spent medium over a range of temperatures. If a temperature optimum exists, at least part of the activity in the cell-free medium can be attributed to extracellular enzymes.

7.7 STUDY OF REACTION PRODUCTS OF A GEOMICROBIAL TRANSFORMATION

Ideally, the products of geomicrobial transformation, if they are precipitates, should be studied not only in respect to chemical composition but also in respect to mineralogical properties through one or more of the following techniques: scanning (SEM) and transmission (TEM) electron microscopy, including energy dispersive X-ray measurements (EDX), microprobe examination, X-ray photoelectron spectroscopy (XPS), infrared spectrometry, X-ray diffraction, and mineralogic and crystallographic examination. Similar studies should ideally be undertaken on the substrate if it is an insoluble mineral or mineral complex, to be able to detect any mineralogical changes it may have undergone with time during a geomicrobial transformation.

Studies of geomicrobial phenomena require ingenuity in the application of standard microbiological, chemical, and physical techniques and often require collaboration among microbiologists, biochemists, geochemists, mineralogists, and other specialists to unravel a problem.

7.8 SUMMARY

Geomicrobial phenomena can be studied in the field and in the laboratory. Direct observation may involve microscopic examination and chemical and physical measurements. Very specific and sensitive molecular biological techniques can be applied to determine the number and identity of microbes in field samples. Laboratory study may involve an artificial reconstruction of a geomicrobial process. Special methods have been devised for sampling, for direct observation, and for laboratory manipulation. The latter two categories include the use of fluorescence microscopy, radioactive tracers, and mass spectrometry for observing microorganisms in situ, for measuring process rates, and for measuring microbial isotope fractionation, respectively. The chemostat principle has been applied in the field to measure natural growth rates. It has also been used under simulated conditions for determining limiting substrate concentrations. Growth on particle surfaces may require special experimental approaches.

REFERENCES

Amann, R. I., L. Krumholz, and D. A. Stahl. 1990. Fluorescent-oligonucleotide probing of whole cells for determinative, phylogenetic, and environmental studies. J. Bacteriol. 172:762–770.

Amann, R. I., B. Zarda, D. A. Stahl, and K.-H. Schleifer. 1992. Identification of individual prokaryotic cells using enzyme-labeled, rRNA-targeted oligonucleotide probes. Appl. Environ. Microbiol. 58: 3007–3011.

Amann, R. I., W. Ludwig, and K.-H. Schleifer. 1995. Phylogenetic identification and in situ detection of individual microbial cells without cultivation. Microbiol. Rev. 59:143–169.

Beeman, R. E., and J. M. Suflita. 1989. Evaluation of deep subsurface sampling procedures using serendipitous microbial contaminants as tracer organisms. Geomicrobiol. J. 7:223–233.

Bien, E., and W. Schwartz. 1965. Geomikrobiologische Untersuchungen. IV. Ueber das Vorkommen konservierter toter and lebender Bakterienzellen in Salzgesteinen. Z. Allg. Mikrobiol. 5:185–105.

Bohlool, B. B., and T. D. Brock. 1974. Immunofluorescence approach to the study of the ecology of *Thermoplasma acidophilium* in coal refuse material. Appl. Microbiol. 28:11–16.

Braun-Howland, E. B., S. A. Danielsen, and S. A. Nierzwicki-Bauer. 1992. Development of a rapid method for detecting bacterial cells in situ using 16S rRNA-targeted probes. Biotechniques 13:928–934.

Braun-Howland, E. G., P. A. Vescio, and S. A. Nierzwicki-Bauer. 1993. Use of a simplified cell blot technique and 16S rRNA-directed probes for identification of common environmental isolates. Appl. Environ. Microbiol. 59:3219–3224.

Brock, T. D. 1978. Thermophilic Microorgansism and Life at High Temperatures. Springer-Verlag, New York.

Brock, T. D., and M. L. Brock. 1968. Measurement of steady-state growth rates of a thermophilic alga directly in nature. J. Bacteriol. 95:811–815.

Bromfield, S. M. 1956. Oxidation of manganese by soil microorganisms. Aust. J. Biol. Sci. 9:238–252.

Bromfield, S. M., and V. B. D. Skerman. 1950. Biological oxidation of manganese in soils. Soil Sci. 69;337–348.

Burdige, D. J., and P. E. Kepkay. 1983. Determination of bacterial manganese oxidation rates in sediments using an in-situ dialysis technique, I. Laboratory studies. Geochim. Cosmochim. Acta 47:1907–1916.

Casida, L. E. 1962. On the isolation and growth of individual microbial cells from soil. Can. J. Microbiol. 8:115–119.

Casida, L.E. 1965. Abundant microorganisms in soil. Appl. Microbiol. 13:327–334.

Casida, L. E. 1971. Microorganisms in unamended soil as observed by various forms of microscopy and staining. Appl. Microbiol. 21:1040–1045.

DeLong, E. F., G. S. Wickham, and N. R. Pace. 1989. Phylogenetic stains: ribosomal RNA-based probes for the identification of single cells. Science 243:1360–1363.

Dubinina, G. A. 1970. Untersuchungen ueber die Morphologie und die Beziehung zu *Mycoplasma*. Z. Allg. Mikrobiol. 10:309–320.

Emerson, S., S. Kalhorn, L. Jacobs, B. M. Tebo, K. H. Nealson, and R. A. Rosson. 1982. Environmental oxidation rate of manganese(II): bacterial catalysis. Geochim. Cosmochim Acta 46:1073–1079.

Emiliani, C., J. Hudson, E. A. Shinn, and R. Y. George. 1978. Oxygen and carbon isotopic growth record in a reef coral from the Florida Keys and a deep-sea coral from the Blake Plateau. Science 202: 627–629.

Eren, J., and D. Pramer. 1966. Application of immunofluorescent staining to studies of the ecology of soil microorganisms. Soil Sci. 101:39–49.

Espejo, R. T., and P. Romero. 1987. Growth of *Thiobacillus ferrooxidans* on elemental sulfur. Appl. Environ. Microbiol. 53:1907–1912.

Estep, M. F., and T. C. Hoering. 1980. Biogeochemistry of the stable hydrogen isotopes. Geochim. Cosmochim. Acta 44:1197–1206.

Gerhardt, P., R. G. E. Murray, R. N. Costilow, E. W. Nester, W. A. Wood, N.R. Krieg, and G. B. Phillips, eds. 1981. Manual of Methods for General Bacteriology. American Society for Microbiology. Washington, DC.

Gerhardt, P., R. G. E. Murray, W. A. Wood, and N. R. Krieg, eds. 1993. Methods for General and Molecular Bacteriology. ASM Press, Washington, DC.

Ghiorse, W. C., and D. L. Balkwill. 1983. Enumeration and morphological characterization of bacteria indigenous to subsurface environments. Dev. Indust. Microbiol. 24:213–224.

Giovannoni, S. J., E. F. DeLong, G. J. Olsen, and N. R. Pace. 1988. Phylogenetic goup-specific oligonucleotide tide probes for identification of single microbial cells. J. Bacteriol. 170:720–726.

Hornor, S. G. 1984. Microbial leaching of zinc concentrate in fresh-water microcosms: Comparison between aerobic and oxygen-limited conditions. Geomicrobiol. J. 3:359–371.

Huber, H., G. Huber, and K. O. Stetter. 1985. A modified DAPI fluorescence staining procedure suitable for the visualization of lithotrophic bacteria. Syst. Appl. Microbiol. 6:105–106.

Ivanov, M. V. 1968. Microbiological processes in the formation of sulfur deposits. Published by the U.S. Department of Agriculture and the National Science Foundation, Washington, DC (Israel Program for Scientific translations).

Jannasch, H. W. 1967. Growth of marine bacteria at limiting concentrations of organic carbon in seawater. Limnol. Oceanogr. 12:264–271.

Jannasch, H. W. 1969. Estimation of bacterial growth in natural waters. J. Bacteriol. 99:156–160.

Jannasch, H. W., and M. J. Mottl. 1985. Geomicrobiology of deep-sea hydrothermal vents. Science 229:717–725.

Jannasch, H. W., and C. O. Wirsen. 1981. Morphological survey of microbial mats near deep-sea thermal vents. Appl. Environ. Microbiol. 41:528–538.

Jones, G. E., and R. L. Starkey. 1957. Fractionation of stable isotopes of sulfur by microorganisms and their role in deposition of native sulfur. Appl. Microbiol. 5:111–118.

Jones, J. G., S. Gardener, and B. M. Simon. 1983. Bacterial reduction of ferric iron in a stratified eutrophic lake. J. Gen. Microbiol. 129: 131–139.

Jones, J. G., S. Gardener, and B. M. Simon. 1984. Reduction of ferric iron by heterotrophic bacteria in lake sediment. J. Gen. Microbiol. 130:45–51.

Jurtshuk, R. J., M. Blick, J. Bresser, G. E. Fox, and P. Jurtshuk, Jr. 1992. Rapid in situ hybridization technique using 16S rRNA segments for detecting and differentiating the closely related Gram-positive organisms *Bacillus polymyxa* and *Bacillus macerans*. 58:2571–2578.

Kemper, R. L. Jr., and J. R. Pratt. 1994. Use of fluorochromes for direct enumeration of total bacteria in environmental samples: Past and present. Microbiol. Rev. 58:603–615.

Kieft, T. L., and D. E. Caldwell. 1984. Weathering of calcite, pyrite, and sulfur by *Thermothrix thiopara* in a thermal spring. Geomicrobiol. J. 3:201–215.

Konishi, Y., Y. Takasaka, and S. Asai. 1994. Kinetics of growth and elemental sulfur oxidation in batch culture of *Thiobacillus ferrooxidans*. Biotechnol. Bioeng. 44:667–673.

LaRock, P. A. 1969. The bacterial oxidation of manganese in a freshwater lake. PhD. thesis. Rensselaer Polytechnic Institute, Troy, NY.

LaRock, P. A., and H. L. Ehrlich. 1975. Observations of bacterial microcolonies on the surface of ferromanganese nodules from Blake Plateau by scanning electron microscopy. Microb. Ecol. 2:84–96.

Mosser, J. L., B. B. Bohlol, and T. D. Brock. 1974. Growth rates of *Sulfolobus acidocaldarius* in nature. J. Bacteriol. 118:1075–1081.

Muyzer, G. A. C. de Bruyn, D. J. M. Schmedding, P. Bos, P. Westbroek, and G. J. Kuenen. 1987. A combined immunofluoresence-DNA fluorescence staining technique for enumeration of *Thiobacillus ferrooxidans* in a population of acidophilic bacteria. Appl. Environ. Microbiol. 53:660–664.

Perfil'ev, B. V., and D. R. Gabe. 1965. The use of the microbial-landscape method to investigate bacteria which concentrate manganese and iron in bottom deposits, pp. 9–54. *In*: B. V. Perfil'ev, D. R. Gabe, A. M. Gal'perina, V. A. Rabinovich, A. A. Sapotnitskii, E. E. Sherman, and E. P. Troshanov, eds. Applied Capillary Microscopy: The Role of Microorganisms in the Formation of Iron- and Manganese Deposits. Consultants Bureau, New York.

Perfil'ev, B. V., and D. R. Gabe. 1969. Capillary Methods of Studying

Microorganisms. Engl. transl. by J. Shewan. University of Toronto Press, Toronto.

Phelps, T. J., and B. F. Russell. 1990. Drilling and coring deep subsurface sediments for microbiological investigations. *In*: C.B. Fliermans and T.C. Hazen, eds. Proceedings of the First International Symposium on Microbiology of the Deep Subsurface. WSRC Information Services Section Publication Group. pp. 2–35 to 2–47.

Sand, M. D., P. A. LaRock, and R. E. Hodson. 1975. Radioisotope assay for the quantification of sulfate reducing bacteria in sediment and water. Appl. Microbiol. 29:626–634.

Saylor, G., and A. C. Layton. 1990. Environmental application of nucleic acid hybridization. Annu. Rev. Microbiol. 44:625–648.

Schmidt, E. L., and R. O. Bankole. 1965. Specificity of immunofluorescent staining for study of *Aspergillus flavus* in soil. Appl. Microbiol. 13:673–679.

Schweisfurth, R. 1969. Mangan-oxydierende Pilze. Z. Parasitenk., Infektionskr., Hyg. Abt. 1, Orig. 212:486–491.

Sieburth, J. McN. 1975. Microbial Seascapes. A Pictorial Essay on Marine Microorganisms and Their Environments. University Park Press, Baltimore, MD.

Skerman, V.B.D. 1967. A Guide to The Identification of the Genera of Bacteria. With Methods and Digests of Generic Characteristics. 2nd ed. Williams & Wilkins, Baltimore.

Tsien, H. C., B. J. Bratina, K. Tsuji, and R. S. Hanson. 1990. Use of oligonucleotide signatures probes for identification of physiological groups of methyltrophic bacteria. Appl. Environ. Microbiol. 56: 2858–2865.

Tuovila, B. J., and P. A. LaRock. 1987. Occurrence and preservation of ATP in Antarctic rocks and its implication in biomass determinations. Geomicrobiol. J. 5:105–118.

Ward, D. M., R. Weller, and M. M. Bateson. 1990. 16S rRNA sequences reveal numerous uncultured microorganisms in a natural community. Nature (Lond.) 345:63–65.

Weirich, G., and R. Scweisfurth. 1985. Extraction and culture of microorganisms from rock. Geomicrobiol. J. 4:1–20.

Wellman, R. P., F. D. Cook, and H. R. Krouse. 1968. Nitrogen-15: Microbiological alterations of abundance. Science 161:269–270.

Zavarzin, G. A. 1961. Symbiotic culture of a new microorganism oxidizing manganese. Mikrobiologiya 30:393–395 (Engl. transl. pp. 343–345).

Zavarzin, G. A. 1962. Symbiotic oxidation of manganese by two species of *Pseudomonas*. Mikrobiologiya 31:586–588 (Engl. transl. pp. 481–482).

8

Microbial Formation and Degradation of Carbonates

8.1 DISTRIBUTION OF CARBON IN THE EARTH'S CRUST

Carbon is an element central to all life on Earth. Even though it is one of the less abundant elements in the crust (320 ppm; Weast and Astle, 1982), it is widely but unevenly distributed (Fig. 8.1). In some places, it occurs at high concentration as nonliving matter. Much of the carbon on the surface of the Earth is tied up inorganically in the form of carbonates such as limestone and dolomite, amounting to approximately 1.8×10^{22} g of carbon. Much is also trapped as aged organic matter such as bitumen and kerogen, and as coal, shale organic matter, natural gas, and petroleum. This carbon amounts to about 2.5×10^{22} g as compared with around 3.5×10^{18} g of carbon in unaged, dead organic matter in soils and sediments, and around 8.3×10^{17} g of carbon in living matter (estimates from Fenchel and Blackburn, 1979; Bowen, 1979). The atmosphere around the Earth holds about 6.4×10^{17} g of carbon as CO_2 (Bolin, 1970; Fenchel and Blackburn, 1979). From the quantities of carbon in each of these compartments it is seen that the carbon in living matter represents only a small fraction of the total carbon as does the carbon in unaged, dead organic matter and atmospheric carbon. The carbon in limestone and dolomite

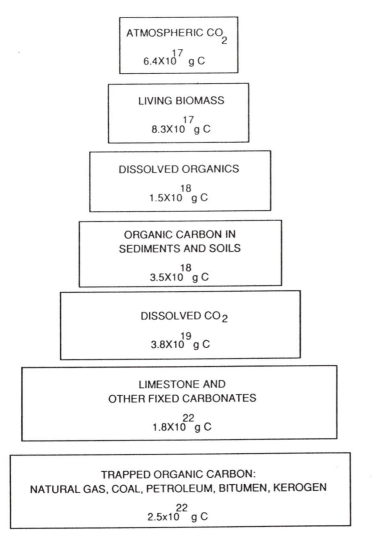

Figure 8.1 Distribution of carbon in the lithosphere of the earth. (Estimates from Fenchel and Blackburn, 1979; Bowen, 1979.)

and in aged organic matter, insofar as it is not mined as fossil fuel and combusted by human beings, is not readily available for assimilation by living organisms. Therefore, living systems have to depend on unaged, dead organic matter and atmospheric carbon. In order not to exhaust this carbon, it has to be recycled by biological mineralization of organic matter (see Chapter 6). In the recycling process, some of the carbon enters the atmosphere as CO_2, some is trapped in carbonate deposits, and some is reassimilated by the living organisms. In the absence of human interference, **homeostasis** is assumed to operate insofar as transfer of carbon among the compartments representing living and dead organic matter and the atmosphere is concerned. Present fears are that human interference is increasing the size of the atmospheric reservoir of carbon through the combustion of fossil fuels because the remaining reservoirs cannot accommodate the extra CO_2 from this combustion. The consequence of this CO_2 build-up in the atmosphere is a blockage of heat radiation into space and an overall warming in the Earth's climate.

8.2 BIOLOGICAL CARBONATE DEPOSITION

Some of the CO_2 generated in biological respiration can be fixed in insoluble carbonates. Indeed, a significant portion of the insoluble carbonate at the Earth's surface is of biogenic origin, but another portion is the result of magmatic and metamorphic activity (e.g., Bonatti, 1966: Skirrow, 1975; Berg, 1986). Biological fixation of carbon in carbonates is carried out by some bacteria, fungi, and algae as well as by some metazoa. The carbonates can be deposited extra- or intracellularly. The bacteria, including cyanobacteria, as well some fungi that are involved, deposit calcium carbonate extracellularly (Bavendamm, 1982; Monty, 1972; Krumbein, 1974, 1979; Morita, 1980; Verrecchia et al., 1990; Chafetz and Buczynski, 1992). The bacterium *Achromatium oxaliferum* seems to be an exception. It has been reported to deposit calcium carbonate intracellularly (Buchanan and Gibbons, 1974). Some algae, including certain green, brown, and red algae, and chrysophytes such as coccolithophores (Lewin, 1965) deposit calcium carbonate as surface structures of their cells, while some protozoa lay it down as tests or shells (foraminifera). Calcium carbonate is also incorporated into skeletal support structures of certain sponges and invertebrates such as coelenterates (corals), echinoderms, bryozoans, brachiopods, and mollusks. In arthropods it is associated with their chitinous exoskeleton. The function of the structural calcium carbonate in each of these organisms is to serve as support and/or protection. In all these cases, calcium and some magnesium ions are combined with carbonate ions of biogenic origin (Lewin 1965). Figure 8.2 illustrates a massive biogenic carbonate deposit in the form of chalk, the White Cliffs of Dover, England.

Historical Perspective of the Study of Carbonate Deposition The beginnings of the study of microbial calcium carbonate precipitation go back

Figure 8.2 White Cliffs of Dover, England; a foraminiferal chalk deposit. (Courtesy of the British Tourist Authority, 40 West 57th Street, New York, NY 10019.)

to the late nineteenth century when Murray and Irvine (1889/1890) and Steinmann (1899/1901) reported formation of $CaCO_3$ in conjunction with urea decomposition and putrefaction in seawater media due to microbial activity (as cited by Bavendamm, 1932). These investigations and others have elicited some controversy.

G. A. Nadson in 1899 and 1903 (see Nadson, 1928) presented the first extensive evidence that bacteria could precipitate $CaCO_3$. Nadson studied the process in Lake Veisovoe in Kharkov, southern Ukraine. He described this lake as shallow with a funnellike deepening to 18 m near its center and as resembling the Black Sea physicochemically and biologically. He found the lake bottom to be covered by black sediment with a slight admixture of calcium carbonate, and he noted that the lake water contained $CaSO_4$, with which it was saturated at the bottom. He found that in winter, the lake water was clear to a depth of 15 m, and from there to the bottom it was turbid, owing to a suspension of elemental sulfur

(S°) and $CaCO_3$. The clear water revealed a varied flora and fauna and a microbial population resembling that of the Black Sea. Gorlenko et al. (1974) demonstrated that a significant part of the sulfur was the result of H_2S oxidation by photosynthetic bacteria, especially the green sulfur bacterium *Pelodictyon phaeum*, along with thiobacilli and other colorless sulfur bacteria. These investigators also reported that the H_2S used by these bacteria arises from bacterial sulfate reduction.

When Nadson incubated black sediment from Lake Veisovoe covered with water from the lake in a test tube open to air, he noted the appearance with time of crystalline $CaCO_3$ in the water above the sediment and the development of a rust color on the sediment surface itself due to oxidized iron (Fig. 8.3). Sterilized sediment did not undergo these changes

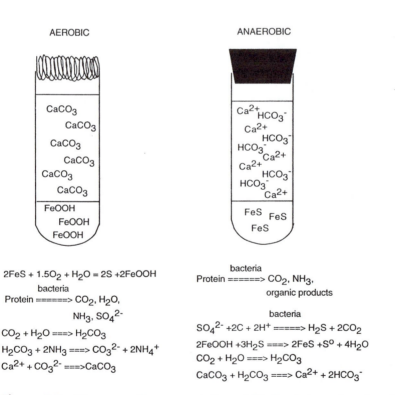

Figure 8.3 Diagrammatic representation of Nadson's experiments with mud from Lake Veisovoe. Contents of the tubes represent the final chemical state after microbial development. Chemical equations describe important reactions leading to the final chemical state.

unless inoculated with a small portion of unsterilized sediment. When access of air was blocked after $CaCO_3$ and ferric oxide development and the test tube reincubated, the reactions were reversed. The $CaCO_3$ dissolved and the sediment again turned black (Fig. 8.3). Nadson isolated a number of organisms from the black lake sediment: *Proteus vulgaris*, *Bacillus mycoides*, *B. salinus* n.sp., *Actinomyces albus* Gasper, *A. verrucosis* n.sp., and *A. roseolus* n.sp. He did not report the presence of sulfate reducing bacteria, which were still unknown at that time.

When in a separate experiment he incubated a pure culture of *P. vulgaris* in a test tube containing sterilized, dried lake sediment and a 2% peptone solution prepared in distilled water, he noted the development of a pellicle of silicic acid which yellowed in time and then became brown and opaque from ferric hydroxide deposition. At the same time he noted that $CaCO_3$ was deposited in the pellicle. None of these changes occurred in sterile, uninoculated medium.

In other experiments, Nadson observed $CaCO_3$ formation by *B. mycoides* in nutrient broth, nutrient agar, and gelatin medium. He did not identify the source of calcium in these instances. He also noted that bacterial decomposition of dead algae and invertebrates in seawater led to $CaCO_3$ precipitation. He further reported the formation of ooliths (i.e., shell-like deposits around normal and involuted cells) by *Bacterium helgolandium* in 1% peptone-seawater broth. He had isolated the bacterium from decomposing *Alcyonidium*, an alga. Lastly, Nadson claimed to have observed precipitation of magnesian calcite or dolomite in two experiments, one involving $3\frac{1}{2}$ years of incubation of black mud from a salt lake near Kharkov, and a second involving $1\frac{1}{2}$ years of incubation of a culture or *P. vulgaris* growing in a mixture of sterilized sediment from Lake Veisovoe and seawater enriched with 2% peptone.

Nadson's pioneering experiments showed that $CaCO_3$ precipitation was not brought about by any special group of bacteria, but depended on appropriate conditions. On the other hand, Drew (1911, 1914), who was apparently not aware of Nadson's studies, concluded from laboratory experiments that he had performed with samples from the tropical seas around the Bahamas that denitrifying bacteria, which he found to be present in significant numbers in his samples, were involved in $CaCO_3$ precipitation in this region. One of these organisms he named *Bacterium calcis*. Although many geologists of his day apparently accepted this interpretation without question, it is not accepted today.

Lipman (1929) rejected Drew's conclusion on the basis of some studies of Pacific waters off the island of Tutuila (American Samoa) and the Tortugas. In his work, Lipman confined himself mainly to an examination of water samples and only a few sediment samples. Using various media, he was able to demonstrate $CaCO_3$ precipitation in the laboratory by a variety of bacteria, not just *B. calcis*, as Drew had suggested. However, since the number of viable organisms in the water samples and in the few sediment samples examined by Lipman seemed small to him, he felt that these organisms were not important in $CaCO_3$ precipitation in the sea. He probably should have examined the sediments more extensively.

Bavendamm (1932) reintroduced Nadson's concept of microbial $CaCO_3$ precipitation as an important geochemical process that is not a specific activity of any special group of bacteria. As a result of his microbiological investigations in the Bahamas, he concluded that this precipitation occurs chiefly in sediments of shallow bays, lagoons, and swamps. He isolated and enumerated heterotrophic and autotrophic bacteria, including sulfur bacteria, photosynthetic bacteria, agar-, cellulose-, and urea-hydrolyzing bacteria, nitrogen fixing bacteria, and sulfate-reducing bacteria, all of which he implicated in an ability to precipitate $CaCO_3$. He rejected the idea of significant participation of cyanobacteria in $CaCO_3$. As we know now, however, some cyanobacteria may cause significant $CaCO_3$ deposition (Golubic, 1973). There is still no complete consensus about the origin of the calcium carbonate suspended in the seas around the Bahamas. Some geochemists favor an inorganic mechanism of formation (Skirrow, 1975). They feel that a finding of supersaturation with $CaCO_3$ of the waters around the Bahamas and an abundant presence of nuclei for $CaCO_3$ deposition explain the $CaCO_3$ precipitation in the simplest way. Others, such as Lowenstam and Epstein (1957), favor algal involvement based on a study of $^{18}O/^{16}O$ ratios of the precipitated carbonates. McCallum and Guhathakurta (1970) isolated a number of different bacteria from sediments from Bimini, Brown's Cay, and Andros Island in the Bahamas with a capacity to precipitate calcium carbonate in most cases as aragonite under laboratory conditions. They concluded that the naturally formed calcium carbonate in the Bahamas is the results of a combination of biological and chemical factors. This also appears from the work of Buczynski and Chafetz (1991), who illustrated in elegant scanning electron photomicrographs the ability of marine bacteria to induce calcium carbonate precipitation in the form of calcite and aragonite (Fig. 8.4).

Despite the now numerous reports on calcium carbonate precipitation by bacteria other than cyanobacteria (e.g., Krumbein, 1974, 1979; Shinano and Sakai, 1969; Shinano, 1972a,b; Ashirov, and Sazonova, 1962; Greenfield, 1963; Abd-el-Malek and Rizk, 1963a; Roemer and Schwartz, 1965; McCallum and Guhathakurta, 1970; Morita, 1980; del Moral et al., 1987; Ferrer et al., 1988a; Ferrer et al., 1988b; Thompson and Ferris, 1990; Rivadeneyra et al., 1991; Buczynski and Chafetz, 1991), the significance of many of them has been questioned by Novitsky (1981). He feels that in most instances, environmental conditions do not meet the requirements

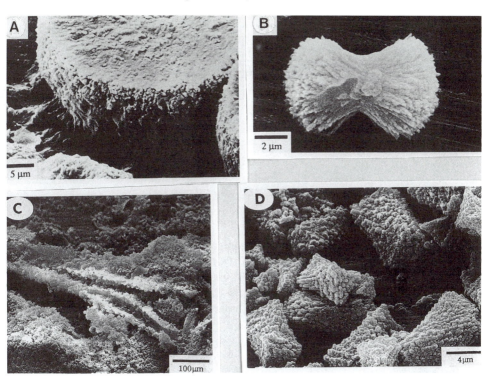

Figure 8.4 Bacterial precipitation of calcium carbonate as aragonite and calcite. Scanning electron micrographs of (A) a crust of a hemisphere of aragonite precipitate formed by bacteria in liquid culture in the laboratory; (B) an aragonite dumbbell precipitated by bacteria in liquid medium in the laboratory; (C) crystal bundles of calcite encrusting dead cyanobacterial filaments that were placed in gelatinous medium inoculated with bacteria from Baffin Bay. The specimen was treated with 30% H_2O_2 to remove organic matter. It should be noted that the crystal aggregates have been cemented to form a rigid crust that does not depend on the organic matter for support; (D) higher magnification view of some of the crystal aggregates from (C) [area of close-up not from field of view in (C)]. These aggregates sometimes resemble rhombohedra, tetragonal disphenoids, or tetragonal dipyramids. [(A) and (B) from Buczinski and Chafetz, 1991, by permission; (C) and (D) from Chafetz and Buczinski, 1992, by permission.]

(especially pH values > 8.3) essential for calcium carbonate precipitation in laboratory experiments. He was unable to demonstrate $CaCO_3$ precipitation with active laboratory isolates at in situ pH in water samples from around Bermuda or from Halifax, Nova Scotia, harbor. Novitsky did not, however, acknowledge the laboratory observation of bacterial $CaCO_3$ pre-

Table 8.1 Solubility Products of Some Carbonates

Compound	Solubility constant (K_{sol})	See footnote
$CaCO_3$	$10^{-8.32}$	a
$MgCO_3 \cdot 3H_2O$	10^{-5}	a
$MgCO_3$	$10^{-4.59}$	b
$CaMg(CO_3)_2$	$10^{-16.7}$	c

[a] Latimer and Hildebrand (1942). Note that Stumm and Morgan (1981) give a solubility product in fresh water of $10^{-8.42}$ (25°C) for calcite and $10^{-8.22}$ (25°C) for aragonite.
[b] Weast and Astle (1982).
[c] Stumm and Morgan (1981).

cipitation by a number of different organisms in a calcium acetate–containing seawater medium with added KNO_3 in which the pH fell to about 6.0 (McCallum and Guhathakukrta, 1970). The authors of this study stressed that the form of the calcium carbonate that precipitated in the presence of the bacteria was aragonite, as is generally observed in situ in marine environments. When calcium carbonate precipitated in any of their sterile controls, it seemed to be in a different mineral form.

Basis for Microbial Carbonate Deposition   Carbonate compounds are relatively insoluble. Table 8.1 lists the solubility constants for several different, geologically important carbonates. Because of the relative insolubility of the carbonates, they are readily precipitated from solution at relatively low carbonate- and counter-ion concentrations. The following will illustrate the fact.

In a solution containing $10^{-4.16}$ M Ca^{2+},* the calcium will be precipitated by CO_3^{2-} at a concentration in excess of $10^{-4.16}$ M. This is because the product of the concentrations of the two ions will exceed the solubility product of $CaCO_3$, which is

$$[Ca^{2+}][CO_3^{2-}] = K_{sol} = 10^{-8.32} \tag{8.1}$$

In general, Ca^{2+} will be precipitated as $CaCO_3$ when the carbonate-ion concentration is in excess of the ratio $10^{-8.32}/[Ca^{2+}]$ [see Eq. (8.1)].

Carbonate ion in an unbuffered aqueous solution undergoes hydrolysis. This process causes a solution of 0.1 M Na_2CO_3 to develop a hydroxyl-ion concentration of 0.004 M. The following reactions explain this phenomenon.

---

* The ion concentrations here refer really to ion activities.

$$Na_2CO_3 \Leftrightarrow 2Na^+ + CO_3^{2-} \tag{8.2}$$

$$CO_3^{2-} + H_2O \Leftrightarrow HCO_3^- + OH^- \tag{8.3}$$

$$HCO_3^- + H_2O \Leftrightarrow H_2CO_3 + OH^- \tag{8.4}$$

Of these three reactions, the third can be considered negligible. Since bicarbonate dissociates according to the reaction

$$HCO_3^- = CO_3^{2-} + H^+ \tag{8.5}$$

whose dissociation constant (K_2) is

$$\frac{[CO_3^{2-}][H^+]}{[HCO_3^-]} = 10^{-10.33} \tag{8.6}$$

and since the ionization constant for water (K_w) at 25°C is

$$[H^+][OH^-] = 10^{-14} \tag{8.7}$$

the equilibrium constant (K_{equil}) for the hydrolysis of CO_3^{2-} [Reaction (8.3)] is

$$\frac{[HCO_3^-][OH^-]}{[CO_3^{2-}]} = \frac{10^{-14}}{10^{-10.33}} = 10^{-3.67} \tag{8.8}$$

At pH 7.0, where the hydroxyl concentration is 10^{-7} (see equation 8.7), the ratio of bicarbonate to carbonate is therefore

$$\frac{[HCO_3^-]}{[CO_3^{2-}]} = \frac{10^{-3.67}}{10^{-7}} = 10^{3.33} \tag{8.9}$$

This means that at pH 7.0, the bicarbonate concentration is $10^{3.33}$ times greater than the carbonate concentration in an aqueous solution, assuming an equilibrium exists between the CO_2 of the atmosphere in contact with the solution and CO_2 in the solution.

From Eq. (8.1) we can predict that at a freshwater concentration of 0.03 g of Ca^{2+} per liter of solution (i.e., at $10^{-3.14}$ M Ca^{2+}), an excess of $10^{-5.18}$ M carbonate ion would be required to precipitate the calcium as calcium carbonate, because from Eq. (8.1) it follows that

$$[CO_3^{2-}] = \frac{10^{-8.32}}{10^{-3.14}} = 10^{-5.18} \tag{8.10}$$

Now, at pH 7.0, $10^{-5.18}$ M carbonate would be in equilibrium with $(10^{-5.18})$ multiplied by $(10^{3.33})$ or $10^{-1.85}$ M HCO_3^- according to equation 8.9. This amount of bicarbonate plus carbonate is equivalent to approximately 0.6 g of CO_2 per liter of solution.

Similarly, from equation 8.1 we can predict that at the calcium-ion concentration in normal seawater, which is 10^{-2} M, a carbonate concentration in excess of $10^{-6.32}$ M would be required to precipitate it. At pH 8, this amount of carbonate ion would be expected to be in equilibrium with approximately $10^{-3.99}$ M bicarbonate ion, which is equivalent to about 0.0045 g of CO_2 per liter of solution. Assuming the combined concentration of carbonate and bicarbonate ions in seawater to be 2.8×10^4 μg of carbon per liter (Marine Chemistry, 1971), we can calculate that the carbonate concentration at pH 8.0 must be about $10^{-4.97}$ M and the bicarbonate concentration about $10^{-2.64}$ M. Since the product of the carbonate concentration ($10^{-4.97}$ M) and that of the calcium concentration (10^{-2}) is $10^{-6.97}$, which is greater than the solubility product for $CaCO_3$ ($10^{-8.32}$), seawater is saturated with respect to calcium carbonate. In reality, this appears to be true only for marine surface waters (Schmalz, 1972). (Mg^{2+} ion in seawater is not readily precipitated as $MgCO_3$ because of the relatively high solubility of $MgCO_3$).

A quantity of 0.6 g of CO_2 can be derived from the complete oxidation of 0.41 g of glucose according to the equation

$$C_6H_{12}O_6 + CO_2 \rightarrow 6CO_2 + 6H_2O \tag{8.11}$$

Similarly, 0.0045 g of CO_2 can be derived from the complete oxidation of 0.003 g of glucose. Such quantities of glucose are readily oxidized by appropriate populations of bacteria or fungi in relatively short time.

Conditions for Extracellular Microbial Carbonate Precipitation   As already mentioned, some bacteria including cyanobacteria, and fungi precipitate $CaCO_3$ or other insoluble carbonates extracellularly under various conditions. Let us define the conditions under which this can take place. They include:

1.   Aerobic and anaerobic oxidation of carbon compounds consisting of carbon and hydrogen with or without oxygen, for example carbohydrates, organic acids, and hydrocarbons. If such oxidations occur in a *well-buffered neutral or alkaline environment* containing adequate amounts of calcium or other appropriate cations, at least some of the CO_2 that is generated will be transformed into carbonate, which will then precipitate with appropriate cations

$$CO_2 + H_2O \Leftrightarrow H_2CO_3 \tag{8.12}$$

$$H_2CO_3 + OH^- \Leftrightarrow HCO_3^- + H_2O \tag{8.13}$$

$$HCO_3^- + OH^- \Leftrightarrow CO_3^{-2} + H_2O \tag{8.14}$$

Calcium carbonate precipitation under these conditions has been illustrated by the formation of aragonite and other calcium carbonates by bacteria and fungi in seawater media containing organic matter at concentrations of 0.01 to 0.1% (Krumbein, 1974). The organic matter in different texperiments consisted of glucose, sodium acetate, or sodium lactate. The aragonite precipitated on the surface of the bacteria or fungi after 36 h of incubation. Between 36 and 90 h, the cells in the $CaCO_3$ precipitate were still viable (although deformed), but after 4–7 days they were nonviable. Phosphate above a critical concentration can interfere with calcite formation by soil bacteria (Rivadeneyra, 1985).

Verrecchia et al. (1990) have suggested that in semiarid regions, the role of fungi is to immobilize Ca^{2+} with oxalate that is a product of their metabolism. Upon the death of the fungi, bacteria convert the calcium oxalate (whewellite) to secondary calcium carbonate by mineralizing the oxalate.

2. Aerobic or anaerobic oxidation of organic nitrogen compounds with the release of NH_3 and CO_2 in unbuffered environments containing sufficient amounts of calcium, magnesium, or other appropriate cations. The NH_3 is formed especially by bacteria in the deamination of amines, amino acids, purines, pyrimidines, and other nitrogen-containing compounds. In water, the NH_3 hydrolyzes to NH_4OH, which dissociates partially to NH_4^+ and OH^-, thereby raising the pH of the environment to the point where at least some of the CO_2 produced may be transformed into carbonate. $CaCO_3$ precipitation under these conditions has been illustrated by the formation of aragonite and other calcium carbonates by bacteria and fungi in seawater media containing nutrients such as asparagine or peptone in a concentration from 0.01 to 0.1%. or homogenized cyanobacteria, as observed by Krumbein (1974). Moderately halophilic bacteria have been shown to precipitate $CaCO_3$ as calcite, aragonite, or vaterite, depending on culture conditions in the laboratory (del Moral et al., 1987; Ferrer et al., 1988a; Rivadeneyra et al., 1991). Other examples are the precipitation of calcium carbonate by various species of *Micrococcus* and a gram-negative rod in peptone media made up in natural seawater and in Lyman's artificial seawater (Shinano and Sakai, 1969; Shinano, 1972a,b). The organisms in this case came from inland seas of the North Pacific and from the western Indian Ocean. Lithification of beachrock along the shores of the Gulf of Aqaba (Sinai) is an example of in situ formation of $CaCO_3$ by heterotrophic bacteria in their mineralization of cyanobacterial remains (Krumbein, 1979).

3. The reduction of $CaSO_4$ to CaS by sulfate-reducing bacteria such as *Desulfovibrio* spp., *Desulfotomaculum* spp., and others using organic carbon, indicated by the formula (CH_2O) in equation 8.15, as their source of reducing power. The CaS formed by these organisms hydrolyzes readily to H_2S, which has a small dissociation constant ($K_1 = 1.1 \times 10^{-7}$; $K_2 = 1 \times 10^{-15}$). The Ca^{2+} then reacts with CO_3^{2-} derived from the CO_2

produced in the oxidation of the organic matter by the sulfate-reducing bacteria. The following equations summarize the chemical sequence of reactions:

$$CaSO_4 + 2(CH_2O) \xrightarrow[\text{reducers}]{\text{sulfate}} CaS + 2CO_2 + 2H_2O \qquad (8.15)$$

$$CaS + H_2O \longrightarrow Ca(OH)_2 + H_2S \qquad (8.16)$$

$$CO_2 + H_2O \longrightarrow H_2CO_3 \qquad (8.17)$$

$$Ca(OH)_2 + H_2CO_3 \longrightarrow CaCO_3 + 2H_2O \qquad (8.18)$$

In the foregoing reactions it should be noted that 2 moles of CO_2 are formed for every mole of SO_4^{2-} reduced. Yet, only 1 mole of CO_2 is required to precipitate the Ca^{2+} ions. Hence, $CaCO_3$ precipitation under these conditions depends on one of three conditions, namely the loss of CO_2 from the environment, the presence of a suitable buffer system, or the development of alkaline conditions. Evidence for $CaCO_3$ deposition linked to bacterial sulfate reduction is found in the work of Abd-el-Malek and Rizk (1963a). They demonstrated the formation of $CaCO_3$ during bacterial sulfate reduction in experiments using fertile clay-loam soil enriched with starch and sulfate, or sandy soil enriched with sulfate and plant matter. Other evidence of microbial carbonate formation during sulfate reduction is found in the work of Ashirov and Sazonova (1962) and Roemer and Schwartz (1965). Ashirov and Sazonova showed that secondary calcite was deposited when an enrichment of sulfate-reducing bacteria was grown in quartz sand bathed in Shturm's medium: $(NH_4)SO_4$, 4 g; $NaHPO_4$, 3.5 g; KH_2PO_4, 1.5 g; $CaSO_4$, 0.5 g; $MgSO_4 \cdot 7H_2O$, 1.0; NaCl, 20 g; $(NH_4)_2Fe(SO_4)_2 \cdot 6H_2O$, 0.5 g; Na_2S, 0.030 g; $NaHCO_3$, 0.5 g; distilled water, 1 liter. The hydrogen donor was either gaseous hydrogen, calcium lactate plus acetate (the acetate probably acted as a carbon source), or petroleum. Petroleum may have first been broken down to usable hydrogen donors for sulfate reduction by other organisms in the mixed culture (e.g., Nazina et al., 1985) which the investigators used as inoculum, before the sulfate reducers carried out their activity. The results from these experiments have lent support to the notion that incidents of sealing of some oil deposits by $CaCO_3$ may be due to the activity of sulfate reducing bacteria at the petroleum/water interface of an oil reservoir.

Roemer and Schwartz (1965) showed that sulfate reducers were able to form calcite ($CaCO_3$) from gypsum ($CaSO_4 \cdot 2H_2O$) and from anhydrite ($CaSO_4$). Their cultures also formed strontianite ($SrCO_3$) from celestite ($SrSO_4$) and witherite ($BaO \cdot CO_2$) from barite ($BaSO_4$), but they formed aluminum hydroxide rather than aluminum carbonate from aluminum sulfate.

Still another example of calcium carbonate formation as a result of bacterial sulfate reduction is the deposition of secondary calcite in cap rock of salt domes, which has been inferred from a study of $^{13}C/^{12}C$ ratios of samples taken from these deposits (Feeley and Kulp, 1957) (see also Chapter 17).

4. The hydrolysis of urea leading to the formation of ammonium carbonate

$$NH_2(CO)NH_2 + 2H_2O \rightarrow (NH_4)_2CO_3 \qquad (8.19)$$

This reaction causes precipitation of Ca^{2+}, Mg^{2+} or other appropriate cations when present at suitable concentrations. Urea is an excretory product of ureotelic animals, including adult amphibians and mammals. This hydrolytic reaction was first observed in experiments by Murray and Irvine (see Bavendamm, 1932). They believed it to be important in the marine environment. However, they did not implicate bacteria in urea hydrolysis whereas Steinmann (1899, 1901), working independently, did (as cited by Bavendamm, 1932). Bavendamm (1932) observed extensive $CaCO_3$ precipitation by urea-hydrolyzing bacteria from the Bahamas. As it is now perceived, this is the least important mechanism of microbial carbonate deposition in nature because urea is not a widely distributed compound.

5. Removal of CO_2 from a bicarbonate-containing solution. Such removal will cause an increase in carbonate ion concentration according to the relationship

$$2HCO_3^- \Leftrightarrow CO_2 + H_2O + CO_3^{2-} \qquad (8.20)$$

In the presence of an adequate supply of Ca^{2+}, $CaCO_3$ will precipitate.

An important process of CO_2 removal is photosynthesis, in which CO_2 is assimilated as the chief source of carbon. Some chemolithotrophs, as long as they do not generate acids in the oxidation of their inorganic energy source, may also promote $CaCO_3$ precipitation because such organisms also rely on CO_2 as their carbon source. Examples of microbial organisms that precipitate $CaCO_3$ around them as a result of their photosynthetic activity include certain filamentous cyanobacteria associated with stromatolites (Monty, 1972; Golubic, 1973; Walter, 1976; Krumbein and Giele, 1979; Wharton et al., 1982; Nekrasova et al., 1984) (also see below). In Flathead Lake delta, Montana, calcareous nodules and crusts are deposited on subaqueous levees by cyanobacteria and algae. The calcium carbonate deposition in the outer portions of the nodules and concretions may result in a local increase in pH, which promotes the dissolution of silica of diatom frustules also found on the nodules and concretions. The dissolved silica is reprecipitated with calcium carbonate in interior zones of these structures. The source of the calcium and organic carbon from which CO_2 is generated by mineralization is not the lake, which is oligotrophic, but the Flathead River feeding into the lake at the site of deposition. Deposition of the concretions and crusts coincides with periods of high productivity (Moore, 1983).

This mechanism may also function in the deposition of structural carbonate. Examples are found in some of the green, brown and red algae, and some of the chrysophytes, which are all known to deposit calcium carbonate in their walls (e.g., Lewin, 1965a; Friedmann et al., 1972). Not all calcareous algae form their $CaCO_3$ as a result of photosynthetic removal of CO_2, however; some form it from respiratory CO_2. In any case, photosynthetic removal of CO_2 is probably one of the most important mechanisms of biogenic $CaCO_3$ formation in the open, aerobic environment.

Carbonate Deposition by Cyanobacteria   Carbonate deposition by cyanobacteria has been described by Golubic (1973) and Pentecost and Bauld (1988). In this process, a distinction must be made between cyanobacteria that entrap and agglutinate *preformed* calcium carbonate in their thalli and those that precipitate it in their thalli as a result of their photosynthetic activity. Preformed calcium carbonate used in the entrapment and agglutination process is formed at a site other than the site of deposition, whereas the calcium carbonate deposited as a result of photosynthetic activity of the cyanobacteria is being formed at the site of deposition (e.g., Krumbein and Potts, 1979; Pentecost and Bauld, 1988). Calcium carbonate associated with stromatolite structures (special types of cyanobacterial mats) may be a result of deposition by entrapment and agglutination or of deposition through cyanobacterial photosynthesis. In the cases of *Homoeothrix crustacea* (Pentecost, 1988) and *Lyngbya aestuarii* and *Scytonema myochrous* it is due to photosynthesis (Pentecost and Bauld, 1988).

Calcium carbonate associated with travertine, lacustrine carbonate crusts, and nodules can result from photosynthetic activity of cyanobacteria in freshwater environments. Travertine, a porous limestone, is formed from rapid calcium carbonate precipitation due to cyanobacterial photosynthesis in waterfalls and streambeds of fast-flowing rivers, which tend to bury the cyanobacteria. By outward growth or movement, the cyanobacteria avoid being trapped and contribute to the porosity of the deposit (Golubic, 1973). Lacustrine carbonate crusts are formed by benthic cyanobacteria attached to rocks or sediment, which deposit calcium carbonate through their photosynthetic activity in shallow portions of lakes with carbonate-saturated waters (Golubic, 1973). Calcareous nodules are also formed around rounded rocks and pebbles or shells to which calcium carbonate-depositing cyanobacteria are attached and which are rolled by water currents, thus exposing different parts of their surface to sunlight at different times and promoting photosynthetic activity and calcium carbonate precipitation by the attached cyanobacteria (Golubic, 1973). High-magnesium calcite precipitated in sheaths of certain cyanobacteria such as *Scytonema* may be related to the ability of the sheaths to concentrate magnesium three to five times over magnesium concentrations in seawater (Monty, 1967; see also discussion by Golubic, 1973).

Spiro and Pentecost (1991) observed that at Waterfall Beck in Yorkshire, England, the calcite deposited by cyanobacteria was richer in ^{13}C than the nearby travertine, suggesting that the travertine was formed by a different mechanism than the calcite formed on the cyanobacteria. Outgassing of CO_2 from the streamwater may have played a role in travertine formation (Pentecost and Spiro, 1990).

Thompson and Ferris (1990) demonstrated the ability of *Synechococcus* from Green Lake, Lafayette, New York, to precipitate calcite, gypsum, and probably magnesite from filter-sterilized water from the lake in laboratory simulations (Fig. 8.5). The lake has an average depth of about 28 m (52.5 m max.). It is meromictic with a distinct, permanent chemocline at a depth of about 18 m (Brunskill and Ludlam, 1969). Its water is naturally alkaline (pH around 7.95), contains in the order of 11 mM Ca^{2+}, 2.8 mM Mg^{2+}, and 10 mM SO_4^{2-}, and has an ionic strength of around 54.1 and an alkalinity of around 3.24 (Thompson and Ferris, 1990). Gypsum crystals developed on the surface of *Synechococcus* cells before calcite crystals, but the calcite deposit became more massive and less prone to being shed by the cells on division than the gypsum deposit. Calcite deposition coincided with a rise in pH in the immediate surround of the cells due to their photosynthetic activity. Gypsum deposition occurred in the dark as well as in light and hence was not driven by photosynthesis as was calcite deposition. Indeed, Thompson and Ferris suggest that calcite may replace gypsum in the developing carbonate bioherm (a microbialite) in the lake. Calcite is deposited on the *Synechococcus* cells as a result of interaction between calcium ions bound at the cell surface and carbonate ions generated as a result of the photosynthetic activity of the cells. The cell-bound calcium ions also capture sulfate ions to form gypsum. This activity explains the origin of the marl and the calcified bioherms that are found in Green Lake (Thompson et al., 1990).

The calcium carbonate formed in the lithification of some cyanobacterial mats was found not to originate during photosynthesis of the cyano-

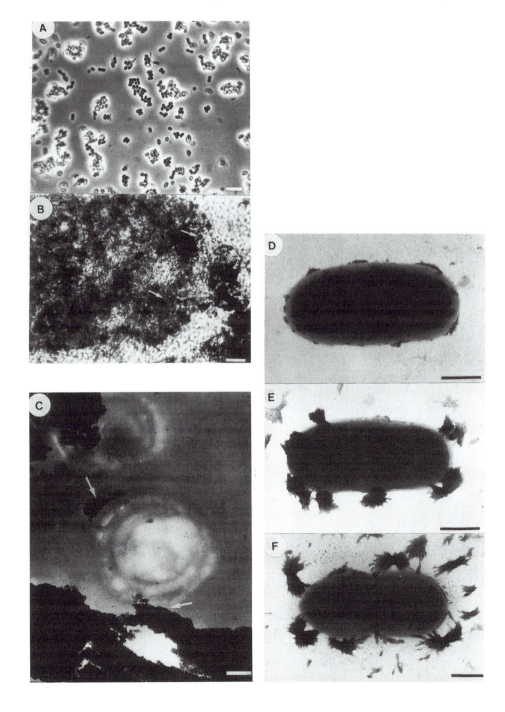

bacteria but was the result of the activity of bacteria associated with the cyanobacteria (Chafetz and Buczynski, 1992). According to Chafetz and Buczynski, cyanobacterial stromatolites may thus owe their existence to bacterial $CaCO_3$ precipitation rather than cyanobacterial photosynthesis.

A Possible Model for Oolite Formation A process resembling structural carbonate deposition to a degree (see below) is that involving the laying down of carbonate on the cell surface of a marine pseudomonad, strain MB-1. It adsorbs calcium and magnesium ions on its cell surface (cell wall-membrane complex) (Greenfield, 1963). Living or dead cells of this organism have this capacity, and hence the process of $CaCO_3$ deposition by this prokaryote is not directly dependent on the living state of the organism. The dead cells adsorb calcium ions more extensively than they adsorb magnesium ions. Carbonate in the medium derives mostly from respiratory CO_2 produced by living cells of the organism. The conversion of CO_2 to carbonate is brought about by hydrolysis of ammonia produced from organic nitrogen compounds by actively metabolizing cells. It results in aragonite ($CaCO_3$) formation on the cells. These cells then serve as nuclei for further calcium carbonate precipitation (Fig. 8.6). Recent observations by Buczynski and Chafetz (1991) strongly support this model. A similar phenomenon was also shown with a marine yeast (Buck and Greenfield, 1964, as cited by McCallum and Guhathakurta, 1970). These phenomena may represent a model for oolite formation.

Figure 8.5 Calcite and gypsum precipitation by *Synechococcus* sp. isolated from Fayetteville Green Lake, New York. (A) Phase contrast photomicrograph of *Synechococcus* in laboratory culture. (B) Petrographic thin-section photomicrograph of calcite crystal from Green Lake showing evidence of occlusion of numerous small bacterial cells within calcite grain (arrows). Note similar size of *Synechococcus* cells in (A) and occluded bacterial cells in (B). Scale bars in (A) and (B) equal 5 μm. (C) Thin-section transmission electron micrograph of two *Synechococcus* cells and calcite from a 72-h culture (cell is represented by white oval area between arrows). Arrows point to calcite (electron-dense material) on the cell surface. Cells are unstained to avoid dissolution of calcite by acidic heavy-metal stains used to provide contrast to biological specimens. Scale bar equals 200 nm). (D–F) Series of transmission electron micrographs showing progression of gypsum precipitation on cell wall of *Synechococcus* (whole mounts): (D) Initiation of numerous nucleation sites on cell surface. (E) Gypsum precipitation spreading away from cell. Gypsum still appears to be covered by thin layer of bacterial slime. (F) Dividing *Synechococcus* cell shedding some of the precipitated gypsum. Scale bars equal 500 nm. (Courtesy of J. B. Thompson from Thompson and Ferris, 1990, by permission.)

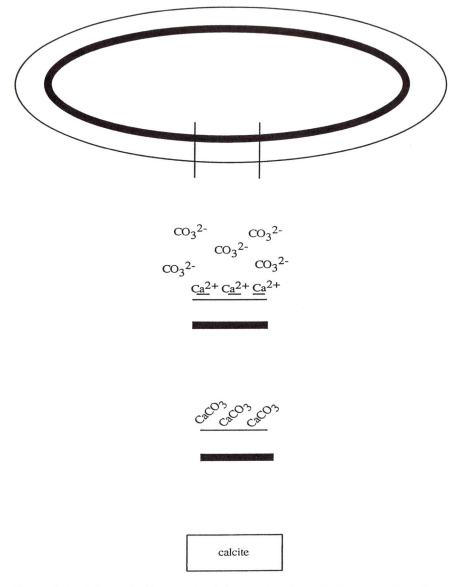

Figure 8.6   Schematic diagram of calcium carbonate deposition on cell surface of a prokaryote, resulting in calcite or aragonite formation. See text for details.

Structural or Intracellular Carbonate Deposition by Microbes In general, morphological and physiological studies have shown that whereas bacteria, including cyanobacteria, cause precipitation of $CaCO_3$ mostly in the immediate environment around their cells, some eukaryotic algae and protozoans lay down $CaCO_3$ as cell-support structures on their cells. Examples of such eukaryotic organisms are green algae (Chlorophyceae) such as *Acetabularia, Chara, Nitella, Penicillus*, and *Padina* (Wefer, 1980); coccolithophores (Fig. 8.7) such as *Scyphosphaera, Rhabdosphaera*,and *Calciococcus*; coralline algae such as *Arthrocardia silvae* and *Amphiroa beauvoisii* (Fig. 8.8); and foraminifera such as *Heterostegina* and *Globig-*

3 μm

Figure 8.7 Coccoliths, the calcareous skeletons of an alga belonging to the class Chrysophyceae (see sketch in Fig. 5.9). These specimens were found residing on the surface of a ferromanganese nodule from Blake Plateau of the Atlantic coast of the United States. (From LaRock, P. A., and H. L. Ehrlich, 1975, Observations of bacterial microcolonies on the surface of ferromanganese nodule from Blake Plateau by scanning electron microscopy. Microb. Ecol. 2:64–96. Copyright ©1975 Springer Verlag, New York, by permission.)

(a)

(b)

Figure 8.8   Articulated coralline (calcareous) algae. (a) *Arthrocardia silvae*, Johansen, California. Scale mark = 1 cm. (b) *Amphiroa beauvoisii*, Lamouroux, Gulf of California. Scale mark = 1.5 cm. (Courtesy of H. William Johansen.)

erina (Fig. 8.9). The details of the biochemical mechanism by which they deposit $CaCO_3$ are largely unknown, although it has been assumed that they derive the carbonate externally as a result of photosynthetic activity or internally as a result of their respiratory activity. In eukaryotic algae, structural calcium carbonate is laid down either as calcite or as aragonite (Lewin, 1965). The amount of carbon incorporated into carbonate as a result of algal photosynthesis may be a significant portion of the total carbon assimilated (Jensen et al., 1985). Wefer (1980) measured in situ $CaCO_3$ production by *Halimeda incrassata, Penicillus capitatus and Padina sanctae-crucis* in Harrinton Sound, Bermuda, of approximately 50, 30 and 240 g m^{-2} y^{-1}, respectively. Gelatinous or mucilaginous substances in association with the cell walls of these algae may be involved in the deposition and organization of the crystalline mineral. If the CO_2 for forming the algal carbonate depended on photosynthesis and came from seawater, it is likely to be enriched in ^{13}C relative to seawater CO_2, whereas if the CO_2 had a respiratory origin, the algal carbonate is likely to be enriched in ^{12}C relative to seawater (Lewin, 1965). The basis for the morphogenesis of structures as intricate as those found associated with calcareous foraminifera and coccolithophores remains to be explained in detail. In coccolithophores, special intracellular vesicles are the site of coccolith formation (De Jong et al., 1983). In the coccolithophore *Emiliania huxleyi* strains A92, L92, and 92D, acidic Ca^{2+}-binding polysaccharides have been found to inhibit precipitation of $CaCO_3$ in vitro (Borman et al., 1982, 1987) and are thought to regulate $CaCO_3$ precipitation in the intracellular coccolith-forming vesicles (see De Jong, et al., 1983).

A Model for Skeletal Carbonate Formation   A possible clue to the biochemical mechanism of $CaCO_3$ deposition in cell structures, especially in protozoa (heterotrophs), may be found in an observation by Berner (1968). He noted that during bacterial decomposition of butterfish and smelts in seawater in a sealed jar, calcium was precipitated not as $CaCO_3$ but as calcium soaps or adipocere (calcium salts of fatty acids) in spite of the presence of HCO_3^- and CO_3^{2-} species and an alkaline pH in the reaction mixture. The prevailing fatty acid concentration favored calcium soap formation over calcium carbonate formation. Berner suggested that in nature such soaps could later be transformed into $CaCO_3$ through mineralization of the fatty acid ligand.

McConnaughey (1991) has proposed that *Chara corallina*, a calcareous alga, lays down $CaCO_3$ pericellularly by a process involving an ATP-driven H^+/Ca^{2+} exchange. In this process protons produced in the reaction

$$Ca^{2+} + CO_2 + H_2O \Rightarrow CaCO_3 + 2H^+ \tag{8.21}$$

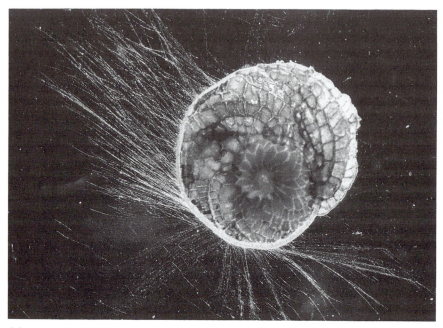

(a)

Figure 8.9 Foraminifera. (a) A living foraminiferan specimen, *Heterostegina depressa*, from laboratory cultivation. Note the multichambered test and the fine protoplasmic threads projecting from the test. Test diameter, 3 mm. (Courtesy of R. Röttger. See also Röttger, 1974). (b) Foraminiferan test (arrow) in Pacific sediment: *Globigerina* (?) ($\times 2430$).

assumed to occur on the $CaCO_3$ surface facing the cell, are exchanged for Ca^{2+}. The protons then react with HCO_3^- in the cell to generate CO_2 for photosynthesis:

$$HCO_3^- + H^+ \Rightarrow CO_2 + H_2O \qquad (8.22)$$

It seems possible that in some living calcareous organisms, calcium intended for calcium carbonate structures is first localized by formation of an organic calcium salt at the site of calcium deposition (the plasma membrane?) and is then converted to $CaCO_3$ in the presence of carbonic anhydrase, an enzyme that can promote the conversion of dissolved metabolic CO_2 to bicarbonate and carbonate in a reversible reaction. Carbonic anhydrase activity has been detected in some microalgae and cyanobacteria (Aizawa and Miyachi, 1986). While Aizawa and Miyachi considered

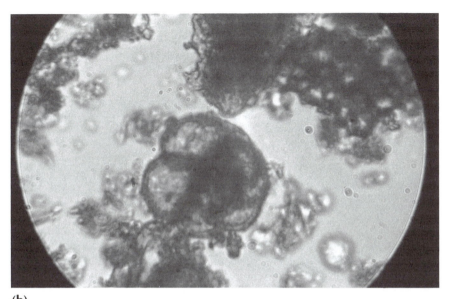

(b)

Figure 8.9 Continued.

the activity of the carbonic anhydrase from the standpoint of its role in CO_2 assimilation in photosynthesis, it may very well also play a role in CO_2 conversion to carbonate under different physiological conditions. *Hyphenomonas*, a coccolithophorid, has been said to utilize "a hydroxyproline-proline-rich peptide and sulfated polysaccharide moieties" in $CaCO_3$ deposition (Isenberg and Lavine, 1973). Microscopically, vesicles derived from the Golgi apparatus of the cell and intracellular fibers produced by them play a role in the mineral deposition process in this organism. Degens (1979) has related this model to calcium transport mechanisms in cells.

Microbial Formation of Carbonates Other than Those of Calcium
Sodium Carbonate   Carbonate may occur in solid phases not only as calcium or calcium plus magnesium salts but also as a sodium salt (natron, $Na_2CO_3 \cdot 10H_2O$). In at least one instance, the Wadi Natrun in the Libyan desert, such a deposit has been clearly associated with the activity of sulfate-reducing bacteria (Abd-el-Malek and Rizk, 1963b). As the authors described it, the wadi (a channel of a watercourse that is dried up except during rainfall; an arroyo) in this case contains a chain of small lakes, 23 m below sea level. The smallest of the lakes dries up almost completely during the summer, and the larger ones dry out partially at that time.

Natron is in solution in the water of all the lakes and in solid form on the bottom of some of the lakes. The water feeding the lakes is supplied partly by springs and partly by streamlets which probably derive their water from the nearby Rosetta branch of the Nile. On its way to the lakes, the surface water passes through cypress swamps. The authors found sulfate and carbonate in high concentration in the lakes (189–204 mEq of carbonate per liter and 324–1107 mEq of sulfate per liter) and at low concentration in the cypress swamp (0 mEq to traces of carbonate per liter and 2–13 mEq of sulfate per liter). Bicarbonates occurred in significant amounts in lakes and swamps (22–294 meq and 11–16 meq liter^{-1}, respectively). Soluble sulfides predominated in the lakes and swamps (7–13 and 1–4 mEq liter^{-1}, respectively). Considerable numbers of sulfate-reducing bacteria (1×10^6 to 8×10^6 per ml) were detected in the swamps and in the soil at a distance of 150 m from the lakes, but not elsewhere (see Table 8.2 for more detailed data). Sulfate reduction was inferred to occur chiefly in the cypress swamps because of the significant presence of sulfate reducers and the readily available organic nutrient supply at those sites. The sulfate reduction leads to the production of bicarbonate, as follows:

$$SO_4^{2-} + 2(CH_2O) = H_2S + 2CO_2 + 2OH^- \tag{8.23}$$

$$2CO_2 + 2OH^- = 2HCO_3^- \tag{8.24}$$

Most of the soluble products of sulfate reduction, HCO_3^- and S^{2-} were found to be washed into the lakes where they became concentrated and partially precipitated as carbonate and sulfide salts on evaporation. Some of the sulfide produced in the swamps was found to combine with iron to form ferrous sulfide, which imparts a characteristic black color to the swamps. The carbonate in the lakes resulted from the loss of CO_2 from the water to the atmosphere, promoted by the warm water temperatures, especially in summer (CO_2 solubility in water decreases with increase in temperature) [Eq. (8.20)].

Manganese and Iron Carbonates   Carbonate may also combine with manganese and iron to form some deposits in nature. The origin of at least some of these deposits has been ascribed to microbial activity. An example is the occurrence of rhodochrosite ($MnCO_3$) and siderite ($FeCO_3$) in Punnus-Ioki Bay of Lake Punnus Yarvi on the Karelian peninsula in the former U.S.S.R. (Sokolova-Dubinina and Deryugina, 1967a). Lake Punnus-Yarvi is 7 km long and 1.5 km wide at its broadest part. Its greatest depth is 14 m. It is only slightly stratified thermally. The oxygen concentration in its surface water was reported by the authors to range from 11.8 to 12.1 mg liter^{-1} and in its bottom water 5.7 to 6.6 mg liter^{-1}. The pH of

Table 8.2 Chemical and Bacteriological Analyses of Water and Soil Samples from Wadi Natrûn

| Type and source of sample | pH range | Milliequivalents[a] of: | | | | | Total soluble salts (g liter^{-1}) | Organic matter (%) | Viable counts of sulfate reducers[b] (10^6 ml^{-1}) |
		CO_3^{2-}	HCO_3^{-}	SO_4^{2-}	Cl^{-}	S^{2-}			
Water									
Artesian wells	7.4–7.8	0	2–5	9–13	18–30	0.2–0.5	2–3[c]		d
Burdi swamps	6.8–7.2	0–trace	11–61	2–13	1–7	1–4	1–8[c]		1–5
Lakes	9.5–10.1	189–240	22–294	324–1107	107–210	7–13	180–239[c]		d
Soil									
Newly cultivated uplands	7.0–7.6	0	1–2	14–18	11–19[c]	c	c	0.2–0.5	d
About 150 m from lakes	7.2–7.5	0	2–3	12–23	1–6[c]	c	c	0.1–5.2	5–8
Swamps	7.4–7.8	Trace	3–11	4–7	3–8[c]	d	c	3.4–7.8	0.8–2

[a] Milliequivalents per liter of water and per 100 g of soil.
[b] Counts per milliliter of water and per gram of soil.
[c] Not determined.
[d] Not detected.
Source: Abd-el-Malek and Rizk (1963b), by permission.

its water was given as slightly acid (pH 6.3 to 6.6). The Mn^{2+} concentration ranged from 0.09 mg liter^{-1} in its surface water to 0.02–0.2 mg liter^{-1} in its bottom water (1.4 mg liter^{-1} in winter). The lake is fed by the Suantaka-Ioki and Rennel rivers and by 24 small streams that drain surrounding swampland. The lake, in turn, feeds into the Punnus-Yarvi river. It is estimated that 48% of the water in the lake is exchanged each year. The manganese and iron in the lake are derived from surface and underground drainage from Cambrian and Quaternary glacial deposits, carrying 0.2 to 0.8 mg of manganese liter^{-1} and 0.4 to 2 mg of iron liter^{-1}. The oxidized forms of manganese and iron are incorporated into silt, where they are reduced and subsequently concentrated by upward migration to the sediment-water interface and reoxidation into lake ore. This occurs mostly in Punnus-Ioki Bay, which has oxide deposits on the sediment at water depths down to 5–7 m. The oxide layer is 5–7 cm thick.

All sediment and ore samples taken from the lake (mainly Punnus-Ioki Bay) contained manganese-reducing bacteria. They were concentrated chiefly in the upper sediment layer. They included an unidentified, nonsporeforming bacillus in addition to *Bacillus circulans* and *B. polymyxa*. Limited numbers of sulfate-reducing bacteria, which the investigators (Sokolova-Dubinina and Deryugina, 1967a) attributed to the lack of extensive accumulation of organic matter, were recovered from the ore. They were associated with hydrotroilite (FeS·nH_2O). Carbonates of calcium and manganese at most stations in the lake and bay were reported of low occurrence (0.1%, calculated on the basis of CO_2). However, in a limited area near the center of Punnus-Ioki Bay, ore contained as much as 4.7% calcite, 5.96% siderite, and 4.99% rhodochrosite, together with 15.8% hydrogoethite and 38.9% barium psilomelanes and wads (complex oxides of manganese) (Sokolova-Dubinina and Deryugina, 1967a). The relatively localized concentration of carbonates was related by the investigators to the localized availability of organic matter and its attack by heterotrophs. It was the ultimate source of CO_2-CO_3^{2-} and the cause of the essential low E_h. It was noted that the decaying remains of the plant life on the lakeshore accumulated in sufficient quantities in the Punnus-Ioki Bay area only where extensive carbonate ores were found. The weak sulfate-reducing activity at this location may explain the low iron and manganese sulfide formation and the significant carbonate formation.

Siderite (ferrous carbonate) beds in the Yorkshire Lias in England are thought to have resulted from Fe_2O_3 reduction and subsequent reaction with HCO_3^-. The bicarbonate could have resulted from microbial mineralization of organic matter. Formation of siderite at this spot can only be explained by the exclusion of sulfate by an overlying clay layer, which blocked the entry of sulfate to the site of siderite deposition. If sulfate had entered the site, it would have been bacterially reduced to sulfide and led to the formation of iron sulfide instead of siderite (see Sellwood, 1971).

On the other hand, siderite is currently forming in rapidly accreting tidal marsh sediments at very shallow depths on the Norfolk coast, England, where extensive bacterial decay of organic matter is occurring at low interstitial sulfate and sulfide concentrations, according to Pye (1984; Pye et al., 1990). Scanning electron microscopic examination and X-ray powder diffraction analysis of siderite concretions from this location revealed that siderite formed a void-filling cement and coating around quartz grains. Traces of greigite, iron monosulfides, and calcite were also detected (Pye, 1984). Carbon-isotope-fractionation studies supported a microbiological role in the origin of the siderite (Pye et al., 1990). Although sulfate reducers were present and active at this site, the investigators explained the simultaneous formation of siderite and ferrous sulfide as resulting from a faster rate of ferric iron than sulfate reduction. This was confirmed by Coleman et al. (1993), who demonstrated that the reduction of ferric iron in this case was the result of anaerobic bacterial respiration, probably by the sulfate reducers, in which Fe(III) was a more effective terminal electron acceptor than sulfate, and that it was not the result of chemical reduction by H_2S formed in bacterial sulfate reduction.

Mixed deposits including managanous and ferrous carbonates, manganous and iron sulfides, and manganous and calcium-iron phosphates have been found in the sediments of the Landsort Deep in the central Baltic Sea. The site is anoxic. The minerals are thought to have formed authigenically as a result of microbial mineralization of organic matter (Suess, 1979). Here metal carbonate and sulfide deposition appeared to be compatible, perhaps because iron was available in a nonlimiting supply.

Bacterial rhodochrosite (manganous carbonate) formation in the reduction of Mn(IV) oxide has been observed in pure culture under laboratory conditions, using the isolate GS-15, now known as *Geobacter metallireducens* (Lovley and Phillips, 1988).

8.3 BIODEGRADATION OF CARBONATES

Carbonates in nature may be readily degraded as a direct or indirect result of biological activity, especially microbiological activity (Golubic and Schneider, 1979). A chemical basis for this decomposition is the instability of carbonates in acid solution. For instance,

$$CaCO_3 + H^+ \rightarrow Ca^{2+} + HCO_3^- \tag{8.25}$$

$$HCO_3^- + H^+ \rightarrow H_2CO_3 \tag{8.26}$$

$$H_2CO_3 \rightarrow H_2O + CO_2 \tag{8.27}$$

Since $Ca(HCO_3)_2$ is very soluble compared to $CaCO_3$, the $CaCO_3$ begins to dissolve even in weak acid solution. In stronger acid solution, $CaCO_3$ dissolves more rapidly because, as equation 8.27 shows, the carbonate is likely to be lost from solution as CO_2. Therefore, from a biochemical standpoint, any organism that generates acid metabolic wastes is capable of dissolving insoluble carbonates. Even the mere metabolic generation of CO_2 during respiration may have this effect, because

$$CO_2 + H_2O \rightarrow H_2CO_3 \tag{8.28}$$

and

$$H_2CO_3 + CaCO_3 \rightarrow Ca^{2+} + 2\ HCO_3^- \tag{8.29}$$

Thus, it is not surprising that various kinds of CO_2- and acid-producing microbes have been implicated in the breakdown of insoluble carbonates in nature.

Biodegradation of Limestone   Breakdown of lime in the cement of reservoir walls and docksides was attributed in part to bacterial action as long ago as 1899 (Stutzer and Hartleb, 1899). However, extensive investigation into microbial decay of limestone was first undertaken three decades later by Paine et al. (1933). These workers showed that both sound and decaying limestones usually carried a sizable bacterial flora, the numbers ranging from 0 to over 18×10^6 per gram. The size of the population in a particular sample seemed to depend in part on the environment around the stone. As one might expect, the surface of the limestone was generally more densely populated than was the interior of the stone. The authors suggested that in many limestones, the bacteria were unevenly distributed, inhabiting pockets and interstices in the limestone structures. The kinds of bacteria found in limestones that they examined included gram-variable and gram-negative rods and cocci. Sporeformers, such as *Bacillus mycoides*, *B. megaterium*, and *B. mesentericus*, appeared to have been rare or absent. They performed an experiment to estimate the rate of limestone decay through bacterial action under controlled conditions. They employed a special apparatus in which evolved CO_2 was trapped in barium hydroxide solution. They found 0.18 mg of CO_2 per hour per 350 g of stone to be evolved in one case, and 59 mg of CO_2 per hour per 350 g of stone in another. In the latter case they calculated from the data that 28 g of CO_2 would have been evolved from 1 kg of stone in 1 year. As expected, the rate of CO_2 evolution from decaying stone was greater than from sound stone. Although in these experiments, organic acids and CO_2 from heterotrophic metabolism of organic matter were the cause of the observed limestone decay, autotrophic nitrifying and sulfur-oxidizing bacteria were also shown to be able to promote limestone decay through the production of nitric and sulfuric acids from ammonia and reduced sulfur, respectively. Nitrifying and sulfur-oxidizing bacteria were actually detected in some limestone samples by the authors.

In a much later study variable numbers of fungi, algae, and ammonifying, nitrifying and sulfur-oxidizing bacteria were found on the surface of some limestones in Germany (Krumbein, 1968). Krumbein found that the number of detectable organisms depended on the type of stone, the length of elapsed time since the collection of the stone from a natural site, the surface structure of the stone (i.e., the degree of weathering), the cleanliness of the stone, and the climatic and microclimatic conditions prevailing at the collection site. In the case of strongly weathered stone, the bacteria had sometimes penetrated the stone to a depth of 10 cm. Ammonifying bacteria were generally most numerous. Nitrogen-fixing bacteria were few and denitrifiers were absent. The number of the organisms was not directly related to the presence of lichens or algae. The greatest numbers of ammonifiers were found on freshly collected and weathered stone. This was related to the pH of the stone surface (pH 8.1–8.3 in water extract). Sulfur oxidizers were more numerous in city environments, where the atmosphere contains more oxidizable sulfur compounds than in the countryside. The concentration of nitrifiers on limestone could not be correlated with city and country atmospheres. The number of ammonifiers on limestone was also found to be greater in stone exposed to city air than to country air. This can be explained on the basis that city air contains more organic carbon that can serve as nutrient to these bacteria than does country air. Laboratory experiments by Krumbein confirmed the weathering activity of limestone by its natural flora. The ammonifiers in these observations were less directly responsible for the weathering of the limestone than they were in generating ammonia from which the nitrifying bacteria could form nitric acid, which then corroded the limestone.

In yet another important study of the decomposition of limestone by microbes, numerous bacteria and fungi were isolated from a number of samples (Wagner and Schwartz, 1965). The active organisms appeared to weather the stone through production of oxalic and gluconic acids. The investigators also noted the presence of nitrifying bacteria and thiobacilli in their samples and suggested that the corresponding mineral acids produced by these organisms also contributed to the weathering of the stone.

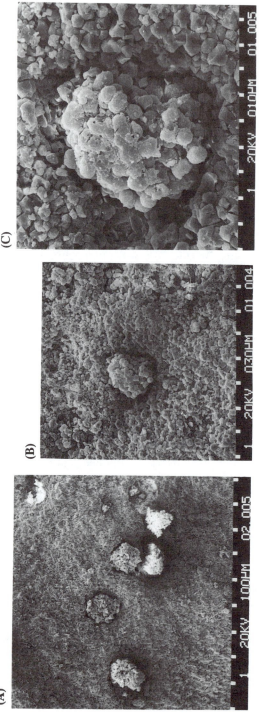

Figure 8.10 SEM photomicrographs (A) to (C) showing a section of marble from the Dionysos Theater, Acropolis, in Athens, Greece, at three different magnifications (note scale and scale marks on the bottom of each photograph). The marble has been extensively corroded by "biopitting." Deep holes of different sizes (between 2 μm and 5 mm in diameter and depth) were incised chemically (etched by metabolically produced substances) and physically (mechanical action) by black yeasts and meristematically growing yeastlike fungi. The microcolonial fungi can be confused with algae in SEM micrographs. The fungi have a cell size very similar to that of the marble grain. The chemical and physical corrosive action of these fungi have been demonstrated in laboratory experiments (Anagnostidis et al., 1992). (Courtesy of Wolfgang E. Krumbein).

Marble, a metamorphic CaCO₃-type of rock, can also be attacked by microorganisms. Figure 8.10 shows the effect of microcolonial fungi. *Micrococcus halobius* was shown to colonize surfaces on Carrara marble slabs, forming biofilm and producing gluconic, lactic, pyruvic, and succinic acids from glucose that etched the marble surfaces (Urzi et al., 1991). This organism also caused a discoloration of the marble surface suggesting that natural patina on marble structures may have a microbial origin. Black fungi of the Dematiaceae have also been implicated in the destruction of some marble and limestone (Gorbushina et al, 1994). However, instead of attacking the marble through formation of corrosive metabolic products such as acids, the fungus appears to attack the marble by exerting physical pressure in pores and crevices in its growth and by changing water activity in its superficial polymers and around the cells in the stone. The melanin pigment produced by the fungus has been implicated in the blackening of surfaces of marble structures.

Cyanobacteria, Algae, and Fungi that Bore into Limestone   Endolithic cyanobacteria, algae, and fungi have been found to cause local dissolution of limestone thereby forming tubular passages in which they can grow. (Fig. 8.11) (Golubic,1975). The kinds of limestone they attack in nature include coral reefs, beach rock, and other types. Active algae include some green, brown and red algae (Golubic, 1969). The mechanism by which any of these organism bore into limestone is not understood. Some filamentous boring cyanobacteria possess a terminal cell that is directly responsible for the boring action, presumably dissolution of the calcium carbonate (Golubic, 1969). Different boring microorganisms form tunnels of characteristic morphology (Golubic et al., 1975). In a pure mineral such as iceland spar, boring follows the planes of crystal twinning, diagonal to the main cleavage planes (Golubic et al, 1975). The depth to which cyanobacteria and algae bore into limestone is limited by light penetration in the rock, since they need light for photosynthesis. Boring cyanobacteria may have unusually high concentration of phycocyanin to compensate for the low light intensity in the limestone. In contrast, boring fungi are not limited by light penetration. Being incapable of photosynthesis, they have no need for light.

Endolithic fungi and cyanobacteria or algae in limestone may form a special symbiotic relationship in the form of lichens (Fig. 8.12). The cyanobacteria or algae in these associations share the carbon which they fix with the fungi while the fungi share minerals which they mobilize and some other less well defined functions with the cyanobacteria or algae. These lichens, although growing within the limestone may serve as food to some snails, as has been reported from the Negev Desert in Israel (Shachak et al., 1987). The snails have a tonguelike organ in their mouth,

(a)

Figure 8.11 Microorganisms that bore into limestone. (a) Limestone sample experimentally recolonized by the cyanobacterium *Hyella balani* Lehman ($\times$234). The exposed tunnels are the result of boring by the cyanobacterium. (b) Casts of the borings of the green alga *Eugamantia sacculata* Kormann (larger filaments) and the fungus *Ostracoblabe implexa* Bornet and Flahault in calcite spar ($\times$2,000). The casts were made by infiltrating a sample of fixed and dehydrated bored mineral with synthetic resin and then etching sections of the embedded material (e.g., with dilute mineral acid) to expose the casts of the organism. (Courtesy of S. Golubic.)

the **radula**, which has embedded in it toothlike structures, frequently consisting of the iron mineral hematite, useful for abrading. With their radula they scrape the surface of limestone beneath which the lichens grow in order to feed on the lichens, their preferred food. They ingest some of the pulverized limestone with the lichen and leave behind a trail of this limestone powder as they move over the limestone surface. It has been estimated that in the Negev Desert this weathering affects 0.7 to 1.1 metric tons of rock per hectare per year (Shachak et al., 1987). A similar biological weathering phenomenon was previously noted on some reef structures in Bermuda (see Golubic and Schneider, 1979).

Cyanobacteria and fungi inhabit not only limestone but also other porous rocks. To distinguish among rock inhabiting microbes, the term

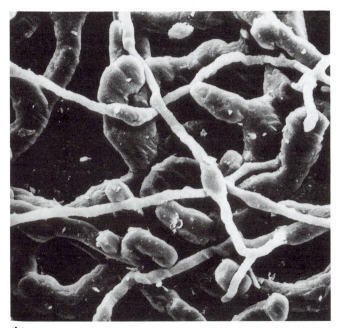

(b)
Figure 8.11 Continued.

euendoliths has been coined for true boring microbes in limestone. Opportunists which invade preexistent pores of rock where they may cause alteration of the rock in the area that they inhabit are classified as **cryptoendoliths**. Those that invade preexistent cracks and fissures without altering the rock structure are classified as **chasmoendoliths** (Golubic et al., 1981).

In some extreme environments, such as Antarctic dry valleys, cryptoendolithic cyanobacteria (Friedmann and Ocampo, 1976) and lichens (in this instance a symbiotic association of a green alga and a filamentous fungus)(Friedmann, 1982; Friedmann and Ocampo-Friedmann, 1984) inhabit superficial cavities in sandstone, 1 to 2 mm below the surface (Fig. 8.9). The near-surface locale in the sandstone (ortho-quartzite) is the major habitat for microorganisms in this inhospitable environment (Friedmann, 1982). These cryptoendolithic microorganisms appear to differ from the boring microbes in that they are not directly responsible for cavity formation in the sandstone but instead invade preexistent pores. But they promote exfoliation of the sandstone surface, a physical process facilitated through solubilization of the cementing substance ($CaCO_3$?), which holds the quartz grains together. (Friedmann, 1977; Friedmann and Weed, 1987). The lichen activity may manifest itself visibly by mobilization of iron in the region of lichen activity. Exfoliation may expose the lichens

Figure 8.12 Cryptoendolithic microorganisms in vertically fractured Beacon sandstone: (a) lichen (small black bodies between rock particles), (b) zone of fungus filaments, (c) yellowish green zone of unicellular eucaryotic alga, (d) blue-green zone of unicellular cyanobacterium. The color difference between (c) and (d) is not apparent in this black-and-white photograph. Sample A 76-77/36, north of Mount Dido, elevation 1750 m, magnification $\times 4.5$. [From E. I. Friedmann, Microorganisms in antarctic desert rock from dry valleys and Dufek Massif, Antarctic Journal of the United States, courtesy of the author.]

to the external environment from which they, if they survive the exposure, may reinvade the standstone through pores. Fissures in granite and grandiorite in the Antarctic dry valleys may also be inhabited by lichens and coccoid cyanobacteria (Friedmann, 1982). These organisms are clearly not boring microbes.

Cryptoendoliths are not unique to the Antarctic Dry Valleys but are also found in hot desert environments (e.g., Mojave Desert, California; Sonoran Desert, Mexico; Negev, Israel) (Friedmann, 1980). Since the hot deserts like the cold Antarctic deserts represent extreme environments in which moisture is a limiting factor to survival, rock interiors afford protection and permit life to persist.

8.4 BIOLOGICAL CARBONATE FORMATION AND DEGRADATION AND THE CARBON CYCLE

The biological aspect of the carbon cycle involves chiefly the fixation of inorganic carbon (CO_2 or its equivalents) as organic carbon, and the remineralization of some of this organic carbon to inorganic carbon (Fig.

8.13). The carbon cycle operates in such a way that a certain portion of the biologically fixed carbon is transitionally immobilized as **standing crop**. However, the standing crop undergoes continual turnover at such a rate that its size does not markedly change unless it is subjected to some major environmental change.

Some sediments may become a permanent sink of organic matter trapped in it. Through burial, such organic matter is removed from the carbon cycle and may be transformed into coal, petroleum, bitumen, kerogen, and/or natural gas (see Chapter 20), or very rarely graphite or diamonds. The carbon in these substances reenters the carbon cycle with few exceptions only through human intervention (the exploitation of fossil fuels).

As this chapter has demonstrated, inorganic carbon may also be removed from the carbon cycle through biotransformation into inorganic carbonates. Such carbon when trapped in sediments may not reenter the active carbon cycle for long geologic time intervals. Reentry, of carbon into the cycle, as this chapter has shown, is under extensive biological control. Indeed, the biological component of the carbon cycle as a whole exerts the chief directing influence over it.

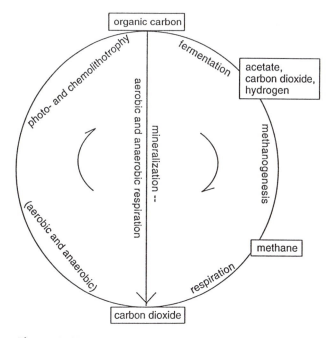

Figure 8.13 The carbon cycle.

For more detailed discussion of the carbon cycle on Earth the following references should be examined (Bolin, 1970; Golubic, Krumbein, and Schneider, 1979; Smith, 1981; Krumbein and Swart, 1983; Post et al., 1990).

8.5 SUMMARY

Carbon dioxide is trapped on Earth mainly as calcium and calcium-magnesium carbonates, but also to a much lesser extent as iron, manganese, and sodium carbonates. In many cases, these carbonates are of biogenic origin. Calcium or calcium-magnesium carbonates may be precipitated by bacteria, cyanobacteria, and fungi by (a) aerobic or anaerobic oxidation of carbon compounds devoid of nitrogen in neutral or alkaline environments with a supply of calcium or magnesium counterions, (b) aerobic or anaerobic oxidation of organic nitrogen compounds in unbuffered environments with a supply of calcium or magnesium counterions, (c) $CaSO_4$ reduction by sulfate-reducing bacteria, (d) the hydrolysis of urea in the presence of Ca and Mg counter ions, or (e) photo- and chemosynthetic autotrophy in the presence of Ca and Mg counterions.

The complete mechanism of calcium carbonate precipitation by living organisms is not understood in all cases, although it always depends on metabolic CO_2 production or CO_2 consumption and on the relative insolubility of calcium carbonate. Bacteria and fungi may deposit it extracellularly. Bacterial cells may also serve as nuclei around which calcium carbonate is laid down. Some algae and protozoans form structural carbonate. Its deposition may involve local fixation of calcium prior to its reaction with metabolically produced carbonate.

Microbial calcium carbonate precipitation has been observed in soil and in freshwater and marine environments. Sodium carbonate (natron) deposition associated with microbial sulfate-reducing activity has been noted in the hot climate of the Wadi Natrun in the Libyan Desert. Ferrous carbonate deposition associated with microbial activity has been noted in some special coastal marine environments. Manganese carbonate deposition associated with microbial activity has been noted in a freshwater lake. Both ferrous and manganous carbonate formation have been observed in the laboratory with pure bacterial cultures.

Insoluble carbonates may be broken down by microbial attack. This is usually the result of organic and inorganic acid formation by the microbes but may also involve physical processes. A variety of bacteria, fungi, and even algae have been implicated. Such activity is particularly evident on limestones and marble used in building construction, but it is also evident in natural limestone formations, such as coral reefs, where limestone-boring algae and fungi are active in the breakdown process. Bacteria and fungi contribute to discoloration of structural limestone and marble and may be the cause of patina.

REFERENCES

Abd-el-Malek, Y., and S. G. Rizk. 1963a. Bacterial sulfate reduction and the development of alkalinity. II. Laboratory experiments with soils. J. Appl. Bacteriol. 26:14–19.

Abd-el-Malek, Y, and S. G. Rizk. 1963b. Bacterial sulfate reduction and the development of alkalinity. III. Experiments under natural conditions in the Wadi Natrun. J. Appl. Bacteriol. 26:20–26.

Aizawa, K., and S. Miyachi. 1986. Carbonic anhydrase and CO_2 concentration mechanism in microalgae and cyanobacteria. FEMS Microbiol. Rev. 39;215–233.

Anagnostidis, K., C. K. Gehrmann, M. Gross, W. E. Krumbein, S. Lisi, A. Pantazidou, C. Urzi, and M. Zagari. 1992. Biodeterioration of marbles of the Parthenon and Propylaea, Acropolis, Athens—Associated organisms, decay, and treatment suggestions. *In*: D. Decrouez, J. Chamay, and F. Zezza, eds. II. Int. Symp. for the Conservation of Monuments in the Mediterranean Basin. Musée d'Histoire Naturelle, Genève, Switzerland, pp. 305–325.

Ashirov, K. B., and I. V. Sazonova. 1962. Biogenic sealing of oil deposits in carbonate reservoirs. Mikrobiologiya 31:680–683 (Engl. transl. pp. 555–557).

Bavendamm, W. 1932. Die mikrobiologische Kalkfaellung in der tropischen See. Arch. Mikrobiol. 3:205–276.

Berg, G. W. 1986. Evidence for carbonate in the mantle. Nature (Lond.) 324:50–51.

Berner, R. A. 1968. Calcium carbonate concretions formed by the decomposition of organic matter. Science 159:195–197.

Bolin, B. 1970. The carbon cycle. Sci. Am. 223 (3):124–132.

Bonatti, E. 1966. Deep-sea authigenic calcite and dolomite. Science 152: 534–537.

Borman, A. H., E. W. de Jong, M. Huizinga, D. J. Kok, and L. Bosch. 1982. The role in $CaCO_3$ crystallization of an acid Ca^{2+}-binding polysaccharide associated with coccoliths of *Emiliania huxleyi*. Eur. J. Biochem. 129:179–183.

Borman, A. H., E. W. de Jong, R. Thierry, P. Westbroek, and L. Bosch. 1987. Coccolith-associated polysaccharides from cells of *Emiliania huxleyi* (Haptophyceae). J. Phycol. 23:118–123.

Bowen, H. J. M. 1979. Environmental Chemistry of the Elements. Academic Press, London.

Brunskill, G. J., and S. D. Ludlam. 1969. Fayetteville Green Lake, New York. I. Physical and chemical limnology. Limnol. Oceanogr. 14: 817–829.

Buchanan, R. E., and N. E. Gibbons, eds. 1974. Bergey's Manual of Determinative Bacteriology. 8th ed. Williams & Wilkins, Baltimore, MD.

Buck, J. D., and L. J. Greenfield. 1964. Calcification in marine-occurring yeast. Bull. Mar. Sci. Gulf Carib. 14:239–245.

Buczynski, C., and H. S. Chafetz. 1991. Habit of bacterially induced precipitates of calcium carbonate and the influence of medium viscosity on mineralogy. J. Sedim. Petrol. 61:226–233.

Chafetz, H. S., and C. Buczynski. 1992. Bacterially induced lithification of microbial mats. Palaios 7:277–293.

Coleman, M. L., D. F. Hedrick, D. R. Lovley, D. C. White, and K. Pye. 1993. Reduction of Fe(III) in sediments by sulfate-reducing bacteria. Nature (Lond.) 361:436–438.

Degens, E. T. 1972. Geochemistry of sediments: ancient. *In*: The Encyclopedia of Geochemistry and Environmental Sciences. Encyclopedia of Earth Sciences Series, Vol IVA. R. W. Fairbridge, ed. Van Nostrand Reinhold, New York, pp. 417–428.

Degens, E. T. 1979. Biogeochemistry of stable isotopes, pp. 304–329. *In*: E. Eglinton and M. T. J. Murphy, eds. Organic Geochemistry: Methods and Results. Springer-Verlag, New York.

De Jong, E. W., P. van der Wal, A. H. Borman, J. P. M. De Vrind, P. Van Emburg, P. Westbroek, and L. Bosch. 1983. Calcification in coccolithophorids. *In*: P. Westbroek and E. W. De Jong, eds. Biomineralization and Biological Metal Accumulation. D. Reidel, Dordrecht, Holland, pp. 291–301.

del Moral, A., E. Roldan, J. Navarro, M. Monteoliva-Sanchez, and A. Ramos-Cormenzana. 1987. Formation of calcium carbonate crystals by moderately halophilic bacteria. Geomicrobiol. J. 5:79–87.

Drew, G. H. 1911. The action of some denitrifying bacteria in tropical and temperate seas, and the bacterial precipitation of calcium carbonate in the sea. J. Mar. Biol. Assoc. U.K. 9:142–155.

Drew, G. H. 1914. On the precipitation of calcium carbonate in the sea by marine bacteria, and on the action of denitrifying bacteria in tropical and temperate seas. Carnegie Inst. Washington, Publ. 182, 5: 7–45.

Feeley, H. W., and J. L. Kulp. 1957. Origin of Gulf Coast salt-dome sulfur deposits. Bull. Am. Assoc. Petrol. Geol. 41:1802–1853.

Fenchel, T., and T. H. Blackburn. 1979. Bacteria and Mineral Cycling. Academic Press, London.

Ferrer, M. R., J. Quevedo-Sarmiento, V. Bejar, R. Delgado, A. Ramos-Cormenzana, and M. A. Rivadeneyra. 1988a. Calcium carbonate formation by *Deleya halophila*: Effect of salt concentration and incubation temperature. Geomicrobiol. J. 6:49–57.

Ferrer, M. R. J. Quevedo-Sarmiento, M. A. Rivadeneyra, V. Bejar, R. Delgado, and A. Ramos-Cormenzana. 1988b. Calcium carbonate precipitation by two groups of moderately halophilic microorganisms at different temperatures and salt concentrations. Curr. Microbiol. 17:221–227.

Friedmann, E. I. 1977. Microorganisms in antarctic desert rocks from dry valleys and Dufek Massif. Antarctic J. U.S. 12:26–30.

Friedmann, E. I. 1980. Endolithic microbial life in hot and cold deserts. Orig. Life 10:223–235.

Friedmann, E. I. 1982. Endolithic microorganisms in the Antarctic cold desert. Science 215:1045–1053.

Friedmann, E. I., and R. Ocampo. 1976. Endolithic blue-green algae in the Dry Valleys: Producers in the Antarctic desert ecosystem. Science 193:147–1249.

Friedmann, E. I., and R. Ocampo-Friedmann. 1984. Endolithic microorganisms in extreme dry environments: Analysis of a lithobiontic microbial habitat. pp. 177–185. *In*: M.J. Klug and C.A. Reddy, eds. Current Perspectives in Microbial Ecology. Proc. Third Int. Symp. Microb. Ecol. American Society for Microbiology, Washington, DC.

Friedmann, E. I., and R. Weed. 1987. Microbial trace-fossil formation, biogenous, and abiotic weathering in the Antarctic Cold Desert. Science 236:703–705.

Friedmann, E. I., W. C. Roth, J. B. Turner, and R. S. McEwen. 1972. Calcium oxalate crystals in the aragonite-producing green alga *Penicillus* and related genera. Science 177. 891–893.

Golubic, S. 1969. Distribution, taxonomy, and boring patterns of marine endolithic algae. Am. Zool. 9:747–751.

Golubic, S. 1973. The relationship between blue-green algae and carbonate deposits. pp. 434–472. *In*: N. G. Carr and B. A. Whitton, eds., The Biology of Blue-Green Algae. Blackwell Scientific Publications, Oxford.

Golubic, S., and J. Schneider. 1979. Carbonate dissolution. pp. 107–129. *In*· P. A. Trudinger and D. J. Swaine, eds. Biogeochemical Cycling of Mineral-Forming Elements. Elsevier, Amsterdam.

Golubic, S., R. D. Perkins, and K. J. Lukas. 1975. Boring microorganisms and microborings in carbonate substrates, pp. 229–259. *In*: R. W. Frey, ed. The Study of True Fossils. Springer-Verlag, New York.

Golubic, S., W. Krumbein, and J. Schneider. 1979. The carbon cycle. pp. 29–45. *In*: P.A. Trudinger and D.J. Swaine, eds. Biogeochemical Cycling of Mineral-Forming Elements. Elsevier, Amsterdam.

Golubic, S., I. Friedmann, and J. Schneider. 1981. The lithobiontic ecological niche, with special reference to microorgansisms. J. Sed. Petrol. 51:475–478.

Gorbushina, A. A., W. E. Krumbein, C. H. Hamman, L. Panina, S. Soukharjevski, and U. Wollenzien. 1994. Role of black fungi in color change and biodetrioration of antique marbles. Geomicrobiol. J. 11: 205–211.

Gorlenko, V. M., E. N. Chebotarev, and V. I. Kachalkin. 1974. Microbial oxidation of hydrogen sulfide in Lake Veisovo (Slavyansk Lake). Mikrobiologiya 43:530–534 (Engl. transl. pp. 450–453).

Greenfield, L. J. 1963. Metabolism and concentration of calcium and magnesium and precipitation of calcium carbonate by a marine bacterium. Ann. NY Acad. Sci. 109:23–45.

Isenberg, H. D., and L. S. Lavine. 1973. Protozoan calcification, pp. 649–686. *In*: I. Zipkin, ed. Biological Mineralization. Wiley, New York.

Jensen, P. R., R. A. Gibson, M. M. Littler, and D. S. Littler. 1985. Photosynthesis and calcification in four deep-water *Halimeda* species (Chlorophyceae, Caulerpales). Deep-Sea Res. 32:451–464.

Krumbein, W. E. 1968. Zur Frage der biologischen Verwitterung: Einfluss der Mikroflora auf die Bausteinverwitterung und ihre Abhaengigkeit von adaphischen Faktoren. Z. Allg. Mikrobiol. 8; 107–117.

Krumbein, W. E. 1973. Ueber den Einfluss von Mikroorganismen auf die Bausteinverwitterung—Eine oekoligische Studie. Deutsch. Kunst. Denkmalpfl. 31:54–71.

Krumbein, W. E. 1974. On the precipitation of aragonite on the surface of marine bacteria. Naturwissenschaften 61:167.

Krumbein, W. E. 1979. Photolithotrophic and chemoorganotrophic activity of bacteria and algae as related to beachrock formation and degradation (Gulf of Aqaba, Sinai). Geomicrobiol. J. 1:139–203.

Krumbein, W. E., and C. Giele. 1979. Calcification in a coccoid cyanobacterium associated with the formation of desert stromatolites. Sedimentology 26:593–604.

Krumbein, W. E., and M. Potts. 1979. *Girvanella*-like structures formed by *Plectonema gloeophilum* (Cyanophyta) from the Borrego Desert in Southern California. Geomicrobiol. J. 1:211–217.

Krumbein, W. E., and P. K. Swart. 1983. The microbial carbon cycle.

pp. 5–62. W.E. Krumbein, ed. Microbial Geochemistry. Blackwell Scientific Publications, Oxford, UK.

LaRock, P. A., and H. L. Ehrlich. 1975. Observation of bacterial micro-colonies on the surface of ferromanganese nodules from Blake Plateau by scanning electron microscopy. Microb. Ecol. 2:84–96.

Latimer, W. M., and J. H. Hildebrand. 1942. Reference Book of Inorganic Chemistry. Rev. ed. Macmillan, New York.

Lewin, J. 1965. Calcification, pp. 457–465. *In*: R. A. Lewin, ed. Physiology and Biochemistry of Algae. Academic Press, New York.

Lipman, C. B. 1929. Further studies on marine bacteria with special reference to the Drew hypothesis on $CaCO_3$ precipitation in the sea. Carnegie Inst. Washington, Publ. 391, 26:231–248.

Lovley, D. R., and E. J. P. Phillips. 1988. Novel mode of microbial energy metabolism: organic carbon oxidation coupled to dissimilatory reduction of iron and manganese. Appl. Environ. Microbiol. 54: 1472–1480.

Lowenstam, H. A., and S. Epstein. 1957. On the origin of sedimentary aragonite needles of the Great Bahama Bank. J. Geol. 65:364–375.

Marine Chemistry. 1971. A report of the Marine Chemistry Panel of the Committee of Oceanography, National Academy of Sciences, Washington, DC.

McCallum, M. F., and K. Guhathakurta. 1970. The precipitation of calcium carbonate from seawater by bacteria isolated from Bahama Bank sediments. J. Appl. Bact. 33:649–655.

McConnaughey, T. 1991. Calcification in *Chara corallina*: CO_2 hydroxylation generates protons for bicarbonate assimilation. Limnol. Oceanogr. 36:619–628.

Monty, C. L. V. 1967. Distribution and structure of recent stromatolitic algal mats, Eastern Andros Island, Bahamas. Ann. Soc. Geol. Belgique 90:55–99

Monty, C. V. L. 1972. Recent algal stromatolitic deposits, Andros Island, Bahamas. Preliminary Report. Geol. Rundschau 61:742–783.

Morita, R. Y. 1980. Calcite precipitation by marine bacteria. Geomicrobiol. J. 2:63–82

Moore, J. N. 1983. The origin of calcium carbonate nodules forming on Flathead Lake delta, northwestern Montana. Limnol. Oceanogr. 28: 646–654.

Murray, J., and R. Irvine. 1889/1890. On coral reefs and other carbonate of lime formations in modern seas. Proc. Roy. Soc. Lond., A17: 79–109.

Nadson, G. A. 1903. Die Mikroorganismen als geologische Faktoren. I. Ueber die Schwefelwasserstoffgaehrung und ueber die Beteiligung der Mikroorganismen bei der Bildung des schwarzen Schlammes (Heilschlammes). Arb. d. Comm. f. d. Erf. d. Mineralseen bei Slaw-jansk, St. Petersburg. 98 pp.

Nadson, G. A. 1928. Beitrag zur Kenntniss der bakteriogenen Kalkablag-erung. Arch. Hyrobiol. 191:154–164.

Nazina, T. N., E. P. Rozanova, and S. I. Kuznetsov. 1985. Microbial oil transformation processes accompanied by methane and hydrogen-sulfide formation. Geomicrobiol. J. 4:103–130.

Nekrasova, V. K., L. M. Gerasimenko, and A. K. Romanova. 1984. Pre-cipitation of calcium carbonate in the presence of cyanobacteria. Mikrobiologiya 53:833–836 (Engl. transl. pp. 691–694).

Novitsky, J. A. 1981. Calcium carbonate precipitation by marine bacteria. Geomicrobiol. J. 2:375–388.

Paine, S. G., F. V. Lingood, F. Schimmer, and T. C. Thrupp. 1933. IV. The relationship of microorganisms to the decay of stone. Roy. Soc. Phil. Trans. 222B:97–127.

Pentecost, A. 1988. Growth and calcification of the cyanobacterium *Ho-moeothrix crustacea*. J. Gen. Microbiol. 134:2665–2671.

Pentecost, A., and J. Bauld. 1988. Nucleation of calcite on the sheaths of cyanobacteria using a simple diffusion cell. Geomicrobiol. J. 6: 129–135.

Pentecost, A., and B. Spiro. 1990. Stable carbon and oxygen isotope com-position of calcites associated with modern freshwater cyanobacteria and algae. Geomicrobiol. J. 8:17–26.

Post, W. M., T-H. Peng, W. E. Emanuel, A. W. King, V. H. Dale, and D. L. DeAngelis. 1990. The global carbon cycle. Am. Sci. 78: 310–326.

Pye, K. 1984. SEM analysis of siderite cements in intertidal marsh sedi-ments, Norfolk, England. Mar. Geol. 56:1–12.

Pye, K., J. A. D. Dickson, N. Schiavon, M. L. Coleman, and M. Cox. 1990. Formation of siderite-Mg-calcite-iron sulfide concretions in in-tertidal marsh and sandflat sediments, north Norfolk, England. Sedi-mentology 37:325–343.

Rivadeneyra, M. A., R. Delgado, E. Quesada, and A. Ramos-Cormenza. 1991. Precipitation of calcium carbonate by *Deleya halophila* in media containing NaCl as sole salt. Curr. Microbiol. 22:185–190.

Rivadeneyra, M. A., I. Perez-Garcia, V. Salmeron, and A. Ramos-Cor-menzana. 1985. Bacterial precipitation of calcium carbonate in pres-ence of phosphate. Soil Biol. Biochem. 17:171–172.

Roemer, R., and W. Schwartz, 1965. Geomikrobiologische Untersu-chungen. V. Verwertung von Sulfatmineralien und Schwermetallen. Tolleranz bei Desulfizierern. Z. Allg. Mikrobiol. 5:122–135.

Röttger, R. 1974. Larger foraminifera: Reproduction and early stages of development in *Heterostegina depressa*. Marine Biol. (Berlin) 26: 5–12.

Schmalz, R. F. 1972. Calcium carbonate: Geochemistry, p. 110. *In*: R. W. Fairbridge, ed. The Encyclopedia of Geochemistry and Environmental Sciences. Encyclopedia of Earth Sciences Series, Vol. IVA. Van Nostrand Reinhold, New York.

Sellwood, R. W. 1971. The genesis of some sideritic beds in the Yorkshire Lias (England). J. Sed. Petrol. 41:854–858.

Shachak, M., C. G. Jones, and Y. Granot. 1987. Herbivory in rocks in the weathering of a desert. Science 236:1098–1099.

Shinano, H. 1972a. Studies on marine microorganisms taking part in the precipitation of calcium carbonate. II. Detection and grouping of the microorganisms taking part in the precipitation of calcium carbonate. Bull. Jpn. Soc. Sci. Fisheries 38:717–725.

Shinano, H. 1972b. Microorganisms taking part in the precipiation of calcium carbonate. III. A taxonomic study of marine bacteria taking part in the precipitation of calcium carbonate. Bull. Jpn. Soc. Sci. Fisheries 38:727–732.

Shinano, H., and M. Sakai. 1969. Studies of marine bacteria taking part in the precipitation of calcium carbonate. I. Calcium carbonate deposited in peptone medium prepared with natural seawater and artificial seawater. Bull. Jpn. Soc. Sci. Fisheries 35:1001–1005.

Skirrow, G. 1975. The dissolved gases—carbon dioxide, pp. 144–152. *In*: J. R. Riley and G. Skirrow. Chemical Oceanography. Vol. 2. 2nd ed. Academic Press, London.

Smith, S. V. 1981. Marine macrophytes as a global carbon sink. Science 211:838–840.

Sokolova-Dubinina, G. A., and Z. P. Deryugina. 1967. On the role of microorganisms in the formation of rhodochrosite in Punnus-Yarvi Lake. Mikrobiologiya 36:535–542 (Engl. transl. pp. 445–451).

Spiro, B., and A. Pentecost. 1991. One day in the life of a stream—a diurnal inorganic carbon mass balance for a travertine-depositing stream (Waterfall Beck, Yorkshire). Geomicrobiol. J. 9:1–11.

Steinmann, G. 1899/1901. Ueber die Bildungsweise des dunklen Pigments bei den Mollusken nebst Bemerkungen ueber die Entstehung von Kalkcarbonat. Ber. Naturf. Ges. Freiburg i.B. 11:40–45.

Stumm, W., and J. J. Morgan. 1981. Aquatic Chemistry. An Introduction Emphasizing Chemical Equilibria in Natural Waters. Wiley, New York.

Stutzer, A. and R. Hartleb. 1899. Die Zersetzung von Cement unter dem Einfluss von Bakterien. Z. Angew. Chem. 12:402 (cited by Paine et al., 1933)

Suess, E. 1979. Mineral phases formed in anoxic sediments by microbial decomposition of organic matter. Geochim. Cosmochim. Acta 43: 339–352.

Taylor, W. R. 1950. Plants of Bikini and Other Northern Marshall Islands. Michigan University Press, Ann Arbor.

Thompson, J. B., and F. G. Ferris. 1990. Cyanobacterial precipitation of gypsum, calcite and magnesite from natural alkaline lake water. Geology 18:995–998.

Thompson, J. B., F. G. Ferris, and D. A. Smith. 1990. Geomicrobiology and sedimentology of the mixolimnion and chemocline in Fayetteville Green Lake, New York. Palaios 5:52–75.

Urzi, C., S. Lisi, G. Criseo, and A. Pernice. 1991. Adhesion to and degradation of marble by a *Micrococcus* strain isolated from it. Geomicrobiol. J. 9:81–90.

Verrecchia, E. P., J.-L. Dumant, and K. E. Collins. 1990. Do fungi building limestone exist in semi-arid regions? Naturwissenschaften 77: 584–586.

Wagner, E., and W. Schwartz. 1965. Geomikrobiologische Untersuchungen. VIII. Ueber das Verhalten von Bakterien auf der Oberflaeche von Gesteinen und Mineralien und ihre Rolle bei der Verwitterung. Z. Allg. Mikrobiol. 7:33–52.

Walter, M. R., ed. 1976. Stromatolites. Developments in Sedimentalogy 20. Elsevier, Amsterdam.

Weast, R. C., and M. J. Astle, eds. 1982. CRC Handbook of Chemistry and Physics. 63rd ed. CRC Press, Boca Raton, FL.

Wefer, G. 1980. Carbonate production by algae *Halimeda*, *Penicillus*, and *Padina*. Nature (Lond.) 285:323–324.

Wharton, R. A. Jr., B. C. Parker, G. M. Simmons Jr., K. S. Seaburg, and F. G. Love. 1982. Biogenic calcite structures in Lake Fryxell, Antarctica. Nature (Lond.) 295:403–405.

9

Geomicrobial Interactions with Silicon

9.1 DISTRIBUTION AND SOME CHEMICAL PROPERTIES

The element silicon is one of the most abundant in the Earth's crust, ranking second only to oxygen. Its estimated crustal abundance is 27.7% (wt/wt) whereas that of oxygen is 46.6% (wt/wt) (Mitchell, 1955). In nature it occurs generally in the form of silicates and silicon dioxide (silica). It is found in primary and secondary minerals. It can be viewed as an important part of the backbone of rock mineral structure. In silicates, silicon is usually surrounded by four oxygen atoms in tetrahedral fashion (see Kretz, 1972) whereas aluminum in aluminosilicates is coordinated with oxygens in tetrahedral or octahedral fashion, depending on the mineral (see Tan, 1986). In silicate minerals, silicon tetrahedra may be linked by sharing oxygen atoms or through ionic bridging with cations. Similarly in aluminosilicates other than clays, silicon tetrahedra and aluminum tetrahedra may be linked by sharing oxygen atoms or through ionic bridging with cations (Tan, 1986). In clays, which result from weathering of primary aluminosilicates, silica-tetrahedral sheets and aluminum-hydroxide-octahedral sheets are layered in different ways depending on clay type. In

montmorillonite-type clays, structural units consisting of single aluminum-hydroxide octahedral sheets are sandwiched between two silica-tetrahedral sheets. The units are interspaced with layers of water molecules of variable thickness into which other polar molecules, including some organic ones, can enter. This variable water layer allows montmorillonite-type clays to exhibit swelling. The structural units of **illite-type clays** resemble montmorillonite type clays but differ from them in that Al replaces some of the Si in the silica-tetrahedral sheets. These substituting Al atoms impart extra charges, which are neutralized by potassium ions between the structural units that act as bridges and prevent the swelling exhibited by montmorillonite in water. In **kaolinite clays**, structural units consist of silica-tetrahedral sheets alternating with aluminum-hydroxide-octahedral sheets joined to one another by oxygen bridges (see Toth, 1955).

Si-O bonds of **siloxane linkages** (Si-O-Si) in silicates and aluminosilicates are very strong (energy of formation ranges from 3,110 to 3,142 kcal mol^{-1}) whereas Al-O bonds are somewhat weaker (energy of formation ranges from 1,793 to 1,878 kcal mol^{-1}). Cationic-oxygen bonds are the weakest (energies of formation ranges from 299 to 919 kcal ml^{-1}) (values cited by Tan, 1986). The strength of these bonds determines their susceptibility to weathering. Thus Si-O bonds are relatively resistant to acid hydrolysis (e.g., Karavaiko et al., 1985), unlike Al-O bonds. Cationic oxygen bonds are readily broken by protonation or cation exchange.

Silicate in solution at pH 2–9 exists in the form of undissociated, monosilicic acid (H_4SiO_4) whereas at pH 9 and above, it transforms into silicate ions (see Hall, 1972). Monosilicic acid polymerizes at a concentration of 2×10^{-3} M and above forming oligomers of polysilicic acids (Iler, 1979), a reaction which appears to be favored around neutral pH (Avakyan et al., 1985). Polymerization of monosilicate can be viewed as a removal of water between adjacent silicates to form a siloxane linkage. **Silica** can be viewed as an anhydride of **silicic acid**:

$$H_4SiO_4 \Rightarrow SiO_2 + 2H_2O \tag{9.1}$$

Dissociation constants for silicic acid are as follows (see Anderson, 1972):

$$H_4SiO_4 \Rightarrow H^+ + H_3SiO_4^- \quad (K_1 = 10^{-9.5}) \tag{9.2}$$

$$H_3SiO_4^- \Rightarrow H^+ + H_2SiO_4^- \quad (K_2 = 10^{-12.7}) \tag{9.3}$$

Silica can exist in partially hydrated form called **metasilicic acid** (H_2SiO_3) or fully hydrated form called **orthosilicic acid** (H_4SiO_4). Each of these hydrated forms can be polymerized, the ortho acid forming, for instance, $H_3SiO_4 \cdot H_2SiO_3 \cdot H_3SiO_3$ (Latimer and Hildebrand, 1940). The polymers may exhibit colloidal properties, depending on size and other

Table 9.1 Abundances of Silicon on the Earth's Surface

Phases	Concentration	Reference
Granite	336,000 ppm	Bowen, 1979
Basalt	240,000 ppm	Bowen, 1979
Shale	275,000 ppm	Bowen, 1979
Limestone	32,000 ppm	Bowen, 1979
Sandstone	327,000 ppm	Bowen, 1979
Soils	330,000 ppm	Bowen, 1979
Seawater	3×10^3 μg L^{-1}	Marine Chemistry, 1971
Fresh water	7 ppm	Bowen, 1979

factors. Colloidal forms of silica seem to exist only locally at high concentrations or at saturation levels and are favored by acid conditions (Hall, 1972).

Common silicon-containing minerals include quartz (SiO_2), olivine [$(Mg,Fe)_2SiO_4$], orthopyroxene ($Mg,FeSiO_3$), biotite [$K(Mg,Fe)_3AlSi_3$-$O_{10}(OH)_2$], orthoclase ($KAlSi_3O_8$), plagioclase [$(Ca,Na)(Al,Si)AlSi_2O_8$], kaolinite [$Al_4Si_4O_{10}(OH)_8$], and others. The concentration of silicon in various components of the Earth's surface is listed in Table 9.1.

Silica and silicates form an important buffer system in the oceans (Garrels, 1965), together with carbon dioxide. The $CO_2/HCO_3^-/CO_3^{2-}$ system, because of its rapid rate of reaction, is a short-range buffer, whereas silica and silicates, owing to their slow rate of reaction, are part of a long-range buffer system (Sillen, 1967).

Aluminosilicates in the form of clay also perform a buffering function in mineral soils. This is because of their ion-exchange capacity, their net electronegative charge, and their adsorption powers. Their ion-exchange capacity and adsorption powers, moreover, make them important reservoirs of cations and organic molecules. Montmorillonite exhibits the greatest ion-exchange capacity, illites exhibit less, and kaolinites least (Dommergues and Mangenot, 1970, p. 469).

9.2 BIOLOGICALLY IMPORTANT PROPERTIES OF SILICON AND ITS COMPOUNDS

Silicon is taken up and concentrated in significant quantities by certain forms of life. These include microbial forms, such as diatoms and other chrysophytes; silicoflagellates and some xanthophytes; radiolarians and actinopods; plant forms, such as horsetails, ferns, grasses, and some flowers and trees; and some animal forms, such as sponges, insects, and even vertebrates. Some bacteria (Heinen, 1960) and fungi (Heinen, 1960; Holzapfel and Engel, 1945a,b) have also been reported to take up silicon to a limited extent. According to Bowen (1966), diatoms may contain from 1,500 to 20,000 ppm, land plants from 200 to 5,000 ppm, and marine animals from 120 to 6,000 ppm of silicon.

While the function of silicon in most higher forms of life, animals and plants, is not presently apparent, it is clearly structural in some microorganisms, such as diatoms, actinopods and radiolarians. In diatoms, silicon also seems to play a metabolic role in the synthesis of chlorophyll (Werner, 1966, 1967), the synthesis of DNA (Darley and Volcani, 1969; Reeves and Volcani, 1984), and the synthesis of DNA polymerase and thymidylate kinase (Sullivan, 1971); Sullivan and Volcani, 1973).

Silicon compounds in the form of clays (aluminosilicates) may exert an effect on microbes in soil. They may stimulate or inhibit microbial metabolism, depending on conditions (e.g., Marshman and Marshall, 1981a,b; Weaver and Dugan, 1972; see also the discussion of Marshall, 1971). These effects of clays are mostly indirect, i.e., clays tend to modify the microbial habitat physicochemically, thereby eliciting a physiological response by the microbes (Stotzky, 1986). For beneficial effect, clays may buffer the soil environment and help to maintain a favorable pH, thereby promoting growth and metabolism of some microorganisms which otherwise might be slowed or stopped if the pH became unfavorable (Stotzky, 1986). Certain clays have been found to enable some bacteria that were isolated from marine ferromanganese nodules or associated sediments to oxidize Mn^{2+}. The microorganisms can oxidize Mn^{2+} if it is bound to bentonite (a montmorillonite-type of clay) or kaolinite but not to illite if each has been pretreated with ferric iron. They cannot oxidize Mn^{2+} free in solution (Ehrlich, 1982). Cell-free preparation of these bacteria oxidize Mn^{2+} bound to bentonite and kaolinite even without ferric iron pretreatment, although the manganese-oxidizing activity of the cell extracts is greater when the clays are pretreated with ferric iron (Ehrlich, 1982). Like intact cells, the cell-free extract cannot oxidize Mn^{2+} free in solution (Ehrlich, 1982). Clays may also enhance activity of some enzymes such as catalase when binding them to their surface (see Stotzky, 1982, p. 404).

By contrast, clays may suppress microbial growth and metabolism by adsorbing microbial nutrients, thereby making them less available to microbes. Clays may also adsorb microbial antibiotics and thereby lower the inhibitory activity of these agents (see Stotzky, 1986). The result may be that an antibiotic producer is outgrown by organisms it normally keeps in check with help of the antibiotic. These effects of clays can be explained, at least in part, by the strength of binding to a negatively charged clay surface and an inability of many microbes to attack adsorbed nutrients, or by the inability of adsorbed antibiotics to inhibit susceptible microbes (see, for instance, Dashman and Stotzky, 1986). High concentrations of clay may interfere with diffusion of oxygen by increasing the viscosity of a solution, which can have a negative effect on aerobic microbial respiration (see Stotzky, 1986). Clays may also modulate other interactions between different microbes and between microbes and viruses in soil, and they may affect the pathogenicity of disease-causing soil microbes (see Stotzky, 1986).

9.3 BIOCONCENTRATION OF SILICON

Bacteria Some bacteria have been shown to accumulate silicon. A soil bacterium B_2 (Heinen, 1960) and a strain of *Proteus mirabilis* (Heinen, 1968) have been found to take up limited amounts of silicon when furnished in the form of silica gel, quartz, or sodium silicate. Sodium silicate was taken up most easily, quartz least easily. The silicon seemed to substitute partially for phosphorus in phosphorus-deficient media (Fig. 9.1). This substitution reaction was reversible (Heinen, 1962). The silicon taken up by the bacteria appeared to be organically bound in a metabolizable form (Heinen, 1962). Sulfide, sulfite, and sulfate were found to affect phosphate-silicate exchange in different ways, depending on concentration, whereas KCl and NaCl were without effect. NH_4Cl, NH_4NO_3, and $NaNO_3$ stimulated the formation of adaptive enzymes involved in the phosphate-silicate exchange (Heinen, 1963a). The presence of sugars such as glucose, fructose, or sucrose, and of amino acids such as alanine, cysteine, glutamine, methionine, asparagine, and citrulline, as well as of metabolic intermediates pyruvate, succinate, and citrate, stimulated silicon uptake, whereas acetate, lactate, phenylalanine, peptone, and wheat germ oil inhibited it. Glucose at an initial concentration of 1.2 mg per milliliter of medium stimulated silicon uptake maximally (Fig. 9.2). Higher concentrations of glucose caused the formation of particles of protein, carbohydrates, and silicon outside the cell. $CdCl_2$ inhibited the stimulatory effect of glucose on silicon uptake, but 2,4-dinitrophenol was without effect. The simultaneous presence of $NaNO_3$ and KH_2PO_4 lowered the stimulatory effect of glucose but did not eliminate it (Heinen, 1963b). The silicon that was fixed in bacteria was readily displaced by phosphate in the absence of external glucose. In cells that were preincubated in a glucose-silicate solution, only a small portion of the silicon was released by glucose-phosphate, but all of the silicon was releasable on incubation in

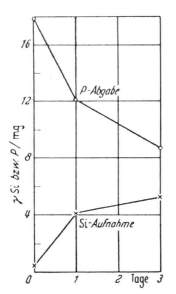

Figure 9.1 Relationship between Si uptake and P_i release during incubation of resting cells of strain B_2 in silicate solution (80 µg of Si per milliliter). (From W. Heinen, Silicium-Stoffwechsel bei Mikroorganismen. II. Mitteilung. Beziehung zwischen dem Silica und Phosphat-Stoffwechsel bei Bakterien. Arch. Mikrobiol. 41:229–246, Copyright © 1962 Springer-Verlag.)

glucose-carbonate solution. Some of the silicon taken up by the bacterial cells appeared to be tied up by labile ester bonds (C-O-Si) whereas other silicon appeared to be tied up in more stable bonds (C-Si) (Heinen, 1963c, 1965). Studies of intact cells and cell extracts of *Proteus mirabilis* after the cells were incubated in the presence of silicate suggested that the silicate taken up was first accumulated in the cell walls and then slowly transferred to the interior of the cell (Heinen, 1965). The silicon was organically bound in the wall and intracellularly. A particulate fraction from *P. mirabilis* bound silicate organically in an oxygen-dependent process (Heinen, 1967).

Fungi Some fungi have also been reported to accumulate silicon (Holzapfel and Engel, 1954 a,b). When growing on a silicate-containing agar medium, the vegetative mycelium of such fungi exhibits an induction period of 5–7 days before taking up silicon. When the silicon in the medium is in the form of galactose or glucose-quartz complexes (Holzapfel, 1951),

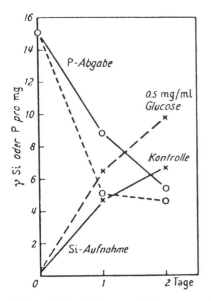

Figure 9.2 Influence of glucose on silicate uptake (x—x, without glucose, control; x--x, with glucose), and phosphate release (o—o, without glucose, control; o--o, with glucose). (From W. Heinen, Silicium-Stoffwechsel bei Mikroorganismen. IV. Mitteilung. Die Wirkung zwischen organischer Verbindungen, insbesondere Glucose, auf den Silicium-Stoffwechsel bei Baketerien. Arch Mikrobiol. 45:162–171, Copyright © 1963 Springer-Verlag.)

silicon uptake by vegetative mycelium can occur within 12–18 h. Evidently inorganic silicates have to be transformed to organic complexes during the prolonged induction period before the silicon is taken into the fungal cells.

Diatoms   Among eukaryotic microorganisms that take up silicon, the diatoms have been the most extensively studied with respect to this property (Fig. 9.3) (Lewin, 1965). Their silicon uptake ability affects redistribution of silica between fresh and marine waters. In the Amazon River estuary, for instance, they remove 25% of the dissolved silica in the river water. Their frustules are not swept oceanward upon their death, but are transported coastward and incorporated into dunes, mud, and sand bars (Millman and Boyle, 1975).

   Diatoms are unicellular algae, enclosed in a wall of silica consisting of two valves in pillbox arrangement. The valves are usually perforated

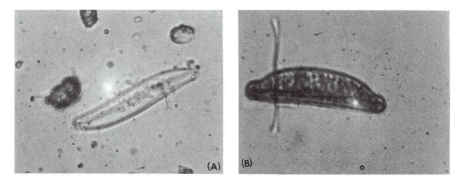

Figure 9.3   Diatoms. (A) *Gyrosigma*, from fresh water (× 1944). (B) *Cymbella*, from fresh water (× 1944). (C) *Fragellaria*, from fresh water (× 1864). (D) Ribbon of diatoms, from fresh water (× 1990). (E) Marine diatom frustule, from Pacific Ocean sediment (× 1944).

plates which may have thickened ribs. Their shape may be pennate or centric. In cell division, each daughter cell receives one of the two valves and synthesizes the other de novo to fit into the one already present. To prevent excessive reduction in size of daughter diatoms which receive the smaller of the two valves after each cell division, a special reproductive step called **auxospore formation** returns the organism to maximum size. It occurs when a progeny cell has reached a minimum size after repeated divisions. Auxospore formation is a sexual reproductive process in which the cells escape from their frustules and increase in size in their zygote membrane, which may become weakly silicified. After a time, the proto-plast in the zygote membrane contracts and forms the typical frustules of the parent cell (Lewin, 1965).

The silica walls of diatoms consist of hydrated, amorphous silica, a polymerized silicic acid (Lewin, 1965). The walls of marine diatoms may contain as much as 96.5% SiO_2 but only 1.5% Al_2O_3 or Fe_2O_3 and 1.9% water (Rogall, 1939). In clean, dried frustules of freshwater *Navicula pelliculosa*, 9.6% water has been found (Lewin, 1957). Thin parts of diatom frustules reveal a foamlike substructure under the electron microscope, suggesting silica gel (Helmcke, 1954), which may account for the adsorptive power of such frustules. The silica gel may be viewed as arranged in small spherical particles about 22 μm in diameter (Lewin, 1965). Because of the low degree of solubility of amorphous silica at the pH of most natural waters, frustules of living diatoms do not dissolve readily (Lewin, 1965). Yet it has been reported that at pH 8, 5% of the silica in the cell walls of *Thalassiosira nana* and *Nitschia linearis* dissolves, and at pH 10,

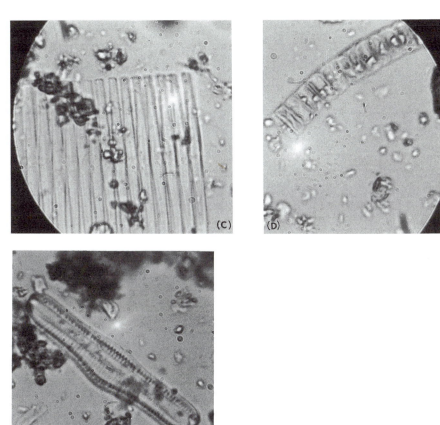

Figure 9.3 Continued.

20% of the silica in the frustules of *N. linearis* and all of the silica in the frustules of *T. nana* dissolves (Jorgensen, 1955). This silica dissolution may reflect the state of integration of newly assimilated silica in the diatom frustule. Some bacteria naturally associated with diatoms have been shown to accelerate dissolution of silica in frustules by an unknown mechanism(s) (Patrick and Holding, 1985). Diatom frustules of living organisms are to some extent protected against dissolution by an organic film, when present, and their rate of dissolution has been shown to observe temperature dependence (Katami, 1982). After the death of diatoms, their frustules may dehydrate to form more crystalline SiO_2, which is much less soluble in alkali and may account for the accumulation of diatomaceous ooze.

Rates of silicic acid uptake and incorporation by diatoms can be easily measured with radioactive [^{68}Ge]-germanic acid as tracer (Azam et al., 1973; Azam,1974, Chisholm et al., 1978). At low concentration (molar ratio of Ge/Si of 0.01), germanium, which is chemically similar to silicon, is apparently incorporated together with silicon into the silicic acid polymer of the frustule. At higher concentration (molar ratio of Ge/Si of 0.1), germanium is toxic to diatoms (Azam et al., 1973).

How silica is deposited in the wall structure of diatoms and arranged in a predetermined, characteristically intricate geometric design is still very poorly understood. Important aspects have, however, been revealed (Lewin, 1965). Diatoms take up silicon as contained in orthosilicate (H_4SiO_4). More highly polymerized forms of silica are apparently not taken up by them unless they are first depolymerized, as by some bacteria (Lauwers and Heinen, 1974). Organic silicates are not available to them either, nor can Ge, C, Sn, Pb, P, As, B, Al Mg, or Fe replace silicon extensively, if at all (Lewin, 1965). The concentration of silicon in a diatom depends to a degree on its concentration in the growth medium and on the rate of cell division (the faster the cells divide, the thinner their frustules). Silicon is essential for cell division, but resting cells in a medium in which silica is not at a limiting concentration continue to take up silica (Lewin, 1965). Synchronously growing cells of *Navicula pelliculosa* take up silica at a constant rate during the cell-division cycle (Lewin, 1965). Silica uptake appears dependent on energy yielding processes (Lewin, 1965; Azam et al., 1974; Azam and Volcani, 1974) and seems to involve intracellular receptor sites (Blank and Sullivan, 1979). Uncoupling of oxidative phosphorylation stops silica uptake by *N. pelliculosa* and *Nitschia angularis*. Starved cells of *N. pelliculosa* show an enhanced silicon uptake rate when fed glucose or lactate in the dark, or when returned to the light, where they can photosynthesize (Healy et al., 1967). Respiratory inhibitors prevent Si and Ge uptake by *Nitschia alba* (Azam et al., 1974; Azam and Volcani, 1974). It has also been noted that total uptake of phosphorus and carbon is decreased during silica starvation of *N. pelliculosa*. Upon restoration of silica to the medium, the total uptake of phosphorus and carbon is again increased (Coombs and Volcani, 1968). Sulfhydryl groups appear to be involved in silica uptake (Lewin, 1965). The cell membrane probably plays a fundamental role in the laying down of new silica to form a frustule (Lewin, 1965), but the details of this process are not yet understood.

9.4   BIOMOBILIZATION OF SILICON

Some bacteria and fungi play an important role in the mobilization of silica and silicates in nature. Part of this microbial involvement is manifest in the weathering of rock silicates and aluminosilicates. The solubilizing action may involve the cleavage of Si-O-Si (siloxane) or Al-O framework bonds or the removal of cations from the crystal lattice of silicate causing the subsequent collapse of the silicate lattice structure. The mode of attack may be by (1) microbially produced ligands of cations, (2) microbially produced organic or inorganic acids, (3) microbially produced alkali (ammonia or amines), or (4) microbially produced extracellular polysaccharide acting at acid pH (glycocalyx).

Solubilization by Ligands Microbially produced ligands of divalent cations have been shown to cause dissolution of calcium containing silicates. For instance, a soil strain of *Pseudomonas* that produced 2-ketogluconic acid from glucose dissolved synthetic silicates of calcium, zinc, and magnesium and the minerals wollastonite ($CaSiO_3$), apophyllite [$KCa_4Si_8O_{20}$ (F,OH)·$8H_2O$], and olivine [$(Mg,Fe)_2SiO_4$] (Webley et al., 1960). The demonstration consisted of culturing the organism for 4 days at 25°C on separate agar media, each containing 0.25% (wt/vol) of one of the synthetic or natural silicates. A clear zone was observed around the bacterial colonies when the silicate was dissolved (Fig. 9.4). A similar silicate-dissolving action with a gram-negative bacterium, strain D_{11}, which resembled *Erwinia* spp., and with *Bacterium* (now *Erwinia*) *herbicola* or with some *Pseudomonas* strains, all of which were able to produce 2-ketogluconate from glucose, has also been shown (Duff et al., 1963). The action of these bacteria was tested in glucose-containing basal medium: KH_2PO_4, 0.54 g; $MgSO_4·7H_2O$, 0.25 g; $(NH_4)_2SO_4$, 0.75g; $FeCl_3$, trace; Difco yeast extract, 2 g; glucose 40 g; Distilled water, 1 liter; and 5–500 mg of pulverized mineral per 5–10 ml of medium. It was found that the dissolution of silicates in this case resulted from the complexation of their cationic components by 2-ketogluconate. The complexes were apparently more stable than the silicates. For example,

$$CaSiO_3 \Leftrightarrow Ca^{2+} + SiO_3^{2-} \tag{9.4}$$

$$Ca^{2+} + \text{2-ketogluconate} \Leftrightarrow Ca(\text{2-ketogluconate}) \tag{9.5}$$

The structure of 2-ketogluconic acid is

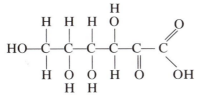

The silicon that was liberated or released in these experiments and subsequently transformed took three forms: (1) low-molecular weight or ammo-

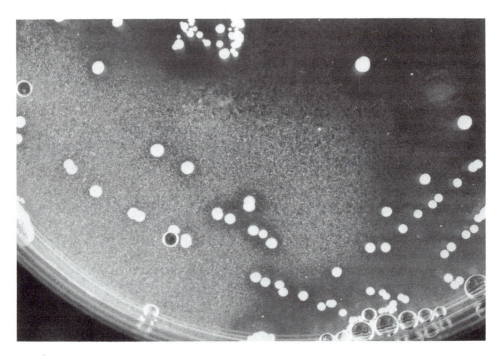

Figure 9.4   Colonies of bacterial isolate C-2 from a sample of weathered rock on synthetic Ca-silicate selection medium showing evidence of Ca-silicate dissolution around colonies. Basal medium was prepared by aseptically mixing 10 ml of sterile solution A (3 g dextrose or 3 g levulose in 100 ml dist. water), 10 ml of sterile solution B (0.5 g $(NH_4)_2SO_4$, 0.5 g $MgSO_4 \cdot 7H_2O$, 0.5 g Na_2HPO_4, 0.5 g KH_2PO_4, 2 g yeast extract, 0.05 $MnSO_4 \cdot H_2O$ in 500 ml dist. water) and 20 ml of sterile 3% agar and distributing the mixture in Petri plates. Capping agar was prepared by mixing 10 ml of a sterile synthetic $CaSiO_3$ suspension with 7.5 ml of sterile solution A, 7.5 ml of solution B, and 15 ml of sterile 3% agar and distributing 3 ml of this mixture aseptically over the surface of the solidified basal agar in the plates.

nium molybdate-reactive silicate (monomeric?), (2) a colloidal-polymeric silicate of higher molecular weight which did not react with dilute hydrofluoric acid, and (3) an amorphous form which could be removed from solution by centrifugation and dissolved in cold 5% aqueous sodium carbonate (Duff et al., 1963). Polymerized silicate can be transformed by bacteria into monomeric silicate, as has been shown in studies with *Proteus mirabilis* and *B. caldolyticus* (Lauwers and Heinen, 1974). The *Pro-*

teus culture is able to assimilate some of the monomeric silicate. The mechanism of depolymerization has not been elucidated.

Gluconic acid produced from glucose by several different types of bacteria was recently shown to solubilize bytownite, albite, kaolinite, and quartz at near-neutral pH (Vandevivere et al., 1994). The activity around neutral pH suggests that the mechanism of action of the gluconate involves chelation.

Quartz (SiO_2) has been shown to be subject to slow dissolution by organic acids such as citric, oxalic, pyruvic, and humic acids (Bennett et al., 1988), all of which can be formed by fungi or bacteria. In a pH range from 3 to 7, the effect was greatest at pH 7, indicating that the mechanism of action was not by acidulation but by chelation. Bennett et al. (1988) suggest that the chelation involves an electron-donor acceptor system. Acetate, fumarate, and tartrate were ineffective in dissolving silica complex.

Solubilization by Acids   The effect of the action of acids in solubilizing silicates has been noted in various studies. Waksman and Starkey (1931) cited the action of CO_2 on orthoclase:

$$2KAlSi_3O_8 + 2H_2O + CO_2 \Rightarrow H_4Al_2Si_2O_9 + K_2CO_3 + 4SiO_2$$

$$(9.6)$$

The CO_2 is, of course, very likely to be a product of respiration or fermentation. Another example of a silicate attacked by acid is that involving spodumene ($LiAlSi_2O_6$) (Karavaiko et al., 1979). An in-situ correlation was observed between the degree of alteration of a spodumene sample and the acidity it produced in aqueous suspension. Unweathered spodumene generated a pH in the range of 5.1 to 7.5, whereas altered spodumene generated a pH in the range of 4.2 to 6.4. Non-sporeforming heterotrophs were found to predominate in weathered spodumene. They included bacteria such as *Arthrobacter pascens, A. globiformis* and *A. simplex* as well as *Nocardia globerula, Pseudomonas fluorescens, P. putida, and P. testosteronii*, and fungi such as *Trichoderma lignorum, Cephalosporum atrum* and *Penicillium decumbens*. Acid decomposition of spodumene may be formulated as follows (Karavaiko et al., 1979):

$$4LiAlSi_2O_6 + 6H_2O + 2H_2SO_4 \Rightarrow 2Li_2SO_4 + Al_4[Si_4O_{10}](OH)_8$$

$$+ 4H_2SiO_3 \qquad (9.7)$$

The aluminosilicate product in this reaction is kaolinite.

Further investigation into microbial spodumene degradation (Karavaiko et al., 1980) revealed that among the most active microorganisms are the fungi *Penicillium notatum* and *Aspergillus niger*, thionic bacteria like *Thiobacillus thiooxidans*, and the slime-producing bacterium *Bacillus mucilaginosus* var. *siliceus* (Karavaiko et al., 1980; Avakyan et al., 1986). The fungi and *T. thiooxidans*, by producing acid, were most effective in solubilizing Li and Al. *B. mucilaginosus* was effective in solubilizing Si in addition to Li and Al by reaction of its extracellular polysaccharide with the silicate of spodumene.

Solubilization of silicon along with other constituents in the primary minerals amphibolite, biotite, and orthoclase by acids (presumably citric and oxalic acids) formed by several fungi and yeasts at the expense of glucose has also been demonstrated (Eckhardt, 1980). These findings of silicon mobilization are similar to those in earlier studies on the action of the fungi *Botritis*, *Mucor*, *Penicillium* and *Trichoderma* from rock surfaces and weathered stone. In these experiments, citric and oxalic acids produced by the fungi solubilized Ca, Mg, and Zn silicates (Webley et al., 1963). In other experiments, *Aspergillus niger* released Si from apophyllite, olivine, saponite, vermiculite, and wollastonite, but not augite, garnet, heulandite, hornblende, hypersthene, illite, kaolinite, labradorite, orthoclase or talc (Henderson and Duff, 1963); *Penicillium simplicissimus* released Si from basalt, granite, granodiorite, rhyolite, andesite, peridotite, dunite, and quartzite with metabolically produced citric acid (Silverman and Munoz, 1970); and *Penicillium notatum* and *Pseudomonas* sp. released Si from plagioclase and nepheline with acids formed by these organisms (Aristovskaya and Kutuzova, 1968; Kutuzova, 1969). In contrast to Henderson and Duff (1963), Berner et al. (1980) have found in laboratory experiments that augite, hypersthene, hornblende, and diopside in soil samples were subject to weathering by soil acids, presumably of biological origin.

In a study of weathering of three different plagioclase specimens, all Ca-Na feldspars, by organic acids that can be metabolic end products of microbial metabolism and by inorganic acids, it was found that steady-state dissolution rates were highest at approximately pH 3 and decreased as the pH was increased toward neutrality (Welch and Ullman, 1993). Polyfunctional acids, including oxalate, citrate, succinate, pyruvate, and 2-ketoglutarate, were the most effective, whereas acetate and propionate were less effective. However, all the organic acids were more effective than the inorganic acids HCl and HNO_3. The polyfunctional acids acted mainly as acidulants at very acid pH and mainly as chelators near neutral pH. Ca and Na were rapidly released in these experiments. Those organic acids that preferentially chelated Al were the most effective in enhancing plagioclase dissolution. Although the products of dissolution of feldspars are usually considered to include separate aluminum and silicate species, soluble aluminum-silicate complexes may be intermediates (Browne and Driscoll, 1992).

The practical effect of acid attack of aluminosilicates can be seen in the corrosion of concrete sewer pipes. Concrete is formed from a mixture of cement (heated limestone, clay and gypsum) and sand. On setting, the cement includes the compounds Ca_3SiO_5 and $Ca_3(AlO_3)_2$, which hold the sand in their matrix. H_2S produced by microbial mineralization of organic sulfur compounds and by bacterial sulfate reduction can itself corrode the concrete. But it is much more effective after it is microbially oxidized to sulfuric acid by thiobacilli (*T. neapolitanus*, *T. intermedius*, *T. novellus*, *T. thiooxidans*) and attacks the concrete (Parker, 1949; Milde et al., 1983; Sand et al. 1984).

Groundwater pollution with biodegradable substances has been found to result in silicate weathering of aquifer rock. Products of microbial biodegradation cause the weathering. This was observed in an oil-polluted aquifer near Bemidji, Minnesota (Hiebert and Bennet, 1992). Microcosm experiments of 14 months duration in the aquifer with a mixture of crystals of albite, anorthite, anorthoclase, microcline, all feldspar minerals, and quartz revealed microbial colonization of the mineral surfaces by individual cells and small clusters and intense etching of the feldspar minerals and light etching of the quartz at or near where the bacteria were seen. Such aquifer rock weathering can affect water quality.

Solubilization by Alkali   Alkaline conditions are very conducive to mobilizing silicon, be that from silicates, aluminosilicates, or even quartz (Karavaiko et al., 1984). This is because not only the Al-O bond but the Si-O bond is very labile under these conditions since both are susceptible to nucleophilic attack (see discussion by Karavaiko et al., 1985). *Sarcina ureae* growing in a peptone-urea broth released silicon readily from nepheline, plagioclase, and quartz (Aristovskaya and Kutuzova, 1968; Kutuzova, 1969). In this instance, ammonia resulting from the hydrolysis of urea was the source of the alkali. In microbial spodumene degradation, alkaline pH also favors silicon release (Karavaiko et al., 1980).

Solubilization by Extracellular Polysaccharide   Extracellular polysaccharide has been claimed to play an important role in silicon release, especially in the case of quartz. Such polysaccharide, whether of bacterial origin [e.g., from *Bacillus mucilaginosus* var. *siliceus* (Avakyan et al., 1986), or unnamed organisms in a microbial mat on rock around a hot spring (Heinen and Lauwers, 1988)] or of fungal origin (e.g., from *Aspergillus niger*; Holzapfel and Engel, 1954), is able to react with siloxanes to form organic siloxanes. The reaction appears not to be enzyme-catalyzed since polysaccharide from which the cells have been removed is reactive. Indeed, such organic silicon-containing compounds can be formed with reagent-grade organics (Holzapfel, 1951; Weiss et al., 1961) and have been isolated from various biological sources other than microbes (Schwarz, 1973). With polysaccharide from *B. mucilaginosus*, the reaction appears to be favored by acid metabolites (Malinovskaya et al., 1990). Welch and Vandevivere (1995) found that polysaccharides from different sources either had no effect or interfered with solubilization of plagioclase by gluconate at a pH between 6.5 and 7.0.

Depolymerization of Polysilicates   Since silicate can exist in monomeric form as well as polymeric form (metasilicates, siloxanes), and since silicon uptake by microbes depends on the monomeric form (ortho-silicate), depolymerization of siloxanes is important. *Proteus mirabilis* as well as *Bacillus caldolyticus* have the capacity to promote this process. In the case of *B. caldolyticus*, it appears to be growth-dependent even though the organism does not assimilate Si (Lauwers and Heinen, 1974). In the weathering of quartz, the degradation of silicon polymer to the monomeric stage appears to proceed through an intermediate oligomeric stage. Organosilicates may also be formed transitionally (Avakyan et al., 1985). The detailed mechanisms by which these transformations proceed are not known. It is clear, however, that these biodegradative processes are of fundamental importance to the biological silica cycle.

9.5   ROLE OF MICROBES IN THE SILICON CYCLE

As the foregoing discussion shows, microbes [even some plants and animals (Drever, 1994)] have a significant influence on the distribution and form of silicon in the biosphere. Those organisms which assimilate silicon, clearly act as concentrators of it. Those which degrade silica, silicates or aluminosilicates act as agents of silicon dispersion. They are an important source of the orthosilicate on which the concentrators depend, and they are also important agents of rock weathering. Comparative electron microscopic studies have provided clues to the extent of microbial weathering action (see, for instance, Berner et al., 1980; Callot et al., 1987).

It has been argued that biological silicate-liberating reactions by acids and complexing agents under laboratory conditions occurred at higher glucose concentrations than might be encountered in nature and may therefore be laboratory artifacts. A counterargument can be made, however, that microenvironments exist in soil and sediment that have appropriately high concentrations of utilizable carbohydrates, nitrogenous compounds, and other needed nutrients from the excretory products and the dead remains of organisms from which appropriate bacteria and/or fungi can form the compounds which can promote the breakdown of quartz, silicates, and aluminosilicates. Indeed, fungal hyphae in the litter zone and A horizon of several different soils have been shown by scanning electron microscopy to carry calcium oxalate crystals attached to them, which is evidence for extensive in situ production of oxalate by the fungus (Graustein et al., 1977). The basidiomycete *Hysterangium crassum* was actually shown to weather clay in situ with the oxalic it produced (Cromack et al., 1979). Lichens, a symbiotic association between fungi and algae, show evidence observable in situ of extensive rock weathering activity (Jones et al., 1981) [see, however, disclaimers by Ahmadjian (1967) and Hale (1967)].

Silica degradation in nature is usually a slower process than in the laboratory because the conditions in natural environments are usually less favorable. If this were not so, rock in the biosphere would be a very unstable substance.

Thus, silicon in nature may follow a series of cyclical biogeochemical transformations (Kuznetsov, 1975; Harriss, 1972; Lauwers and Heinen, 1974). Silica, silicates, and aluminosilicates in rocks are subject to the weathering action of biological, chemical, and physical agents. The extent of the contribution of each of these agents must depend on the particular environmental circumstances. Silicon liberated in these processes as soluble silicate may be leached away by surface or groundwater, and from these waters it may either be removed by chemical or biological precipitation at new sites, or it may be swept into bodies of fresh water or the sea. There, silicate will tend to be removed by biological agents. Upon their death, these biological agents will release their silicon back into solution or their siliceous remains will be incorporated into the sediment (e.g., Allison, 1981; Patrick and Holding, 1985; Weaver and Wise, 1974), where some or all of it may later be returned to solution. The sediments of the ocean appear to be a sink for excess silica swept into the oceans since the silica concentration of seawater tends to remain relatively constant. But over geologic time, even this silicon is not permanently immobilized. Plate tectonics will ultimately cause even this silicon to be recycled.

9.6 SUMMARY

The environmental distribution of silicon is significantly influenced by microbial activity. Certain microorganisms assimilate it and build it into cell-support structures. They include diatoms and some other chryso-

phytes, silicoflagellates, some xanthophytes, radiolarians, and actinopods. Silicon uptake rates by diatoms have been measured, but the mechanism by which silicon is assimilated is still only partially understood.
Certain silicon-incorporating bacteria may provide a biochemical model
for silicon assimilation in general. Silicon-assimilating microorganisms
such as diatoms and radiolaria are important in the formation of siliceous
oozes in oceans, and diatoms are important in forming such oozes in lakes.

Some bacteria and fungi are able to solubilize silicates and silica.
They accomplish this by forming chelators, acids or bases, which react
with silicates, and exopolysaccharides, which react with silica and silicates. These actions are important in the weathering of rock and in cycling
silicon in nature.

REFERENCES

Ahmadjian, V. 1967. The Lichen Symbiosis. Blaidsell Publishing Co., Div.
of Ginn and Co., Waltham, MA.

Allison, C. W. 1981. Siliceous microfossils from the Lower Cambrian of
northwest Canada: Possible source for biogenic chert. Science 211:
53–55.

Anderson, G. M. 1972. Silica solubility, pp. 1085–1088. In: R. W. Fair-
bridge, ed. The Encyclopedia of Geochemistry and Environmental
Sciences. Encyclopedia of Earth Sciences Series, Vol. IVA. Van
Nostrand Reinhold, New York.

Aristovskaya, T. V., and R. S. Kututzova. 1968. Microbiological factors
in the mobilization of silicon from poorly soluble natural compounds.
Pochvovedenie (12):59–66.

Avakyan, Z. A., N. P. Belkanova, G. I. Karavaiko, and V.P.Piskunov.
1985. Silicon compounds in solution during bacterial quartz degradation. Mikrobiologiya 54:301–307 (Engl. transl. pp. 250–256).

Avakyan, Z. A., T. A. Pivovarova, and G.I. Karavaiko. 1986. Properties
of a new species, *Bacillus mucilaginosus*. Mikrobiologiya 55:
477–482 (Engl. transl. pp. 369–374).

Azam, F. 1974. Silicic-acid uptake in diatoms studied with [⁶⁸]germanic
acid as tracer. Planta 121:205–212.

Azam, F., and B. E. Volcani. 1974. Role of silicon in diatom metabolism.
VI. Active transport of germanic acid in the heterotrophic diatom
Nitschia alba. Arch. Microbiol. 101:1–8.

Azam, F., B. B. Hemmingsen, and B. E. Volcani. 1973. Germanium incorporation into the silica of diatom cell walls. Arch. Mikrobiol. 92:
11–20.

Azam, F., B. B. Hemmingsen, and B. E. Volcani. 1974. Role of silicon in diatom metabolism. V. Silicic acid transport and metabolism in the heterotrophic diatom *Nitschia alba.* Arch. Microbiol. 97:103–114.

Bennett, P. C., M. E. Melcer, D. I. Siegel, and J. P. Hassett. 1988. The dissolution of quartz in dilute aqueous solutions of organic acids at 25°C. Geochim. Cosmochim. Acta 52:1521–1530.

Berner, R. A., E. L. Sjoeberg, M. A. Velbel, and M. D. Krom. 1980. Dissolution of pyroxenes and amphiboles during weathering. Science 207:1205–1206.

Blank, G. S., and C. W. Sullivan. 1979. Diatom mineralization of silicic acid. III. Si(OH)$_4$ binding and light dependent transport in *Nitschia angularis.* Arch. Microbiol. 123:157–164.

Bowen, H. J. M. 1966. Trace Elements in Biochemistry. Academic Press, London.

Bowen, H. J. M. 1979. Environmental Chemistry of the Elements. Academic Press, London.

Browne, B. A., and C. T. Driscoll. 1992. Soluble aluminum silicates: Stoichiometry, stability, and implications for environmental geochemistry. Science 256:1667–1670.

Bunt, J. S., and A. D. Rovira. 1955. Microbiological studies of some subantarctic soils. J. Soil Sci. 6:119–128.

Callot, G., M. Maurette, L. Pottier, and A. Dubois. 1987. Biogenic etching of microfractures in amorphous and crystalline silicates. Nature (Lond.) 328:147–149.

Chisholm, S. W., F. Azam, and R. W. Eppley. 1978. Silicic acid incorporation in marine diatoms on light-dark cycles: Use as an assay for phased cell division. Limnol. Oceanogr. 23:518–529.

Coombs, J., and B. E. Volcani. 1968. The biochemistry and fine structure of silica shell formation in diatoms. Silicon induced metabolic transients in *Navicula pelliculosa.* Planta 80:264–279.

Cromack, K. Jr., P. Sollins, W. C. Graustein, K. Speidel, A.W. Todd, G. Spycher, C. Y. Li, and R. L. Todd. 1979. Calcium oxalate accumulation and soil weathering in mats of the hypogeous fungus *Hysterangium crassum.* Soil Biol. Biochem. 11:463–468.

Darley, W. M., and B. E. Volcani. 1969. Role of silicon in diatom metabolism. A silicon requirement for deoxyribonucleic acid synthesis in the diatom *Cylindrotheca fusiformis* Reimann and Lewin. Exp. Cell Res. 58:334–343.

Dashman, T. and G. Stotzky. 1986. Microbial utilization of amino acids and a peptide bound on homoionic montmorillonite and kaolinite. Soil Biol. Biochem. 18:5–14.

Dommergue, Y., and F. Mangenot. 1970. Ecologie Microbienne du Sol. Masson, Paris.

Drever, J. I. 1994. The effect of land plants on weathering of silicate minerals. Geochim. Geophys. Acta 58:2325–2332.

Duff, R. B., D. M. Webley, and R. O. Scott. 1963. Solubilization of minerals and related materials by 2-ketogluconic acid-producing bacteria. Soil Sci. 95:105–114.

Eckhardt, F. E. W. 1980. Microbial degradation of silicates. Release of cations from aluminosilicate minerals by yeasts and filamentous fungi. pp.107–116. In: T. A. Oxley, G. Becker, and D. Allsopp, eds. Biodeterioration. The Proceedings of the Fourth International Biodeterioration Symposium. Berlin. Pitman, London.

Ehrlich, H. L. 1982. Enhanced removal of Mn^{2+} from seawater by marine sediment and clay minerals in the presence of bacteria. Can. J. Microbiol. 28:1389–1395.

Garrels, R. M. 1965. Silca: Role in the buffering of natural waters. Science 148:69.

Graustein, W. C., K. Cromack Jr., and P. Sollins. 1977. Calcium oxalate: occurrence in soils and effect on nutrient and geochemical cycles. Science 198:1252–1254.

Hale, M. E. Jr. 1967. The Biology of Lichens. Edward Arnold (Publishers), London.

Hall, F. R. 1972. Silica cycle, pp. 1082–1085. In: R. W. Fairbridge, ed. The Encyclopedia of Geochemistry and Environmental Sciences. Encyclopedia of Earth Sciences Series, Vol. IVA. Van Nostrand Reinhold, New York.

Harriss, R. C. 1972. Silica - Biogeochemical cycle, 1082–1085. In: R.W. Fairbridge, ed. The Encyclopedia of Geochemistry and Environmental Sciences. Encyclopedia of Earth Sciences Series, Vol. IVA. Van Nostrand Reinhold, New York.

Healy, F. P, J. Combs, and B. E. Volcani. 1967. Changes in pigment content of the diatom Navicula pelliculosa in silicon starvation synchrony. Arch. Mikrobiol. 59:131–142.

Heinen, W. 1960. Silicium-Stoffwechsel bei Mikroorganismen. I. Mitteilung. Aufnahme von Silicium durch Bakterien. Arch. Mikrobiol. 37:199–210.

Heinen, W. 1962. Siliciumstoffwechsel bei Mikroorganismen. II. Mitteilung. Beziehung zwischen dem Silicat- und Phosphat-Stoffwechsel bei Bakterien. Arch. Mikrobiol. 41:229–246.

Heinen, W. 1963a. Silicium-Stoffwechsel bei Mikroorganismen. III. Mitteilung. Einfluss verschiedener Anionen auf den bakteriellen Si-Stoffwechsel. Arch. Mikrobiol. 45:145–161.

Heinen, W. 1963b. Silicium-Stoffwechsel bei Mikroorganismen. IV. Mitteilung. Die Wirkung organischer Verbindungen, insbesondere Glucose, auf den Silicium-Stoffwechsel bei Bakterien. Arch. Mikrobiol. 45:162–171.

Heinen, W. 1963c. Silicium-Stoffwechsel bei Mikroorganismen. V. Mitteilung. Untersuchungen zur Mobilitaet der inkorporierten Kieselsaeure. Arch. Mikrobiol. 45:172–178.

Heinen, W. 1965. Time-dependent distribution of silicon in intact cells and cell-free extracts of *Proteus mirabilis* as a model of bacterial silicon transport. Arch. Biochem. Biophys. 110:137–149.

Heinen, W. 1967. Ion accumulation in bacterial systems. III. Respiration-dependent accumulation of silicate by a particulate fraction from *Proteus mirabilis* cell-free extract. Arch. Biochem. Biophys. 120: 101–107.

Heinen, W. 1968. The distribution and some properties of accumulated silicate in cell-free bacterial extracts. Acta Bot. Neerl. 17:105–113.

Heinen, W., and A. M. Lauwers. 1988. Leaching of silica and uranium and other quantitative aspects of the lithobiontic colonization in a radioactive thermal spring. Microb. Ecol. 15:135–149.

Helmcke, J-G. 1954. Die Feinstruktur der Kieselsaeure und ihre physiologische Bedeutung in Diatomenschalen. Naturwissenschaften 11: 254–255.

Henderson, M. E. K., and R. B. Duff. 1963. The release of metallic and silicate ions from minerals, rocks and soils by fungal activity. J. Soil Sci. 14:236–246.

Hiebert, F. K., and P. C. Bennett. 1992. Microbial control of silicate weathering in organic-rich ground water. Science 258:278–281.

Holzapfel, L. 1951. Siliziumverbindungen in biologischen Systemen. Organ. Kieselsaeureverbindungen. XX. Mitteillung. Z. Elektrochem. 55:577–580.

Holzapfel, L., and W. Engel, 1954a. Ueber die Abhaengigkeit der Wachstumsgeschwindigkeit von *Aspergillus niger* in Kieselsaeureloesungen bei O_2-Belueftung. Naturwissenschaften 41:191–192.

Holzapfel, L., and W. Engel, 1954b Der Einfluss organischer Kieselsaeureverbindungen auf das Wachstum von *Aspergillus niger* und *Triticum*. Z. Naturforsch. 9b:602–606.

Iler, R. 1979. Chemistry of Silica. Wiley & Sons, New York.

Jones, D., M. J. Wilson, and W. J. McHardy. 1981. Lichen weathering of rock-forming minerals: application of scanning electron microscopy and microprobe analysis. J. Microsc. 124 (Pt. 1):95–104.

Jorgensen, E.G. 1955. Solubility of the silica in diatoms. Physiol. Plant. 8:846–851.

Karavaiko, G. I., Z. A. Avakyan, V. S. Krutsko, E. O. Mel'nikova, A. V. Zhdanov, and V. P. Piskunov. 1979. Microbiological investigations on a spodumene deposit. Mikrobiologiya 48:502–508 (Engl. transl. pp. 383–398).

Karavaiko, G. I., V. S. Krutsko, E. O. Mel'nikova, Z. A. Avakyan, and Yu.I. Ostroushko. 1980. Role of microorganisms in spodumene degradation. Mikrobiologiya 49:547–551 (Engl. transl. pp. 402–406).

Karavaiko, G. I., N. P. Belkanova, V. A. Eroshchev-Shak, and Z. A. Avakyan. 1985. Role of microorganisms and some physicochemical factors of the medium in quartz destruction. Mikrobiologiya 53: 976–981 (Engl. transl. pp. 795–800).

Katami, A. 1982. Dissolution rates of silica from diatoms decomposing at various temperatures. Marine Biol. (Berlin) 68:91–96.

Kretz, R. 1972. Silicon: Element and geochemistry, pp. 1091–1092. *In*: R. W. Fairbridge. The Encyclopedia of Geochemistry and Environmental Sciences. Encyclopedia of Earth Sciences Series, Vol. IVA. Van Nostrand Reinhold, New York.

Kutuzova, R. S. 1969. Release of silica from minerals as result of microbial activity. Mikrobiologiya 38:714–721 (Engl. transl. pp. 596–602).

Kuznetsov, S. I. 1975. The role of microorganisms in the formation of lake bottom deposits and their diagenesis. Soil Sci. 119:81–88.

Latimer, W. M., aand J. H. Hildebrand. 1940. Reference Book of Inorganic Chemistry. Rev. ed. Macmillan, New York.

Lauwers, A. M., and W. Heinen. 1974. Bio-degradation and utilization of silica and quartz. Arch. Microbiol. 95:67–78.

Lewin, J. C. 1957. Silicon metabolism in diatoms. IV. Growth and frustule formation in *Navicula pelliculosa*. J. Gen. Physiol. 39:1–10.

Lewin, J. C. 1965. Silification, pp. 447–455. *In*: R. A. Lewin, ed. Physiology and Biochemistry of Algae. Academic Press, New York.

Malinovskaya, I. M., L. V. Kosenko, S. K. Votselko, and V. S. Podgorskii. 1990. Role of *Bacillus mucilaginosus* polysaccharide in degradation of silicate minerals. Mikrobiologiya 59:70–78 (Engl. transl. pp. 49–55).

Marine Chemistry. 1971. A report of the Marine Chemistry Panel of the Committee of Oceanography. National Academy of Sciences, Washington, D.C.

Marshall, K. C. 1971. Sorption interactions between soil particles and microorganisms, pp. 409–445. *In*: A. D. McLaren and J. Skujins, eds. Soil Biochemistry, Vol. 2. Marcel Dekker, New York.

Marshman, N. A., and K. C. Marshall. 1981a. Bacterial growth on proteins in the presence of clay minerals. Soil Biol. Biochem. 13: 127–134.

Marshman, N. A., and K. C. Marshall. 1981b. Some effect of montmorillonite on the growth of mixed microbial cultures. Soil Biol. Biochem. 13:135–141.

Milde, K., W. Sand, W. Wolff, and E. Bock. 1983. Thiobacilli of the corroded concrete walls of the Hamburg sewer system. J. Gen. Microbiol. 129:1327–1333.

Millman, J. D., and E. Boyle. 1975. Biological uptake of dissolved silica in the Amazon River estuary. Science 189:995–997.

Mitchell, R. L. 1955. Trace elements, pp. 253–285. *In*: F. E. Bear, ed. Chemistry of the Soil. Reinhold, New York.

Parker, C. D. 1947. Species of sulfur bacteria associated with corrosion of concrete. Nature (Lond.)159:439.

Patrick, S., and A. J. Holding. 1985. The effect of bacteria on the solubilization of silica in diatom frustules. J. Appl. Bacteriol. 59:7–16.

Reeves, C. D., and B. E. Volcani. 1984. Role of silicon in diatom metabolism. Patterns of protein phosphorylation in *Cylindrotheca fusiformis* during recovery from silicon starvation. Arch. Microbiol. 137: 291–294.

Rogall, E. 1939. Ueber den Feinbau der Kieselmembran der Diatomeen. Planta 29:279–291.

Sand, W., and E. Bock. 1984. Concrete corrosion in the Hamburg sewer system. Env. Technol. Lett. 5:517–528.

Schwarz, K. 1973. A bound form of silicon in glycosaminoglycans and polyuronides. Proc. Natl. Acad. Sci. USA 70:1608–1612.

Sillen, L. G. 1967. The ocean as chemical system. Science 156:1189–1197.

Silverman, M. P., and E. F. Munoz. 1970. Fungal attack on rock: solubilization and altered infrared spectra. Science 169:985–987.

Stotzky, G. 1986. Influences of soil mineral colloids on metabolic processes, growth, adhesion, and ecology of microbes and viruses. pp. 305–428. *In*: P. M. Huang and M. Schnitzer. Interactions of Soil Minerals With Natural Organics and Microbes. Soil Science Society of America, Madison, WI.

Sullivan, C. W. 1971. A silicic acid requirement for DNA polymerase, thymidylate kinase, and DNA synthesis in the marine diatom *Cylindrica fusiformis*. Ph.D. thesis, University of California.

Sullivan, C. W., and E. B. Volcani. 1973. Role of silicon in diatom metabolism. The effects of silicic acid on DNA polymerase, TMP kinase and DNA synthesis in *Cyclotheca fusiformis*. Biochim. Biophys. Acta 308:212–229.

Tan, K. H. 1986. Degradation of soil minerals by organic acids. pp. 1–27. *In*: P. M. Huang and M. Schnitzer, eds. Interactions of Soil Minerals with Natural Organics and Microbes. SSSA Special Publication Number 17. Soil Science Society of America, Madison, WI.

Toth, S. J. 1955. Colloid chemistry of soils. pp. 85–106. *In*: F.E. Bear, ed. Chemistry of the Soil. Reinhold, New York.

Vandevivere, P., S. A. Welch, W. J. Ullman, and D. L. Kirchman. 1994.

Enhanced dissolution of silicate minerals by bacteria at near-neutral pH. Microb. Ecol. 27:241–251.

Waksman, S. A., and R. L. Starkey. 1931. The Soil and the Microbe. Wiley, New York.

Weaver, F. M., and S. W. Wise, Jr. 1974. Opaline sediments of the Southeastern Coastal Plain and horizon A: biogenic origin. Science 184: 899–901.

Weaver, T. L., and P. R. Dugan. 1972. Enhancement of bacterial methane oxidation by clay minerals. Nature (Lond.) 237:518.

Webley, D. M., R. B. Duff, and W. A. Mitchell. 1960. A plate method for studying the breakdown of synthetic and natural silicates by soil bacteria. Nature (Lond.) 188:766–767.

Webley, D. M., M. E. F. Henderson, and I. F. Taylor. 1963. The microbiology of rocks and weathered stones. J. Soil Sci. 14:102–112.

Weiss, A., G. Reiff, and A. Weiss. 1961. Zur Kenntnis wasserbestaendiger Kieselsaeureester. Z. Angorg. Chem. 311:151–179.

Welch, S. A., and W. J. Ullman. 1993. The effect of organic acids on plagioclase dissolution rates and stoichiometry. Geochim. Cosmochim. Acta 57:2725–2736.

Welch, S. A., and P. Vandevivere. 1995. Effect of microbial and other naturally occurring polymers on mineral dissolution. Geomicrobiol. J. 12:227–238.

Werner, D. 1966. Die Kieselsaeure im Stoffwechsel von *Cyclotella cryptica* Reimann, Lewin and Guillard. Arch. Mikrobiol. 55:278–308.

Werner, D. 1967. Hemmung der Chlorophyllsynthese und des NADP$^+$-abhaengigen Glyzerinaldehyd-3-phosphat-dehydrogenase durch Germaniumsaeure bei *Cyclotella cryptica*. Arch. Mikrobiol. 57: 51–60.

10

Geomicrobial Interactions with Phosphorus

10.1 BIOLOGICAL IMPORTANCE OF PHOSPHORUS

Phosphorus is an element fundamental to life, being a structural and functional component of all organisms. It is found universally in such vital cell constituents as nucleic acids, nucleotides, phosphoproteins, and phospholipids, and in teichoic and teichuronic acids in gram-positive bacteria, and phytins (also known as inositol phosphates) in plants. In many types of bacteria and some yeasts it may also be present intracellulary as polyphosphate granules. In the pentavalent state, the element is capable of forming anhydrides in the form of organic and inorganic pyrophosphates (see Fig. 6.2). In this valence state, phosphorus is also capable of forming anhydrides with organic acids and amines, and with sulfate (as in adenosine phosphosulfate). The phosphate anhydride bond serves to store biochemically useful energy. For example, a free energy change ($\Delta G°$) of -7.3 kcal (-30.6 kJ) per mole of adenosine 5'-triphosphate (ATP) is associated with the hydrolysis of its terminal anhydride bond, yielding adenosine 5'-diphosphate (ADP) + P_i. Unlike many anhydrides, some of those of phosphate, such as ATP, are unusually resistant to hydrolysis in the aqueous environment (Westheimer, 1987). Chemical hydrolysis of these bonds require 7 min of heating in dilute acid (e.g., 1 N HCl) at the temperature of boiling water (Lehninger, 1970, p. 290). At more neutral pH and physiological temperature, hydrolysis proceeds at an optimal rate only in the presence of appropriate enzymes (e.g., ATPase). The relative resistance of phosphate anhydride bonds to hydrolysis is attributable to the negative charges on the phosphates at neutral pH (Westheimer, 1987) and is the probable reason why ATP came to be selected in the evolution of life as a universal transfer agent of chemical energy in biological systems.

10.2　OCCURRENCE IN EARTH'S CRUST

Phosphorus is found in all parts of the biosphere. It occurs mostly in inorganic phosphates and in organic phosphate derivatives. Total phosphorus concentrations in mineral soil range from 35 to 5300 mg kg^{-1} (average 800 mg kg^{-1}) (Bowen, 1979). An average concentration in fresh water is 0.02 mg kg^{-1} (Bowen, 1979) and in seawater 0.09 mg L^{-1} (*Marine Chemistry*, 1971). The ratio of organic to inorganic phosphorus (P_{org}/P_i) varies widely in these environments. In mineral soil, P_{org}/P_i may range from 1:1 to 2:1 (Cosgrove, 1967, 1977). In lake water, as much as 50% of the organic fraction may be phytin (i.e., hydrolyzable by phytase) (Herbes et al., 1975). The organic phosphorus in such lake water may constitute 80 to 99% of the total soluble phosphorus. In the particular example cited by Herbes et al., the total organic phosphorus amounted to 40 μg phosphate per liter. They speculated that phosphatase-hydrolyzable compounds were largely absent because they are much more labile than phytins. Readily measurable phosphatase activity was detected in Tokyo, Sagami and Suruga Bays by Kobori and Taga (1979) and Taga and Kobori (1978).

10.3　CONVERSION OF ORGANIC INTO INORGANIC PHOSPHORUS AND THE SYNTHESIS OF PHOSPHATE ESTERS

An important source of free organic phosphorus compounds in the biosphere is the breakdown of animal and vegetable matter, as well as its extraction by living microbial cells (Shapiro, 1967) and by animals. Organically bound phosphorus is for the most part not directly available to living organisms because it cannot be taken into the cell in this form. To be taken up, it must first be freed from organic combination through mineralization. This is accomplished through hydrolytic cleavage catalyzed by phosphatases. In soil as much 70–80% of the microbial population may be able to participate in this process (Dommergues and Mangenot, 1970,

p. 266). Active organisms include bacteria such as *Bacillus megaterium, B. subtilis, B. malabarensis, Serratia* spp., *Proteus* spp., *Arthrobacter* spp., *Streptomyces* spp., and fungi such as *Aspergillus* sp., *Penicillium* sp., *Rhizopus* sp., *Cunninghamella* sp. (Dommergues and Mangenot, 1970, p. 266). These organisms secrete or liberate upon their death, phosphatases with greater or lesser substrate specificity (Skujins, 1967). Such activity has also been noted in the marine environment (Ayyakkannu and Chandramohan, 1971).

Phosphate liberation from phytin generally requires the enzyme phytase:

$$\text{Phytin} + 6H_2O \Rightarrow \text{inositol} + 6P_i \tag{10.1}$$

Phosphate liberation from nucleic acid requires the action of nucleases, which yield nucleotides, followed by the action of nucleotidases, which yield nucleosides and inorganic phosphate:

$$\text{Nucleic acid} \xrightarrow[+\ H_2O]{\text{nuclease}} \text{nucleotides} \xrightarrow[+\ H_2O]{\text{nucleotidase}} \text{nucleoside} + P_i$$

$$\tag{10.2}$$

Phosphate liberation from phosphoproteins, phospholipids, ribitol and glycerol phosphates requires phosphomono- and phosphodiesterases. The phosphodiesterases attack diester linkages while the phosphomonoesterases (phosphatases) attack monoester linkages (Lehninger, 1975, pp. 184, 323–325),

$$
R\!-\!O\!-\!\overset{\displaystyle O}{\underset{\displaystyle OH}{\overset{\|}{P}}}\!-\!O\!-\!R' \xrightarrow[H_2O]{\text{diesterase}} ROH
$$

$$
+ \; HO\!-\!\overset{\displaystyle O}{\underset{\displaystyle OH}{\overset{\|}{P}}}\!-\!O\!-\!R' \xrightarrow[H_2O]{\text{phosphatase}} HO\!-\!\overset{\displaystyle O}{\underset{\displaystyle OH}{\overset{\|}{P}}}\!-\!OH + HDR' \tag{10.3}
$$

Synthesis of organic phosphates (monomeric phosphate esters) is an intracellular process and normally proceeds through a reaction between a carbinol group and ATP in the presence of an appropriate kinase. For example,

$$\text{Glucose} + \text{ATP} \xrightarrow{\text{glucokinase}} \text{glucose 6-phosphate} + \text{ADP} \tag{10.4}$$

Phosphate esters in cells may also arise through phosphorolysis of certain polymers, such as starch or glycogen:

$$(Glucose)_n + H_3PO_4 \xrightarrow{\text{phosphorylase}} (glucose)_{n-1} \tag{10.5}$$
$$+ \text{ glucose 1-phosphate}$$

$$\text{Glucose 1-phosphate} \xrightarrow{\text{phosphoglucomutase}} \text{glucose 6-phosphate} \tag{10.6}$$

10.4 ASSIMILATION OF PHOSPHORUS

Adenosine 5′-triphosphate (ATP) may be generated from ADP by adenylate kinase:

$$2ADP \Rightarrow ATP + AMP \tag{10.7}$$

or by substrate-level phosphorylation, as in the reaction

$$\text{3-phosphoglyceraldehyde} + NAD^+$$
$$+ P_i \xrightarrow[\text{dehydrogenase}]{\text{triosephosphate}} \text{1,3-diphosphoglycerate} + NADH + H^+ \tag{10.8}$$

$$\text{1,3-diphosphoglycerate} + ADP \xrightarrow{\text{ADP kinase}} \text{3-phosphoglycerate} + ATP \tag{10.9}$$

It may also be generated by oxidative phosphorylation:

$$ADP + P_i \xrightarrow[\text{ATPase}]{\text{electron transport system}} ATP \tag{10.10}$$

or by photophosphorylation,

$$ADP + P_i \xrightarrow[\text{light, ATPase}]{\text{photosynthetic system}} ATP \tag{10.11}$$

Phosphate polymers are generally produced through reactions such as the following:

$$(Polynucleotide)_{n-1} + \text{nucleotide triphosphate}$$
$$\xrightarrow{\text{polymerase}} (polynucleotide)_n + P{\sim}P \tag{10.12}$$

$$P{\sim}P \xrightarrow{\text{pyrophosphatase}} 2 P_i \tag{10.13}$$

Inorganic pyrophosphate (P~P) was recently reported to be able to serve as an energy source to some bacteria (Liu et al., 1982; Varma et al, 1983). Whereas it is easy to conceive that this ability can be of great importance in the sense of energy conservation for intracellularly formed pyrophosphate in bacteria, it remains to be clarified how important this may be for extracellularly available pyrophosphate in nature. Liu et al. (1982) found gram-positive and gram-negative, motile and nonmotile bacteria in pyrophosphate enrichments from freshwater anaerobic environments which grew at the expense of the pyrophosphate as energy source. Nothing appears to be known about the mechanism of pyrophosphate uptake by these organisms.

Like pyrophosphate, intracellular, inorganic polyphosphate granules formed by some microbial cells are a form of metaphosphate and can represent an energy storage compound (e.g., see van Groenestijn et al., 1987) as well as a phosphate reserve. In the case of the cyanobacterium *Anabaena cylindrica* it may also play a role as a detoxifying agent by combining with aluminum ions that are taken into the cell (Pettersson et al., 1985).

10.5 MICROBIAL SOLUBILIZATION OF PHOSPHATE MINERALS

Inorganic phosphorus may occur in soluble and insoluble form in nature. The most common inorganic form is orthophosphate (H_3PO_4). As an ionic species, the concentration of phosphate is controlled by its solubility in the presence of an alkaline earth cation such as Ca^{2+} and Mg^{2+}, and of metal cations ions such as Fe^{2+}, Fe^{3+} and Al^{3+} at appropriate pH values (see Table 10.1). In seawater, for instance, the soluble phosphate concentration (about 3×10^{-6} M, maximum) is controlled by Ca^{2+} ions (4.1×10^2 mg L^{-1}), which form hydroxyapatite with phosphate in the prevailing pH range of 7.4–8.1.

Insoluble phosphate occurs most commonly in the form of apatite $[Ca_5(PO_4)_3(F,Cl,OH)]$, in which the (F,Cl,OH) radical may be represented exclusively by F, Cl, or OH or any combinations of these. In soil, insoluble

Table 10.1 Solubility Products of Some Phosphate Compounds

Compound	K_2	Reference
$CaHPO_4 \cdot 2H_2O$	2.18×10^{-7}	Kardos (1955), p. 185
$Ca_{10}(PO_4)_6(OH)_2$	1.53×10^{-112}	Kardos (1955), p. 188
$Al(OH)_2HPO_4$	2.8×10^{-29}	Kardos (1955) p. 184
$FePO_4$	1.35×10^{-18}	From ΔG of formation

phosphate may also occur as an aluminum salt (e.g., variscite, Al-$PO_4\cdot2H_2O$) or the iron salts strengite ($FePO_4\cdot2H_2O$) or vivianite $[Fe_3(PO_4)_2\cdot8H_2O]$.

Insoluble forms of inorganic phosphorus (calcium, aluminum and iron phosphates) may be solubilized through microbial action. The mechanism by which the microbes accomplish this solubilization varies. It may be by (1) the production of inorganic or organic acids that attack the insoluble phosphates; (2) the production of chelators, such as gluconate and 2-ketogluconate (Duff and Webley, 1959; Banik and Dey, 1983; Babu-Khan et al., 1995; see also Chapter 9), citrate, oxalate, and lactate, all of which can complex the cationic portion of the insoluble phosphate salts and thus force their dissociation; (3) the reduction of iron in ferric phosphate (strengite) to ferrous iron by dissimilatory iron reduction in sediment-water systems (Jansson, 1987); and (4) the production of hydrogen sulfide (H_2S) which can react with the iron in iron phosphate and precipi-

Table 10.2 Some Organisms Active in Phosphate Solubilization

Organism	Mechanism solubilization	References
B. megatherium	H_2S production FeS precipitation	Sperber (1958b); Swaby and Sperber (1958)
Thiobacillus sp.	H_2SO_4 production from sulfur	Lipman and McLean (1916)
Nitrifying bacteria	NH_3 oxidation to HNO_3	Dommergues and Mangenot (1970), p. 263
Pseudomonads, Arthrobacter, and *Erwinia*like bacterium	Chelate production; glucose converted to gluconate or 2-ketogluconate	Duff et al. (1963); Sperber (1958a); Dommergues and Mangenot (1970), p. 262; Babu-Khan et al. (1995)
Sclerotium	?	Dommergues and Mangenot (1970), p. 262
A. niger, A. flavus, Sclerotium rolfsii, Fusarium oxysporum, Cylindrosporium sp., and *Penicillium* sp.	organic acid production (e.g., citric acid)	Agnihotri (1970)

tate it as iron sulfide, thereby liberating phosphate, as in the reaction

$$2FePO_4 + 3H_2S \Rightarrow 2FeS + 2H_3PO_4 + S° \qquad (10.14)$$

Table 10.2 lists some organisms active in phosphate solubilization.

Solubilization of phosphate minerals has been noted directly in soil (Alexander, 1977; Babenko et al., 1984; Chatterjee and Nandi, 1964; Dommergues and Mangenot, 1970; Patrick et al., 1973). Indeed, soil containing significant amounts of immobilized calcium, aluminum or iron phosphates has been thought to benefit from inoculation with phosphate mobilizing bacteria (see, for instance, discussion by Dommergues and Mangenot, 1970, p. 262). Important microbial phosphate solubilizing activity in soil occurs in rhizospheres (Alexander, 1977), probably because root secretions allow phosphate-solubilizing bacteria to generate sufficient acid or ligands to effect dissolution of calcium and other insoluble phosphates. Phosphate-deficient soil may be profitably fertilized with insoluble inorganic phosphate rather than soluble phosphate salts because the former will be solublized slowly and thus will be better conserved than soluble phosphate salts, which can be readily leached. Soluble phosphate in soil may consist not only of orthophosphate but also of pyrophosphate (metaphosphate). The latter is readily hydrolyzed by pyrophosphatase, especially in flooded soil (Racz and Savant, 1972).

10.6 MICROBIAL PHOSPHATE IMMOBILIZATION

Microorganisms can cause fixation or immobilization of phosphate, either by promoting the formation of inorganic precipitates or by assimilation into organic cell constituents or intracellular polyphosphate granules. The latter two processes have been called transitory immobilization by Dommergues and Mangenot (1970) because of the ready resolubilization through mineralization upon death of the cell. In soil and freshwater environments, transitory immobilization is often the more important, although fixation of phosphate by Ca^{2+}, Al^{3+}, or Fe^{3+} is recognized. In a few marine environments (coastal waters or shallow seas) where phosphorite deposits occur, the precipitation mechanism may be more important.

Phosphorite Deposition Phosphorite (Fig. 10.1) in nature may form authigenically or diagenetically. In the first case, the phosphorite forms as a result of a reaction of soluble phosphate with calcium ions forming corresponding insoluble calcium phosphate compounds. In diagenesis, phosphate may replace carbonate in calcareous concretions. The role of microbes in these processes may be one or more of the following: (1) making

Figure 10.1 Micronodules of phosphorite (phosphatic pellets) from the Peru shelf. The average diameter of such pellets is 0.25 mm. According to Burnett (personal communication), such pellets are more representative of what is found in the geologic record than the larger phosphorite nodules. (Courtesy of William C. Burnett.)

reactive phosphate available, (2) making reactive calcium available, or (3) generating or maintaining the pH and redox conditions which favor phosphate precipitation.

Authigenic Formations   Models of authigenic phosphorite genesis assume mineralization of organic phosphorus where biologically productive waters occur, such as at ocean margins overlying shallow continental slopes, shelf areas, or plateaus. Here detrital accumulations may be mineralized at the sediment/water interface and in interstitial pore waters, liberating phosphate, some of which may then interact chemically with calcium in seawater to form phosphorite grains. These may subsequently be redistributed within the sediment units (e.g., Riggs, 1984; Mullins and Rasch, 1985). Dissolution of fish debris (bones) has also been considered an important source of phosphate in authigenic phosphorite genesis (Suess, 1981). Upwelling probably plays an important role in many cases of authigenic phosphorite formations at latitudes on western continental margins in both the northern and southern hemispheres, where prevailing winds (e.g., trade winds) cause upwelling (see, for instance, discussion by Burnett et al., 1982; Jahnke et al., 1983; Riggs, 1984). Nathan et al. (1993) cite evidence that in the southern Benguela upwelling system (Cape Peninsula, western coast of South Africa) during nonupwelling periods in winter, phosphate-sequestering bacteria of the oxidative genera *Pseudomonas* and *Acinetobacter* become dominant in the water column, whereas the fermentative *Vibrio* and Enterobacteriacea are dominant during upwelling in summer. It has been suggested that *Pseudomonas* and *Acinetobacter*, which sequester phosphate as polyphosphate under aerobic conditions and hydrolyze the polyphosphate under anaerobic conditions to obtain energy of maintenance and to sequester volatile fatty acids for polyhydroxybutyrate formation, contribute to authigenic phosphorite formation. Locally elevated, excreted orthophosphate becomes available for precipitation as phosphorite by reacting with seawater calcium. In the northern Benguela upwelling system off the coast of Namibia where upwelling occurs year-around, Nathan et al. (1993) found that phosphate-sequestering cocci occurred in the water column and suggest that these organisms, like *Pseudomonas* and *Acinetobacter*, may release sequestered phosphate when they reach waters with low oxygen concentration below 10-m water depth and thereby contribute to phosphorite formation.

Authigenic phosphorite formation at some eastern continental margins, where upwelling, if it occurs at all, is a weak and intermittent process, may have been formed more directly as a result of intracellular bacterial phosphate accumulation, which became transformed into carbonate fluorapatite upon death of the cells and accumulated in sediments in areas where the sedimentation rate was very low (O'Brien and Veeh, 1980; O'Brien et al., 1981). Ruttenberg and Berner (1993) concluded that carbonate fluorapatite accumulations in Long Island Sound and Mississippi Delta sediments are the result of mineralization of organic phosphorus. These accumulations increased as organic phosphorus concentrations decreased with depth. Thus, important phosphorus sinks occur in sediments of continental margins outside upwelling regions.

Youssef (1965) has proposed that phosphorite could be formed in a marine setting through mineralization of phytoplankton remains which have settled into a depression on the sea floor, leading to localized accumulation of dissolved phosphate, which then precipitated as a result of reaction with calcium in seawater. Piper and Codespoti (1975) have proposed that carbonate fluorapatite $[Ca_{10}(PO_4,CO_3)_6F_{2-3}]$ precipitation in the marine environment may be dependent on bacterial denitrification in the oxygen minimum layer of the ocean as it intersects with the ocean floor. A loss of nitrogen due to denitrification means lowered biological activity and can lead to excess accumulation of phosphate in this zone. The more acid conditions (pH 7.4–7.9) in the deeper waters keep phosphate dissolved and allow for its transport by upwelling to regions where phosphate precipitation is favored (pH > 8) (Fig. 10.2). This model takes into account the conditions of marine apatite formation described by Gul-

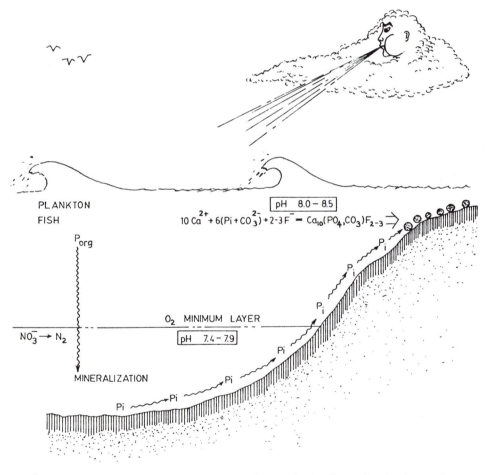

Figure 10.2   Schematic representation of phosphorite formation in the marine environment according to the model of Piper and Codispoti (1975). Note that the rising of P_i upslope is due to upwelling.

brandsen (1969) and helps to explain the occurrence of probable contemporary formation of phosphorite in regions of upwelling such as the continental margin of Peru (Veeh et al., 1973; see however, Suess, 1981) and on the continental shelves of southwestern Africa (Baturin, 1973; Baturin et al., 1969). To explain the more extensive ancient phosphorite deposits, a periodic warming of the ocean can be invoked, which would reduce oxygen solubility and favor more intense denitrification in deeper waters resulting in temporarily lessened biological activity and a consequent increase in phosphate concentration leading to phosphate precipitation (Piper and Codespoti, 1975). Mullins and Rasch (1985) proposed an oxygen-depleted, sedimentary environment for biogenic apatite formation along the continental margin of central California during the Miocene. They view oolitic phosphorite as having resulted from organic matter mineralization by sulfate reducers in sediments in which dolomite was concurrently precipitated. The phosphate, according to their model, then tended to precipitate interstitially as phosphorite, in part around bacterial nuclei. Mullins and Rasch (1985) found fossilized bacteria in the phosphorite. O'Brien et al (1981) had previously reported the discovery of fossilized bacteria in a phosphorite deposit on the East Australian margin.

Diagenetic Formation Models of phosphorite formation through diagensis generally assume an exchange of phosphate for carbonate in calcium carbonate accretions in the form of calcite or aragonite. The role of bacteria in this process is to mobilize phosphate by mineralizing detrital organic matter. This has been demonstrated in marine and freshwater environments under laboratory conditions (Lucas and Prevot, 1984; Hirschler et al., 1990a,b). Adams and Burkhart (1967) propose that diagenesis of calcite to form apatite explains the origin of some deposits in the North Atlantic. Phosphorite deposits off Baja California and in a core from the eastern Pacific Ocean seem to have formed as a result of partial diagenesis (d'Anglejan, 1967, 1968).

Occurrences of Phosphorite Deposits Sizable phosphorite deposits are associated with only six brief geological intervals: the Cambrian, Ordovician, Devonian-Mississippian, Permian, Cretaceous, and Cenozoic eras. Since in many instances these phosphorite deposits are associated with black shales and contain uranium in reduced form (Altschuler et al., 1958), they are presumed to have accumulated under reducing conditions. Apatite appears to be forming at the present time in the sediments at the Mexican continental margin (Jahnke et al., 1983) and in the deposits off the coast of Peru (e.g., Burnett et al., 1982; Suess, 1981; Veeh et al., 1973).

Deposition of Other Phosphate Minerals   Microbes can also play a role
in the authigenic or diagenetic formation of other phosphate minerals such
as vivianite, strengite, and variscite. In these instances the bacteria may
generate orthophosphate by degrading organic phosphate in detrital matter
and/or by mobilizing iron or aluminum in minerals. Authigenic formation
of such phosphate minerals is probably most common in soil. A case of
diagenetic formation of vivianite from siderite ($FeCO_3$) in the North Atlantic coastal plain has been proposed by Adams and Burkhart (1967). Microbial control of pH and E_h can influence the stability of these phosphates
(Patrick et al.. 1973; Williams and Patrick, 1971).

   Citrobacter sp. has been reported to form metal phosphate precipitates, e.g., Cd phosphate ($CdHPO_4$) and uranium phosphate (UO_2HPO_4),
that encrust the cells (Macaskie et al., 1987, 1992). The precipitates form
as a result of the action of a cell-bound, metal-resistant phosphatase on
organophosphates such as glyccerol-2-phophate, liberating orthophosphate (HPO_4^{2-}) that reacts in the immediate surround of the cells with
metal cations to form corresponding metal phosphates. The metal phosphates deposit on the cell surface.

   Some microbes, such as certain strains of *Arthrobacter, Flavobacterium, Listeria, and Pseudomonas*, can cause struvite ($MgNH_4PO_4 \cdot 6H_2O$)
formation, at least under laboratory conditions (Rivadeneyra et al., 1983,
1992). A major, but not necessarily exclusive, microbial contribution to
this process appears to be ammonia formation (Rivadeneyra et al., 1992).
Struvite forms in seawater solutions in which the NH_4^+ concentration is
0.01 M when orthophosphate is added at pH 8.0 (Handschuh and Orgel,
1973). The presence of calcium ion in sufficient quantity can suppress
struvite formation and promote apatite formation instead (Rivadeneyra et
al., 1983). Although struvite formation is probably of little significance in
nature today, it may have been in the primitive world of Precambrian times
if NH_4^+ was present at concentrations as high as 10^{-2} M (Handschuh and
Orgel, 1973).

10.7   MICROBIAL REDUCTION OF OXIDIZED FORMS OF PHOSPHORUS

Phosphorus may also undergo redox reactions, some or all of which may
be catalyzed by microbes. Of biogenic interest are the $+5$, $+3$, $+1$, and
-3 oxidation states, as in orthophosphate (H_3PO_4), orthophosphite
(H_3PO_3), hypophosphite (H_3PO_2), and phosphine (PH_3), respectively. Reduction of phosphate to phosphine by soil bacteria has been reported
(Rudakov, 1927; Tsubota, 1959, Devai et al., 1988). Mannitol appeared

to be a suitable electron donor in the reaction described by Rudakov (1927) and glucose in the experiments described by Tsubota (1959). Phosphite and hypophosphite were claimed to be intermediates in the reduction (Rudakov, 1927; Tsubota, 1959). Devai et al. (1988) detected phosphine evolution in anaerobic sewage treatment in Imhoff tanks in Hungary and confirmed it in anaerobic laboratory experiments. Iron phosphide (Fe_3P_2) is reported to have been formed when a cell-free preparation of *Desulfovibrio* was incubated in the presence of steel in a yeast extract broth under hydrogen gas (Iverson, 1968). Hydrogenase from *Desulfovibrio* may have been responsible for the formation of phosphine from phosphate contained in the yeast extract. The phosphine could then have reacted with ferrous iron to form the Fe_3P_2 (Iverson, 1968).

Questions have been raised about the ability of microbes to reduce phosphate. Liebert (1927) showed that on the basis of thermodynamic calculations using heats of formation, the reduction of phosphate to phosphite by mannitol is an energy-consuming process and could therefore not serve a respiratory function. He calculated a heat-of-reaction value (ΔH) of $+20$ kcal on the basis of the following equation:

$$C_6H_{14}O_6 + 13Na_2HPO_4 \Rightarrow 13Na_2HPO_3 + 6CO_2 + 7H_2O$$

$$\text{316 kcal} \qquad \text{5,390 kcal} \qquad \text{4,460 kcal} \qquad \text{566 kcal} \qquad \text{478 kcal}$$

$$(10.15)$$

He also calculated a ΔH of $+483$ kcal for the reduction of phosphate to hypophosphite and a ΔH of $+1,147$ kcal for the reduction of phosphate to phosphine. These same conclusions can also be reached when free-energy changes (ΔG) are considered. In 1962, Woolfolk and Whiteley reported that they were unable to reduce phosphate with hydrogen in the presence of an extract of *Veillonella alcalescens* (formerly *Micrococcus lactilyticus*), even though this extract could catalyze the reduction of some other oxides with hydrogen. Skinner (1968) also questioned the ability of bacteria to reduce phosphate. He could not find such organisms in soils he tested. However, Burford and Bremner (1972), while unable to demonstrate phosphine evolution from water logged soils, were not able to rule out microbial phosphine genesis because they found that soil constituents can adsorb phosphine. Thus, unless bacteria form phosphine in excess of the adsorption capacity of a soil, phosphine detection in the gas phase may not be possible.

Interestingly, Barrenscheen and Beckh-Widmanstetter (1923) reported the production of hydrogen phosphide (phosphine, PH_3) from organically bound phosphate during putrefaction of beef blood. Much more recently, Metcalf and Wanner (1991) presented evidence supporting the existence of a C-P lyase in *Escherichia coli* that catalyzes the reductive cleavage of compounds such as methylphosphonate to phosphite and methane,

$$
\underset{P\,(+5)}{\overset{O}{\underset{\parallel}{\underset{OH}{\overset{\parallel}{HO-P-CH_3}}}}} + 2(H) \xrightleftharpoons[\text{C-P lyase}]{} \underset{P\,(+3)}{\overset{O}{\underset{\parallel}{\underset{OH}{\overset{\parallel}{HO-P-H}}}}} + CH_4 \qquad (10.16)
$$

This enyme activity was previously studied in *Agrobacterium radiobacter* (Wackett et al., 1987), although it was described by these authors as a hydrolytic enzyme. Phosphonolipids are known to exist in organisms from bacteria to mammals (Hilderbrand and Henderson, 1989, cited by Metcalf and Wanner, 1991). Thus biochemical mechanisms for synthesizing organophosphonates exist, and it is therefore highly likely that an organophosphonate such as methyl or ethylphosphonate that requires C-P lyase to release the phosphorus (Metcalf and Wanner, 1991) is an intermediate in the conversion of orthophosphate to orthophosphite, and that the C-P lyase activity represents the reductive step in this transformation. This needs further investigation.

10.8   MICROBIAL OXIDATION OF REDUCED FORMS OF PHOSPHORUS

Reduced forms of phosphate may be aerobically and anaerobically oxidized by bacteria. Thus *Bacillus caldolyticus*, a thermophile, can oxidize hypophosphite to phosphate aerobically (Heinen and Lauwers, 1974). The active enzyme system consists of an $(NH_4)_2SO_4$-precipitable protein fraction, NAD, and respiratory chain components. The enzyme system does not oxidize phosphite. Adams and Conrad (1953) first reported the aerobic oxidation of phosphite to phosphate by bacteria and fungi from soil. All phosphite that was oxidized by these strains was assimilated. None was released into the medium before death of the organisms. Phosphate added to the medium inhibited phosphite oxidation. Active organisms included *P. fluorescens*, *P. lachrymans*, *Aerobacter* (now known as *Enterobacter*) *aerogenes*, *Erwinia amylovora*, *Alternaria*, *Aspergillus niger*, *Chaetomium*, *Penicillium notatum*, and some actinomycetes. In later studies, Casida (1960) found that a culture of *P. fluorescens* strain 195 formed orthophosphate aerobically from orthophosphite in excess of its needs and released phosphate into the medium. The culture was heterotrophic, and its phosphite-oxidizing activity was inducible and stimulated by yeast extract. The enzyme system involved in phosphite oxidation was an ortho-phosphite-nicotinamide adenine dinucleotide oxidoreductase, which was inactive on arsenite, hypophosphite, nitrite, selenite or tellurite and was inhibited by sulfite (Malacinski and Konetzka, 1966, 1967).

Oxidation of reduced phosphorus compounds can also occur anaerobically. A soil bacillus has been isolated which is capable of anaerobic oxidation of hypophosphite and phosphite to phosphate (Foster et al., 1978). In a mixture of phosphite and hypophosphite, phosphite was oxidized first. Phosphate inhibited the oxidation of either phosphite or hypophosphite. The organism did not release phosphate into the medium.

Since phosphite and hypophosphite have not been reported in detectable quantities in the natural environment, it has been suggested that microbial ability to utilize the compounds, especially anaerobically, may be a vestigial property developed at a time when the Earth had a reducing atmosphere surrounding it which favored the occurrence of phosphite (Foster et al., 1978).

10.9 MICROBIAL ROLE IN THE PHOSPHORUS CYCLE

In many ecosystems, phosphorus availability may determine the extent of microbial growth and activity. The element follows cycles in which it

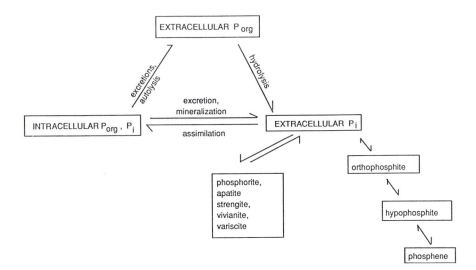

Figure 10.3   The phosphorus cycle.

finds itself alternately outside and inside living cells, in organic and inorganic form, free or fixed, dissolved or precipitated. Microbes play a central role in these changes of state, as outlined in Figure 10.3 and as discussed in this chapter.

10.10 SUMMARY

Phosphorus is a very important element for all forms of life. It is used in cell structure as well as cell function. It plays a role in transducing biochemically useful energy. When free in the environment, it occurs primarily as organic phosphate esters and as inorganic phosphates. Some of the latter, such as calcium, aluminum, and iron phosphates, are very insoluble at neutral or alkaline pH. To be nutritionally utilizable, organic phosphates have to be enzymatically hydrolyzed to liberate the orthophosphate. Microbes play a central role in this process. Microbes may also free orthophosphate from insoluble inorganic phosphates by production of organic or mineral acids or chelators, or, in the case of iron phosphates, the production of H_2S. Under some conditions, microbes may promote the formation of insoluble inorganic phosphates, such as those of calcium, aluminum, or iron. They have been implicated in phosphorite formation in the marine environment.

Microbes have been implicated in the reduction of pentavalent phosphorus to lower valence states. The experimental evidence for this is somewhat equivocal, however. It is likely that organic phosphorus compounds are intermediates in these reductions. Microbes have also been implicated in the oxidation of reduced forms of phosphorus to phosphate. The experimental evidence in this case is strong. It includes demonstration of enzymatic involvement. The geomicrobial significance of these redox reactions in nature is not clearly understood at present.

REFERENCES

Adams, J. K., and B. Burkhart. 1967. Diagenetic phosphates from the northern Atlantic coastal plain. Abstr., Annu. Geophys. Soc. Amer. and Assoc. Soc., Joint Meet., New Orleans, LA, Nov. 20 to 22, 1967. Program, p. 2.

Adams, F., and J. P. Conrad. 1953. Transition of phosphite to phosphate in soil. Soil Sci. 75:361–371.

Agnihotri, V. P. 1970. Solubilization of insoluble phosphates by some soil fungi isolated from nursery seedbeds. Can. J. Microbiol. 16:877–880.

Alexander, M. 1977. Introduction to Soil Microbiology. 2nd ed. Wiley, New York.

Ayyakkannu, K., and D. Chandramohan. 1971. Occurrence and distribution of phosphate solubilizing bacteria and phosphatase in marine sediments at Porto Nova. Mar. Biol. 11:201–205.

Babenko, Yu. S., G. I. Tyrygina, E. F. Grigor'ev, L. M. Dolgikh, and T. I. Borisova. 1984. Biological activity and physiological-biochemical properties of phosphate dissolving bacteria. Mikrobiologiya 53: 533–539 (Engl. transl. pp. 427–433).

Banik, S., and B. K. Dey. 1983. Phosphate solubilizing potentiality of the microorganisms capable of utilizing aluminum phosphate as a sole phosphate source. Z. Mikrobiol. 138:17–23.

Barrenscheen, H. K., and H. A. Beckh-Widmanstetter. 1923. Ueber bakterielle Reduktion organisch gebundener Phopshorsaeure. Biochem. Z. 140:279–283.

Babu-Khan, S., T. C. Yeo, W. L. Martin, M. R. Duran, R. D. Rogers, and A. H. Goldstein. 1995. Cloning of a mineral phosphate-solubilizing gene from *Pseudomonas cepacia*. Appl. Environ. Microbiol. 61: 972–978.

Baturin, G.N. 1973. Genesis of phosphorites of the Southwest African shelf. Tr. Inst. Okeanol. Akad. Nauk SSSR 95:262–267.

Baturin, G. N., K. I. Merkhulova, and P. I. Chalov. 1969. Radiometric evidence for recent formation of phosphatic nodules in marine shelf sediments. Mar. Geol. 13:M37-M41.

Bowen, H. J. M. 1979. Environmental Chemistry of the Elements. Academic Press, London.

Burford, J. R., and J. M. Bremner. 1972. Is phosphate reduced to phosphine in waterlogged soils? Soil Biol. Biochem. 4:489–495.

Burnett, W. C., M. J. Beers, and K. K. Roe. 1982. Growth rates of phosphate nodules from the continental margin off Peru. Science 215: 1616–1618.

Casida, L. E., Jr. 1960. Microbial oxidation and utilization of orthophosphite during growth. J. Bacteriol. 80:237–241.

Chatterjee, A. K., and P. Nandi. 1964. Solubilization of insoluble phosphates by rhizosphere fungi of legumes (*Arachis cyanopsis, Desmodium*). Trans. Bose Res. Inst. (Calcutta) 27:115–120.

Cosgrove, D. J. 1967. Metabolism of organic phosphates in soil. pp. 216–228. *In*: A. D. McLaren and G. H. Petersen, eds. Soil Biochemistry. Vol. 1. Marcel Dekker, New York.

Cosgrove, D. J. 1977. Microbial transformations in the phosphorus cycle. Adv. Microb. Ecol. 1:95–134.

d'Anglejan, B. F. 1967. Origin of marine phosphates off Baja California, Mexico. Marine Geol. 5:15–44.

d'Anglejan, B. F. 1968. Phosphate diagenesis of carbonate sediments as a mode of *in situ* formation of marine phosphorites: observation in a core from the Eastern Pacific. Can. J. Earth Sci. 5:81–87.

Devai, I., L. Feldoeldy, I. Wittner, and S. Plosz. 1988. Detection of phosphine: new aspects of the phosphorus cycle in the hydrosphere. Nature (Lond.) 333:343–345.

Dommergues, Y., and F. Mangenot. 1970. Ecologie Microbienne du Sol. Masson, Paris.

Duff, R. B., and D. M. Webley. 1959. 2-Ketoglucuronic acid as a natural chelator produced by soil bacteria. Chem. Ind., pp. 1376–1377.

Duff, R. B., D. M. Webley, and R. O. Scott. 1963. Solubilization of minerals and related materials by 2-ketogluconic acid-producing bacteria. Soil Sci. 95:105–114.

Foster, T.L., L. Winans Jr., and J. S. Helms. 1978. Anaerobic utilization of phosphite and hypophosphite by *Bacillus* sp. Appl. Environ. Microbiol. 35:937–944.

Gulbrandsen, R. A. 1969. Physical and chemical factors in the formation of marine apatite. Econ. Geol. 64:365–382.

Handschuh, G. J., and L. E. Orgel. 1973. Struvite and prebiotic phosphorylation. Science 179:483–484.

Heinen, W., and A. M. Lauwers. 1974. Hypophosphite oxidase from *Bacillus caldolyticus*. Arch. Microbiol. 95:267–274.

Herbes, S. E., H. E. Allen, K. H. Mancy. 1975. Enzymatic characterization of soluble organic phosphorus in lake water. Science 187: 432–434.

Hilderbrand, R. L., and T. O. Henderson. 1989. Phosphonic acids in nature. *In*: R. L. Hilderbrand, ed. The Role of Phosphonates in Living Systems. CRC Press, Boca Raton, FL, pp. 6–25.

Hirschler, A., J. Lucas, and J-C. Hubert. 1990a. Bacterial involvement in apatite genesis. FEMS Microbiol. Ecol. 73:211–220.

Hirschler, A., J. Lucas, and J-C. Hubert. 1990b. Apatite genesis: A biologically induced or biologically controlled mineral formation process? Geomicrobiol. J. 7:47–57.

Iverson, W. P. 1968. Corrosion of iron and formation of iron phosphide by *Desulfovibrio desulfuricans*. Nature (Lond.) 217:1265.

Jahnke, R. A., S. R. Emerson, K. K. Roe, and W. C. Burnett. 1983. The present day formation of apatite in Mexican continental margin sediments. Geochim. Cosmochim. Acta 47:259–266.

Jansson, M. 1987. Anaerobic dissolution of iron-phosphorus complexes in sediment due to the activity of nitrate-reducing bacteria. Microb. Ecol. 14:81–89.

Kardos, L. T. 1955. Soil fixation of plant nutrients, pp. 177–199. *In*: F.E. Bear, ed. Chemistry of the Soil. Reinhold, New York.

Kobori, H., and N. Taga. 1979. Phosphatase activity and its role in the mineralization of organic phosphorus in coastal sea water. J. Exp. Mar. Biol. Ecol. 36:23–39.

Lehninger, A. L. 1970. Biochemistry. 1st ed. Worth Publishers, New York.

Lehninger, A. L. 1975. Biochemistry. 2nd ed. Worth Publishers, New York.

Liebert, F. 1927. Reduzieren Mikroben Phosphate? Z. Bakt. Parasitenk. Infrektionskr. Hyg. 72:369–374.

Lipman, J. G., and H. McLean. 1916. The oxidation of sulfur soils as a means of increasing the availability of mineral phosphates. Soil Sci. 1:533–539.

Liu, C. L., N. Hart, and H. D. Peck, Jr. 1982. The utilization of inorganic pyrophosphate as a source of energy for the growth of sulfate reducing bacteria, *Desulfotomaculum* sp. Science 217:363–364.

Lucas, J., and L. Prévot. 1984. Synthese de l'apatite par voie bactérienne a partir de matière organique phosphatée et de divers carbonates de calcium dans des eaux douce et marine naturelles. Chem. Geol. 42: 101–118.

Macaskie, L. E., A. C. R. Dean, A. K. Cheetham, R. J. B. Jakeman, and A. J. Skarnulis. 1987. Cadmium accumulation by a *Citrobacter* sp.: The chemical nature of the accumulated metal precipitate and its location on the bacterial cells. J. Gen. Microbiol. 133:539–544.

Macaskie, L. E., R. M. Empson, A. K. Cheetham, C. P. Grey, and A. J. Skarnulis. 1992. Uranium bioaccumulation by a *Citrobacter* sp. as a result of enzymatically mediated growth of polycrystalline HUO_2PO_4. Science 257:782–784.

Malacinski, G., and W. A. Konetzka. 1966. Bacterial oxidation of orthophosphite. J. Bacteriol. 91:578–582.

Malacinski, G., and W. A. Konetzka. 1967. Orthophosphite-nicotinamide adenine dinucleotide oxidoreductase from *Pseudomonas fluorescens*. J. Bacteriol. 93:906–1910.

Marine Chemistry. 1971. A report of the Marine Chemistry Panel of the Committee on Oceanography. National Academy of Sciences, Washington, DC.

McConnell, D. 1965. Precipitation of phosphates in sea water. Econ. Geol. 60:1059–1062.

Metcalf, W. W., and B. L. Wanner. 1991. Involvement of the *Escherichia coli phn (psiD)* gene cluster in assimilation of phosphorus in the form of phosphonates, phosphite, P_i esters, and P_i. J. Bacteriol. 173: 587–600.

Mullins, H. T., and R. F. Rasch. 1985. Sea-floor phosphorites along the Central California continental margin. Econ. Geol. 80:696–715.

Nathan, Y., J. M. Bremner, R. E. Loewenthal, and P. Monteiro. 1993. Role of bacteria in phosphorite genesis. Geomicrobiol. J. 11:69–76.

O'Brien, G. W. O., and H. H. Veeh. 1980. Holocene phosphorite on the East Australian margin. Nature (Lond.) 288:690–692.

O'Brien, G. W. O., J. R. Harris, A. R. Milnes, and H. H. Veeh. 1981. Bacterial origin of East Australian continental margin phosphorites. Nature (Lond.) 294:442–444.

Patrick, W. H. Jr., S. Gotoh, and B. G. Williams. 1973. Strengite dissolution in flooded soils and sediment. Science 179:564–565.

Pettersson, A., L. Kunst, B. Bergman, and G.M. Roomans. 1985. Accumulation of aluminum by *Anabaena cyclindrica* into polyphosphate granules and cell walls: an X-ray energy-dispersive microanalysis study. J. Gen. Microbiol. 131:2545–2548.

Piper, D. Z., and L. A. Codespoti. 1975. Marine phosphorite deposits and the nitrogen cycle. Science 179:564–565.

Racz, G. J., and N. K. Savant. 1972. Pyrophosphate hydrolysis in soil as influenced by flooding and fixation. Soil Sci. Soc. Am. Proc. 36: 678–682.

Riggs, S. R. 1984. Paleooceanographic model of neogene phosphorite deposition, U.S. Atlantic Continental Margin. Science 223:123–131.

Rivadeneyra, M. A., A. Ramos-Cormenzana, and A. Garcia-Cervigon. 1983. Bacterial formation of struvite. Geomicrobiol. J. 3:151–163.

Rivadeneyra, M. A., I. Perez-Garcia, and A. Ramos-Cormenzana. 1992a. Struvite precipitation by soil and fresh water bacteria. Curr. Microbiol. 24:343–347.

Rivadeneyra, M. A., I. Perez-Garcia, and A. Ramos-Cormenzana. 1992b. Influence of ammonium ion on bacterial struvite production. Geomicrobiol. J. 10:125–137.

Rudakov, K. J. 1927. Die Reduktion der mineralischen Phosphate auf biologischem Wege. II. Mitteilung. Z. Bakt. Parasitenk., Infektionskr. Hyg. Abt. II 79:229–245.

Ruttenberg, K. C., and R. A. Berner. 1993. Authigenic apatite formation and burial in sediments from non-upwelling, continental margin environments. Geochim. Cosmochim. Acta 57:991–1007.

Shapiro, J. 1967. Induced rapid release and uptake of phosphate by microorganisms. Science 155:1269–1271.

Skinner, F. A. 1968. The anaerobic bacteria of soil, pp. 573–592. *In*: T. R. G. Gray and D. Parkinson, eds. The Ecology of Soil Bacteria. University of Toronto Press, Toronto, Ontario, Canada.

Skukjins, J. J. 1967. Enzymes in soil. *In*: A. D. McLaren and G. H. Peterson, eds. Soil Biochemistry. Vol. 1. Marcel Dekker, New York, pp. 371–414.

Sperber, J. I. 1958a. The incidence of apatite-solubilizing organisms in the rhizosphere and soil. Aust. J. Agric. Res. 9:778–781.

Sperber, J. I. 1958b. Release of phosphates from soil minerals by hydrogen sulfide. Nature (Lond.) 181:934.

Suess, E. 1981. Phosphate regeneration from sediments of the Peru continental margin by dissolution of fish debris. Geochim. Cosmochim. Acta 45:577–588.

Swaby, R. J., and J. Sperber. 1958. Phosphate-dissolving microorganisms in the rhizosphere of legumes. *In*: E. G. Hallsworth, ed. Nutrition of Legumes, Proc. Univ. Nottingham Easter School Agric. Sci., 5th. Academic Press, London, pp. 289–294.

Taga, N., and H. Kobori. 1978. Phosphatase activity in eutrophic Tokyo Bay. Marine Biol. (Berl.) 49:223–229.

Tsubota, G. 1959. Phosphate reduction in the paddy field. I. Soil Plant Food 5:10–15.

Van Groenestijn, J. W., M. H. Deinema, and A. J. B. Zehnder. 1987. ATP production from polyphosphate in *Acinetobacter* strain 210A. Arch. Microbiol. 148:14–19.

Varma, A. K., W. Rigsby, and D. C. Jordan. 1983. A new inorganic pyrophosphate utilizing bacterium from a stagnant lake. Can. J. Microbiol. 29:1470–1474.

Veeh, H. H., W. C. Burnett, and A. Soutar. 1973. Contemporary phosphorites on the continental margin of Peru. Science 181:844–845.

Wackett, L. P., S. L. Shames, C. P. Venditti, and C. T. Walsh. 1987. Bacterial carbon-phosphorus lyase: products, rates, and regulation of phosphonic and phosphinic acid metabolism. J. Bacteriol. 169:710–717.

Westheimer, F. H. 1987. Why nature chose phosphates. Science 235:1173–1178.

Williams, B. G., and W. H. Patrick, Jr. 1971. Effect of E_h and pH on the dissolution of strengite. Nature (Phys. Sci.) 234:16–17.

Woolfolk, C. A., and H. R. Whiteley. 1962. Reduction of inorganic compounds with molecular hydrogen by *Micrococcus lactilyticus*. I. Stoichiometry with compounds of arsenic, selenium, tellurium, transition and other elements. J. Bacteriol. 84:647–658.

Youssef, M.I. 1965. Genesis of bedded phosphates. Econ. Geol. 60:590–600.25

11

Geomicrobially Important Interactions with Nitrogen

11.1 NITROGEN IN THE BIOSPHERE

Nitrogen is an element essential to all life. It is abundant in the atmosphere, mostly as dinitrogen (N_2), representing roughly four-fifths by volume or three-fourths by weight of the total gas, the rest being mainly oxygen plus lesser amounts of CO_2 and other common gases. Small amounts of oxides of nitrogen are also present. In soil, sediment, and fresh and ocean water, nitrogen exists in both inorganic and organic forms. Geomicrobially important inorganic forms include ammonia and ammonium ion, nitrite, nitrate, and gaseous oxides of nitrogen. Table 11.1 summarizes the abundances of some of these forms in the environment (see also Stevenson, 1972). Geomicrobially important organic nitrogen compounds include humic and fulvic acids, proteins, peptides, and amino acids, purines, pyrimidines, pyridines, other amines, and amides. Table 11.1 also summarizes the abundance of organic nitrogen in the environment.

Chemically, nitrogen occurs in the oxidation states -3 (e.g., NH_3), -2 (e.g., N_2H_4), -1 (e.g., NH_2OH), 0 (N_2), $+1$; (N_2O), $+2$ (e.g., NO), $+3$ (e.g., HNO_2), $+4$ (e.g., N_2O_4) and $+5$ (e.g., HNO_3). Of these, the -3, -1, 0, $+1$, $+2$, $+3$, and $+5$ oxidation states have biological significance

Table 11.1 Abundance of Nitrogen in the Biosphere

Biosphere compartment	Form of nitrogen	Quantity of nitrogen (kg N)
Atmosphere	N_2	3.9×10^{18}
Oceans	Organic N	9×10^{14}
	NH_4^+, NO_2^-, NO_3^-	10^{14}
Land	Organic N	8×10^{14}
	NH_4^+, NO_2^-, NO_3^-	1.4×10^{14}
Sediments	Total N	4×10^{17}
Rocks	Total N	1.9×10^{20}
Living biomass	Total N	1.3×10^{13}

Estimates from Fenchel and Blackburn (1979) and Brock and Madigan (1988).

because they can be enzymatically altered. Although nitrogen compounds with nitrogen in the oxidation state of +4 are not metabolized by microbes, nitrite formed from the dismutation of NO_2 after it is absorbed by soil can be oxidized to nitrate by as yet unidentified organisms (Ghiorse and Alexander, 1978).

Generally, inorganic nitrogen compounds exist in nature either as gases in the atmosphere and dissolved in water, or as compounds in aqueous solution. Exceptions are small deposits of nitrates of sodium, potassium, calcium, magnesium, or ammonium known as guano or cave, playa or caliche nitrates (Lewis, 1965). In some cases these nitrate deposits were apparently formed by bacteriological transformation of organic nitrogen formed by nitrogen-fixing bacteria, including cyanobacteria. In other cases these deposits arose from bacterial transformation of organic nitrogen in animal droppings such as those of birds or bats. The organic nitrogen was released as ammonia and then oxidized to nitrate by a consortium of bacteria (Ericksen, 1983) (see section 11.2 for discussion of the reactions). Deposits of this type in Chile, which were probably formed mainly from the microbiota in playa lakes (Ericksen, 1983), are commercially exploited.

Organic nitrogen in nature may exist dissolved in an aqueous phase or in a solid state, in the latter case usually in polymers (e.g., certain proteins like keratin). Insofar as is known, organic nitrogen is usually metabolizable by microbes.

Nitric acid formed through bacterial nitrification can be an important agent in weathering of rocks and minerals (see Chapters 4, 6, 8, 9, and 10).

11.2 MICROBIAL INTERACTIONS WITH NITROGEN

Ammonification Most plants derive the nitrogen that they assimilate from soil. This nitrogen is in most instances in the form of nitrate. The anion nitrate is much less readily bound by mineral soil particles, especially clays that have a net negative charge, than the cation ammonium. The nitrate supply in soil depends on recycling of spent organic nitrogen (plant excretions and remains, animal excretions and remains, microbial excretions and remains). The first step in this recycling is **ammonification**, in which the organic nitrogen is transformed into ammonia. An example of ammonification is the deamination of amino acids:

$$R-\underset{\underset{NH_2}{|}}{C}HCOOH + NAD^+ \Rightarrow \underset{\underset{NH}{\|}}{R}CCOOH + NADH_2 \qquad (11.1)$$

$$\underset{\underset{NH}{\|}}{R}CCOOH + H_2O \Rightarrow \underset{\underset{O}{\|}}{R}CCOOH + NH_3 \qquad (11.2)$$

The NH_3 reacts with water to form ammonium hydroxide, which dissociates:

$$NH_3 + H_2O \Rightarrow NH_4OH \Rightarrow NH_4^+ + OH^- \qquad (11.3)$$

In the laboratory it is commonly observed that when heterotrophic bacteria grow in a proteinaceous medium in which the organic nitrogen serves as energy, carbon, and nitrogen source, the pH rises with time due to the liberation of ammonia and its hydrolysis to ammonium ion. Indeed, ammonification is always an essential first step when an amino compounds such as amino acid serves as an energy source.

Ammonia is also formed as a result of urea hydrolysis catalyzed by the enzyme urease:

$$NH_2CONH_2 + 2H_2O \Rightarrow 2NH_4^+ + CO_3^{2-} \qquad (11.4)$$

Urea is a nitrogen waste product excreted in the urine of many mammals. Although urease is produced by a variety of prokaryotic and eukaryotic microbes, a few prokaryotic soil microbes, e.g., *Bacillus pasteurii* and *Bacillus freudenreichii*, seem to be specialists in degrading urea. They prefer to grow at an alkaline pH such as is generated when urea is hydrolyzed (see Alexander, 1977).

Nitrification Plants can readily assimilate ammonia. But the ammonia produced in ammonification in aqueous systems at neutral pH exists as a positively charged ammonium ion (NH_4^+) due to protonation, which is adsorbed by clays and then not readily available to plants. Thus it is important that ammonia be converted into an anionic nitrogen species, which is not readily adsorbed by clays and thus more readily utilizable by plants. **Nitrifying bacteria** play a central role in this conversion. The majority of nitrifying bacteria are autotrophs and can be divided into two groups: one includes those bacteria that oxidize ammonia to nitrous acid (e.g., *Nitrosomonas, Nitrocystis*), and the second includes those bacteria that oxidize nitrite to nitrate (e.g., *Nitrobacter, Nitrococcus*). All the members of these two groups of nitrifying bacteria are obligate autotrophs except for a few *Nitrobacter winogradskyi* strains, which appear to be facultative. They are all aerobes. Representatives are found in soil, freshwater, and seawater (for further characterization see Alexander, 1977; Buchanan and Gibbons, 1974).

Although ammonification is the major source of ammonia in soil and sediments, a special anaerobic respiratory process in which nitrate as the terminal electron acceptor is reduced to ammonia via nitrite may also be a significant source of ammonia in some environments (Jørgensen and Sørensen, 1985, 1988; Binnerup et al., 1992):

$$NO_3^- + 8(H) + H^+ \Rightarrow NH_3 + 3H_2O \tag{11.5}$$

This process is known as **nitrate ammonification** and can be carried on by a number of different facultative and strictly anaerobic bacteria (see, e.g. review by Ehrlich, 1993, pp. 232–233; Dannenberg et al., 1992; Sørensen, 1987). A variety of organic compounds, H_2, and inorganic sulfur compounds can serve as electron donors (H) in this reaction (Dannenberg et al., 1992).

Ammonia oxidation by the ammonia oxidizers involves hydroxylamine as an intermediate (see review by Wood, 1988). The formation of hydroxylamine is catalyzed by an oxygenase:

$$NH_3 + \tfrac{1}{2}O_2 \Rightarrow NH_2OH \tag{11.6}$$

This reaction does not yield biochemically useful energy. Chemoautotrophic ammonia oxidizers obtain their energy from the subsequent oxidation of hydroxylamine to nitrite. The overall reaction of oxidation of hydroxylamine to HNO_2 can be summarized as

$$NH_2OH + O_2 \Rightarrow HNO_2 + H_2O \tag{11.7}$$

but actually involves some intermediate steps (Hooper, 1984). It is the source of energy (ATP) for the organism through chemiosmotic coupling, i.e., oxidative phosphorylation. The oxygenase which converts ammonia to hydroxylamine appears to be non-specific. It can also catalyze the oxygenation of methane to methanol (Jones and Morita, 1983):

$$CH_4 + \tfrac{1}{2}O_2 \Rightarrow CH_3OH \tag{11.8}$$

This does not mean, however, that ammonia oxidizers can use methanol as an energy source. They lack the ability to oxidize methanol. Ammonia oxidizers can also form some NO and N_2O in side reactions, especially under oxygen limitation when nitrite may replace oxygen as terminal electron acceptor (see discussions by Knowles, 1985 and Davidson, 1993). This is an important observation because it means that biogenically formed N_2O and NO is not solely the result of denitrification (see next section).

The nitrite oxidizers convert nitrite to nitrate:

$$NO_2^- + \tfrac{1}{2}O_2 \Rightarrow NO_3^- \tag{11.9}$$

They generate useful energy by coupling the process chemiosmotically to ATP generation (Aleem and Newell, 1984; Wood, 1988).

Ammonia may also be heterotrophically converted to nitrate, although the importance of this process is probably of minor importance in nature in most instances. Rates of heterotrophic nitrification measured under laboratory conditions so far are significantly slower than those of autotrophic nitrification. The organisms capable of heterotrophic nitrification include both bacteria, such as *Arthrobacter* sp., and fungi, such as *Aspergillus flavus*. They gain no energy from the conversion. The pathway from ammonia to nitrate may involve intermediates such as hydroxylamine, nitrite, and 1-nitrosoethanol in the case of bacteria, and 3-nitropropionic acid in the case of fungi (see Alexander, 1977).

Evidence has now been presented that one or more bacteria, as yet unidentified, can oxidize ammonium anaerobically to dinitrogen (N_2). A disproportionation reaction with nitrates is probably involved (Van de Graaf et al., 1995),

$$5NH_4^+ + 3NO_3^- \Rightarrow 4N_2 + 9H_2O + 2H^+ \tag{11.10}$$

The standard free energy change of this reaction is -354.6 kcal ($-1,483.5$ kJ). As the discoverers indicate, this is a large enough energy yield for some of the energy to be conserved by the active organism.

Denitrification   Nitrate, nitrite, and nitrous and nitric oxides can serve as electron acceptors in microbial respiration, usually under anaerobic conditions. The transformation of nitrate to nitrite is called **dissimilatory nitrate reduction**, and the reduction of nitrate to nitric oxide (NO), nitrous oxide (N_2O), and/or dinitrogen is called **denitrification**. **Assimilatory nitrate reduction** is the first step in a processes in which nitrate is reduced to ammonia for the purpose of assimilation. Only as much nitrate is consumed in this process as is needed for assimilation. It is not a form of respiration and is performed by many organisms that cannot use nitrate for respiration. Some nitrate-respiring bacteria are only capable of nitrate reduction, lacking the enzymes for reduction of nitrite, while others are capable of reducing nitrite to ammonia instead of dinitrogen in a process which has been called **nitrate ammonification** by Sørensen (1987)(see also disccusion in previous section). All the nitrate respiratory processes have been found to operate to varying degrees in terrestrial, freshwater, and marine environments and represent an important part of the nitrogen cycle favored by anaerobic conditions.

The following half-reaction describes nitrate reduction:

$$NO_3^- + 2H^+ + 2e \Rightarrow NO_2^- + H_2O \tag{11.11}$$

The electron donor may be any one of a variety of organic metabolites or reduced sulfur such as H_2S or $S°$. The enzyme catalyzing reaction (11.09) is called **nitrate reductase** and is an iron-molybdo-protein. It is not only capable of catalyzing nitrate reduction but may also catalyze reduction of ferric to ferrous iron (see Chapter 14) and reduction of chlorate to chlorite. Nitrate can affect ferric iron reduction by nitrate reductase competitively (Ottow, 1969).

Nitrite may be reduced to dinitrogen by the following series of half-reactions, with organic metabolites or reduced sulfur acting as electron donor:

$$NO_2^- + 2H^+ + e \Rightarrow NO + H_2O \tag{11.12}$$

$$2NO + 2H^+ + 2e \Rightarrow N_2O + H_2O \tag{11.13}$$

$$N_2O + 2H^+ + 2e \Rightarrow N_2 + H_2O \tag{11.14}$$

The reduction of nitrite to ammonia may be summarized by the equation

$$NO_2^- + 7H^+ + 6e \Rightarrow NH_3 + 2H_2O \tag{11.15}$$

The electron donor may be one of a variety of organic metabolites.

Although it has been generally believed that these reactions can only occur at low oxygen tension or in the absence of oxygen, recent evidence indicates that in some exceptional organisms like *Thiosphaera pantotropha*, the reaction takes place at near normal oxygen tension (Robertson and Kuenen, 1984a,b; but see also disagreement by Thomsen et al., 1993). This organism can actually use oxygen and nitrate simultaneously as terminal electron acceptor. The explanation is that in *Tsa. pantotropha* the enzymes of denitrification are produced aerobically as well anaerobically, whereas in oxygen-sensitive denitrifiers they are only produced at low oxygen tension or anaerobically. Nitrate reductase in *Tsa. pantotropha* is constitutive whereas in many anaerobic denitrifiers it is inducible. Moreover, nitrate reductase of *Tsa. pantotropha* is not inactivated by oxygen, as in some anaerobic denitrifiers. Finally, oxygen does not repress formation of the denitrifying enzymes in *Tsa. pantotropha* as it does in some anaerobic denitrifiers. Aerobic denitrification in this organism appears to be linked to heterotrophic nitrification (Robertson et al., 1988; Robertson and Kuenen, 1990). The organism seems to use denitrification as a means of disposing of excess reducing power because its cytochrome system is insufficient for this purpose. Oxygen tolerance in denitrification has also been observed with some other bacteria (Hochstein et al., 1984; Davies et al., 1989; Bonin et al., 1989).

For a more complete discussion of denitrification the reader is referred to a monograph (Payne, 1981) and a review article by the same author (Payne, 1983).

Nitrogen Fixation   If nature had not provided for a microbial mechanism, **nitrogen fixation**, to reverse the effect of microbial depletion of fixed nitrogen from soil or water as volatile nitrogen oxides or dinitrogen by denitrification, life on Earth would not have long continued after the process of denitrification first evolved. Nitrogen fixation is dependent on a special enzyme, **nitrogenase**, which is found only in prokaryotic organisms, including aerobic and anaerobic photosynthetic and non-photosynthetic eubacteria and archaebacteria. Nitrogenase is an oxygen-sensitive enzyme, usually a combination of an iron- and a molybdo-protein (Eady and Postgate, 1974; Orme-Johnson, 1992), but in some cases (e.g., *Azotobacter chroococcum*) may also be a combination of an iron- and vanado-protein (Robson et al., 1986; Eady et al., 1987), and in another case (*Azotobacter vinelandii*) may be a combination of two iron proteins (Chiswell et al, 1988). Nitrogenase catalyzes the reaction:

$$N_2 + 6H^+ + 6e \Rightarrow 2NH_3 \tag{11.16}$$

The enzyme is not specific for dinitrogen. It can also catalyze the reduction of acetylene (CH≡CH), hydrogen cyanide (HCN), cyanogen (NCCN), hydrogen azide (HN$_3$), hydrogen thiocyanate (HNCS), protons (H$^+$), carbon monoxide (CO), and some other compounds (Smith, 1983).

The reducing power needed for dinitrogen reduction is provided by reduced ferredoxin and in heterotrophs may be formed from oxidized ferredoxin in a reaction in which pyruvate is oxidatively decarboxylated (Lehninger, 1975):

$$CH_3COCOOH + NAD^+ + CoASH$$
$$\Rightarrow CH_3CO{\sim}SCoA + CO_2 + NADH + H^+ \tag{11.17}$$

$$NADH + H^+ + (ferredoxin)_{ox} \Rightarrow NAD^+ + (ferredoxin)_{red} + 2H^+$$
$$(11.18)$$

In phototrophs, the reduced ferredoxin is produced as part of the photophosphorylation mechanism (see Chapter 6).

Nitrogen fixation is a very energy intensive reaction, consuming as many 16 moles of ATP per mole of dinitrogen in its reduction to ammonia (Newton and Burgess, 1983).

Nitrogen fixation may proceed symbiotically or nonsymbiotically. **Symbiotic nitrogen fixation** requires that the nitrogen-fixing bacterium associate with a specific plant (e.g., legumes, several nonleguminous angiosperms, the water fern *Azolla*), fungi (certain lichens) or, in rare cases, with an animal host in order to carry out nitrogen fixation. Even then, dinitrogen will be fixed only if the fixed-nitrogen level in the surrounding environment of the plant or the diet of the animal is deficient. In some plants (e.g., legumes or the alder), the nitrogen fixer may be localized in cells of the cortical root tissue which are transformed into nodules. Invasion of the plant tissue may have been via root hairs. In some other plants the nitrogen fixers may be localized in special leaf structures (e.g., in *Azolla*). In animals, the nitrogen fixer may be found to be a member of the flora of the digestive tract (Knowles, 1978). The plant host in symbiotic nitrogen fixation provides the energy source required by the nitrogen fixer for generating ATP. The energy source may take the form of compounds such as succinate, malate and fumarate. The plant also provides an environment in which access to oxygen is controlled so that nitrogenase in the nitrogen fixer is not inactivated. In root nodules of legumes, leghemoglobin is involved in this control of oxygen. The nitrogen fixer shares the ammonia which it forms from dinitrogen with its plant or animal host.

Symbiotic nitrogen-fixing bacteria include members of the genus *Rhizobium, Bradyrhizobium, Frankia, and Anabaena*. Some strains of *B. japonicum* have been found to be able to grow autotrophically on hydrogen as energy source because they possess uptake hydrogenase. They can couple hydrogen oxidation to ATP synthesis which they can use in CO_2 assimilation via the ribulose-bisphosphate carboxylase system. In nitrogen fixation, the ability to couple hydrogen oxidation to ATP synthesis may represent an energy conservation system since the nitrogenase generates a significant amount of hydrogen during nitrogen fixation, the energy content of which would otherwise be lost to the system.

Relatively recently, a special symbiotic nitrogen-fixing relationship was discovered in Brazil between certain cereal grasses, such as maize, and nitrogen fixing spirilla, such as *Azospirillum lipoferum* (Day et al., 1975, Smith et al., 1976; Von Buelow and Doebereiner, 1975). In these symbioses, the bacterium does not invade the plant roots but lives in close association with them in the **rhizosphere**. Apparently the plants excrete compounds which the bacterium can use as energy sources and which enables it to fix nitrogen that it can share with the plant if the soil is otherwise deficient in fixed nitrogen.

In **nonsymbiotic nitrogen fixation**, the active organisms are free-living in soil or water and fix nitrogen if fixed nitrogen is limiting. Their nitrogenase is not distinctly different from that of symbiotic nitrogen fixers. Unlike the symbiotic nitrogen fixers, non-symbiotic nitrogen-fixers appear to be able to maintain an intracellular environment in which nitrogenase is not inactivated by oxidizing conditions. The capacity for nonsymbiotic nitrogen fixation is widespread among prokaryotes. The best known examples and the most efficient include the aerobes *Azotobacter* and *Beijerinckia* and the anaerobe *Clostridium pasteurianum*, but many other aerobic and anaerobic genera include species with nitrogen-fixing capacity, even some photo- and chemolithotrophs. Most of the nitrogen fixers are only active at environmental pH values between about 5 and 9, but some strains of the acidophile *Thiobacillus ferrooxidans* have been shown to fix nitrogen at a pH as low as 2.5.

For a more detailed discussion of nitrogen fixation the reader is referred to Alexander (1984), Broughton (1983), Newton and Orme-Johnson (1980), and Quispel (1974).

11.3 MICROBIAL ROLE IN THE NITROGEN CYCLE

Owing to their special capacities for transforming inorganic nitrogen compounds that plants and animals lack, microbes, especially prokarkyotes and certain fungi, play a central role in the nitrogen cycle (Fig. 11.1). Many reactions of the cycle are entirely dependent on them; nitrogen fixation is restricted to prokaryotes. The direction of transformations in the cycle is determined by environmental conditions, especially the availability of oxygen, but also by the supply of particular nitrogen compounds. Anaerobic conditions may encourage denitrification and thus cause nitrogen limitation unless the process is counteracted by anaerobic nitrogen fixation. Limitations in the supply of organic or ammonia nitrogen affect the availability of nitrate. Fixed nitrogen is frequently a growth limiting factor in the marine environment but infrequently in fresh water, where phosphate is more likely to limit productivity. Fixed nitrogen may be a limiting factor in soil, especially agriculturally exploited soils.

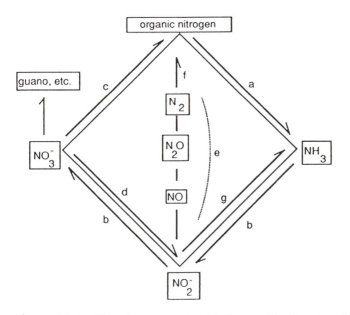

Figure 11.1 The nitrogen cycle. (a) Ammonification (aerobic and anaerobic), (b) autotrophic nitrification (strictly aerobic), (c) nitrate assimilation (aerobic or anaerobic), (d) nitrate reduction [usually anaerobic, but see (e)], (d,e) denitrification (usually anaerobic but sometimes aerobic; see text), (f) nitrogen fixation (aerobic and anaerobic), (d,g) nitrate ammonification.

11.4 SUMMARY

Nitrogen is essential to all forms of life. It is assimilated by cells in the form of ammonia and/or nitrate. It is released from organic combination in the form of ammonia. The latter process is called ammonification and occurs both aerobically and anaerobically. Ammonia is an energy-rich compound and can be oxidized to nitrate by way of nitrite by some aerobic, autotrophic eubacteria (nitrifiers). It can also be converted to nitrate by some heterotrophic bacteria and some fungi in a non-energy-yielding process, but this is much less common process. The conversion of ammonia to nitrate is important in soil and sediments because negatively charged clay particles can adsorb ammonia, making it unavailable to plants. Under reducing conditions, nitrate can be transformed by anaerobic respiration to nitrite, nitrous and nitric oxides, dinitrogen, or ammonia by appropriate bacteria. In soil this can have the effect of lowering its fertility. Depletion of soil nitrogen through dinitrogen evolution can, however, be biologically reversed by symbiotic and nonsymbiotic nitrogen-fixing bacteria which are able to reduce dinitrogen to ammonia. These various biological interactions are part of a cycle which is essential to sustenance of life on Earth.

REFERENCES

Aleem, M. I. H., and D. L. Sewell. 1984. Oxidoreductase systems in *Nitrobacter agilis. In*: W. R. Strohl and O. H. Tuovinen, eds. Microbial Chemoautotrophy. Ohio State University Press, Columbus, pp. 185–210.

Alexander, M. 1977. Introduction to Soil Microbiology. 2nd ed. Wiley, New York.

Alexander, M., ed. 1984. Biological Nitrogen Fixation. Ecology, Technology, and Physiology. Plenum Press, New York.

Binnerup, S. J., K. Jensen, N. P. Revsbech, M. H. Jensen, and J. Sørensen. 1992. Denitrification, dissimilatory reduction of nitrate to ammonium, and nitrification in a bioturbated estuarine sediment as measured with ^{15}N and microsensor techniques. Appl. Environ. Microbiol. 58:303–313.

Bonin, P., M. Gilewicz, and J. C. Bertrand. 1989. Effects of oxygen on each step of denitrification on *Pseudomonas nautica*. Can. J. Microbiol. 35:1061–1064.

Brock, T. D., and M. T. Madigan. 1988. Biology of Microorganisms. 5th ed. Prentice-Hall, Englewood Cliffs, NJ.

Broughton, W. J., ed. 1983. Nitrogen Fixation. Vol. I, Ecology; Vol. 2, Rhizobium; Vol. 3, Legumes. Clarendon Press, Oxford.

Buchanan, R. E., and N. E. Gibbons, eds. 1974. Bergey's Manual of Determinative Bacteriology. 8th ed. Williams & Wilkins, Baltimore.

Chiswell, J. R., R. Premakumar, and P. E. Bishop. 1988. Purification of a second alternative nitrogenase from a *nif*HDK deletion strain of *Azotobacter vinelandii*. J. Bacteriol. 170:27–33.

Dannenberg, S., M. Kroder, W. Dilling, and H. Cypionka. 1992. Oxidation of H_2, organic compounds and inorganic sulfur compounds coupled to reduction of O_2 or nitrate by sulfate-reducing bacteria. Arch. Microbiol. 158:93–99.

Davidson, E. A. 1993. Soil water content and the ratio of nitrous to nitric oxide emitted from soil. *In*: R. S. Oremland, ed. Biogeochemistry of Global Change. Radiatively Active Trace Gases. Chapman and Hall, New York, pp. 369–386.

Davies, K. J. P., D. Lloyd, and L. Boddy. 1989. The effect of oxygen on denitrification in *Paracoccus denitrificans* and *Pseudomonas aeruginosa*. J. Gen. Microbiol. 135:2445–2451.

Day, J. M., M. C. P. Neves, and J. Doebereiner. 1975. Nitrogenase activity on the roots of tropical forage grasses. Soil Biol. Biochem. 7: 107–112.

Eady, R.R., and J.R. Postgate. 1974. Nitrogenase. Nature (Lond.) 249: 805–810.

Eady, R. R., R. L. Robson, T. H. Richardson, R. W. Miller, and M. Hawkins. 1987. The vanadium nitrogenase of *Azotobacter chroococcum*. Purification and properties of of the vanadium-iron protein. Biochem. J. 244:197–207.

Ehrlich, H. L. 1993. Bacterial mineralization of organic matter under anaerobic conditions. *In*: J-M. Bollag and G. Stotzky. Soil Biochemistry, Vol. 8. Marcel Dekker, New York, pp. 219–238.

Ericksen, G.E. 1983. The Chilean nitrate deposits. Am. Sci. 71:366–374.

Fenchel, T., and T. H. Blackburn. 1979. Bacteria and Mineral Cycling. Academic Press, London.

Ghiorse, W. C., and M. Alexander. 1978. Nitrifying populations and the destruction of nitrogen dioxide in soil. Microb. Ecol. 4:233–240.

Hochstein, L. I., M. Betlach, and K. Kritikos. 1984. The effect of oxygen on denitrification during steady-state growth of *Paracoccus halodenitrificans*. Arch. Microbiol. 137:74–78.

Hooper, A. B. 1984. Ammonia oxidation and energy transduction in the nitrifying bacteria. pp. 133–167. *In*: W. R. Strohl and O. H. Tuovinen, eds. Microbial Chemoautotrophy. Ohio State University Press, Columbus.

Jones, R. D. and R. Y. Morita. 1983. Methane oxidation by *Nitrosococcus oceanus* and *Nitrosomonas europaea*. Appl. Environ. Microbiol.45: 401–410.

Jørgensen, B. B., and J. Sørensen. 1985. Seasonal cycles of O_2, NO_3^- and SO_4^- reduction in estuarine sediments: the significance of an NO_3^- reduction maximum in the spring. Mar. Ecol. Prog. Ser. 24: 65–74.

Jørgensen, B. B., and S. Sørensen. 1988. Two annual maxima of nitrate reduction and denitrification in estuarine sediment (Norsminde Fjord, Denmark). Mar. Ecol. Prog. Ser. 48:147–154.

Knowles, R. 1978. Free-living bacteria. pp. 25–40. *In*: J. Doebereiner, R. H. Burris, A. Hollaender, A. A. Franco, C. A. Neyra, and D. B. Scott, eds. Limitations and Potentials for Biological Nitrogen Fixation in the Tropics. Plenum Press, New York.

Knowles, R. 1985. Microbial transformations as sources and sinks of nitrogen oxides. pp. 411–426. *In*: D. E. Caldwell, J. A. Brierley, and C. L. Brierley, eds. Planetary Ecology. Van Nostrand Reinhold, New York.

Lehninger, A. L. 1975. Biochemistry. 2nd ed. Worth Publishers, New York.

Lewis, R. W. 1965. Nitrogen, pp. 621–629. *In*: Mineral Facts and Problems. Bull. 630. U.S. Bureau of Mines, Washington, DC.

Newton, W. E., and B. K. Burgess. 1983. Nitrogen fixation: Its scope and importance. *In*: A. Mueller and W. E. Newton. Nitrogen Fixation. The Chemical-Biochemical-Genetic Interface. Plenum Press, New York, pp. 1–19.

Newton, W. E., and W. H. Orme-Johsnon, eds. 1980. Nitrogen Fixation. University Park Press, Baltimore.

Orme-Johnson, W. H. 1992. Nitrogenase structure: Where to now? Science 257:1639–1640.

Ottow, J. C. G. 1969. Der Einfluss von Nitrat, Chlorat, Sulfat, Eisenoxydform und Wachstumsbedingungen auf das Ausmass der bakteriellen Eisenreduktion. Z. Pflanzenernaehr. Duengung Bodenk. 124:238–253.

Payne, W. J. 1981. Denitrification. Wiley, New York.

Payne, W. J. 1983. Bacterial denitrification: asset or defect. BioScience 33: 319–325.

Quispel, A., ed. 1974. The Biology of Nitrogen Fixation. North-Holland Publishing Company, Amsterdam.

Robertson, L. A., and J. G. Kuenen. 1984a. Aerobic denitrification: A controversy revived. Arch. Microbiol. 139:351–354.

Robertson, L. A., and J. G. Kuenen. 1984b. Aerobic denitrification—Old wine in new bottles? Antonie van Leeuwenhoek 50:525–544.

Robertson, L. A., and J. G. Kuenen. 1990. Combined heterotrophic nitrification and aerobic denitrification in *Thiosphaera pantotropha* and other bacteria. Antonie van Leeuwenhoek 57:139–152.

Robertson, L. A., E. W. J. van Niel, R. A. M. Torremans, and J. G. Kuenen. 1988. Simultaneous nitrification and denitrification in aerobic chemostat cultures of *Thiosphaera pantotropha*. Appl. Environ. Microbiol. 54:2812–2818.

Robson, R. L., R. R. Eady, T. H. Richardson, R. W. Miller, M. Hawkins, and J. R. Postgate. 1986. The alternative nitrogenase of *Azotobacter chroococcum* is a vanadium enzyme. Nature (Lond.) 322:388–390.

Smith, B. E. 1983. Reactions and physicochemical properties of the nitrogenase MoFe proteins. *In*: A. Mueller and W. E. Newton, eds. Nitrogen Fixation. The Chemical-Biochemical-Genetic-Interface. Plenum Press, New York, pp. 23–62.

Smith, R. L., J. H. Bouton, S. C. Schank, K. H. Quesenberry, M. E. Tyler, J. R. Milam, M. H. Gaskins, and R. C. Littell. 1976. Nitrogen fixation in grasses inoculated with *Spirillum lipoferum*. Science 193: 1003–1005.

Sørensen, J. 1987. Nitrate reduction in marine sediment: pathways and interactions with iron and sulfur cycling. Geomicrobiol. J. 5: 401–421.

Stevenson, F. J. 1972. Nitrogen: Element and geochemistry. *In*: R. W. Fairbridge, ed. The Encyclopedia of Geochemistry and Environmental Sciences. Encyclopedia of Earth Sciences Series, Vol. IVA. Van Nostrand Reinhold, New York, pp. 795–801.

Thomsen, J. K., J. J. L. Iversen, and R. P. Cox. 1993. Interactions between respiration and denitrification during growth of *Thiosphaera pantotropha* in continuous culture. FEMS Microbiol. Lett. 110: 319–324.

Van de Graaf, A. A., A. Mulder, P. de Bruijn, M. S. M. Jetten, L. A. Robertson, and J. G. Kuenen. 1995. Anaerobic oxidation of ammonium is a biologically mediated process. Appl. Environ. Microbiol, 61:1246–1251.

von Buelow, J. F. W., and J. Doebereiner. 1975. Potential for nitrogen fixation in maize genotypes in Brazil. Proc. Natl. Acad. Sci. USA 72:2389–2393.

Wood, P. 1988. Chemolithotrophy. *In*: C. Anthony. Bacterial Energy Transduction. Academic Press, London, pp. 183–230.

12

Geomicrobial Interactions with Arsenic and Antimony

Although arsenic and antimony are generally toxic to life, some microorganisms exist which can metabolize forms of these elements. In some cases they can use them as energy sources, in others as electron acceptors, and in still others they metabolize them to detoxify them. These reactions are important from a geomicrobial standpoint because they indicate that some microbes contribute to arsenic and antimony mobilization or immobilization in the environment.

12.1 ARSENIC

Distribution Arsenic is widely distributed in the upper crust of the Earth, mostly at very low concentration (Carapella, 1972). It rarely occurs in elemental form. More often it is found combined with sulfur, as in orpiment (As_2S_3) or realgar (AsS); with selenium, as in As_2Se_3; with tellurium, as in As_2Te; or as sulfosalts, as in enargite (Cu_3AsS_4) or arsenopyrite (FeAsS). It is also found as arsenides of heavy metals such as iron (loellingite, $FeAs_2$), copper (domeykite, Cu_3As), nickel (nicolite, NiAs), and cobalt (Co_2As). Sometimes the element occurs in the form of arsenite minerals (arsenolite or claudetite, As_2O_3) or in the form of arsenate minerals [erythrite, $Co_3(AsO_4)_2 \cdot 8H_2O$; scorodite, $FeAsO_4 \cdot 2H_2O$; olivenite, Cu_2 $(AsO_4)(OH)$]. Arsenopyrite is the most common and widespread mineral form of arsenic; orpiment and realgar are also fairly common. The ultimate source of arsenic on the Earth's surface is igneous activity. On weathering of arsenic-containing rocks, which may harbor 1.8 ppm of the element, the arsenic is dispersed through the upper lithosphere and the hydrosphere.

Arsenic concentration in soil may range from 0.1 to more than 1,000 ppm. The average concentration in seawater is given as 3.7 μg liter^{-1} and in fresh water as 0.5 μg liter^{-1}. In air, arsenic may be found in the concentration range of 1.5–53 ng m^{-3} (Bowen, 1979). Some living organisms may concentrate arsenic manyfold over its level in the environment. Thus, some algae have been found to accumulate arsenic 200–3,000 times in excess of its concentration in their growth medium (Lunde, 1973). Man may artificially raise the arsenic concentration in soil and water through the introduction of sodium arsenite (Na_2AsO_3) or cacodylic acid [$(CH_3)_2$ $AsO \cdot OH$] as herbicide.

The valence states in which arsenic is usually encountered in nature include -3, 0, $+2$, $+3$, and $+5$.

Toxicity Arsenic compounds are toxic to most living organisms. Arsenite (AsO_3^{2-} or AsO_2^{-}) has been shown to inhibit dehydrogenases such as pyruvate, α-ketoglutarate, and dihydrolipoate dehydrogenases (Mahler and Cordes, 1966). Arsenate uncouples oxidative phosphorylation; i.e., it inhibits ATP synthesis by chemiosmosis (Da Costa, 1972).

Both the uptake of arsenate and the inhibitory effect of arsenate on metabolism can be modified by phosphate (Button et al., 1973; Da Costa, 1971, 1972). A common transport mechanism for phosphate and arsenate seems to exist in the membranes of some organisms, although a separate transport mechanism for phosphate may also exist (Bennett and Malamy, 1970). In the latter case, phosphate uptake does not affect arsenate uptake, nor does arsenate uptake affect phosphate uptake. In one reported case, that of a fungus, *Cladosporium herbarium*, arsenite toxicity could also be ameliorated by the presence of phosphate. In that instance, prior oxidation of arsenite to arsenate by the fungus appeared to be the cause of the effect (Da Costa, 1972). In growing cultures of *Candia humicola*, phosphate can inhibit the formation of trimethylarsine from arsenate, arsenite, and monomethylarsinate, but not from dimethyl arsinate (Cox and Alexander, 1973). In similar cultures, phosphite can also suppress trimethylarsine production from monomethylarsinate but not from arsenate or dimethylarsinate, and hypophosphite can cause temporary inhibition of the conversion of arsenate, monomethylarsonate, and dimethylarsinate (Cox and Alexander, 1973). High antimonate concentrations lower the rate of conversion of arsenate to trimethylarsine by resting cells of the fungus (Cox and Alexander, 1973).

Bacteria may develop genetically determined resistance to arsenic (Ji and Silver, 1992; Ji et al., 1993). This resistance may be determined by a locus on a plasmid, for example in *Staphylococcus aureus* (Dyke et al., 1978) and *Escherichia coli* (Hedges and Baumberg, 1973). The mechanism of resistance in these species is a special pumping mechanism that expels the arsenic taken up as arsenate by the cells (Silver and Keach, 1982). It involves reduction of arsenate to arsenite intracellulary followed by efflux of the arsenite promoted by an oxyanion translocating ATPase (Broeer et al., 1993; Ji et al., 1993). Some of the resistant organisms have the capacity to oxidize reduced forms of arsenic to arsenate and others to reduce oxidized forms of arsenic to reduced forms (see below).

Microbial Oxidation of Reduced Forms of Arsenic

Dissolved Arsenic Species   Bacterial oxidation of arsenite to arsenate was first reported by Green in 1918. He discovered an organism with this ability in arsenical cattle-dipping solution used for the protection against insect bites. He named the organism *Bacillus arsenoxydans*. Quastel and Scholefield (1953) observed arsenite oxidation in perfusion experiments in which they passed 2.5×10^{-3} M sodium arsenite solution through columns charged with Cardiff soil. They did not isolate the organism or organisms responsible but showed that 0.1% solution of NaN_3 inhibited the oxidation. The onset of arsenite oxidation in their experiments occurred after a lag. The length of this lag was increased when sulfanilamide was added with the arsenite. A control of pH was found important for sustained bacterial activity. An almost stoichiometric O_2 consumption was observed with arsenite oxidation.

Further investigations of arsenical cattle-dipping solutions led to the isolation of 15 arsenite-oxidizing strains of bacteria (Turner, 1949, 1954). These organisms were assigned to the genera *Pseudomonas*, *Xanthomonas*, and *Achromobacter*. *Achromobacter arsenoxydans-tres* is considered synonymous with *Alcaligenes faecalis* (Hendrie et al., 1974). This organism is described in the eighth edition of *Bergey's Manual of Determinative Bacteriology* (Buchanan and Gibbons, 1974) as frequently having the capacity for arsenite oxidation.

Of the 15 isolates, *P. arsenoxydans-quinque* was studied in detail with respect to arsenite oxidation. Resting cells of this culture oxidized arsenite at an optimum pH of 6.4 and at an optimum temperature of 40°C (Turner and Legge, 1954). Cyanide, azide, fluoride, and pyrophosphate inhibited the activity. Under anaerobic conditions, 2,6-dichlorophenol indophenol, *m*-carboxyphenolindo-2,6-dibromophenol, and ferricyanide could act as electron acceptors. Pretreatment of the cells with toluene and acetone, or by desiccating them, rendered them incapable of oxidizing arsenite in air. The arsenite-oxidizing enzyme was "adaptable." Examination of arsenite oxidation by cell-free extracts of *P. arsenoxydans-quinque* suggested the presence of soluble "dehydrogenase" activity which under anaerobic conditions conveyed electrons from arsenite to 2,6-dichlorophenol indophenol (Legge and Turner, 1954). The activity was partly inhibited by 10^{-3} M *p*-chloromercurybenzoate. The entire arsenite-oxidizing system was believed to consist of "dehydrogenase" and an oxidase (Legge, 1954).

An arsenite-oxidizing soil strain of *Alcaligenes faecalis* was isolated in 1973 whose arsenite-oxidizing ability was found to be inducible by arsenite and arsenate (Osborne and Ehrlich, 1976). It oxidized arsenite stoichiometrically to arsenate (Table 12.1):

$$AsO_2^- + H_2O + \tfrac{1}{2}O_2 \Rightarrow AsO_4^{3-} + 2H^+ \tag{12.1}$$

Inhibitor and spectrophotometric studies suggested that the oxidation involved an oxidoreductase with a bound flavin that passed electrons from arsenite to oxygen by way of cytochrome c and cytochrome oxidase (Osborne and Ehrlich, 1976).

Anderson et al. (1992) isolated an inducible arsenite-oxidizing enzyme that was located on the outer surface of the plasma membrane of *A. faecalis* strain NCIB 8687. The enzyme location suggests that in intact cells of this organism, arsenite oxidation occurs in its periplasmic space. Biochemical characterization showed the enzyme to be a molybdenum-containing hydroxylase consisting of a monomeric 85-kDa peptide with a pterin cofactor and one atom of molybdenum, five or six atoms of iron,

Table 12.1 Stoichiometry of Oxygen Uptake by *Alcaligenes faecalis* on Arsenite Based on the Reaction of $AsO_2^- + H_2O + \tfrac{1}{2}O_2 \rightarrow AsO_4^{3-} + 2H^+$

NaAsO$_2$ added (μmol)	Oxygen uptake		
	Theoretical	Experimental	Percent of theoretical
19.25	9.63	8.79	91.3
38.50	19.25	18.48	96.0
57.75	28.88	27.05	93.7
77.00	38.50	37.05	96.2

Source: Osborne (1973), by permission.

and inorganic sulfide. Both azurin and cytochrome c from *A. faecalis* served as electron acceptors in arsenite oxidation catalyzed by this enzyme.

A strain of *A. faecalis* similar to that isolated by Osborne and Ehrlich (1976) was independently isolated and characterized by Phillips and Taylor (1976). Neither their strain nor that of Osborne and Ehrlich oxidized arsenite strongly until the late exponential phase or stationary phase of growth was reached in batch culture (Phillips and Taylor, 1976; Ehrlich, 1978). Other heterotrophic arsenite-oxidizing bacteria that have been identified more recently include *Pseudomonas putida* strain 18 and *Alcaligenes eutrophus* strain 280, both of which were isolated from gold-arsenic deposits (Abdrashitova et al., 1981).

The observation by Osborne and Ehrlich (1976) that their strain passes electrons from arsenite to oxygen via an electron transport system which involves c-type cytochrome and cytochrome oxidase suggested that their organism may be able to couple this oxidation to ATP synthesis, i.e., to derive energy from the process. Indeed, indirect evidence indicates that the organism may be able to derive maintenance energy from arsenite oxidation (Ehrlich, 1978). Much stronger evidence that arsenite can be used as an energy source by some bacteria was presented by Ilyaletdinov and Abdrashitova (1981). They isolated a culture, *Pseudomonas arsenitoxidans* from a gold-arsenic ore deposit which could grow autotrophically with arsenite as sole source of energy. It was also reported to attack the arsenic in arsenopyrite. It appeared to be an obligate autotroph, being unable to use organic carbon and energy sources.

Arsenite [As(III)] may under some conditions also be subject to abiotic oxidation by Mn(IV) but apparently to a much lesser extent or not at all by Fe(III) (Oscarson et al., 1981).

Arsenic-Containing Minerals Arsenic combinations with iron, copper, and sulfur are also oxidized by bacteria. The simplest of these compounds, orpiment (As_2S_3), was found to be oxidized by *Thiobacillus ferrooxidans* TM with the production of arsenite and arsenate in a mineral salts solution (9K medium without iron; see Silverman and Lundgren, 1959) to which the mineral had been added in pulverized form (Ehrlich, 1963) (Fig. 12.1). The initial pH was 3.5 and dropped to 2.0 in 35 days. In contrast, in an uninoculated control, in which orpiment oxidized spontaneously but slowly, the pH rose from 3.5 to 5. Realgar (AsS) was not attacked by *T. ferrooxidans* TM.

Arsenopyrite (FeAsS) and enargite (Cu_3AsS_4) were also oxidized by an iron-oxidizing *Thiobacillus* culture under the same test conditions as used with orpiment (Ehrlich, 1964). During growth on arsenopyrite, the

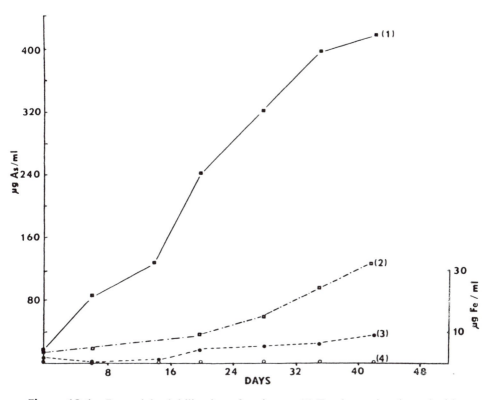

Figure 12.1   Bacterial solubilization of orpiment. (1) Total arsenic released with bacteria (*T. ferrooxidans*); (2) total arsenic released without bacteria; (3) total iron released with bacteria (*T. ferrooxidans*); (4) total iron released without bacteria. (From Ehrlich, 1963, by permission.)

arsenic was transformed to arsenite and arsenate. At least some of the iron was oxidized and extensively precipitated as iron arsenite and arsenate. The pH dropped from 3.5 to 2.5 in both inoculated and uninoculated flasks. Oxidation of arsenopyrite in the absence of bacteria was significantly slower (Fig. 12.2). In a later study in stirred-tank reactors, Carlson et al. (1992) identified the minerals jarosite and scorodite as the solid products formed in an oxidation of arsenic-containing pyrite by a mixed culture of moderately thermophilic acidophiles. In another study of *T. ferrooxidans* at 22°C and a moderately thermophilic mixed culture at 45°C on arsenopyrite, the mixed culture oxidized the mineral nearly completely

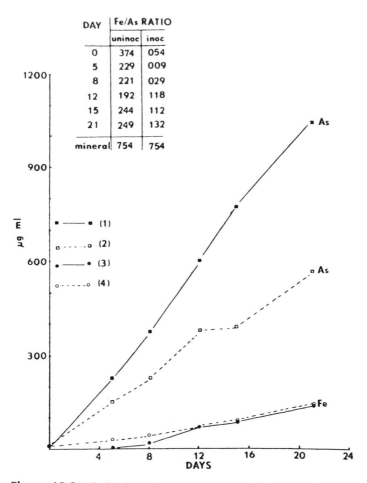

Figure 12.2 Oxidation of arsenopyrite by *T. ferrooxidans*. Curves (1) and (3) represent changes in inoculated flasks. Curves (2) and (4) represent changes in uninoculated flasks. (From Ehrlich, 1964, by permission.)

whereas *T. ferrooxidans* did not (Tuovinen et al., 1994). Jarosite was a major sink for ferric iron in both cultures, but some amorphous ferric arsenate and S° were also formed. Because of insufficient arsenopyrite oxidation of scorodite did not form. The extremely thermophilic archaebacterium *Sulfolobus* was also found to oxidize arsenopyrite (Groudeva et al., 1986; Ngubane and Baecker, 1990).

Although *T. ferrooxidans* appears not to be able to oxidize arsenite to arsenate, the thermophilic archaebacterium *Sulfolobus acidocaldarius* strain BC is able to do so (Sehlin and Lindstroem, 1992).

On enargite *T. ferrooxidans* released cupric copper and arsenate into solution together with some iron that was present as an impurity in the mineral (Ehrlich, 1964). With active bacteria, a pH drop from 3.5 to between 2 and 2.5 was observed, whereas in the absence of bacteria, a pH rise to 4.5 was usually observed. Some copper and arsenic remained out of solution in the experiments. The rate of enargite oxidation without bacteria was significantly slower and may have followed a different course of reaction on the basis of the different rates of Cu and As solubilization.

The ability of *T. ferrooxidans* and other acidophilic, iron-oxidizing bacteria to oxidize arsenopyrite is now beginning to be industrially exploited in beneficiation of precious metal–containing, sulfidic ores (Olson, 1994; Pol'kin et al., 1973; Pol'kin and Tanzhnyanskaya, 1968). The process involves the selective removal of arsenopyrite as well as pyrite by bacterial oxidation (bioleaching) before gold or silver is chemically (nonbiologically) extracted by cyanidation (complexing with cyanide). Arsenopyrite and pyrite interfere with cyanidation for several reasons. They encapsulate finely disseminated gold and silver, blocking access of cyanide reagent used to solubilize the gold and silver. They consume oxygen that is needed to convert metallic gold to aurous (Au^+) and auric (Au^{3+}) gold before it can be complexed by cyanide. Finally, the oxidation products of arsenopyrite and pyrite consume cyanide by forming thiocyanate and iron-cyanide complexes from which cyanide cannot be regenerated (Livesey-Goldblatt et al., 1983; Karavaiko et al., 1986). This increases the amount of cyanide reagent required for precious metal recovery.

In addition to plain sulfidic gold ores, carbonaceous gold ores containing as much as 6% arsenic have been reported to be beneficiated by *T. ferrooxidans* (Kamalov et al., 1973). As much as 90% of the arsenic in the ore was removed in 10 days under some circumstances. In another study, it was found possible to accelerate leaching of arsenic from a copper-tin-arsenic concentrate 6–7 times, using an adapted strain of *T. ferrooxidans* (Kulibakin and Laptev, 1971).

Microbial Reduction of Oxidized Arsenic Species Some bacteria, fungi, and algae are able to reduce arsenic compounds. One of the first reports on arsenite reduction involved fungi (Gosio, 1897). Although originally the product of this reduction was thought to be diethylarsine (Bignelli, 1901), it was later shown to be trimethylarsine (Challenger et al., 1933). A bacterium from cattle-dipping tanks which reduced arsenate to arsenite was also described quite some time ago (Green, 1918). More recently, it has been reported that a strain of the green alga *Chlorella* can reduce a part of the arsenate it absorbs from the medium to arsenite (Blasco et al., 1971).

The fungus *Scopulariopsis brevicaulis* is able to convert arsenite to trimethylarsine by a mechanism that includes the following steps (Challenger, 1951):

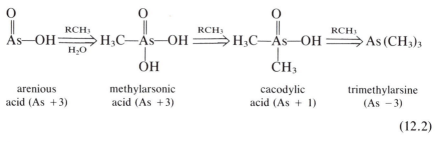

arenious acid (As +3)	methylarsonic acid (As +3)	cacodylic acid (As + 1)	trimethylarsine (As −3)

$$(12.2)$$

Methionine and other methyl compounds are the methyl donors (RCH_3). Besides *S. brevicaulis*, other fungi, such *Aspergillus, Mucor, Fusarium, Paecilomyces*, and *Candida humicola*, have also been found active in such reductions (Alexander, 1977; Cox and Alexander, 1973; Pickett et al., 1981). Trimethylarsine oxide was found to be an intermediate in trimethylarsine formation by *C. humicola* (Pickett et al., 1981).

Ahmann et al. (1994) isolated a comma-shaped motile rod, strain MIT-13, from arsenic-contaminated sediment from the Aberjona watershed in Massachusetts. Anaerobically growing cultures of this organism respired on arsenate using lactate but not acetate as electron donor and producing stoichiometric amounts of arsenite. Sulfate was used as an alternative electron acceptor with lactate but did not inhibit arsenate reduction when the two were present together. Arsenate was the preferential electron acceptor. Molybdate inhibited growth on arsenate transiently.

Bacterial reduction of arsenate to arsenite by H_2 has been demonstrated with cell extracts of *Micrococcus (Veillonella) lactilyticus* and whole cells of *M. aerogenes* (Woolfolk and Whiteley, 1962). The active enzyme was hydrogenase. Arsine (AsH_3) was not formed in these reactions. *Pseudomonas* sp. and *Alcaligenes* sp., on the other hand, are able to reduce arsenate and arsenite anaerobically to arsine (Cheng and Focht, 1979). Dimethylarsine was produced from arsenate with H_2 by whole cells and cell extracts of the strict anaerobe *Methanobacterium* M.O.H. (McBride and Wolfe, 1971). Studies with extracts of this organism revealed that methylcobalamin (CH_3B_{12}) was the immediate methyl donor. The reaction, moreover, required the consumption of ATP. 5-CH_3-tetrahydrofolate or serine could not replace CH_3B_{12}, although CO_2 could when tested by isotopic tracer technique. The reaction sequence was:

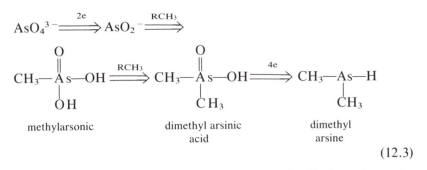

$$AsO_4^{3-} \xrightarrow{2e} AsO_2^- \xrightarrow{RCH_3}$$

methylarsonic   dimethyl arsinic   dimethyl
                       acid   arsine

(12.3)

In an excess of arsenite, methylarsonic acid was the final product, the supply of CH_3B_{12} being limiting. In an excess of CH_3B_{12}, a second methylation step yielding dimethyl arsinic acid (cacodylic acid) followed the first one. The last step was shown to occur in the absence of CH_3B_{12}. All steps were enzymatic. Cell extracts of *Desulfovibrio vulgaris* were also found to produce a volatile arsenic derivative, presumably an arsine (McBride and Wolfe, 1971).

Only a few direct observations of microbial reduction of arsenic compounds in nature have been reported. One was the bacterial reduction of arsenate to arsenite in seawater (Johnson, 1972). In this study, bacteria from phytoplankton samples from Narragansett Bay and from Sargasso Sea water were able to reduce arsenate added to Sargasso Sea water. An arsenate reduction rate of 10^{-11} μmol cell^{-1} min^{-1} was measured after 12 h of incubation at 20–22°C. Arsenic was not accumulated by the bacteria, and none was lost from the medium through volatilization. These observations may help to explain why the observed As^{+5}/As^{+3} ratio was 10^{-1} to 10 instead of 10^{26}, as predicted under equilibrium conditions in oxygenated seawater at pH 8.1 (Johnson, 1972). Another observation was the oxidation of arsenite to arsenate in activated sludge, as reported by Myers et al. (1973). Over a longer period, arsenate was reduced stepwise to lower oxidation states in the activated sludge. *P. fluorescens* was an active reducer in this system under aerobic conditions (Myers et al., 1973). Cheng and Focht (1979) reported that bacteria can reduce arsenate and arsenite to arsine under anaerobiosis in soils. They indicated that unlike fungi, the bacteria produce mono- and dimethylarsine only when methyl arsonate or dimethyl arsinate is available.

Figure 12.3 summarizes those reactions involving arsenic compounds that are catalyzed by microorganisms. The oxidation of methylated arsine, although not indicated in the diagram, has been suggested by Cheng and Focht (1979). The arsenic cycle in natural fresh waters has been discussed by Ferguson and Gavis (1972). In marine waters the cycle has been examined by Andreae (1979) and Scudlark and Johnson (1982).

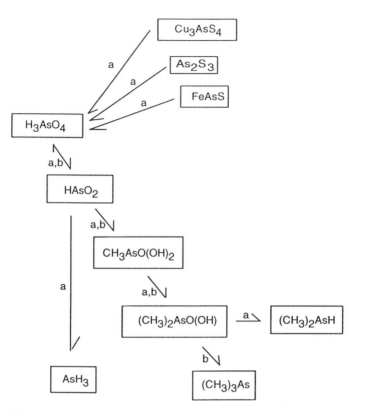

Figure 12.3 Summary of possible microbial interactions with arsenic compounds: (a) performed by bacteria, (b) performed by fungi.

12.2 ANTIMONY

Antimony Distribution in the Earth's Crust Antimony is a rare element. Its average concentration in igneous rocks is 0.2 ppm, in shales 1.5 ppm, in limestone 0.3 ppm, in sandstone 0.05 ppm, and in soil 1 ppm; in seawater the average concentration is 0.24 ppb and in fresh water 0.2 ppb (Bowen, 1979). As a mineral it may occur as stibnite (Sb_2S_3), hermesite (Sb_2S_2O), senarmontite (Sb_2O_3), jamesonite ($2PbS \cdot Sb_2S_3$), barlangevite ($5PbS \cdot 2Sb_2S_3$), as sulfantimonides of copper, silver, and nickel, and sometimes as elemental antimony (Gornitz, 1972). Stibnite is the most common mineral. Antimony can exist in the oxidation states -3, 0, $+3$, and $+5$. Like arsenic compounds, antimony compounds tend to be toxic to most living organisms. The basis for this toxicity has not been clearly established.

Microbial Oxidation of Antimony Compounds   Among the earliest reports of the oxidation of antimony compounds is one by Bryner et al (1954) in which the oxidation of the mineral tetrahedrite ($4Cu_2S \cdot Sb_2S_3$) by *Thiobacillus ferrooxidans* is recorded. Lyalikova in 1961 reported the oxidation of antimony trisulfide (Sb_2S_3) by *T. ferrooxidans*. In both these instances the oxidation proceeded under acid conditions (pH 2.45)(Kuznetsov et al., 1963). The oxidation state of the solubilized antimony was not determined in either case. More recently, Silver and Torma (1974) reported on the oxidation of synthetic antimony sulfides by *T. ferrooxidans*, and Torma and Gabra (1977) on the oxidation of stibnite (Sb_2S_3) by the organism. The latter authors suspected that *T. ferrooxidans* oxidized trivalent antimony [Sb(III)] to pentavalent antimony [Sb(V)] but offered no proof. Lyalikova et al. in 1972 reported on the bacterial oxidation of of Sb-Pb sulfides, Sb-Pb-Te sulfides, and Sb-Pb-As sulfides with the formation of such minerals as anglesite ($PbSO_4$) and valentinite (Sb_2O_3). The formation of valentinite suggests that the antimony in the minerals was solubilized but not oxidized. Apparently most or all of the oxidation involved the sulfide in the minerals.

Trivalent antimony is, however, susceptible to direct oxidation by *Stibiobacter senarmontii* (Lyalikova, 1972, 1974; Lyalikova et al., 1976). This organism, which was isolated from a sample from a Yugoslavian mine drainage, oxidizes senarmontite (Sb_2O_3) or Sb_2O_4 to Sb_2O_5, deriving useful energy from the process. It is a gram-positive, motile rod (0.5–1.8 $\times$ 0.5 μm) with a single polar flagellum and has the ability to form rudimentary mycelium in certain stages of development. It grows at neutral pH and generates acid when oxidizing Sb_2O_3 (pH range 7.5–5.5). When grown on reduced antimony oxide (senarmontite), the organism possesses the enzyme ribulose bisphosphate carboxylase, indicating that it has chemolithotrophic propensity (Lyalikova et al., 1976). Antimony sulfide ores can thus be oxidized completely in two steps:

$$Sb_2S_3 \xrightarrow[\text{(1)}]{O_2} Sb_2O_3 \xrightarrow[\text{(2)}]{O_2} Sb_2O_5 \qquad (12.4)$$

The first step is catalyzed by an organism such as *Thiobacillus* Y or *T. ferrooxidans* [see (1) in reaction 12.4], and the second by *Stibiobacter senarmontii* [see (2) in reaction 12.4] (Lyalikova et al., 1974).

Microbial Reduction of Oxidized Antimony Minerals   No evidence exists that microbes are able to reduce oxidized antimony compounds, whether as terminal electron acceptor or for other purposes (see, for instance, Iverson and Brinckman, 1978). This lack of evidence should not, however, be taken to mean that it does not occur, but rather that it has so far not been observed.

12.3   SUMMARY

Although arsenic and antimony compounds are toxic to most forms of life, some microbes metabolize them. Arsenite has been found to be enzymatically oxidized by several different bacteria. The enzyme system is inducible. In laboratory experiments, *Alcaligens faecalis* oxidizes arsenite most intensely only after having gone through active growth. The organism probably can derive maintenance energy from arsenite oxidation. *Pseudomonas arsenitoxidans* can grow autotrophically with arsenite as sole source of energy. Simple and compound arsenic sulfides are oxidized by *Thiobacillus ferrooxidans*. No evidence has been obtained, however, that this organism can oxidize trivalent arsenic enzymatically to pentavalent arsenic.

Arsenite and arsenate have also been shown to be reduced by certain bacteria and fungi. When reducing the arsenic to the -3 oxidation state, the bacteria produce arsine and/or dimethyl arsine, whereas the fungi produce trimethylarsine. All of these arsines are volatile. Some bacteria may merely reduce arsenate to arsenite.

Antimony-containing compounds have also been shown to be microbially oxidized. *T. ferrooxidans* has been shown to attack a variety of antimony-containing sulfides. Although enzymatic oxidation of Sb(III) to Sb(V) has been claimed in the oxidation of Sb_2S_3 by *T. ferrooxidans*, clear proof is lacking. Generally, only the sulfide moiety and ferrous iron, if present in the mineral, are oxidized by this organism. *Stibiobacter senarmontii*, an autotroph that was isolated from an ore deposit in Yugoslavia, on the other hand, oxidizes the antimony in Sb_2O_3 or Sb_2O_4 to Sb_2O_5. Microbial reduction of oxidized antimony compounds has not been reported so far.

REFERENCES

Abdrashitova, S. A., B. N. Mynbaeva, and A. N. Ilyaletdinov. 1981. Oxidation of arsenic by the heterotrophic bacteria *Pseudomonas put-*

ida and *Alcaligenes eutrophus*. Mikrobiologiya 50:41–45 (Engl. transl. pp. 28–31).

Ahmann, D., A. L. Roberts, L. R. Krumholz, and F. M. M. Morel. 1994. Microbe grows by reducing arsenic. Nature (Lond.) 371:750.

Alexander, M. 1977. Introduction to Soil Microbiology. Wiley, New York.

Anderson, G. L., J. Williams, and R. Hille. 1992. The purification and characterization of arsenite oxidase from *Alcaligenes faecalis*, a molybdenum-containing hydroxylase. J. Biol. Chem. 267:23674–23682.

Andreae, M. O. 1979. Arsenic speciation in seawater and interstitial waters: The influence of biological-chemical interactions on the chemistry of a trace element. Limnol. Oceanogr. 24:440–452.

Bennett, R. L., and M. H. Malamy. 1970. Arsenate-resistant mutants of *Escherichia coli* and phosphate transport. Biochem. Biophys. Res. Commun. 40:496–503.

Bignelli, P. 1901. Gazz. Chim. Ital. 31:58 (as cited by Challenger, 1951).

Blasco, F., C. Gaudin, and R. Jeanjean. 1971. Absorption of arsenate ions by *Chlorella*. Partial reduction of arsenate to arsenite. C.R. Acad. Sci. (Paris) Ser. D 273:812–815.

Bowen, H. J. M. 1979. Environmental Chemistry of the Elements. Academic Press. London

Broeer, S., G. Ji, A. Broeer, and S. Silver. 1993. Arsenic efflux governed by the arsenic resistance determinant of *Staphylococcus aureus* plasmid pI258. J. Bacteriol. 175:3480–3485.

Bryner, L. C., J. V. Beck, B. B. Davis, and D. G. Wilson. 1954. Microorganisms in leaching sulfide minerals. Ind. Eng. Chem. 46:2587–2592.

Buchanan, R.E., and N.E. Gibbons. 1974. Bergey's Manual of Determinative Bacterology. Williams & Wilkins, Baltimore.

Button, D. K., S. S. Dunker, and M. L. Moore. 1973. Continuous culture of *Rhodotorula rubra*: kinetics of phosphate-arsenate uptake, inhibition and phosphate-limited growth. J. Bacteriol. 113:599–611.

Carapella, S. C. Jr. 1972. Arsenic: element and geochemistry, *In*: R. W. Fairbridge, ed. The Encyclopedia of Geochemistry and Environmental Sciences. Encyclopedia of Earth Sciences Series IVA. Van Nostrand Reinhold, New York, pp.31–42.

Carlson, L., E. B. Lindstroem, K. B. Hallberg, and O. H. Tuovinen. 1992. Solid-phase products of bacterial oxidation of arsenic pyrite. Appl. Environ. Microbiol. 58:1046–1049.

Challenger, F. 1933. The formation of organo-metalloidal compounds by microorganisms. Part II. Trimethylarsine and dimethylarsine. J. Chem. Soc., pp. 95–101.

Challenger, F. 1951. Biological methylation. Adv. Enzymol. 12:429–491.

Cheng, C-N., and D.D.Focht. 1979. Production of arsine and methylar-
sines in soil and in culture. Appl. Environ. Microbiol. 38:494–498.

Cox, D. P., and M. Alexander. 1973. Effect of phosphate and other anions
on trimethylarsine formation by *Candida humicola*. Appl. Microbiol.
25:408–413.

Da Costa, E. W. B. 1971. Suppresion of the inhibitory effects of arsenic
compounds by phosphate. Nature (New Biol.) (Lond.) 231:32.

Da Costa, E. W. B. 1972. Variation in the toxicity of arsenic compounds
to microorganisms and the suppression of the inhibitory effects of
phosphate. Appl. Microbiol. 23:46–53.

Dyke, K. G. H., M. T. Parker, and M. H. Richmond. 1970. Penicillinase
production and metal-ion resistance in *Staphylococcus aureus* cultures isolated from hospital patients. J. Med. Microbiol. 3:125–136.

Ehrlich, H. L. 1963. Bacterial action on orpiment. Econ. Geol. 58:
991–994.

Ehrlich, H. L. 1964. Bacterial oxidation of arsenopyrite and enargite.
Econ. Geol. 59:1306–1312.

Ehrlich, H. L. 1978. Inorganic energy sources for chemolithotrophic and
mixotrophic bacteria. Geomicrobiol. J. 1:65–83.

Ferguson, J. F., and J. Gavis. 1972. A review of the arsenic cycle in
natural waters. Water Res. 6:1259–1274.

Gornitz, V. 1972. Antimony; Element and geochemistry. *In*: R. W. Fair-
banks, ed. The Encyclopedia of Geochemistry and Environmental
Sciences. Encylopedia of Earth Sciences Series IVA. Van Nostrand
Reinhold, New York, pp. 33–36.

Gosio, B. 1897. Ber. [Dtsch. Chlem. Ges.) 30:1024 (as cited by Challenger,
1951).

Green, H.H. 1918. Description of a bacterium which oxidizes arsenite to
arsenate, and of one which reduces arsenate to arsenite, isolated
from a cattle-dipping tank. South Afr. J. Sci. 14:465–467.

Groudeva, V. I., S. N. Groudev, and K. I. Markov. 1986. A comparison
between mesophilic and thermophilic bacteria with respect to their
ability to leach sulfide minerals. *In*: R. W. Lawrence, R.M.R. Branion, and H.G. Ebner, eds. Fundamental and Applied Biohydrometallurgy. Elsevier, Amsterdam, pp. 484–485.

Hedges, R. W., and S. Baumberg. 1973. Resistance to arsenic compounds
conferred by a plasmid transmissible between strains of *Escherichia
coil*. J. Bacteriol. 115:459–460.

Hendrie, M. S., A. J. Holding, and J. M. Shewan. 1974. Emended description of the genus *Alkaligenes* and of *Alkaligenes faecalis* and a proposal that the generic name of *Achromobacter* be rejected: status of

the named species of *Alkaligenes* and *Achromobacter*. Int. J. Syst. Bacteriol. 24:534–550.

Ilyaletdinov, A. N., and S. A. Abdrashitova. 1981. Autotrophic oxidation of arsenic by a culture of *Pseudomonas arsenitoxidans*. Mikrobiologiya 50:197–204 (Engl. transl. pp. 135–140).

Iverson, W. P., and F. E. Brinckman. 1978. Microbial metabolism of heavy elements. *In*: Ralph Mitchell, ed. Water Pollution Microbiology. Vol. 2. Wiley, New York, pp. 201–232.

Ji, G., and S. Silver. 1992. Regulation and expression of the arsenic resistance operon from *Staphylococcus aureus* plasmid pI258. J. Bacteriol. 174:3684–3694.

Ji, G., S. Silver, E. A. E. Garber, H. Ohtake, C. Cervantes, and P. Corbisier. 1993. Bacterial molecular genetics and enzymatic transformations of arsenate, arsenite, and chromate. *In*: A. E. Torma, M. L. Apel, and C. L. Brierley, eds. Biohydrometallurgical Technologies. Vol. 2. Fossil Energy Materials Bioremediation, Microbial Physiology. Minerals, Metals, and Materials Society, Warrendale, PA, pp. 529–539.

Johnson, D. L. 1972. Bacterial reduction of arsenite in seawater. Nature (Lond.) 240:44–45.

Kamalov, M. R., G. I. Karavaiako, A. N. Ilyaletdinov, and S.A. Abdrashitova. 1973. The role of *Thiobacillus ferrooxidans* in leaching arsenic from a concentrate of carbonaceous gold-containing ore. Izv. Akad. Nauk. Kaz. SSR Ser. Biol. 11:37–44.

Karavaiko, G. I., L. K. Chuchalin, T. A. Pivovarova, B. A. Yemel'yanov, and A.G. Forofeyev. 1986. Microbiological leaching of metals from arsenopyrite containing concentrates. *In*: R. W. Lawrence, R. M. R. Branion, and H. G. Ebner. 1986. Fundamental and Applied Biohydrometallurgy. Elsevier, Amsterdam, pp. 115–126.

Kulibakin, V. G., and S. F. Laptev. 1971. Effect of adaptation of *Thiobacillus ferrooxidans* to a copper-arsenic-tin concentrate on the arsenic leach rate. Sb. Tr. Tsent. Nauk-Issled. Inst. Olomyan Prom., No. 1, pp. 75–76.

Kuznetsov, S. I., M. V. Ivanov, and N. N. Lyalikova. 1963. *Introduction to Geological Microbiology*. English translation. McGraw-Hill, New York.

Legge, J.W. 1954. Bacterial oxidation of arsenite. IV. Some properties of the bacterial cytochromes. Aust. J. Biol. Sci. 7:504–514.

Legge, J.W., and A.W. Turner. 1954. Bacterial oxidation of arsenite. III. Cell-free arsenite dehydrogenase. Aust. J. Biol. Sci. 7:496–503.

Livesey-Goldblatt, E., P. Norman, and D.R. Livesey-Goldblatt. 1983.

Gold recovery from arsenopyrite/pyrite ore by bacterial leaching and cyanidation. *In*: G. Rossi and A. E. Torma, eds. Recent Progress in Biohydrometallurgy. Associazione Mineraria Sarda, 09016 Iglesias, Italy, pp. 627–641.

Lunde, G. 1973. Synthesis of fat and water soluble arseno organic compounds in marine and limnetic algae. Acta Chem. Scand. 27: 1586–1594.

Lyalikova, N. N. 1961. The role of bacteria in the oxidation of sulfide ores. Tr. In-ta Mikrobiol. AN SSSR, No. 9 (as cited by Kuznetsov et al., 1963).

Lyalikova, N.N. 1972. Oxidation of trivalent antimony to higher oxides as an energy source for the development of a new autotrophic organism *Stibiobacter* gen. n. Dokl. Akad. Nauk SSSR Ser. Biol. 205: 1228–1229.

Lyalikova, N. N. 1974. *Stibiobacter senarmontii*—A new antimony-oxidizing microorganism. Mikrobiologiya 43:941–948 (Engl. transl. pp.799–805).

Lyalikova, N. N. L. B. Shlain, O. G. Unanova, and L. S. Anisimova. 1972. Transformation of products of compound antimony and lead sulfides under the effect of bacteria. Izv. Akad. Nauk SSSR Ser. Biol. No. 4, 564–567.

Lyalikova, N. N., L. B. Shlain, and V. G. Trofimov. 1974. Formation of minerals of antimony(V) under the effect of bacteria. Izv. Akad. Nauk SSSR Ser. Biol. No. 3, 440–444.

Lyalikova, N. N., I. Ya. Vedenina, and A. K. Romanova. 1976. Assimilation of carbon dioxide by a culture of *Stibiobacter senarmontii*. Mikrobiologiya 45:552–554 (Engl. transl. pp. 476–477).

Mahler, H. R., and E. H. Cordes. 1966. Biological Chemistry. Harper & Row, New York.

McBride, B. C., and R. S. Wolfe. 1971. Biosynthesis of dimethyl arsine by *Methanobacterium*. Biochemistry 10:4312–4317.

Myers, D. J., M. E. Heimbrook, J. Osteryoung, and S. M. Morrison. 1973. Arsine oxidation state in the presence of microorganisms. Examination by differential pulse polarography. Environ. Lett. 5: 53–61.

Ngubane, W. T., and A. A. W. Baecker. 1990. Oxidation of gold-bearing pyrite and arsenopyrite by *Sulfolobus acidocladarius* and *Sulfolobus* BC in airlift bioreactors. Biorecovery 1:255–269.

Olson, G. J. 1994. Microbiological oxidation of gold ores and gold bio-leaching. FEMS Microbiol. Lett. 119:1–6.

Osborne, F. H. 1973. Arsenite oxidation by a soil isolate of *Alcaligenes*. Ph.D. thesis. Rensselaer Polytechnic Institute, Troy, NY.

Osborne, F. H., and H. L. Ehrlich. 1976. Oxidation of arsenite by a soil isolate of Alcaligenes. J. Appl. Bacteriol. 41:295–305.

Oscarson, D. W., P. M. Huang, C. Defosse, and A. Herbillon. Oxidative power of Mn(IV) and Fe(III) oxides with respect to As(III) in terrestrial and aquatic environments. 1981. Nature (Lond.):291:50–51.

Phillips, S. E., and M. L. Taylor. 1976. Oxidation of arsenite to arsenate by *Alkaligenes faecalis*. Appl. Environ. Microbiol. 32:392–399.

Pickett, A. W., B. C. McBride, W. R. Cullen, and H. Manji. 1981. The reduction of trimethylarsine oxide by *Candida humicola*. Can. J. Microbiol. 27:773–778.

Pol'kin, S. I., and Z. A. Tanzhnyanskaya. 1968. Use of bacterial leaching for ore enrichment. Izv. Vyssh. Ucheb. Zaved. Tsvet. Met. 11: 115–121.

Pol'kin, S. I., N. Yudina, V. V. Nanin, and D. Kh. Kim. 1973. Bacterial leaching of arsenic from an arsenpyrite gold-containing concentrate in a thick pulp. Nauchno-Issled. Geologorazved. Inst. Tsvetn. Blagorodn. Metall. no. 107, pp. 34–41.

Quastel, J. H., and P. G. Scholefield. 1953. Arsenite oxidation in soil. Soil Sci. 75:279–285.

Scudlark, J. R., and D. L. Johnson. 1982. Biological oxidation of arsenite in seawater. Estuarine, Coastal and Shelf Science 14:693–706.

Sehlin, H. M., and E. B. Lindstroem. 1992. Oxidation and reduction of arsenic by *Sulfolobus acidocaldarius* strain BC. FEMS Microbiol. Lett. 93:87–92.

Silver, M. and A. E. Torma. 1974. Oxidation of metal sulfides by *Thiobacillus ferrooxidans* grown on different substrates. Can. J. Microbiol. 20:141–147.

Silver, S., and D. Keach. 1982. Energy-dependent arsenate flux: the mechanism of plasmid-mediated resistance. Proc. Natl. Acad. Sci. USA 79:6114–6118.

Silverman, M. P., and D. G. Lundgren. 1959. Studies on the chemoautotrophic iron bacterium *Ferrobacillus ferrooxidans*. I. An improved medium and a harvesting procedure for securing high cell yields. J. Bacteriol. 77:642–647.

Torma, A. E., and G. G. Gabra. 1977. Oxidation of stibnite by *Thiobacillus ferrooxidans*. Antonie v. Leeuwenhoek 43:1–6.

Tuovinen, O. H., T. M. Bhatti, J. M. Bigham, K. B. Hallberg, O. Garcia, Jr., and E. B. Linstroem. 1994. Oxidative dissolution of arsenopyrite by mesophilic and mederately thermophilic acidophiles. Appl. Environ. Microbiol. 60:3268–3274.

Turner, A. W. 1949. Bacterial oxidation of arsenite. Nature (Lond.) 164: 76–77.

Turner, A. W. 1954. Bacterial oxidation of arsenite. I. Description of bacteria isolated from arsenical cattle-dipping fluids. Aust. J. Biol. Sci. 7:452–478.

Turner, A. W., and J. W. Legge. 1954. Bacterial oxidation of arsenite. II. The activity of washed suspensions. Aust J. Biol. Sci. 7:479–495.

Woolfolk, C. A., and H. R. Whiteley. 1962. Reduction of inorganic compounds with molecular hydrogen by *Micrococcus lactilyticus*. I. Stoichiometry with compounds of arsenic, selenium, tellurium, transition and other elements. J. Bacteriol. 84:647–658.15

13

Geomicrobiology of Mercury

The element mercury has been known as a specific chemical from at least as far back as 1500 B.C. The physician Paracelsus (A.D. 1493–1541) attempted to cure syphilis by administering metallic mercury to sufferers of the disease. His treatment was probably based on intuitive or empirical knowledge that at an appropriate dosage, mercury was more toxic to the cause of the disease than to the patient. The true etiology of syphilis was, however, unknown to him. The extent of mercury toxicity for human beings and other animals became very apparent only in recent times as a consequence of environmental pollution by mercury compounds. The toxicity manifests itself in major physical impairments and death due to intake of the compounds in food and water. Incidents of mercury poisoning in Japan (Minamata disease), Iraq, Pakistan, Guatemala, and the United States drew special attention to the problem. In some of these cases, food made from seed grain treated with mercury compounds to inhibit fungal damage before planting was consumed. The seed grain had not been intended for food use. In other cases, food such as meat had become tainted because the animals yielding the meat drank water that had become polluted by mercury compounds or they had eaten mercury-tainted feed. Tracing of the fate of mercury introduced into the environment has revealed an intimate role of microbes in the interconversion of some mercury compounds.

13.1 DISTRIBUTION OF MERCURY IN THE EARTH'S CRUST

The concentration of mercury in the Earth's crust has been reported as 0.08 ppm (Jonasson, 1970). Its concentration in uncontaminated soils has been given as 0.07 ppm (Jonasson, 1970). Its concentration in fresh waters may range from 0.01 to 10 ppb, although concentrations as high as 1,600 ppb have been measured in waters in contact with copper deposits in the southern Urals (Jonasson, 1970). The maximum permissible level in potable waters in the United States has been set at 5 ppb. The average mercury concentration in seawater has been reported at 0.2 ppb (Marine Chemistry, 1971).

Mercury can exist in nature as mercury metal or as mercury compounds. The metal is liquid at ambient temperature and has a significant vapor pressure (1.2×10^{-3} mm Hg at 20°C) and a heat of vaporization of 14.7 cal mol^{-1} at 25°C (Vostal, 1972). The most prevalent mineral of mercury is cinnabar (HgS). It is found in highest concentrations in volcanically active zones, such as the circum-Pacific volcanic belt, the East Pacific Rise, and the Mid-Atlantic Ridge. The occurrence of mercury metal is rarer. In water, inorganic mercury may exist as aquo, hydroxo, halido, and bicarbonate complexes of mercuric ion, but the mercuric ion may also be adsorbed to particulate or colloidal materials in suspension (Jonasson, 1970). In soil, inorganic mercury may exist in the form of elemental mercury vapor that may be adsorbed to soil matter, at least in part. It can also exist as mercuric humate complexes at pH 3–6 or as Hg(OH)$^+$ and Hg(OH)$_2$ in the pH range 7.5–8.0. The latter two species may be adsorbed to soil particles (Jonasson, 1970). Mercury in soil and water may also exist as methylmercury [(CH$_3$)Hg$^+$] which may be adsorbed by negatively charged particles such as clays.

13.2 ANTHROPOGENIC MERCURY

The local mercury level may be affected by human activity. Some industrial operations, such as the synthesis of certain chemicals like vinyl chloride and acetaldehyde, which employ inorganic mercury compounds as catalysts, or the electrolytic production of chlorine gas and caustic soda, which employs mercury electrodes, or the manufacture of paper pulp, which makes use of phenylmercuric acetate as a slimicide (Jonasson,

1970), may pollute the environment. In agriculture, organic compounds used as fungicides to prevent fungal attack of seed may pollute the soil. In mining, the exposure of mercury ore deposits and other deposits in which mercury is only a trace component leads to weathering and resultant solubilization and spread of some of the mercury into the environment.

13.3 MERCURY IN THE ENVIRONMENT

As Jonasson (1970) has pointed out, inorganic mercury compounds were considered less toxic in the past than organic mercury compounds, but since the discovery that inorganic mercury compounds can be converted into organic ones (e.g., methyl mercury), this is no longer considered to be true. Living tissue has a high affinity for methylmercury [$(CH_3)Hg^+$]. Fish have been found to concentrate it up to 3,000 times over the concentration found in water. This is because methylmercury is fat- as well as water-soluble and is more readily taken up by living cells than mercuric ion. Owing to its lipid solubility, nervous tissue, especially the brain, has a high affinity for methylmercury. It is also bound by inert matter, especially negatively charged particles such as clays.

Dimethylmercury [$(CH_3)_2Hg$] is volatile. It can thus enter the atmosphere from soil or water phases. The ultraviolet component of sunlight can, however, dissociate dimethylmercury into volatile elemental mercury, methane, and ethane.

Microorganisms have in recent years been shown to be intimately involved in interconversions of inorganic and organic mercury compounds. The initial discoveries of the microbial activities were those of Jensen and Jerneloev (1969), who demonstrated the production of methylmercury from mercuric chloride ($HgCl_2$) added to lake sediment samples and incubated for several days in the laboratory. They also noted the production of dimethylmercury [$(CH_3)_2Hg$] from decomposing fish tissue, containing methylmercury or supplemented with Hg^{2+}, and incubated for several weeks. Later work established that methylation could be brought about by bacteria and fungi (see below).

Most microbial interactions with mercury are **detoxification** reactions. By forming volatile elemental mercury or dimethylmercury, neither of which is water soluble, the microbes ensure removal of mercury from their environment into the atmosphere. Even microbial methylmercury formation may be a form of mercury detoxification, because methylmercury can be immobilized in sediment or soil by adsorption to negatively charged clay particles, which removes it as a toxicant from the microbial environment. Similarly, the precipitation of HgS by reaction of Hg^{2+} with biogenic H_2S is a type of mercury detoxification since the solubility of HgS is very low (K_{sol}, 10^{-49}). Of all these detoxification mechanisms, formation of volatile metallic mercury appears to be the predominant microbial detoxification mechanism (Robinson and Tuovinen, 1984). Baldi et al. (1987) demonstrated the presence of a significant number of mercury-resistant bacteria that could reduce Hg^{2+} to Hg° but not methylate it at sites surrounding natural mercury deposits situated in Tuscany, Italy. Baldi et al. (1989) also isolated 37 strains of aerobic, mercury-resistant bacteria from Fiora River in southern Tuscany, which receives drainage from a cinnabar mine. All 37 strains were able to reduce Hg^{2+} to Hg° but only three were also able to degrade methylmercury. None were able to generate methylmercury.

13.4 SPECIFIC MICROBIAL INTERACTIONS WITH MERCURY

Microbial Methylation of Mercury   An early study of the biochemistry of microbial methylation of mercury involved the use of cell extract of a methanogenic culture in the presence of low concentrations of Hg^{2+}, which caused formation of $(CH_3)_2Hg$ but little methane (CH_4), through preferential interaction between methylcobalamin and Hg^{2+} (Wood et al., 1986). Although the production of methylcobalamin in this instance depended on enzymatic catalysis, the production of $(CH_3)_2Hg$ from the reaction of Hg^{2+} with methylcobalamin did not. This nonenzymatic nature of mercury methylation by methylcobalamin was confirmed by Bertilsson and Neujahr (1971), Imura et al. (1971), and Schrauzer et al. (1971) (see, however, the more recent observations with *Desulfovibrio desulfuricans* described below). The mechanism of mercury methylation by methylcobalamin can be summarized as follows (DeSimone et al., 1973):

$$Hg^{2+} \xrightarrow{\text{CH}_3\text{B}_{12}} (CH_3)Hg^+ \xrightarrow{\text{CH}_3\text{B}_{12}} (CH_3)_2Hg \tag{13.1}$$

The initial methylation step of Hg^{2+} in this reaction sequence proceeds 6,000 times as fast as the second methylation step (Wood, 1974).

Subsequent studies have revealed that mercury can be methylated by microbes other than methanogens, including both anaerobes and aerobes (see review by Robinson and Tuovinen, 1984). Among anaerobes, *Clostridium cochlearium* was shown to methylate mercury contained in HgO, $HgCl_2$, $Hg(NO_3)_2$, $Hg(CN)_2$, $Hg(SCN_2$, and $Hg(OOCCH_3)_2$ (Yamada and Tonomura, 1972a,b).

Among aerobes, *Pseudomonas* spp., *Bacillus megaterium*, *Escherichia coli*, *Enterobacter aerogenes*, and others have been implicated (see summary by Robinson and Tuovinen, 1984). Even fungi such as *Aspergillus niger*, *Scopulariopsis brevicaulis*, and *Saccharomyces cerevisiae* have been found capable of mercury methylation (see Robinson and Tuovinen, 1984).

Sulfate reducers such as *Desulfovibrio desulfuricans* appear to be the principal methylators of mercury in some anoxic estuarine sediments when sulfate is limiting and fermentable organic energy sources are available (Compeau and Bartha, 1985). The methyl group was shown to originate from carbon-3 of the amino acid serine, which was transformed to methyl carbon via tetrahydrofolate and transferred to mercury via cobalt prophyrin (methylcobalamin) (Berman et al., 1990; Choi and Bartha, 1993). In *D. desulfuricans* the methyl transfer from methylcobalamin to mercury was shown to be enzymatically catalyzed. It involves a methylcobalamin-protein complex, the protein being the catalyst of the methyl transfer (Choi et al., 1994a,b). The methylation of the cobalamin-protein complex results from a methyl group transfer from methyl tetrahydrofolate (methyl THF). The methyl THF is thought to derive its methyl group from serine via serine hydroxymethyl transferase activity and subsequent reduction of the methylene group on the THF to a methyl group. Methyl THF may also arise from formate via formyl THF, methenyl THF, and methylene THF. The normal role of the methylcobalamin-protein complex in *D. desulfuricans* is to provide the methyl group in acetate synthesis from CO_2 by the acetyl~ScoA pathway (see Chapters 6 and 17). Hg^{2+} evidently acts as a competing methyl acceptor in acetate formation in the organism (Choi et al., 1994b). The methyl transfer from the methylcobalamin protein complex to mercury follows Michaelis-Menten kinetics and is 600 times faster at pH 7 than from methyl cobalamin alone. These findings raise a question about the abiotic mercury methylation ascribed to other unrelated bacteria and other microorganisms to which reference was made above.

Besides mercury, bacteria and other microbes have been found to be able to methylate other metals. Among these metals are cadmium, lead and tin, and the metalloids selenium and tellurium (Brinckman et al., 1976; Guard et al., 1981; Wong et al., 1975; review by Summers and Silver, 1978). Methylation of some metals may occur as a result of nonbiological transmethylation by biogenic methylated donor compounds such as trimethyltin and methyl iodide (Brinckman and Olson, 1986). Methyl halides, including methyl iodides (White, 1982; Brinckman and Olson, 1986) are produced by some marine algae and also by associated microorganisms (White, 1982; Manley and Dastoor, 1988) and fungi (Harpur, 1985). They can readily react nonenzymatically with some metal salts (Brinckman and Olson, 1986). Trimethyl tin yields carbanions which can methylate metal ions such as those of palladium, thallium, platinum and gold, forming unstable methylated intermediates which undergo reductive demethylation to yield the metal in the elemental state. Mercuric ion reacting with trimethyl tin forms, however, stable methylmercury (Brinckman and Olson, 1986).

The fungus *Neurospora crassa* uses a still different reaction for methylating mercury (Landner, 1971). This organism first complexes mercuric ion with homocysteine or cysteine nonenzymatically, and then, with the help of a methyl donor and the enzyme transmethylase, cleaves (CH_3Hg^+) from this complex. The following illustrates this reaction with homocysteine:

$$
\begin{array}{ccc}
\text{SH} & \text{SHg}^+ & \text{SH} \\
| & | & | \\
\text{CH}_2 & \text{CH}_2 & \text{CH}_2 \\
| & | & | \\
\text{CH}_2 & \text{CH}_2 & \text{CH}_2 \\
| & | & \xrightarrow[\text{transmethylase}]{\text{methyl donor}} | \\
\text{Hg}^{2+} + \text{CHNH}_2 \longrightarrow \text{CHNH}_2 & & \text{CHNH}_2 + (\text{CH}_3)\,\text{Hg}^+ \\
| & | & | \\
\text{COOH} & \text{COOH} & \text{COOH}
\end{array}
$$

(13.2)

Suitable methyl donors are betaine or choline but not methylcobalamin.

Although in the laboratory, bacterial methylation of mercury appears to be favored by anaerobic conditions, partially aerobic conditions are needed in nature. This is because under in situ anaerobic conditions, biogenic H_2S may prevail and, as a result, mercuric mercury will exist most probably as HgS (Fagerstrom and Jerneloev, 1971; Vostal 1972). HgS cannot be methylated without prior conversion to a soluble Hg^{2+} salt or HgO (Yamada and Tonomura, 1972c). It has been shown that H_2S, which in nature is frequently of biogenic origin, can transform preexisting methylmercury into dimethylmercury (Craig and Bartlett, 1978).

Microbial Diphenylmercury Formation   A case of microbial conversion of phenylmercuric acetate to diphenylmercury has been reported (Matsumara et al., 1971). The reaction can be summarized as follows:

$$2\phi Hg^+ \rightarrow \phi\text{-Hg-}\phi + \text{unknown Hg compound and trace of } Hg^{2+}$$

(13.3)

Microbial Reduction of Mercuric Ion   Mercuric ion is reduced to volatile metallic mercury by a number of bacteria and fungi. Active organisms include strains of *Pseudomonas* spp., enteric bacteria, *Staphylococcus aureus*, *Thiobacillus ferrooxidans*, group B *Streptococcus*, *Streptomyces*, and *Cryptococcus* (Brunker and Bott, 1974; Komura et al., 1970; Nelson et al., 1973; Olson et al., 1981; Nakahara et al., 1985; Summer and Lewis, 1973). On the basis of mercury-resistance, which correlates with Hg^{2+}-reducing ability, *Bacillus*, *Vibrio*, coryneform bacteria, *Cytophaga*, *Flavobacterium*, *Achromobacter*, *Alcaligenes*, and *Acinetobacter* (Nelson and Colwell, 1974) should also be included. A significant number of *Pseudomonas* strains having Hg^{2+}-reducing ability were found unable to utilize glucose (Nelson and Colwell, 1974).

The enzyme system in Hg^{2+} reduction to $Hg°$ involves a soluble, NADPH-dependent, cytoplasmic flavoprotein which is active in the presence of excess exogenously supplied thiols. The thiol compounds (in the lab they may be mercaptoethanol, dithiothreitol, glutathione, or cysteine) insure the reduced state of mercuric reductase and the formation of a mercapto derivative of mercuric mercury, which is the actual substrate for mercuric reductase. The reaction catalyzed by mercuric reductase may be summarized as follows (Fox and Walsh, 1982; Foster, 1987):

$$NADPH + H^+ + RS\text{-}Hg\text{-}SR \rightarrow NADP^+ + Hg° + 2RSH.$$

$$(13.4)$$

The dimercaptyl derivative of Hg^{2+} may be replaced by a monomercaptyl or an ethylenediamine (EDTA) derivative in some reactions. NADH can be substituted for NADPH with enzyme preparations from some organisms, but it is less active. The kinetics for the purified enzyme are biphasic. [See review by Robinson and Tuovinen (1984) for a more detailed discussion]. Although the reaction occurs under reducing conditions, it is performed by many obligate and facultative aerobes (e.g., Nelson et al., 1973; Spangler et al., 1973a,b).

Mercuric reductase activity is not entirely substrate specific. Besides mercuric ion, the enzyme can also reduce ionic silver and ionic gold to corresponding metal colloids (Summers and Sugarman, 1974; Summers, 1986). Silver or gold resistance in bacteria is, however, not related to this enzyme activity (Summers, 1986).

Not all reduction of mercuric mercury observed in nature is biological (Nelson and Colwell, 1975). Chemical reduction may occur as a result of interaction with humic acid (Alberts et al., 1974).

Microbial Decomposition of Organomercurials   Phenyl- and methylmercury can be microbially converted to volatile $Hg°$ by bacteria in lake and estuarine sediments and in soil (Nelson et al., 1973; Spangler et al., 1973a,b; Tonamura et al., 1968). The bacteria which seem most frequently involved are mercury resistant strains of *Pseudomonas*. Although mercury-resistant strains of other genera, such as *Escherichia* and *Staphylococcus*, also exhibit this activity, they seem to be much less active (Nelson and Colwell, 1975). The removal of phenyl or methyl groups linked to mercuric mercury is catalyzed by a class of enzymes called mercuric lyases. They catalyze the cleavage of carbon-mercury bonds and in laboratory demonstrations require the presence of an excess of reducing agent such as L-cysteine. They release Hg^{2+}, which is then reduced to $Hg°$ by mercuric reductase (Furukawa and Tonomura, 1971, 1972a,b; Tezuka and Tonomura, 1976, 1978; Robinson and Tuovinen, 1984). Phenylmercuric lyase can be inducible (Nelson et al., 1973; Robinson and Tuovinen, 1984). The overall reactions can be summarized as follows:

$$\phi Hg^{+} + H^{+} + 2e \rightarrow Hg° + \phi H \tag{13.5}$$

$$(CH_3)Hg^{+} + H^{+} + 2e \rightarrow Hg° + CH_4 \tag{13.6}$$

Besides methyl- and phenylmercury compounds, some bacteria are able to decompose one or more of the following organomercurials: ethylmercuric chloride (EMC), fluorescein mercuric acetate, parahydroxymercuribenzoate (pHMB), thimerosal, and merbromin (see review by Robinson and Tuovinen, 1984).

Oxidation of Metallic Mercury   Elemental mercury ($Hg°$) has been reported to be oxidizable to mercuric ion in the presence of certain bacteria (Holm and Cox, 1975). Whereas strains of *P. aeruginosa*, *P. fluorescens*, *E. coli* and *Citrobacter* oxidized only small amounts of $Hg°$, strains of *B. subtilis* and *B. megaterium* oxidized more significant amounts. In none of these cases was methylmercury formed. The observed oxidation was not enzymatic, but was due to reaction with metabolic products which acted as oxidants. Even yeast extract was found to be able to oxidize $Hg°$.

13.5   GENETIC CONTROL OF MERCURY TRANSFORMATIONS

In general, resistance to the toxicity of inorganic mercury compounds in bacteria is attributable to the ability to form mercuric reductase, and for certain organomercurials, mercuric lyase. However, in a strain of *Enterobacter aerogenes*, bacterial resistance to some organomercurials may be due to membrane impermeability (Pan Hou et al., 1981), and in *Clostridium cochlearium* it is due to demethylation followed by precipitation with H_2S generated by the organism (Pan Hou and Imura, 1981).

The genes encoding mercuric reductase and mercuric lyase are usually found on plasmids; i.e., they are R- or sex-factor-linked (Belliveau and Trevors, 1990; Komura and Izaki, 1971; Loutit, 1970; Novice, 1967; Richmond and John, 1964; Smith, 1967; Schottel et al., 1974; Summers and Silver, 1972). Except in *Thiobacillus ferrooxidans*, the mercury resistance genes (*mer*) in all bacteria so far tested are expressed only in the presence of mercury compounds; i.e., the enzymes they code for are *inducible* (Robinson and Tuovinen, 1984). Depending on the organisms, induction of the two enzymes, mercuric lyase and mercuric reductase, may be **coordinated**; i.e., the two genes are under common regulatory control. In such an instance, an organomercurial would induce both the lyase and the mercuric reductase (see Robinson and Tuovinen, 1984). In *T. ferrooxidans* the mercuric resistance (Hg^r) trait appears to be constitutive (Olson et al., 1982). In a number of instances the mercury resistance trait has been found to be **transposable**; i.e., the gene complex may move to other positions on the same or other plasmids or the bacterial chromosome within a given cell (Foster, 1987). The genetic determinants in *C. cochlearium* for demethylation of methyl mercury and mercury precipitation as a sulfide are also plasmid-encoded (Pan Hou and Imura, 1991).

Many gram-positive bacteria that are sensitive to mercury have been found to possess the gene (*merA*) that codes for the mercuric reductase enzyme. Because synthesis of mercuric reductase is inducible in these strains, their mercury sensitivity is apparently due to the lack of a determinant(s) for a functional mercury transport mechanism into the cell (Bogdanova et al., 1992). The needed components of the mercury-resistance gene complex (operon) have been determined by Hamlett et al. (1992).

In nature, plasmid-determined mercury resistance in bacteria can be transferred from resistant to susceptible cells through conjugation or phage transduction among gram-negative organisms, and through phage transduction among gram-positive organisms (Summers and Silver, 1972). In other words, a mercury sensitive strain can become mercury resistant by acquiring a plasmid containing the *mer* genes from a Hg^r bacterium.

In the fungus, *Neurospora crassa*, in which Hg^{2+} methylation is the basis for mercury tolerance, the mercury-resistant strain isolated by Landner (1971) appears to be a constitutive mutant that has lost control over one of the last enzymes in methionine biosynthesis, so that methylation of Hg^{2+} no longer competes with methionine synthesis, as it does in the wild-type parent strain. The resistant strain could tolerate 225 mg Hg liter^{-1}.

13.6   ENVIRONMENTAL SIGNIFICANCE OF MICROBIAL MERCURY TRANSFORMATIONS

The enzymic attack of mercury compounds is not for the dissipation of excess reducing power or respiration, nor is it for the production of useful metabolites. Its function is detoxification. This is well illustrated in experiments with mercury-sensitive and -resistant strains of *Thiobacillus ferrooxidans* (Baldi and Olson, 1987). The mercury-sensitive strain could not oxidize pyrite admixed with cinnabar (HgS) in oxidation columns, whereas two resistant strains could. Yet, even the resistant strains could not use the cinnabar as an energy source. The mercury-resistant strains volatilized the mercury as Hg°.

Volatile mercury (Hg°) and dimethylmercury, owing to their high volatility and low water solubility, are readily lost to the atmosphere from the normal habitat of microbes and other creatures. Methylmercury because of its positive charge can become immobilized by adsorption to negatively charged clay particles in soil and sediment. The metabolic transformation to volatile forms of mercury or to methylmercury which may become fixed in soil or sediment protects not only the organisms actively involved in the conversion but also co-inhabitants which lack this ability and are more susceptible to mercury poisoning. By contrast, development of mercury resistance in microbes which is due to lessened permeability to a mercury compound benefits only those organisms which have acquired this trait.

13.7 A MERCURY CYCLE

On the basis of the interactions of microbes with mercury compounds described in the foregoing sections, it is apparent that microbes play an important role in the movement of mercury in nature, i.e., in soil, sediment, and aqueous environments. One of the main results of microbial action on mercury, whether in the form of mercuric ion or alkyl- or arylmercury ions, seems to be its volatilization as Hg°. Methylmercury ion may also be an important end-product of microbial action on mercuric ion, but it can also be an intermediate in the formation of volatile dimethylmercury. Methylmercury is more toxic to susceptible forms of life than mercuric ion, owing to the greater lipid solubility combined with its positive charge of the former.

A mercury cycle is outlined in Figure 13.1. Mercuric sulfide of volcanic origin slowly autoxidizes to mercuric sulfate on exposure to air and moisture and may become disseminated in soil and water through groundwater movement. Bacteria and fungi may reduce the Hg^{2+} to $Hg°$, as may humic substances. The volatile mercury (Hg°) may be adsorbed by soil, sediment, and humic substances, or lost to the atmosphere. Some of the Hg^{2+} may also become methylated through the action of bacteria and fungi. Some of the positively charged mercuric ions, methyl mercury

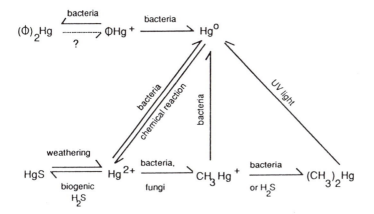

Figure 13.1 Mercury transformations by microbes and by chemical and physical agents.

ions as well as phenyl mercuric ions, may be fixed by negatively charged soil and sediment particles and by humic matter and thereby become immobilized. This may explain why mercury concentrations in soil are often higher in top soil than in subsoil (Anderson, 1967). Those ions that are not fixed can be further disseminated by water movement.

Mercury methylation and demethylation has been observed to occur seasonally in the surficial sediments of the deep parts of an oligotrophic lake, with measurable amounts of the methyl mercury passing into the overlying water column to be subsequently demethylated and reduced to volatile metallic mercury (Korthals and Winfrey, 1987). Methylmercury in the surficial sediments was also demethylated. During the year, methylation was most intense from mid-July through September, whereas demethylation was most active from spring to late summer (Korthals and Winfrey, 1987).

Methylation and demethylation of mercury together with mercury reduction has also been observed in estuarine sediments (Compeau and Bartha, 1984), sulfate reducers being the principal agents of methylation when the supply of sulfate is limiting (Compeau and Bartha, 1985).

Some methylmercury ions may be further methylated to volatile dimethylmercury, which readily escapes into the atmosphere from soil and water. In the atmosphere it may be decomposed into Hg°, methane and ethane by ultraviolet radiation from the Sun. Phenylmercury, which is usually of manmade origin, may be reduced to volatile mercury by bacteria in soil, but it may also be converted to diphenylmercury. Mercuric ion may be converted to mercuric sulfide by bacterially generated H_2S and thereby be immobilized. Only the formation of mercuric sulfide is strictly dependent on anaerobiosis; methylation or reduction of mercuric mercury can occur aerobically or anaerobically.

13.8 SUMMARY

Mercuric ion (Hg^{2+}) may be methylated by bacteria and fungi to methylmercury [$(CH_3)Hg^+$]. In some bacteria this methylation is enzymatic. At least some fungi use a methylation mechanism different from bacteria. Methylmercury is water-soluble and more toxic than mercuric ion. In nature, some methylmercury may, however, be adsorbed by sediment, which removes it as a toxicant as long as it remains adsorbed. Some bacteria can methylate methylmercury, forming volatile dimethylmercury. This may be a form of detoxification when it occurs in soil or sediment because this compound is water-insoluble and can escape into the atmosphere. Methylmercury as well as phenylmercury can be enzymatically reduced to volatile metallic mercury ($Hg°$) by some bacteria. Phenylmercury can also be microbially converted to diphenylmercury. Most importantly, mercuric ion can be enzymatically reduced to metallic mercury by bacteria and fungi. Reduction of Hg^{2+} to $Hg°$ is a more important detoxification reaction in nature than its methylation. This is a form of detoxification, because the product is volatile. Metallic mercury may also be oxidized to mercuric mercury by bacteria. However, this reaction is not enzymatic but rather the result of interaction with metabolic by-products. Microbes metabolizing mercury are generally resistant to its toxic effects.

The mercury cycle in nature is under the influence of microorganisms.

REFERENCES

Alberts, J. J., J. E. Schindler, R. W. Miller, and D. E. Nutter, Jr. 1974. Elemental mercury evolution mediated by humic acid. Science 184: 895–897.

Anderson, A. 1967. Mercury in soil. Grundfoerbaettring 20:95–105.

Baldi, F., and J. G. Olson. 1987. Effects of cinnabar on pyrite oxidation by *Thiobacillus ferrooxidans* and cinnabar mobilization by a mercury-resistant strain. Appl. Environ. Microbiol. 53:772–776.

Baldi, F., G. J. Olson, and F. E. Brinckman. 1987. Mercury transformations by heterotrophic bacteria isolated from cinnabar and other metal sulfide deposits in Italy. Geomicrobiol. J. 5:1–16.

Baldi, F., M. Filippelli, and G. J. Olson. 1989. Biotransformation of mercury by bacteria isolated from a river collecting cinnabar mine waters. Microb. Ecol. 17:263–274.

Belliveau, B. H., and J. T. Trevors. 1990. Mercury resistance determined by a self-transmissible plasmid in *Bacillus cereus* 5. Biol. Metals 3: 188–196.

Berman, M., T. Chase, Jr., and R. Bartha. 1990. Carbon flow in mercury biomethylaltion by *Desulfovibrio desulfuricans*. Appl. Environ. Microbiol. 56:298–300.

Bertilsson, L., and H. Y. Neujahr. 1971. Methylation of mercury compounds by methylcobalamin. Biochemistry 10:2805–2808.

Bogdanova, E. S., S. Z. Mindlin, E. Pakrova, M. Kocur, and D. A. Rouch. 1992. Mercuric reductase in environmental gram-positive bacteria sensitive to mercury. FEMS Microbiol. Lett. 97:95–100.

Brinckman, F. E., and G. J. Olson. 1986. Chemical principles underlying bioleaching of metals from ores and solid wastes, and bioaccumulation of metals from solution. *In*: H. L. Ehrlich and D. S. Holmes, eds. Workshop on Biotechnology for the Mining, Metal-Refining, and Fossil Fuel Processing Industries. Biotech. Bioeng. Symp. 16. Wiley, New York, pp. 35–44.

Brinckman, F. E., W. P. Iverson, and W. Blair. 1976. Approaches to the study of microbial transformations of metals. *In*: J. M. Sharpley and A. M. Kaplan, eds. Proceedings of the Third International Biodegradation Symposium. Applied Science Publishers, pp. 919–936.

Brunker, R. L., and T. L. Bott. 1974. Reduction of mercury to the elemental state by a yeast. Appl. Microbiol. 27:870–873.

Compeau, G., and R. Bartha. 1984. Methylation and demethylation of mercury under controlled redox, pH, and salinity conditions. Appl. Environ. Microbiol. 48:1203–1207.

Compeau, G, and R. Bartha. 1985. Sulfate-reducing bacteria: principal methylators of mercury in anoxic estuarine sediments. Appl. Environ. Microbiol. 50:498–502.

Choi, S-C., and R. Bartha. 1993. Cobalamin-mediated mercury methylation by *Desulfovibrio desulfuricans* LS. Appl. Environ. Microbiol. 59:290–295.

Choi, S-C., T. Chase, Jr., and R. Bartha. 1994a. Enzymatic catalysis of mercury methylation by *Desulfovibrio desulfuricans* LS. Appl. Environ. Microbiol. 60:1342–1346.

Choi, S-C., T. Chase, Jr., and R. Bartha. 1994b. Metabolic pathways leading to mercury methylation in *Desulfovibrio desulfuricans* LS. Appl. Environ. Microbiol. 60:4072–4077.

Craig, P. J., and P. D. Bartlett. 1978. The role of hydrogen sulfide in environmental transport of mercury. Nature (Lond.) 275:635–637.

DeSimone, R. E., M. W. Penley, L. Charbonneau, S. G. Smith, J. K. M. Wood, H. A. O. Hill, J. M. Pratt, S. Ridsdale, and J. P. Williams. 1973. The kinetics and mechanism of cobalamin-dependent methyl and ethyl transfer to mercuric ion. Biochim. Biophys. Acta 304: 851–863.

Fagerstrom, T., and A. Jerneloev. 1971. Formation of methyl mercury from pure mercuric sulfide in aerobic organic sediment. Water Res. 5:121–122.

Foster, T. J. 1987. The genetics and biochemistry of mercury resistance. CRC Crit. Rev. Microbiol. 15:117–140.

Fox, B., and and C. T. Walsh. 1982. Mercuric reductase: purification and characterization of a transposon-encoded flavoprotein containing an oxidation-reduction-active disulfide. J. Biol. Chem. 257:2498–2503.

Furukawa, K., and K. Tonomura. 1971. Enzyme system involved in decomposition of phenyl mercuric acetate by mercury-resistant *Pseudomonas*. Agric. Biol. Chem. 35:604–610.

Furukawa, K., and K. Tonomura. 1972a. Metallic mercury-releasing enzyme in mercury-resistant *Pseudomonas*. Agric. Biol. Chem. 36: 217–226.

Furukawa, K., and K. Tonomura. 1972b. Induction of metallic-mercury-releasing enzyme in mercury-resistant *Pseudomonas*. Agric. Biol. Chem. 36(Suppl. 13):2441–2448.

Guard, H. E., A. B. Cobet, and W. M. Coleman III. 1981. Methylation of trimethyltin compounds by estuarine sediments. Science 213: 770–771.

Hamlett, N. V., E. C. Landsdale, B. H. Davis, and A. O. Summers. 1992. Roles of the Tn*21*, *merT*, *merP*, *and merC* gene products in mercury resistance and mercury binding. J. Bacteriol. 174:6377–6385.

Harper, D. B. 1985. Halomethane from halide ion—A highly efficient fungal conversion of environmental significance. Nature (Lond.) 315:55.

Holm, H. W., and M. F. Cox. 1975. Transformation of elemental mercury by bacteria. Appl. Microbiol. 29:491–494.

Imura, N., E. Sukegawa, S-K. Pan, K. Nagao, J-Y. Kim, T. Kwan, and T. Ukita. 1971. Chemical methylation of inorganic mercury with methyl-cobalamin, a vitamin B_{12} analog. Science 172:1248–1249.

Jensen, S., and A. Jerneloev. 1969. Biological methylation of mercury in aquatic organisms. Nature (Lond.) 223:753–754.

Jonasson, I. P. 1970. Mercury in the natural environment: a review of recent work. Geol. Surv. Can., Paper 1970, 70–57, 39 pp.

Komura, I., and K. Izaki. 1971. Mechanism of mercuric chloride resistance in microorganisms. I. Vaporization of a mercury resistant compound from mercuric chloride by multiple drug resistant strains of *Escherichia coli*. J. Biochem. (Tokyo) 70:885–893.

Komura, I., K. Izaki, and H. Takahashi. 1970. Vaporization of inorganic mercury by cell-free extracts of drug-resistant *Escherichia coli*. Agric. Biol. Chem. 34:480–482.

Korthals, E. T., and M. R. Winfrey. 1987. Seasonal and spatial variations in mercury methylation and demethylation in an oligotrophic lake. Appl. Environ. Microbiol. 53:2397–2404.

Landner, L. 1971. Biochemical model for the biological methylation of mercury suggested from methylation studies in vivo with *Neurospora crassa*. Nature (Lond.) 230:452–454.

Loutit, J. S. 1970. Mating systems of *Pseudomonas aeruginosa* strain I. VI. Mercury resistance associated with the sex factor (FP). Genet. Res. 16:179–184.

Manley, S. L., and M. N. Dastoor. 1988. Methyl iodide (CH_3I) production by kelp and associated microbes. Mar. Biol. (Berl.) 98:477–482.

Marine Chemistry. 1971. A Report of the Marine Chemistry Panel of the Committee on Oceanography. National Academy of Sciences, Washington, DC.

Matsumara, F., Y. Gotoh, and G. M. Brush. 1971. Phenyl mercuric acetate: Metabolic conversion by microorganisms. Science 173:49–51.

Nakahara, H., J. L. Schottel, T. Yamada, Y. Miyagawa, M. Asakawa, J. Harville, and S. Silver. 1985. Mercuric reductase enzymes from *Streptomyces* species and group B *Streptococcus*. J. Gen. Microbiol. 131:1053–1059.

Nelson, J. D. Jr., and R. R. Colwell. 1974. Metabolism of mercury compounds by bacteria in Chesapeake Bay. *In*: R. F. Acker, B. F. Brown, and J. R. De Palma, eds. Proc. Int. Congr. Mar. Corros. Fouling, 3rd, 1972. Northwestern University Press, Evanston, IL, pp. 767–777.

Nelson, J. D., and R. R. Colwell. 1975. The ecology of mercury-resistant bacteria in Chesapeake Bay. Microb. Ecol. 1:191–218.

Nelson, J. D., W. Blair, F. E. Brinckman, R. R. Colwell, and W. P. Iverson. 1973. Biodegradation of phenylmercuric acetate by mercury-resistant bacteria. Appl. Microbiol. 26:321–326.

Novick, R. P. 1967. Penicillinase plasmids of *Staphylococcus aureus*. Fed. Proc. 26:29–38.

Olson, G. J., W. P. Iverson, and F. E. Brinckman. 1981. Volatilization of mercury by *Thiobacillus ferrooxidans*. Curr. Microbiol. 5:115–118.

Olson, G. J., F. D. Porter, J. Rubinstein, and S. Silver. 1982. Mercuric reductase enzyme from a mercury-volatilizing strain of *Thiobacillus ferrooxidans*. J. Bacteriol. 151:1230–1236.

Pan Hou, H. S. K., and N. Imura. 1981. Role of hydrogen sulfide in mercury resistance determined by plasmid of *Clostridium cochlearum* T-2. Arch. Microbiol. 129:49–52.

Pan Hou, H. S., M. Nishimoto, and N. Imura. 1981. Possible role of membrane proteins in mercury resistance of *Enterobacter aerogenes*. Arch. Microbiol. 130:93–95.

Richmond, M. H., and M. John. 1964. Cotransduction by a staphylococcal phage of the genes responsible for penicillinase sysnthesis and resistance to mercury salts. Nature 202:1360–1361.

Robinson, J. B., and O. H. Tuovinen. 1984. Mechanism of microbial resistance and detoxification of mercury and organomercury compounds: Physiological, biochemical, and genetic analyses. Microbiol. Rev. 48:95–124.

Schottle, J. A., D. Mandel, S. Clark, S. Silver, and R. W. Hodges. 1974. Volatilization of mercury and organomercurials determined by inducible R-factor systems in enteric bacteria. Nature (Lond.) 251: 335–337.

Schrauzer, G. N., J. H. Weber, T. M. Beckham, and R. K. Y. Ho. 1971. Alkyl group transfer from cobalt to mercury: reaction of alkyl cobalamins, alkyl cobaloximes, and of related compounds with mercury acetate. Tetrahedron Lett. 3:275–277.

Smith, D. H. 1967. R. Factors mediate resistance to mercury, nickel, and cobalt. Science 156:1114–1116.

Spangler, W. A., J. L. Spigarelli, J. M. Rose, and H. M. Miller. 1973a. Methylmercury: bacterial degradation in lake sediments. Science 180:192–193.

Spangler, W. J., J. L. Spigarelli, J. M. Rose, R. S. Fillipin, and H. H. Miller. 1973b. Degradation of methylmercury by bacteria isolated from environmental samples. Appl. Microbiol. 25:488–493.

Steffan, R. J., E. T. Korthals, and M. R. Winfrey. 1988. Effects of acidification on mercury methylation, demethylation, and volatilization in sediments form an acid-suceptible lake. Appl. Environ. Microbiol. 54:2003–2009.

Summers, A. O., 1986. Genetic mechanisms of heavy-metal and antibiotic resistances. *In*: D. Carlisle, W. L. Berry, I. R. Kaplan, and J. R. Watterson, eds. Mineral Exploration: Biological Systems and Organic Matter. Rubey Vol. 5. Prentice-Hall, Englewood Cliffs, NJ, pp. 265–281.

Summers, A. O., and E. Lewis. 1973. Volatilization of mercuric chloride by mercury-resistant plasmid-bearing strains of *Escherichia coli*, *Staphylococcus aureus* and *Pseudomonas aeruginosa*. J. Bacteriol. 113:1070–1072.

Summers, A. O., and S. Silver. 1972. Mercury resistance in a plasmid-bearing strain of *Escherichia coli*. J. Bacteriol. 112:1228–1236.

Summers, A. O., and S. Silver, 1978. Microbial transformations of metals. Annu. Rev. Microbiol. 32:637–672.

Summers, A. O., and L. I. Sugarman. 1974. Cell-free mercury(II)-reducing activity in a plasmid-bearing strain of *Escherichia coli*. J. Bacteriol. 119:242–249.

Tezuka, T., and K. Tonomura. 1976. Purification and properties of an enzyme catalyzing the splitting of carbon-mercury linkages from mercury resistant *Pseudomonas* K 62-strain. J. Biochem. (Tokyo) 80:79–87.

Tezuka, T., and K. Tonomura. 1978. Purification and properties of a second enzyme catalyzing the splitting of carbon-mercury linkages from mercury-resistant *Pseudomonas* K-62. J. Bacteriol. 135:138–143.

Tonamura, K., T. Makagami, F. Futai, and D. Maeda. 1968. Studies on the action of mercury-resistant microorganisms on mercurials. I. The isolation of a mercury-resistant bacterium and the binding of mercurials to the cells. J. Ferment. Technol. 46:506–512.

Trevors, J. T. 1986. Mercury methylation by bacteria. J. Basic Microbiol. 26:499–504.

Vostal, J. 1972. Transport and transformation of mercury in nature and possible routes of exposure. *In*: L. Friberg and J. Vostal, eds. Mercury in the Environment. An Epidemiological and Toxicological Appraisal. CRC Press, Cleveland, pp. 15–27.

White, R. H. 1982. Analysis of dimethyl sulfonium compounds in marine algae. J. Mar. Res. 40:529–536.

Wong, P. T. S., Y. K. Chau, and P. L. Luxon. 1975. Methylation of lead in the environment. Nature (Lond.) 253:263–264.

Wood, J. M. 1974. Biological cycles of toxic elements in the environment. Science 183:1049–1052.

Wood, J. M., F. S. Kennedy, and C. G. Rosen. 1968. Synthesis of methylmercury compounds by extracts of a methanogenic bacterium. Nature (Lond.) 220:173–174.

Yamada, M., and K. Tonomura. 1972a. Formation of methylmercury compounds from inorganic mercury by *Clostridium cochlearium*. J. Ferment. Technol. 50:159–166.

Yamada, M., and K. Tonomura. 1972b. Further study of formation of methylmercury from inorganic mercury by *Clostridium cochlearium* T-2. J. Ferment. Technol. 50:893–900.

Yamada, M., and K. Tonomura. 1972c. Microbial methylation of mercury in hydrogen sulfide-evolving environments. J. Ferment. Technol. 50:901–909.

14

Geomicrobiology of Iron

14.1 IRON DISTRIBUTION IN THE EARTH'S CRUST

Iron is the fourth most abundant element in the Earth's crust, and is the most abundant element in the Earth as a whole (see Chapter 2). Its average concentration in the crust has been estimated to be 5.0% (Rankama and Sahama, 1950). It is found in a number of minerals in rocks, soil and sediments. Table 14.1 lists mineral types in which Fe is a major or a minor structural component.

The primary source of iron accumulated at the Earth's surface is volcanic activity. However, weathering of iron-containing rocks and minerals is often an important phase in the formation of local iron accumulations, including sedimentary ore deposits.

14.2 GEOCHEMICALLY IMPORTANT PROPERTIES

Iron is a very reactive element. Its common oxidation states are 0, +2, and +3. In a moist environment exposed to air or in aerated solution at pH values greater than 5, its ferrous form (+2) readily autoxidizes to the ferric form (+3). In the presence of an appropriate reducing agent, i.e.,

Table 14.1 Iron-Containing Minerals

Igneous (primary) minerals	Secondary minerals	Sedimentary minerals
Pyroxenes	Montmorillonite[a] $(OH_4Si_8Al_4O_{20}\cdot nH_2O$	Siderite $(FeCO_3)$
Amphiboles		Goethite $(Fe_2O_3\cdot H_2O)$
Olivines	Illite $(OH)_4K_y(Al_4Fe_4Mg_4Mg_6)$- $(Si_{8-y}Al_y)O_{20}$	Limonite $(Fe_2O_3\cdot nH_2O$ or $FeOOH)$
Micas		Hematite (Fe_2O_3)
		Magnetite (Fe_3O_4)
		Pyrite, marcasite (FeS_2)
		Pyrrhotite (Fe_nS_{n+1}); $n = 5-6$
		Ilmenite $(FeO\cdot TiO_2)$

[a] Montmorillonite contains iron by lattice substitution for aluminum.

under reducing conditions, ferric iron is readily reduced to the ferrous state. In dilute acid solution, metallic iron readily oxidizes to ferrous iron with the production of hydrogen:

$$Fe^\circ + 2H^+ \rightarrow Fe^{2+} + H_2 \tag{14.1}$$

Ferric iron precipitates as a hydroxide or oxide in neutral to slightly alkaline solution, but is soluble as Fe^{3+} in acid solution. It dissolves in strongly alkaline solution because of the amphoteric nature of $Fe(OH)_3$, which allows it to form an oxyanion. In an aqueous environment, a mixture of ferric iron and metallic iron undergoes a dismutation reaction resulting in the formation of ferrous iron:

$$Fe^\circ + 2Fe^{3+} \rightarrow 3Fe^{2+} \tag{14.2}$$

Hydrogen sulfide reduces ferric iron readily to ferrous iron and precipitates it as a sulfide, or disulfide in an excess of hydrogen sulfide:

$$2\,Fe^{3+} + H_2S \rightarrow 2Fe^{2+} + 2H^+ + S^\circ \tag{14.3}$$

$$Fe^{2+} + H_2S \rightarrow FeS + 2H^+ \tag{14.4}$$

$$2Fe^{3+} + 2H_2S \rightarrow FeS_2 + Fe^{2+} + 4H^+ \tag{14.5}$$

Reactions 14.3 to 14.5 are important microbiologically (see also Chapter 18).

14.3   BIOLOGICAL IMPORTANCE OF IRON

Function of Iron in Cells   All organisms, prokaryotic and eukaryotic, single-celled and multicellular, require iron nutritionally. A small group of homolactic fermenting bacteria consisting of the lactic streptococci are the only known exception. The other organisms need iron in some enzymatic processes that involve the transfer of electrons, as in aerobic and anaerobic respiration in which cytochromes (their heme iron) and other nonheme iron-sulfur proteins play a role in the transfer of electrons to a terminal acceptor. Photosynthesizers also need iron for ferredoxin, a nonheme iron-sulfur protein and some cytochromes that are part of the photosynthetic system. Some cells also employ iron in some of their superoxide dismutases, which convert superoxide to hydrogen peroxide by a dismutation reaction (see Chapter 3). Most aerobic cells produce the heme enzymes catalase and peroxidase as a means of decomposing the toxic hydrogen peroxide to water and oxygen. Prokaryotes capable of nitrogen fixation employ ferredoxin and another nonheme iron-sulfur protein (component II of nitrogenase) as well as an iron-molybdo, an iron-vanado, or an iron-iron protein (component I) of nitrogenase.

As will be discussed in section 14.4, ferrous iron may also serve as a major energy source to certain bacteria. Ferric iron may serve as a terminal electron acceptor under other conditions for the same or different bacteria. As discussed in Chapter 3, ferrous iron may have served as an important reductant during the evolutionary emergence of oxygenic photosynthesis by scavenging the toxic oxygen produced in the process until the appearance of oxygen-protective superoxide dismutase. Cloud (1973) has pointed to the Banded Iron Formations, that arose in the sedimentary record from 3.3 to 2 billion years ago, as evidence for this oxygen-scavenging action by ferrous iron. Banded iron formations consist of cherty magnetite (Fe_3O_4) and hematite (Fe_2O_3) (see discussion in section 14.9). From a biogeochemical viewpoint, large-scale microbial iron oxidation is important because it leads to extensive iron precipitation, and microbial iron reduction is important because it leads to extensive iron solubilization. In some anaerobic environments, microbial iron reduction plays an important role in mineralization of organic carbon (iron respiration).

Iron Assimilation by Microbes   Since in aerobic environments of neutral pH, ferric iron precipitates readily from solution and iron oxides and oxyhydroxides do not readily dissolve, and since iron cannot be readily taken into cells in insoluble form, a number of microorganisms have acquired the ability to synthesize chelators which help to keep ferric iron in solution or which may return it to solution in sufficient quantities to be nutritionally available. Examples of such chelators, known collectively as **siderophores**, are enterobactin or enterochelin, a catechol derivative from *Salmonella typhimurium* (Pollack and Neilands, 1970); aerobactin, a hydroxamate derivative produced by *Enterobacter aerogenes* (formerly *Aerobacter aerogenes*) (Gibson and Magrath, 1969); and rhodotorulic acid, a hydroxamate derivative produced by the yeast *Rhodotorula* (Neilands, 1974) (Fig. 14.1). The chelated iron is usually taken up by first binding to ferri-siderophore-specific receptors at the cell surface of the microbial species which produced the siderophore. In some cases, it may also bind to surface receptors of certain species of microorganisms incapable of synthesizing

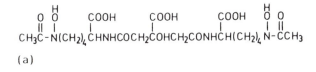

(a)

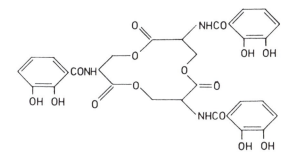

(b)

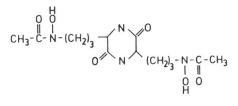

(c)

Figure 14.1 Examples of siderophores: (a) aerobactin (bacterial), (b) enterochelin (bacterial), and (c) rhodotorulic acid (fungal).

the desferrisiderophore. After transport into a cell, the chelated ferric iron is usually enzymatically reduced to ferrous iron and rapidly released by the siderophore, which has only a low affinity for it (Brown and Ratledge, 1975; Cox, 1980; Ernst and Winkelmann, 1977; Tait, 1975). The desferrisiderophore may then be excreted again for further scavenging of iron. In some instances the ferrisiderophore is degraded first before the ferric iron is reduced to ferrous iron. The desferrisiderophore in that instance is not recyclable. In still another few instances, liganded ferric iron is exchanged with another ligand at the initial binding site at the cell surface before being transported into the cell and reduced to ferrous iron. The ferrous iron is immediately assimilated into heme protein or nonheme iron-sulfur protein.

14.4 IRON AS ENERGY SOURCE FOR BACTERIA

Acidophiles Microbes may promote iron oxidation, but this does not mean that the oxidation is always enzymatic. Because of a tendency of ferrous iron to autoxidize in aerated solution at pH values above 5, it is difficult to demonstrate enzyme catalyzed iron oxidation in near neutral solutions. At this time, the strongest evidence for enzyme catalyzed iron oxidation involving bacteria has been amassed at pH values below 5.

Thiobacillus ferrooxidans

General traits The most widely studied acidophilic, iron-oxidizing bacterium is *Thiobacillus ferrooxidans*, which was first isolated by Colmer et al. (1950) and named and characterized as a chemolithotroph by Temple and Colmer (1951). It is a gram-negative, motile rod (0.5 × 1.0 μm) (Fig. 14.2) which derives energy and reducing power from the oxidation of ferrous iron, reduced forms of sulfur (H_2S, $S°$, $S_2O_3^{2-}$), metal sulfides, H_2, and formate. It gets its carbon from CO_2, and it derives its nitrogen preferentially from NH_3-N but can also use NO_3-N (Temple and Colmer, 1951; Lundgren et al., 1964). Some strains can fix N_2 (Mackintosh, 1978; Stevens et al., 1986.)

Morphologically, the cells of *T. ferrooxidans* exhibit the multilayered cell wall typical of gram-negative bacteria (Avakyan and Karavaiko, 1970; Remsen and Lundgren, 1966) (Fig. 14.3). They do not contain special internal membranes like those found in nitrifiers and methylotrophs. Cell division is mostly by constriction, but occasionally also by partitioning (Karavaiko and Avakyan, 1970).

T. ferrooxidans is acidophilic. The original isolate by Temple and Colmer (1951) grew and oxidized iron in a pH range of 2.0–2.5, but no

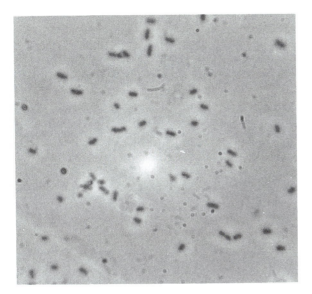

Figure 14.2 *Thiobacillus ferrooxidans* ($\times 5,170$). Cell suspension viewed by phase contrast.

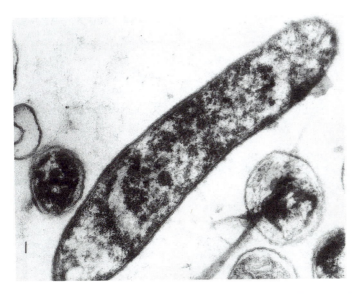

Figure 14.3 *Thiobacillus ferrooxidans* ($\times 30,000$). Electron photomicrograph of a thin section. (Courtesy of D. G. Lundgren.)

overall growth range was reported. The TM strain used by Silverman and Lundgren (1959a) oxidized iron optimally in a pH range of 3.0 to 3.6, whereas a strain (NCIB No. 9490) used by Landesman et al. (1966) oxidized iron optimally at pH 1.6. Razzell and Trussell (1963) gave the optimum pH for growth on iron as 2.5, and Jones and Kelly (1983) gave the pH range for growth on iron as 1 to 4.5. Differences in pH optima for iron oxidation and growth on iron can be attributed to strain differences and differences in experimental conditions. In general, all strains grow and oxidize iron around pH 2. Growth can be observed over a range of initial pH of 1.5 to 6.0. Ferric iron produced by the organism precipitates above pH 1.9 (Buchanan and Gibbons, 1974).

T. ferrooxidans is mesophilic; i.e., its growth range on iron is generally from 15 to 42°C with an optimum in the range of 30–35°C (Silverman and Lundgren, 1959a; Ahonen and Tuovinen, 1989; Niemelae et al., 1994). Ahonen and Tuovinen (1989) found acidophilic, iron-oxidizing cultures resembling *T. ferrooxidans* to grow from water samples from a Finnish copper mine at incubation temperatures of 4, 7, 10, 13,16,19, 28, and 37°C but not at 46°C. Optimum rates of iron oxidation by resting cells of *T. ferrooxidans* have been observed around 40°C with at least some strains (Silverman and Lundgren, 1959b; Tuttle and Dugan, 1976). A linear increase in iron oxidation rates has been measured between 4 and 28°C with the Finnish cultures (Ahonen and Tuovinen, 1989).

Iron is not the only energy source for *T. ferrooxidans*. As its generic name *Thiobacillus* implies, it can also use reduced forms of sulfur such as H_2S, $S°$, $S_2O_3^{2-}$, and metal sulfides as sole sources of energy (see Chapters 17 and 18). Recently it was discovered that *T. ferrooxidans* can also use hydrogen as a sole source of energy with oxygen as terminal electron acceptor (Drobner et al., 1990) and formate with oxygen or ferric iron as terminal electron acceptor (Pronk et al., 1991a,b). The key enzyme hydrogenase that enables utilization of hydrogen as energy source is inducible. When hydrogen is utilized, the optimal pH range for growth is 3.0–5.8 as opposed to an optimum with Fe^{2+} of pH 2.0.

Some strains of *T. ferrooxidans* were originally named *Ferrobacillus ferrooxidans* (Leathen et al., 1956) and *F. sulfooxidans* (Kinsel, 1960). These have since been considered synonymous with *T. ferrooxidans* (Unz and Lundgren, 1961; Ivanov and Lyalikova, 1962; Hutchinson et al., 1966; Kelly and Tuovinen, 1972; Buchanan and Gibbons, 1974).

Laboratory cultivation   Laboratory study of *T. ferrooxidans* depends on ease of culturing. Table 14.2 list ingredients of four liquid media, of which the 9K and T&K media are the most suitable. Cultivation on solid media gelled with a grade of agar commonly used in bacteriology laborato-

Table 14.2 Media for Cultivating *T. ferrooxidans*

Ingredients	Quantity of ingredients (g liter^{-1})			
	9K[a]	T and K[b]	L[c]	T and C[d]
$(NH_4)SO_4$	3.0	0.4	0.15	0.5
KCl	0.1	—	—	—
K_2HPO	0.5	0.4	0.05	—
$MgSO_4 \cdot 7H_2O$	0.5	0.4	0.5	1.0
$Ca(NO_3)_2$	0.01	—	0.01	—
$FeSO_4 \cdot 7H_2O$	44.22	33.3	1.0	129.1

[a] 9K medium of Silverman and Lundgren (1959a). $FeSO_4$ is dissolved in 300 ml of distilled water and filter-sterilized. The remaining salts are dissolved in 700 ml of distilled water and autoclaved after adjustment of the pH with 1 ml of 10 N H_2SO_4.
[b] Medium of Tuovinen and Kelly (1973). The salts are dissolved in 0.11 N H_2SO_4. The medium is sterilized by filtration.
[c] Medium of Leathen et al. (1956).
[d] Medium of Temple and Colmer (1951). The pH of this medium is adjusted to 2.0–2.5 with H_2SO_4. The medium is sterilized by filtration.

ries or with silica gel, has had limited success. Tuovinen and Kelly (1963) have suggested that galactose from agar may be inhibitory to the organism. Growing the organism on solid media is essential for culture purification and cloning for genetic studies as well as for estimating population size by viable counts. Recently, it was found that most strains will form rust-colored colonies of varying morphology, depending on the strain, in one to two weeks on a modified 9K medium solidified with agarose or purified agar (Manning, 1975; Mishra and Pradosh, 1979; Mishra et al., 1983; Holmes et al., 1983). Another approach, which currently is not extensively used, is to grow *T. ferrooxidans* cells on membrane filters (Sartorius or Millipore types) on T&K medium solidified with 0.4% of Japanese agar at an initial pH of 1.55 and incubating at 20°C (Tuovinen and Kelly, 1973). Rust-colored colonies of 1.0–1.5-mm diameter develop after 2 weeks.

Johnson et al. (1987) proposed the following solid medium for con-current isolation and enumeration of *T. ferrooxidans* and acidophilic heterotrophic bacteria that usually accompany *T. ferrooxidans* in nature. The medium contains 1 part by volume of a 20% ferrous sulfate solution, 14 parts by volume of a basal salts/tryptone soya broth solution consisting of (in g/L) $(NH_4)_2SO_4$ 1.8, $MgSO_4 \cdot 7H_2O$ 0.7, tryptone soya broth (TSB, Oxoid Ltd., U.K.) 0.35, and 5 parts by volume of 0.7% agarose type 1

(Sigma Ltd., U.S.A.). The best final pH values for optimal recovery of *T. ferrooxidans* were 2.7 and 2.3. A modification of this medium that also supported growth of moderately thermophilic acidophilic, iron-oxidizing bacteria was developed by Johnson and McGinness in 1991.

Facultative heterotrophy A number of claims have appeared in the literature that some strains of *T. ferrooxidans* can be adapted to grow heterotrophically, using glucose instead of iron as sole energy and carbon source (Lundgren et al., 1964; Shafia and Wilkinson, 1969; Shafia et al, 1972; Tabita and Lundgren, 1971a; Tuovinen and Nicholas, 1977; Sugio et al., 1981). Some of these strains could be reverted to iron-dependent autotrophy while others could not. The discovery that some cultures of *T. ferrooxidans* contained heterotrophic satellite organisms (Zavarzin, 1972; Guay and Silver, 1975; Macintosh, 1978; Arkesteyn, 1979; Harrison, 1981, 1984; Johnson and Kelso, 1983; Lobos et al., 1986; Wichlacz et al., 1986) has cast doubt on the existence of facultative autotrophy in any *T. ferrooxidans* strains.

Consortia with T. ferrooxidans The existence of satellite organisms that appear to live in close association with *T. ferrooxidans* was first reported by Zavarzin (1972). As noted above, confirmation of their existence soon followed. Indeed, taxonomically different organisms were isolated from different *T. ferrooxidans* consortia. Zavarzin (1972) isolated his organism in modified Leathen's medium in which yeast extract (0.01–0.02%) was substituted for $(NH_4)_2SO_4$ and adjusted to pH 3–4. The organism was morphologically distinct from *T. ferrooxidans*, being a single rod. It could not oxidize iron even though it required its presence at high concentration in the medium. It was acidophilic (optimum pH range 2–3). It required yeast extract but did not grow in an excess of it. Glucose, sucrose, fructose, ribose, maltose, xylose, citric, succinic and fumaric acids, mannitol and ethanol could serve as energy sources. The organism was originally isolated from peat in an acid bog. It resembled *Acetobacter xylinum* but grew at lower pH than the latter and exhibited only weak acetic acid- and ethanol-forming ability.

Guay and Silver (1975) derived a satellite organism from *T. ferrooxidans* strain TM by subculturing in 9K medium with increasing amounts of glucose concentrations from 0.1 to 1.0% and concomitant decreases in Fe^{2+} from 9,000 ppm to 10 ppm in four steps. A pure culture of the satellite organism, which they named *Thiobacillus acidophilus*, was isolated on 9K basal salts–glucose agar containing 1 ppm Fe at pH 4.5 and 25°C. The pure culture consisted of a gram-negative, motile rod (0.5–0.8 μm × 1.0–1.5 μm) and was thus morphologically not very distinct from *T*.

ferrooxidans. Physiological study of the organism by Guay and Silver (1975) and subsequently by Norris et al. (1986), Mason et al. (1987), and Mason and Kelly (1988), have shown the organism to be capable of hetero- trophic growth in suitable medium and also capable of mixotrophic growth on tetrathionate plus glucose. In addition it was found to grow autotrophi- cally in media with elemental sulfur (S°), thiosulfate, trithionate or tetrathi- onate as energy source and CO_2 as carbon source. The organism is incapa- ble of oxidizing Fe^{2+} or metal sulfides. It is able to use D-ribose, D-xylose, L-arabinose, D-glucose, D-fructose, D-galactose, D-mannitol, sucrose, cit- rate, malate, dl-aspartate, and dl-glutamate as carbon and energy sources in 9K basal medium without Fe^{2+}. It cannot not use D-mannose, L-sor- bose, L-rhamnose, ascorbic acid, lactose, D-maltose, cellobiose, treha- lose, D-melibiose, raffinose, acetate, lactate, pyruvate, glyoxalate, fumar- ate, succinate, mandelate, cinnamate, phenylacetate, salicylate, phenol, benzoate, phenylalanine, tryptophane, tyrosine, or proline.

Like *T. ferrooxidans*, *T. acidophilus* is acidophilic (pH range 1.5–6.0; opt. pH 3.0). Its DNA has a GC ratio of 62.9–63.2 mol%, which is distinctly different from that of *T. ferrooxidans*, whose GC ratio is 56.1 mol%. Differences in key enzymes such as glucose 6-phosphate dehydro- genase, 6-phosphogluconate dehydrogenase, fructose 1,6-diphosphate al- dolase, isocitric dehydrogenase, 2-ketogluconate dehydrogenase, NADH: acceptor oxidoreductase, thiosulfate oxidizing enzyme, and rhodanese in *T. ferrooxidans* and *T. acidophilus* were also noted. Its ribulose 1,5-bis- phosphate carboxylase level for CO_2 fixation depends on whether the organism is growing mixotrophically or autotrophically. Glucose in mixo- trophic culture can partially repress the enzyme (Mason and Kelly, 1988). What is puzzling about *T. acidophilus* is that it was carried for years as a satellite of *T. ferrooxidans* in 9K medium without an organic supplement. The explanation may be as given by Arkesteyn and DeBont (1980), that *T. acidophilus* is oligotrophic and, in coculture with *T. ferrooxidans*, lives at the expense of organic excretions (organic acids, alcohols, amino acids) from *T. ferrooxidans* that inhibit the growth of the latter organism if they build up in the medium.

Other satellite organisms found associated with some *T. ferroxidans* cultures include *Acidiphilium cryptum* (Harrison, 1981), *A. organovorum* (Lobos et al., 1986), *A. angustum*, *A. facilis*, and *A. rubrum* (Wichlacz et al., 1986). *A. cryptum* is a heterotrophic, motile, gram-negative rod (0.3–0.4 μm × 0.5–0.8 μm) that grows in very dilute organic media in a pH range of 1.9–5.9. The organism is an oligotroph whose growth is inhib- ited at high organic nutrient concentrations. Its GC content is 68–70 mol%, which clearly distinguishes it from *T. ferrooxidans* and *T. acidophilus*. It cannot use sulfur or ferrous iron as an energy source and is inhibited by as little as 0.01% acetate. *A. organovorum* differs from *A. cryptum* by a lower GC ratio (64 mol%) and especially by its facultative oligotrophy. The *Acidiphilium* species isolated by Wichlacz et al. (1986) differed from *A. cryptum* in terms of cell size, pigment production, nutritional traits and genome homology. In support of the theory of Arkesteyn and DeBont (1980) concerning the natural role of *T. acidophilus* in relation to its association with *T. ferrooxidans*, Harrison (1984) showed that growth of *T. ferrooxidans* is enhanced in the presence of *A. cryptum* when 0.004% pyruvate is present in the medium. Pyruvate can be used by *A. cryptum* as carbon and energy source but inhibits the growth of *T. ferrooxidans* at a critical concentration even though *T. ferrooxidans* excretes pyruvate during its growth (Schnaitman and Lundgren, 1965).

Genetics Understanding of the genetics of *T. ferrooxidans* has lagged behind that of many other bacteria. One reason for this has been the past difficulty of culturing the organism on solid media to obtain pure cultures and to select different strains. Another reason has been that the acidophily of *T. ferrooxidans* made the application of typical molecular biological techniques difficult.

With the development of new culture techniques, significant advances in unraveling the genetics of *T. ferrooxidans* have been made in the last 10 years or so. For instance, in the older literature, reports appeared of "training" of *T. ferroxidans* to tolerate high concentrations of toxic metal ions such as those of Cu, Ni, Co, Pb, and so forth, by serial subculture in the presence of increasing amounts of the toxicants (e.g., Marchlewitz et al., 1961; Tuovinen et al., 1971; Kamalov, 1972; Tuovinen and Kelly, 1974). Many interpreted this phenomenon on the basis of spontaneous mutation and selection. However, considering the relatively rapid rate at which these cultures developed tolerance for these metals and their limited growth yields in appropriate media, this interpretation was questioned by some. Pure cultures of rapidly growing heterotrophs acquire resistance to a toxicant by a spontaneous mutation and selection process more gradually. The recent discovery of mobile genetic elements in the genome of many *T. ferrooxidans* strains offers a more plausible explanation (Holmes et al., 1987; Yates and Holmes, 1987; Schrader and Holmes, 1988).

Like most bacteria, *T. ferrooxidans* contains most of its genetic information on a **chromosome**, and like most bacteria, most strains of *T. ferrooxidans* also contain some nonessential genetic information on extrachromosomal DNA called **plasmids** (e.g., Mao et al., 1980; Martin et al., 1981; Rawlings et al., 1986). Some of the chromosomal and plasmid genes have been mapped and characterized (Rawlings and Kusano, 1994). These include genes of nitrogen metabolism, such as glutamine synthase (see Rawlings and Kusano, 1994) and nitrogenase (e.g., Pretorius et al., 1986, 1987; Rawlings and Kusano, 1994), genes involved in CO_2 fixation and in energy conservation (see Rawlings and Kusano, 1994), genes responsible for resistance to mercury, and other genes (see Rawlings and Kusano, 1994). The genomes of different strains of *T. ferrooxidans* and *T. thiooxidans* have been compared (Harrison, 1982, 1986). The 13 strains of *T. ferrooxidans* that were examined fell into seven different DNA homology groups. Though genetically related, they were not identical. None showed significant DNA homology with six strains of *T. thiooxidans*, indicating little genetic relationship. The *T. thiooxidans* strains fell into two different DNA homology groups. On the basis of 16S rRNA relationships, *T. ferrooxidans* belongs to the Proteobacteria, most to the beta-subgroup, but at least one strain to the gamma-subgroup (see Rawlings and Kusano, 1994).

After many previous failures, it has recently become possible under laboratory conditions to introduce genes into strains of *T. ferrooxidans* lacking them, using bacterial transformation technique via electroporation and bacterial conjugation and having them expressed by the recipient strains. Thus, the mercury resistance gene complex (*mer* operon) was introduced into a mercury-sensitive strain of *T. ferrooxidans* by bacterial transformation via electroporation (Kusano et al., 1992). Heterogeneous arsenic resistance genes on plasmids were transferred from a strain of *Escherichia coli* (heterotroph) to an arsenic-sensitive strain of *T. ferrooxidans* by filter mating (Peng et al., 1994). To what extent gene transfer occurs among *T. ferrooxidans* strains in nature is still unknown.

The energetics of ferrous iron oxidation   Oxidation of ferrous iron does not furnish much energy on a molar basis, if compared, for instance, to oxidation of glucose. In the past, estimates of the ΔG for iron oxidation have ranged around 10 kcal mol^{-1}, as calculated for example by Baas Becking and Parks (1927) from the equation

$$4FeCO_3 + O_2 + 6H_2O$$

$$\rightarrow 4Fe(OH)_3 + 4CO_2 \ (\Delta G_{298} = -40 \ kcal) \tag{14.6}$$

A more recent examination of the question of the free-energy yield from iron oxidation by Lees et al. (1969), who assumed that the reaction proceeded as follows:

$$Fe^{2+} + H^+ + \tfrac{1}{4}O_2 \rightarrow Fe^{3+} + \tfrac{1}{2}H_2O \tag{14.7}$$

and who took into account effects of pH and ferric iron solubility upon the reaction, led to the following equation for calculating the molar free energy ΔG between pH 1.5 and 3:

$$\Delta G = -1.3(7.7 - pH - 0.17) \tag{14.8}$$

From this equation it may be calculated that the ΔG at pH 2.5 is 6.5 kcal mol^{-1}, barely enough for the synthesis of 1 ml of ATP (which requires about 7 kcal mol^{-1}). If we assume that it takes 120 kcal of energy to assimilate 1 mol of carbon at 100% efficiency (Silverman and Lundgren, 1959b), approximately 18.5 mol of Fe would have to be oxidized to incorporate this much carbon. However, *T. ferrooxidans* is not 100% efficient in using energy available from iron oxidation. An early experimental estimate of the true efficiency of carbon assimilation at the expense of Fe^{2+} was 3.2% (Table 14.3) (Temple and Colmer, 1951). At that efficiency level, it would take 577 mol of Fe. Taking the efficiency determination of Silverman and Lundgren (1959b) given in Table 14.3, a consumption of 90.1 mol of Fe^{2+} to assimilate 1 mol of carbon would be predicted. This is greater than the consumption of 50 mol observed by Silverman and Lundgren (1959b) and raises the question as to whether the assumption that it takes 120 kcal in these cells to assimilate 1 mol of carbon is correct. However, this estimate of 90.1 mol of Fe^{2+} per mole of carbon approaches the results of another experiment (Beck, 1960), in which about 100 mol of Fe^{2+} had to be oxidized to fix 1 mol of CO_2 by a strain of *T. ferrooxidans*. No matter what the actual efficiency of energy utilization, the observed molar ratios of Fe^{2+} oxidized to CO_2 assimilated illustrate that large amounts of iron have to be oxidized to satisfy the energy requirements for growth of these organisms.

Iron-oxidizing enzyme system in T. ferrooxidans   Significant advances in elucidating the enzymatic mechanism of iron oxidation by *T. ferrooxidans* have been made. Kinetic studies with whole cells of the

Table 14.3   Estimates of Free-Energy Efficiency of Carbon Assimilation by Iron-Oxidating Thiobacilli

Efficiency (%)	Age of cells	Reference
3.2	17 days	Temple and Colmer (1951)
30	?	Lyalikova (1958)
4.8–10.6	?	Beck and Elsden (1958)
20.5 ± 4.3	Late log phase	Silverman and Lundgren (1959b)

organism yielded apparent K_m values of 2.2 mM and for cell-free preparations of 5.6 mM (pH optimum 3.5) (Ingledew, 1982).*

Ferric iron resulting from Fe^{2+} oxidation, causes product inhibition and limits growth of *T. ferrooxidans* as it builds up in the medium (Kelly and Jones, 1978; Jones and Kelly, 1983; Kovalenko et al., 1982). The inhibitory effect can be modulated by change in temperature. Increasing temperature decreases the inhibitory effect. Physiological age also has an effect on ferric iron susceptibility, lag phase cells being more sensitive to inhibition than log phase cells (Kovalenko et al., 1982). Ferrous iron may cause substrate inhibition in chemostat growth (Jones and Kelly, 1983).

A requirement for sulfate ions by the iron oxidizing system of *T. ferrooxidans* was established by Lazaroff (1963). Changes in SO_4^{2-} concentration as well as pH affect the values of V_{max} but not K_m. Even when cells were adapted to the presence of Cl^-, Cl^- could not fully replace SO_4^{2-} (Lazaroff, 1963; see also Kamalov, 1967; Vorreiter and Madgwick, 1982). In at least one instance, SO_4^{2-} could be partially replaced by HPO_4^{2-} or $HAsO_4^{2-}$ but not by BO_3^-, MoO_4^{2-}, NO_3^-, or Cl^- ions. Formate and MoO_4^{2-} ions inhibited iron oxidation (Schnaitman et al., 1969). Selenate could replace sulfate in iron oxidation by *T. ferrooxidans* but did not permit its growth. In the presence of sulfate or selenate, iron oxidation was enhanced by tellurate, tungstate, arsenate, or phosphate (Lazaroff, 1977). A role of sulfate in iron oxidation by *T. ferrooxidans* appears to be to stabilize the hexa-aquated complex of Fe(II) which serves as substrate for its iron-oxidizing enzyme system. Selenate can replace sulfate completely as anionic stabilizer, and tellurate, tungstate, arsenate, and phosphate can replace it partially (Lazaroff, 1983). The extensive formation of jarosite, a crystalline basic iron sulfate, in the presence but not in the absence of *T. ferrooxidans* suggests that in nonbiological oxidation of Fe(II), water displaces sulfate in the ferric product (Lazaroff et al., 1982, 1985).

One currently accepted model for the iron-oxidizing system in *T. ferroxidans* is illustrated in Figure 14.4a. According to this model, ferrous iron is oxidized at the outer surface of the outer membrane of the cell envelope with the electron of the substrate Fe^{2+} reducing iron in the +3 oxidation state that is structurally bound in the outer membrane, also called polynuclear iron (Ingledew, 1986). The resultant bound ferrous iron in the outer membrane gives up its electron to the copper protein rusticyanin catalyzed by an as unidentified catalytic component (X). The rusticya-

---

* K_m is defined by the Michaelis-Menton equation: $v = (V_{max}[S])/([S] + K_m)$, where v is the reaction velocity, V_{max} the maximal velocity, $[S]$ the intitial substrate concentration, and K_m is a constant (Segel, 1975).

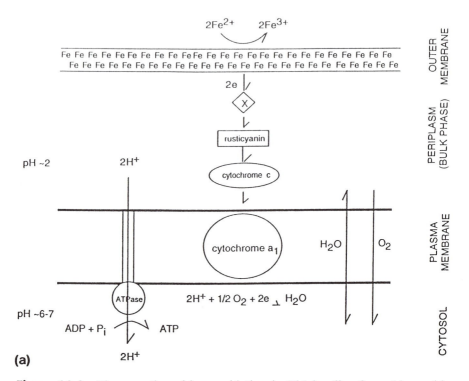

Figure 14.4 Bioenergetics of iron oxidation in *Thiobacillus ferooxidans*. (a) Model as proposed by Ingledew [modified from Ingledew et al. (1977)]; for discussion see text. (b) Model as proposed by Blake et al. (1992); for discussion see text. The periplasmic cytochrome c in Blake's model could be a component of the outer membrane, where it would function in place of the polynuclear iron in the model of Ingledew et al. (1977) in transferring electrons from Fe^{2+} or other electron donors; or it could serve as an electron shuttle between the outer membrane and the periplasm.

nin in turn yields its acquired electrons to periplasmic cytochrome c. The reduced cytochrome c then binds to the outer surface of the plasma membrane, allowing for the transfer of the electrons across the membrane to cytochrome oxidase (cytochrome a_1) located on the inside surface of the plasma membrane. The reduced cytochrome oxidase reacts with O_2 leading to the formation of water. The inclusion of the catalytic component (X) in the model in Figure 14.4a is necessary because the kinetics for electron transfer from Fe(II) to periplasmic cytochrome c are otherwise too slow to explain the observed rate of iron oxidation by intact cells

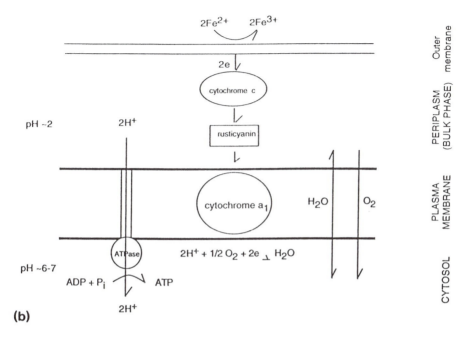

(b)

(Ingledew, 1986; Cox and Boxer, 1986; Blake and Shute, 1987). Fry et al. (1986) believe that they found evidence for such an enzyme in the form of a nonheme iron protein. The suggestion by the model that outer-membrane-bound iron may be the initial electron acceptor is based on a proposal by Dugan and Lundgren (1965) that some iron is bound in the cell envelope of *T. ferrooxidans* and on the finding by Agate and Vishniac (1970) that phosphatidylserine in the membrane may be the site of this iron binding.

Alternative models for the iron-oxidizing system have been proposed (Blake et al., 1992). The most plausible of the alternative models is one based on direct observations by Blake et al. (1992) of rapid electron transfer from a molar excess of Fe^{2+} to rusticyanin catalyzed by a partially purified iron-rusticyanin oxidoreductase that appeared to consist of some form of cytochrome c (Fig. 14.4b). The reaction depended on the presence of sulfate ions, and the kinetics of iron oxidation were consistent with those observed with intact cells. Polynuclear iron in the outer membrane and the postulated catalyst X are not required by this model. Evidence for a satisfactory rate of electron transfer from rusticyanin to cytochrome

a_1 (38 s^{-1}) was previously presented by Yamanaka et al. (1991), cited by Blake et al. (1992).

Although evidence for the involvement of the cytochrome system in the oxidation of iron by *T. ferrooxidans* was first presented by Vernon et al., 1960 and subsequently in a more detailed analysis by Tikhonova et al. (1967), the basis for the currently most plausible models of iron oxidation emerged only with the discovery of the copper protein rusticyanin in the periplasm (Cobley and Haddock, 1970; Cox and Boxer, 1978; Ingledew et al., 1977; Ingledew and Houston, 1986; Cox and Boxer, 1986). If the model of Ingledew et al. (1977) is representative of the actual process of iron oxidation in intact cells of *T. ferrooxidans*, the Fe(II):Fe(III) oxido-reductase still remains to be clearly identified. Blaylock and Nason (1963) separated an iron-cytochrome c reductase from a particulate iron oxidase preparation from *Ferrobacillus* (*Thiobacillus*) *ferrooxidans*. Yates and Nason (1966) believed this reductase to be a DNA-containing enzyme protein (Yates and Nason, 1966a,b) but Din et al. (1967) showed it to be a RNA-containing enzyme. The location of this enzyme in the intact cell remains unclear. If its location is in the periplasm of *T. ferrooxidans*, and if it can react with rusticyanin, it may be the missing enzyme which catalyzes the transfer of electrons from the outer-membrane-bound iron to rusticyanin in the model of Ingledew et al. (1977). On the other hand, Fukumori et al. (1988) recently isolated an Fe(II) oxidizing enzyme from a *T. ferrooxidans* strain which conveyed electrons from Fe(II) to cytochrome c-552 but not to rusticyanin. Whether it is related to Din's iron-cytochrome c reductase is not known. Thus, major questions remain to be answered about Fe(II) oxidation of *T. ferrooxidans*.

Energy coupling in iron oxidation   Energy coupling in iron oxidation by *T. ferrooxidans* is best understood in terms of a chemiosmotic mechanism (Ingledew, 1982; Ingledew et al., 1977). Such a mechanism implies that a proton motive force is set up across the plasma membrane owing to charge separation across the two sides of the membrane. Proton motive force results from a pH gradient generated from the higher proton concentration in the acid periplasm relative to the near-neutral cytoplasm of the active *T. ferrooxidans* cell, and from a transmembrane electrical potential. The transfer of electrons to O_2 via the electron transport system results in pumping of protons from the cytoplasm into the periplasm which together with protons formed from the hydrolysis of ferric iron generated in the oxidation are a cause of the proton gradient. Energy coupling, i.e., ATP synthesis, results from the fact that the plasma membrane is impermeable to protons in the periplasm except at the sites where adenosine 5'-triphosphatase (ATPase) is anchored in the membrane. ATPase contains a proton channel which allows passage of protons in the direction of the cytoplasm and as a result of their movement causes the synthesis of ATP by the ATPase in the reaction

$$ADP + P_i \rightarrow ATP \tag{14.9}$$

Stoichiometrically, only one ATP can be synthesized per electron pair passed to oxygen for every two Fe^{2+} oxidized at 100% efficiency. Since, as already discussed, the efficiency of energy coupling is much less than 100%, much ferrous iron needs to be oxidized to meet the energy demand of the *T. ferrooxidans* cell.

For a discussion of a molecular biochemical analysis of the ATPase of *T. ferrooxidans*, the reader is referred to Rawlings and Kusano (1994).

Reverse electron transport The assimilation of CO_2 by any autotroph requires a source of reducing power. When *T. ferrooxidans* grows on iron, this source is Fe^{2+}. Ferrous iron thus has a dual function in the nutrition of this organism, namely as a source of energy and as a source of reducing power. To reduce fixed CO_2, electrons from Fe^{2+} are shunted to $NADP^+$ via the cytochrome system against an electropotential gradient by expenditure of energy (consumption of ATP) in a process called **reverse electron transport** (Aleem et al., 1963; Ingledew, 1982). The expenditure of energy is necessary because the cytochrome c couple (the level at which electrons from iron enter the membrane-bound electron transport system) has a much higher E_h' ($+245$ mV) than does the NAD/NADH couple (E_h' -320 mV) or the NADPH couple (E_h' -324 mV). In one strain of *T. ferrooxidans*, cytochrome c, cytochromes c_1 and b (bc_1 complex), and a flavin have been identified as participants in the reverse electron transport system (Tikhonova et al., 1967). As expected, arsenate, an uncoupler of oxidative phosphorylation, and amytal, an inhibitor of flavin reduction, blocked this system.

Carbon assimilation The major mechanism of CO_2 assimilation in *T. ferrooxidans* involves the Calvin-Benson cycle (Din et al., 1967b; Gale and Beck, 1967; Maciag and Lundgren, 1964). A minor CO_2 fixation mechanism involving phosphoenolpyruvate carboxylase also exists in the organism (Din et al., 1967b). The latter enzyme system is needed for the formation of certain amino acids. In the Calvin-Benson cycle, CO_2 is fixed by ribulose 1,5-bisphosphate obtained from ribulose 5-phosphate as follows (see also Chapter 6).

$$Ribulose\ 5\text{-phosphate} + ATP \xrightarrow{\text{phosphoribulokinase}}$$

$$ribulose\ 1,5\text{-bisphosphate} + ADP \tag{14.10}$$

$$\text{Ribulose 1,5-bisphosphate } + \text{ CO}_2 \xrightarrow{\underset{\text{carboxylase}}{\text{ribulose bisphosphate}}}$$

$$2(3\text{-phosphoglycerate}) \quad (14.11)$$

Each 3-phosphoglycerate is then reduced to 3-phosphoglyceraldehyde:

$$3\text{-Phosphoglycerate } + \text{ NADPH } + \text{ H}^+ + \text{ ATP} \xrightarrow{\text{dehydrogenase}}$$

$$3\text{-phosphoglyceraldehyde } + \text{ NADP}^+ + \text{ ADP } + \text{ P}_i \quad (14.12)$$

The 3-phosphoglyceraldehyde is converted in a series of steps to various cell constituents as well as to catalytic amounts of ribulose 5-phosphate to keep the Calvin-Benson cycle operating.

Phosphoenolpyruvate carboxylase catalyzes the fixation of CO_2 by phosphoenolpyruvate, which is formed from 3-phosphoglycerate as follows:

$$3\text{-Phosphoglycerate } \xrightarrow{\text{phosphoglyceromutase}} 2\text{-phosphoglycerate}$$
$$(14.13)$$

$$2\text{-Phosphoglycerate } \xrightarrow{\text{enolase}} \text{phosphoenolpyruvate} \quad (14.14)$$

The phosphoenolpyruvate is then combined with CO_2:

$$\text{Phosphoenolpyruvate } + \text{ CO}_2 \xrightarrow{\text{PEP carboxylase}} \text{oxalacetate } + \text{ P}_i$$
$$(14.15)$$

Whether a functional tricarboxylic acid (TCA) cycle exists in *T. ferrooxidans* when growing autotrophically on iron is presently unclear. Anderson and Lundgren (1969) reported experimental evidence for it in iron grown cells, but Tabita and Lundgren (1971b) found it only in glucose grown cells. The strain used in these studies may have been mixed with a satellite organism.

Other Acidophilic Iron-Oxidizing Bacteria

Mesophiles   Other acidophilic bacteria capable of oxidizing ferrous iron enzymatically have been discovered in more recent years. Many of these are quite unrelated to *T. ferrooxidans* (Harrison, 1984; 1986). Some of them are mesophilic, others are thermophilic, and most are chemolithotrophic (Norris, 1990; Pronk and Johnson, 1992). Some can oxidize only iron while others can also oxidize reduced sulfur and metal sulfides.

Among the mesophiles that can oxidize iron are *Thiobacillus prosperus* (Huber and Stetter, 1989), *Leptospirillum ferrooxidans* and similar bacteria (Markosyan 1972; Balashova et al., 1974; Pivovarova et al.,1981; Harrison, 1984; Harrison and Norris, 1985, 1990), a strain of *Metallogenium* (Walsh and Mitchell, 1972), and an unnamed isolate from England (Cameron et al., 1981). *T. prosperus* resembles *T. ferrooxidans* closely except that it can grow in the presence of up to 6% NaCl, whereas *T. ferroxidans* cannot; chloride is toxic to it.

Typical *L. ferrooxidans* was originally isolated from a copper deposit in Armenia. It is a vibrioid cell with a polar flagellum about 25 nm in diameter. Involution cells may have a spiral shape. Tori form at pH values of less than 2. In the laboratory, organisms have been grown in Leten's (Leathen's) medium at pH 2–3 (Kuznetsov and Romanenko, 1963) and in 9K medium (Silverman and Lundgren, 1959a) at pH 1.5. The organism oxidizes iron for energy but cannot oxidize reduced sulfur and is incapable of growth in organic media. Its cells have been shown to contain an active ribulose 1,5-bisphosphate carboxylase, typical of chemolithotrophs (autotrophs), which use the Calvin-Benson cycle for CO_2 assimilation. Not all morphologically similar acidophilic iron oxidizers should, however, be assigned to the genus *Leptospirillum* (Harrison and Norris, 1985).

An acid-tolerant *Metallogenium* was reported from mesoacidic, iron-bearing groundwaters (Walsh and Mitchell, 1972). It is a filamentous organism consisting of branching filaments (0.1–0.4 μm and >1 μm long), usually encrusted with iron. The organisms tolerates a pH range 3.5–6.8, with an optimum at pH 4.1. In the laboratory, the addition of 0.024 M phthalate at pH 4.1, but not of acetate, citrate, or phosphate, was important for observing iron oxidation and growth at initial Fe^{2+} concentrations greater than 100 mg L^{-1}. The growth medium was constituted as follows; $(NH_4)_2SO_4$, 0.1%; KH_2PO_4, 0.001%; $CaCO_3$, 0.01%; $MgSO_4 \cdot 7H_2O$, 0.02%; $FeSO_4 \cdot 7H_2O$, 9%; distilled water; pH adjustment to 4.1. It is not clear whether this organism is an autotroph or a heterotroph. Care has to be taken in identifying it. Some inorganic precipitates may resemble this type of *Metallogenium* morphologically (Ivarson and Sojak, 1978).

An unnamed acidophilic bacterium isolated from a stream in England oxidized Fe^{2+} optimally at pH 2.5 in an inorganic medium in the laboratory (Cameron et al., 1981). It could not oxidize Mn^{2+}, thiosulfate, or sulfide, nor could it grow on complex media. It formed long, unbranched, iron-encrusted filaments.

Thermophiles Thermophilic iron-oxidizing bacteria can be divided into moderate and extreme thermophiles. To the former group belong an incompletely characterized assemblage of *Thiobacillus*-like bacteria (LeRoux et al., 1977; Brierley and Lockwood, 1977; Brierley, 1978; Brierley et al., 1978; Norris and Barr, 1985; Norris, 1990; Norris et al., 1986) and *Sulfobacillus thermosulfidooxidans* (Golovacheva and Karavaiko, 1978; Norris 1990). Their relationship to *Thiobacillus* is uncertain. At least some like *S. thermosulfidooxidans* do not belong to the genus *Thiobacillus* at all. In general they oxidize iron at 50 and 55°C, and in some cases at 60°C. Although they grow chemolithotrophically at the expense of iron, they frequently require the presence of small amounts of yeast extract, cysteine or glutathione for growth. Some cannot use reduced inorganic sulfur compounds as an energy source.

Sulfobacillus thermosulfidooxidans is a gram-positive, nonmotile, sporeforming (most strains), rod-shaped eubacterium with tapered ends (0.6–0.8 × 1.0–1.3 μm, sometimes as long as 6 μm). The GC content of its DNA is 53.6–53.9 mol%. Its temperature range for growth is 28–60°C, with an optimum around 50°C. It can grow autotrophically on Fe(II), S°, or metal sulfides as energy source (Golovacheva and Karavaiko, 1978). It cannot use sulfate as a source of sulfur and consequently must be supplied with reduced sulfur (Norris and Barr, 1985). Autotrophic growth is stimulated by a trace (0.01–0.05%) yeast extract, but 0.1% yeast extract is inhibitory. Its pH range for growth is 1.9–3.0, the optimum range being pH 1.9–2.4 (Golovacheva and Karavaiko, 1978).

Well-studied examples of extreme thermophiles among acidophilic, iron-oxidizing bacteria are *Sulfolobus acidocaldarius* (Brock et al., 1972, 1976) and *Acidianus* (formerly *Sulfolobus*) *brierleyi*. Both genera belong to the archaebacteria. At this time it is unclear whether *S. solfataricus* and *S. ambivalens* can oxidize iron (Zillig et al., 1980; 1985). All four organisms can oxidize S°. *A. brierleyi* is a nonsporeforming, nonmotile, pleomorphic organism (1–1.5 μm in diameter)(Brierley and Brierley, 1973; Brierley and Murr, 1973; McClure and Wyckoff, 1982; Segerer et al., 1986)(Fig. 14.5). The GC content of its DNA is 57 $\pm$ 3 mol%, compared to *S. acidocaldarius*, which has a GC content of 60–68 mol%. The cells of neither organism contain peptidoglycan in their cell wall nor do they feature an outer membrane characteristic of gram-negative eubacteria (Berry and Murr, 1980). Instead the cells are surrounded by a protein layer called the S-layer (Taylor et al., 1982). In laboratory media, the organisms grow best at a pH around 3.0. Their temperature range of growth is between 55 and 90°C with an optimum in the range of 70–75°C. The organisms can grow autotrophically with Fe(II) or S° as energy source. Their autotrophic growth is stimulated by a trace (0.02%) yeast extract. They do not assimilate CO_2 via the Calvin-Benson cycle but more likely by a reverse tricarboxylic acid (TCA) cycle mechanism (Brock and Madigan, 1988). They also can grow heterotrophically with yeast extract at high concentration.

Neutrophiles

Gallionella   Although, as already shown, convincing evidence for enzymatic iron oxidation by iron bacteria at acid pH exists, unequivocal evidence for enzymatic iron oxidation by iron bacteria at neutral pH is mostly lacking. The strongest evidence to date for enzymatic iron oxidation at neutral pH is that associated with *Gallionella*, especially *G. ferruginea*. This organism was first described by Ehrenberg in 1836. In its most easily recognizable form it appears as a bean-shaped cell with a lateral stalk of twisted bundles of fibrils (Fig. 14.6; see also Figures in Hanert, 1981, and Ghiorse, 1984). The bean-shaped cell on the lateral stalk was first recognized as an integral part of the organism by Cholodny (1924). The stalks may branch dichotomously, each carrying a bean-shaped cell at the tip of each branch. The stalk is usually anchored to a solid surface. It may be heavily encrusted with ferric hydroxide. The cells, which may form one or two polar flagellae, may detach from their stalk, swim away as swarmers, and seek a new site for attachment and a stalked growth habit.

In culture experiments in the laboratory under aerobic gradient conditions in which *G. ferruginea* was growing exponentially, the cells were free-living and did not form stalks but were motile with a single flagellum. Stalks began to be formed only in stationary phase when the cell population exceeded 6×10^5 ml^{-1} at a pH above 6. No stalks were formed when ferrous iron did not autoxidize, conditions that include microaerobiosis (e.g., pH 6.5, $E_h < -40$ mV) (Hallbeck and Pedersen, 1990). Hallbeck and Pedersen (1990) suggest that the stalk formation protects against "increasing reducing capacity of ferrous iron as it becomes unstable in an environment that becomes oxidizing."

Gallionella has been generally described as a so-called **gradient organism**; i.e., it grows best under low oxygen tension (0.1–1 mg of O_2 per liter) and in an E_h range of $+200$ to $+320$ mV (Kucera and Wolfe, 1957; Hanert, 1981). It is a mesophile growing optimally at 20°C (Hallbeck and Pedersen, 1990), although growth in nature of some strains has been observed up to 47°C (Hanert, 1981). It prefers a pH range of 6.0–7.6 (Hanert, 1981). Its low oxygen requirement explains why this organism can catalyze Fe^{2+} oxidation at neutral pH. Under these partially reduced conditions, the iron autoxidizes only slowly (Wolfe, 1964).

Gallionella can grow autotrophically and mixotrophically. Hallbeck and Pedersen (1991) demonstrated that the organism obtained all its carbon from CO_2 when growing in a mineral salts medium in an aerobic gradient with iron sulfide as energy source. The same investigators showed that glucose, fructose, and sucrose could meet part of the energy

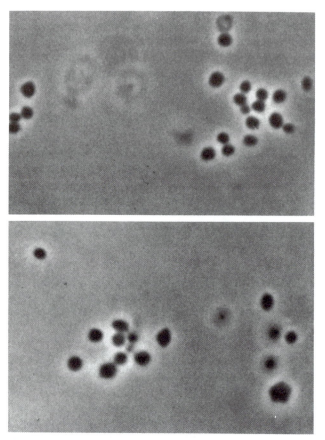

(A)

Figure 14.5   Iron- and sulfur-oxidizing archaebacteria. (A) *Sulfolobus acidocaldarius*. (Reproduced from Brock et al., Sulfolobus: A new genus of sulfur–oxidizing bacteria living at low pH and high temperature ($\times 3,450$). Arch. Mikrobiol. 84: 54. Copyright © 1972 by Springer-Verlag, by permission.) (B) *Acidianus* (formerly *Sulfolobus*) sp. ($\times 28,760$). (Courtesy of J. A. Brierley and C. L. Brierley.)

requirement and part or all of the carbon requirement, depending on the concentration of respective sugars. Hanert (1968) previously claimed that *Gallionella* will not grow without oxidizable iron in the medium. It may well be that strictly autotrophic as well as facultatively autotrophic strains exist. The organism appears to fix CO_2 via ribulose 1,5-bisphosphate carboxylase/oxidase (Luetters and Hanert, 1989; Hallbeck and Pedersen, 1991).

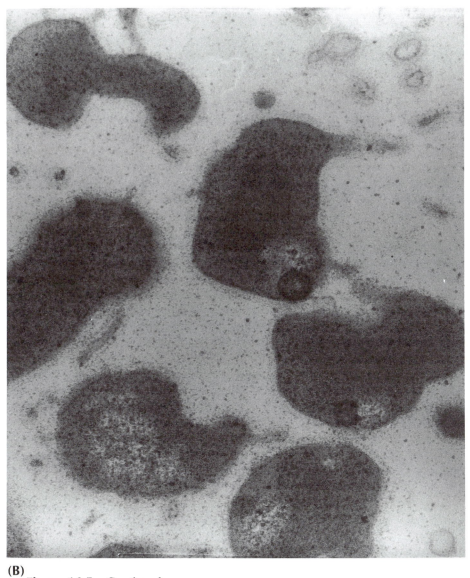

(B)
Figure 14.5 Continued.

According to Luetters-Czekalla (1990), *G. ferruginea* strain BD was able to grow using sulfide and thiosulfate as energy sources and electron donors but not sulfur or tetrathionate at the interface of the oxidizing and reducing zones of a microgradient culture. Addition of organic carbon did not stimulate growth. Under these growth conditions, the organism did

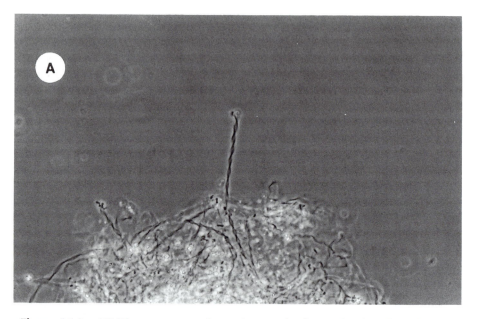

Figure 14.6 (A) Phase-contrast photomicrograph of a tangle of *Gallionella fer-ruginea*. Note the small cell at the tip of the twisted stalk projecting from the tangle ($\times 576$). (B) Electron photomicrograph of unstained and unshadowed *Gallionella* cell showing fibrillar nature of lateral twisted stalk ($\times 65,160$). (Courtesy of W. C. Ghiorse.)

not form the characteristic stalk it forms when growing on Fe(II). Luetters-Czekalla (1990) took this to indicate that the stalk is a product of iron oxidation. The culture, nevertheless, excreted a significant amount of an unidentified, extracellular polymeric material. Another strain of *G. ferruginia*, isolated by Hallbeck (1993), was not able to grow with sulfide or thiosulfate as sole sources of energy and reducing power.

Isolation and propagation methods for *Gallionella* have been summarized by Hanert (1981). They are adaptations of the method originally described by Kucera and Wolfe (1957), using the medium formulated by these authors, The medium consists of NH_4Cl, 0.1%; K_2HPO_4, 0.05%; $MgSO_4$, 0.02%; and freshly prepared ferrous sulfide suspension making up 10% of the total volume of medium. Tap water or natural water are used in making up the medium because distilled water apparently does not supply a required component. A small amount of CO_2 is bubbled

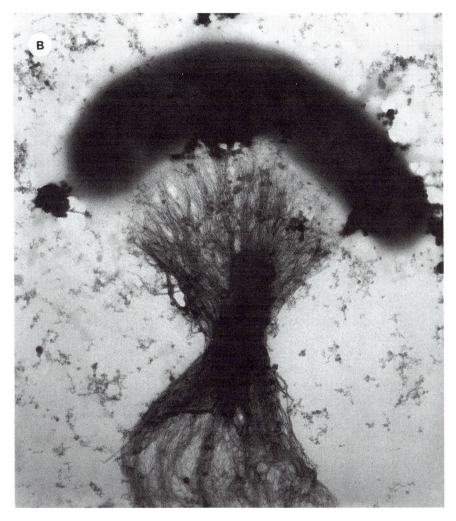

Figure 14.6 Continued.

through the salts solution prior to the addition of ferrous sulfide. The medium is placed in test tubes which are stoppered to prevent loss of CO_2. A redox gradient is established in the culture medium as oxygen from the air diffuses into it and reacts with some of the FeS. *Gallionella*, when growing in this medium, occupies the lower two-thirds of the volume above the ferrous sulfide. The organism will not grow anaerobically, whether nitrate is added or not.

Owing to the complex growth habit of *Gallionella* under some culture conditions, determination of its growth rate can be a problem. Individual development of the stalked organism can be followed quantitatively in microculture by microscopically following stalk elongation and twisting (Hanert, 1974a). In other words, growth is measured in terms of increase in mass of the organism. An elongation rate in the first generation of 40–50 μm h^{-1} has been obtained. This is 2 to 4 times faster than in natural environments. Stalks do not elongate further after three or four divisions of the apical cell. Stalk lengthening occurs at the tip where the apical cell is attached. The stalk twists as it lengthens, owing to the rotation of the apical cell. The rotation occurs at a constant rate. In a natural environment, the rate of growth of *Gallionella* is measured in terms of the rate of attachment to a solid surface such as a submerged microscope slide and the rate of stalk elongation, as shown in the relationship (Hanert, 1973)

$$V_t = \frac{b_v l_v}{2} t^2 \tag{14.16}$$

Here b_v is the average rate of attachment, l_v the average rate of stalk elongation, and t the length of the growth period, which should not be longer than 10 h if this relationship is to hold. V_t is a measure of the amount of growth at time t.

Growth can also be followed by determining viable counts by a most-probable-numbers method, and total counts by a direct counting method with an epifluorescence microscope after staining of the cells with acridine orange (Hallbeck and Pederson, 1990). These latter methods have to be employed for measuring growth when no stalks are formed.

The rate of iron oxidation by *Gallionella* in the natural environment may be measured by submerging a microscope slide at a site of *Gallionella* development for a desired length of time, then removing the slide and measuring the amount of iron deposited on it (Hanert, 1974b). Iron deposition can be expressed in terms of the amount of iron per unit surface area of the slide which was submerged and on which iron was laid down.

Other Iron Bacteria   Other bacteria that have been associated by some investigators with iron oxidation include sheathed bacteria (Fig. 14.7) such as *Sphaerotilus*, *Leptothrix* spp., *Crenothrix polyspora*, *Clonothrix* sp., and *Lieskeella bifida* and some encapsulated bacteria like the Siderocapsaceae group. Many of these are more likely iron-depositing bacteria; i.e. they bind preoxidized iron at their cell surface (Ghiorse, 1984; Ghiorse and Ehrlich, 1992).

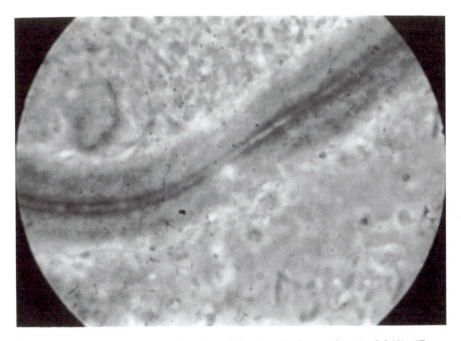

Figure 14.7 *Leptothrix* spp. Portion of the sheathed organism (× 5,240). (Courtesy of E. J. Arcuri.)

Ultrastructural examination of the sheath of *Leptothrix discophora* SP-6 showed it to be a tubular structure of condensed fibrils (6.5 nm in diameter) overlaid by a somewhat diffuse capsular layer (Emerson and Ghiorse, 1993a). The fibrillar part of the sheath was anchored by bridges to the outer membrane of the gram-negative cells in the sheath. The capsular layer had a net negative charge. Purified sheaths contained 34–35% polysaccharide consisting of a 1:1 mixture of uronic acids and galactosamine, 23–25% protein enriched in cysteine, 8% lipid, and 4% inorganic ash. The cysteine in the sheath protein is thought to be important in the maintenance of integrity of the sheath (Emerson and Ghiorse, 1993b).

Leptothrix spp., sometimes classed with *Sphaerotilus* (Pringsheim, 1949; Stokes 1954; Hoehnl, 1955), have been examined on several occasions for enzymatic iron oxidation. Winogradsky (1988) first reported that *L. ochracea* (probably *L. discophora*, according to Cholodny, 1926) could oxidize iron. He found that he could grow the organism in hay infusion only if he added ferrous carbonate. The iron was oxidized and deposited in its sheath. He inferred from this observation that the organism was an autotroph. Molisch (1910) and Pringsheim (1949) disagreed with Winogradsky's conclusion about iron oxidation by *L. ochracea*, believing that the organism merely deposited autoxidized iron in its sheath. However, Lieske (1919) confirmed Winogradsky's observations of growth on ferrous carbonate in very dilute organic solution and suggested that the organism might be mixotrophic. Cholodny (1926), Sartory and Meyer (1947), and Praeve (1957) also made observations similar to those of Winogradsky and Lieske. Most claims for enzymatic oxidation by *Leptothrix* have been based mainly on the growth requirement of ferrous iron in dilute medium and on the oxidation of this ferrous iron during growth. However, Praeve (1957) also showed a stimulation of oxygen uptake by organisms when Fe^{2+} was present as the only exogenous, oxidizable substrate in Warburg respiration experiments. Significantly, he found that empty sheaths were unable to take up oxygen on Fe^{2+}. Most recently, De Vrind-De Jong et al. (1990) reported iron-oxidizing activity in spent medium from a culture of *L. discophora* SS-1. Corstjens et al. (1992) related this iron-oxidizing activity to a 150-kDa protein. It behaved like an enzyme and was not produced by a spontaneous mutant strain that lacked iron-oxidizing activity. The factor was distinct from the manganese-oxidizing protein excreted by the wild-type strain described by Adams and Ghiorse (1987).

Dubinina (1978a,b) more recently reported that *L. pseudoochracea*, *Metallogenium*, and *Arthrobacter siderocapsulatus* oxidized Fe^{2+} with metabolically produced H_2O_2 through catalysis by catalase of the organism. It may be that different mechanisms of enzymatic iron oxidation exist.

For all other sheathed bacteria, enzymatic iron oxidation is at most presumptive, based on gross morphological similarities with *Leptothrix* and the observation of oxidized iron deposits on their sheaths or capsules. It is quite likely that most or all of these organisms merely deposit preoxidized iron on their sheaths or capsules.

A wall-less bacterium, *Mycoplasma laidlawii* (now known as *Acholeplasma laidlawii*) has been reported to oxidize iron (Balashova and Zavarzin, 1972). The organism was cultured in a salt-free, meat-peptone medium containing iron wire or powder. Ferric iron was formed during active growth and, in part, precipitated on the cells of the organism. Catalase was found to depress ferric oxide production, suggesting that H_2O_2 played a role in the oxidation process. It is interesting that in this instance, catalase did not accelerate iron oxidation as found by Dubinina (1978a). It is not clear from the report whether other enzymes played a direct role in the oxidation of iron in this instance.

Although up to now, bacterial oxidation of ferrous iron has been generally assumed to require oxygen as terminal electron acceptor, an exception that occurs around neutral pH has been found. Two anaerobic, phototrophic strains resembling *Rhodomicrobium vannielii* and *R. palustris*, respectively, and one strain resembling *Thiodictyon* spp. morphologically were shown to oxidize Fe(II) to Fe(III) oxides in the light. These bacteria used the reducing power generated in the oxidation of Fe(II) to Fe(III) oxide for CO_2 fixation (Widdel et al., 1993). No iron oxidation was observed anaerobically in the dark or in the light in the absence of the bacteria. No growth occurred in the test media in the absence of added iron. Widdel et al. (1993) suggest that this anaerobic oxidation of Fe(II) by phototrophic bacteria may have contributed to the early stages of banded-iron formations that arose in Archean times. This bacterial process also permits the completion of an iron cycle under anaerobic conditions.

14.5 IRON(III) AS TERMINAL ELECTRON ACCEPTOR IN BACTERIAL RESPIRATION

Ferric iron in nature may be microbiologically reduced to ferrous iron. The ferric iron may be in dissolved form or as an insoluble mineral like limonite, goethite, hematite, and so forth. As in the case of iron oxidation, this reduction may be enzymatic or nonenzymatic. Enzymatic ferric iron reduction may manifest itself as a form of respiration, mostly anaerobic, in which ferric iron serves as **dominant** or **exclusive** terminal electron acceptor, or it may accompany fermentation, in which ferric iron serves as a **supplementary**, as opposed to dominant or exclusive, terminal electron acceptor. Both of these ferric-iron-reducing processes are forms of **dissimilatory iron reduction**.

When ferric iron is reduced during uptake for incorporation into specific cellular components, the process represents **assimilatory iron reduction**. Relatively large quantities of iron are consumed in dissimilatory reduction whereas only very small quantities are consumed in assimilatory reduction. The ferric iron in assimilatory reduction when acquired at approximately neutral pH is usually complexed by siderophores and may be reduced in this complexed form or after release from the ligand in the cell envelope (discussed earlier).

The emphasis in the following sections will be on dissimilatory iron reduction.

Bacterial Ferric Iron Reduction Accompanying Fermentation For some time, ferric iron has been known to influence fermentative metabolism of bacteria as a result of its ability to act as terminal electron acceptor. Roberts (1947) showed a change in fermentation balance when comparing the action of *Bacillus polymyxa* on glucose anaerobically in the presence and absence of iron (Table 14.4). The ferric iron in these experiments was supplied as freshly precipitated, dialyzed ferric hydroxide suspension obtained in a reaction of ferric chloride and an excess of potassium hydroxide. The suspension had a pH of 7.8. The ferric iron seemed to act as a supplementary electron acceptor in the fermentation and in this way changed the quantities of certain products formed from glucose. Thus, in the presence of iron, less H_2 and CO_2, more ethanol, and less 2,3-butylene glycol were formed in either organic or synthetic medium than in the absence of iron. Also, more glucose was consumed in the presence of iron than in its absence in either medium.

Bromfield (1954a) showed that besides *B. polymyxa*, growing cultures of *B. circulans* can also reduce ferric iron. Depending on the me-

Table 14.4 Fermentation Balances for *Bacillus polymyxa* Growing in Two Different Media in the Presence and Absence of Ferric Hydroxide[a]

Products	Synthetic medium[b] (mol/100 mol glucose)		Organic medium[c] (mol/100 mol glucose)	
	$-Fe(OH)_3$	$+Fe(OH)_3$	$-Fe(OH)_3$	$+Fe(OH)_3$
CO_2	199	170	186	178
H_2	51	31	53	33
HCOOH	11	12	9	12
Lactic acid	17	19	14	7
Ethanol	72	82	78	94
Acetoin	0.5	1	1	2
2,3-Butylene glycol	64	51	49	44
Acetic acid	0	0	0	0
Iron reduced	—	42	—	61
Glucose fermented (mg/100 ml)	1,029	2,333	1,334	2,380
C recovery (%)	112.1	101.8	98.8	97.2
O/R index	1.06	1.0	1.06	1.03

[a] Incubation was for 7 days at 35°C.
[b] Glucose, 2.4%; asparagine, 0.5%; K_2HOP_4, 0.08%; KH_2PO_4, 0.02%; KCl, 0.02%; $MgSO_4 \cdot 7H_2O$, 0.5%
[c] Glucose, 2.5%; peptone, 1%; K_2HPO_4, 0.08%; KH_2PO_4, 0.02%; KCl, 0.02%; $MgSO_4 \cdot 7H_2O$, 0.5%
Source: Roberts (1947), by permission.

dium, he found that even *Escherichia freundii*, *Aerobacter* (now *Enterobacter*) sp. and *Paracolobactrum* (now probably *Citrobacter*) can do so. However, he inferred from his results that the reduction of iron was not directly involved in the oxidation of the substrate (energy source), which is at variance with his results with resting cells (Bromfield, 1954b). He found that completely anaerobic conditions were not required to obtain ferric iron reduction by the bacteria. But when the level of aeration of the cultures was increased, ferrous iron became reoxidized due to autoxidation.

Bromfield (1954b) also showed that washed cells of *B. circulans*, *B. megaterium*, and *E. aerogenes* reduced ferric iron of several ferric compounds [$FeCl_3$, $Fe(OH)_3$, $Fe(lactate)_3$] in the presence of such suitable electron donors such as glucose, succinate, and malate. He was able to inhibit the reduction by boiling the cells or by adding chloroform or toluene to the reaction mixture, but he did not observe inhibition with either azide or cyanide. He interpreted his findings as indicating that ferric iron reduction was associated with dehydrogenase activity. He thought, however, that the reduction of insoluble ferric iron (e.g., ferric hydroxide) could only have occurred in the presence of a complexing agent. From more recent studies, it has become clear that although some complexing agents may speed up the rate of reduction, as did α, α-dipyridyl in his experiments, it is not essential (e.g., De Castro and Ehrlich, 1970).

Some other bacteria have since been shown to be able to reduce ferric iron in conjunction with a fermentative process. They include aerobes such as *Pseudomonas* spp., *Vibrio* sp., anaerobes such as *Clostridium* spp., *Bacteroides hypermegas*, *Desulfovibrio desulfuricans*, and *Desulfotomaculum nigrificans* (see review by Lovley, 1987). The iron-reducing activity has been viewed in several of these instances as a sink for reducing power that does not furnish energy to the cells (see reviews by Lovley, 1987, 1991). However, this explanation holds only if it can be demonstrated that iron reduction is not accompanied by energy conservation. *Pseudomonas ferrireductans* (now *Shewanella putrefaciens* strain 200) appears to contain both constitutive and inducible ferric iron reductase. The constitutive enzyme is involved in ferric iron respiration, and the inducible enzyme, produced at lower oxygen tension, is involved in electron scavenging without energy conservation (electron sink) (Arnold et al., 1986a).

Even some fungi have been implicated in Fe(III) reduction (e.g., Ottow and von Klopotek, 1969). However, their ability to reduce ferric iron is not likely to involve anaerobic respiration but instead involves either assimilatory iron reduction or the production of metabolic product(s) that act as chemical reductants of the iron.

Ferric Iron Respiration   Typical heterotrophic ferric iron respirers include both strictly anaerobic and facultative bacteria. Examples of strict anaerobes include *Geobacter metallireducens* (formerly strain GS-15)(Lovley et al., 1993), *G. sulfurreducens* (Caccavo et al., 1994), and various sulfate reducers such as *Desulfuromonas acetoxidans* (Roden and Lovley, 1993) and *Desulfovibrio desulfuricans* (Coleman et al., 1993) in the absence of sulfate as terminal electron acceptor. Examples of facultative bacteria include various strains of *Shewanella* (formerly *Pseudomonas, Alteromonas*) *putrefaciens* (Obuekwe and Westlake, 1982; Arnold et al., 1986a,b; Myers and Nealson, 1988; Lovley et al., 1989) and *Pseudomonas* sp. (Balashova and Zavarzin, 1979). The facultative bacteria reduce iron only anaerobically.

Some autotrophs also can respire with iron as terminal electron acceptor. *T. thiooxidans*, *T. ferrooxidans*, and *Sulfolobus* spp. can reduce ferric iron with $S°$ as electron donor (Brock and Gustafson, 1976; Pronk et al., 1992). *T. thiooxidans* can bring this reduction about aerobically because the ferrous iron it produces at acid pH (around pH 2.5) does not autoxidize readily. On the other hand, *T. ferrooxidans* forms Fe^{2+} only anaerobically because it reoxidizes the Fe^{2+} aerobically. *S. acidocaldarius* can form Fe^{2+} microaerophilically at 70°C because under this limited oxygen availability, it does not reoxidize Fe^{2+}. Some growing cultures of *T. ferrooxidans* appear to use a branched pathway when oxidizing sulfur aerobically in the presence of ferric iron (Corbett and Ingledew, 1987) in which electrons from sulfite can pass to iron(III) or O_2 via a bc_1 complex. The reaction can be summarized as follows:

$$SO_3^{2-} + H_2O \Rightarrow SO_4^{2-} + 2H^+ + 2e \qquad (14.17)$$

$$2Fe^{3+} + 2e \Rightarrow 2Fe^{2+} \qquad (14.18)$$

$$\tfrac{1}{2}O_2 + 2H^+ + 2e \Rightarrow H_2O \qquad (14.19)$$

The sulfite is a metabolic intermediate in the oxidation of $S°$. Its formation involves an oxygenation and thus requires oxygen (see Chapter 17).

T. ferrooxidans strain AP19–3 appears to reduce ferric iron both aerobically and anaerobically with sulfur by an enzyme system that includes a sulfide:Fe(III)- and a sulfite:Fe(III)-oxidoreductase (Sugio et al., 1989; Sugio et al., 1992a). Other strains of *T. ferrooxidans* and *Leptospirillum ferrooxidans*, which was not previously known to oxidize sulfur ($S°$), also appear to possess this enzyme system (Sugio et al., 1992b). In this process, the bacteria first transform elemental sulfur to sulfide in the presence of reduced glutathione. The reactions involved in sulfur oxidation with Fe(III) by *T. ferrooxidans* AP19–3 as presently understood can be summarized as follows.

$$S° + 2GSH \Rightarrow H_2S + GSSG \tag{14.20}$$

$$H_2S + 6Fe^{3+} + 3H_2O \Rightarrow SO_3^{2-} + 6Fe^{2+} + 8 H^+ \tag{14.21}$$

$$SO_3^{2-} + 2Fe^{3+} + H_2O \Rightarrow SO_4^{2-} + 2Fe^{2+} + 2H^+ \tag{14.22}$$

Growth on sulfur in the presence of Fe(III) by *T. ferrooxidans* AP19-3 occurs only aerobically (Sugio et al., 1988a,b). *L. ferrooxidans* has so far not been grown on sulfur (Sugio et al., 1992b). It seems to conserve energy only from iron oxidation, not sulfur oxidation by ferric iron. Chemical oxidation of intermediate sulfite by ferric iron in the periplasm of *T. ferrooxidans* AP19-3 also occurs (Sugio et al., 1985). Sulfite oxidation by ferric iron in *T. ferrooxidans* strain AP19-3 is viewed as a mechanism of detoxification since sulfite is toxic if allowed to accumulate (Sugio et al., 1988b).

Dissimilatory iron reduction in the form of anaerobic respiration has now been recognized as an important means of mineralization of organic matter in environments in which sulfate or nitrate occurs in amounts insufficient to sustain sulfate or nitrate respiration, respectively (Lovley, 1987, 1991; Nealson and Saffarini, 1994). The process can operate with various organic acids, including volatile and fatty acids, and with aromatic compounds as electron donors. Furthermore, it can displace methanogenesis by outcompeting for H_2 and acetate (Lovley, 1991).

Metabolic Evidence for Enzymatic Ferric Iron Reduction Most of the evidence for dissimilatory ferric iron reduction rests on observations with growing cultures. Troshanov (1968, 1969), following up on Bromfield's earlier observations, demonstrated that *B. circulans*, *Pseudomonas liquefaciens*, *Bacillus mesentericus*, *B. cereus*, *B. centrosporus*, *B. mycoides*, *B. polymyxa*, and *Micrococcus* sp., which he isolated from sediment from several lakes on the Karelian peninsula in the former U.S.S.R., could reduce ferric iron to varying degrees. He found that all his cultures that reduced iron could also reduce manganese, but the reverse was not true. The effect of oxygen on iron reduction in his experiments depended on the culture he tested. Some organisms, such as *B. circulans*, reduced iron more readily anaerobically; others, including *B. polymyxa*, did not. Troshanov noted that the form in which ferric iron was presented to his cultures affected the rate of its reduction. Insoluble ferric iron in bog ore was reduced more slowly than soluble $FeCl_3$. Cultures also varied in their ability to reduce insoluble ferric iron. Troshanov found that *B. circulans* actively reduced bog iron ore whereas *B. polymyxa* did not.

Similar findings were made independently with soil bacteria by Ottow and collaborators (Ottow, 1969a; Ottow, 1971; Ottow and Glathe,

1971; Hammann and Ottow, 1974; Munch et al., 1978; Munch and Ottow, 1980). Support for the notion that ferric iron reduction in these instances was enzymatic was gained from the observation that nitrate and chlorate can reversibly inhibit ferric iron reduction by members of the genera *Enterobacter* and *Bacillus* as well as *Pseudomonas* and *Micrococcus* (Ottow, 1969b; Ottow, 1970a). Since all of these organisms possess dissimilatory nitrate reductase, which catalyzes reduction of nitrate and chlorate (Pichinoty, 1963), it was inferred that in these organisms iron reductase is the same enzyme as nitrate reductase. However, this cannot be the case in all iron reducers, since some other bacteria (e.g., *B. pumilus*, *B. sphaericus*, *Clostridium saccharobutylicum*, and *C. butyricum*) that lack nitrate reductase activity can nevertheless reduce ferric iron (Ottow, 1970a; Munch and Ottow, 1977). These organisms must possess another kind of iron reductase. This was suggested by observations of ferric iron reduction by mutants lacking nitrate reductase (Nit⁻), isolated from wild-type strains possessing this enzyme (Nit⁺). Iron reduction with these strains was found to be insensitive to inhibition by nitrate or chlorate (Ottow, 1970a). Most of these Nit⁻ mutants reduced iron less rapidly than did their wild-type parent, but a Nit⁻ mutant of *B. polymyxa* reduced iron more intensely than its wild-type parent. Consistent with the observation that some dissimilatory iron reducers use an enzyme other than nitrate reductase in Fe(III) reduction, Ottow and Ottow (1970) noted that the size of the soil microflora capable of reducing iron is usually greater than the size of the microflora capable of reducing nitrate. The list of inhibitors of bacterial ferric iron reduction has been extended to include permanganate, dichromate, sulfite, thiosulfate, and the redox dyes methylene blue, indochlorophenol, and phenazine methosulfate when *Shewanella* (*Pseudomonas*) *putrefaciens* strain 200 is the test organism (Obuekwe and Westlake, 1982).

Nitrate inhibition of ferric iron reduction need not be due to competitive inhibition of iron reductase (nitrate reductase) (Obuekwe et al., 1981). Nitrate was found to stimulate ferric iron reduction by *S. putrefaciens* (*Pseudomonas* sp.) during short-term incubation but to depress it during long-term incubation. The inhibitory effect of nitrate in this instance was related to chemical oxidation of ferrous iron to ferric iron by nitrite (see also discussion by Sørensen, 1987). Nevertheless, the authors found that when *S. putrefaciens* (*Pseudomonas* sp.) was preinduced by nitrate, the resultant induced strain reduced Fe(III) faster than the uninduced strain, suggesting that nitrate reductase can catalyze ferric iron reduction by this organism. Obuekwe and Westlake (1982) explained this effect as merely reflecting a better physiological state of induced than uninduced cells. Nitrate may also act as a noncompetitive inhibitor of ferric iron reduction, as in *Staphylococcus aureus* (Lascelle and Burke, 1978).

Other evidence for an enzymatic nature of Fe(II) reduction is the observation of De Castro and Ehrlich (1970) that a cell extract from marine *Bacillus* strain 29A, whose intact cells actively reduced ferric iron, could reduce the iron in the mineral limonite. This activity was partially destroyed by heating and inhibited by mercuric chloride and *para*-chloromercuribenzoate. Lascelle and Burke (1978) detected ferric iron reduction by a membrane fraction from *Staphylococcus aureus* which could also reduce nitrate. They found evidence for involvement of a branched electron transport pathway in ferric iron reduction by *Staphylococcus aureus*. By use of selective inhibitors they showed that this branched pathway passed electrons to nitrate via a cytochrome-b-requiring branch whereas it passed electrons to ferric iron via a branch ahead of cytochrome b. Nitrate is thought to inhibit ferric iron reduction by this organism because nitrate accepts electrons more readily than ferric iron in this system. Obuekwe (1986) demonstrated that the ferric reductase in *S.* (*Pseudomonas*) *putrefaciens*) strain 200 was inducible and that ferric iron reduction was inhibited by sodium amytal, $2n$-heptyl-4-hydroxyquinoline-*N*-oxide (HQNO), and sodium cyanide in whole cells, suggesting involvement of a cytochrome pathway in transferring electrons from donor to ferric iron. He was, however, unable to demonstrate ferric iron reducing activity with cell membranes, or periplasmic or cytoplasmic content of the cells. A kinetic study of ferric iron reduction by *S. putrefaciens* (*Pseudomonas*) strain 200 indicated that the form of soluble ferric iron species determines reaction rate (Arnold et al., 1986b).

The most direct evidence for enzymatically catalyzed ferric iron reduction comes from studies with *Geobacter metallireducens* (strain GS-15), *G. sulfurreducens*, three strains of *Shewanella putrefaciens*, and *Desulfuromonas acetoxidans*. Fe(III) reduction as a form of anaerobic respiration is readily demonstrable with growing and resting cultures of *Geobacter metallireducens* (formerly strain GS-15), a strict anaerobe isolated from freshwater sediment, (Lovley and Phillips, 1988a), *G. sulfurreducens* isolated from a hydrocarbon contaminated ditch (Caccavo et al., 1994), and the facultative anaerobe *Shewanella* (formerly *Alteromonas, Pseudomonas*) *putrefaciens* strains MR-1, 200 and ATCC 8071 (Myers and Nealson, 1988; Obuekwe et al., 1981; Lovley and Phillips, 1989). Strain MR-1 was isolated from a lake and strain 200 from a Canadian oil pipeline. All were found to use Fe(III) or Mn(IV) as terminal electron acceptor for growth.

G. metallireducens reduced amorphous iron oxide to magnetite (Mn_3O_4) in laboratory experiments, but it did not readily reduce crystalline iron oxides (Lovley and Phillips, 1986b). It can use acetate (Fig. 14.8), ethanol, butyrate, and propionate as electron donors. It oxidizes the acetate to CO_2 in the process. In studies with intact cells (Gorby and Lovley,

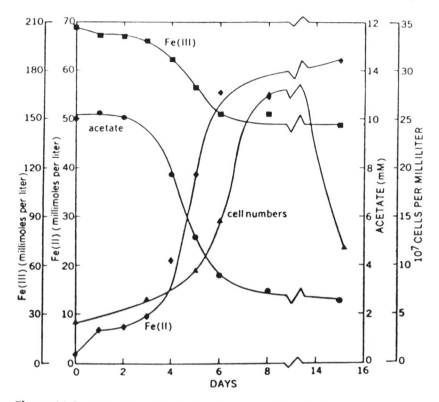

Figure 14.8 Reduction of ferric iron by acetate through the mediation of anaerobic bacterial strain GS-15 *Geobacter metallireducens*. This family of curves illustrates the reduction of oxalate-extractable Fe(III) to Fe(II) at the expense of acetate consumption during growth of culture GS-15 in FWA medium containing amorphic Fe(III) oxide. (Reproduced from Lovley and Phillips, 1988, by permission.)

1991), the organism conveyed reducing power from an appropriate electron donor to ferric iron via a membrane-bound electron transport system that included b- but not c-type cytochrome and Fe(III) reductase. The electron transport inhibitors 2-heptyl-4-hydroxyquinoline N-oxide, sodium azide, and sodium cyanide had no effect on the system. It oxidizes acetate via the tricarboxylic (citric) acid cycle (Champine and Goodwin, 1991).

G. *sulfurreducens*, unlike G. *metallireducens*, can use either H_2 or acetate as electron donor in iron(III) reduction. Figure 14.9 shows diagrammatically how the organism uses H_2 in this process. Elemental sulfur, Co(III)-EDTA, fumarate, and malate can serve as alternative sole electron acceptors, but not Mn(IV) or U(VI). Unlike G. *metallireducens*, the organ-

$$0.5\ H_2 + 2H^+ + FeOOH \Longrightarrow Fe^{2+} + 2H_2O$$

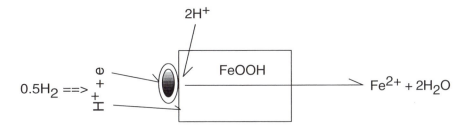

Figure 14.9 Diagrammatic representation of the reduction of iron(III) in the form of FeOOH with H_2 by an organism like *Geobacter sulfurreducens* or *Shewanella putrefaciens*.

ism cannot use propionate, butyrate, benzoate, or phenol as electron donors, nor can it use some other organic compounds as reductants of Fe(III). The organism cannot reduce nitrate, sulfate sulfite, or thiosulfate with acetate as electron donor (Caccavo et al., 1994).

Shewanella putrefaciens can use H_2 (Fig. 14.9) as well as formate, lactate, and pyruvate as electron donors in the anaerobic reduction of Fe(III), but the last two donors are only incompletely oxidized to acetate and CO_2 (Lovley et al., 1989; Myers and Nealson, 1988, 1990). Myers and Nealson (1990) found that energy was conserved when *S. putrefaciens* strain MR-1 reduced ferric iron with lactate as electron donor anaerobically by demonstrating respiration-linked proton translocation, which was completely inhibited by 20 μM carbonyl cyanide *m*-chlorophenylhydrazone and partially inhibited to completely inhibited by 50 μM 2-*n*-heptyl-4-hydroxyquinoline *N*-oxide. When grown anaerobically, *S. putrefaciens* MR-1 has 80% of its cytochrome complement (mostly the c type) in the outer membrane of its cell envelope, with the rest (c and b types) being mostly or entirely associated with the plasma membrane. When grown aerobically, the same organism contains a major portion of its cytochrome complement in its plasma membrane, which, insofar as studied, is the more common location of cytochromes in other bacteria, except for periplasmic c-type cytochrome (Myers and Myers, 1992a). Slightly more than 50% of formate-dependent ferric reductase activity of anaerobically grown *S. putrefaciens* was found associated with the outer membrane, the rest with the plasma membrane. The membranes of aerobically grown cells were devoid of this activity (Myers and Myers, 1993).

Most of the formate dehydrogenase activity was soluble (Myers and Myers, 1993), suggesting a possible periplasmic location, as with fumarase reductase in anaerobically grown cells of this organism (Myers and Myers, 1992b). Addition of nitrate, nitrite, fumarate, or trimethylamine *N*-oxide as alternative electron acceptors did not inhibit ferric reductase activity; NADH could replace formate as electron donor in experiments with ferric-reductase-containing membrane fraction (Myers and Myers, 1993).

S. putrefaciens strain 200 forms a constitutive and inducible Fe(III) reductase. The latter is only induced anaerobically. Both reductases are responsible for rapid Fe(III) reduction under anaerobic conditions. The constitutive reductase will reduce Fe(III) at low O_2 tension, but the rate of Fe(III) reduction is very slow. Apparently competition with O_2 is responsible for the slow rate of reduction. A branched respiratory pathway is postulated in which one branch leads to O_2 and the other to Fe(III) (Arnold et al., 1990).

The iron reductase activity detected in *Pseudomonas aeruginosa* (Cox, 1980) evidently represents an assimilatory iron reductase system because it only reduced complexed iron such as ferripyrochelin and ferric citrate (pyrochelin and citrate are known siderophores). The ferric citrate reductase had a cytoplasmic location, whereas the ferripyrochelin reductase had a periplasmic and cytoplasmic location. The activities appear to be pyridine nucleotide-linked, although reduced glutathione could also serve as electron donor. Similarly, the iron(III) reduction observed with *Escherichia coli* K12 (Williams and Poole) involved an assimilatory iron reductase. Like *Pseudomonas aeruginosa*, *E. coli* K12 reduced ferric citrate directly with reduced pyridine nucleotide as electron donor without involvement of the respiratory chain in the membrane of the organism. Adenosine 5′-triphosphate (ATP) and cyanide were found to stimulate ferric citrate reduction, possibly by forming complexes with Fe(III).

Bioenergetics of Dissimilatory Iron Reduction   Because proton translocation has been demonstrated during Fe(III) reduction by *S. putrefaciens* MR-1 (Myers and Nealson, 1990), it can be inferred that this organism conserves energy chemiosmotically in this process. The other strains of *S. putrefaciens*, such as those studied by Obuekwe (1986), Arnold et al. (1986a,b), and Lovley et al. (1989), probably conserve energy by the same mechanism as strain MR-1. At least, they grow anaerobically when Fe(III) is the only terminal electron acceptor but not in its absence or that of any alternative acceptor. *Geobacter metallireducens* most likely conserves energy chemiosmotically in Fe(III) respiration because, as previously mentioned, electrons from a suitable energy source are transported to Fe(III) by a membrane-bound system that includes b-type cytochrome and an Fe(III) reductase (Gorby and Lovley, 1991).

Free-energy calculations show that when acetate is the electon donor (reductant), the standard free-energy change at pH 7 ($\Delta G_r^{o'}$) when Fe^{3+} is the electron acceptor (oxidant) is -193.4 kcal mol^{-1} (-808.4 kJ mol^{-1}), which is close to that when O_2 is the reductant (-201.8 kcal mol^{-1} or -843.5 kJ mol^{-1}). However, when $Fe(OH)_3$ is the oxidant of acetate, the standard free-energy change at pH 7 is only -5.5 kcal mol^{-1} (-23.0 kJ mol^{-1}) (Ehrlich, 1993). This indicates that undissolved iron oxides or hydroxides are poor electron acceptors at neutral or alkaline pH and do not allow for effective biochemical energy conservation. They are better acceptors and more effective for biochemical energy conservation below pH 7.

It has been suggested that not all iron(III)-reducing bacteria gain energy in the process. For instance, Ghiorse (1988) and Lovley (1991) propose that a number of bacteria reduce ferric iron merely to dispose of excess reducing power via secondary respiratory pathways without conserving energy, or that the iron reduction that was observed with these orgranisms was part of their iron assimilation process. An absence of growth stimulation under anaerobic conditions when an organism reduces Fe(III) with stoichiometric release of extracellular Fe^{2+} is not necessarily an indication that energy is not conserved in the reduction. The Fe(III) reduction process may simply yield insufficient energy for growth if it is the only source of energy and may have to be accompanied by an additional respiratory or fermentative process to meet the full energy demand.

The recent observation by Lovley et al. (1995) that *Pelobacter carbinolicus* can catabolize 2,3-butanediol with the production of ethanol and acetate by fermentation as well as by reduction of Fe(III) or S° provides new insight into this puzzle. The authors noted the ratio of ethanol to acetate produced from 2,3-butanediol was significantly greater in the absence of Fe(III) than in its presence. Since *P. carbinolicus* lacks c-type cytochromes, it may be using the Fe(III) merely as an electron sink in this case. On the other hand, the authors also found that the organisms can grow on H_2 as sole energy source in a medium in which acetate is the sole carbon source (it cannot use acetate as energy source) and Fe(III) is the only terminal electron acceptor. Under these conditions the organism must respire and conserve energy chemiosmotically. This raises a question whether when growing on 2,3-butanediol, the organism uses Fe(III) when present merely as an electron sink and does not conserve energy from its reduction.

Reduction of Ferric Iron by Fungi Some fungi seem to be able to reduce ferric iron. Lieske reported such a phenomenon (see Starkey and Halvorson, 1927). Ottow and von Klopotek (1969) observed reduction of iron in hematite (Fe_2O_3) by *Alternaria tenuis, Fusarium oxysporum,* and *F. so-*

lani, all of which possessed an inducible nitrate reductase. Fungi incapable of reducing nitrate in this study were also incapable of reducing ferric iron. These investigators concluded, therefore, that nitrate reductase functions in fungi in the reduction of ferric iron as well as nitrate, a mechanism thus similar to that in many bacteria. Their conclusion raises a question, however, as to where in the fungi the nitrate reductase is located to be able to act on hematite and whether this represents dissimilatory or assimilatory iron reduction. The latter seems more likely.

Types of Ferric Compounds Attacked via Dissimilatory Iron Reduction   The ease with which ferric iron is reduced by bacteria depends in part on the form in which they encounter it. In the case of insoluble forms, the order of decreasing solubilization in one study (Ottow, 1969b) was $FePO_4 \cdot 4H_2O$ > $Fe(OH)_3$ > lepidochrocite (gamma-$FeOOH$) > goethite (α-$FeOOH$) > hematite (α-Fe_2O_3). In another study (De Castro and Ehrlich, 1970), using glucose as electron donor, marine *Bacillus* 29A was found to solubilize larger amounts of Fe from limonite and goethite than from hematite. In initial stages of the reduction, the order of decreasing activity was goethite > limonite > hematite. The organism did not reduce ferric iron extensively when it occurred in ferromanganese nodules from the deep sea (Ehrlich et al., 1973). Because of the insoluble nature of the iron in the nodules, the aerobic conditions, and in some cases the circumneutral pH at which these observations were made, it seems unlikely that the absence of measurable iron reduction was due to the manganese inhibition phenomenon described by Lovley and Phillips (1988b). In a study of anaerobic reduction of pedogenic iron oxides by *Clostridium butyricum*, Munch and Ottow (1980) found that amorphous oxides were more readily attacked than crystalline iron oxides. This was also observed by Lovley and Phillips (1986b) in anaerobic experiments with an enrichment culture from Potomac River sediment tested on various forms of Fe(III).

Since oxides of ferric iron are very insoluble at near neutral pH, the question arises how bacteria can attack these compounds. The tendency of the oxides to dissociate into soluble species at near neutral pH is simply too small to explain the phenomenon. Physical contact between the active cell and the surface of the oxide particle appears essential. This was clearly demonstrated by Munch and Ottow (1982) in experiments in which they placed iron oxide inside a dialysis bag whose pores did not permit the passage of bacterial cells but did permit ready diffusion of culture medium with its dissolved inorganic and organic chemical species. No iron reduction occurred when bacteria (*Clostridium butyricum* or *Bacillus polymyxa*) were placed in medium outside the bag, but it did occur when bacteria and iron oxide were both present inside the bag. Interestingly, metabolites produced by the bacteria, including acids, were not able to dissolve significant amounts of the oxide. Furthermore, the lowered E_h (redox potential) generated by the bacteria, whether inside or outside the dialysis bag was also unable to cause reduction of ferric iron in the oxide.

14.6 NONENZYMATIC OXIDATION AND REDUCTION OF FERRIC IRON BY MICROBES

Nonenzymatic Oxidation Many different kinds of microorganisms can promote iron oxidation indirectly (i.e., nonenzymatically). They can accomplish this by affecting the redox potential of the environment or, more precisely, by generating oxidants that chemically oxidize Fe(II). They can also accomplish this by generating a pH above 5 at which Fe^{2+} is oxidized by oxygen in the air (**autoxidation**). Among the first to recognize indirect iron oxidation were Harder (1919), Winogradsky (1922), and Cholodny (1926). Starkey and Halvorson (1927) demonstrated indirect oxidation of iron in laboratory experiments with bacteria and explained their findings in terms of reaction kinetics involving the oxidation of ferrous iron and the solubilization of ferric iron by acid. From their work it can be inferred that any organism that raises the pH of a medium by forming ammonia from protein-derived material [Eq. (14.16)] or by consuming salts of organic acids [Eq. (14.24)] can promote ferrous iron oxidation in an aerated medium:

$$\underset{\substack{| \\ NH_2}}{RCHCOOH} + H_2O + \tfrac{1}{2}O_2 \longrightarrow \underset{\substack{|| \\ O}}{RCCOOH} + NH_4^+ + OH^-$$

amino acid   keto-acid   (14.23)

$$C_3H_5O_3^- + 3O_2 \rightarrow 3CO_2 + 2H_2O + OH^- \qquad (14.24)$$
lactate

A more specialized case of indirect microbial iron oxidation is that associated with cyanobacterial and algal photosynthesis. The photosynthetic process may promote ferrous iron oxidation by creating conditions that favor autoxidation in two ways: (1) by raising the pH of the waters in which they grow and (2) by raising the oxygen level in the waters around them. The rise in pH is explained by Eqs. (14.25) and (14.26):

$$2HCO_3^- \rightarrow CO_3^{2-} + CO_2 + H_2O \qquad (14.25)$$

$$CO_3^{2-} + H_2O \rightarrow HCO_3^- + OH^- \qquad (14.26)$$

Reaction (14.25) is promoted by CO_2 assimilation in photosynthesis, the reaction of which may be summarized as follows:

$$CO_2 + H_2O \xrightarrow[\text{light}]{\text{chl}} CH_2O + O_2 \qquad (14.27)$$

Reaction (14.27) also yields much of the oxygen needed for autoxidation of the iron. Its genesis causes a rise of E_h because of increased saturation or even supersaturation of the water.

Ferrous iron may be protected from chemical oxidation at elevated pH and E_h by chelation with oxalate, citrate, humic acids, and tannins. In that instance, bacterial breakdown of the ligand will free the ferrous iron, which then autoxidizes to ferric iron. This has been demonstrated in the laboratory with a *Pseudomonas* and a *Bacillus* strain (Kullman and Schweisfurth, 1978). These cultures do not derive any energy from iron oxidation but rather from the oxidation of the ligand.

The production of ferric iron from the oxidation of ferrous iron at pH values above 5 usually leads to precipitation of the iron. But the presence of chelating agents such as humic substances, citrate, and the like can prevent the precipitation. Unchelated ferric iron tends to hydrolyze at higher pH values and may form compounds such as ferric hydroxide:

$$Fe^{3+} + 3H_2O \rightarrow Fe(OH)_3 + 3H^+ \qquad (14.28)$$

Ferric hydroxide is relatively insoluble and will settle out of suspension. It may crystallize and dehydrate, forming FeOOH, goethite ($Fe_2O_3 \cdot H_2O$), or hematite (Fe_2O_3).

Nonenzymatic Reduction   Starkey and Halvorson (1927) tried to explain ferric iron reduction in nature as an indirect effect of microbes. They argued that by causing a drop in pH and a lowering of oxygen tension, ferric iron would be changed to ferrous iron according to the relationship

$$4Fe^{2+} + O_2 + 10H_2O \Leftrightarrow 4Fe(OH)_3 + 8H^+ \qquad (14.29)$$

in which Fe^{3+} was considered to be an insoluble phase [$Fe(OH)_3$]. From this chemical equation they derived the relationship

$$Fe^{2+} = K \frac{[H^+]^2}{[O_2]^{1/4}} \qquad (14.30)$$

But as the experiments of Munch and Ottow (1982) have convincingly shown, this mode of iron reduction is usually not very significant in nature. This was also shown in a study by Lovley et al. (1991). Iron oxide minerals are relatively stable in the absence of oxygen when a strong reducing agent is not present. But in the presence of such reducing agents, chemical Fe(III) reduction does occur. For instance, H_2S produced by sulfate-reducing bacteria may reduce ferric to ferrous iron before precipitating ferrous sulfide (Berner, 1962).

$$2HFeO_2 + 3H_2S \xrightarrow{\text{(at pH 7–9)}} 2FeS + S^\circ + 4H_2O \qquad (14.31)$$
$$\text{goethite}$$

$$2HFeO_2 + 3H_2S \xrightarrow{\text{(at pH 4)}} FeS + FeS_2 + 4H_2O \qquad (14.32)$$

Marine bacteria that disproportionate elemental sulfur into H_2S and sulfate have also been shown to chemically reduce Fe(III) and Mn(IV) anaerobically (Thamdrup et al., 1993):

$$4S° + 4H_2O \Rightarrow 3H_2S + SO_4^{2-} + 2H^+ \qquad (14.33)$$

Formate produced by a number of bacteria (e.g., *Escherichia coli*) can reduce Fe(III).

$$2Fe^{3+} + HCOOH \rightarrow 2Fe^{2+} + 2H^+ + CO_2 \qquad (14.34)$$

Some other metabolic products can also act as reductant of ferric iron (see discussion by Ghiorse, 1988). In all cases reduction is favored by acid pH.

14.7 MICROBIAL PRECIPITATION OF IRON

Enzymatic Processes   The clearest example of enzymatic iron(III) precipitation is that of *Gallionella ferruginea*. The ferric iron it produces in its oxidation of ferrous iron is deposited in its stalk fibrils by a mechanism that is completely unknown at this time (see discussion by Ghiorse, 1984; Ghiorse and Ehrlich, 1992). Enzymatic iron reduction by other organisms can also result in precipitation of magnetite (Fe_3O_4) (Lovley et al., 1987) and in precipitation of Fe(II) as siderite ($FeCO_3$) (see Chapter 8).

Nonenzymatic Processes   Some bacteria may deposit ferric oxides which they formed nonenzymatically on their cells. This nonenzymatic oxidation may involve ligand destruction of iron chelates (e.g., Aristovskaya and Zavarzin, 1971). Such organisms may include sheathed bacteria such as *Leptothrix* spp. (Fig. 14.7), *Siderocapsa* (Fig. 14.10), *Naumanniella*, *Ochrobium*, *Siderococcus*, *Pedomicrobium*, *Herpetosyphon*, *Seliberia* (Fig. 14.11), *Toxothrix* (Krul et al., 1970), *Acinetobacter* (MacRae and Celo, 1975) and *Archangium*.

Bioaccumulation of Iron   Most or all of the bacteria listed in the previous section may also accumulate ferric iron produced by other organism either enzymatically or nonenzymatically. Usually the iron is passively collected on the cell surface by reacting with acidic exopolymer (glycocalyx) which exposes negative charges. The exopolymer may be organized in the form of a sheath, a capsule or slime (see discussion by Ghiorse, 1984). Some protozoans such as *Anthophysa, Euglena* (Mann et al., 1987), *Bikosoeca*, and *Siphomonas* are also known to deposit iron on their cells.

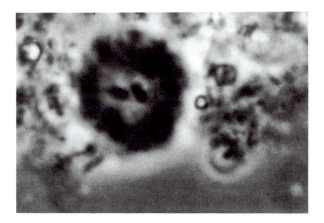

Figure 14.10 *Siderocapsa geminata*, Skuja (1956) (×7,000). Specimen from filtered water from Pluss See, Schleswig-Holstein, Germany. Note the capsule surrounding the pair of bacterial cells. (Courtesy of W. C. Ghiorse and W.-D. Schmidt.)

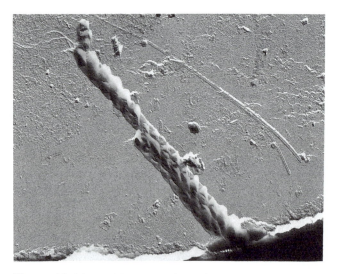

Figure 14.11 *Seliberia* sp. from forest pond neuston (×8,200). (Courtesy of W. C. Ghiorse.)

14.8 THE CONCEPT OF IRON BACTERIA

A survey of the literature on bacteria which interact with iron shows that the term **iron bacteria** is differently defined by different authors. Some have included in this term any bacteria which precipitate iron, whether by

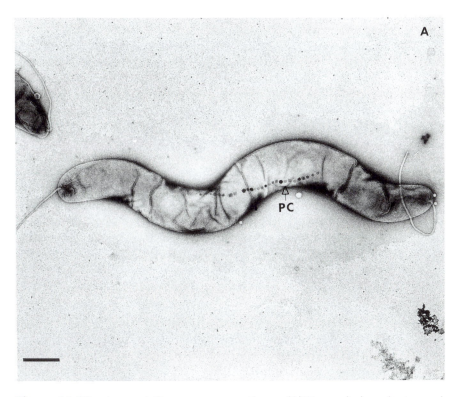

Figure 14.12 *Aquaspirillum magnetotacticum.* (A) Transmission electron micrograph of a negatively stained cell, showing electron dense particle chain (PC) of magnetosomes (bar = 500 nm). (Reproduced from Balkwill, D. L., D. Maratea, and R. P. Blakemore. 1980. Ultrastructure of a Magnetotactic Spirillum, J. Bacteriol. 141:1399–1408, by permission.) (B) Transmission electron micrograph of thin sections showing the trilaminate structure of the membranous vesicles (MV), which lie along the same axis as complete magnetosomes (bar = 250 nm). (Reproduced from Gorby, Y. A., T. J. Beveridge, and R. P. Blakemore. 1988. Characterization of the Bacterial Magnetosome Membrane. J. Bacteriol. 170:834–841, by permission.)

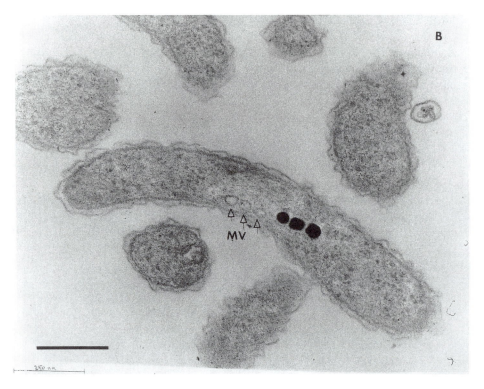

Figure 14.12 Continued.

active (enzymatic) oxidation or by passive accumulation of ferric oxides or hydroxides about their cells, whether they possess cellular structures with specific affinity for ferric iron or not. As Starkey (1945) suggested, the term **iron bacteria** is best reserved for those bacteria which oxidize iron enzymatically. Bacteria that passively accumulate iron should be called **iron accumulators**, among which **specific** and **nonspecific** accumulators could be recognized as subgroups.

Magnetotactic bacteria occupy a special position among iron bacteria (Blakemore, 1982) (Fig. 14.12). They have the capacity to take up complexed ferric iron and to transform it into magnetite Fe_3O_4 by a mechanism which may involve reduction and partial reoxidation (Frankel et al., 1983). The magnetite is deposited as crystals in membrane-bound structures called **magnetosomes** (Fig. 14.12) (Blakemore, 1982). The magnetite crystals are single-domain magnets which aid the bacteria which form them in their orientation in the environment by aligning with the Earth's magnetic field, which is inclined downward in the northern hemisphere and upward in the southern hemisphere. This aids them in seeking out their preferred habitat, a partially reduced environment. Upon the death of magnetotactic bacteria, the magnetite crystals in them are liberated and in nature may become incorporated in sediment. It has been suggested that remanent magnetism detected in some rocks may be due to magnetite residues from magnetotactic bacteria (Blakemore, 1982; Karlin et al., 1987). However, magnetite can also be formed extracellularly by some non-magnetotactic bacteria (Lovley et al., 1987) (see Section 14.10).

14.9 SEDIMENTARY IRON DEPOSITS OF PUTATIVE BIOGENIC ORIGIN

Sedimentary iron deposits, many representing large ore formations, may feature iron in the form of oxides, sulfides, or carbonates. Many of each of these types may have been formed as a result of specific microbial activity. Microbial participation in the biogenic formation of any of these deposits may be inferred from (a) the presence of fossilized microbes with imputed iron-oxidizing or accumulating potential in some ancient deposits, (b) the presence of living iron-oxidizing bacteria or iron accumulating bacteria in currently forming ferric oxide (Fe_2O_3) deposits; the presence of Fe^{3+} reducers in currently forming magnetite deposits (Fe_3O_4); or the presence of sulfate reducing bacteria in the case of currently forming iron sulfide deposits, or (c) the inference of probable environmental conditions prevailing at the time of formation of the deposit which would favor biogenic iron deposition. Identification of microfossils in an iron deposit may also permit deduction of prevailing environmental conditions from knowledge of environmental requirements and activities of modern microorganisms which resemble the microfossils. In this section only iron oxide deposits will be considered. Biogenic iron sulfide formation will be discussed in Chapter 18. Biogenic iron carbonate (siderite) formation was briefly treated in Chapter 8.

Among the most ancient iron deposits in the formation of which microbes may have played a central role are the banded-iron formations (BIF) formed mostly in the Precambrian in the time interval roughly between about 3.3 and 1.8 eons ago. The major deposits were formed between 2.2 and 1.9 eons ago (see discussions by Cloud, 1973; Lundgren and Dean, 1979; Walker et al. 1983; Nealson and Myers, 1990). These formations have been found in various parts of the world and in many places are extensive enough to be economically exploitable as iron ore.

They are characterized by alternating layers rich in chert, a form of silica (SiO_2), and layers rich in iron minerals such as hematite (Fe_2O_3), magnetite (Fe_3O_4), the iron silicate chamosite, and even siderite ($FeCO_3$). Ferric iron predominates over ferrous iron in the iron-rich layers. Average thickness of the layers is 1–2 cm, but they may be thinner or thicker. Because the most important BIF were formed during that part of the Precambrian when the atmosphere changed from a reducing to an oxidizing one due to the emergence of oxygenic photosynthesis, Cloud (1973) has argued that the alternating layers reflect episodic deposition of iron oxides, involving seasonal, annual, and longer-period cycles. In the reducing atmosphere of the Archean, iron in the Earth's crust was mostly in the ferrous form and thus could act as a scavenger of the oxygen initially produced by oxygenic photosynthesis. The scavenging reaction involved autoxidation of the iron. Since autoxidation of iron is a very rapid reaction at near-neutral pH and higher, the supply of ferrous iron could have been periodically depleted and further oxygen scavenging would have had to wait until the supply of dissolved ferrous iron was replenished by leaching from rock or by hydrothermal emissions (Holm, 1987). Alternatively, if much of the ferrous iron for scavenging O_2 was formed by ferric-iron-respiring bacteria while consuming the organic carbon produced by seasonally growing photosynthesizers, periodic depletion of organic carbon rather than ferrous iron could have been the origin of the iron-poor layers (Nealson and Myers, 1990). Bacterial reduction of the oxidized iron could also explain the origin of magnetite in the iron-rich layers of BIF, since Lovley and collaborators (1987) and Lovley and Phillips (1988) found that in their laboratory experiments, *Geobacter metallireducens* precipitated magnetite in reducing Fe^{3+}. Previously it was proposed that the magnetite in the iron-rich layers may have resulted from partial reduction of hematitic iron by organic carbon from biological activity (Perry et al., 1973).

If iron was, indeed, the scavenger of the oxygen of this early oxygenic photosysnthesis, it would mean that the sediments became increasingly more oxidizing relative to the atmosphere due to the accumulation of the oxidized iron (Walker, 1987). Only when free oxygen began to accumulate would the atmosphere have become oxidizing relative to the sediments. Chert was deposited in BIF because no silicate-depositing microorganism (e.g., diatoms, radiolarians) had yet evolved. As more oxygen was generated by the photosynthesizers, ferrous iron reserves became largely depleted, permitting the oxygen content of the atmosphere to increase to levels we associate with an oxidizing atmosphere. This, in turn, would have restricted Fe(III) reduction to the remaining and ever-shrinking oxygen-depleted (anaerobic) environments. Although an extensive array of microfossils has been found in the cherty layers of various BIF

(Fig. 14.13) (e.g., Gruner, 1923; Cloud and Licari, 1968), not all sedimentologists agree that the origin of BIF depended on biological activity (see discussion by Walker et al., 1983).

The discovery of Fe(II) oxidation by photosynthesizing anaerobic bacteria (Widdel et al., 1993) suggests that some BIF may have been formed before oxygenic photosynthesis arose and/or that anaerobic photo-

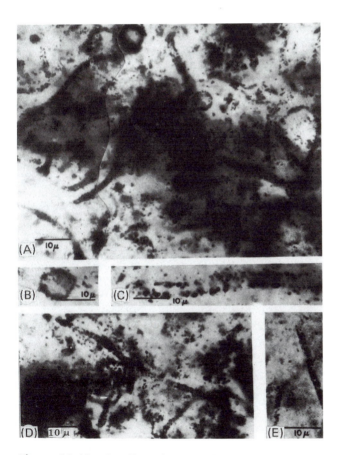

Figure 14.13 Gunflint microbiota in the stromatolites of the Biwabik Iron Formation, Corsica Mine, Minnestoa. (A) *Gunflintia* filaments and *Huroniospora spheroides* replaced by hematite are abundant in some of the dark laminations in stromatolitic rocks at the Corsica Mine. (B) *Huroniospora*, Corsica Mine. (C) Wide filament, Corsica Mine. (D,E) *Gunflintia* filaments, Corsica Mine. (Reproduced from Cloud and Licari, 1968, by permission.)

synthesizers initiated BIF from the abundant Fe(II) when the Earth's atmosphere was still reducing. Anbar and Holland (1992) previously suggested that BIF may have begun to form in anoxic Precambrian oceans as a result of abiotic photochemical oxidation of Mn^{2+} to a manganese(IV) oxide like birnessite, which then served as oxidant in an oxidation of ferrous iron hydroxide complexes to γ-FeOOH or amorphous ferric hydroxide.

Among modern iron deposits in whose formation microorganisms are or have been involved are ochre deposits, bog and lake iron ores, and others. Formation of ochre deposits, consisting of amorphous iron oxides (Ivarson and Sojak, 1978), is a common observation in field drains. *Gallionella*, sometimes in association with *Leptothrix*, is usually the causal organism. Hanert (1974b) has measured rates of ochre deposition in field drains in terms of ferric iron deposition on submerged microscope slides. He found 4.0, 8.8 and 20.2 μg of iron deposited per square centimeter in 1, 2, and 3 days, respectively. Biogenic ochre deposits have also been reported to be forming in a bay (the caldera of Santorini) of the Greek island Palaea Kameni in the Aegean Sea (Puchelt et al., 1973; Holm, 1987). *Gallionella ferruginea* has been identified as the source of this ochre. The source of the ferrous iron which is oxidized by *Gallionella* in this instance is of hydrothermal origin.

Bog iron deposition is currently noted in the Pine Barrens of southern New Jersey, U.S.A. (Crerar et al., 1979; Madsen et al., 1986). The depositional process seems to depend on several different kinds of bacteria: *Thiobacillus ferrooxidans*, *Leptothrix ochracea*, *Crenothrix polyspora*, *Siderocapsa geminata*, and an iron-oxidizing *Metallogenium* (sensu Walsh and Mitchell, 1972a). According to Crerar et al. (1979), the iron is oxidized and precipitated by the iron-oxidizing bacteria from acid surface waters, which exhibit a pH range of 4.3–4.5 in summer. It accumulates as limonite (FeOOH) impregnating sands and silts. The source of the iron is glauconite and to a much lesser extent pyrite in underlying sedimentary formations, from which it is released into groundwater—whether with microbial help has not yet been ascertained. The iron is brought to the surface by groundwater base flow which feeds local streams. Biocatalysis of iron oxidation seems essential to account for the rapid rate of iron oxidation in the acid waters. However, *T. ferrooxidans* is probably least important because it is only infrequently encountered, perhaps because the environmental pH is above its optimum. Dissolved ferric iron in the streams in the Pine Barrens can be photoreduced (Madsen et al., 1986).

Trafford et al. (1973) describe an example in which *T. ferrooxidans* appeared responsible for ochre formation in field drains from pyritic soils.

It is surprising that no jarosite was reported to be formed under these conditions (Silver et al., 1986).

14.10 MICROBIAL MOBILIZATION OF IRON FROM MINERALS IN ORE, SOIL, AND SEDIMENTS

Bacteria and fungi are able to mobilize major quantities of iron from minerals in ore, soil and sediment in which the iron is a major constituent. The minerals include carbonates, oxides, and sulfides. Attack of iron sulfides will be discussed in Chapter 18; attack of carbonates was briefly discussed in Chapter 8. In this section the stress will be on iron oxides.

As indicated in Section 14.5, iron reduction has been observed with bacteria from soil and aquatic sources. In recent years a more systematic study of the importance of this activity in soils and sediments has been undertaken (see, for instance, review of activity in marine sediments by Burdige, 1993). Ferric iron reduction (iron respiration) can be an important form of anaerobic respiration in environments in which nitrate or sulfate is present in insufficient quantities as terminal electron acceptors and in which methanogenesis is not occurring (Lovley, 1987, 1991; Sørensen and Jørgensen, 1987). Indeed, the presence of bioreducible ferric iron appears to inhibit sulfate reduction and methanogenesis when electron donors like acetate or hydrogen are limiting (Lovley and Phillips, 1986a, 1987). Caccavo et al. (1992) found a facultative anaerobe, strain BrY, that oxidized hydrogen anaerobically using Fe(III) as electron acceptor in the Great Bay Estuary, New Hampshire, U.S.A. Evidence that some ferric iron-reducing bacteria can use hydrogen as electron donor was first reported by Balashova and Zavarzin (1979). The form in which ferric iron occurs also determines whether bacterial iron respiration will occur. In sediments from a freshwater site in the Potomac River Estuary in Maryland, U.S.A., only amorphous FeOOH present in the upper 4 cm was bacterially reducible. Mixed Fe(II)-Fe(III) compounds, which were present in deeper layers, were not attacked (Lovely and Phillips, 1986b). Local E_h and pH conditions can determine in what form the iron from bioreduction of FeOOH will occur. A mixed culture from sediment of Contrary Creek in central Virginia, U.S.A., produced magnetite (Fe_3O_4) in the laboratory when the culture medium was allowed to go alkaline (pH 8.5 in the absence of glucose) but produced Fe^{2+} when the medium was allowed to go acid (to pH 5.5 in the presence of glucose)(Bell et al., 1987). E_h values dropped from a range of 0 to -100 mV to less than -200 mV at the end of the experiments. A growing, pure culture of a strict anaerobe, *Geobacter metallireducens* strain GS-15, produced magnetite from amorphous iron oxide directly with acetate as reductant at pH 6.7–7.0 and 30–35°C (Lovley and Phillips, 1988a; Lovley et al., 1987) (see also section 14.5).

In soil, the phenomenon of **gleying** has come to be associated with bacterial reduction of iron oxides. It is a process which occurs under anaerobic conditions such as result from water-logging. The affected soil becomes sticky and takes on a gray or light greenish-blue coloration (Alexander, 1977, p. 377). Although once attributed to microbial sulfate reduction, this is no longer considered to be the primary cause of gleying (Bloomfield, 1950). Waterlogged soil has been observed to bleach before sulfate reduction is detectable (Takai, and Kamura, 1966). Bloomfield (1951) suggested that gleying was at least in part due to plant degradation products, although he had earlier demonstrated gleying under artificial conditions in a sugar-containing medium that was allowed to ferment. A microbial reduction of iron in which ferric iron is reduced by anaerobic bacterial respiration is now favored as an explanation of gleying (Ottow, 1969a, 1970b, 1971). The reaction which iron undergoes in this microbial reduction can be summarized as follows (Ottow, 1971):

$$\text{Energy source} \xrightarrow[\text{dehydrogenase}]{\text{substrate}} e^- + H^+ + ATP + \text{end products}$$

$$(14.35)$$

$$Fe(OH)_3 + 3H^+ + e^- \longrightarrow Fe^{2+} + 3H_2O \qquad (14.36)$$

$$2Fe(OH)_3 + Fe^{2+} + 2OH^- \longrightarrow \underset{\substack{\text{gray-green mixed} \\ \text{oxide of gley}}}{Fe_3(OH)_8} \qquad (14.37)$$

14.11   MICROBES AND THE IRON CYCLE

Microbial transformations of iron play an important role in the cycling of iron in nature (Fig. 14.14). Weathering of iron-containing minerals in rocks, soils and sediments introduces iron into the cycle. This weathering action is partly promoted by bacterial action and partly by chemical activity (Bloomfield, 1953a,b). The microbial action involves in many cases the interaction of the minerals with metabolic end products (Berthelin and Kogblevi, 1972, 1974); Berthelin and Dommergues, 1972; Berthelin et al., 1974). The mobilized iron, if ferrous, may be biologically or non-biologically oxidized to ferric iron at a pH above 5 under partially or fully aerobic conditions independent of light, and biologically in light under anaerobic conditions. At a pH below 4 it is oxidized mainly biologically. The oxidation may be followed by immediate precipitation of the iron as a hydroxide, oxide, phosphate, or sulfate. If complexing agents such as humic

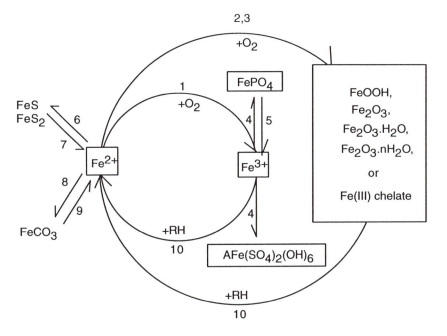

NUMERICAL CODE:
1. microbially at acid pH
2. microbially at neutral pH, in the dark when O_2 tension is low or
   in light when O_2 is absent (in anoxygenic photosynthesis)
3. chemically at neutral pH when O_2 tension is high
4. chemically
5. H^+, microbially
6. H_2S, often of microbial origin
7. $+O_2$, microbially or chemically
8. $+CO_3^{2-}$
9. H^+, microbially or chemically
10. microbially or chemically

Figure 14.14 The iron cycle.

substances abound, the ferric iron may be converted to soluble complexes and be dispersed from its site of formation. In podzolic soils (spodosols) in temperate climates, for instance, this complexed iron may be transported from the upper A horizons to the B horizons. In hot, humid climates, ferric iron is more likely to precipitate at the site of its release from iron-containing soil minerals, owing to intense microbial action that rapidly and fairly completely mineralizes available organic matter. This, in many instances, prevents extensive formation of soluble ferric iron complexes. The iron precipitates tend to cement soil particles together in a process known as **laterization**. Aluminum hydroxide liberated in the weathering process may also be precipitated and contribute to laterization (Brooks and Kaplan, 1972, p. 75; Merkle, 1955, p. 204).

In freshwater environments, mobilization of iron may be prominent in sediments when the hypolimnion is oxygen-depleted. In sediment from Blelham Tarn in the English Lake District, bacteria occur which mineralize organic matter by iron respiration (Jones et al., 1983, 1984). Formation of Fe(II) by ferric iron reduction by a *Vibrio* of the bacterial flora was stimulated by the addition of malate and inhibited by nitrate, MnO_2, and Mn_2O_3. H_2 could serve as a reductant (energy source) for the organism. Mineralization of organic matter by iron respiration may be very important for the carbon cycle in some reducing environments (Lovley and Phillips, 1986a; 1987; 1988a). Mobilized ferrous iron appears to be oxidized by *Gallionella* in the metalimnion of Hortlandsstemma, a small eutrophic lake in Norway where the dissolved oxygen concentration was only 0.005 mmol L^{-1} at a pH of 6.5 (Heldal and Tumyr, 1983). At higher oxygen concentrations the iron would have been autoxidized.

The microbial genesis of iron sulfides, including pyrite, at near neutral pH and the biological oxidation of iron sulfides at acid pH make very important contributions to the cycling of iron in environments such as salt marshes and sulfide ore deposits, respectively (see Chapter 18).

14.12 SUMMARY

Iron may be enzymatically oxidized by some bacteria for generating useful energy. Best evidence for this comes from the study of acidophilic iron oxidizers, like *Thiobacillus ferrooxidans*, *Sulfolobus* spp., *Acidianus brierleyi*, *Sulfobacillus thermosulfidooxidans*, and *Leptospirillum ferrooxidans*. The first of these has been most extensively studied. The reason why the most convincing evidence of microbial iron oxidation has come from the study of acidophiles is that ferrous iron is least susceptible to autoxidation below pH 5. Some bacteria growing near neutral pH can, however, also oxidize iron. The stalked bacterium *Gallionella ferruginea* uses it chemolithotrophically, whereas the sheathed bacteria *Leptothrix* spp. may use it mixotrophically. These organisms require a partially reduced environment for this activity. *Leptothrix* spp. may also precipitate iron passively on their sheath.

Iron may also be oxidized nonenzymatically by microorganisms when they raise the redox potential and/or pH of their environment, thereby favoring autoxidation. The rise in redox potential results from the buildup of oxidants in their metabolism. The rise in pH is the consequence of ammonia production, the consumption of organic salts, some of which may chelate ferrous iron, or photosynthesis.

Ferric iron precipitation need not involve iron oxidation. Ferric iron can be stabilized in solution by chelation. Naturally produced chelators that may solubilize extensive amounts of ferric iron include microbially produced oxalate, citrate, humic acids, and tannins. Ferric iron precipitation may result from microbial destruction (mineralization) of these chelators. A number of bacteria have been found active in this manner. Ferric iron may be locally concentrated by adsorption to the cell surface. A variety of microorganisms, including bacteria and protozoans, have been found capable of this.

Microbial formation and accumulation of iron in aqueous environments may cause concurrent accumulation of other heavy metal ions by coprecipition and adsorption to hydrous iron oxides (e.g., Gunkel, 1986). The adsorbed metals may be remobilized by acidification without reduction of ferric adsorbent (Tipping et al., 1986).

Ferric iron can be enzymatically reduced to ferrous iron with a suitable electron donor. Nitrate reductase A is one enzyme that may be involved. However, at least one other enzyme must also be independently active, especially in organisms that cannot synthesize nitrate reductase. Both bacteria and fungi have been implicated in this reaction. Not all microbial ferric iron reduction is enzymatic. Some may be the result of reaction with metabolic products such as H_2S or formate.

Both oxidative and reductive reactions of iron brought about by microbes play important roles in the iron cycle in nature. They affect the mobility of iron as well as local accumulations of iron. The formation of some sedimentary iron deposits have been directly attributed to microbes acting on iron. Ochre formation has been associated with bacterial iron oxidation, and gleying has been associated with bacterial iron reduction. Bacterial iron reduction as a form of respiration by heterotrophic bacteria in reducing environments can make a significant contribution to the carbon cycle by promoting anaerobic mineralization.

REFERENCES

Adams, L. F., and W. C. Ghiorse. 1987. Characterization of extracellular Mn^{2+}-oxidizing activity and isolation of Mn^{2+}-oxidizing protein from *Leptothrix discophora* SS-1. J. Bacteriol. 169:1279–1285.

Agate, A. D., and W. Vishniac. 1970. Transport characteristics of phospholipids from thiobacilli. Bacteriol. Proc., GP14, p. 50.

Ahonen, L., and O. H. Tuovinen. 1989. Microbial oxidation of ferrous iron at low temperature. Appl. Environ. Microbiol. 55:312–316.

Aleem, M. I. H., H. Lees, and D. J. D. Nicholas. 1963. Adenosine triphosphate-dependent reduction of nicotinamide adenine dinucleotide by ferro-cytochrome c in chemoautotrophic bacteria. Nature (Lond.) 200:759–761.

Alexander, M. 1977. Introduction to Soil Microbiology. 2nd ed. Wiley, New York.

Anbar, A. D., and H. D. Holland. 1992. The photochemistry of manganese and the origin of banded iron formations. Geochim. Cosmochim. Acta 56:2595–2603.

Anderson, K. J., and D. G. Lundgren. 1969. Enzymatic studies of the iron-oxidizing bacterium, *Ferrobacillus ferrooxidans*: Evidence for a glycolytic pathway and Krebs cycle. Can. J. Microbiol. 15:73–79.

Aristovskaya, T. V., and G. A. Zavarzin. 1971. Biochemistry of iron in soil. *In*: A. D. McLaren and J. Skujins, eds. Soil Biochemistry. Vol. 2. Marcel Dekker, New York, pp. 385–408.

Arkesteyn, G. J. M. W. 1979. Pyrite oxidation by *Thiobacillus ferrooxidans* with special reference to the sulfur moiety of the mineral. Antonie v. Leeuwenhoek 45:423–435.

Arkesteyn, G. J. M. W., and J. A. M. DeBont. 1980. *Thiobacillus acidophilus*: a study of its presence in *Thiobacillus ferrooxidans* cultures. Can. J. Microbiol. 26:1057–1065.

Arnold, R. G., T. J. DiChristina, and M. R. Hoffmann. 1986a. Inhibitor studies of dissimilative Fe(III) reduction by *Pseudomonas* sp. strain 200 ("*Pseudomonas ferrireductans*"). Appl. Environ. Microbiol. 52: 281–289.

Arnold, R. G., T. M. Olson, and M. R. Hoffmann. 1986b. Kinetics and mechanism of dissimilative Fe(III) reduction by *Pseudomonas* sp. 200. Biotech. Bioeng. 28:1657–1671.

Arnold, R. G., M. R. Hoffmann, T. J. DeChristina, and F. W. Picardal. 1990. Regulation of dissimilatory Fe(III) reduction activity in *Shewanella putrefaciens*. Appl. Environ. Microbiol. 56:2811–2817.

Avakyan, A. A., and G. I. Karavaiko. 1970. Submicroscopic organization of *Thiobacillus ferrooxidans*. Mikrobiologiya 39:855–861 (Engl. transl., pp. 744–751).

Baas Becking, L. G. M., and G. S. Parks. 1927. Energy relations in the metabolism of autotrophic bacteria. Physiol. Rev. 7:85–106.

Balashova, V. V., and G. A. Zavarzin. 1972. Iron oxidation by *Myco-*

plasma laidlawii. Mikrobiologiya 41:909–911 (Engl. transl. pp. 808–810).

Balashova, V. V., and G. A. Zavarzin. 1979. Anaerobic reduction of ferric iron by hydrogen bacteria. Mikrobiologiya 48:773–778 (Engl. transl. pp. 635–639).

Balashova, V. V., I. Ya. Vedina, G. E. Markosyan, and G. A. Zavarzin. 1974. *Leptospirillum ferrooxidans* and aspects of its autotrophic growth. Mikrobiologiya 43:581–585 (Engl. transl. pp. 491–494).

Balkwill, D. L., D. Maratea, and R. P. Blakemore. 1980. Ultrastructure of a magnetotactic spirillum. J. Bacteriol. 141:1399–1408.

Beck, J. V. 1960. A ferrous-ion oxidizing bacterium. I. Isolation and some general physiological characteristics. J. Bacteriol. 79:502–509.

Beck, J. V., and S. R. Elsden. 1958. Isolation and some characteristics of an iron-oxidizing bcterium. J. Gen. Microbiol. 19:i.

Bell, P. E., A. L. Mills, and J. S. Herman. 1987. Biogeochemical conditions favoring magnetite formation during anaerobic iron reduction. Appl. Environ. Microbiol. 53:2610–2616.

Berner, R. A. 1962. Experimental studies of the formation of sedimentary iron sulfides. *In*: M. L. Jensen, ed. Biogeochemistry of Sulfur Isotopes. Yale University Press, New Haven, pp. 107–120.

Berry, V. K., and L. E. Murr. 1980. Morphological and ultrastructural study of the cell envelope of thermophilic and acidophilic microorganisms as compared to *Thiobacillus ferrooxidans*. Biotech. Bioeng. 22:2543–2555.

Berthelin, J., and Y. Dommergues. 1972. Rôle de produits du métabolisme microbien dans la solubilization des minéraux d'une arène granitique. Rev. Ecol. Biol. Sol 9:397–406.

Berthelin, J., and A. Kogblevi. 1972. Influence de la stérilization partielle sur la solubilization microbienne des minéraux dans les sols. Rev. Ecol. Biol. Sol 9:407–419.

Berthelin, J., and A. Kogblevi. 1974. Influence de l'engorgement sur l'altération microbienne des minéraux dans les sols. Rev. Ecol. Biol. Sol 11:499–509.

Berthelin, J., A. Kogblevi, and Y. Dommergues. 1974. Microbial weathering of a brown forest soil: Influence of partial sterilization. Soil Biol. Biochem. 6:393–399.

Blake, R. C., and E. A. Shute. 1987. Respiratory enzymes of *Thiobacillus ferrooxidans*. J. Biol. Chem. 262:14983–14989.

Blake, R., II, J. Waskovsky, and A. P. Harrison, Jr. 1992. Respiratory components in acidophilic bacteria that respire on iron. Geomicrobiol. J. 10:173–192.

Blakemore, R. P. 1982. Magnetotactic bacteria. Annu. Rev. Microbiol. 36:217–238.

Blaylock, B. A., and A. Nason. 1963. Electron transport systems of the chemoautotroph *Ferrobacillus ferrooxidans*. 1. Cytochrome c–containing iron oxidase. J. Biol. Chem. 238:3453–3462.

Bloomfield, C. 1950. Some observations on gleying. J. Soil Sci. 1:205–211.

Bloomfield, C. 1951. Experiments on the mechanism of gley formation. J. Soil Sci. 2:196–211.

Bloomfield, C. 1953a. A study of podzolization. Part I. The mobilization of iron and aluminum by Scots pine needles. J. Soil Sci. 4:5–16.

Bloomfield, C. 1953b. A study of podzolization. Part II. The mobilization of iron and aluminum by the leaves and bark of *Agathis australis* (Kauri). J. Soil Sci. 4:17–23.

Brierley, C. L., and J. A. Brierley. 1973. A chemoautotrophic microorganism isolated from an acid hot spring. Can. J. Microbiol. 19:183–188.

Brierley, C. L., and L. E. Murr. 1973. Leaching: Use of a thermophilic and chemoautotrophic microbe. Science 179:488–499.

Brierley, J. A. 1978. Thermophilic iron oxidizing bacteria found in copper leaching dumps. Appl. Environ. Microbiol. 36:523–525.

Brierley, J. A., and S. J. Lockwood. 1977. The occurrence of thermophilic iron-oxidizing bacteria in a copper leaching system. FEMS Microbiol. Lett. 2:163–165.

Brierley, J. A., P. R. Norris, D. P. Kelly, and N. W. LeRoux. 1978. Characteristics of a moderately thermophilic and acidophilic iron-oxidizing *Thiobacillus*. Eur. J. Appl. Microbiol. 5:291–299.

Brock, T. D., and J. Gustafson. 1976. Ferric iron reduction by sulfur- and iron-oxidizing bacteria. Appl. Environ. Microbiol. 32:567–571.

Brock, T. D., and M. T. Madigan. 1988. Biology of Microorganisms. Prentice-Hall, Englewood Cliffs, NJ.

Brock, T. D., K. M. Brock, R. T. Belly, and R. L. Weiss. 1972. *Sulfolobus*: A new genus of sulfur-oxidizing bacteria living at low pH and high temperature. Arch. Microbiol. 84:54–68.

Brock, T. D., S. Cook, S. Peterson, and J. L. Mosser. 1976. Biogeochemistry and bacteriology of ferrous iron oxidation in geothermal habitats. Geochim. Cosmochim. Acta 40:493–500.

Bromfield, S. M. 1954a. The reduction of iron oxide by bacteria. J. Soil Sci. 5:129–139.

Bromfield, S. M. 1954b. Reduction of ferric compounds by soil bacteria. J. Gen. Microbiol. 11:1–6.

Brooks, R. R., and I. R. Kaplan. 1972. Biogeochemistry, pp. 74–82. *In*: The Encyclopedia of Geochemistry and Environmental Sciences.

Encyclopedia of Earth Sciences Series. Vol. IVA. Van Nostrand Reinhold, New York.

Brown, K. A., and C. Ratledge. 1975. Iron transport in *Mycobacterium smegmatis*: Ferrimycobactin reductase (NAD(P)H:ferrimycobactin oxidoreductase), the enzyme releasing iron from its carrier. FEBS Lett. 53:262–266.

Buchanan, R. E., and N. E. Gibbons, eds. 1974. Bergey's Manual of Determinative Bacteriology. 8th ed. Williams & Wilkins, Baltimore.

Burdige, D. J. 1993. The biogeochemistry of magnanese and iron reduction in marine sediments. Earth-Sci. Rev. 35:249–284.

Caccavo, F., Jr., R. P. Blakemore, and D. R. Lovley. 1992. A hydrogen-oxidizing, Fe(III)-reducing microorganism from the Great Bay Estuary, New Hampshire. Appl. Environ. Microbiol. 58:3211–3216.

Caccavo, F., Jr., D. J. Lonergan, D. R. Lovley, M. Davis, J. F. Stolz, and M. J. McInerney. 1994. *Geobacter sulfurreducens* sp. nov., hydrogen- and acetate-oxidizing dissimilatory metal-reducing microorganism. Appl. Environ. Microbiol. 60:3752–3759.

Cameron, F. J., C. Edwards, and M. V. Jones. 1981. Isolation and preliminary characterization of an iron-oxidizing bacterium from an ochre-polluted stream. J. Gen. Microbiol. 124:213–217.

Champine, J. E., and S. Goodwin. 1991. Acetate catabolism in the dissimilatory iron-reducing isolate GS-15. J. Bacteriol. 173:2704–2706.

Cholodny, N. 1924. Zur Morphologie der Eisenbakterien *Gallionella* und *Spirophyllum*. Ber. dtsch. bot. Ges. 42:35–44.

Cholodny, N. 1926. Die Eisenbakterien. Beitraege zu einer Monographie. Verlag von Gustav Fischer, Jena.

Cloud, P. E. Jr. 1973. Paleoecological significance of the banded-iron-formations. Econ. Geol. 68:1135–1143.

Cloud, P. E. Jr., and G. R. Licari. 1968. Microbiotas of the banded iron formations. Proc. Natl. Acad. Sci. USA 61:779–786.

Cobley, J. G., and B. A. Haddock. 1975. The respiratory chain of *Thiobacillus ferrooxidans*: The reduction of cytochromes by Fe^{2+} and the preliminary characterization of rusticyanin, a novel 'blue' copper protein. FEBS Lett. 60:33.

Coleman, M. L., D. B. Hedrick, D. R. Lovley, D. C. White, and K. Pye. 1993. Reduction of Fe(III) in sediments by sulfate-reducing bacteria. Nature (Lond.) 361:436–438.

Colmer, A. R., K. L. Temple, and H. E. Hinkle. 1950. An iron-oxidizing bacterium from the acid drainage of some bituminous coal mines. J. Bacteriol. 59:317–328.

Corbett, C. M., and W. J. Ingledew. 1987. Is Fe^{3+}/Fe^{2+} cycling an intermediate in sulfur oxidation by Fe^{2+}-grown *Thiobacillus ferrooxidans*? FEMS Microbiol. Lett. 41:1–6.

Corstjens, P. L. A. M., J. P. M. De Vrind, P. Westbroek, and W. E. De Vrind-De Jong. 1992. Enzymatic iron oxidation by *Leptothrix discophora*: Identification of an iron-oxidizing protein. Appl. Eviron. Microbiol. 58:450–454.

Cox, C. D. 1980. Iron reductases from *Pseudomonas aeruginosa*. J. Bacteriol. 141:199–204.

Cox, J. C., and D. H. Boxer. 1978. The purification and some properties of rusticyanin, a blue copper protein involved in iron(II) oxidation from *Thiobacillus ferrooxidans*. Biochem. J. 174:497–502.

Cox, J. C., and D. H. Boxer. 1986. The role of rusticyanin, a blue copper protein, in the electron transport chain of *Thiobacillus ferrooxidans* grown on iron or thiosulfate. Biotech. Appl. Biochem. 8: 269–275.

Crerar, D. A., G. W. Knox, and J. L. Means. 1979. Biogeochemistry of of bog iron in the New Jersey Pine Barrens. Chem. Geol. 24:111–135.

De Castro, A. F., and H. L. Ehrlich. 1970. Reduction of iron oxide minerals by a marine *Bacillus*. Antonie v. Leeuwenhoek 36:317–327.

De Vrind-De Jong, E. W., P. L. A. M. Corstjens, E. S. Kempers, P. Westbroek, and J. P. M. De Vrind. 1990. Oxidation of manganese and iron by *Leptothrix discophora*: Use of N,N,N',N'-tetramethyl-p-phenylenediamine as an indicator of metal oxidation. Appl. Environ. Microbiol. 56:3458–3462.

Din, G. A., I. Suzuki, and H. Lees. 1967a. Ferrous iron oxidation by *Ferrobacillus ferrooxidans*. Purification and properties of Fe^{++}-cytochrome c reductase. Can. J. Biochem. 45:1523–1546.

Din, G. A., I. Suzuki, and H. Lees. 1967b. Carbon dioxide fixation and phosphoenolpyruvate carboxylase in *Ferrobacillus ferrooxidans*. Can. J. Microbiol. 13:1413–1419.

Drobner, E., H. Huber, and K. O. Stetter. 1990. *Thiobacilllus ferrooxidans*, a facultative hydrogen oxidizer. Appl. Environ. Microbiol. 56: 2922–2923.

Dubinina, G. A. 1978a. Mechanism of the oxidation of divalent iron and manganese by iron bacteria growing at a neutral pH of the medium. Mikrobiologiya 47:591–599 (Engl. transl. 471–478).

Dubinina, G. A. 1978b. Functional role of bivalent iron and manganese oxidation in *Leptothrix pseudoochracea*. Mikrobiologiya 47:783–789 (Engl. transl. 631–636).

Dugan, P. R., and D. G. Lundgren. 1965. Energy supply for the chemoautotroph *Ferrobacillus ferrooxidans*. J. Bacteriol. 89:825–834.

Ehrenberg, C. G. 1836. Vorläufige Mitteilungen über das wirkliche Vorkommen fossiler Infusorien und ihre grosse Verbreitung. Poggendorf's Ann. Phys. Chem. 38:213–227.

Ehrlich, H. L. 1993. Bacterial mineralization of organic carbon under anaerobic conditions. *In*: J-M. Bollag and G. Stotzky, eds. Soil Biochemistry, Vol. 8. Marcel Dekker, New York, pp. 219–247.

Ehrlich, H. L., S. H. Yang, and J. D. Mainwaring. 1973. Bacteriology of manganese nodules. VI. Fate of copper, nickel and cobalt, and iron during bacterial and chemical reduction of the manganese (IV). Z. Allg. Mikrobiol. 13:39–48.

Emerson, D., and W. C. Ghiorse. 1993a. Ultrastructure and chemical composition of the sheath of *Leptothrix discophora* SP-6. J. Bacteriol. 175:7808–7818.

Emerson, D., and W. C. Ghiorse. 1993b. Role of disulfide bonds in maintaining the structural integrity of the sheath of *Leptothrix discophora* SP-6. J. Bacteriol. 175:7819–7827.

Ernst, J. F. and G. Winkelmann. 1977. Enzymatic release of iron from sideramines in fungi. NADH:sideramine oxidoreductase in *Neurospora crass*. Biochim. Biophys. Acta 500:27–41.

Frankel, R. B., G. C. Papaefthymiou, R. P. Blakemore, and W. O'Brien. 1983. Fe_3O_4 precipitation in magnetotactic bacteria. Biochim. Biophys. Acta 763:147–159.

Fry, I. V., N. Lazaroff, and L. Packer. 1986. Sulfate-dependent iron oxidation by *Thiobacillus ferrooxidans*: Characterization of a new EPR detectable electron transport component on the reducing side of rusticyanin. Arch. Biochem. Biophys. 246:650–654.

Fukumori, Y., T. Yano, A. Sato, and T. Yamanaka. 1988. Fe(II)-oxidizing enzyme purified from *Thiobacillus ferrooxidans*. FEMS Microbiol. Lett. 50:169–172.

Gale, N. L., and J. V. Beck. 1967. Evidence for the Calvin cycle and hexose monophosphate pathway in *Thiobacillus ferrooxidans*. J. Bacteriol. 94:1052–1059.

Ghiorse, W. C. 1984. Biology of iron- and manganese-depositing bacteria. Annu. Rev. Microbiol. 38:515–550.

Ghiorse, W.C. 1988. Microbial reduction of manganese and iron. *In*: A. J. B. Zehnder, ed. Biology of Anaerobic Microorganisms. Wiley, New York, pp. 305–331.

Ghiorse, W. C., and H. L. Ehrlich. 1992. Microbial biomineralization of iron and manganese. *In*: H. C. W. Skinner and R. W. Fitzpatrick (eds.). Biomineralization. Process of Iron and Manganese. Modern and Ancient Environments. Catena Supplement 21. Catena Verlag, Cremlingen, Germany, pp. 75–99.

Gibson, F., and D. I. Magrath. 1969. The isolation and characterization of a hydroxamic acid (aerobactin) formed by *Aerobacter aerogenes* 62-1. Biochim. Biophys. Acta 192:175–184.

Golovacheva, R. S., and G. I. Karavaiko. 1978. A new genus of thermophilic spore-forming bacteria, *Sufobacillus*. Mikrobiologiya 47: 815–822 (Engl. transl. pp. 658–665.

Gorby, Y. A., and D. R. Lovley. 1991. Electron transport in the dissimilatory iron reducer, GS-15. Appl. Environ. Microbiol. 57:867–870.

Gorby, Y. A., T. J. Beveridge, and R. P. Blakemore. 1988. Characterization of the bacterial magnetosome membrane. J. Bacteriol. 170: 834–841.

Gruner, J. W. 1923. Algae, believed to be Archean. J. Geol. 31:146–148.

Guay, R., and M. Silver. 1975. *Thiobacillus acidophilus* sp. nov., isolation and some physiological characteristics. Can. J. Microbiol. 21: 281–288.

Gunkel, G. 1986. Studies of the fate of heavy metals in lakes. I. The role of ferrous oxidizing bacteria in coprecipitation of heavy metals. Arch. Hydrobiol. 105:489–515.

Hallbeck, L. 1993. On the biology of the iron-oxidizing and stalk-forming bacterium *Gallionella ferruginea*. Ph.D. thesis. University of Goeteborg, Sweden.

Hallbeck, L., and K. Pedersen. 1990. Culture parameters regulating stalk formation and growth rate of *Gallionella ferruginea*. J. Gen. Microbiol. 136:1675–1680.

Hallbeck, L., and K. Pedersen. 1991. Autotrophic and mixotrophic growth of *Gallionella ferruginea*. J. Gen. Microbiol. 137:2657–2661.

Hammann, R., and J. C. G. Ottow. 1974. Reductive dissolution of Fe_2O_3 by saccharolytic Clostridia and *Bacillus polymyxa* under anaerobic conditions. Z. Pflanzenernaehr. Bodenk. 137:108–115.

Hanert, H. 1968. Untersuchungen zur Isolierung, Stoffwechselphysiologie und Morphologie von *Gallionella ferruginea* Ehrenberg. Arch Mikrobiol. 60:348–376.

Hanert, H. 1973. Quantifizierung der Massenentwicklung des Eisenbakteriums *Gallionella ferruginea* unter natuerlichen Bedingungen. Arch. Microbiol. 88:225–243.

Hanert, H. 1974a. Untersuchungen zur individuellen Entwicklungskinetik von *Gallionella ferruginea* in statischer Mikrokultur. Arch. Mikrobiol. 96:59–74.

Hanert, H. 1974b. In situ-Untersuchungen zur Analyse und Intensitaet der Eisen(III)-Faellung in Draenungen. Z. Kulturtech. Flurberein. 15:80–90.

Hanert, H. H. 1981. The genus *Gallionella*. pp. 509–515. *In*: M. P. Starr,

H. Stolp, H. G. Trueper, A. Balows, and H. G. Schlegel. The Prokaryotes. A Handbook on Habitats, Isolation, and Identification of Bacteria. Springer-Verlag, Berlin.

Harder, E. C. 1919. Iron depositing bacteria and their geologic relations. U.S. Geol. Surv. Prof. Pap. 113, 89 pp.

Harrison, A. P. Jr. 1981. *Acidiphilium cryptum* gen. nov., sp. nov., heterotrophic bacterium from acidic mineral environments. Int. J. Syst. Bacteriol. 31:327–332.

Harrison, A. P. Jr. 1982. Genomic and physiological diversity amongst strains of *Thiobacillus ferrooxidans*, and genomic comparison with *Thiobacillus thiooxidans*. Arch. Microbiol. 131:68–76.

Harrison, A. P. Jr. 1984. The acidophilic thiobacilli and other acidophilic bacteria that share their habitat. Annu. Rev. Microbiol. 38:265–292.

Harrison, A.P. Jr. 1986. The phylogeny of iron-oxidizing bacteria. *In*: H. L. Ehrlich and D. S. Holmes, eds. Workshop on Biotechnology for the Mining, Metal-Refining and Fossil Fuel Processing Industries. Wiley, New York, pp. 311–317.

Harrison, A. P. Jr., and P. R. Norris. 1985. *Leptospirillum ferrooxidans* and similar bacteria: some characteristics and genomic diversity. FEMS Microbiol. Lett. 30:99–102.

Heldal, M., and O. Tumyr. 1983. *Gallionella* from metalimnion in an eutrophic lake: morphology and X-ray energy-dispersive microanalysis of apical cells and stalks. Can. J. Microbiol. 29:303–308.

Hoehnl, G. 1955. Ein Beitrag zur Physiologie der Eisenbakterien. Vom Wasser 22:176–193.

Holm, N. G. 1987. Biogenic influence on the geochemistry of certain ferruginous sediments of hydrothermal origin. Chem. Geol. 63: 45–57.

Holmes, D. S., J. H. Lobos, L. H. Bopp, and G.C. Welch. 1983. Setting up a genetic system de novo for studying the acidophilic thiobacillus *T. ferrooxidans*. *In*: G. Rossi and A. E. Torma, eds. Recent Progress in Biohydrometallurgy. Associazione Mineraria Sarda, Iglesias, Italy, pp. 541–554.

Holmes, D. S., J. R. Yates, and J. Schrader. 1989. Mobile repeated DNA sequences in *Thiobacillus ferrooxidans* and their significance for biomining. *In*: P. R. Norris and D. P. Kelly, eds. Biohydrometallurgy. Science and Technology Letters, Kew Surrey, U.K., pp. 153–160.

Huber, H., and K. O. Stetter. 1989. *Thiobacillus prosperus* sp. nov., represents a new group of halotolerant metal-mobilizing bacteria isolated from a marine geothermal field. Arch. Microbiol. 151:479–485.

Hutchinson, M., K. I. Johnstone, and D. White. 1966. Taxonomy of the acidophilic Thiobacilli. J. Gen. Microbiol. 44:373–381.

Ingledew, W. J. 1982. *Thiobacillus ferrooxidans*. The bioenergetics of an acidophilic chemolithotroph. Biochim. Biophys. Acta 683:89–117.

Ingledew, W. J. 1986. Ferrous iron oxidation by *Thiobacillus ferrooxidans*. *In*: H. L. Ehrlich and D. S. Holmes, eds. Workshop on Biotechnology for the Mining, Metal-Refining and Fossil Fuel Processing Industries. Biotech. Bioeng. Symp. 16. Wiley, New York, pp. 23–33.

Ingledew, W. J., and A. Houston. 1986. The organization of the respiratory chain of *Thiobacillus ferrooxidans*. Biotech. Appl. Biochem. 8: 242–248.

Ingledew, W. J., J. C. Cox, and P. J. Halling. 1977. A proposed mechanism for energy conservation during Fe^{2+} oxidation by *Thiobacillus ferrooxidans*: Chemiosmotic coupling to net H^+ influx. FEMS Microbiol. Lett. 2:193–197.

Ivanov, M. V., and N. N. Lyalikova. 1962. Taxonomy of iron-oxidizing thiobacilli. Mikrobiologiya 31:468–469 (Engl. transl. pp. 382–383).

Ivarson, K. C, and M. Sojak. 1978. Microorganisms and ochre deposition field drains of Ontario. Can. J. Soil Sci. 58:1–17.

Johnson, D. B., and W. I. Kelso. 1983. Detection of heterotrophic contaminants in cultures of *Thiobacillus ferrooxidans* and their elimination by subculturing in media containing copper sulfate. J. Gen. Microbiol. 129:2969–2972.

Johnson, D. B., and S. McGinness. 1991. A highly efficient and universal solid medium for growing mesosphilic and moderately thermophilic, iron-oxidizing acidophilic bacteria. J. Microbiol. Methods 13: 113–122.

Johnson, D. B., J. H. M. Macvicarl, and S. Rolfe. 1987. A new solid medium for the isolation and enumeration of *Thiobacillus ferrooxidans* and acidophilic heterotrophic bacteria. J. Microbiol. Methods 7:9–18.

Jones, C. A., and D. P. Kelly. 1983. Growth of *Thiobacillus ferrooxidans* on ferrous iron in chemostat culture: Influence of product and substrate inhibition. J. Chem. Tech. Biotechnol. 33B:241–261.

Jones, J. G., S. Gardener, and B. M. Simon. 1983. Bacterial reduction of ferric iron in a stratified eutrophic lake. J. Gen. Microbiol. 129: 131–139.

Jones, J. G., S. Gardener, and B. M. Simon. 1984. Reduction of ferric iron by heterotrophic bacteria in lake sediments. J. Gen. Microbiol. 130:45–51.

Kamalov, M. R. 1967. Oxidation of divalent iron by a *Thiobacillus ferrooxidans* culture in the presence of chloride and sulfate ions. Izv. Akad. Nauk. Kaz. SSR Ser. Biol. 5:47–50.

Kamalov, M. R. 1972. Adaptation of *Thiobacillus ferrooxidans* cultures to increased amounts of copper, zinc, and molybdenum in an acid medium. Izv. Akad. Nauk Kaz. SSR Ser. Biol. 10:39–44.

Karavaiko, G. I., and A. A. Avakyan. 1970. Mechanisms of *Thiobacillus ferrooxidans* multiplication. Mikrobiologiya 39:950–952 (Engl. transl. pp. 833–836).

Karlin, R., M. Lyle, and G. R. Heath. 1987. Authigenic magnetite formation in suboxic marine sediments. Nature (Lond.) 326:490–493.

Kelly, D. P., and C. A. Jones. 1978. Factors affecting metabolism and ferrous iron oxidation in suspensions and batch cultures of *Thiobacillus ferrooxidans*: relevance to ferric iron leach solution regeneration. *In*: L. E. Murr, A. E. Torma, and J. A. Brierley, eds. Metallurgical Applications of Bacterial Leaching and Related Microbiological Phenomena. Academic Press, New York, pp. 19–44.

Kelly, D. P., and O. H. Tuovinen. 1972. Recommendation that the names *Ferrobacillus ferrooxidans* Leathen and Braley and *Ferrobacillus sulfooxidans* Kinsel be recognized as synonymous of *Thiobacillus ferrooxidans* Temple and Colmer. Int. J. Syst. Bacteriol. 22:170–172.

Kinsel, N. 1960. New sulfur oxidizing iron bacterium: *Ferrobacillus sulfooxidans* sp.n. J. Bacteriol. 80:628–632.

Kovalenko, T. V., G. I. Karavaiko, and V. P. Piskunov. 1982. Effect of Fe^{3+} ions in the oxidation of ferrous iron by *Thiobacillus ferrooxidans* at various temperatures. Mikrobiologiya 51:156–160 (Engl. transl. pp. 142–146).

Krul, I. M., P. Hirsch, and J. T. Staley. 1970. *Toxothrix trichogenes* (Mol) Berger and Bingmann: The organism and its biology. Antonie v. Leeuwenhoek 36:409–420.

Kucera, S., and R. S. Wolfe. 1957. A selective enrichment method for *Gallionella ferruginea*. J. Bacteriol. 74:344–349.

Kullmann, K-H., and R. Schweisfurth. 1978. Eisenoxydierende, staebchenfoermige Bakterien. II. Quantitative Untersuchungen zum Stoffwechsel und zur Eisenoxydation mit Eisen(II)-oxalat. Z. Allg. Mikrobiol. 18:321–327.

Kusano, T., K. Sugawara, C. Inoue, T. Takeshima, M. Numata, and T. Shiratori. 1992. Electrotransformation of *Thiobacillus ferrooxidans* with plasmids containing a *mer* determinant as the selective marker by electroporation. J. Bacteriol. 174:6617–6623.

Kuznetsov, S. I., and V. I. Romanenko. 1963. The Microbiological Study of Inland Waters. Izvddatel'stvo Akad. Nauk SSSR, Moscow-Leningrad.

Landesman, J., D. W. Duncan, and C. C. Walden. 1966. Iron oxidation by washed cell suspensions of the chemoautotroph, *Thiobacillus fer-rooxidans*. Can. J. Microbiol. 12:25–33.

Lascelle, J. and K. A. Burke. 1978. Reduction of ferric iron by L-lactate and DL-glycerol-3-phosphate in membrane preparations from *Staph-ylococcus aureus* and interactions with the nitrate reductase system. J. Bacteriol. 134:585–589.

Lazaroff, N. 1963. Sulfate requirement for iron oxidation by *Thiobacillus ferrooxidans*. J. Bacteriol. 78–83.

Lazaroff, N. 1977. The specificity of the anionic requirement for iron oxidation by *Thiobacillus ferrooxidans*. J. Gen. Microbiol. 101: 85–91.

Lazaroff, N. 1983. The exclusion of D_2O from the hydration sphere of $FeSO_4.7H_2O$ oxidized by *Thiobacillus ferrooxidans*. Science 222: 1331–1334.

Lazaroff, N., W. Sigal, and A. Wasserman. 1982. Iron oxidation and pre-cipitation of ferric hydroxysulfates by resting *Thiobacillus ferrooxi-dans* cells. Appl. Environ. Microbiol. 43:924–938

Lazaroff, N., L. Melanson, E. Lewis, N. Santoro, and C. Pueschel. 1985. Scanning electron microscopy and infrared spectroscopy of iron sed-iments formed by *Thiobacillus ferrooxidans*. Geomicrobiol. J. 4: 231–268.

Leathen, W. W., N. A. Kinsel, and S. A. Braley, Sr. 1956. *Ferrobacillus ferrooxidans*: a chemosynthetic autotrophic bacterium. J. Bacteriol. 72:700–704.

Lees, H., S. C. Kwok, and I. Suzuki. 1969. The thermodynamics of iron oxidation by ferrobacilli. Can. J. Microbiol. 15:43–46.

Le Roux, N. W., D. S. Wakerley, and S. D. Hunt. 1977. Thermophilic *Thiobacillus*-type bacteria from Icelandic thermal areas. J. Gen. Mi-crobiol. 100:197–201.

Lieske, R. 1919. Zur Ernaehrungsphysiologie der Eisenbakterien. Z. Bakt. Parasitenk. Infektionskr. Hyg. Abt. II, 49:413–425.

Lobos, J. H., T. E. Chisholm, L. H. Bopp, and D. S. Holmes. 1986. *Acidiphilium organovorum* sp. nov., an acidophilic heterotroph iso-lated from a *Thiobacillus ferrooxidans* culture. Int. J. Syst. Bacte-riol. 36:139–144.

Lovley, D. R. 1987. Organic matter mineralization with the reduction of ferric iron: a review. Geomicrobiol. J. 5:375–399.

Lovley, D. R. 1991. Dissimilatory Fe(III) and Mn(IV) reduction. Micro-biol. Rev. 55:259–287.

Lovley, D. R., and E. J. P. Phillips. 1986a. Organic matter mineralization with the reduction of ferric iron in anaerobic sediments. Appl. Envi-ron. Microbiol. 51:683–689.

Lovley, D. R., and. E. J. P. Phillips. 1986b. Availability of ferric iron for microbial reduction in bottom sediments of the freshwater tidal Potomac River. Appl. Environ. Microbiol. 52:751–757.

Lovley, D. R., and E. J. P. Phillips. 1987. Competitive mechanisms for inhibition of sulfate reduction and methane production in the zone of ferric iron reduction in sediments. Appl. Environ. Microbiol. 53: 2636–2641.

Lovley, D. R., and E. J. P. Phillips. 1988a. Novel mode of microbial energy metabolism: organic carbon oxidation coupled to dissimilatory reduction of iron or manganese. Appl. Environ. Microbiol. 54: 1472–1480.

Lovley, D. R., and E. J. P. Phillips. 1988b. Manganese inhibition of microbial iron reduction in anaerobic sediments. Geomicrobiol. J. 6: 145–155.

Lovley, D. R., J. F. Stolz, G. L. Nord, and E. J. P. Phillips. 1987. Anaerobic production of magnetite by a dissimilatory iron-reducing microorganism. Nature (Lond.) 330:252–254.

Lovley, D. R., E. J. P. Phillips, and D. J. Lonergan. 1989. Hydrogen and formate oxidation coupled to dissimilatory reduction of iron and manganese by *Alteromonas putrefaciens*. Appl. Environ. Microbiol. 55:700–706.

Lovley, D. R., E. J. P. Phillips, and D. J. Lonergan. 1991. Enzymatic versus nonenzymatic mechanisms for Fe(III) reduction in aquatic sediments. Environ. Sci. Technol. 25:1062–1067.

Lovley, D. R., E. J. P. Phillips, D. J. Lonergan, and P. K. Widman. 1995. Fe (III) and S° reduction by *Pelobacter carbinolicus*. Appl. Environ. Microbiol. 61:2132–2138.

Lovley, D. R., S. J. Giovannoni, D. C. White, J. E. Champine, E. J. P. Phillips, Y. A. Gorby, and S. Goodwin. 1993. *Geobacter metallireducens* gen. nov. sp. nov., a microorganism capable of coupling the complete oxidation of organic compounds to the reduction of iron and other metals. Arch. Microbiol. 159:336–344.

Luetters, S., and H. H. Hanert. 1989. The ultrastructure of chemolithoautotrophic *Gallionella ferruginea* and *Thiobacillus ferrooxidans* as revealed by chemical fixation and freeze-etching. Arch. Microbiol. 151:245–251.

Luetters-Czekalla, S. 1990. Lithoautotrophic growth of the iron bacterium *Gallionella ferruginea* with thiosulfite or sulfide as energy source. Arch. Microbiol. 154:417–421.

Lundgren, D. G., and W. Dean. 1979. Biogeochemistry of iron. *In*: P. A. Trudinger and D. J. Swaine, eds. Biogeochemical Cycling of Mineral-Forming Elements. Elsevier, Amsterdam, pp. 211–251.

Lundgren, D. G., K. J. Andersen, C. C. Remsen, and R. P. Mahoney. 1964. Culture, structure, and physiology of the chemoautotroph *Ferrobacillus ferrooxidans*. Dev. Indust. Microbiol. 6:250–259.

Lyalikova, N. N. 1958. A study of chemosynthesis in *Thiobacillus ferrooxidans*. Mikrobiologiya 27:556–559.

Maciag, W. J., and D. G. Lundgren. 1964. Carbon dioxide fixation in the chemoautotroph, *Ferrobacillus ferrooxidans*. Biochem. Biophys. Res. Commun. 17:603–607.

Mackintosh, M. E. 1978. Nitrogen fixation by *Thiobacillus ferrooxidans*. J. Gen. Microbiol. 105:215–218.

MacRae, I. C., and J. S. Celo. 1975. Influence of colloidal iron on the respiration of a species of the genus *Acinetobacter*. Appl. Microbiol. 29:837–840.

Madsen, E. L., M. D. Morgan, and R. E. Good. 1986. Simultaneous photoreduction and microbial oxidation of iron in a stream in the New Jersey Pinelands. Limnol. Oceanogr. 31:832–838.

Mann, H., T. Tazaki, W. S. Fyfe, T. J. Beveridge, and R. Humphrey. 1987. Cellular lepidocrocite precipitation and heavy-metal sorption in *Euglena* sp. (unicellular alga): implications for biomineralization. Chem. Geol. 63:39–43.

Manning, H. L. 1975. New medium for isolating iron-oxidizing and heterotrophic acidophilic bacteria from acid mine drainage. Appl. Microbiol. 30:1010–1016.

Mao, M. W. H., P. R. Dugan, P. A. W. Martin, and O. H. Tuovinen. 1980. Plasmid DNA in chemoorganotrophic *Thiobacillus ferrooxidans* and *T. acidophilus*. FEMS Microbiol. Lett. 8:121–125.

Marchlewitz, B., D. Hasche, and W. Schwartz. 1961. Untersuchungen über das Verhalten von Thiobakterien gegenüber Schwermetallen. Z. Allg. Mikrobiol. 1:179–191.

Markosyan, G.E. 1972. A new iron-oxidizing bacterium *Leptospirillum ferrooxidans* gen. et sp. nov. Biol. Zh. Arm. 25:26.

Martin, P. A. W., P. R. Dugan, and O. H. Tuovinen. 1981. Plasmid DNA in acidophilic, chemolithotrophic thiobacilli. Can. J. Microbiol. 27: 850–853.

Mason, J., and D. P. Kelly. 1988. Mixotrophic and autotrophic growth of *Thiobacillus acidophilus*. Arch. Microbiol. 149:317–323.

Mason, J., D. P. Kelly, and A. P. Wood. 1987. Chemolithotrophic and autotrophic growth of *Thermothrix thiopara* and some thiobacilli on thiosulfate and polythionates, and a reassessment of the growth yields of *Thx. thiopara* in chemostat culture. J. Gen. Microbiol. 133: 1249–1256.

McClure, M. A., and R. W. G. Wyckoff. Ultrastructural characteristics of *Sulfolobus acidocalderius*. J. Gen. Microbiol. 128:433–437.

Merkle, F. G. 1955. Oxidation-reduction processes in soils. *In*: F. E. Bear, ed. Chemistry of the Soil. Reinhold, New York, pp. 200–218.

Mishra, A. K., and P. Roy. 1979. A note on the growth of *Thiobacillus ferrooxidans* on solid medium. J. Appl. Bacteriol. 47:289–292.

Mishra, A. K., P. Roy, and S. S. Roy Mahapatra. 1983. Isolation of *Thiobacillus ferrooxidans* from various habitats and their growth pattern on solid medium. Curr. Microbiol. 8:147–152.

Molisch, H. 1910. Die Eisenbakterien. Gustav Fischer Verlag, Jena.

Munch, J. C., and J. C. G. Ottow. 1977. Modelluntersuchungen zum Mechanismus der bakteriellen Eisenreduktion in Hydromorphen Boeden. Z. Pflanzenaehr. Bodenk. 140:549–562.

Munch, J. C., and J. C. G. Ottow. 1980. Preferential reduction of amorphous to crystalline iron oxides by bacterial activity. Soil Sci. 129: 1–21.

Munch, J. C., and J. C. G. Ottow. 1982. Einfluss von Zellkontakt und Eisen(III)-Oxidform auf die Bakterielle Eisenreduktion. Z. Pflanzenernaehr. Bodenk. 145:66–77.

Munch, J. C., Th. Hillebrand, and J. C. G. Ottow. 1978. Transformations in the Fe_o/Fe_d ratio of pedogenic iron oxides affected by iron-reducing bacteria. Can. J. Soil Sci. 58:475–486.

Myers, C. R., and J. M. Myers. 1992a. Localization of cytochromes to the outer membrane of anaerobically grown *Shewanella putrefaciens* MR-1. J. Bacteriol. 174:3429–3438.

Myers, C. R., and J. M. Myers. 1992b. Fumarate reductase is a soluble enzyme in anaerobically grown *Shewanella putrefaciens* MR-1. FEMS Microbiol. Lett. 98:13–20.

Myers, C. R., and J. M. Myers. 1993. Ferric reductase is associated with the membranes of anaerobically grown *Shewanella putrefaciens* MR-1. FEMS Microbiol. Lett. 108:15–22.

Myers, C. R., and K. H. Nealson. 1988. Bacterial manganese reduction and growth with manganese oxide as sole electron acceptor. Science 240:1319–1321.

Myers, C. R., and K. H. Nealson. 1990 Respiration-linked proton translocation coupled to anaerobic reduction of manganese(IV) and iron(III) in *Shewanella putrefaciens* MR-1. J. Bacteriol. 172:6232–6238.

Nealson, K. H., and D. Saffarini. 1994. Iron and manganese in anaerobic respiration: environmental significance, physiology, and rergulation. Annu. Rev. Microbiol. 48:311–343.

Neilands, J. B., ed. 1974. Microbial Iron Metabolism. Academic Press, New York.

Niemelae, S. I., C. Sivalae, T. Luoma, and O. H. Tuovinen. 1994. Maximum temperature limits for acidophilic, mesophilic bacteria in biological leaching systems. Appl. Environ. Microbiol. 60:3444–3446.

Norris, P. R. 1990. Acidophilic bacteria and their activity in mineral sulfide oxidation. *In*: H. L. Ehrlich and C. L. Brierley (eds.), Microbial Mineral Recovery. McGraw-Hill, New York, pp. 3–27.

Norris, P. R., and D. W. Barr. Growth and iron oxidation by acidophilic moderate thermophiles. FEMS Microbiol. Lett. 28:221–224.

Norris, P. R., R. M. Marsh, and E. B. Lindstrom. 1986a. Growth of mesophilic and thermophilic acidophilic bacteria on sulfur and tetrathionate. Biotechnol. Appl. Biochem. 8:313–329.

Norris, P. R., L. Parrott, and R. M. Marsh. 1986b. Moderately thermophilic mineral-oxidizing bacteria. *In*: H. L. Ehrlich and D. S. Holmes. Workshop on Biotechnology for the Mining,Metal-Refining, and Fossil-Fuel Processing Industries. Biotech. Bioeng. Symp. No. 16. Wiley, New York, pp. 254–262.

Obuekwe, C. O. 1986. Studies on the physiological reduction of ferric iron by *Pseudomonas* species. Microbios. Lett. 33:81–97.

Obuekwe, C. O., and D. W. S. Westlake. 1982. Effect of reducible compounds (potential electron acceptors) on reduction of ferric iron by *Pseudomonas* species. Microbios. Lett. 19:57–62.

Obuekwe, C. O., D. W. S. Westlake, and F. D. Cook. 1981. Effect of nitrate on reduction of ferric iron by a bacterium isolated from crude oil. Can. J. Microbiol. 27:692–697.

Ottow, J.C.G. 1969a. The distribution and differentiation of iron-reducing bacteria in gley soils. Zbltt. Bakteriol. Parasitenk. Infektionskr. Hyg. Abt. 2 123:600–615.

Ottow, J. C. G. 1969b. Der Einfluss von Nitrat, Chlorat, Sulfat, Eisoxidform und Wachstumsbedingungen auf das Ausmass der bakteriellen Eisenreduktion. Z. Pflanzenernaehr. Bodenk. 124:238–253.

Ottow, J. C. G. 1970a. Selection, characterization and iron-reducing capacity of nitrate reductaseless (nit⁻) mutants of iron- reducing bacteria. Z. Allg. Mikrobiol. 10:55–62.

Ottow, J. C. G. 1970b. Bacterial mechanism of gley formation in artificially submerged soil. Nature (Lond.) 225:103.

Ottow, J. C. G. 1971. Iron reduction and gley formation by nitrogen-fixing Clostridia. Oecologia 6:164–175.

Ottow, J. C. G., and H. Glathe. 1971. Isolation and identification of iron-reducing bacteria from gley soils. Soil Biol. Biochem. 3:43–55.

Ottow, J. C. G., and H. Ottow. 1970. Gibt es eine Korrelation zwischen der eisenreduzierenden und der nitratreduzierenden Flora des Bodens? Ztbltt. Bakteriol. Parasitk. Infektionskr. Hyg. Abt. 2 124: 314–318.

Ottow, J. C. G., and A. von Klopotek. 1969. Enzymatic reduction of iron oxide by fungi. Appl. Microbiol. 18:41–43.

Peng, J.-B., W.-M. Yan, and X. Z. Bao. 1994. Expression of heterogenous arsenic resistance genes in the obligately autotrophic biomining bacterium *Thiobacillus ferrooxidans*. Appl. Environ. Microbiol. 60: 2653–2656.

Perry, E. C. Jr., F. C. Tan, and G. B. Morey. 1973. Geology and stable isotope geochemistry of Biwabik Iron Formation, northern Minnesota. Econ Geol. 68:1110–1125.

Pichinoty, F. 1963. Récherches sur la nitrate réductase d'*Aerobacter aerogenes*. Ann. Inst. Pasteur (Paris) 104:394–418.

Pivovarova, T. A., G. E. Markosyan, and G. I. Karavaiko. 1981. Morphogenesis and fine structure of *Leptospirillum ferrooxidans*. Mikrobiologyia 50:482–486 (Engl. transl. pp. 339–344).

Pollack, J. R., and J. B. Neilands. 1970. Enterobactin, an iron transport compound. Biochem. Biophys. Res. Commun. 38:989–992.

Praeve, P. 1957. Untersuchungen ueber die Stoffwechselphysiologie des Eisenbakteriums *Leptothrix ochracea* Kuetzing. Arch. Mikrobiol. 27:33–62.

Pretorius, I. M., D. E. Rawlings, and D. R. Woods. 1986. Identification and cloning of *Thiobacillus ferrooxidans* structural *nif*-genes in *Escherichia coli*. Gene 45:59–65.

Pretorius, I. M., D. E. Rawlings, E. G. O'Neill, W. A. Jones, R. Kirby, and D. R. Woods. 1987. Nucleotide sequence of the gene encoding the nitrogenase iron protein of *Thiobacillus ferrooxidans*. J. Bacteriol. 169:367–370.

Pringsheim, E. G. 1949. The filamentous bacteria *Sphaerotilus, Leptothrix* and *Cladothrix* and their relations to iron and manganese. Phil. Trans. Roy. Soc. Lond. Ser. B Biol. Sci 233:453–482.

Pronk, J. T., and D. B. Johnson. 1992. Oxidation and reduction of iron by acidophilic bacteria. Geomicro. J. 10:149–171.

Pronk, J. T., W. M. Meijer, W. Hazeu, J. P. van Dijken, P. Bos, and J. G. Kuenen. 1991a. Growth of *Thiobacillus ferooxidans* on formic acid. Appl. Environ. Microbiol. 57:2057–2062.

Pronk, J. T., K. Liem, P. Bos, and J. G. Kuenen. 1991b. Energy transduction by anaerobic ferric iron respiration in *Thiobacillus ferrooxidans*. Appl. Environ. Microbiol. 57:2063–2068.

Pronk, J. T., de Bruyn, P. Bos, and J. G. Kuenen. 1992. Anaerobic growth of *Thiobacillus ferooxidans*. Appl. Environ. Microbiol. 58: 2227–2230.

Puchelt, H., H. H. Schock, Schroll, and H. Hanert. 1973. Rezente marine Eisenerze auf Santorin, Griechenland. Geol. Rundschau 62:786–812.

Rankama, K., and T. G. Sahama. 1950. Geochemistry. University of Chicago Press, Chicago, pp. 657–676.

Rawlings, D. E., and T. Kusano. 1994. Molecular genetics of *Thiobacillus ferrooxidans*. Microbiol. Rev. 58:39–55.

Rawlings, D. E., I. M. Pretorius, and D. R. Woods. 1986. Expression of *Thiobacillus ferrooxidans* plasmid functions and the development of genetic systems for the thiobacilli. *In*: H. L. Ehrlich and D. S. Holmes, eds. Workshop on Biotechnology for the Mining, Metal-Refining and Fossil Fuel Processing Industries. Wiley, New York, pp. 281–287.

Razzell, W. E., and P. C. Trussell. 1963. Isolation and properties of an iron-oxidizing *Thiobacillus*. J. Bacteriol. 85:595–603.

Remsen, C., and D. G. Lundgren. 1966. Electron microscopy of the cell envelope of *Ferrobacillus ferrooxidans* prepared by freeze-etching and chemical fixation techniques. J. Bacteriol. 92:1765–1771.

Roberts, J. L. 1947. Reduction of ferric hydroxide by strains of *Bacillus polymyxa*. Soil Sci. 63:135–140.

Roden, E. E., and D. R. Lovley. 1993. Dissimilatory Fe(III) reduction by the marine microorganism *Desulfuromonas acetoxidans*. Appl. Environ. Microbiol. 59:734–742.

Sartory, A., and J. Meyer. 1947. Contribution à l'étude du métabolisme hydrocarboné des bactéries ferrugineuses. C.R. Acad. Sci. (Paris) 225:541–542.

Schnaitman, C. A., and D. G. Lundgren. 1965. Organic compounds in the spent medium of *Ferrobacillus ferrooxidans*. Can. J. Microbiol. 11:23–27.

Schnaitman, C. A., M. S. Korczinski, and D. G. Lundgren. 1969. Kinetic studies of iron oxidation by whole cells of *Ferrobacillus ferrooxidans*. J. Bacteriol. 99:552–557.

Schrader, J., and D. S. Holmes. 1988. Phenotypic switching of *Thiobacillus ferrooxidans*. J. Bacteriol. 170:3915–3923.

Segel, I. H. 1975. Enzyme Kinetics. Behavior and Analysis of Rapid Equilibrium and Steady-State Enzyme Systems. Wiley, New York.

Segerer, A., A. Neuner, J. K. Kristjansson, and K. O. Stetter. 1986. *Acidianus infernus* gen. nov., sp. nov., and *Acidianus brierleyi* comb. nov.: Facultatively aerobic, extremely acidophilic thermophilic sulfur-metabolizing archaebacteria. Int. J. Syst. Bacteriol. 36: 559–564.

Shafia, F., and R. F. Wilkinson. 1969. Growth of *Ferrobacillus ferrooxidans* on organic matter. J.Bacteriol. 97:256–260.

Shafia, F., K. R. Brinson, M. W. Heinzman, and J. M. Brady. 1972. Transition of chemolithotroph *Ferrobacillus ferrooxidans* to obligate organotrophy and metabolic capabilities of glucose-grown cells. J. Bacteriol. 111:56–65.

Short, K. A., and R. P. Blakemore. 1986. Iron respiration-driven proton translocation in aerobic bacteria. J. Bacteriol. 167:729–731.

Silver, M., H. L. Ehrlich, and K. C. Ivarson. 1986. Soil mineral transformation mediated by soil microbes. *In*: P. M. Huang and M. Schnitzer, eds. Interactions of soil minerals with natural organics and microbes. SSSA Special Publication Number 17. Soil Science Society of America, Madison, WI, pp. 497–519.

Silverman, M. P., and D. G. Lundgren. 1959a. Studies on the chemoautotrophic iron bacterium *Ferrobacillus ferrooxidans*. I. An improved medium and a harvesting procedure for securing high cell yields. J. Bacteriol. 59:642–647.

Silverman, M. P., and D. G. Lundgren. 1959b. Studies on the chemoautotrophic iron bacterium *Ferrobacillus ferrooxidans*. II. Manometric studies. J. Bacteriol. 78:326–331.

Sorensen, J. 1987. Nitrate reduction in marine sediment:pathways and interactions with iron and sulfur cycling. Geomicrobiol. J. 5: 401–421.

Sørensen, J., and B. B. Jørgensen. 1987. Early diagenesis in sediments from Danish coastal waters: microbial activity and Mn-Fe-S geochemistry. Geochim. Cosmochim. Acta 51:1583–1590.

Starkey, R. L. 1945. Precipitation of ferric hydrate by iron bacteria. Science 102:532–533.

Starkey, R. L., and H. O. Halvorson. 1927. Studies on the transformations or iron in nature. II. Concerning the importance of microorganisms in the solution and precipitation of iron. Soil Sci. 24:381–402.

Stevens, C. J., P. R. Dugan, and O. H. Tuovinen. 1986. Acetylene reduction (nitrogen fixation) by *Thiobacillus ferrooxidans*. Biotechnol. Appl. Biochem. 8:351–359.

Stokes, J. L. 1954. Studies on the filamentous sheathed iron bacterium *Sphaerotilus natans*. J. Bacteriol. 67:278–291.

Sugio, T., Y. Anzai, T. Tano, and K. Imai. 1981. Isolation and some properties of an obligate and a facultative iron-oxidizing bacterium. Agric. Biol. Chem. 45:1141–1151.

Sugio, T., C. Domatsu, O. Munakata, T. Tano, and K. Imai. 1985. Role of a ferric ion reducing system in sulfur oxidation of *Thiobacillus ferrooxidans*. Appl. Environ. Microbiol. 49:1401–1406.

Sugio, T., K. Wada, M. Mori, K. Inagaki, and T. Tano. 1988a. Synthesis of an iron-oxidizing system during growth of *Thiobacillus ferrooxidans* on sulfur-basal salts medium. Appl. Environ. Microbiol. 54: 150–152.

Sugio, T., T. Katagiri, M. Moriyama, Y. L. Zhen, K. Inagaki, and T. Tano. 1988b. Existence of a new type of sulfite oxidase which utilizes ferric ions as an electron acceptor in *Thiobacillus ferrooxidans*. Appl. Environ. Microbiol. 54:153–157.

Sugio, T., T. Katagiri, K. Inagaki. and T. Tano. 1989. Actual substrate for elemental sulfur oxidation by sulfur:ferric ion oxidoreductase purified from *Thiobacillus ferrooxidans*. Biochim. Biophys. Acta 973:250–256.

Sugio, T., T. Hirose, Y. Li-Zhen, and T. Tano. 1992a. Purification and some properties of sulfite:ferric ion oxidoreductase from *Thiobacillus ferrooxidans*. J. Bacteriol. 174:4189–4192.

Sugio, T. , K.J. White, E. Shute, D. Choate, and R.C. Blake II. 1992b. Existence of a hydrogen sulfide:ferric ion oxidoreductase in iron-oxidizing bacteria. Appl. Environ. Microbiol. 58:431–433.

Tabita, R., and D. G. Lundgren. 1971a. Utilization of glucose and the effect of organic compounds on the chemolithotroph *Thiobacillus ferrooxidans*. J. Bacteriol. 108:328–333.

Tabita, R., and D. G. Lundgren. 1971b. Heterotrophic metabolism of the chemolithotroph *Thiobacillus ferrooxidans*. J. Bacteriol. 108: 334–342.

Tait, G. H. 1975. The identification and biosynthesis of siderochromes formed by *Micrococcus denitrificans*. Biochem. J. 146:191–204.

Takai, Y., and T. Kamura. 1966. The mechanism of reduction in water-logged paddy soil. Folia Microbiol. 11:304–315.

Taylor, K. A., J. F. Deatherage, and L. A. Amos. 1982. Structure of the S-layer of *Sulfolobus acidocaldarius*. Nature (Lond.) 299:840–842.

Temple, K. L., and A. R. Colmer. 1951. The autotrophic oxidation of iron by a new bacterium: *Thiobacillus ferrooxidans*. J. Bacteriol. 62: 605–611.

Thamdrup, Bo, K. Finster, J. Wuergler Hansen, and F. Bak. 1993. Bacterial disproportionation of elemental sulfur coupled to chemical reduction of iron or manganese. Appl. Environ. Microbiol. 59: 101–108.

Tikhonova, G. V., L. L. Lisenkova, N. G. Doman, and V. P. Skulachev. 1967. Electron transport pathways in *Thiobacillus ferrooxidans*. Mikrobiologiya 32:725–734 (Engl. transl. 599–606.

Tipping, E., D. W. Thompson, M. Ohnstad, and N. B. Hetherington. 1986. Effects of pH on the release of metals from naturally occurring oxides of manganese and iron. Environ. Technol. Lett. 7:109–114.

Trafford, B. D., C. Bloomfield, W. I. Kelso, and G. Pruden. 1973. Ochre formation in field drains in pyritic soils. J. Soil Sci. 24:453–460.

Troshanov, E. P. 1968. Iron- and manganese-reducing microorganisms in ore-containing lakes of the Karelian isthmus. Mikrobiologiya 37: 934–940 (Engl. transl. pp. 786–791).

Troshanov, E. P. 1969. Conditions affecting the reduction of iron and manganese by bacteria in the ore-bearing lakes of the Karelian isthmus. Mikrobiologiya 38:634–643 (Engl. transl. pp. 528–535).

Tuovinen, O. H., and D. P. Kelly. 1973. Studies on the growth of *Thiobacillus ferrooxidans*. I. Use of membrane filters and ferrous iron agar to determine viable numbers, and comparison with $^{14}CO_2$-fixation and iron oxidation as measures of growth. Arch. Mikrobiol. 88:285–298.

Tuovinen, O. H., and D. P. Kelly. 1974. Studies on the growth of *Thiobacillus ferrooxidans*. II. Toxicity of uranium to growing cultures and tolerance conferred by mutation, other metal cations and EDTA. Arch. Microbiol. 95:153–164.

Tuovinen, O. H., and D. J. D. Nicholas. 1977. Transition of *Thiobacillus ferrooxidans* KG-4 from heterotrophic growth on glucose to autotrophic growth on ferrous-iron. Arch. Microbiol. 114:193–195.

Tuovinen, O. H., S. I. Niemelae, and H.G. Gyllenberg. 1971. Tolerance of *Thiobacillus ferrooxidans* to some metals. Antonie v. Leeuwenhoek 37:489–496.

Tuttle, J. H., and P. R. Dugan. 1976. Inhibition of growth, iron, and sulfur oxidation by *Thiobacillus ferrooxidans* by simple organic compounds. Can. J. Microbiol. 22:719–730.

Unz, R. F. and D. G. Lundgren, 1961. A comparative nutritional study of three chemoautotrophic bacteria: *Ferrobacillus ferrooxidans*, *Thiobacillus ferrooxidans* and *Thiobacillus thiooxidans*. Soil Sci. 92:302–313.

Vernon, L. P., J. H. Mangum, J. V. Beck, and F. M. Shafia. 1960. Studies on a ferrous-ion-oxidizing bacterium. II. Cytochrome composition. Arch. Biochem. Biophys. 88:227–231.

Vorreiter, L., and J. C. Madgwick. 1982. The effect of sodium chloride on bacterial leaching of low-grade copper ore. Proc. Australas. Inst. Min. Metall., No. 284, pp. 63–66.

Walker, J. C. G. 1987. Was the Archean biosphere upside down? Nature (Lond.) 329:710–712.

Walker, J. C. G., C. Klein, M. Schidlowski, J. W. Schopf, D. J. Stevenson, and M. R. Walter. 1983. Environmental evolution of the Archean-Early Proterozoic Earth. *In*: J. W. Schopf, ed. Earth's Earliest Biosphere. Its Origin and Evolution. Princeton University Press, Princeton, NJ, pp. 260–290.

Walsh, F., and R. Mitchell. 1972. An acid-tolerant iron-oxidizing Metallogenium. J. Gen. Microbiol. 72:369–376.

Wichlacz, P. L., R. F. Unz, and T. A. Langworthy. 1986. *Acidiphilium angustum* sp. nov., *Acidiphilium facilis* sp. nov., and *Acidiphilium rubrum* sp. nov.: acidophilic heterotrophic bacteria isolated from acidic coal mine drainage. Int. J. Syst. Bacteriol. 36:197–201.

Widdel, F., S. Schnell, S. Heising, A. Ehrenreich, B. Assmus, and B. Schink. 1993. Ferrous iron oxidation by anoxygenic phototrophic bacteria. Nature (Lond.) 362:834–836.

Winogradsky, S. 1888. Ueber Eisenbakterien. Bot. Ztg. 46:261–276.

Winogradsky, S. 1922. Eisenbaketerien als Anorgoxydanten. Z. Bakteriol. Parasitenk. Infektionskr. Hyg. Abt. II 57:1–21.

Wolfe, R. S. 1964. Iron and manganese bacteria. *In*: H. Heukelekian and N. Dondero, eds. Principles and Applications in Aquatic Microbiology. Wiley, New York, pp. 82–97.

Yamanaka, T., T. Yano, M. Kai, H. Tamegai, A. Sato, and Y. Fukumori. 1991. The electron transfer system in an acidophilic iron-oxidizing bacterium. *In*: Y. Mukohata (Ed.), New Era of Bioenergetics. Academic Press, Tokyo, pp. 223–246.

Yates, J. R., and D. S. Holmes. 1987. Two families of repeated DNA sequences in *Thiobacillus ferrooxidans*. J. Bacteriol. 169:1861–1870.

Yates, M. G., and A. Nason. 1966a. Enhancing effect of nucleic acids and their derivatives in the reduction of cytochrome c by ferrous ions. J. Biol. Chem. 241:4861–4871.

Yates, M. G., and A. Nason. 1966b. Electron transport systems of the chemoautotroph *Ferrobacillus ferrooxidans*. II. Purification and properties of a heat-labile iron–cytochrome c reductase. J. Biol. Chem. 241:4872–4880.

Zavarzin, G. A. 1972. Heterotrophic satellite of *Thiobacillus ferrooxidans*. Mikrobiologiya 41:369–370 (Engl. transl. pp. 323–324.

Zillig, W., K. O. Stetter, S. Wunderl, W. Schulz, H. Priess, and I. Scholz. 1980. The *Sulfolobus*-"*Caldariella*" group: Taxonomy on the basis of the structure of DNA-dependent RNA polymerases. Arch. Microbiol. 125:259–269.

Zillig, W., S. Yeats, I. Holz, A. Boeck, F. Gropp, M. Rettenberger, and S. Lutz. 1985. Plasmid-related anaerobic autotrophy of the novel archaebacterium *Sulfolobus ambivalens*. Nature (Lond.) 313:789–791.

15

Geomicrobiology of Manganese

15.1 OCCURRENCE OF MANGANESE IN THE EARTH'S CRUST

The abundance of manganese in the Earth's crust is 0.1% (Alexandrov, 1972, p. 670). The element is, therefore, only 1/50 as plentiful as iron in this part of the Earth. Its distribution in the crust is by no means uniform. In soils, for instance, its concentration can range from 0.002 to 10% (Goldschmidt, 1954). An average concentration in fresh water has been reported to be 8 μg kg^{-1} (Bowen, 1979). Concentrations slightly in excess of 1 mg kg^{-1} can be encountered in anoxic hypolimnia of some lakes. In seawater, an average concentration has been reported to be 0.2 μg kg^{-1} (Bowen, 1979), but concentrations more than 3 orders of magnitude greater may be encountered near active hydrothermal vents on midocean spreading centers.

Manganese is found as a major or minor component in more than 100 naturally occurring minerals (Mineral Facts and Problems, 1965, p. 556). Major accumulations of manganese occur in the form of oxides, carbonates, and silicates. Among the oxides, psilomelane [Ba,Mn^{2+}Mn$_8$$^{4+}$ O$_{16}$(OH)$_4$], birnessite (δMnO$_2$), pyrolusite and vernadite (MnO$_2$), manganite [MnOOH], todorokite [(Mn^{2+},Mg^{2+},Ba^{2+},Ca^2,K$^+$,Na$^+$)$_2$Mn$_5$$^{4+}$

$O_{12} \cdot 3H_2O$], and hausmannite (Mn_3O_4) are important examples. Among the carbonates, rhodochrosite ($MnCO_3$) is important, and among the silicates, rhodonite ($MnSiO_3$) and braunite [$(Mn,Si)_2O_3$] are important. The oxides, carbonates, and silicates of magnanese originated mostly as secondary minerals from weathering of manganese-containing primary minerals or as authigenic minerals from the precipitation of dissolved manganese. Minerals that contain manganese as a minor constituent include ferromagnesian minerals such as pyroxenes and amphiboles (Trost, 1958), or micas such as biotite (Lawton, 1955, p.59). They are all of igneous origin.

15.2 GEOCHEMICALLY IMPORTANT PROPERTIES OF MANGANESE

Manganese is one of the elements of the first transition series, which includes, in order of increasing atomic number from 21 to 29, the elements Sc, Ti, V, Cr, Mn, Fe, Co, Ni, and Cu. Electronically, these elements differ mostly in the degree to which their d orbitals are filled (Latimer and Hildebrand, 1940). Their increasing oxidation states are attributed to removal of 4s and 3d electrons (Sienko and Plane, 1966).

Manganese can exist in the oxidation states 0, +2, +3, +4, +6, and +7. However, in nature only the +2, +3, and +4 oxidation states are commonly found, Of the three naturally occurring oxidation states, only manganese in the +2 oxidation state can occur as a free ion in solution. It may also occur as a soluble inorganic or organic complex. Manganese in the +3 oxidation state can occur in solution only when it is complexed. The free +3 ion tends to disproportionate (dismutate) into the +2 and +4 oxidation states:

$$2Mn^{3+} + 2H_2O \Leftrightarrow Mn^{2+} + MnO_2 + 4H^+ \qquad (15.1)$$

The +4 oxides of manganese are insoluble in water. They have amphoteric properties (Latimer and Hildebrand, 1940) that account for their affinity for various cations, especially for heavy metals such as Co, Ni, and Cu. Mn(IV) oxides have long been known as scavengers of metallic cations (Geloso, 1927; Goldberg, 1954). They are frequently associated with ferric iron in nature.

Manganous ion is more stable than ferrous ion under similar conditions of pH and E_h. Based on equilibrium computations, manganese should exist predominantly as Mn^{2+} below pH 5.5 and 3.8×10^{-4} atm of CO_2, and as Mn(IV) above pH 5.5 if the E_h is approximately 800 mV and the Mn^{2+} activity is 0.1 ppm. At an E_h of 500 mV and below, Mn^{2+} at an activity as high as 10 ppm may dominate up to pH 7.8–8.0 (Hem, 1963).

Although in theory, 0.1 ppm of Mn^{2+} ions in aqueous solution should readily autoxidize when exposed to air at pH values above 4, they usually do not do so until the pH exceeds 8. Apart from Mn^{2+} concentration and E_h effects, one explanation for this resistance of Mn^{2+} ions to oxidation is the high energy of activation requirement of the reaction (Crerar and Barnes, 1974). Another explanation is that the Mn^{2+} may be extensively complexed and thereby stabilized by such inorganic ions as Cl^-, SO_4^{2-}, and HCO_3^- (Hem, 1963; Goldberg and Arrhenius, 1958), or by organic compounds such as amino acids, humic acids, and others (Graham, 1959; Hood, 1963; Hood and Slowey, 1964).

15.3 BIOLOGICAL IMPORTANCE OF MANGANESE

Manganese is an important trace element in biological systems. It is essential in microbial, plant, and animal nutrition. It is required as an activator by a number of enzymes such as isocitric dehydrogenase or malic enzyme, and may replace Mg^{2+} ion as an activator, for example in enolase (Mahler and Cordes, 1966). It is also required in oxygenic photosynthesis, where it functions in the production of oxygen from water by photosystem II (e.g., Klimov, 1984). In section 15.5, the ability of Mn(II) to serve as an energy source to some bacteria will be discussed, and in section 15.6, the ability of Mn(III) and Mn(IV) to serve as terminal electron acceptors in respiration by some bacteria will be examined. As in the case of iron, the most important geomicrobial interactions with manganese are those which lead to precipitation of dissolved manganese in an insoluble phase (immobilization) or solubilization of insoluble forms of manganese. These reactions frequently but not always involve oxidations or reductions of manganese.

15.4 MANGANESE-OXIDIZING AND -REDUCING BACTERIA AND FUNGI

Manganese-Oxidizing Bacteria and Fungi   Jackson (1901a,b) and Beijerinck (1913) were the first to describe the existence of manganese-oxidizing bacteria. Since their discovery, a number of other bacteria, many of which are taxonomically unrelated, have been reported to oxidize manganese. Among these, some promote the oxidation enzymatically, others nonenzymatically. In the case of still some other manganese-oxidizing bacteria, it is as yet unclear whether they oxidize manganese enzymatically or nonenzymatically. As a group, manganese-oxidizing bacteria include gram-positive and gram-negative forms. They are represented by sporeforming and nonsporeforming rods, cocci, vibrios, spirilla, and sheathed and appendaged bacteria. To date, all recognized manganese-oxidizing bacteria seem to be aerobic eubacteria. Among all the known manganese oxidizers, only two may be able to grow autotrophically on manganese (Ali and Stokes, 1971; Kepkay and Nealson, 1987). In most other instances, growth in the presence of manganese is either mixotrophic (manganese oxidation supplies some or all of the energy; the carbon source is organic) or heterotrophic (manganese oxidation furnishes no useful energy). Table 15.1 lists examples of bacteria that have been shown to oxidize manganese enzymatically or nonenzymatically. They have been detected in very diverse environments, such as "desert varnish" on rock surfaces, in soil, in the water column and in sediments of freshwater lakes and streams, in ocean water and sediments, and on and in ferromanganese concretions from the ocean floor.

Some mycelium-forming fungi have also been found to promote manganese oxidation, at least under laboratory conditions. In most instances this oxidation is probably nonenzymatic and due to interaction with a metabolic product (e.g., hydroxy acid) or a fungal cell component. However, in the case of some fungi, such as the white rot fungus *Phanerochaete chrysosporium*, oxidation may be the result of an **extracellular**, Mn(II)-dependent peroxidase which oxidizes Mn(II) to Mn(III) (see section 15.52 below) (Glenn and Gold, 1985; Glenn et al., 1986). This reaction is similar to one observed with peroxidase in plant extracts (horse radish, turnip) (Kenten and Mann, 1949, 1950). The Mn(III) is stabilized as a complex $[Mn^{3+}]$, e.g., lactate complex, which may then react with an organic compound (YH) such as a lignin component that reduces the Mn(III) back to Mn(II) and is itself oxidized (Fig. 15.1) (Glenn et al., 1986; Paszczynski et al, 1986). This rereduction of Mn(III) to Mn(II) makes manganese merely an electron shuttle in the peroxidase system and would not have a geochemical impact insofar as manganese redistribution is concerned. However, Kenten and Mann (1949) proposed that at very low H_2O_2 concentration, oxidized manganese may accumulate because rereduction of the oxidized manganese would be negligible. In the peroxidase M2 reaction in the absence of YH, Glenn et al. (1986) observed formation of a brown precipitate which was a manganese oxide (tentatively identified as MnO_2, but could have been Mn_2O_3 or Mn_3O_4) in a laboratory experiment in which H_2O_2 slowly diffused into a weakly buffered solution of enzyme and Mn(II).

Manganese-Reducing Bacteria and Fungi A number of different, taxonomically unrelated bacteria have been found to reduce manganese either enzymatically or non-enzymatically. The bacteria which reduce man-

Table 15.1 Some Bacteria that Oxidize Manganese

A. Attack of dissolved Mn^{2+} enzymatically
 1. Derive useful energy:
  Marine strains SSW_{22}, S_{13}, HCM-41, and E_{13} (all are gram-negative rods)
  (Ref. 1; unpublished results)
  Hyphomicrobium manganoxidans (Ref. 2)
  Pseudomonas strain S-36 (Ref. 3)
 2. Do not derive useful energy:
  Arthrobacter siderocapsulatus (Ref. 4)
  Leptothrix discophora (Ref. 17)
  Leptothrix pseudoochracea (Ref. 4)
  Metallogenium (Ref. 4)
  Strain FMn-1 (Ref. 5)
 3. Not known if able to derive useful energy:
  Arthrobacter B (Ref. 6)
  Arthrobacter citreus (Ref. 7)
  Arthrobacter globiformis (Ref. 7)
  Arthrobacter simplex (Ref. 7)
  Citrobacter freundii E_4 (Ref. 8)
  Hyphomicrobium T37 (Ref. 9)
  Pedomicrobium (Ref. 10)
  Pseudomonas E_1 (Ref. 8)
  Pseudomonas spp. (Ref. 11)
B. Attack of Mn^{2+} prebound to Mn(IV) oxide or some clays enzymatically
 1. Derive useful energy:
  Arthrobacter 37 (Ref. 12)
  Oceanospirillum (Ref. 13)
  Vibrio (Ref. 13)
  Marine strain CFP-11 (Ref. 1)
C. Attack Mn^{2+} nonenzymatically
 Pseudomonas manganoxidans (Ref. 14)
 Streptomyces sp. (Ref. 15)
 Bacillus (Ref. 16; but see also Ref. 18)

References: (1) Ehrlich, 1983, 1985; (2) Eleftheriadis, 1976; (3) Kepkay and Nealson, 1987; (4) Dubinina 1978a,b; (5) Zindulis and Ehrlich, 1983; (6) Bromfield, 1956; Bromfield and David, 1976; (7) Dubinina and Zhdanova, 1975; (8) Douka, 1977; (9) Tyler and Marshall, 1967b; (10) Aristovskaya, 1961; Ghiorse and Hirsch, 1979; (11) Zavarzin, 1962; (12) Ehrlich, 1969; Arcuri, 1978; (13) Ehrlich, 1976, 1978a; (14) Jung and Schweisfurth, 1979; (15) Bromfield, 1978, 1979; (16) De Vrind et al., 1986b; (17) Adams and Ghiorse, 1987; (18) Van Wassbergen et al., 1993.

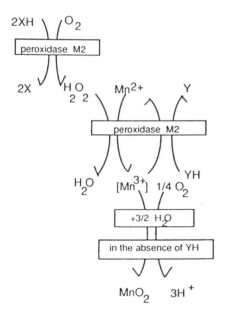

Figure 15.1 Interaction of extracellular peroxidase M2 from *Phanerochaete chrysosporium* with manganese. (Based on description by Glenn et al., 1986, Pacsczynski et al., 1986.)

Manganese-Reducing Bacteria and Fungi A number of different, taxonomically unrelated bacteria have been found to reduce manganese either enzymatically or nonenzymatically. The bacteria which reduce manganese enzymatically may do so as a form of respiration in which the oxidized manganese serves as terminal electron acceptor and is reduced to Mn(II). Some bacteria can reduce the oxidized manganese aerobically or anaerobically, whereas others can reduce it only anaerobically. Thus, manganese-reducing bacteria include aerobes, and strict and facultative anaerobes. In some cases, manganese may be reduced to satisfy a nutritional need for soluble Mn(II) (see De Vrind et al., 1986; also discussion by Ehrlich, 1987) or to scavenge excess reducing power, as in some cases of NO_3^- reduction (Robertson et al., 1988) and ferric iron reduction (Ghiorse, 1988; Lovley, 1991).

Bacteria that reduce manganese oxides include gram-positive and gram-negative forms, rods, cocci and vibrios and appear to be eubacteria. The majority of the bacteria studied to date that can respire with manganese oxide as terminal electron acceptor seem to require reduced carbon as electron donor (reductant), but a few exceptional types appear to be able to use hydrogen. As in the case of the manganese oxidizers (section

15.1), they have been found in very diverse environments, including soil, freshwater, and marine habitats. Most of the manganese-reducing bacteria described to date do not seem to have the ability to oxidize Mn(II) as well, but a few exceptions have been reported. Representative examples of Mn reducers are listed in Table 15.2.

Some mycelial fungi have been found to reduce manganese oxides, at least under laboratory conditions. As in the case of Mn(II) oxidation by fungi, reduction of manganese oxides by them must be largely nonenzymatic in most if not all cases, although experimental proof for any specific mechanism is generally lacking. Nonenzymatic reduction of manganese oxides by fungi is best explained in terms of production of metabolic products by them that act as reductants.

Table 15.2   Some Mn(IV)-Reducing Bacteria

A.   Gram-positive bacteria
    Bacillus 29 (Ref. 1)
    Coccus 32 (Ref. 1)
    Bacillus SG-1 (Ref. 2)
    Bacillus circulans (Ref. 3)
    Bacillus polymyxa (Ref. 3)
    Bacillus mesentericus (Ref. 3)
    Bacillus mycoides (Ref. 3)
    Bacillus cereus (Ref. 3)
    Bacillus centrosporus (Ref. 3)
    Bacillus filaris (Ref. 3)
    Arthrobacter strain B (Ref. 4)
    Bacillus GJ33 (Ref. 5)
B.   Gram-negative bacteria
    Geobacter metallireducens (formerly strain GS-15) (Ref. 6)
    Geobacter sulfurreducens (Ref. 7)
    Shewanella (formerly *Alteromonas*) *putrefaciens* (Ref. 8)
    Pseudomonas liquefaciens (Ref. 3)
    Strain BIII 32 (Ref. 9)
    Strain BIII 41 (Ref. 9)
    Strain BIII 88 (Ref. 11)
    Acinetobacter calcoaceticus (Ref. 10)

References: (1) Trimble and Ehrlich, 1968; (2) De Vrind et al., 1986a; (3) Troshanov, 1968, 1969; (4) Bromfield and David, 1976; (5) Ehrlich et al., 1973; (6) Lovley and Phillips, 1988; (7) Caccavo, et al., 1994; (8) Myers and Nealson, 1988; (9) Ehrlich, 1980b; (10) Karavaiko et al., 1986; (11) Ehrlich, 1993a.

15.5 BIOOXIDATION OF MANGANESE

Like iron oxidation, manganese oxidation by microbes may be enzymatic or nonenzymatic. However, unlike enzymatic iron oxidation, it has not been reported to occur at very acid pH. Enzymatic iron oxidation in air-saturated solutions or solutions close to air saturation is favored by very acid pH because at that pH iron does not autoxidize at a significant rate. Enzymatic manganese oxidation in solutions at similar air saturation levels is not favored at pH values much lower than neutrality because, in contrast to iron, the free energy change when manganese is oxidized with oxygen as electron acceptor decreases steadily until it assumes a positive value near pH 1.0 (see free energy values at pH 0 and 7 listed by Ehrlich, 1978). One instance of nonenzymatic manganese biooxidation at a pH of around 5.0 has been documented (Bromfield, 1978, 1979).

Among the first to report on bacterial manganese oxidation were Jackson (1901a,b) and Beijerinck (1913). The latter suggested that it may be associated with autotrophic growth. Lieske (1919) and Sartory and Meyer (1947) suggested that manganese oxidation may also be associated with mixotrophic growth. Either way, an enzymatic process was implied. The first experiments to demonstrate Mn^{2+} oxidation by resting cells were performed by Bromfield (1956). He showed that stationary phase cells of *Arthrobacter* strain B (formerly called *Corynebacterium* strain B) cells oxidized Mn^{2+} in an 0.005% $MnSO_4 \cdot 4H_2O$ solution at 40° but not above 45°C. The oxidation was inhibited by copper and mercury salts, and by azide and cyanide. In a later study (Bromfield, 1974), he showed that organic substrate concentration and composition affected manganese oxidation by *Arthrobacter* strain B. Subsequently, Bromfield and David (1976) found that cells of *Arthrobacter* strain B rapidly adsorbed Mn^{2+} ions from solution besides oxidizing them. The adsorbed but unoxidized Mn^{2+} could be displaced by addition of 5 mM Cu^{2+}. Oxidation kinetics of this organisms are shown in Figure 15.2. Bromfield and David (1976) also found that this organism could reduce oxidized manganese. An impression left by most studies to date is that **enzymatic** manganese oxidation is restricted mainly to bacteria. An exceptional case of enzymatic manganese oxidation by a fungus is discussed in section 15.4.

Enzymatic Manganese Oxidation   When considering all reports of enzymatic manganese(II) oxidation by bacteria, it becomes evident that not all bacteria use the same mechanism of oxidation. The mechanisms, insofar as they have been studied to date, can be divided into three major groups, at least one of which can be further divided into several subgroups.

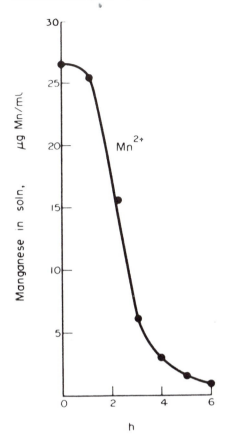

Figure 15.2 Oxidation of 0.5 mM Mn^{2+} by bacterial cell suspensions of *Arthrobacter* strain B (formerly *Corynebacterium* strain B) at pH 7.0. The mixture (80 ml) contained 4 mg cells per milliliter and 27 μg of Mn per milliliter. (Reprinted with permission from Soil Biol. Biochem., 8:37–43, Bromfield, S. M., and D. J. David, Sorption and oxidation of manganous ions and reduction of manganese oxide by cell suspensions of a manganese oxidizing bacterium, Copyright 1976, Pergamon Press plc, by permission.)

Group I Manganese Oxidizers   This group includes those bacteria that oxidize *dissolved* manganese species, e.g., Mn^{2+}, using oxygen as terminal electron acceptor. Some bacteria can gain useful energy from this reaction (Subgroup Ia), while others do not (Subgroup Ib). The overall oxidation can be summarized by the following reaction:

$$Mn^{2+} + \tfrac{1}{2}O_2 + H_2O \Rightarrow MnO_2 + 2H^+ \tag{15.2}$$

A number of bacteria in Group I whose manganese-oxidizing system has been examined so far appear to oxidize manganese in their cell envelope external to the plasma membrane, i.e., in the periplasm when gram-negative bacteria are involved or the wall region [periplasmic equivalent (Hobot, 1984)] when gram-positive bacteria are involved (Ehrlich, 1983; Ehrlich and Zapkin, 1985). In **Subgroup Ia** bacteria, this oxidation can be coupled to ATP synthesis, as has been demonstrated directly with everted membrane vesicles from a marine bacterium, strain SSW_{22} (Ehrlich, 1983; Ehrlich and Salerno, 1990) or by uncoupling of ATP synthesis in this organism with 2,4-dinitrophenol (Ehrlich and Salerno, 1988, 1990). Synthesis of ATP in strain SSW_{22} appears to be very tightly coupled to manganese oxidation since ADP stimulates manganese oxidation by everted membrane vesicles from it (Ehrlich and Salerno, 1990). A periplasmic factor is required for Mn^{2+} oxidation by membranes from strain SSW_{22} (Clark and Ehrlich, 1988; Clark, 1991), but the exact function of this factor (electron transfer or chelation of Mn^{2+}) in the system has yet to be established. A diagram illustrating the current working model to explain coupling of ATP synthesis to manganese oxidation in Group Ia bacteria is shown in Figure 15.3. The model assumes that chemiosmosis is the underlying energy transducing mechanism, namely that the manganese-oxidizing half-reaction occurs external to the plasma membrane in the periplasm or its equivalent. It also assumes that oxygen reduction occurs on the inner surface of the plasma membrane. Proton generation by the manganese oxidizing half-reaction and proton pumping linked to the passage of electrons from manganese to oxygen via the electron transport system in the membrane result in a proton gradient across the membrane (outside acid relative to the cytoplasm). This gradient contributes to the proton motive force needed for ATP synthesis via ATPase. ADP stimulation of Mn^{2+} oxidation by everted membrane vesicles from strain SSW_{22} is explained by a mass action effect involving vectorial and scalar proton consumption in ATP synthesis. The everted membrane vesicles appeared to become freely permeable to Mn^{2+} in their preparation.

A *Pseudomonas* strain S-36 (Nealson, 1978) should also be assigned to Subgroup Ia bacteria. This organism binds and oxidizes Mn^{2+} (Kepkay et al., 1984), appears to derive energy from the oxidation of manganese, and may use some of this energy to fix CO_2 (Kepkay and Nealson, 1987). Nealson et al. (1988) did not rule out, however, the possibility, even though remote, that manganese stimulated carbon utilization in this organism by a mechanism other than oxidation.

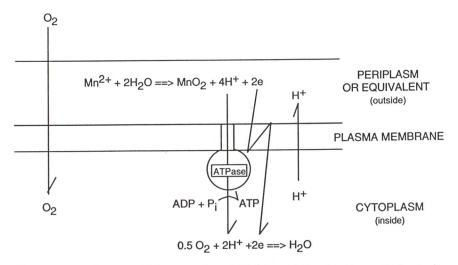

Figure 15.3 Proposed bioenergetics of Mn^{2+} oxidation by Group Ia bacteria. Note that manganese oxidation occurs in the periplasm in this scheme, whereas oxygen reduction takes place at the inner surface of the plasma membrane. Protons pumped into the periplasm from the cytoplasm by the electron transport system in the plasma membrane (details not shown) together with protons from the oxidative half-reaction of Mn^{2+} contribute to the proton gradient across the membrane and therefore to ATP synthesis.

In **Subgroup Ib** bacteria [only one example is known so far (Zindulis and Ehrlich, 1983)], manganese oxidation is not coupled to ATP synthesis. Manganese oxidation appears to occur external to the plasma membrane in the periplasm. Mn^{2+} is oxidized in this instance by an as-yet-unidentified periplasmic factor, which behaves like an oxidant and not like an enzyme. This factor, which is reduced by its interaction with manganese, is enzymatically reoxidized on the inside of the cell by oxygen without essential involvement of the electron transport system in the plasma membrane (Zindulis and Ehrlich, 1983). Manganese oxidation by Group 1b bacteria thus does not generate useful energy but may be a mechanism of detoxification aimed at Mn^{2+} in the environment. A diagram representing our current understanding of manganese oxidation of Group 1b bacteria is shown in Figure 15.4.

Sheathed bacteria, in particular *Leptothrix* spp., should be assigned to **Subgroup Ic** of Group I manganese oxidizers. Early work with *L. disco-*

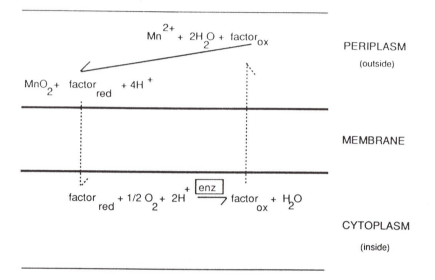

Figure 15.4   Current model for manganese oxidation by a Group Ib bacterium. The electron transport system of the plasma membrane is not directly involved. Manganese oxidation is the result of interaction with an as-yet-unidentified factor, which is reduced. The reduced factor is reoxidized in the cytoplasm. Whether the protons from the manganese oxidation in the periplasm, if formed as shown, contribute to ATP synthesis is not known.

phora led to contradictory findings. Resting cell suspensions of *Leptothrix discophora* (also called *Sphaerotilus discophora*) were shown to oxidize Mn^{2+} if the culture had been grown in a medium supplemented with manganese but not without manganese supplementation (Johnson and Stokes, 1966). A subsequent study with different strains of *L. discophora* (Van Veen, 1972), however, failed to confirm inducibility of manganese-oxidizing activity in the organism. Strain difference might have been the cause of this difference in results. In another strain, manganese-oxidizing activity in *L. discophora* has been associated with a particulate cell fraction that sediments at $48,000 \times g$. It appeared to be coupled to cytochrome oxidase because it was inhibited by 10^{-5} M cyanide and 10^{-4} M azide, but coupling to b- and c-type cytochromes was not observed (Hogan, 1973; Mills, 1979). *L. discophora* has been reported to be able to grow autotrophically and mixotrophically on Mn(II) as an energy source (Ali and Stokes, 1971), but these observations were not confirmed (Hajj and Makemson, 1976).

Recent studies with a freshly isolated strain of *L. discophora* (SS-1; ATCC 43182) have called the earlier findings into question. Strain SS-1, which has lost the ability to form a sheath but which continues to be able to oxidize Mn^{2+} (Adams and Ghiorse, 1986), releases a Mn^{2+}-oxidizing protein, having an apparent molecular weight of 100,000–110,000, into liquid culture medium. A polysaccharide moiety appears to be conjugated with the protein. The manganese-oxidizing activity of the protein exhibits a pH optimum of around 7.5 and a temperature optimum of 25°C and is inhibited by cyanide, *o*-phenanthroline, and $HgCl_2$ but not by azide. The oxidized manganese becomes associated with the protein and can be removed by reduction with ascorbate, thereby restoring the manganese-oxidizing activity of the protein (Adams and Ghiorse, 1987; Boogerd and De Vrind, 1987). These observation are consistent with a claim by Mulder (1972) and Van Veen (1972) that *Leptothrix* spp. excrete proteins that promote manganese oxidation. The protein acts as a catalyst in association with acidic exopolymer in oxidizing Mn(II) to Mn(III) (Adams and Ghiorse, 1988). The production of Mn in the $+3$ oxidation state suggests that the oxidation is probably not a source of energy for the organism because it yields barely enough energy (~ 7 kcal mol^{-1}) at pH 7 under standard conditions when contrasted with oxidation of Mn(II) to Mn(IV), which yields ~ 16 kcal mol^{-1} at pH 7 under standard conditions. Excess concentrations of Mn^{2+} inhibit the growth of *L. discophora* SS-1, probably because the manganese oxide (MnO_2) that it deposits on the cells in the absence of a sheath deprives it of essential iron by binding it (Adams and Ghiorse, 1985).

Based on the manganese-oxidizing activity of its mature, dormant spores, marine *Bacillus* SG-1 should also be assigned to a subgroup of Group I manganese oxidizers. Its vegetative cells do not oxidize Mn(II) (Nealson and Ford, 1980; Rosson and Nealson, 1982; De Vrind et al., 1986). Studies to date suggest an assignment to **Subgroup 1c**. Spores of *Bacillus* SG-1 bind and oxidize Mn^{2+} (Rosson and Nealson, 1982). The spores, whose formation by the vegetative cells is enhanced by solid surfaces in dilute seawater medium (Kepkay and Nealson, 1982), bind Mn^{2+} to their surface, and a protein component of the spore coat oxidizes it. The protein has a MW of 205,000 and may be a glycoprotein (Tebo et al, 1988). Two regions, given the label Mnx, on the chromosome of *Bacillus* SG-1 code for factors required for manganese oxidation by the spores (Van Waasbergen et al., 1993). The initial product of manganese oxidation by the spores was first thought to be Mn_3O_4 (hausmannite)(Hastings and Emerson, 1986), but was later identified as Mn(IV) oxide (10Å manganate) (Mandernak, 1992). A strain of *Bacillus megaterium* (strain BC1) has been isolated more recently from a microbial mat at Laguna Figueroa, Baja

California, Mexico, whose spores oxidize Mn^{2+}, presumably like those of *Bacillus* SG-1 (Gong-Collins, 1986).

Whereas vegetative cells of *Bacillus* SG-1 cannot oxidize Mn^{2+}, De Vrind et al. (1986) discovered that they can reduce MnO_2 at low oxygen tension. The electron donor in these organisms is an unidentified intracellular compound; externally supplied glucose or succinate did not act as electron donor. MnO_2 reduction involved a branched, membrane-bound, electron transport pathway in which oxygen at normal tension competed with MnO_2 as terminal acceptor. The MnO_2-reducing activity in this organism is thought to supply needed Mn^{2+} for sporulation in environments where the supply of Mn^{2+} is limited or absent, but where manganese oxide is available (De Vrind et al., 1986).

The oxidation of Mn(II) by spores but not vegetative cells of members of the genus *Bacillus* is by no means a universal phenomenon. Ehrlich and Zapkin (1985) and Vojak et al. (1984) isolated bacilli whose oxidation of Mn(II) depended on vegetative cells, in the latter case cells that were in the process of sporulating. These organisms should be assigned to Subgroup Ia based on knowledge to date.

Two strains of bacteria, *Pseudomonas* strain E_1 and *Citrobacter freundii* strain E_4, isolated from manganese concretions found in an alfisol soil in West Peleponnesus, Greece, must be assigned to Group I of enzymatic manganese oxidizers because they attack unbound Mn^{2+}. However, based on information to date about their Mn oxidase, they do not seem to belong to Subgroup Ia, Ib, or Ic, but should be assigned to **Subgroup Id** (Douka, 1977, 1980; Douka and Vaziourakis, 1982). Oxidation was demonstrated with intact cells and cell extract. The activity was heat-sensitive and inhibited by $HgCl_2$. Cell extracts exhibited a temperature optimum at 34°C. Whether these bacteria can derive energy from manganese oxidation remains to be determined.

Group II Manganese Oxidizers   This group includes those bacteria that oxidize Mn^{2+} only when it is prebound external to the cell to one of several inorganic solids. Like Group I bacteria, they also use oxygen as terminal electron acceptor. At least some bacteria using this mechanism can gain useful energy from the oxidation. The oxidative reaction catalyzed by these bacteria when Mn^{2+} is bound to a hydrated Mn(IV) oxide like $MnO_2 \cdot H_2O$ (H_2MnO_3) can be summarized as follows:

$$MnMnO_3 + \tfrac{1}{2}O_2 + 2H_2O \Rightarrow 2H_2MnO_3 \qquad (15.3)$$

The $MnMnO_3$ results from a reaction in which Mn^{2+} displaces the protons of H_2MnO_3:

$$Mn^{2+} + H_2MnO_3 \Rightarrow MnMnO_3 + 2H^+ \qquad (15.4)$$

Those Group II organisms in which manganese oxidation has been examined in some detail oxidize Mn(II) in their periplasm (gram-negative eubacteria) or their periplasmic equivalent (gram-positive eubacteria) like Group 1, Subgroup 1a organisms. They differ from Subgroup 1a organisms by attacking only Mn^{2+} that is prebound to a solid substrate like manganese(IV) oxide, ferromanganese, or clays like montmorillonite or kaolinite but not illite (Arcuri, 1978; Arcuri and Ehrlich, 1979; Ehrlich, 1982; Ehrlich, 1984). In the case of oxidation of Mn^{2+} bound to clays, pretreatment of the clays overnight with ferric chloride in the absence of added Mn^{2+} was essential for whole cells to oxidize manganese subsequently bound to the clay (Ehrlich, 1982). Such pretreatment was not necessary when cell extracts were used, but the amount of manganese oxidized in 4 h was less without than with ferric chloride pretreatment. The iron was not a factor required for bacterial activity because when ferric chloride and manganese were added simultaneously to the reaction mixture, bacteria did not oxidize the bound manganese. The iron seems to play a steric role in making bound Mn^{2+} more accessible to oxidation by the cells or cell extract. So far it is not clear how the Mn^{2+} bound to an appropriate inorganic phase outside the cell is transported into its periplasm or its equivalent. At least one periplasmic factor has been detected that is essential to manganese oxidation by Group II organisms (Arcuri and Ehrlich, 1979, 1980), but whether it is involved in transferring Mn^{2+} into the periplasm or whether it is involved in electron transfer to the electron transport system in the membrane remains to be established. Like members of Group Ia, Group II manganese-oxidizing bacteria can couple ATP synthesis to Mn(II) oxidation (Ehrlich, 1976) using a respiratory chain in their plasma membrane (Arcuri and Ehrlich, 1979). A diagram of a working model for manganese oxidation of Group II bacteria is shown in Figure 15.5.

Group III Manganese Oxidizers   This group includes those bacteria that oxidize dissolved manganese(II) with H_2O_2 using catalase as the enzyme. The reaction catalyzed by these bacteria can be summarized as follows:

$$Mn^{2+} + H_2O_2 \Rightarrow MnO_2 + 2H^+ \qquad (15.5)$$

Group III bacteria so far include *Metallogenium*, *Leptothrix pseudoochracea*, and *Arthrobacter siderocapsulatus* (Dubinina, 1978a). These organisms appear to generate H_2O_2 faster in their metabolism than their catalase can destroy it by the reaction

$$2H_2O_2 \Rightarrow 2H_2O + O_2 \qquad (15.6)$$

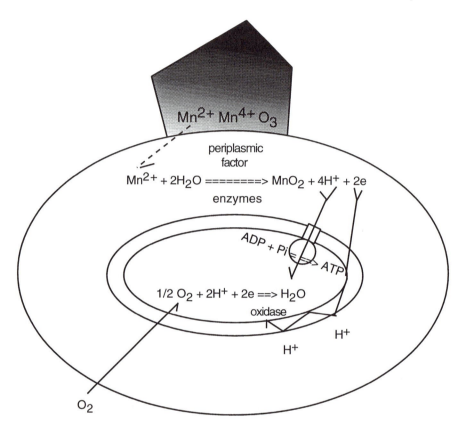

Figure 15.5   Proposed bioenergetics of $MnMnO_3$ oxidation by Group II bacteria. Note that this scheme is similar to that for Group Ia bacteria, except that the Mn^{2+} derives from bound Mn(II).

In the presence of Mn^{2+} at neutral pH, their catalase can function as a peroxidase and use H_2O_2 as an oxidant to generate MnO_2 according to Reaction (15.4) (Dubinina, 1978b). The reaction can be reproduced in the laboratory by generating H_2O_2 from glucose oxidation by glucose oxidase and then causing the H_2O_2 to oxidize Mn^{2+} with a commercial preparation of catalase (Dubinina, 1978a). The oxidation product was identified as a Mn(IV) product, but since it was complexed by pyrophosphate it is more likely that it was Mn(III). Fe^{2+} can replace Mn^{2+} in this reaction (Dubinina, 1978b). At acid pH, H_2O_2 can reduce MnO_2 to Mn^{2+} chemically according to the reaction

$$MnO_2 + H_2O_2 + 2H^+ \Rightarrow Mn^{2+} + 2H_2O + O_2 \qquad (15.7)$$

This reaction proceeds without catalase.

Nonenzymatic Manganese Oxidation   Autoxidation of Mn^{2+} is promoted when environmental conditions feature an E_h greater than $+500$ mV, a pH greater than 8, and a Mn^{2+} concentration greater than 0.01 ppm. These pH-E_h limits are much narrower than those required for autoxidation of iron (see Chapter 14). Nonenzymatic manganese oxidation may also be promoted through production of one or more metabolic end products which cause chemical oxidation of Mn^{2+}. According to Soehngen (1914), a large number of microorganisms can cause such manganese oxidation when in the presence of hydroxycarboxylic acids such as citrate, lactate, malate, gluconate, or tartrate. He indicated that metabolic utilization of a part of such acids causes a rise in pH of the culture medium (if unbuffered), and that when the pH turns alkaline, residual hydroxycarboxylic acid catalyzes the oxidation of Mn^{2+}. Such hydroxycarboxylic acids can be formed by both bacteria and fungi. In apparent agreement, Van Veen (1973) found that with *Arthrobacter* 216, hydroxycarboxylic acids are required for Mn^{2+} oxidation. However, he felt that other microorganisms with a more specific manganese-oxidizing capacity play a more important role in manganese oxidation in soil.

Another example of nonenzymatic manganese oxidation is furnished by an actinomycete, *Streptomyces* sp., found in Australian soil. In the laboratory it causes Mn^{2+} to be oxidized in a pH range of 5–6.5 when growing on soil agar (Bromfield, 1978, 1979). The actinomycete produces a water-soluble, extracellular compound which is responsible for the oxidation. The oxidation product is thought to protect the organism from inhibition by Mn^{2+} ions (Bromfield, 1978).

Pseudomonas MnB1 has been reported to form a Mn^{2+}-oxidizing protein which oxidizes Mn(II) intracellularly and is consumed in the process. The protein is therefore not an enzyme. The oxidation does not require oxygen. It proceeds optimally at pH 7.0. The formation of the protein depends on cessation of growth by the culture at the end of its exponential phase but does not require added Mn^{2+} in the medium to stimulate the synthesis of the protein (Jung and Schweisfurth, 1979).

Zygospores of the alga *Chlamydomonas* from soil have been reported to become encrusted with Mn(IV) oxide, presumably through oxidation of Mn^{2+} (Schulz-Baldes and Lewin, 1975), but how the manganese came to be oxidized was not explained. It could have been a case of passive accumulation of manganese oxide (see section 15.7).

15.6 BIOREDUCTION OF MANGANESE

Microbial reduction of oxidized manganese may be enzymatic or non-enzymatic, as in the case of microbial oxidation of manganese. The occurrence of microbial reduction of manganese oxide was suggested as far back as 1894 by Adeny who found that manganese dioxide, formed when he added potassium permanganate to sewage, was reduced to manganous carbonate. He attributed this reduction to the action of bacteria and thought it analogous to bacterial nitrate reduction. Mann and Quastel (1946) reported the microbial reduction of biogenically formed manganese oxide in a soil perfusion column by feeding it with glucose solution. They were able to inhibit this reaction by feeding the column with azide. They suggested that the oxides of manganese acted as hydrogen acceptors. Troshanov in 1968 isolated a number of Mn(IV)-reducing bacteria from reduced horizons of several Karelian lakes (U.S.S.R.). His isolates included *Bacillus circulans*, *B. polymyxa*, *Pseudomonas liquefaciens*, *B. mesentericus*, *B. mycoides*, *Micrococcus* sp., *B. cereus*, *B. centrosporus*, and *B. filaris* among others. Some of these strains could reduce both manganese and iron oxides, but strains that could reduce only manganese oxides were encountered most frequently. Nitrate did not inhibit Mn(IV) reduction the way it did Fe(III) reduction by *B. circulans*, *B. polymyxa*, *P. liquefaciens*, and *B. mesentericus* (Troshanov, 1969). Anaerobic conditions stimulated Mn(IV) reduction by *B. centrosporus*, *B. mycoides*, B. filaris, and *B. polymyxa* but not by *B. circulans*, *B. mesentericus* or *Micrococcus albus*. All carbohydrates tested were not equally good sources of reducing power for a given organism (Troshanov, 1969). Although a significant part of the Mn(IV) seemed to be reduced enzymatically, some of it appeared also to be reduced chemically since some reduction of oxidized manganese in lake ores by glucose and xylose and some organic acids (e.g., acetic and butyric acids) was observed at pH of 4.3–4.4 under sterile conditions (Troshanov, 1968). Such reaction of glucose and probably other organic reducing agents with Mn(IV) oxides is, however, dependent on the form of the oxide; hydrous Mn(IV) oxides react readily, whereas crystalline MnO_2 does not (see, for instance, Ehrlich, 1984; Nealson and Saffarini, 1994).

Bromfield and David (1976) reported that *Arthrobacter* strain B, which can oxidize manganese aerobically, is able to reduce manganese oxides at lowered oxygen tension. Ehrlich (1988) reported that several gram-negative marine bacteria that can oxidize Mn(II) and thiosulfate aerobically and reduce tetrathionate anaerobically (Tuttle and Ehrlich, 1986) can also reduce MnO_2 aerobically and anaerobically.

Clearly a variety of microbes have the capacity to reduce oxidized manganese. Some bacteria can reduce Mn(IV) oxide aerobically or anaerobically (e.g., Ehrlich, 1966; Di-Ruggiero and Gounot, 1990; Troshanov, 1968). Others reduce it only anaerobically (e.g., Burdige and Nealson, 1985; Lovley and Phillips, 1988a; Myers and Nealson, 1988). Among these latter bacteria, some, like *Shewanella* (*Alteromonas, Pseudomonas*) *putrefaciens*, are facultative but nevertheless reduce Mn(IV) oxide only anaerobically (Myers and Nealson, 1988), while others, like *Geobacter metallireducens* (formerly strain GS-15), are strict anaerobes (Lovley and Phillips, 1988a).

Aerobic and Anaerobic Enzymatic Reduction of Manganese Oxides   As in the case of enzymatic oxidation of Mn(II) by bacteria, enzymatic reduction of oxidized manganese by bacteria does not follow the same electron transport pathway in all organisms. Growing cultures of *Bacillus* 29, of marine origin, are able to reduce MnO_2 aerobically and anaerobically using glucose as electron donor. However, a component of the electron transport system involved in MnO_2 reduction in this culture is only aerobically inducible. Both Mn^{2+} and MnO_2 can serve as inducers. Because of the aerobic requirement for induction, uninduced *Bacillus* 29 cannot reduce MnO_2 anaerobically unless an artificial electron carrier such as ferricyanide is added (Ehrlich, 1966; Trimble and Ehrlich, 1968, 1970). Once initiated by a growing culture of *Bacillus* 29, MnO_2 reduction is quickly inhibited by addition of $HgCl_2$ to a final concentration of 10^{-3} M. During MnO_2 reduction with glucose as electron donor, the organism forms more lactate and pyruvate from glucose than in the absence of MnO_2. It also consumes more glucose in the presence of MnO_2 than in its absence (Table 15.3). Calculations from the results on day 11 in this table show that *Bacillus* 29 had formed 7.9 nmol pyruvate with MnO_2 but only 3.1 nmol without it per 100 nmol glucose consumed. Likewise, it had formed 20.4 nmol lactate with MnO_2 and 24.2 nmol without it per 100 nmol glucose consumed. Glucose consumption and total product formation qualitatively similar to those with growing cultures of *Bacillus* 29 occurred with marine coccus 32 (Trimble and Ehrlich, 1968).

Aerobic MnO_2 reduction has also been studied with resting-cell cultures and with cell extracts. In the case of *Bacillus 29*, uninduced resting cells were able to reduce MnO_2 only in the presence of dilute ferri- or ferrocyanide, whereas induced resting cells were able to reduce MnO_2 without added ferri- or ferrocyanide. The ferri- or ferrocyanide substituted for a missing component in the electron transport system to MnO_2 in the uninduced cells (Ehrlich 1966; Ehrlich and Trimble 1970). Substrate level concentrations ferrocyanide can serve as electron donor in place of glucose in the reduction of MnO_2 by *Bacillus* 29 (Ghiorse and Ehrlich, 1976). Mann and Quastel (1946) had previously observed that pyocyanin could

Table 15.3 Chemical and Physical Changes During Growth[a] of *Bacillus* 29 and

Culture	Time (days)	Glucose consumed (nmol/ml)		Manganese released (nmol/ml)	Pyruvate produced (nmol/ml)	
		+ MnO$_2$	− MnO$_2$	+ MnO$_2$	+ MnO$_2$	− MnO$_2$
29	0	0	0	0	0	0
	1	1,720	655	25	43	32
	3	3,060	1,239	240	190	153
	4	3,686	2,342	420	438	330
	7	6,120	3,635	1,060	705	243
	8	8,750	5,286	1,260	879	330
	11	11,893	7,281	1,830	936	225
32	0	0	0	0	0	0
	1	480	0	0	50	52
	2	1,440	1,440	0	84	75
	3			130	138	79
	7	3,000	2,650	490	1,025	575
	9	4,900	3,125	780	1,650	980
	11	5,550	3,600	1,020	2,625	1,120

[a] Culture medium: For *Bacillus* 29, 0.48% glucose and 0.048% peptone in seawater; for coccus 32, 0.60% glucose and 0.048% peptone in seawater. One-gram portions of reagent grade MnO$_2$ were added to 20 ml of medium, as needed. Incubation was at 25°C.
Source: From Trimble and Ehrlich (1968), by permission.

serve as electron carrier in MnO$_2$ reduction by bacteria. But they believed that such a carrier was essential for such reduction to occur.

Induction of the missing electron transport component involves protein synthesis, since inhibitors of protein synthesis prevent its induction by manganese (Trimble and Ehrlich, 1970). Reduction of MnO$_2$ by cell extracts from *Bacillus* 29 proceeds in the same way as with resting cells (Ghiorse and Ehrlich, 1976). Extracts from cells previously grown without added MnSO$_4$ do not reduce MnO$_2$ in the absence of added small amounts of ferri- or ferrocyanide, whereas extracts from cells previously grown in the presence of added MnSO$_4$ do. A spectrophotometric study of the electron transport components in plasma membrane of *Bacillus* 29 and inhibition tests of MnO$_2$ reduction by cell extracts of the organism initially led to the inference that electron transfer from electron donor to MnO$_2$ in induced cells involved a branched pathway consisting of flavoproteins, c-type cytochrome, and two metallo-proteins, one an MnO$_2$ reductase and the other a conventional oxidase (Fig. 15.6a). The existence of two MnO$_2$-reducing metalloproteins was initially postulated on the basis of stimulation of MnO$_2$-reducing activity by sodium azide at 1.0 to 10 mM concentra-

Coccus 32 in the Presence and Absence of MnO₂

Lactate produced (nmol/ml)		E_h (mv)		pH		Cells/ml ($\times 10^7$)	
$+MnO_2$	$-MnO_2$	$+MnO_2$	$-MnO_2$	$+MnO_2$	$-MnO_2$	$+MnO_2$	$-MnO_2$
0	0	469	474	7.3	7.2	1.6	1.6
461	449	528	482	6.6	6.7	3.8	4.0
1,635	1,612	502	512	5.4	5.6	4.5	4.8
1,430	1,955	573	514	5.7	5.5	4.5	5.0
2,080	1,387	615	594	5.7	5.5	4.6	5.2
2,040	1,705	617	589	5.5	4.9	4.8	5.0
2,430	1,761	602	589	5.8	5.1	5.0	5.0
0	0	426	426	7.7	7.7	2.5	2.5
0	0	394	419	7.0	6.9	4.6	5.8
9	5	519	490	6.8	6.5	6.2	7.0
		491	496	6.6	6.4	6.2	7.0
174	68	484	495	5.7	4.7	6.0	4.8
250	124	514	549	5.4	4.5	5.8	4.2
250	123	475	509	5.6	4.3	5.4	4.8

tions. The oxidase was presumed to be more azide sensitive than the MnO₂ reductase (Ghiorse and Ehrlich, 1976). Subsequent observations, however, have led to the conclusion that the azide stimulation in *Bacillus* 29 was the result of inhibition of catalase activity with a resultant reduction of some MnO₂ by accumulating H₂O₂ (see reaction 15.7) (Fig. 15.6b) (Ghiorse, 1988). This reduction of MnO₂ by metabolically produced H₂O₂ is similar to that previously reported by Dubinina (1979a,b) with *Leptothrix pseudoochracea*, *Arthrobacter siderocapsulatus*, and *Metallogenium*.

The organization of the MnO₂-reducing system in *Bacillus 29* is not universal in manganese-reducing bacteria. In other marine MnO₂-reducing bacteria so far examined, electron transport appears to involve more conventional electron transport carriers and enzymes, as suggested by electron transport inhibitor studies. These inhibitor studies, moreover, indicated that the pathway appears not to be exactly the same in each of the organisms tested (Ehrlich, 1980, 1993b). H₂O₂ appears to play no significant role in MnO₂ reduction by these organisms because 1 mM azide did not stimulate MnO₂ reduction. Ghiorse (1988) has suggested that in *Bacillus 29*, MnO₂ reduction may serve as a means of disposing of excess reducing power without energy coupling. In strains involving more conventional electron transport components for the reduction of MnO₂, such reduction in air may be a supplemental source of energy.

How is it that MnO_2 can partially replace oxygen as terminal electron acceptor for some aerobic bacteria? *Bacillus* 29 was found to consume less oxygen in the presence of MnO_2 than in its absence when respiring on glucose (Trimble, 1967; Ehrlich, 1981). MnO_2 reduction by *Bacillus GJ33* was accelerated in rotary-shake culture in air at 200 rpm but depressed at 300 rpm (Ehrlich, 1988). This can be explained thermodynamically (Ehrlich, 1987). Considering that the free energy change at pH 7.0 for the half-reaction

$$MnO_2 + 4H^+ + 2e^- \Rightarrow Mn^{2+} + 2H_2O \tag{15.8}$$

is only -18.5 kcal and that for the half-reaction

$$\tfrac{1}{2}O_2 + 2H^+ + 2e^- \Rightarrow H_2O \tag{15.9}$$

at pH 7.0 is -37.6 kcal, it might be concluded that oxygen should be the better electron acceptor and therefore MnO_2 should not depress its consumption. However, the free energy values above are for standard conditions at pH 7.0 (1 molar solute concentration, except for H^+ and 1 atm for gases). These conditions do not apply to laboratory experiments or in nature. Oxygen concentration in pure water at atmospheric pressure and 25°C is less than $10^{-2.89}$ M, whereas the concentration of Mn(IV) in MnO_2 which the bacteria encounter suspended in water is orders of magnitude greater than 1 M because MnO_2 is insoluble in water. Because of its insolubility, the bacteria must be in physical contact with the MnO_2 (see discussion by Ghiorse, 1988). The specific mechanism by which bacteria are able to transfer electrons to insoluble MnO_2 remains to be worked out, but a possible model is presented below. The problem must be similar to that encountered when bacteria oxidize metal sulfides by direct action (see Chapter 18).

MnO_2 reducing ability has been found inducible in all marine cultures tested so far by Ehrlich (1973). Despite differences in electron pathway from donor to recipient, the overall reaction involving MnO_2 reduction appears to be the same in all marine organisms studied by Ehrlich and collaborators (Ehrlich et al., 1972; Ehrlich, 1973). It may be summarized as follows:

$$\text{Glucose} \xrightarrow{\text{bacteria}} n e^- + nH^+ + \text{end products} \tag{15.10}$$

$$\frac{n}{2} MnO_2 + n e^- + nH^+ \xrightarrow[\substack{\text{or uninduced bacteria} \\ \text{+ ferri- or ferrocyanide}}]{\text{induced bacteria}} \frac{n}{2} Mn(OH)_2 \tag{15.11}$$

$$\frac{n}{2} Mn(OH)_2 + nH^+ \Rightarrow \frac{n}{2} Mn^{2+} + nH_2O \tag{15.12}$$

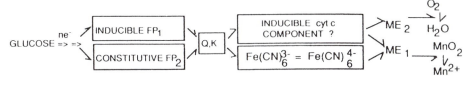

(a)

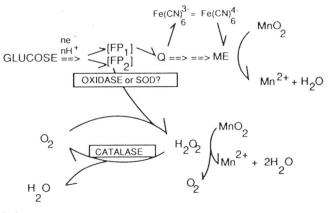

(b)

Figure 15.6 Electron transfer pathways in MnO_2 reduction by *Bacillus* 29. (a) Original proposal. (Redrawn from Ghiorse and Ehrlich, 1976, by permission.) FP_1, flavoprotein (1); FP_2, flavoprotein (2); Q,K, coenzyme Q; ME_1, metalloprotein (1) (MnO_2 reductase); ME_2, metalloprotein (2) (an oxidase such as cytochrome oxidase). (b) Amended proposal. (Modified from Ghiorse, 1988.) FP_1, flavoprotein; FP_2, glucose oxidase; SOD, superoxide dismutase (converts superoxide radicals formed during oxygen reduction to H_2O_2 and O_2); Q, coenzyme Q; ME, metalloprotein (MnO_2 reductase). (By permission from W. C. Ghiorse.)

The reason for representing the direct product of reduction of MnO_2 by the bacteria as $Mn(OH)_2$ rather than Mn^{2+} is that in resting cell experiments or in experiments with cell extract which are run for only 3–4 h, it is necessary to acidify the reaction mixture to about pH 2 on completion of incubation in order to bring Mn^{2+} into solution. No manganese is solublized upon acidification at time 0 from the MnO_2 employed in these experiments Such acidification is not necessary in growth experiments where acid production from glucose by the bacteria or complexation by medium constituents bring Mn^{2+} into solution (Trimble and Ehrlich, 1968).

The enzyme directly responsible for MnO_2 reduction by organisms that do it aerobically or anaerobically has so far not been isolated and characterized. In the case of *Acinetobacter calcoaceticus*, assimilatory nitrate reductase has, however, been implicated (Karavaiko et al., 1986). MnO_2 reduction by cell extract from this organism was found to be inhibited by NH_4^+ and NO_3^-. It was stimulated if the cells had been cultured in the presence of added molybdenum, or if molybdate was added to the reaction mixture in which MnO_2 was being reduced with extract from cells grown in the absence of molybdenum-supplemented medium (Karavaiko et al, 1986). MnO_2-reducing activity by *A. calcoaceticus* appears to be assisted by the simultaneous production of organic acids by *A. calcoaceticus* (Yurchenko et al., 1987).

Ehrlich (1993a,b) has proposed a model (Fig. 15.7) to explain how bacteria are able to reduce insoluble MnO_2 either aerobically or anaerobically. This model requires that the bacteria are in direct physical contact with MnO_2 particles. Mn^{2+} bound in the outer cell envelope enters into a disproportionation reaction [Reaction (15.1)] with the oxidation state of the bound Mn^{2+} being raised to Mn^{3+}. Reducing power removed from electron donors by the bacteria is transferred by enzymes and carriers to the Mn^{3+} in the outer cell envelope, re-reducing it to Mn^{2+}. Most of this reduced Mn^{2+} is released into the surround of the cell but some remains bound in the outer cell envelope to continue the disproportionation of the MnO_2. The model is based on the following experimental observations with marine pseudomonad BIII 88 (Ehrlich, 1993a).

The plasma membrane of marine pseudomonad BIII 88 contains a respiratory system that includes b and c cytochromes. Aerobically, MnO_2 reduction with acetate by intact cells was stimulated by electron transport inhibitors antimycin A and 2-heptyl-4-hydroxyquinoline N-oxide (HQNO), suggesting a branched respiratory pathway to the terminal electron acceptors MnO_2 and O_2, the branch to O_2 being more sensitive to the inhibitors than the branch to MnO_2. Anaerobically, both inhibitors caused a decrease in MnO_2 reduction. Oxidative phosphorylation uncouplers 2,4-dinitrophenol and carbonyl cyanide m-chlorophenylhydrozone (CCCP) stimulated MnO_2 reduction at critical concentration aerobically and anaerobically, indicating that energy is conserved in MnO_2 reduction. Induced cells contained significantly more manganese in their cell envelopes than uninduced cells. The extent of their MnO_2-reducing activity was strongly correlated with the manganese concentration in their cell envelope. Cell envelopes of marine bacterial strains that could oxidize Mn^{2+} or could neither oxidize Mn^{2+} nor reduce MnO_2, contained less manganese and exhibited marginal or no MnO_2 reducing activity.

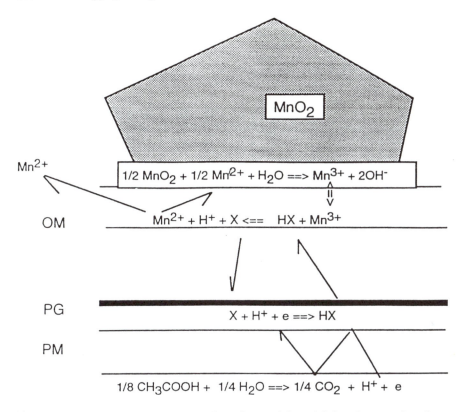

Figure 15.7 Schematic representation of a model explaining the transfer of reducing power (electrons) across the interface between the cell surface of marine pseudomonad BIII88 and the surface of an MnO_2 particle with which the bacterium is in physical contact. OM, outer cell membrane; PG, peptidoglycan layer; PM, plasma membrane; X, hypothetical carrier of reducing power in the cell envelope. (From Ehrlich, 1993b, with permission.)

This model helps explain why marine pseudomonad BIII 88, unlike *Geobacter metallireducens* or *Shewanella putrefaciens*, can reduce MnO_2 aerobically as well as anaerobically. It depends on the stability of Mn^{2+} in air at circumneutral and weakly acid pH in the absence of a catalyst, as opposed to Fe^{2+}, which would rapidly autoxidize. Bacteria that can reduce MnO_2 only anaerobically are postulated to use an oxygen-sensitive Fe(II)/Fe(III) shuttle between the surface of MnO_2 and the cell envelope.

The above discussion centered around enzymatic attack of Mn(IV) as terminal electron acceptor. Since Mn(III)-containing minerals like hausmannite exist, their possible function as terminal electron acceptors in bacterial respiration in nature needs to be considered. Very limited evidence has been obtained to date. Gottfreund and Schweisfurth (1983) reported the reduction of Mn(III), when complexed by pyrophosphate, to Mn(II) by bacterial strain Red 16 using glucose, glycerol, or fructose as electron donor aerobically and glucose, fructose, glycerol, lactate, succinate, and acetate anaerobically. However, strain Red 16 could not grow on acetate aerobically or anaerobically. The authors proposed that this organism produces Mn(III) in complexed form as a detectable intermediate in the reduction of Mn(IV) to Mn(II). Another study (Gottfreund et al., 1985) indicated the existence of bacteria that need not attack Mn(IV) but can produce a ligand that binds Mn(III). The Mn(III) is believed to be derived by nonenzymatic reduction from Mn(IV). The bacteria then reduce the chelated Mn(III) to Mn(II). The authors suggested that Mn(III) is an intermediate in Mn(IV) reduction as well as Mn(II) oxidation. However, whether Mn(III) is an obligatory intermediate in manganese oxidation or reduction has yet to be clearly established (see, e.g., Ehrlich, 1988, 1993a,b). The study of Mn(III) as an intermediate is fraught with difficulty because of the ease with which Mn(II) can react with some forms of Mn(IV) in a dismutation reaction to form Mn(III) in the presence of a ligand for Mn(III) such as pyrophosphate (e.g., Ehrlich, 1964).

Enzymatic Reduction of Manganese Oxides Restricted to Anaerobic Conditions Bacterial reduction of manganese oxides such as MnO_2 which occurs *only* anaerobically has now been well documented (Burdige and Nealson, 1985; Lovley and Phillips, 1988a; Lovley and Goodwin, 1988; Myers and Nealson, 1988; Lovley et al., 1989; Lovley et al., 1993). In at least three instances, the organisms, *Shewanella* (formerly *Alteromonas, Pseudomonas*) *putrefaciens* (Myers and Nealson, 1988), *Bacillus polymyxa* D1 and *Bacillus* MBX1 (Rusin et al., 1991a; 1991b), are facultative anaerobes. In a number of other instances, the organisms, *Geobacter metallireducens* (formerly strain GS-15)(Lovley and Phillips, 1988a), *G. sulfurreducens* (Caccavo et al., 1994), *Desulfovibrio desulfuricans, Desulfomicrobium baculatum, Desulfobacterium autotrophicum*, and *Desulfuromonas acetoxidans* (Lovley and Phillips, 1994a) are obligate anaerobes. The organisms are quite versatile in regard to the electron donors they can utilize for MnO_2 reduction. For instance, *G. metallireducens* can use butyrate, propionate, lactate succinate and acetate, and several aromatic compounds which are all completely oxidized to CO_2 (Lovley and Philips, 1988a; Lovley, 1991). *S. putrefaciens*, on the other hand, can utilize lactate and pyruvate as electron donors but oxidizes these only to acetate. It can, however, use H_2 and formate as electron donors whereas *G. metallireducens* cannot (Lovley et al., 1989). *G. sulfurreducens* can also use H_2 as an electron donor (Caccavo et al., 1994). As reported in Chapter 14, these organisms are not restricted to MnO_2 as terminal electron acceptor anaerobically. Depending on which organism is considered, Fe(III), nitrate, fumarate, glycine, iodate, elemental sulfur, sulfite, sulfate, and thiosulfate may serve. In the case of *S. putrefaciens*, oxygen, nitrate, and fumarate inhibited MnO_2 reduction but sulfate, sulfite, molybdate, nitrite, or tungstate did not (Myers and Nealson, 1988).

Anaerobic utilization of acetate, which is unfermentable except in methanogenesis, as electron donor in reduction of MnO_2 is proof of anaerobic respiration. Since acetate can also be respired in air where oxygen can serve as terminal electron acceptor, it is interesting that gram-negative, marine strain SSW_{22} can reduce MnO_2 with acetate, succinate and glucose aerobically as well as anaerobically to the same degree (Ehrlich, 1988; Ehrlich, et al., 1991; Ehrlich, unpublished data). This raises a question about the differences in electron transport pathways between those organisms in which electron transport from acetate to MnO_2 is blocked by oxygen and those in which it is not. In at least some instances it may have to do with the nature of the manganese reductase. As discussed in the previous section, in marine pseudomonad BIII 88, which can reduce MnO_2 aerobically and anaerobically, the manganese reductase is proposed to act on Mn(III) produced by reaction between cell-envelope-bound Mn^{2+} and external MnO_2. In *S. putrefaciens*, c-type cytochrome in the outer membrane of anaerobically but not aerobically grown cells is probably directly or indirectly involved in MnO_2 reduction as it is in dissimilatory Fe(III) reduction (Myers and Myers, 1992; 1993). Cytochrome c_3 may be involved in MnO_2 reduction by *Desulfovibrio* spp., as it was shown to be in the reduction of chromate (Lovley and Phillips, 1994b) and of uranium(VI) (Lovley et al., 1993). The lesser sensitivity of Mn(II) than that of cytochromes to autoxidation may explain, at least in part, why some bacteria can reduce MnO_2 only anaerobically.

Nonenzymatic Reduction of Manganese Oxides   Some bacteria and most of those fungi that reduce Mn(IV) oxides such as MnO_2, reduce them indirectly (nonenzymatically). A likely mechanism of reaction is the production of metabolic products that are strong enough reductants for Mn(IV) oxides. *Escherichia coli*, for instance, produces formic acid from glucose which is capable of reacting nonenzymatically with MnO_2:

$$3H^+ + HCOO^- + MnO_2 \Rightarrow Mn^{2+} + CO_2 + 2H_2O \qquad (15.13)$$

Pyruvate, which is a metabolic product of some bacteria can also react nonenzymatically with manganese oxides at acid pH (Stone, 1987). Sulfate reducers produce H_2S from sulfate anaerobically and *Shewanella putrefaciens* can produce H_2S and sulfite from thiosulfate and Fe^{2+} from Fe(III) in their respiratory processes. Such biogenic H_2S, sulfite, and Fe^{2+} can nonenzymatically reduce some manganese oxides readily (Burdige and Nealson, 1986; Mulder, 1972; Lovley and Phillips, 1988; Myers and Nealson, 1988b; Nealson and Saffarini, 1994).

Many fungi produce oxalic acid, which can also reduce MnO_2 nonenzymatically (e.g., Stone, 1987):

$$4H^+ + {}^-OOCCOO^- + MnO_2 \Rightarrow Mn^{2+} + 2CO_2 + 2H_2O \quad (15.14)$$

Since the electron transport mechanism in fungi, which are eukaryotic organisms, is located in mitochondrial membranes and not in the plasma membrane, as in prokaryotic cells, fungi cannot be expected to reduce MnO_2 enzymatically (Ehrlich, 1978a), except possibly by extracellular enzymes. MnO_2 reduction in acid solution under anoxic conditions by reversal of the reaction

$$Mn^{2+} + \tfrac{1}{2}O_2 + H_2O \Rightarrow MnO_2 + 2H^+ \quad (15.15)$$

is not very likely because of the high energy of activation requirement.

The fact that MnO_2 is reduced indirectly by fungi can be readily demonstrated on a glucose-containing agar medium into which MnO_2 has been incorporated (e.g., Schweisfurth, 1968; Tortoriello, 1971). Fungal colonies growing on such a medium develop a halo (clear zone) around them in which MnO_2 has been dissolved (reduced). Since enzymes do not work across a separation between them and their substrate, the formation of the halo can be explained either on the basis that a reducing compound was released by the fungus, which then reacted with the MnO_2, or on the basis that acid produced by the fungus lowered the pH to a range where a residual compound like glucose, if a constituent of the medium, reduced MnO_2. The only exception would be the release of an enzyme into the medium by the fungus, which would then catalyze the reduction. But the existence of such an enzyme has so far not been reported.

15.7 BIOACCUMULATION OF MANGANESE

Just as microorganisms exist that accumulate ferric oxide passively, so do microorganisms that accumulate manganese oxides. In a number of cases, the same organism may accumulate either metal oxide or both. As with iron oxides, they normally deposit manganese oxides on the surface

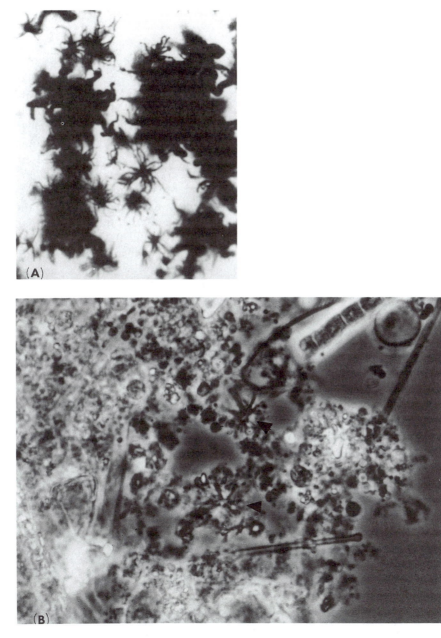

Figure 15.8 *Metallogenium*. (A) Microaccretions of *Metallogenium* from Lake Ukshezero in a split peloscope after 50 days; growth in central part of microzone (×900). (From Perfil'ev and Gabe, 1965, by permission.) (B) *Metallogenium* from filtered water from Pluss See, Schleswig-Holstein, Germany (×2,450). (Courtesy of W. C. Ghiorse and W.-D. Schmidt.)

of their cells or on some structure surrounding the cell such as a slime layer or a sheath, but not intracellularly (see reviews by Ghiorse, 1984; Ghiorse and Ehrlich, 1992). In some instances, the accumulation may not be passive but may follow prior oxidation of Mn^{2+} by the organism. Prokaryotes that accumulate manganese oxides include sheathed bacteria, especially *Leptothrix, Metallogenium* (Fig. 15.8), *Caulococcus, Kusnezovia, Pedomicrobium* (Fig. 15.9) (Ghiorse and Hirsch, 1979; Ghiorse and Ehrlich, 1992), *Hyphomicrobium* (Fig. 15.10), *Siderocapsa* (Fig.14.9) (now believed to belong to the genus *Arthrobacter*), *Naumanniella, Planctomyces* (Schmidt et al., 1981, 1982), and Mn^{2+}-oxidizing fungi (Fig. 15.11). Representative bacteria, actinomycetes, and fungi from soil of the Chiatura manganese biogeochemical province of the former Georgian S.S.R. were found to accumulate between 0.04 and 0.3 mg of Mn per gram dry weight of biomass in Czapek's medium unsupplemented with manganese, and from 9.2 to 101 mg/g of biomass in Czapek's medium supplemented with 0.1% Mn (Letunova et al, 1978). The organisms ana-

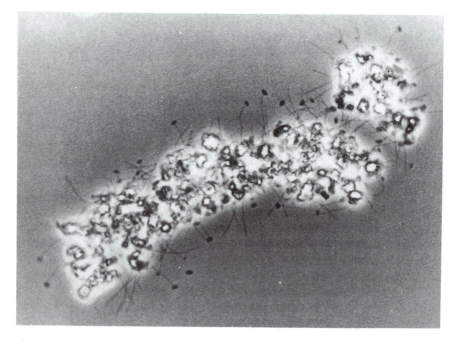

Figure 15.9 *Pedomicrobium* in association with manganese oxide particles (×2,800). (Courtesy of W. C. Ghiorse.)

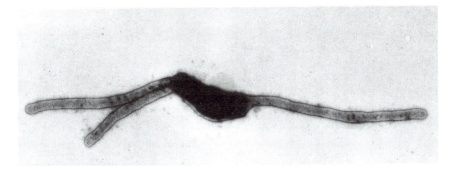

Figure 15.10 Manganese-oxidizing *Hyphomicrobium* sp. isolated from a Baltic Sea iron-manganese crust ($\times$15,600). (Courtesy of W. C. Ghiorse.)

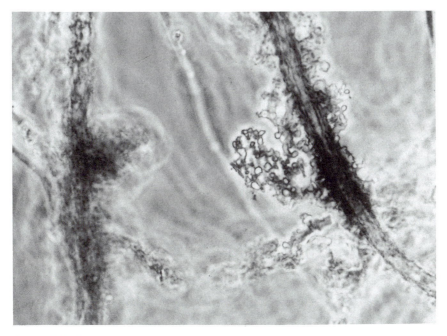

Figure 15.11 Manganese oxide deposition on fungal hyphae ($\times$990). (Courtesy of R. Schweisfurth.)

lyzed in this study included *Bacillus megaterium, Actinomyces violaceus, A. indigocolor, Cladosporium herbarium*, and *Fusarium oxysporum*. The manganese concentration in the soil samples from which these organisms were isolated ranged from 1.05 to 296 g kg^{-1}. Development of the actinomycetes and fungi but not the bacteria appeared to be stimulated by high manganese levels in their native soil habitat.

In nature, bacterial manganese and/or iron deposition may be transient. With *Leptothrix* spp., it may be influenced by temperature because temperature influences the size of the biomass. Ghiorse and Chapnick (1983) noted that a *Leptothrix* isolated from water from a swamp (Sapsucker Woods, Ithaca, New York) grew optimally between 20 and 30°C in the laboratory and poorly at 4 and 35°C. This correlated with observation that *Leptothrix* and particulate iron and manganese were most abundant in the surface waters of the swamp when the temperature in the surface water was in the range 20–30°C.

15.8 MICROBIAL MANGANESE DEPOSITION IN SOIL AND ON ROCKS

Soil Manganese is an essential micronutrient for plants. Lack of sufficiently available manganese in soil can lead to manganese deficiency in plants, which may manifest itself in the form of gray-speck disease, for instance. Such lack of availability may be due to lack of manganese mobility in a soil and can be the result of activity of manganese-oxidizing microbes.

Microorganisms can have a profound effect on the distribution of manganese in soil. Oxidizers of Mn(II) will tend to immobilize it and make it less available or unavailable to plants by converting it to an insoluble oxide. Mobile forms of manganese include Mn^{2+} ions, complexes of Mn^{2+}, especially complexes such as those of humic and tannic acids, and possibly complexes of Mn(III). Immobile forms of manganese include Mn^{2+} ions adsorbed to clays or manganese or ferromanganese oxides, $Mn(OH)_2$, and insoluble salts of Mn(II) such as $MnCO_3$, $MnSiO_3$, and MnS, and various Mn(III)- and Mn(IV)-oxides. Oxidized manganese can accumulate as concretions in some soils, often accompanied by iron and various other trace elements (Dosdorff and Nikiforoff, 1940; Taylor et al., 1964; Taylor and McKenzie, 1966; McKenzie, 1967; McKenzie, 1970). The formation of these nodules may be microbially mediated (Drosdoff and Nikiforoff, 1940; Douka, 1977; see also sections on bacterial manganese oxidation). The insolubility of various immobile manganese compounds is governed by prevailing pH and E_h conditions (Collins and Buol,

1970). Reducing conditions at near-neutral or alkaline pH are especially favorable for the stability of manganous compounds, while oxidizing conditions at near-neutral or alkaline pH are especially favorable for the stability of Mn(III) and Mn(IV) oxides. Agriculturally, the most important forms of insoluble manganese include the oxides of Mn(IV) and mixed oxides such as $MnO \cdot MnO_2$ (also written Mn_2O_3) and $2MnO \cdot MnO_2$ or $MnO \cdot Mn_2O_3$ (also written Mn_3O_4). The stability of these oxides may also be affected by the presence of iron (Collins and Buol, 1970). Ferrous iron generated by Fe(III)-reducing bacteria can chemically reduce manganese(IV) oxides (Lovley and Phillips, 1988b; Myers and Nealson, 1988b).

Soil chemists have distinguished between the various forms of manganese in soil in an empirical fashion through the use of different extraction methods (see, e.g., Robinson, 1929; Sherman et al., 1942; Leeper, 1947; Reid and Miller, 1963, Bromfield and David, 1978). Thus Sherman et al. (1942) measured, in successive steps, water-soluble manganese by extracting a soil sample with distilled water, then exchangeable manganese by extracting the residue with with 1 N ammonium acetate at pH 7.0, and, finally, easily reducible manganese by extracting the second residue with 1 N ammonium acetate containing 0.2% hydroquinone. Other investigators have used different extraction reagents to measure exchangeable and easily reducible manganese (e.g., Robinson, 1929; Leeper, 1947).

The soil percolation experiments by Mann and Quastel (1946) clearly showed the role which microbes can play in immobilizing manganese in soil. These investigators showed that when they continuously perfused nonsterile soil columns with 0.02 M $MnSO_4$ solution, the manganese was progressively retained in the columns by being transformed into oxide paralleled by a disappearance of manganous manganese in the effluent. The oxidized state of the manganese was demonstrated by reacting it in situ with hydroxylamine reagent (a reducing agent). This reaction released manganese from the column in soluble form. The oxidation of the perfused manganese in the columns was inferred to have been microbial because poisons such as chloretone, sodium iodoacetate, and sodium azide inhibited it.

Manganese oxidation by soil microbes can also be demonstrated in the laboratory by the method of Gerretsen (1937). It involves the preparation of Petri plates of agar mixed with unsterilized soil. A central core is removed from the agar and replaced with a sterile sandy soil mixture containing 1% $MnSO_4$. As Mn^{2+} diffuses from the core into the agar, some developing bacterial and fungal colonies accumulate brown precipitate of manganese oxide in and around them as a result of enzymatic or nonenzymatic manganese oxidation, if the pH was held below 8 (Fig. 15.12). Controls in which the soil is antiseptically treated before adding it to the agar will not show evidence of manganese oxidation. This method was effectively used by Leeper and Swaby (1940) in demonstrating the presence of Mn^{2+} oxidizing microbes in Australian soils.

Mn(II) oxidation can also be studied in deep cultures of a soil-agar mixture. Uren and Leeper (1978) found that manganese oxidation in deep cultures of soil-agar mixtures (85 ml of 1% soil agar in a 150-ml jar), set up like Gerretsen plates, seemed to occur preferentially at reduced oxygen levels because oxidation was most intense 10–17 mm below the agar surface when air was passed over the surface. Above this zone, sparse manganese oxide was deposited only around bits of organic matter from the soil and fungal hyphae. Raising the agar concentration in the soil agar to 2% caused the intense zone of oxidation to appear 4–11 mm below the surface. When the jar was sealed with an air space above the agar, an

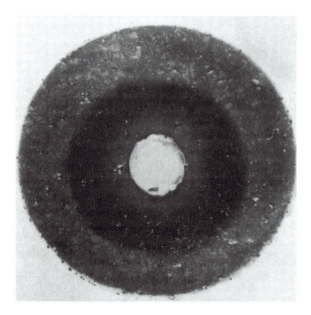

Figure 15.12   Gerretsen plate as modified by Leeper and Swaby (1940), showing manganese oxide deposition (dark halo) as a result of microbial growth around the central MnSO$_4$-containing agar plug. Initial pH 7.3; final pH 6.7. (From G. W. Leeper and R. J. Swaby, The oxidation of manganous compounds by microorganisms in the soil, Soil Science 49:163–169, Copyright © 1940 by Williams and Wilkins; by permission.)

8-mm-thick oxidation zone developed from the surface downward. The investigators found that the factor that controlled where manganese oxidation was most intense in the soil column was available CO_2, since aeration at the surface with a mixture of 97% N_2, 2% O_2, and 1% CO_2 permitted development of an oxidation zone in the top 1 cm while a gas mixture of 98% N_2 and 2% O_2 did not. The observed oxidation was due to microbial activity because incorporation of azide to a final concentration of 1 mM into the soil agar inhibited manganese oxidation. Microbial manganese-oxidizing activity was noted even in soils as acid as pH 5.0 (Uren and Leeper, 1987; Sparrow and Uren, 1987).

In situ manganese oxidation by soil microbes has been observed with pedoscopes (Perifil'ev and Gabe, 1965). This apparatus consists of one or more capillaries with optically flat sides for direct microscopic observation of the capillary content. Inserting a sterilized pedoscope into soil enables development of soil microbes in the lumen of the capillaries under soil conditions. Periodic removal of the pedoscope and examination under a microscope permits visual assessment of the developmental progress of the organism and the deposition, if any, of manganese oxide (see Chapter 7). Manganese oxidizing and/or depositing organisms detected for the first time by use of the pedoscope include *Metallogenium personatum*, *Kusnetsovia polymorpha*, *Caulococcus manganifer*, and others (Perfil'ev and Gabe, 1965).*

Application of the above techniques or others has provided ample evidence of the important role which microbes can play in immobilizing

---

* The status of the genus *Metallogenium* is presently uncertain. Although isolated and cultured repeatedly in the former Soviet Union by a number of different investigators (e.g., Perifil'ev and Gabe, 1965; Zavarzin, 1961, 1964; Mirchink et al., 1970; Dubinina, 1970, 1978b, 1984, 1988), living cultures which oxidize manganese have so far not been isolated outside Russia. Structures resembling *Metallogenium* have been collected by membrane filtration from waters of several lakes, but these structures proved nonviable. Indeed they did not contain any trace of nucleic acid, protein, or membrane lipid (Klaveness, 1977; Gregory et al., 1980; Schmidt and Overbeck, 1984; Maki et al., 1987). Structures interpreted as *Metallogenium symbioticum*, which is said to grow in obligate association with some fungi and bacteria (Zavarzin, 1961, 1964; Dubinina, 1984) may be fibrous manganese oxide produced non-enzymatically by the fungus (Schweisfurth, 1969; Schweisfurth and Hehn, 1972; Schmidt and Overbeck, 1984) as, for instance, in manganese oxidation by exopolymers produced by an arthroconidial anamorph of a basidomycete (Emerson et al., 1989). It is possible that the starlike structures (arais) associated with *Metallogenium*-coenobia are nonliving appendages from which the actual cell which gave rise to the appendages has become detached before or during sample collection. Like the stalks of *Gallionella*, the "appendages" may be structures impregnated with manganese oxide. A recent observation of arais showing "secondary coccoid body formation" as part of a manganese depositing biofilm on the inner walls a water pipe lend support to this view (Sly et al., 1988).

manganese in soil by its oxidation. Thus the method of Gerretsen was adapted to show the role microbes played in manganese oxidation in Australian soils (Leeper and Swaby, 1940; Leeper and Uren, 1978; Sparrow and Uren, 1978), and in soils of south Sakhalin (U.S.S.R.) (Ten Khakmun, 1967). Other investigators who have demonstrated the presence of manganese-oxidizing microbes (bacteria and/or fungi) in soils in various parts of the world by various methods include Timonin (1950a,b) Timonin et al. (1972), Bromfield and Skerman (1950), Bromfield (1956, 1978), Aristovskaya (1961), Aristovskaya and Parinkina (1963), Perfil'ev and Gabe (1965), Schweisfurth (1969, 1971), Van Veen (1973), and others. Among the bacteria identified as active agents of Mn(II) oxidation in these studies, *Arthrobacter* or *Corynebacterium*, *Pedomicrobium*, *Pseudomonas*, and *Metallogenium* were the most frequently mentioned, but a number of other unrelated genera were also identified as being active.

Rocks Manganese oxides are sometimes found in thin, brown to black veneers, and iron oxides in orange veneers (each up to 100 μm thick) covering rock surfaces in semiarid and arid regions of the world. In the brown to black coatings, manganese and/or iron oxides are major components (20–30%) along with about 60% clay and various trace elements (Potter and Rossman, 1977; Dorn, 1991). These manganese- or iron-rich coatings are known as **desert varnish** or **rock varnish**. Manganese-rich coatings have been detected on some rocks in the Sonoran and Mojave Deserts (North America), Negev Desert (Middle East), the Gibson and Victoria Deserts (Western Australia), and the Gobi Desert (Asia). Although they are very likely the result of microbial activity (Krumbein, 1969; Dorn and Oberlander, 1981; Taylor-George et al., 1983; Hungate et al., 1987), this view is not universally shared. Krumbein (1969) and Krumbein and Jens (1981) have implicated cyanobacteria and fungi in the formation of desert varnish in the Negev Desert, whereas Dorn and Oberlander have implicated *Pedomicrobium*- and *Metallogenium*-like bacteria in desert varnish from Death Valley in the Mojave Desert. The latter investigators produced desert varnish–like deposits in the laboratory with their bacterial isolates. Dorn (1991) is of the opinion that the following conditions favor bacterial participation in manganese-rich varnish formation: (1) the varnish is formed on moist rock surfaces, (2) it forms on rock surfaces of low nutrient content that favor bacterial manganese oxidation by mixotrophs, and (3) it forms on rock surfaces exhibiting circum-neutral pH.

Adams et al. (1992) have suggested that in the formation of iron-rich varnish, siderophores produced by bacteria on rock surfaces where the varnish forms mobilize ferric iron from wind-borne dust or the rock surfaces. The mobilized iron is then concentrated as iron oxide or oxyhydroxide on the cell walls of bacteria on the rock surfaces. Upon death of the iron-coated bacteria, the iron oxide becomes part of the varnish.

Microscopic examination of desert varnish from the Sonoran Desert revealed the presence of fungi (dematiaceous hyphomycetes) and bacteria. Manganese oxidizers in samples from this source included *Arthrobacter*, *Micrococcus*, *Bacillus*, *Pedomicrobium*, and possibly *Geodermatophilus* (Taylor-George et al., 1983). Active respiration but little CO_2 fixation was detected in the varnish. The respiration was attributed mainly to the fungi. The absence of significant CO_2 fixation indicated an absence of photosynthetic, i.e., cyanobacterial or algal, activity. Varnish formation on the Sonoran Desert rocks has been postulated to be the result of formation of fungal microcolonies on rock surfaces which trap wind-borne clay and other mineral particles. Upon death of part of the fungi, intermittent, moisture-dependent weathering of minerals is thought to lead to microbial or inorganic concentration of mobilized manganese through oxidation. Dead fungal biomass could provide nutrients to manganese-accumulating bacteria at this stage. Further development of fungal microcolonies and their coalescence was postulated to lead ultimately to the formation of continuous films of varnish. The fungal mycelium in this model seems to serve as an "anchor" for Mn oxide formed at least in part by bacteria, besides contributing to rock weathering (Taylor-George et al., 1983). The MnO_2 in the varnish may act as a screen against ultraviolet radiation and may thus serve a protective function (Taylor-George et al., 1983). Examination of the bacterial flora of desert varnish from the Negev Desert yielded a range of bacterial isolates similar, but not identical, to that from varnish from the Sonoran Desert. Of 79 bacterial isolates, 74 oxidized manganese under laboratory conditions. They were assigned to the genera *Bacillus*, *Geodermatophilus*, *Arthrobacter*, and *Micrococcus* (Hungate et al., 1987). The prevalence of manganese oxidizers in desert varnish samples from geographically very distinct desert sites further supports the idea that they play a role in its formation (Hungate et al, 1983).

Ores   It has been suggested that some sedimentary manganese ore deposits are of biogenic origin. This is based in part on the observation of structures in the ore which have been identified as microfossils by the discoverers. Manganese ore from the Groote Eylandt deposit in Australia has revealed the presence of algal (cyanobacterial?) stromatolite structures, cyanobacterial oncolites, coccoid microfossils, and microfossils enclosed in metacolloidal oxides of manganese (Ostwald, 1981). In this deposit the microbes are viewed as being the main cause of manganese oxide formation and accretion, with subsequent nonbiological diagenetic changes leading to the ultimate form of the deposit. Some Precambrian and a Creta-ceous-Paleogene manganese deposit (Seical deposit on the island of Timor) have been attributed to the activity of *Metallogenium* because coenobial structures resembling modern forms of this organism described by Russian investigators have been found by microscopic study of sections of the respective ores (Crerar et al., 1979). Similar observations were made by Shternberg (1967) on samples of the Oligocene Chiatura man-ganese deposit and the Paleozoic Tetri-Tsarko manganese deposit.

15.9 MICROBIAL MANGANESE DEPOSITION IN FRESHWATER ENVIRONMENTS

Manganese-oxidizing microorganisms have been detected in freshwater environments since the beginning of the twentieth century. Early reports include those by Neufeld (1904), Molisch (1910), and Lieske (1919) as cited by Moese and Brantner (1966), Thiel (1925), von Wolzogen-Kuehr (1927), Zappfe, (1931), and Sartory and Meyer (1947). The organisms found by these investigators were usually detected in sediments, organic debris, or manganiferous crusts. More recent investigations concerned microbial manganese oxidation in springs, lakes, and in water distribution systems.

Bacterial Manganese Oxidation in Springs   Precipitation of manganese hy-droxide (presumably hydrous manganese oxide) which upon aging became transformed into pyrolusite and birnessite in a mineral spring near Ko-maga-dake on the island of Hokkaido, Japan, was attributed to the activity of manganese-oxidizing bacteria, including sheathed bacteria (Hariya and Kikuchi, 1964). The bacteria were detected by filtration of water from the spring through ordinary filter paper and membrane filters and by incubat-ing samples of water in the laboratory. The water contained (in mg L^{-1}): Mn^{2+}, 4.75; K^+, 16; Na^+, 128, Ca^{2+}, 101; Mg^{2+}, 52; Cl^-, 156.2; SO_4^{2-}, 481; HCO_3^- and CO_2, 11, but only traces of Fe^{2+} or Fe^{3+}. The tempera-ture of the water was 23°C and its pH was 6.8. During incubation of water samples in the laboratory, manganese precipitated progressively between 20 and 50 days accompanied by a fall in pH and a gradual rise in E_h after 20 days and an abrupt rise after 45 days. Only 1.18 mg of dissolved Mn^{2+} per liter were left after 50 days.

   Another instance of bacterial manganese oxidation associated with a spring was reported from Squalicum Creek Valley near Bellingham, Washington, U.S.A. (Mustoe, 1981). In this example, a black soil sur-rounded the spring over an area of 5 m by 25 m and to a depth of 30 cm.

It contained 43% MnO_2 and 20–30% iron oxide (calculated as Fe_2O_3). Two pseudomonad strains which could oxidize Mn^{2+} and Fe^{2+} when growing on a medium containing soil organic matter were isolated from it. The oxides were deposited extracellularly. The organisms were considered to be a cause of the manganese and iron deposition in the soil.

Still another instance of bacterial manganese oxidation in spring water and sediment was observed near Ein Feshkha on the western shore of the Dead Sea, Israel (Ehrlich and Zapkin, 1985). Gram-positive, spore-forming rods and gram-negative rods were isolated from water and sediment samples from drainage channels, which oxidized Mn^{2+} enzymatically. They were of the Group I type (see section 15.5). It was postulated that at least some of the manganese oxide formed by the bacteria in the spring water was carried to the Dead Sea and became incorporated as laminations into calcareous crusts found along the shore of the lake.

Bacterial Manganese Oxidation in Lakes   Evidence of bacterial manganese oxidation has been observed in Lake Punnus-Yarvi on the Karelian peninsula in the U.S.S.R. (Sokolova-Dubinina and Deryugina, 1967a,b). The lake is of glacial origin, oligotrophic, and slightly stratified thermally. It is 7 km long, up to 1.5 km wide, and up to 14 m deep. It contains significant amounts of dissolved manganese (0.02–1.4 mg L^{-1}) and iron (0.7–1.8 mg L^{-1}) only in its deeper waters. The lake is fed by 2 rivers and 24 surrounding streams that drain surrounding swamps. The manganese in the lake is supplied by surface- and groundwater drainage containing 0.2–0.8 mg L^{-1} and 0.4–2 mg L^{-1}, respectively. Most of the iron and manganese in the lake are deposited on the northwestern banks of the Punnus-Ioki Bay situated at the outflow from the lake into the Punnus-Ioki River in a deposit 5–7 cm thick at a water depth of up to 5–7 m. The deposit includes hydrogoethite ($nFeO·nH_2O$), wad ($MnO_2·nH_2O$) and psilomelane ($mMnO·MnO_2$). The iron content of the deposit ranges from 18 to 60% and the manganese content from 10 to 58%. The deposit also includes 5–16% SiO_2, and Al, Ba, and Mg in amounts ranging from 0.3 to 0.7%. The bacterium held responsible for manganese oxidation was *Metallogenium*, which was found in all parts of the lake. Manganese deposition was demonstrated in the laboratory by incubating sediment samples at 8°C for several months. Dark brown, compact spots of manganese oxide were then detected, which revealed no characteristic bacterial structures by examination with a light microscope but which did show the presence of *Metallogenium* by electron microscopy which closely resembled *Metallogenium* fossils in Chiatura manganese ore (Dubinina and Deryugina, 1971). *Metallogenium* was cultured from lake samples on manganese acetate agar, but sediment from portions of the lake where no microconcretions of manganese oxide were found did not yield *Metallogenium* cultures. *Metallogenium* development was also studied in peloscopes, showing progressive encrustation with manganese oxide. Manganese deposition in the lake occurred in a redox potential range of 435–720 mV and a pH range of 6.3–7.1, although it was concluded from data gathered by other investigators that ore formation may begin at an E_h as low as 230 mV and at a pH of 6.5. At Mn^{2+} concentrations below 10 mg L^{-1}, autoxidation was not considered likely. Manganese oxide deposition attributed to *Metallogenium* activity was also noted in some other Karelian lakes where a steady supply of dissolved manganese occurred, redox conditions were favorable, and bacterial reducing processes did not occur (Sokolova-Dubinina and Deryugina, 1968). *Metallogenium* has also been reported to occur in some sediment samples from Lake Geneva (Lac Leman) (Jaquet et al., 1982), and is implicated in manganese oxidizing activity in Lake Constance (Stabel and Kleiner, 1983).

In Lake Oneida, near Syracuse, New York, manganese deposition has been studied for many years because of the occurrence of ferromanganese concretions on the sediment surface of shallow, well oxygenated central areas of the lake (Fig. 15.13). The lake has a surface area of 207 km^2 and varies in depth, its average depth being 6.8 m and its maximum depth 16.8 m (Dean and Ghosh, 1978). It never exhibits seasonal thermal stratification. Wind agitation keeps it well aerated to all depth when it is not frozen, although deeper water may have oxygen concentrations 50% of saturation (Dean and Ghosh, 1978). The ferromanganese concretions appear as crust around rocks at the edge of shoals where the water is less than 4.3 m in depth (Dean, 1970), but some have also been recovered from hard sediment at depths of 8 m (Chapnick et al., 1982). The crust may take on a saucer shape, concave upward (Dean 1970). Many exhibit coarse concentric banding of alternating zones rich in manganese and rich in iron (Moore, 1981). At other times the concretions have been described as flattened, disc-shaped structures. Their rate of growth as determined by natural radioisotope analysis has been estimated to vary between >1 mm/100 years at some periods and no growth at other periods (Moore et al., 1980).

The origin of the ferromanganese concretions in Lake Oneida has been a puzzle. Hypotheses of abiogenesis (e.g., Ghosh, 1975) and biogenesis (e.g., Gillette, 1961; Dean, 1970; Dean and Ghosh, 1978; Dean and Greeson, 1979; Chapnick et al., 1982) have been offered at various times. Gillette (1961) and Chapnick et al. (1982) favor a direct bacterial role, whereas Dean (1970) and Dean and Ghosh (1978) favor an indirect role played by algae in the lake. Gillette (1970) recognized bacterial cells in pulverized concretions by microscopic examination and showed that some of the isolates precipitated iron. He speculated that bacterial iron and manganese deposition in the lake could be the result of degradation of organic complexes, but he also suggested that iron may be precipitated from autoxidation with oxygen generated in the photosynthesis of algae. Dean and Ghosh (1978) and Dean and Greeson (1979) favor a primary role for cyanobacteria and algae as producers of chelators for iron and manganese and accumulators of the chelates. These notions are supported by recent observations that cyanobacteria in the lake belonging to the genus *Microcystis* sp. generate microenvironments with strongly alkaline conditions (pH >9) in their mats as a result of photosynthetic activity (Richardson et al., 1988) (see equations 14.25 and 14.26 in Chapter 14 for a chemical explanation). Such alkaline conditions promote autoxidation of Mn^{2+}. The oxides become entrapped in the mats. Upon death, the cyanobacterial and algal biomass is seen as carrying the oxidized iron and manganese to the lake bottom to be released on its decay and somehow incorporated into concretions. Chapnick et al. (1982) have demonstrated that bacteria in bottom water of the lake can catalyze the oxidation of the Mn^{2+} dissolved in it. They also noted some binding of Mn^{2+} without oxidation. Analysis of a manganese budget for the lake supports the notion that most of the dissolved manganese (95%) becomes incorporated into the nodules, with cyanobacteria and algae helping in transporting the manganese from the surface water to the nodule-forming regions, and bacteria participating in oxidation of Mn^{2+} in the bottom waters to an as yet undetermined extent. Bottom water movement carries nodules or fragments of nodules to the surface of deeper anoxic sediments, where they are buried and undergo reduction with resolubilization of manganese. *Shewanella* (formerly *Alternaria*) *putrefaciens* is one organism found in this sediment that has been demonstrated in laboratory experiments to be an effective reducer of Mn(IV) oxide (Myers and Nealson, 1988). Deep water circulation reintroduces the dissolved manganese into the water column and the cycle is repeated. Manganese which is lost in lake outflow (50 tons/year) is more than replaced by manganese by influx (75 tons per year) into the lake (Dean et al., 1981).

Another lake in which ferromanganese concretions have been found is Lake Charlotte, Nova Scotia, first studied by Kindle in 1932 and later by Harriss and Troup (1970). The concretions occur in a shallow region of the Lake that Kindle (1935) named Concretion Cove. It receives runoff from soils rich in metals and organics. Kindle (1932) thought that an indirect mechanism was operating in which removal of CO_2 from the lake water through photosynthesis by diatoms caused a rise in pH which promoted autoxidation of Mn^{2+} to MnO_2. Recent investigations suggest more direct bacterial involvement, however (Kepkay, 1985). Using a special

Figure 15.13   Ferromanganese concretions from the "nodule rich" area in Lake Oneida, central New York State. (A) Two concretions: the one on the right is typical of the flat "pancake" variety, the one on the left is a rock with 2–3 cm of ferromanganese oxide crust. (B) Epifluorescent light photomicrograph of a portion of the surface of the concretion on the left in (A) stained with 0.01% acridine orange. The sample was removed from the surface and then viewed under violet epi-illumination. The brightly fluorescent, filamentous, and coccoid bacteria are present in the ferromanganese-mineralized biofilm on the surface. Scale bar = 10 µm. (C) Transmission electron photomicrograph of an ultrathin section of a piece of material from a concretion similar to those shown in (A). The sample was fixed in glutaraldehyde followed by osmium tetroxide and uranyl acetate, dehydrated in ethanol, and embedded in epoxyresin. A portion of a filamentous bacterial cell approximately 0.6 µm in diameter is surrounded by a dark-colored complex composed of polymer-ferromanganese oxide. Note the smaller encrusted coccoid cell in the upper left surrounded by a clear zone produced by shrinkage during preparation for electron microscopy. The wispy dendritic black material is Fe/Mn oxide. Scale bar = 1 µm. (From Ghiorse and Ehrlich, 1992, by permission.)

device called a peeper (Burdige and Kepkay, 1983) emplaced into sediment in Concretion Cove, Kepkay (1985a,b) demonstrated bacterially dependent oxidation of Mn(II) in the dark, i.e., without direct participation of algal photosynthesis. He also demonstrated the binding of nickel and, to a lesser extent, copper, which was enhanced by microbial oxidation of manganese. In experiments in which filter membranes with 0.2-µm pore size were submerged in Lake Charlotte water, a succession of bacterial

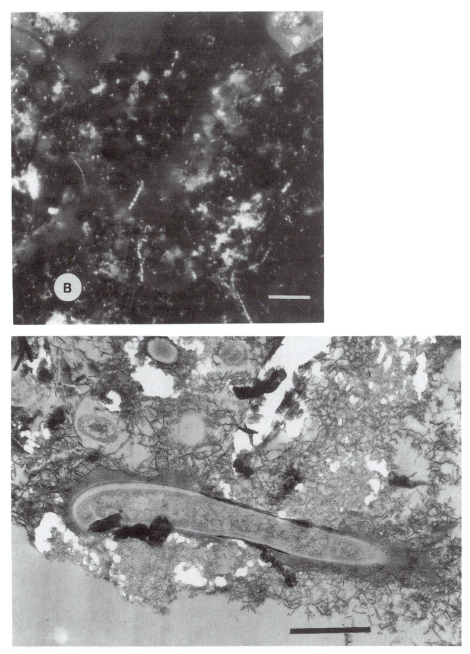

Figure 15.13 Continued.

cell types was observed: cocci developed into rods, suggesting *Arthrobacter*. The coccus-to-rod transformation coincided with a peak in oxygen consumption. CO_2 fixation was detectable as the rods started to bind Mn and Fe to their cell surface. If this metal binding involved at least in part oxidation, especially of manganese, autotrophic growth at the expense of manganese could have been taking place at that point (Kepkay et al., 1986).

Sediment samples from the experimental site in Lake Charlotte where concretions are found revealed the presence of microscopic precipitates and iron (Kepkay, 1985a,b). How these precipitates are related to the concretions has yet to be elucidated. In places where macrophytes (*Eriocaulon sepangulare*) grew in the Concretion Cove area, bacterial manganese oxidation resulted in the formation of a finely disperse oxide within the sediments. In places where macrophytes were absent, ferromanganese concretions were noted at the sediment surface. The macrophyte root system appeared to influence manganese oxidation by somehow promoting downward movement of manganous manganese in the sediments. Manganese oxidation was most active 1–3 cm below the sediment surface where the macrophytes grew but occurred at the sediment surface where macrophytes were absent (Kepkay, 1985c).

Ferromanganese concretions have been reported in other lakes in North America [e.g., Great Lake, Nova Scotia and Mosque Lake, Ontario (Harriss and Troup, 1970), Lake Ontario (Cronan and Thomas, 1970); Lake George, New York (Schoettle and Friedman, 1971), and Lake Michigan (Rossman and Callender, 1968)]. Typical compositions are listed in Table 15.4. In general, concretions form in areas of lakes where the sedimentation rate is low. Growth rates may be very slow, as in Lake Ontario (0.015 mm yr^{-1}), or more rapid, as in Mosque Lake (1.5 mm yr^{-1}). Not all geologists are convinced that microbes play a role in formation of the concretions. Varentsov (1972), for instance, explains the formation of such nodules in Eningi-Lampi Lake, Karelia, Russia, in terms of chemosorption and autocatalytic oxidation.

In the artificial environment created behind a tailings dam in the mining operation of the Mary Kathleen Mine, North Queensland, Australia, bacteria occur in association with the microalga *Chlamydomonas acidophilus*, which oxidize Mn^{2+} to form disordered γ-MnO_2 via manganite (γ-MnOOH) under laboratory conditions (Greene and Madgwick, 1991). An unidentified component of the algal cells stimulated the oxidation process. Intact cells were, however, most effective as stimulants of the oxidation process. This process is evidently an example in which Mn(III) oxide appears to be an intermediate in Mn(IV) oxide formation.

Table 15.4 Composition of Some Lake Ores (Values in Percent)

Mn	Fe	Si	Al	Source	Reference
36.08 (MnO$_2$)	13.74 (Fe$_2$O$_3$), 7.70 (FeO)	12.75 (SiO$_2$)	12.50 (Al$_2$O$_3$)	Ship Harbour Lake	Kindle (1932)
13.4–15.4 (Mn)	19.5–27.5 (Fe)	4–10 (Si)	0.7–0.95 (Al)	Oneida Lake	Dean (1970)
31.7–35.9 (Mn)	14.2–20.9	—	—	Great Lake	Harriss and Troup (1970)
15.7 (Mn)	39.8–40.2 (Fe)	—	—	Mosque Lake	Harriss and Troup (1970)
17.0 (Mn)	20.6 (Fe)	—	—	Lake Ontario	Cronan and Thomas (1970)
3.57 (Mn)	33.2	—	—	Lake George	Schoettle and Friedman (1971)
0.89–22.2 (MnO)	1.34–60.8 (FeO)	—	—	Lake Michigan	Rossman and Callender (1968)

Bacterial Manganese Oxidation in Water Distribution Systems   In 1962, a case of microbial manganese precipitation in a water pipeline connecting a reservoir with a filtration plant of the waterworks in the city of Trier, Germany, was reported (Schweisfurth and Mertes, 1962). The accumulation of precipitate in the pipes caused a loss of water pressure in the line. The sediment in the pipe had a dark brown to black coloration and was rich in manganese but relatively poor in iron content. Microscopic examination of the sediment revealed the presence of cocci and rods after removal of MnO$_2$ with 10% oxalic acid solution. Sheathed bacteria were found only on the rubber-sealed seams, and then always in association with other bacteria. Evidence of fungal mycelia and streptomycetes was also found. In culture experiments, only gram-negative rods and fungi were detected. Chemical examination of the reservoir revealed that the manganese concentration in the bottom water was between 0.25 and 0.5 mg L^{-1} during most of 1960, except in September-October, when it ran as high as 6 mg L^{-1}. The peak in manganese concentration in bottom waters was correlated with water temperature which reached its peak at about the same time as the manganese concentration in the water. The two feed streams into the reservoir did not contribute large amounts of manganese, only about 0.05 mg L^{-1}. The major source of the manganese could have been the manganiferous minerals of the reservoir basin and surrounding watershed.

Similar observations were made in some pipelines connecting water reservoirs with hydroelectric plants in Tasmania, Australia (Tyler and Marshall, 1967a). In this instance, pipelines from Lake King William were found to have heavy deposits of manganic oxide, whereas those from Great Lake did not. The deposits in pipes leading from Lake King William accumulated to a maximum thickness of 7 mm in 6–12 months. The manganic oxide deposition process was reproduced in the laboratory in a recirculatory apparatus (Tyler and Marshall, 1967a). With water from Lake King William, a deposit of a brown manganic oxide formed at the edge of coverslips after 24 h and for 6 days thereafter. Subsequent addition of $MnSO_4$ to the water caused further deposition after 6 days. By contrast, only slight traces of deposit developed with Great Lake water in similar experiments. The difference in manganese deposition from the two lake waters was explained in terms of the difference in manganese content of the two waters. Lake King William water contained 0.01–0.07 ppm of manganese, while Great Lake water contained only 0.001–0.013 ppm. In the laboratory, inoculation of Great Lake water with some Lake King William water did not promote the oxidation unless $MnSO_4$ was also added. It was also found that if Lake King William water was autoclaved or was treated with azide (10^{-3} M final concentration), manganic oxide deposition was prevented, suggesting the participation of a biological agent in the reaction. Inoculation of autoclaved Lake King William water with untreated water caused resumption of the oxidation. The dominant organism involved in the oxidation appeared at be one identified as *Hyphomicrobium* sp. Sheathed bacteria and fungi and possibly *Metallogenium symbioticum* were also encountered in platings from the pipeline deposit but only at low dilutions. It was not resolved whether *Hyphomicrobium* sp. oxidized Mn^{2+} enzymatically or non-enzymatically (Tyler and Marshall, 1967b). Since the publication of the original studies of manganic oxide pipeline deposit in Tasmania, *Hyphomicrobium* has been found in pipeline deposits in other parts of the world (Tyler, 1970). It was not always the only manganese-oxidizing organism present, however.

15.10   MICROBIAL MANGANESE DEPOSITION IN MARINE ENVIRONMENTS

Manganese-oxidizing bacteria have been detected in various parts of the marine environment, including the water column and surface sediments in estuaries, continental shelves and slopes, and abyssal depths. They appear to play an integral role in the manganese cycle in the sea.

Manganese is unevenly distributed in the marine environment. It occurs in greater quantities on and in sediments than in seawater, and in greater quantities in seawater than in the biomass (see Table 15.5 for Mn distribution in the Pacific Ocean). A significant portion of manganese at the sediment water interface at abyssal depths in some parts of the oceans is concentrated in ferromanganese concretions (nodules) and crusts. The concentration of manganese in surface seawater from the Pacific Ocean has been reported to fall in the range of 0.3 to 3.0 nmol kg^{-1} (16.4–164.8 ng kg^{-1}) (Klinkhammer and Bender, 1980; Landing and Bruland, 1980). The concentration in bottom water is generally less than that in surface water. Klinkhammer and Bender (1980) found it to be one-fourth or less at some stations in the Pacific Ocean. Manganese concentration in surface waters over the continental slope near the mouth of major rivers such as the Columbia River on the coast of the state of Washington, U.S.A., may be as high as 5.24 nmol kg^{-1} (164.8 ng kg^{-1}) (Jones and Murray, 1985). Exceptionally high manganese concentrations occur around active hydrothermal vents on midocean spreading centers; manganese concentrations in hydrothermal solution as high as 1002 μmol kg^{-1} (55.05 mg kg^{-1}) have been reported (Von Damm et al., 1985). Although diluted as much as 8500 times within tens of meters from a vent source when the hydrothermal solution mixes with bottom water, plumes containing elevated manganese concentrations may extend for 1 km or more from the vent source (Baker and Massoth, 1986), and in some instances for hundreds of kilometers.

The dominant oxidation state of manganese in seawater is +2 despite the alkaline pH of seawater (7.5–8.3)(Park, 1966) and its E_h of 430 mV (ZoBell, 1946). The stability of the divalent manganese in seawater is attributable to its complexation by the abundant chloride ions (Goldberg and Arrhenius, 1958), by sulfate and bicarbonate ions (Hem, 1963), and by organic substances such as amino acids (Graham (1959). The oxidation

Table 15.5 Manganese Budget for the Pacific Ocean

Total Mn (As MnO) in sediments	1.4 × 10^{15} tons
Total Mn (as MnO) in nodules	3.1 × 10^{11} tons (170 times)[a]
Total Mn (as MnO) in seawater	1.8 × 10^9 tons
Total Mn (as MnO) in biomass	1 × 10^7 tons (0.0055 times)[a]

[a] Relative to manganese in seawater.
Source: Based on data from Poldervaart (1955), Mero (1962), and Bowen (1966).

state of manganese in marine sediments and ferromanganese concretions is mainly $+4$ based on samples taken in the Pacific Ocean north of the equator and in the Central Indian Ocean (Kalhorn and Emerson, 1984; Murray et al., 1984; Pattan and Mudholkar, 1990). A primary source of manganese in the oceans is that injected by hydrothermal solutions at vents on midocean spreading centers. Magma and volcanic exhalations from submarine volcanoes also make a contribution. Other contributions are from continental runoff and from aeolian sources (windblown dust).

Manganese at the low concentrations found in most of the seawater is important biologically as a micronutrient, but it is too dilute to serve as energy source without prior concentration. The manganese assimilated by marine organisms can be viewed as transiently immobilized but is returned to the available manganese pool upon the death of the organisms.

Microbial Manganese Oxidation in Bays, Estuaries, Inlets, etc. Thiel (1925) detected manganese-oxidizing bacteria in marine mud collected in Woods Hole, Massachusetts, U.S.A. He also found that in anaerobic culture, sulfate-reducing bacteria precipitated $MnCO_3$. Krumbein (1971) reported the presence of heterotrophic bacteria and fungi in the Bay of Biscay and in the Heligoland Bight (North Sea) by culturing samples collected at these locations. In the Bay of Biscay the organisms were present in all samples collected in the uppermost millimeters of sediment at depths of 13–180 m, but few (only in one of several samples taken) between 280 and 540 m. In plate counts from sediment and water samples from the estuary of the River Tamar and the English Channel on a $MnSO_4$-containing culture medium prepared in seawater (3.0% sea salt) with peptone and yeast-extract, Vojac et al. (1985a) found that 11–85% of the total bacterial population formed colonies representing manganese oxidizing bacteria. They identified the manganese oxidizing colonies by applying berbelin blue reagent (Krumbein and Altman, 1973), which is turned blue by oxidized manganese. No obvious correlation between total number of bacteria, proportion of manganese oxidizers and particulate load or salinity was noted, nor were any seasonal trends with regard to the distribution of manganese oxidizers observed. Manganese oxidation rates in water from Tamar Estuary amended with 2 mg Mn^{2+} L^{-1} were 3.32 μg L^{-1} h^{-1} in freshwater containing 30 ml L^{-1} suspended matter and 0.7 μg L^{-1} h^{-1} in saline water (32 o/oo salinity) with the same amount of suspended matter. The rate of manganese oxidation was proportional to particulate load up to 100 mg L^{-1}. Oxidation was depressed in the presence of metabolic inhibitors such as chloramphenicol (100 μg ml^{-1}) and $HgCl_2$ (10 μg ml^{-1}). The oxidation had a temperature optimum of 30°C, which was above the in situ temperature (13.5°C) when the water sample was taken (Vojak et al., 1985b).

Saanich Inlet, on the southeast side of Vancouver Island, presents an interesting natural setting in which to study bacterial Mn^{2+} oxidation. As described by Anderson and Devol (1973) and Richards (1965) (see also Tebo et al., 1985), it is a fjord having a maximum water depth of 220 m and features a sill at its mouth at a depth of 70 m. Water behind the sill develops a chemocline in late winter and summer, becoming anoxic below 130 m. The anoxic condition prevails for about 6 months after which it is displaced by dense oxygenated water pushed over the sill as a result of strong coastal upwelling. Bacterial Mn^{2+} oxidizing activity was measured just above the O_2/H_2S interface and found to be O_2-limited and temperature-dependent. Excess concentration of Mn^{2+} inhibited its oxidation (Emerson et al., 1982; Tebo and Emerson, 1985). Manganese removal involved both binding to particulates (bacterial cells and inorganic aggregates) and oxidation of manganese. The oxidation state of particulate manganese in the water samples upon collection was in the range 2.3–2.7, suggesting the possible in situ formation of Mn(III) and mixed Mn(II)-Mn(IV) oxides (Emerson et al., 1982). Mn(II) binding and oxidation was accompanied by Co(II) binding at the chemocline (Tebo et al., 1984).

In Framvaren Fjord, Norway, evidence of bacterial Mn^{2+} oxidation has also been observed above the chemocline, which is located in the euphotic zone at a depth of 17 m and persists all year. The manganese oxidation rate, measured as O_2-dependent ^{54}Mn binding, was greater than in Saanich Inlet (Tebo et al., 1984).

Manganese Oxidation on the Ocean Floor Manganese oxides can be found in large quantities in concretions (nodules) or crusts on the ocean floor at great distances from hydrothermal discharges where the rate of sedimentation is low (Fig. 15.14) (Margolis and Burns, 1976). They have been found in all the oceans of the world (Horn et al., 1972). They may cover vast areas of the ocean floor, as on some parts of the Pacific Ocean floor, or be distributed in patches. Typical composition of such nodules is given in Table 15.6. The chemical components of a nodule are not evenly distributed throughout its mass (Sorem and Foster, 1972). When examined in cross-section, nodules are seen to have developed around a nucleus, which may be a foraminiferal test, a piece of pumice, a shark's tooth, a piece of coral, an ear bone from a whale, an older nodule fragment, etc.

The oxidation state of manganese in nodules is mainly +4 (Murray et al., 1984; Piper et al., 1984; Pattan and Mudholkar, 1990). Manganese occurs as todorokite, birnessite, and δ-MnO_2 (disordered birnessite, ac-

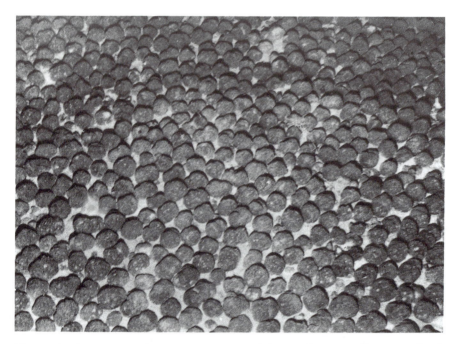

Figure 15.14 A bed of ferromanganese nodules on the ocean floor at a depth of 5,292 m in the southwest Pacific Ocean at 43°01'S and 139°37'W. Nodules may range in size from <1 to 25 cm in diameter. Average size has been given as 3 cm. (From Heezen and Hollister, 1971, by permission.)

cording to the nomenclature of Burns et al., 1974). The Mn(IV) oxides in the nodules have a strong capacity to scavenge cations (e.g., Crerar and Barnes, 1974; Ehrlich et al., 1973; Loganathan and Burau, 1973; Varentsov and Pronina, 1973), particularly Mn^{2+} and other transition metals of the first series. The nodules thus serve as concentrators of divalent manganese and some other ions in seawater and are therefore able to furnish manganese in sufficient concentration to serve as an energy source to Mn(II)-oxidizing bacteria that live on nodules. This makes ferromanganese nodules a selective habitat for these bacteria and other organisms that may depend on them directly or indirectly. Electron microscopic examination and culture tests have shown the presence of various kinds of bacteria on the surface and within nodules (Fig. 15.15) (LaRock and Ehrlich, 1975; Burnett and Nealson, 1981; Ehrlich et al., 1972). Their numbers have been found to range from hundreds to tens of thousands

Table 15.6 Average Concentration of Some
Major Constituents of Manganese Nodules
from the Pacific Ocean (Percent by Weight)

Mn	24.2	Mg	2.7
Fe	14.0	Na	2.6
Co	0.35	Al	2.9
Cu	0.53	Si	9.4
Ni	0.99	L.O.I.[a]	25.8

[a] L.O.I., Loss on ignition.
Source: From Mero (1962).

per gram of nodule, as determined by plate counts on seawater-nutrient agar at 14–18°C and atmospheric pressure (Ehrlich et al., 1972; see also Sorokin, 1972). These numbers are probably underestimates, since the organisms grow in clumps or microcolonies on or in a nodule, from which they cannot be easily dislodged. Indeed, it is necessary to plate suspended, crushed nodule material rather than washings of nodule fragments to make the counts.

The microbial population on nodules includes three types of bacteria when considered in terms of their action on manganese compounds (Ehrlich et al., 1972). The three types are Mn(II) oxidizers, Mn(IV) reducers, and a group that can neither oxidize Mn(II) nor reduce Mn(IV). In the nodules examined by Ehrlich et al. (1972), the Mn(IV) reducers were most numerous. These findings, however, must not be interpreted to mean that the Mn(IV) in the nodules examined was necessarily undergoing active reduction. Neither the Mn(II) oxidizers nor the Mn(IV) reducers are dependent on respective forms of manganese in order to grow. Thus, it it is quite conceivable that the nodules were undergoing net manganese accretion at the time of collection. A plentiful supply of Mn(II) and oxygen as well as a deficit in organic carbon are needed to favor Mn(II) oxidation, whereas an adequate supply of oxidizable organic carbon is needed to favor Mn(IV) reduction. The bacteria that neither oxidize nor reduce manganese may play an important role in keeping the level of oxidizable organic carbon low, thereby favoring the Mn(II)-oxidizing activity by the Mn(II)-oxidizing bacteria. The inability of Sorokin (1972) to culture Mn(II)-oxidizing bacteria from his specimens may have had to do with the culturing method and the method of detection of manganese oxidation he employed.

Most of the organisms found on nodules are gram-negative rods (Ehrlich et al., 1972), although gram-positive bacilli, micrococci, and *Arthro-*

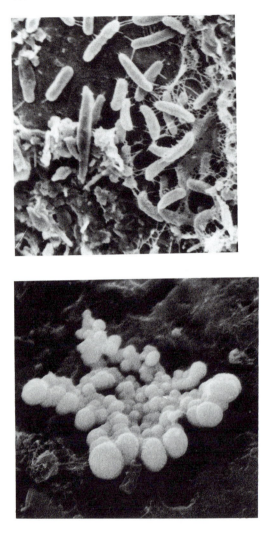

Figure 15.15   Scanning electron photomicrographs of bacteria attached to the surface of ferromanganese nodules from Blake Plateau, off the southern Atlantic coast of the United States. Note the slime strands anchoring the rod-shaped bacteria to the nodule surface. (Reproduced from LaRock, P. A. and H. L. Ehrlich, Observations of Bacterial Microcolonies on the Surface of Ferromanganese Nodules from Blake Plateau by Scanning Electron Microscopy. Microbial Ecology 2: 84. Copyright © 1975 Springer-Verlag, New York, by permission.)

bacter have also have been isolated from them (Ehrlich, 1963). A curious and as yet unexplained finding has been that a significantly greater number of isolates recovered from the central Pacific were unable to grow in freshwater media than from the eastern Pacific.

All Mn(II)-oxidizing bacterial cultures isolated from nodules and associated sediments by Ehrlich and associates are of the Group-II type (see section 15.5). They oxidize Mn^{2+} only if it is first bound to Mn(IV) oxide, ferromanganese, or certain ferric iron-coated sediment particles (Ehrlich, 1978b; 1982; 1984). As previously mentioned, Mn(IV) oxides at neutral to alkaline pH act as scavengers of cations, including Mn^{2+} ions. Given that Mn(IV) oxide is the product of manganese oxidation, the oxidation process generates new scavenging sites, and in this way it keeps nodules growing. The scavenging action of Mn(IV) oxides probably explains how other cationic constituents get incorporated into nodules:

$$H_2MnO_3 + X^{2+} \Rightarrow XMnO_3 + 2H^{2+} \tag{15.16}$$

where X^{2+} represents a divalent cation (e.g., Co^{2+}, Cu^{2+}, Ni^{2+}) (Goldberg, 1964; Ehrlich et al., 1973; Loganathan and Burau, 1973; Varentsov and Pronina, 1973). Iron, if incorporated in this fashion, is probably picked up as Fe(III).

The initial Mn(IV) oxide formed by the bacteria is probably amorphous. The characteristic mineral assemblages identified in nodules by mineralogists are probably formed subsequently by a slow "aging" process (crystallization?) involving structural rearrangements of the nodule components (diagenesis).

The growth rate of manganese nodules in the deep sea is reportedly very slow. Ku and Broecker (1969) and Kadko and Burckle (1980), for instance, have reported rates ranging from 1 to 10 mm/10^6 years, based on radioisotope dating methods. Reys et al. (1982) found a nodule in the Peru Basin whose rate of growth they estimated radiometrically to have been 168 ± 24 mm/10^6 years. These estimates assume a constant rate of growth. Heye and Beiersdorf (1973) found variable growth rates for deep sea nodules that they had examined by a fission track method. They reported rates varying from 0 to 15.1 mm/10^6 years. In other words, the nodules these investigators examined did not grow at constant rates. This suggests that conditions must not be continually favorable for nodule growth and that quiescent periods and, perhaps, even periods of nodule degradation, very likely helped by the Mn(IV)-reducing bacteria, must intervene between growth periods. Bender et al. (1970) estimated manganese accumulation rates from five nodule specimens as ranging from 0.2 to 1.0 mg cm^{-2} 10^{-3} yr^{-1}. If manganese incorporation and therefore nodule growth is an intermittent process, actual growth rates may be somewhat faster than the above estimates from radiodating, but probably by no more than an order of magnitude. Rates of manganese oxidation with bacteria isolated from nodules determined under optimal conditions in the laboratory (15°C, hydrostatic pressure of ~1 atm) were many orders of magnitude faster than any estimated in situ rates of manganese incorporation into nodules (and therefore nodule growth rates). By contrast, growth of nodule bacteria and bacterial manganese oxidation rates are greatly slowed by in-situ hydrostatic pressures (e.g., 300–500 atm) and low temperature (e.g., 4°C) (Ehrlich, 1972). Hence the slow growth rates of nodules are not a sufficient argument against microbial participation in nodule genesis. Conditions appear not to be continually favorable for nodule growth. Quiescent periods and, perhaps, even periods of nodule degradation, which may be the result, at least in part, of the activity of Mn(IV)-reducing bacteria must intervene between nodule growth periods.

Other biogenic mechanisms of nodule growth have been evoked in the past. Butkevich (1928) tried to explain the growth in terms of iron precipitation by *Gallionella* and *Persius marinus*, the latter newly isolated by him but not otherwise known, which he found associated with brown deposits of the Petchora and White Seas. However, Sorokin (1971) has concluded that *Gallionella* does not occur in deep-sea nodules, nor could he find evidence for the presence of *Metallogenium*. The only types of bacteria that he detected in nodules were some other heterotrophs. The Petchora and White Seas are, of course, marginal seas and untypical of the open ocean, which may explain Butkevich's findings. Novozhilova and Berezina (1984) did report finding *Gallionella* in a sample from Atlantis Deep I in the Red Sea. They found *Metallogenium, Caulococcus, Siderococcus* and *Naumanniella* besides some other types of bacteria in samples from the northwestern Indian Ocean. Thus, these bacteria can be isolated from marine settings. It is, however, very likely that the stations from which they collected their samples were subject to continental influences that contributed these bacteria.

Graham (1959) proposed that manganese nodules form as a result of bacterial destruction of organic complexes of Mn(II) in seawater liberating Mn^{2+} ions, which then autoxidize and precipitate, the oxides collecting around foci such as foraminiferal tests. According to Graham, other trace metals could be deposited in a similar manner. He felt that amino acids or peptides were the complexing agents attacked. He detected amino acid-like material in nodules from Blake Plateau. Graham and Cooper (1959) were able to recover foraminiferal tests from Blake Plateau, among which arenaceous ones were coated with a veneer of manganese-rich material containing Cu, Ni, Co, and Fe. Since the manganese-rich coating was on the surface of the tests, the authors reasoned that the manganese was deposited on the foraminifera after their death. They found calcareous foraminiferal tests to be free of this deposit. Graham and Cooper implicated a protein-rich coating on the arenaceous tests as responsible for providing a habitat for organisms (bacteria?) that remove trace-element chelates from seawater.

Kalinenko et al. (1962) visualized the origin of ferromanganese nodules in terms of bacterial colonies growing on ooze particles where they mineralize organic matter coating the particles. In the process, manganese and iron oxides are supposed to be formed and deposited together with other trace elements on the colonies. Slime formed by the bacteria is assumed to cause the deposits to agglutinate, producing micronodules. The investigators studied this process in laboratory experiments by watching bacterial development on glass slides introduced into oozes from the bottom of several Indian Ocean stations. They described the bacteria as the living cement that holds the nodules together.

Ghiorse (1980) examined the surface layers of ferromanganese concretions from the Baltic Sea by scanning and transmission electronmicroscopy and by light microscopy. He noted filamentous structures (possibly fungal hyphae) and single cells, microcolonies or aggregates of bacteria on the outer surface and some rods and cocci within a matrix of polymerlike material and occasionally sheathed bacteria. Budding and star-shaped bacteria and bacteria with internal membranes were also seen (possibly cyanobacteria, purple photosynthetic bacteria, or methanotrophic bacteria). The bacterial assemblage described by Ghiorse reflects a not unexpected terrestrial influence on the development of the Baltic Sea concretions and suggests an involvement of some different microbes in the growth of these nodules.

Burnett and Nealson (1983) examined the surface of some fragments from North Pacific Ocean nodules by energy dispersive X-ray analysis. They found regions of high Mn and Fe concentrations that outlined microorganismlike objects on the nodule surface. They inferred from this observation that the microorganisms were involved in accretion or in removal the metals because they would have expected the Mn and Fe to have been evenly distributed over the surface if a purely physicochemical process had been involved.

Greenslate (1974a) observed manganese deposition in microcavities of planktonic debris, especially diatoms, and proposed that such deposition was the beginning of nodule growth. He also found remains of shelter-building organisms such as benthic foraminifera on nodules, which became encrusted and ultimately buried in the nodule structure and proposed that the skeletal remains provided a framework on which manganese and other nodule components were deposited, perhaps with the help of bacterial

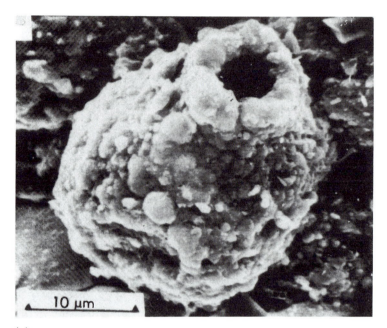

(a)

Figure 15.16 Benthic, test-forming protozoans inhabiting the surface of ferro-manganese nodules. (a) Fresh remains of chambered encrusting protozoans on a nodule surface, showing siliceous biogenic material used in test construction. (b) Surface of a nodule with a partial test of *Saccorhiza ramosa,* the most common and longest of any tubular agglutinating foraminifera yet found on nodules. (c) Test of unidentified form composed almost entirely of manganese micronodules. (From Dugolinsky et al., J. Sediment. Petrol. 47:428–445, 1977, by permission.)

action (Greenslate, 1974b). Others have since reported evidence of traces of such organisms on nodules (Fredericks-Jantzen et al., 1975; Bignot and Dangeard, 1976; Dugolinsky et al, 1977; Harada, 1978; Riemann, 1983) (Fig. 15.16).

The finding that benthic foraminifera and other protozoans are growing, have grown, and may presently be growing on ferromanganese nodules is of significance in explaining the role of Mn(II)-oxidizing bacteria in nodule growth. Since these foraminifera and other protozoans are **phagotrophic** (i.e., they live at the expense of bacterial cells), they probably feed on the Mn(II)-oxidizing bacteria, among other types, on the nodules. To maintain this food supply, uneaten Mn(II)-oxidizing bacteria must, therefore, continue to multiply. Thus, Mn(II)-oxidizing bacteria on nod-

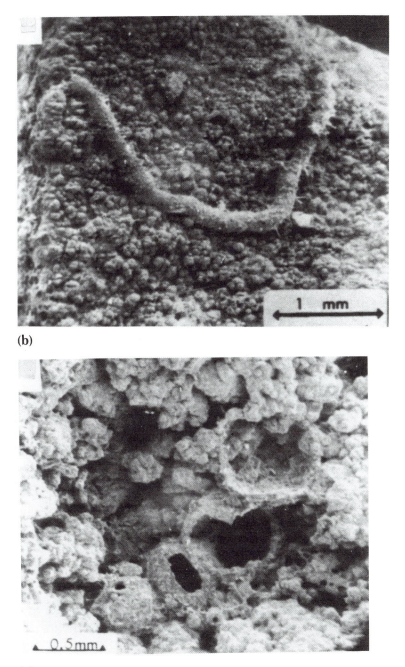

(b)

(c)

ules may play a dual role: (a) to aid in manganese accretion to nodules and (b) to serve as food for phagotrophic protozoans. A somewhat different interpretation of the interrelationship of bacteria, protozoans, and tube-building polychaete worms resident on nodules has been offered by Riemann (1983). He found rhizopods on the nodules to accumulate large volumes of fecal pellets (stercomata) containing biogenic and mineral particles (manganese oxides, etc.) in their tests as a result of feeding on bacteria and detritus and organic primary film on the nodule surface. He proposed that manganese oxides ingested by the protozoa are at least partially reduced, released and reprecipitated, in part by bacterial activity on the nodule surface. He viewed polychaete worms feeding on the rhizopods as further aiding in the concentration of the manganese oxide-containing stercomata.

Although the hypotheses of Graham, Greenslate, and to some extent of Kalinenko et al. may well have some bearing on the initiation of formation of some nodules, it seems doubtful that they apply to the major growth phase of nodules. Graham and Kalinenko et al. assumed that most of the Mn^{2+} in seawater is organically complexed. It is incorporated into nodules as preformed Mn(IV) oxide that resulted from autoxiation of the Mn^{2+} freed from organic complexes through ligand degradation by microbes. Incorporation of Mn(IV) oxide into nodules is also a part of the assumption of Riemann, but his hypothesis does not absolutely depend on it. Sorokin (1972) also assumed the role of heterotrophic bacteria on nodules in nodule genesis to be one of digesting organic manganese complexes, releasing Mn^{2+}. However, he postulated that the Mn^{2+} becomes bound to the surface of nodules and is then abioticallly oxidized. Filter-feeding, benthic invertebrates resident on the nodules were viewed by Sorokin as the source of the organically complexed manganese. None of these proposals give recognition to the fact that most manganese in seawater, as previously pointed out, exists as inorganic complexes from which Mn^{2+} is readily adsorbed by nodules.

Manganese Oxidation Around Hydrothermal Vents   A novel type of biological community driven by geothermal rather than radiant energy from the sun has been discovered around hydrothermal vents situated on mid-ocean spreading centers. Because sunlight does not reach the ocean depths where these communities exist and because introduction of nutrients by thermal convection from overlying surface waters does not furnish sufficient nutrients, the ecosystem depends on primary producers that are chemolithotrophic bacteria. The consumers in the community depend on these chemolithotrophs, directly at the first trophic level and indirectly at the other trophic levels. Indeed, at the first trophic level, many of the consumers have established an intimate symbiotic relationship with some of the primary producers (see Chapter 18 and discussions by Jannasch and Wirsen, 1979; Jannasch, 1984; Jannasch and Mottl, 1985). The available energy sources for the primary producers (chemolithotrophic bacteria) vary, depending on the output of the vents. Potential energy sources include reduced forms of sulfur (H_2S, $S°$, $S_2O_3^{2-}$), H_2, NH_4^+, NO_2^-, Mn^{2+}, CH_4, and CO (Jannasch and Mottle, 1985). Although ferrous iron is frequently a major constituent of hydrothermal fluid from black smokers, it is least likely to serve as energy source because of its great tendency to rapidly precipitate as iron sulfide at the mouth of a vent when the hydrothermal solution meets cold, oxygenated bottom water. Most of any iron which escapes sulfide precipitation quickly autoxidizes. Available electron acceptors for bacteria in the vent community include O_2, NO_3^-, $S°$, SO_4^{2-}, CO_2, and Fe^{3+} (Jannasch and Mottl, 1985), and possibly Mn(IV) oxide.

The hydrothermal solution results from deep penetration (as much as 1–3 km) of seawater into the basalt at midocean ridges. When a magma chamber underlies the region of seawater penetration, the heat diffusing from the chamber (the temperature may be in excess of 350°C) and the high hydrostatic static pressure exerted at this site cause the seawater to react with the basalt. Seawater sulfate may be chemically reduced to sulfide by ferrous iron in the basalt (a rare example of chemical sulfate reduction in nature), bicarbonate may become reduced to methane, Mg^{2+} from seawater may react with silica in the basalt to form talc and protons according to the reaction (Shanks et al., 1981):

$$Fe^{2+} + 2 Mg^{2+} + 4H_2O + 4SiO_2 = 6H^+ + FeMg_2Si_4O_{10}(OH)_2$$
$$(15.17)$$

and render the altered seawater acidic. This acidity (protons) aids in leaching of base metals (Mn^{2+}, Cu^{2+}, Zn^{2+}, Co^{2+}, etc.) from the basalt. Acidity may also be generated in the formation of calcium compounds with basalt constituents and seawater calcium (Bischoff and Rosenbauer, 1983, Seyfried and Janecky, 1985). As a result, the altered seawater (hydrothermal solution) becomes loaded with H_2S, Fe^{2+}, Mn^{2+}, Cu^{2+}, Zn^{2+}, etc. (Jannasch and Mottl, 1985).

Vents on the midocean spreading centers from which the hydrothermal solution enters the ocean are differentiated between so-called white smokers and black smokers. White smokers emit gases and a hydrothermal solution which has a temperature measured between 6 and 23°C at the mouth of the respective vents, whereas black smokers emit gases and a hydrothermal solution at a temperature near 350°C at the mouth of respective vents. In the case of white smokers the upward moving hydrothermal solution meets cold, downward moving seawater and mixes with it before emerging from a vent. As a result of this mixing with seawater, the hydrothermal solution loses a significant portion of its metal charge through precipitation in the rock stratum through which it travels before it emerges from the vent. On emergence it still contains significant amounts of H_2S and Mn^{2+} but little iron. In the case of the black smokers, the upward moving hydrothermal solution does not encounter significant amounts of downward moving cold seawater and issues little altered from vents. On emergence and mixing with cold seawater voluminous amounts of black iron-, copper-, and zinc-sulfides precipitate around the mouth of the vent of black smokers and may be deposited in the form of a chimney along with anhydrite ($CaSO_4$) (Fig. 15.17). Much of the iron in the hydrothermal solution precipitates as sulfide but that which does not, precipitates as iron oxide ($FeOOH$). A major portion of the manganese (Mn^{2+}) remains in solution and may be carried a considerable distance (tens of kilometers) before it is precipitated.

Although H_2S appears to be the most widely used energy source by the primary producers of vent communities at midocean spreading centers, Mn^{2+} because of its relative stability in seawater could also be a potential energy source. Indeed, Mn^{2+} bacteria have been isolated from samples collected around a white smoker (Mussel Bed Vent) on the Galapagos Rift (Ehrlich, 1983) and Black Smokers on the East Pacific Rise at 21° north (Ehrlich, 1985) and 10° north (Ehrlich, unpublished results). Mandernack (1992) measured moderate to high microbial Mn removal (oxidizing) activity at the source of Mn plumes on a hydrothermal vent field on the Galapagos Rift and at about 10 km distance, at between 50 and 100 m over the mouth of a vent. This study included Mussel Bed Vent. Similarly, he measured low to high microbial Mn removal (oxidizing) activity at the source of Mn plumes on the Juan de Fuca hydrothermal vent field on Endeavor Ridge and at about 10 km distance, at between 50 and 200 m over the mouth of a vent.

Two isolates from around Mussel Bed Vent and three from around a vent at 21° north have been studied in some detail. They are all gram-negative bacteria and, depending on the isolate, have the shape of short or curved rods or spirilla (Fig. 15.18). The isolates from around the white smoker grow in a temperature range of 4–37°C and those from around the black smoker in a temperature range of 5–45°C. None grow in nutrient broth prepared in freshwater. The two isolates from Mussel Bed Vent also do not grow in nutrient broth prepared with 3% NaCl. Both isolates from around Mussel Bed Vent and two of the isolates from around the vent at 21° north are examples of Group I manganese oxidizers while the third isolate from around the vent at 21° north is an example of a Group

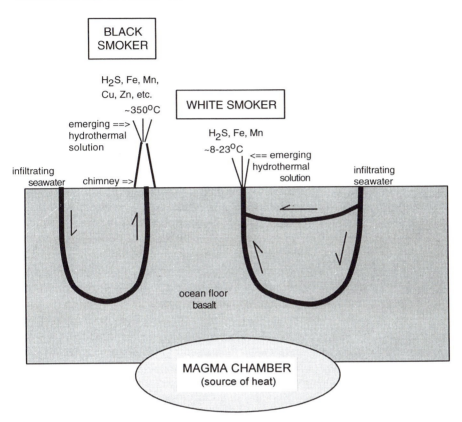

Figure 15.17 Schematic representation of the origin of hydrothermal solutions from black smoker and white smoker vents at midocean rifts. (Adapted from Jannasch and Mottl, 1985.)

II manganese oxidizer. In all, the manganese-oxidizing system is inducible, even in the Group II manganese oxidizer. This is in contrast to all Group II manganese oxidizers isolated from ferromanganese nodules and associated sediments in which the manganese-oxidizing system is constitutive. All five vent isolates appear to be able to obtain energy from Mn(II) oxidation. This conclusion was reached on the basis of inhibition studies of manganese oxidation with electron transport inhibitors, and in the case of the Mussel Bed Vent isolates by direct determination of ATP synthesis coupled to Mn^{2+} oxidation (Ehrlich, 1983, 1985). In the Mussel Bed Vent isolates, ATP synthesis is very tightly coupled to Mn^{2+} oxidation, and

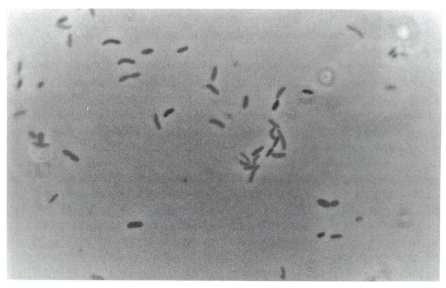

(a)

Figure 15.18 Bacterial strain SSW_{22} isolated from a water sample from near a hydrothermal vent on the Galapagos Rift. (a) Typical morphology in a seawater medium containing tryptone, dextrose, yeast extract and $MnSO_4$. Note range of cell size, typical of growth in this medium ($\times 1,500$). (b) Longitudinal sections of SSW_{22} viewed by electron microscopy ($\times 32,580$). (Courtesy of W. C. Ghiorse.)

growth in batch culture shows significant stimulation by Mn^{2+} in initial stages, which may be attributable to mixotrophy (Ehrlich and Salerno, 1988, 1990).

The two isolates from around Mussel Bed Vent are metabolically very versatile. In addition to being able to oxidize Mn^{2+}, they are also able to reduce MnO_2 aerobically and anaerobically with glucose, succinate, and acetate as electron donors (Ehrlich et al., 1988; unpublished data). Their MnO_2 reductase is inducible. They are also able to oxidize thiosulfate aerobically and reduce tetrathionate anaerobically (Tuttle and Ehrlich, 1986). The ability to use this variety of inorganic electron donors and acceptors suggests that these organisms are opportunists which will use whatever energy source or electron acceptor is in plentiful supply. Mn^{2+} emitted by the vents, thiosulfate and tetrathionate formed from partial chemical or biological oxidation of H_2S emitted by the vents, or MnO_2 from biological or chemical oxidation can all be expected to occur in

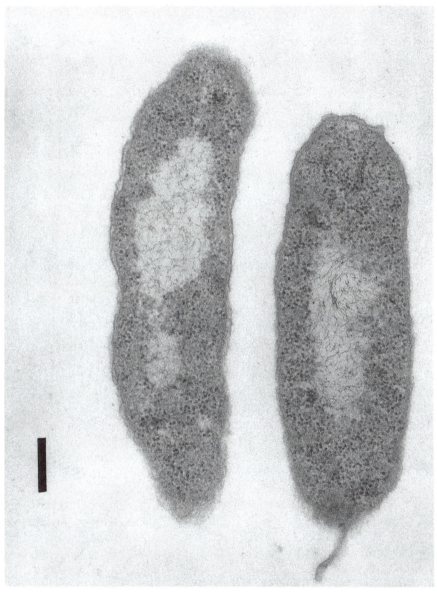

(b)
Figure 15.18 Continued.

various parts of the vent community or beyond at various times and be exploited by these bacteria. It has yet to be determined whether these bacteria may also be able to reduce ferric oxides or oxyanions, such as arsenate or selenate, and oxidize other reduced sulfur species besides thiosulfate.

While the the manganese-oxidizing cultures from around hydrothermal vents in the work cited above were obtained from water samples, Durand et al. (1990) reported finding significant numbers of heterotrophic, gram-negative manganese-oxidizing bacteria associated with the epidermis of polychaete worms and their tubes. The organisms included members of *Aeromonas* and *Pseudomonas*. They oxidized Mn(II) and grew better at 40 than at 20°C. Their significance to the worms and the rest of the vent community remains to be established.

Bacterial Manganese Precipitation in the Seawater Column   Encapsulated bacteria with iron and manganese precipitates in their capsules (presumably their glycocalyx) have been detected in the oceanic water column below 100 m (Fig. 15.19). They were found associated with flocculant amorphous aggregates (marine snow) and occasionally in fecal pellets (Cowen and Silver, 1984). In the eastern subtropical North Pacific, manganese deposits on bacterial capsules were absent in a depth range from 100 to 700 m but became increasingly noticeable below 700 m (Cowen and Bruland, 1985). Such manganese-scavenging activity by encapsulated bacteria was also very prominent in a hydrothermal particle plume (Cowen et al., 1986). A positive effect of increased hydrostatic pressure between 1 and 200 bars (1.01×10^2 to 2.03×10^4 kPa) on Mn^{2+} binding by a natural population of bacteria in such a plume has been observed (Cowen, 1989). The pressure effect may be related to exopolymer production by the bacteria. Since capsules have been found to be abundant in sediment, it has been suggested that encapsulated bacteria play a prominent role in manganese sedimentation in the ocean (Cowen and Bruland, 1985).

In the case of Mn-containing plumes of hydrothermal origin at the cleft segment of the Juan de Fuca Ridge, about 300 km west of the Oregon coast (USA), Cowen et al. (1990) found that the direct biological contribution to dissolved manganese scavenging at plume depth at on-ridge-axis stations was negligible. By contrast, biological manganese scavenging at nonplume depths contributed almost 50% to the total process. At most off-ridge axis stations, most of the dissolved manganese scavenging was almost totally biological at plume depth. Manganese-scavenging rates from a vent plume ranged from 1.7 to 3.4 mM m^{-2} y^{1-}.

Tambiev and Demina (1992) found dissolved manganese to be much more rapidly converted to particulate manganese in the southern trough

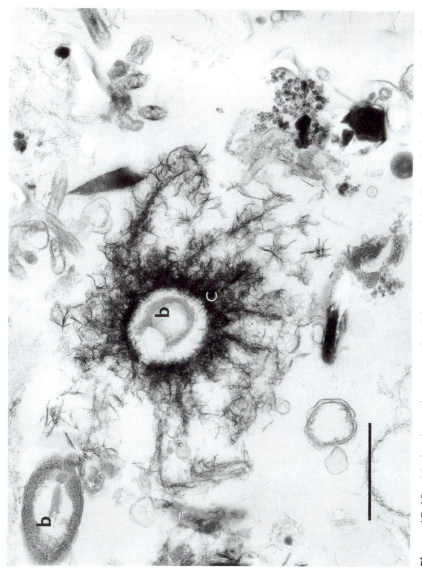

Figure 15.19 Marine bacteria encapsulated in manganese oxide. Transmission electron photomicrograph. b, bacterial cell; c, bacterial capsule. Scale bar = 1 μm. (From Cowen et al., 1986; reprinted by permission from Nature 322:169–171. Copyright © 1986 Macmillan Magazines Ltd., by permission.)

of the Guaymas Basin, Gulf of California (Sea of Cortez), than in the caldera of the Axial Mountain on the Juan de Fuca Ridge. They suggested that the rapid conversion of dissolved manganese in the Guaymas Basin involved the participation of bacteria and that benthopelagic zooplankton grazed on bacteria that had become enriched in manganese, excreting maganese oxide in the form of vernadite (MnO_2). They found particulate manganese at their Juan de Fuca Ridge stations to be associated with bacterialike aggregates.

Of the marine manganese-oxidizing bacteria studied in various laboratories to date, only some accumulate significant amounts of oxidized manganese on their cell envelope.

15.11 MICROBIAL MOBILIZATION OF MANGANESE IN SOILS AND ORES

Soils Because manganese (III) and (IV) oxides are insoluble, their manganese is not directly available as a micronutrient to plants and to soil microbes, which have a need for it in their nutrition. The manganese in these oxides must be reduced to manganese (II) to be available nutritionally. Mn^{2+} that is sorbed by the oxides as well as by clays may be available by ion exchange. Reduction of manganese oxides may be brought about by bacterial respiration in which MnO_2 replaces oxygen as terminal electron acceptor. As explained earlier, this process is not always dependent on anaerobiosis, although bacteria which only reduce MnO_2 anaerobically exist and may dominate. The electron donors are usually organic compounds, but hydrogen may serve under anaerobic conditions (Lovley and Goodwin, 1988).

Under some conditions, Mn(IV) oxides in soil can be solubilized as a result of their reduction by elemental sulfur ($S°$) or thiosulfate ($S_2O_3^{2-}$) in the presence of *Thiobacillus thiooxidans* (Vavra and Frederick, 1952). This was shown in soil perfusion studies in the laboratory. The production of sulfuric acid was not solely responsible for the solubilization of the MnO_2, because in the presence of acid but in the absence of any reduced sulfur, the MnO_2 was not solubilized. Although the greatest quantity of MnO_2 was reduced when *T. thiooxidans* cells were in contact with MnO_2, slightly more than half of the MnO_2 was reduced under the same conditions when the cells were separated from the MnO_2 by a collodion membrane. This may be a case of simultaneous direct and indirect reduction of MnO_2, the indirect reduction being the result of chemical reaction with partially reduced sulfur species generated in the bacterial oxidation of sulfur. A field study in Indiana, U.S.A., confirmed that $S°$ and $S_2O_3^{2-}$ can mobilize fixed manganese in an agriculturally manganese deficient soil (Garey and Barber, 1952).

As previously pointed out, ferrous iron and H_2S produced in microbial iron(III) and sulfate respiration can chemically reduce Mn(IV) oxides in soils and sediments (Lovley and Phillips, 1988b; Myers and Nealson, 1988; Nealson and Saffarini, 1994).

Reduction of manganese oxides of microbial origin in soil can also be brought about chemically by some constituents of root exudates as from oats and vetch especially at acid pH (Bromfield, 1958a,b). Bacteria are apparently not needed to promote Mn(IV)reduction by active root-exudate components (Bromfield, 1958a). Malate is an active root exudate component that has been shown to reduce Mn(IV) oxide chemically (Jaurequi and Reisenauer, 1982). However, in view of the widespread ability of various types of bacteria to reduce MnO_2, bacterial reduction of oxides of Mn(IV) oxides in soil is probably a more important factor in remobilizing fixed manganese in nature.

Enumeration of manganese reducers in soil is an incompletely solved problem. Attempts have been made to use heterotrophic agar medium containing hydrous manganese oxide for differential plate counts (e.g., Schweisfurth, 1968). MnO_2-reducing colonies have been identified as those which develop a clear halo around them, suggesting reduction of the manganese oxide in the area of the halo. The absence of manganese oxide in the halo can be confirmed by applying berbelin blue or leucocrystal violet reagent to the colony (Krumbein and Altmann, 1973; Kessick et al., 1972). While the clear halo indeed indicates manganese reduction, enzymatic reduction cannot be inferred because manganese reductase, which insofar as is known is not an extracellular enzyme, cannot work over a distance but must be in physical contact with any manganese oxide whose reduction it catalyzes. This means that the manganese reducing cell must be in physical contact with it. Reduction in the halo must therefore be interpreted as due to nonenzymatic reaction. If the medium contains glucose, the manganese oxide reduction may be merely the result of reaction with residual glucose at acid pH generated by the colony, but it could also be due to reaction with a metabolic product(s) formed by the bacterial colony from glucose. Potential enzymatic reduction in the same medium may be indicated if an artificial electron shuttle such as ferricyanide is added to the medium at low concentration. In that instance, a manganese oxide-free halo around a colony might indicate enzymatic or nonenzymic reduction or both. If enumeration is done on plates with and without added ferricyanide, a not very discriminating differential count for enzymatic and nonenzymatic organisms is possible (Tortoriello, 1971).

Ores The action of the autotroph *Thiobacillus thiooxidans* on elemental sulfur has been shown to be able to extract manganese from ores (Kazutami and Tano, 1967; Gosh and Imai, 1985a,b). Kazutami and Tano leached an ore containing MnO_2 (10.6%), Fe_2O_3 (25%), SiO_2 (55%), MgO (5.23%), and traces of Ca, Al, and S. The ore was suspended at a concentration (pulp density) of 3% in a culture medium containing K_2HPO_4 (0.4%), $MgSO_4$ (0.03%), $CaCl_2$ (0.02%), $FeSO_4$ (0.001%), $(NH_4)_2SO_4$ (0.2%), and S° (1%). Addition of FeS or $FeSO_4$ stimulated both the growth of *T. thiooxidans* and the solubilization of manganese, while the addition of ZnS stimulated only the solubilization of manganese; $Fe_2(SO_4)_3$ was without effect. The reducing activity of partially reduced sulfur species like thiosulfate or sulfite produced by *T. thiooxidans* in the oxidation of elemental sulfur to sulfuric acid was probably the basis for the solubilization of the MnO_2 in the ore.

Ghosh and Imai (1985a) leached manganese from MnO_2 (>90% pure) with *Thiobacillus ferrooxidans* as well as *T. thiooxidans* in medium containing the mineral salts mixture of 9K medium (see Chapter 14) with 1% elemental sulfur but without ferrous sulfate (Silverman and Lundgren, 1959). As with *T. thiooxidans*, leaching was probably the result of production of partially reduced, soluble sulfur species, such as SO_3^{2-} produced from S° by *T. ferrooxidans* strain AP19–3 (Sugio et al., 1988a,b), which then reduced MnO_2 to Mn^{2+}. They were also able to leach manganese from MnO_2 when S° in their medium was replaced by chalcocite (Cu_2S) or covellite (CuS). Increasing the **pulp density** (particle concentration) of MnO_2 from 0.5 to 5.0% increased the amount of manganese leached from MnO_2 but decreased the amount of Cu leached from chalcocite or covellite (Ghosh and Imai, 1985b).

Heterotrophic leaching of manganese ores was first attempted by Perkins and Novielli (1962). They isolated a *Bacillus* that was able to solubilize manganese from a variety of ores in an organic culture medium which contained molasses as one ingredient. The mineralogy of the ores they tested was not reported, and thus it is not clear whether reduction of manganese was involved or solubilization of manganous minerals. Heterotrophic leaching that did involve reduction of manganese oxides was reported by Trimble and Ehrlich (1968), Ehrlich et al. (1973), Agate and Deshpande (1977), Mercz and Madgwick (1982), Holden and Madgwick (1983), Kozub and Madgwick (1983), Babenko et al. (1983), Silverio (1985), Rusin et al. (1991a,b), and others.

15.12 MICROBIAL MOBILIZATION OF MANGANESE IN FRESHWATER ENVIRONMENTS

Detection of manganese-reducing bacteria in some of the Karalian Lakes (U.S.S.R) has led to the inference that manganese reduction is occurring in these lakes (Sokolova-Dubinina and Deryugina, 1967a; Troshanov,

1968). In Lake Punnus-Yarvi, *Bacillus circulans, B. polymyxa* and an un-identified nonsporeforming rod were thought to be involved in formation of rhodochrosite ($MnCO_3$) from bacteriogenic manganese oxide. The man-ganese-reducing activity that led to $MnCO_3$ formation was found to occur on the shoreward side of a depression in Punnus Ioki Bay and was related to the bacterial degradation of plant debris originating from plant life along the shore of the lake. The manganese oxides appeared to act as terminal electron acceptors. CO_2 from the mineralization of organic matter contrib-uted the carbonate. Limited bacterial sulfate reduction was thought to assist the process by helping to maintain reducing conditions. As much as 5% $MnCO_3$ has been found in the sediment at the active site (see also Chapter 8). Reduction of MnO_2 with acetate by the action of *Geobacter metallireducens* under anaerobic conditions, leading to the production of $MnCO_3$ identified as rhodochrosite, has been demonstrated in the labora-tory (Lovley and Phillips, 1988a). *G. metallireducens* was isolated from Potomac River (U.S.A.) sediment. In these instances Mn was mobilized and at least a portion reimmobilized.

Microbial Mn(IV)-reducing activity has been detected in sediments of Blelham Tarn in the English Lake District. A malate-fermenting *Vibrio* was isolated which could use Mn(IV), Fe(III), and NO_3^- as terminal elec-tron acceptors (Jones et al., 1984). In laboratory experiments, this organ-ism exhibited a 20% greater molar growth yield anaerobically on Mn(IV) oxide than on Fe(III). Mn(IV) oxide and NO_3^- inhibited reduction of iron by the organism. Similar inhibition of iron reduction was observed when a sample of lake sediment was amended with Mn_2O_3 (Jones et al., 1983).

Davison et al. (1982) reported microbial manganese reduction in the deeper sediment in the Esthwaite Water (U.K.). It was intense during winter months when manganese oxides accumulated transiently in the sediment. In the summer months, manganese oxides generated in the water column were reduced in the hypolimnion before reaching the sedi-ment. Myers and Nealson (1988) found a *Shewanella* (formerly *Alteromo-nas) putrefaciens* isolate from Lake Oneida (New York) that, although it is a facultative aerobe, reduces Mn(IV) only anaerobically with a variety of organic electron donors. It is likely one of the organisms responsible for recycling oxidized manganese in the lake. Gottfreund et al. (1983) detected manganese-reducing bacteria in groundwater to which they at-tributed Mn(III)-reducing ability.

15.13 MICROBIAL MOBILIZATION OF MANGANESE IN MARINE ENVIRONMENTS

Mn(IV)-reducing bacteria have been isolated from seawater, marine sedi-ment, and deep-sea ferromanganese concretions (nodules). To date, all isolates tested in the laboratory have been heterotrophs that can use one or more of the following organic electron donors: glucose, lactate, succinate, acetate, or the inorganic donor H_2. In 1988, in situ observations suggested that anaerobically respiring thiobacilli and some other bacteria in anaerobic sediment at the edge of a salt marsh on Skidaway Island, Georgia (U.S.A.), catalyze reduction of Mn(IV) oxides that are in contact with sulfide in a **solid phase** (e.g., FeS). The Mn(IV) oxide was thought to act as electron acceptor when sulfide was oxidized to sulfate. The process was inferred from experiments in which microbial reduction of Mn(IV) oxide by sulfide was inhibited by azide and 2,4-dinitrophenol (Aller and Rude, 1988). A process resembling this has now been demonstrated in the laboratory with elemental sulfur as electron donor (Lovley and Phillips, 1994). Organisms capable of catalyzing this reaction include *Desulfovibrio desulfuricans*, *Desulfomicrobium baculatum*, *Desulfobacterium autotrophicum*, *Desulfuromonas acetoxidans*, and *Geobacter metallireducens*. Stoichiometric transformation according to the reaction

$$S^\circ + 3MnO_2 + 4H^+ \Rightarrow SO_4^{2-} + 3Mn^{2+} + 2H_2O \qquad (15.18)$$

was demonstrated with *D. desulfuricans*. Fe(III) could not replace MnO_2 as electron acceptor in this reaction. This activity is not only of significance for the marine manganese cycle, but it also presents an important mechanism by which sulfate can be regenerated from reduced forms of sulfur anaerobically in the dark.

H$_2$S produced anaerobically by sulfate- reducing bacteria has been shown to be able to reduce Mn(IV) oxide nonenzymatically, with S° being the chief product of sulfide oxidation (Burdige and Nealson, 1986). In addition, H_2S produced by bacterial disproportionation of elemental sulfur into H_2S and SO_4^{2+} has also been shown to reduce Mn(IV) oxide chemically. Indeed, this reaction with the Mn(IV) oxide (or FeIII) appears to be thermodynamically essential in promoting continued bacterial dispr, potionation of the sulfur (Thamdrup et al., 1993).

Most of the bacterial isolates from marine environments studied to date have been aerobes that can reduce MnO_2 aerobically or anaerobically, but some evidence for strictly anaerobic reduction has been obtained (see reviews by Ehrlich, 1987; Burdige 1993). Caccavo et al. (1992) reported the isolation of a facultative anaerobe, strain BrY, from the Great Bay Estuary, New Hampshire, U.S.A., which reduced Mn(IV) oxide only anaerobically with hydrogen or lactate. The extent of in situ activity of marine MnO_2-reducing bacteria has so far not been estimated in any part of the marine environment.

Since Mn(IV) oxides are good scavengers of trace metals such as Cu, Co, and Ni, it is noteworthy that bacterial reduction of the manganese oxide in ferromanganese concretions in laboratory experiments resulted in solubilization of these metals along with manganese (Fig. 15.20) (Ehrlich et al., 1973). Ni and Co solubilization was absolutely dependent on Mn(IV) reduction. Cu release, on the other hand, was only partially dependent on it, being initially solubilized by complexation with peptone in the culture medium. Only in later stages did Cu solubilization show a direct dependence on bacterial action. This finding suggests that Cu may be more loosely bound in the nodule structure than Ni or Co. The need for bacterial action for Cu release appears to arise only when it is encapsulated by Mn(IV) oxides and not in direct contact with the solvent. Also noteworthy in these experiments was the observation that only negligible amounts of iron were solubilized, even though the Mn(IV)-reducing organisms used in these studies also had the capacity to reduce limonite and goethite. Whether this apparent inability to solubilize iron was due to an inability of the bacteria to reduce it in nodules, or to chemical reoxidation of any Fe(II) produced with remaining Mn(IV) oxide in the nodules, or to immedi-

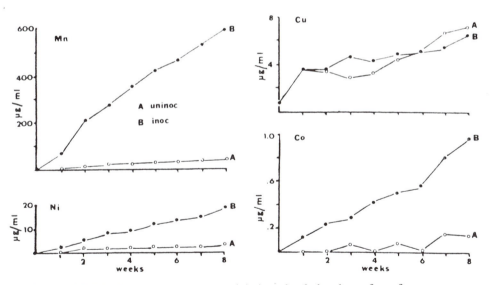

Figure 15.20 Manganese, copper, nickel, and cobalt release from ferromanganese nodule substance by *Bacillus* GJ33 in seawater containing 1% glucose and 0.05% peptone. [From H. L. Ehrlich, S. Y. Yang, and J. D. Mainwaring, Jr. Bacteriology of manganese nodules. VI. Fate of copper, cobalt, and iron during bacterial and chemical reduction of the manganese (IV). Z. Allg. Mikrobiologie 13:39–48. Copyright © 1973; by permission.]

ate autoxidation and precipitation after microbial attack is not known. On the basis of free energy yield from the reduction of iron oxide minerals, the first explanation is the most plausible. The solubilization of the scavenged trace metals by bacterial ferromanganese reduction may be of nutritional importance in the ecology of the marine environment where manganese oxides such as ferromanganese occur.

15.14 MICROBIAL MANGANESE REDUCTION AND MINERALIZATION OF ORGANIC MATTER

As discussed earlier, enzymatic manganese reduction by bacteria is often a form of respiration that can occur aerobically or anaerobically, depending on the type of organism and on prevailing environmental conditions. Organic electron donors in this bacterial process can be any one of several different compounds. Even anaerobically, they may be completely degraded to CO_2 by a single species (Lovley et al., 1993), but often complete degradation requires the successive action of two or more species (see, for instance, Lovley et al., 1989). Thus, manganese oxide respiration can be viewed environmentally as a form of mineralization of organic compounds (Ehrlich, 1993; Nealson and Saffarini, 1994). This form of mineralization is unlike that by bacterial sulfate, and most forms of iron and nitrate respiration because it can occur readily aerobically and anaerobically if the appropriate organisms are present. However, it is probably only anaerobically that manganese respiration has significant impact on the carbon cycle, and then only if no other competing forms of anaerobic respiration occur (Lovley and Phillips, 1988; Ehrlich, 1987).

15.15 MICROBIAL ROLE IN THE MANGANESE CYCLE IN NATURE

It must be inferred from the wide-spread occurrence of manganese-oxidizing and -reducing microorganisms in terrestrial, freshwater and marine environments that they play an important role in the geochemical cycle of manganese. At neutral pH under aerobic conditions, manganese-oxidizing microbes clearly are more important in immobilizing manganese in soils and sediments than are iron-oxidizing microbes in immobilizing iron in view of the relative resistance of Mn(II) to autoxidation at pH values below 8 in contrast to Fe(II) (see, e.g., Diem and Stumm, 1984; also chapter 14 in this book). The manganese-reducing bacteria are important at neutral pH because manganese oxides under reducing conditions in the absence of strong reducing agents such as H_2S or Fe^{2+} are relatively stable. As Figure 15.21 shows, manganese oxidation reactions generally lead to manganese fixation because most Mn(III) products and all Mn(IV) products are usually insoluble. Gottfreund and Schweisfurth (1983) did suggest that soluble Mn(III) complexes may be formed in microbial manganese oxidation. By contrast, reduction of Mn(III) and Mn(IV) oxides generally leads to solubilization of manganese. Under some reducing conditions, the solubilized manganese may be reprecipitated as $MnCO_3$, however (see Dubinina-Sokolova and Deryugina, 1967a; Lovley and Phillips, 1988a; also section 15.9).

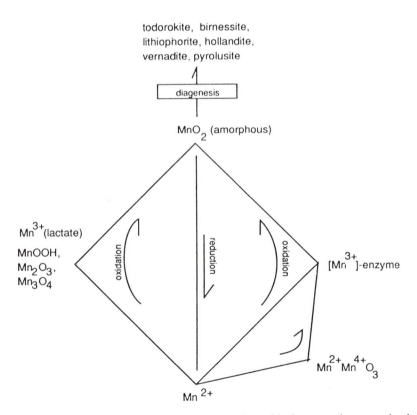

Figure 15.21 The manganese cycle. The oxidation reactions on the left side of the diagram involve a dismutation reaction of Mn(III), whereas the oxidation reactions on the right side of the diagram involve direct oxidation of enzyme-bound Mn(III). Mn(IV) reduction may or may not involve Mn(III) as an intermediate. Mn(III) oxide reduction by bacteria appears to occur but is not indicated in this diagram.

Studies of biogeochemical aspects of the manganese cycle in soil and aquatic environments are still few. One reason for this are difficulties in methodology. Two approaches are being taken. One is the reconstruction of the process in laboratory experiments. Thus, manganese and iron reduction and oxidation cycles in an acidic (pH 5.6), flooded, coastal prairie soil of southwest Louisiana, U.S.A., were reproduced in the laboratory under controlled pH and E_h conditions (Patrick and Henderson, 1981). The results showed that the rate of redox potential change could affect the oxidation and reduction rates differentially if the change was too rapid. Some Mn(IV) reduction in soils may be caused non-microbiologically by interaction with nitrite of microbial origin in which Mn(IV) may be reduced to Mn(II) and/or Mn(III), depending on nitrite concentration relative to Mn(IV) oxide (Bartlett, 1981). At high MnO_2/nitrite ratios, Mn(III) has been found to predominate.

$$NO_2^- + MnO_2 + 2H^+ \Rightarrow Mn^{2+} + NO_3^- + H_2O \qquad (15.19)$$

$$NO_2^- + 2MnO_2 \Rightarrow NO_3^- + Mn_2O_3 \qquad (15.20)$$

The complete reduction to Mn^{2+} is favored by acidic conditions, whereas the reduction to Mn_2O_3 is pH independent.

The second approach to studying biogeochemical aspects of the manganese cycle are in situ investigations (Nealson et al., 1988). Some examples were mentioned earlier in this chapter (Lake Punnus-Yarvi; Lake Oneida; Saanich Inlet; Framvaren Fjord). In the studies of Oneida Lake, Saanich Inlet, and Framvaren Fjord, microbial Mn(II) oxidation was demonstrated by manganese removal in the presence and absence of cell poisons from water samples collected at various water depths and spiked with $^{54}Mn^{2+}$. The metabolic poisons included sodium azide, penicillin G, and tetracycline HCl. A critical evaluation of poisons of potential use as inhibitors of manganese oxidation in environmental samples was made by Rosson et al. (1984). The removal of manganese reflected both binding of manganese(II) to particulates (bacterial cells and organic and inorganic aggregates) and oxidation of manganese(II). Binding was distinguished partially from oxidation by use of formaldehyde. The oxidation state of particulate manganese in the water samples upon collection was in the range of 2.3–2.7, suggesting the possible in situ formation of Mn(III) and mixed Mn(II)-Mn(IV) oxides (Emerson et al., 1982). An even better assessment of manganese oxidation rate can be obtained by measuring Mn(II) removal from solution in the presence and absence of oxygen instead of using metabolic poisons (Tebo and Emerson, 1985). In the absence of oxygen, only Mn(II) binding to surfaces should be observed. Thus the difference between total Mn(II) removal in air and Mn(II) removal in the absence of oxygen should represent the amount of Mn(II) that was oxidized.

An important question in the biogeochemical cycle of manganese is the oxidation state of manganese in the first detectable product of biological and chemical oxidation of Mn(II). Diem and Stumm (1984) and Tipping et al. (1984, 1985) found that Mn(IV) oxide was formed microbiologically as the first recognizable oxidation product in manganese-containing water samples. Tippping et al. (1984, 1985) found the oxide they recovered resembled vernadite (δ-MnO$_2$) and had an oxidation state of 3.5. By contrast, Hastings and Emerson (1986) found that spores of marine *Bacillus* SG-1 transformed Mn^{2+} to a Mn(III) oxide, which they identified as hausmannite (Mn$_3$O$_4$ or MnO$_{1.33}$). However, subsequent study by Mandernack (1992) suggested that the immediate product of manganese oxidation by the spores may be Mn(IV) oxide. The oxidation in this case was caused by a spore coat or exosporium protein (Tebo et al, 1988; Mann et al., 1988). If Hastings and Emerson's oxidation state determination was correct, the first oxidation product formed by the spore coat is an amorphous manganese(III) oxide, which recrystallizes slowly to hausmannite, especially in an excess of Mn(II) (Mann et al., 1983). The first stages of the Mn(II) oxidation by the spore coat protein may be enzymatic, but once the spore-coat protein is covered by the insoluble manganese oxide, further oxidation is abiotic (De Vrind et al., 1986; Mann et al., 1988). Since the first oxidation product of abiotic Mn^{2+} oxidation has been found to be Mn$_3$O$_4$,

$$3Mn^{2+} + 3H_2O + \tfrac{1}{2}O_2 \Rightarrow Mn_3O_4 + 6H^+ \tag{15.21}$$

which subsequently may dismutate at a very slow rate to MnO$_2$ and Mn^{2+} under appropriate conditions of pH,

$$Mn_3O_4 + 4H^+ \Rightarrow MnO_2 + 2Mn^{2+} + 2H_2O \tag{15.22}$$

(Hem, 1981; Hem and Lind, 1983; Hem et al., 1982; Murray et al., 1985), the following interpretation can be offered for the contrasting biological and chemical oxidation results to date. When Mn^{2+} oxidation is enzymatically catalyzed by a manganese oxidase system **with energy conservation**, the reaction may be a two-step sequence, in which the first step involves enzymatic oxidation of Mn^{2+} to Mn(III), with the Mn(III) existing as a bound (possibly enzyme-bound) intermediate:

$$Mn^{2+} + H^+ + \tfrac{1}{4}O_2 \Rightarrow [Mn^{3+}] + \tfrac{1}{2}H_2O \tag{15.23}$$

This is followed by rapid enzymatic oxidation to the final, and at the same time, first detectable product, namely a form of Mn(IV)oxide, here written as MnO$_2$:

$$[Mn^{3+}] + \tfrac{3}{2}H_2O + \tfrac{1}{4}O_2 \Rightarrow MnO_2 + 3H^+ \tag{15.24}$$

In this reaction sequence, all Mn^{2+} that is oxidized is transformed to MnO_2.

When Mn^{2+} is biologically oxidized by a nonenzymatic process or by an enzyme system **without energy conservation** (e.g., with spore-coat protein or by peroxidation with catalase in the presence of H_2O_2), Mn^{2+} is oxidized according to the abiotic mechanism, in which Mn(III) oxide such as Mn_3O_4 is the first detectable product of oxidation, and MnO_2, if it forms at all, is formed very slowly by disproportionation of the Mn_3O_4. The major difference between the reactions catalyzed by manganese oxidase on the one hand, and the reactions involving *Bacillus* SG-1 spores, abiotic oxidation, and probably the peroxidase reaction on the other, is that the manganese oxidase prevents dismutation of Mn(III) by catalyzing its complete conversion to Mn(IV) at a rapid rate, whereas in the other reactions Mn_3O_4 is formed rapidly, and Mn(IV) oxide, if it forms at all, is formed very slowly by abiotic dismutation, with the result that only one-third of the reacting Mn(III) is converted to Mn(IV), the other two-thirds being reconverted to Mn(II) [Reaction (15.22)]. Environmentally it may thus be possible to associate rapidly formed, fresh Mn(IV) oxide deposits with biological catalysis and Mn(III) oxide deposits with abiotic reactions or enzymatic reactions which do not yield useful energy (e.g., like those with *Bacillus* SG-1 spores or peroxidase reactions).

15.16 SUMMARY

Some bacteria can oxidize manganous manganese enzymatically or non-enzymatically. They may oxidize Mn^{2+} in solution with O_2, catalyzed by an inducible or a constitutive oxidase, or they may oxidize it with metabolic H_2O_2, catalyzed by catalase. They may oxidize pre-bound Mn^{2+} with O_2, catalyzed by a constitutive or inducible oxidase. At least some of the oxidase-catalyzed reactions yield energy to the bacteria. Some fungi can oxidize Mn^{2+} with an extracellular peroxidase. Oxidation of Mn^{2+} in solution occurs in soil, and in freshwater and marine environments. Oxidation of pre-bound Mn^{2+} has so far been demonstrated only with bacteria from marine environments.

Some bacteria and fungi may promote nonenzymatic manganese oxidation by utilization of hydroxycarboxylic acids, leading to a rise in pH above neutrality followed by oxidation of Mn(II) catalyzed by residual hydroxycarboxylic acid. Others may produce other metabolic end-products which then act as oxidants of Mn(II). Still other microbes may raise the pH by deamination to a range where Mn(II) autoxidizes. Some cyanobacteria and algae can promote Mn^{2+} autoxidation by raising the pH of their immediate surroundings photosynthetically.

Some bacteria and fungi may precipitate preformed, oxidized manganese through adsorption to their cell surface or to extracellular slime. Some bacteria may reduce Mn(IV) oxides to manganous manganese [Mn(II)] with suitable electron donors. This reduction may be aerobic and/or anaerobic depending on the bacterial strain and environmental conditions. Such reduction has been demonstrated with bacteria from soil and from freshwater and marine environments. Some bacteria and fungi may also reduce Mn(IV) oxides non-enzymatically with metabolic end-products such as H_2S formate, Fe^{2+}, or oxalate.

Manganese-oxidizing and -reducing microorganisms play an important role in the immobilization and mobilization, respectively, of manganese in soil. In some anaerobic environments, manganese-reducing bacteria may play an important role in mobilizing fixed, unavailable manganese and in mineralization of organic matter. Fixed manganese in soil can be concentrated in concretions, or in arid environments as desert varnish. Since manganese is required in plant nutrition, manganese-oxidizing and reducing activity is ecologically very significant.

Manganese-oxidizing and -reducing bacteria can also play a significant role in the manganese cycle in freshwater and marine environments. The manganese oxidizing microbes may contribute to the accumulation of manganese oxides on and in sediments. The oxides they form may sometimes be deposited as concretions formed around a nucleus, such as a sediment grain, a pebble, or a dead biological structure (e.g., mollusk shell, coral fragment, or other debris. Conversely, manganese-reducing microorganisms may mobilize the oxidized or fixed manganese, releasing it into the water column. In reducing freshwater environments, in the presence of abundant plant debris, the microbial reduction of manganese oxides may lead to manganous carbonate formation, a different form of fixed manganese.

Ferromanganese nodules on parts of the ocean floor are inhabited by bacteria. Some of these bacteria can oxidize and others can reduce manganese. Together with benthic foraminifera, which can be expected to feed on the bacteria and which accumulate manganese oxides on their tests and tubes, they appear to constitute a biological system that contributes to nodule formation.

REFERENCES

Adams, J. B., F. Palmer, and J. T. Staley. 1992. Rock weathering in deserts: mobilization and concentration of ferric iron by microorganisms. Geomicrobiol. J. 10:99–114.

Adams, L. F., and W. C. Ghiorse. 1985. Influence of manganese on growth of a sheathless strain of *Leptothrix discophora*. Appl. Environ. Microbiol. 49:556–562.

Adams, L. F., and W. C. Ghiorse. 1986. Physiology and ultrastructure of *Leptothrix discophora* SS-1. Arch. Microbiol. 145:126–135.

Adams, L. F., and W. C. Ghiorse. 1987. Characterization of extracellular Mn^{2+}-oxidizing activity and isolation of an Mn^{2+}-oxidizing protein from *Leptothrix discophora* SS-1. J. Bacteriol. 169:1279–1285.

Adams, L. F., and W. C. Ghiorse. 1988. Oxidation state of Mn in the Mn oxide produced by *Leptothrix discophora* SS-1. Geochim. Cosmochim. Acta 52:2073–2076.

Adeny, W.E. 1894. On the reduction of manganese peroxide in sewage. Proc. Roy. Dublin Soc., Chapter 27, pp. 247–251.

Agate, A. D., and H. A. Deshpande. 1977. Leaching of manganese ores using *Arthrobacter* species. *In*: W. Schwartz, ed. Conference Bacterial Leaching. Gesellschaft fur Biotechnologische Forschung mbH Braunschweig-Stoeckheim. Verlag Chemie, Weinheim, pp. 243–250.

Alexandrov, E. A. 1972. Manganese: element and geochemistry. *In*: R. W. Fairbridge, ed. Encyclopedia of Geochemistry and Environmental Sciences. Encyclopedia of Earth Sciences Series, Vol. IVA. Van Nostrand Reinhold, New York, pp. 670–671.

Ali, S. H. and J. L. Stokes. 1971. Stimulation of heterotrophic and autotrophic growth of *Sphaerotilus discophorus* by manganous ions. Antonie v. Leeuwenhoek 37:519–528.

Aller, R. C., and P. D. Rude. 1988. Complete oxidation of solid phase sulfides by manganese and bacteria in anoxic marine sediments. Geochim. Cosmochim. Acta 52:751–765.

Anderson, J. J., and A. H. Devol. 1973. Deep water renewal in Saanich Inlet, an intermittently anoxic basin. Estuarine Coastal Mar. Sci. 1: 1–10.

Arcuri, E. J. 1978. Identification of the cytochrome complements of several strains of marine manganese oxidizing bacteria and their involvement in manganese oxidation. Ph.D. thesis. Rensselaer Polytechnic Institute, Troy, NY.

Arcuri, E. J., and H. L. Ehrlich. 1979. Cytochrome involvement in Mn(II) oxidation by two marine bacteria. Appl. Environ. Microbiol. 37: 916–923.

Arcuri, E. J., and H. L. Ehrlich. 1980. Electron transfer coupled to Mn(II) oxidation in two deep-sea Pacific Ocean isolates. *In*: P. A. Trudinger, M. R. Walter, and B. J. Ralph, eds. *Biogeochemistry of Ancient and Modern Environments*. Australian Academy of Sciences, Canberra, and Springer-Verlag, Berlin, pp. 339–344.

Aristovskaya, T. V. 1961. Accumulation of iron in breakdown of organo-mineral humus complexes by microorganisms. Dokl. Akad. Nauk SSSR 136:954–957.

Aristovskaya, T. V., and O. M. Parinkina. 1963. A new soil microorganism *Seliberia stellata* n.gen. n. sp. Izv. Akad. Nauk SSSR Ser. Biol. 218:49–56.

Babenko, Yu. S., L. M. Dolgikh, and M. Z. Serebryanaya. 1983. Characteristics of the bacterial breakdown of primarily oxidized manganese ores from the Nikopol' deposit. Mikrobiologiya 52:851–856.

Baker, E. T., and G. J. Massoth. 1986. Hydrothermal plume measurements: a regional perspective. Science 234:980–982.

Balashova, V. V., and G. A. Dubinina. The ultrastructure of *Metallogenium* in axenic culture. Mikrobiologiya 58:841–846 (Engl. transl. pp. 681–685).

Bartlett, R. J. 1981. Nonmicrobial nitrite-to-nitrate transformations in soil. Soil Sci. Soc. Am. J. 45:1054–1058.

Beijerinck, M. W. 1913. Oxydation des Mangancarbonates durch Bakterien und Schimmelpilze. Fol. Microbiol. Hollaend. Beitr. Gesamt. Mikrobiol. Delft 2:123–134.

Bender, M. L., Teh-Lung Ku, and W. S. Broecker. 1970. Accumulation rates of manganese in pelagic sediments and nodules. Earth Planet. Sci. Lett. 8:143–148.

Bignot, G., and L. Dangeard. 1976. Contribution à l'étude de la fraction biogène des nodules polymétalliques des fonds oceaniques actuels. C.R. Somm. Soc. geol. Fr. Fasc. 3, pp. 96–99.

Bischoff, J. L., and R. J. Rosenbauer. 1983. A note on the chemistry of seawater in the range of 350°-500°C. Geochim. Cosmochim. Acta 47:139–144.

Boogerd, F. C., and J. P. M. De Vrind. 1987. Manganese oxidation by *Leptothrix discophora*. J. Bacteriol. 169:489–494.

Bowen, H. J. M. 1966. Trace Elements in Biogeochemistry. Academic Press, London.

Bowen, H. J. M. 1979. Environmental Chemistry of the Elements. Academic Press, London.

Bromfield, S. M. 1956. Oxidation of manganese by soil microorganisms. Aust. J. Biol. Sci. 9:238–252.

Bromfield, S. M. 1958a. The properties of a biologially formal manganese oxide, its availability to oats and its solution by root washings. Plant Soil 9:325–337.

Bromfield, S. M. 1958b. The solution of γ-MnO_2 by substances released from soil and from roots of oats and vetch in relation to manganese availability. Plant Soil 10:147–160.

Bromfield, S. M. 1974. Bacterial oxidation of manganous ions as affected by organic substrate concentration and composition. Soil Biol. Biochem. 6:383–392.

Bromfield, S. M. l978. The oxidation of manganous ions under acid conditions by an acidophilous actinomycete from acid soil. Aust. J. Soil Res. 16:91–100.

Bromfield, S. M. 1979. Manganous ion oxidation at pH values below 5.0 by cell-free substances from *Streptomyces* sp. cultures. Soil Biol. Biochem. 11:115–118.

Bromfield, S. M., and D. J. David. 1976. Sorption and oxidation of manganous ions and reduction of manganese oxide by cell suspensions of a manganese oxidizing bacterium. Soil Biol. Biochem. 8:37–43.

Bromfield, S. M., and D. J. David. 1978. Properties of biologically formed manganese oxide in relation to soil manganese. Aust. J. Soil Res. 16:79–89.

Bromfield, S. M., and V. D. B. Skerman. 1950. Biological oxidation of manganese in soils. Soil Sci. 69:337–348.

Burdige, D. J. (1993). The biogeochemistry of manganese and iron reduction in marine sediments. Earth-Sci. Rev. 35:249–284.

Burdige, D. J., and P. E. Kepkay. 1983. Determination of bacterial manganese oxidation rates in sediments using an in-situ dialysis technique. I. Laboratory studies. Geochim. Cosmochim. Acta 47: 1907–1916.

Burdige, D. J., and K. H. Nealson. 1985. Microbiological manganese reduction by enrichment cultures from coastal marine sediments. Appl. Environ. Microbiol. 50:491–497.

Burdige, D. J., and K. H. Nealson. 1986. Chemical and microbiological studies of sulfide-mediated manganese reduction. Geomicrobiol. J. 4:361–3878.

Burnett, B. R., and K. H. Nealson. 1981. Organic films and microorganisms associated with manganese nodules. Deep-Sea Res. 28A: 637–645.

Burnett, B. R., and K. H. Nealson. 1983. Energy dispersive X-ray analysis of the surface of a deep-sea ferromanganese nodule. Mar. Geol. 53: 313–329.

Burns, R. G., V. M. Burns, and W. Smig. 1974. Ferromanganese nodule mineralogy: suggested terminology of the principal manganese oxide phases. Abstr. Ann. Meet. Geol. Soc. Am.

Butkevich, V. S. 1928. The formation of marine iron-manganese deposits and the role of microorganisms in the latter. Wissenschaft. Meeresinst. Ber. 3:7–80.

Caccavo, F., Jr., R. P. Blakemore, and D. R. Lovley. 1992. A hydrogen-oxidizing, Fe(III)-reducing microorganism from the Great Bay Estuary, New Hampshire. Appl. Environ. Microbiol. 58:3211–3216.

Caccavo, F., Jr., D. J. Lonergan, D. R. Lovley, M. Davis, S. F. Stolz, and M. J. McInerney. 1994. *Geobacter sulfurreducens* sp. nov., a new hydrogen- and acetate-oxidizing dissimilatory metal-reducing microorganism. Appl Environ. Microbiol. 60:3752–3759.

Chapnick, S. D., W. S. Moore, and K. H. Nealson. 1982. Microbially mediated manganese oxidation in a freshwater lake. Limnol. Oceanogr. 27:1004–1014.

Clark, T. R. 1991. Manganese-oxidation by SSW$_{22}$, a hydrothermal vent bacterial isolate. Ph.D. thesis. Rensselaer Polytechnic Institute, Troy, NY.

Clark, T. R., and H. L. Ehrlich. 1988. Manganese oxidation by cell fractions from a bacterial hydrothermal vent isolate. Abstr., Ann. Meet. Am. Soc. Microbiol. K-152, p. 232.

Collins, J. F., and S. W. Buol. 1970. Effects of fluctuations in the E_h-pH environment on iron and/or manganese equilibria. Soil Sci. 110: 111–118.

Cowen, J. P. 1989. Positive effect on manganese binding by bacteria in deep-sea hydrothermal plumes. Appl. Environ. Microbiol. 55: 764–766.

Cowen, J. P., and K. W. Bruland. 1985. Metal deposits associated with bacteria: implications for Fe and Mn marine biogeochemistry. Deep-Sea Res. 32:253–272.

Cowen, J. P., and M. W. Silver. 1984. The association of iron and manganese with bacteria on marine macroparticulate material. Science 224:1340–1342.

Cowen, J. P., G. J. Massoth, and E. T. Baker. 1986. Bacterial scavenging of Mn and Fe in a mid- to far-field hydrothermal particle plume. Nature (Lond.) 322:169–171.

Cowen, J. P., G. J. Massoth, and R. A. Feely. 1990. Scavenging rates of dissolved magnanese in a hydrothermal vent plume. Deep-Sea Res. 37:1619–1637.

Crerar, D. A., and H. L. Barnes. 1974. Deposition of deep-sea manganese nodules. Geochim. Cosmochim. Acta 38:279–300.

Crerar, D. A., A. G. Fischer, and C. L. Plaza. 1979. *Metallogenium* and biogenic deposition of manganese from Precambrian to Recent time. *In*: Geology and Geochemistry of manganese. Vol. III. Publishing House of the Hungarian Academy of Sciences, Budapest, pp. 285–303.

Cronan, D. S., and R. L. Thomas. 1970. Ferromanganese concretions in Lake Ontario. Can. J. Earth Sci. 7:1346–1349.

Davison, W., C. Woof, and E. Rigg. 1982. The dynamics of iron and manganese in a seasonally anoxic lake; direct measurement of fluxes using sediment traps. Limnol. Oceanogr. 27:987–1003.

Dean, W. E. 1970. Fe-Mn oxidate crusts in Oneida Lake, New York. Proc. 13th Conf. Great Lakes Res. Int. Assoc. Great Lakes Res. pp. 217–226.

Dean, W. E., and S. K. Ghosh. 1978. Factors contributing to the formation of ferromanganese nodules in Oneida Lake, New York. J. Res. U.S. Geol. Surv. 6:231–240.

Dean, W. E., and P. E. Greeson. 1979. Influences of algae on the formation of freshwater ferromanganese nodules, Oneida Lake, New York. Arch. Hydrobiol. 86:181–192.

Dean, W. E., W. S. Moore, and K. H. Nealson. 1981. Manganese cycles and the origin of manganese nodules, Oneida Lake, New York, U.S.A. Chem. Geol. 34:53–64.

De Vrind, J. P. M., F. C. Boogerd, and E. W. De Vrind-de Jong. 1986a. Manganese reduction by a marine *Bacillus* species. J. Bacteriol. 167: 30–34.

De Vrind, J. P. M., E. W. De Vrind-de Jong, J-W. H. DeVogt, P. Westbroek, F. C. Boogerd, and R. Rosson. 1986b. Manganese oxidation by spore coats of a marine *Bacillus* sp. Appl. Environ. Microbiol. 52:1096–1100.

Diem, D., and W. Stumm. 1984. Is dissolved Mn^{2+} being oxidized by O_2 in absence of Mn-bacteria or surface catalysts? Geochim. Cosmochim. Acta 48:1571–1573.

Di-Ruggiero, J., and A. M. Gounot. 1990. Microbial manganese reduction by bacterial strains isolated from aquifer sediments. Microb. Ecol. 20:53–63.

Dorn, R. I. 1991. Rock Varnish. Amer. Scientist. 79:542–553.

Dorn, R. I., and T. M. Oberlander. 1981. Microbial origin of desert varnish. Science 213:1245–1247.

Douka, C. E. 1977. Study of bacteria from manganese concretions. Precipitation of manganese by whole cells and cell-free extracts of isolated bacteria. Soil Biol. Biochem. 9:89–97.

Douka, C. E. 1980. Kinetics of manganese oxidation by cell-free extracts of bacteria isolated from manganese concretions from soil. Appl. Environ. Microbiol. 39:74–80.

Douka, C. W., and C. D. Vaziourakis. 1982. Enzymic oxidation of manganese by cell-free extracts of bacteria from manganese concretions from soil. Folia Microbiol. 27:418–422.

Drosdoff, M., and C. C. Nikiforoff. 1940. Iron-manganese concretions in Dayton soils. Soil Sci. 49:333–345.

Dubinina, G. A. 1970. Untersuchungen ueber die Morphologie von *Metallogenium* und die Beziehungen zu *Mycoplasma*. Z. Allg. Mikrobiol. 10:309–320.

Dubinina, G. A. 1978a. Functional role of bivalent iron and manganese oxidation in *Leptothrix pseudoochracea*. Mikrobiologiya 47:783–789 (Engl. transl. pp. 631–636).

Dubinina, G. A. 1978b. Mechanism of the oxidation of divalent iron and manganese by iron bacteria growing at neutral pH of the medium. Mikrobiologiya 47:591–599 (Engl. transl. pp. 471–478).

Dubinina, G. A. 1984. Infection of procaryotic and eukaryotic microorganisms with *Metallogenium*. Curr. Microbiol. 11:349–356.

Dubinina, G. A., and Z. P. Deryugina. 1971. Electron microscope study of iron-manganese concretions from Lake Punnus-Yarvi. Dokl. Akad. Nauk SSSR 201:714–716 (Engl. transl. pp. 738–740).

Dubinina, G. A., and A. V. Zhdanov. 1975. Recognition of the iron bacteria "*Siderocapsa*" as *Arthrobacter siderocapsulatus*. n.sp. Int. J. Syst. Bacteriol. 25:340–350.

Dugolinsky, B. K., S. V. Margolis, and W. C. Dudley. 1977. Biogenic influence on growth of manganese nodules. J. Sediment. Petrol. 47: 428–445.

Durand, P. D. Prieur, C. Jeanthon, and E. Jacq. 1990. Occurrence and activity of heterotrophic manganese oxidizing bacteria associated with Alvinellids (polychaetous annelids) from a deep hydrothermal vent site on the East Pacific Rise. C.R. Acad. Sci. (Paris) 310, series III:273–278.

Ehrlich, H. L. 1963. Bacteriology of manganese nodules. I. Bacterial action on manganese in nodule enrichments. Appl. Microbiol. 11: 15–19.

Ehrlich, H. L. 1964. Microbial transformation of minerals. *In*: H. Heukelekian and N. Dondero, eds. Principles and Applications in Aquatic Microbiology. Wiley, New York, pp. 43–60.

Ehrlich, H. L. 1966. Reactions with manganese by bacteria from marine ferromanganese nodules. Dev. Ind. Microbiol. 7:43–60.

Ehrlich, H. L. 1968. Bacteriology of manganese nodules. II. Manganese oxidation by cell-free extract from a manganese nodule bacterium. Appl. Microbiol. 16:197–202.

Ehrlich, H. L. 1972. Response of some activities of ferromanganese nodule bacteria to hydrostatic pressure. *In*: R. R. Colwell and R. Y. Morita, eds. Effect of the Ocean Environment on Microbial Activities. University Park Press. Baltimore, MD, pp. 208–221.

Ehrlich, H. L. 1973. The biology of ferromanganese nodules. Determination of the effect of storage by freezing on the viable nodule flora, and a check on the reliability of the results from a test to identify MnO₂-reducing cultures. *In*: Interuniversity Program of Research of Ferromanganese Deposits on the Ocean Floor. Phase I Report, Seabed Assessment Program, Interantional Decade of Ocean Exploration. National Science Foundation, Washington, DC, pp. 217–219.

Ehrlich, H. L. 1976. Manganese as an energy source for bacteria. *In*: Environmental Biogeochemistry. Vol. 2. J. O. Nriagu, ed. Ann Arbor Science Publishers, Ann Arbor, MI, pp. 633–644.

Ehrlich, H. L. 1978a. Inorganic energy sources for chemolithotrophic and mixotrophic bacteria. Geomicrobiol. J. 1:65–83.

Ehrlich, H. L. 1978b. Conditions for bacterial participation in the intiation of manganese deposition around marine sediment particles. *In*: W. E. Krumbein, ed. Environmental Biogeochemistry and Geomicrobiology. Vol. 3: Methods, Metals and Assessment. Ann Arbor Science Publishers Inc., Ann Arbor, MI, pp. 839–845.

Ehrlich, H. L. 1980. Bacterial leaching of manganese ores. *In*: P. A. Trudinger, W. R. Walter, and B. J. Ralph, eds. Biogeochemistry of Ancient and Modern Environments. Australian Academy of Science, Canberra, and Springer-Verlag, Berlin, pp. 609–614.

Ehrlich, H. L. 1981. Microbial oxidation and reduction of manganese as aids in its migration in soil. *In*: Colloques Internationaux du C.N.R.S. No. 303—Migrations Organo-Minérales Dans les Sols Temperés. Editions du C.N.R.S., Paris, pp. 209–213.

Ehrlich, H. L. 1982. Enhanced removal of Mn^{2+} from seawater by marine sediments and clay minerals in the presnece of bacteria. Can. J. Microbiol. 28:1389–1395.

Ehrlich, H. L. 1983. Manganese-oxidizing bacteria from a hydrothermally active area on the Galapagos Rift. Ecol. Bull. (Stockh.) 35:357–366.

Ehrlich, H. L. 1984. Different forms of bacterial manganese oxidation. *In*: W. R. Strohl and O. H. Tuovinen. Microbial Chemoautotrophy. Ohio State University Press, Columbus, pp. 47–56.

Ehrlich, H. L. 1985. Mesophilic manganese-oxidizing bacteria from hydrothermal discharge areas at 21° north on the East Pacific Rise. *In*: D. E. Caldwell, J. A. Brierley, and C. L. Brierley, eds. Planetary Ecology. Van Nostrand Reinhold, New York, pp. 186–194.

Ehrlich, H. L. 1987. Manganese oxide reduction as a form of anaerobic respiration. Geomicrobiol. J. 5:423–431.

Ehrlich, H. L. 1988. Bioleaching of manganese by marine bacteria. *In*: Proc. 8th Intern. Biotechnol. Symp., July 17–22, 1988, Paris, France.

Ehrlich, H. L. 1993a. Electron transfer from acetate to the surface of MnO₂ particles by a marine bacterium. J. Indust. Microbiol. 12: 121–128.

Ehrlich, H. L. 1993b. A possible mechanism for the transfer of reducing power to insoluble mineral oxide in bacterial respiration. *In*: A. E. Torma, M. L. Apel, and C. L. Brierley, eds. Biohydrometallurgical Technologies. Vol. II. The Minerals, Metals and Materials Society, Warrendale, PA, pp. 415–422.

Ehrlich, H. L., and J. C. Salerno. 1988. Stimulation by ADP and phosphate of Mn^{2+} oxidation by cell cell extracts and membrane vesicles of induce Mn-oxidizing bacteria. Abst. Ann. Meeting, American Soc. Microbiol. K-151, p. 231.

Ehrlich, H. L., and J. C. Salerno. 1990. Energy coupling in Mn^{2+} oxidation by a marine bacterium. Arch. Microbiol. 154:12–17.

Ehrlich, H. L., and M. A. Zapkin. 1985. Manganese-rich layers in calcareous deposits along the western shore of the Dead Sea may have a bacterial origin. Geomicrobiol. J. 4:207–221.

Ehrlich, H. L., W. C. Ghiorse, and G. L. Johnson II. 1972. Distribution of microbes in manganese nodules from the Atlantic and Pacific Oceans. Dev. Ind. Microbiol. 13:57–65.

Ehrlich, H. L., S. H. Yang, and J. D. Mainwaring Jr. 1973. Bacteriology of manganese nodules. VI. Fate of copper, nickel, cobalt and iron during bacterial and chemical reduction of the manganese (IV). Z. Allg. Mikrobiol. 13:39–48.

Ehrlich, H. L., L. A. Graham, and J. S. Salerno. 1991. MnO_2 reduction by marine Mn^{2+} oxidizing bacteria from around hydrothermal vents a mid-ocean spreading center and from the Black Sea. *In*: J. Berthelin ed., Diversity of Environmental Biogeochemistry. Elsevier, Amsterdam, pp. 217–224.

Eleftheriadis, D. K. 1976. Mangan- und Eisenoxydation in Mineral- und Termal-Quellen-Mikrobiologie, Chemie, und Geochemie. Ph.D thesis. Universitaet des Saarlandes, Saarbruecken, West Germany.

Emerson, D., R. E. Garen, and W. C. Ghiorse. 1989. Formation of *Metallogenium*-like structures by a manganese-oxidizing fungus. Arch. Microbiol. 151:223–231.

Emerson, S., S. Kalhorn, L. Jacobs, B. M. Tebo, K. H. Nealson, and R.A. Rosson. 1982. Environmental oxidation rate of manganese(II): bacterial catalysis. Geochim. Cosmochim. Acta 46:1073–1079.

Fredericks-Jantzen, C. M., H. Herman, and P. Herley. 1975. Microorganisms on manganese nodules. Nature (Lond.) 258:270.

Garey, C. L., and S. A. Barber. 1952. Evaluation of certain factors involved in increasing manganese availability with sulfur. Soil Sci. Soc. Am. Proc. 16:173–175.

Geloso, M. 1927. Adsorption au sein des solutions salines par le bioxide de manganèse et généralités sur ce phenomène. III. Adsorption par le bioxide de manganèse colloidal. Ann. Chim. 7:113–150.

Gerretsen, F. C. 1937. Manganese deficiency of oats and its relation to soil bacteria. Ann. Bot. 1:207–230.

Ghiorse, W. C. 1980. Electron microscopic analysis of metal-depositing microorganisms in surface layers of Baltic Sea ferromanganese concretions. *In*: P. A. Trudinger, M. R. Walter, and B. J. Ralph. Biogeochemistry of Ancient and Modern Enviroments. Australian Academy of Science, Canberra, Springer-Verlag, Berlin, pp. 345–354.

Ghiorse, W. C. 1984. Biology of iron- and manganese-depositing bacteria. Annu. Rev. Microbiol. 38:515–550.

Ghiorse, W. C. 1988. Microbial reduction of manganese and iron. *In*: A. J. B. Zehnder, ed. Biology of Anaerobic Microorganisms. Wiley, New York, pp. 305–331.

Ghiorse, W. C., and S. D. Chapnick. 1983. Metal-depositing bacteria and the distribution of manganese and iron in swamp waters. Ecol. Bull. (Stockh.) 35:367–376.

Ghiorse, W. C., and H. L. Ehrlich. 1976. Electron transpsort components of the MnO_2 reductase system and the location of the terminal reductase in a marine bacillus. Appl. Environ. Microbiol. 31:977–985.

Ghiorse, W. C., and H. L. Ehrlich. 1992. Microbial biomineralization of iron and manganese. *In*: H. C. W. Skinner and R. W. Fitzpatrick, eds. Biomineralization. Process of iron and manganese. Modern and Ancient Environments. Catena Supplement 21. Catena, Verlag, Crelingen, Germany, pp. 75–99.

Ghiorse, W. C., and P. Hirsch. 1979. An ultrastructural study of iron and manganese deposition associated with extracellual polymers of *Pedomicrobium*-like budding bacteria. Arch. Microbiol. 123: 213–226.

Ghosh, S. K. 1976. Origin and geochemistry of ferromanganese nodules in Oneida Lake, New York. Diss. Abstr. Int. 36:4905-B.

Ghosh, J., and K. Imai. 1985a. Leaching of manganese dioxide by *Thiobacillus ferrooxidans* growing on elemental sulfur. J. Ferment. Technol. 63:259–262.

Ghosh, J., and K. Imai. 1985b. Leaching mechanism of manganese dioxide by *Thiobacillus ferrooxidans*. J. Ferment. Technol. 63:295–298.

Gillette, N. J. 1961. Oneida Lake Pancakes. New York State Conservationist, 41 (April-May):21.

Glenn, J. K., and M. H. Gold. 1985. Purification and characterization of an extracellular Mn(II)-dependent peroxidase from the lignin degrading basidiomycete *Phanerochaete chrysosporium*. Arch. Biochem. Biophys. 242:329–341.

Glenn, J. K., L. Akileswaram, and M. H. Gold. 1986. Mn(II) oxidation is the principal function of the extracellular Mn-peroxidase from *Phanerochaete chrysosporium*. Arch. Biochem. Biophys. 251: 688–696.

Goldberg, E. D. 1954. Marine geochemistry. 1. Chemical scavengers of the sea. J. Geol. 62:249–265.

Goldberg, E. D., and G. O. S. Arrhenius. 1958. Chemistry of Pacific pelagic sediments. Geochim. Cosmochim. Acta 13:153–212.

Goldschmidt, V. M. 1954. Geochemisltry, Clarendon Press, Oxford, pp. 621–642.

Gong-Collins, E. 1986. A euryhaline, manganese- and iron-oxidizing *Bacillus megaterium* from a microbial mat at Laguna Figueroa, Baja California, Mexico. Microbios 48:109–126.

Gottfreund, J., and R. Schweisfurth. 1983. Mikrobiologische Oxidation und Reduktion von Manganspecies. Fresenius Z. Anal. Chem. 316: 634–638.

Gottfreund, J., G. Schmitt, and R. Schweisfurth. 1983. Chemische und mikrobiologische Untersuchungen an Komplexverbindungen des Mangans. Mitteilgn. Dtsch. Bodenkundl. Gesellsch. 38:325–330.

Gottfreund, J., G. Schmitt, and R. Schweisfurth. 1985. Wertigkeitswechsel von Manganspecies durch Bakterien in Naehrloesungen und in Lockergestein. Landwirtsch. Forschung 38:80–86.

Graham, J. W. 1959. Metabolically induced precipitation of trace elements from sea water. Science 129:1428–1429.

Graham, J. W., and S. Cooper. 1959. Biological origin of manganese-rich deposits of the sea floor. Nature (Lond.) 183:1050–1051.

Greene, A. C., and J. C. Madgwick. 1991. Microbial formation of manganese oxides. Appl. Environ. Microbiol. 57:1114–1120.

Greenslate, J. 1974a. Manganese and biotic debris associations in some deep-sea sediments. Science 186:529–531.

Greenslate, J. 1974b. Microorganisms participate in the construction of manganese nodules. Nature (Lond.) 249:181–183.

Gregory, E., R. S. Perry, and J. T. Staley. 1980. Characterization, distribution, and significance of *Metallogenium* in Lake Washington. Microb. Ecol. 6:125–140.

Hajj, J., and J. Makemson. 1976. Determination of growth of *Sphaerotilus discophorus* in the presence of manganese. Appl. Environ. Microbiol. 32:699–702.

Harada, K. 1978. Micropaleontologic investigation of Pacific manganese nodules. Mem. Fac. Sci. Kyoto Univ. Ser. Geol. Mineral. 45: 111–132.

Hariya, Y., and T. Kikuchi. 1964. Precipitation of manganese by bacteria in mineral springs. Nature (Lond.) 202:416–417.

Harriss, R. C., and A. G. Troup. 1970. Chemistry and origin of freshwater ferromanganese concretions. Limnol. Oceanogr. 15:702–712.

Hastings, D., and S. Emerson. 1986. Oxidation of manganese by spores of a marine bacillus: kinetic and thermodynamic considerations. Geochm. Cosmochim. Acta 50:1819–1824.

Heezen, B. C., and C. Holister. 1971. The Face of the Deep. Oxford University Press, New York.

Hem, J. D. 1963. Chemical equilibria and rate of manganese oxidation. U.S. Geol. Surv. Water Supply Paper 1667-A.

Hem, J. D. 1981. Rates of manganese oxidation in aqueous systems. Geochim. Cosmochim. Acta 45:1369–1374.

Hem, J. D., and C. J. Lind. 1983. Nonequilibrium models for predicting forms of precipitated manganese oxides. Geochim. Cosmochim. Acta 47:2037–2046.

Hem, J. D., E. Roberson, and R. B. Fournier. 1982. Stability of β-MnOOH and manganese oxide deposition from springwater. Water Resources Res. 18:563–570.

Heye, D., and H. Beiersdorf. 1973. Radioaktive und magnetische Untersuchungen an Manganknollen zur Ermittlung der Wachstumsgeschwindigkeit bzw. zur Altersbestimmung. Z. Geophysik 39:703–726.

Hobot, J. A., E. Carlemalm, W. Villigier, and E. Kellenberger. 1984. Periplasmic gel: new concept resulting from the reinvestigation of bacterial cell envelope ultrastructure by new methods. J. Bacteriol. 160:143–152.

Hogan, V. C. 1973. Electron transport and manganese oxidation in *Leptothrix discophorus*. Ph.D. thesis. Ohio State University, Columbus.

Holden, P. J., and J. C. Madgwick. 1983. Mixed culture bacterial leaching of manganese dioxide. Proc. Australas. Inst. Min. Metall. No. 286, June, pp. 61–63.

Hood, D. W. 1963. Chemical oceanography. Oceanaogr. Mar. Biol. Annu. Rev. 1:129–155.

Hood, D. W., and J. F. Slowey. 1964. Texas A. and M. Univ. Progr. Rept., Proj. 276, AEC Contract No. AT-(40-1)-2799.

Horn, D. R., B. M. Horn, and M. N. Delach. 1972. Distribution of ferromanganese deposits in the world ocean. *In*: D. R. Horn, ed. Ferromanganese Deposits on the Ocean Floor. The Office of the International Decade of Ocean Exploration. National Science Foundation, Washington, DC, pp. 9–17.

Hungate, B., A. Danin, N. B. Pellerin, J. Stemmler, P. Kjellander, J. B. Adams, and J. T. Staley. 1987. Characterization of manganese-oxidizing (MnII $\Rightarrow$ MnIV) bacteria from Negev Desert rock varnish:

implications in desert varnish formation. Can. J. Microbiol. 33: 939–943.

Jackson, D. D. 1901a. A new species of *Crenothrix* (*C. manganifera*). Trans. Am. Microsc. Soc. 23:31–39.

Jackson, D. D. 1901b. The precipitation of iron, manganese, and aluminum by bacterial action. J. Soc. Chem. Ind. 21:681–684.

Jacobs, L., S. Emerson, and J. Skei. 1985. Partitioning and transport of metals across an O_2/H_2S interface in a permanently anoxic basin: Framvaren Fjord, Norway. Geochim. Cosmochim. Acta 49: 1433–1444.

Jannasch, H. W. 1984. Microbial processes at deep sea hydrothermal vents. *In*: P. A. Rona, K. Bostrom, L. Laubier, and K. L. Smith, eds. Hydrothermal Processes at Seafloor Spreading Centers. Plenum Press, New York, pp. 677–709.

Jannasch, H. W., and M. J. Mottl. 1985. Geomicrobiology of Deep-Sea Hydrothermal Vents. Science 229:717–725.

Jannasch, H. W., and C. O. Wirsen. 1979. Chemosynthetic primary production at East Pacific sea floor spreading centers. BioScience 29: 592–598.

Jaquet, J. M., G. Nembrini, J. Garcia, and J. P. Vernet. 1982. The manganese cycle in Lac Léman, Switzerland: The role of *Metallogenium*. Hydrobiology 91:323–340.

Jaureguin, M. A., and H. M. Reisenauer. 1982. Dissolution of oxides of manganese and iron by root exudate components. Soil Sci. Soc. Am. J. 46:314–317.

Johnson, A. H., and J. L. Stokes. 1966. Manganese oxidation by *Sphaerotilus discophorus*. J. Bacteriol. 91:1543–1547.

Jones, C. J., and J. W. Murray. 1985. The geochemistry of manganese in the northeast Pacific Ocean off Washington. Limnol. Oceanogr. 30: 81–92.

Jones, J. G., S. Gardener, and B. M. Simon. 1983. Bacterial reduction of ferric iron in a stratified eutrophic lake. J. Gen. Microbiol. 129: 131–139.

Jones, J. G., S. Gardener, and B. M. Simon. 1984. Reduction of ferric iron by heterotrophic bacteria in lake sediments. J. Gen. Microbiol. 130:45–51.

Jung, W. K., and R. Schweisfurth. 1979. Manganese oxidation by an intracellular protein of a *Pseudomonas* species. Z. Allg. Mikrobiol. 19: 107–115.

Kadko, D, and L. H. Burckle. 1980. Manganese nodule growth rates determined by fossil diatom dating. Nature (Lond.) 287:725–726.

Kalhorn, S., and S. Emerson. 1984. The oxidation state of manganese in surface sediments of the deep sea. Geochim. Cosmochim. Acta 48: 897–902.

Kalinenko,V. O., O. V. Belokopytova, and G. G. Nikolaeva. 1962. Bacteriogenic formation of iron-manganese concretions in the Indian Ocean. Okeanologiya 11:1050–1059 (Engl. transl.).

Karavaiko, G. I., V. A. Yurchenko, V. I. Remizov, and T. M. Klyushnikova. 1986. Reduction of manganese dioxide by cell-free *Acinetobacter calcoaceticus* extracts. Mikrobiologiya 55:709–714.

Kazutami, J., and T. Tano. 1967. Leaching of manganese by *Thiobacillus thiooxidans*. Hakko Kyokaischi 25:166–167.

Kenten, R. H., and P. J. G. Mann. 1949. The oxidation of manganese by plant extracts in the presence of hydrogen peroxide. Biochem. J. 45:225–263.

Kenten, R. H., and P. J. G. Mann. 1950. The oxidation of maganese by peroxidase systems. Biochem J. 50:67–73.

Kepkay, P. E. 1985a. Kinetics of microbial manganese oxidation and trace metal binding in sediments: Results from an in situ dialysis technique. Limnol. Oceanogr. 30:713–726.

Kepkay, P. E. 1985b. Microbial manganese oxidation and trace metal binding in sediments: results from an *in situ* dialysis technique. *In*: D. E. Caldwell, J. A. Brierley, and C. L. Brierley, eds. Planetary Ecology. Van Nostrand Reinhold, New York, pp. 195–209.

Kepkay, P. E. 1985c. Microbial manganese oxidation and nitrification in relation to the occurrence of macrophyte roots in a lacustrine sediment. Hydrobiology 128:135–142.

Kepkay, P. E., and K. H. Nealson. 1982. Surface enhancement of sporulation and manganese oxidation by a marine bacillus. J. Bacteriol. 151: 1022–1026.

Kepkay, P. E., and K. H. Nealson. 1987. Growth of a manganese oxidizing *Pseudomonas* sp. in continuous culture. Arch. Microbiol. 148: 63–67.

Kepkay, P. E., D. J. Burdige, and K. H. Nealson. 1984. Kinetics of bacterial manganese binding and oxidation in the chemostat. Geomicrobiol. J. 3:245–262.

Kepkay, P. E., P. Schwinghamer, T. Willar, and A. J. Bowen. 1986. Metabolism and metal binding by surface colonizing bacteria: results of microgradient measurements. Appl. Environ. Microbiol. 51: 163–170.

Kessick, M. A., J. Vuceta, and J. J. Morgan. 1972. Spectrophotometric determination of oxidized manganese with leuco crystal violet. Environ Sci. Technol. 6:642–644.

Kindle, E. M. 1932. Lacustrine concretions of manganese. Am. J. Sci. 24:496–504.

Kindle, E. M. 1935. Manganese concretions in Nova Scotia lakes. Roy. Soc. Canada Trans. Sec. IV 29:163–180.

Klaveness, D. 1977. Morphology, distribution and significance of the manganese-accumulating micoorganisms *Metallogenium* in lakes. Hydrobiologia 56:25–33.

Klimov, V. V. 1984. Charge separation in photosystem II reaction centers. The role of pheophytin and manganese, Vol. 1. *In*: C. Sybesina, ed. Adv. Photosynth. Res., Proc. Int. Congr. Photosynth. 6th, 1983. Nijhoff, The Hague, The Netherlands, pp. 131–138.

Klinkhammer, G. P., and M. L. Bender. The distribution of manganese in the Pacific Ocean. Earth Planet Sci. Lett. 46:361–384.

Kozub, J. M., and J. C. Madgwick. 1983. Microaerobic microbial manganese dioxide leaching. Proc. Australas. Inst. Min. Metall. No. 288, December, pp. 51–54.

Krumbein, W. E. 1969. Ueber den Einfluss der Mikroflora auf die exogene Dynamik (Verwitterung und Krustenbildung). Geol. Rundschau 58: 333–365.

Krumbein, W. E. 1971. Manganese oxidizing fungi and bacteria in recent shelf sediments of the Bay of Biscay and the North Sea. Naturwissenschaften 58:56–57.

Krumbein, W. E., and H. J. Altmann. 1973. A new method for the detection and enumeration of manganese-oxidizing and reducing compounds. Helgolaender Wissenschaftliche Meeresuntersuchungen 25: 347–356.

Krumbein, W. E., and K. Jens. 1981. Biogenic rock varnishes of the Negev Desert (Israel): an ecological study of iron and manganese transformation by cyanobacteria and fungi. Oecologia (Berl.) 50: 25–38.

Ku, T. L., and W. S. Broecker. 1979. Radiochemical studies on manganese nodules of deep-sea origin. Deep-Sea Res. Oceanogr. Abstr. 16:625–635.

Landing, W. M., and K. Bruland. 1980. Manganese in the North Pacific. Earth Planet Sci. Lett. 49:45–56.

LaRock, P. A., and H. L. Ehrlich. 1975. Observations of bacterial microcolonies on the surface of ferromanganese nodules from Blake Plateau by scanning electron microscopy. Microb. Ecol. 2:84–96.

Latimer, W. M., and J. H. Hildebrand. 1940. Reference Book of Inorganic Chemistry. Rev. ed. Macmillan, New York.

Lawton, K. 1955. Chemical composition of soils. *In*: F. E. Bear, ed. Chemistry of the Soil. Reinhold, New York, pp. 53–54.

Leeper, G. W. 1947. The forms and reactions of manganese in the soil. Soil Sci. 63:79–94.

Leeper, G. W., and R. J. Swaby. 1940. The oxidation of manganous compounds by microorganisms in the soil. Soil. Sci. 49:163–169.

Letunova, S. V., M. V. Ulubekova, and V. I. Shcherbakov. 1978. Manganese concentration by microorganisms inhabiting soils of the manganese biogeochemical province of the Georgian SSR. Mikrobiologiya 47:332–337 (Engl. transl. pp. 273–278.

Lieske, R. 1919. Zur Ernaehrungsphysiologie der Eisenbakterien. Ztbltt. Bakteriol. Parasitenk. Infektionskr. Hyg. Abt. II 49:413–425.

Loganathan, P., and R. G. Burau. 1973. Sorption and heavy metal ions by a hydrous manganese oxide. Geochim. Cosmochim. Acta 37: 1277–1293.

Lovley, D. R. 1991. Dissimilatory Fe(III) and Mn(IV) reduction. Microbiol. Rev. 55:259–287.

Lovley, D. R., and S. Goodwin. 1988. Hydrogen concentrations as an indicator of the predominant terminal electron-accepting reactions in aquatic sediments. Abstr. Ann. Meeting, Am. Soc. Microbiol. NK-30, p. 249.

Lovley, D. R., and E. J. P. Phillips. 1988a. Novel mode of microbial energy metabolism: Organic carbon oxidation coupled to dissimilatory reduction of iron or manganese. Appl. Environ. Microbiol. 54: 1472–1480.

Lovley, D. R., and E. J. P. Phillips. 1988b. Manganese inhibition of microbial iron reduction in anaerobic sediments. Geomicrobiol. J. 6: 145–155.

Lovley, D. R., and. E. J. P. Phillips. 1994a. Novel processes for anaerobic sulfate production from elemental sulfur by sulfate-reducing bacteria. Appl. Environ. Microbiol. 60:2394–2399.

Lovley, D. R., and E. J. P. Phillips. 1994b. Reduction of chromate by *Desulfovibrio vulgaris* and its c_3 cytochrome. Appl. Environ. Microbiol. 60:726–728.

Lovley, D. R., E. J. P. Phillips, and D. J. Lonergan. 1989. Hydrogen and formate oxidation coupled to dissimilatory reduction of iron and manganese by *Alteromonas putrefaciens*. Appl. Environ. Microbiol. 55:700–706.

Lovley, D. R., P. K. Widman, J. C. Woodward, and J. P. Phillips. 1993a. Reduction of uranium by cytochrome c_3 of *Desulfovibrio vulgaris*. Appl. Environ. Microbiol. 59:3572–3576.

Lovley, D. R., S. J. Giovannoni, D. C. White, J. E. Champine, E. J. P. Phillips, Y. A. Gorby, and S. Goodwin. 1993b. *Geobacter metalli-*

reducens gen. nov. sp. nov., a microorganism capable of coupling the complete oxidation of organic compounds to the reduction of iron and other metals. Arch. Microbiol. 159:336–344.

Mahler, H. R., and E. H. Cordes. 1966. Biological Chemistry. Harper & Row, New York.

Maki, J. S., B. M. Tebo, F. E. Palmer, K. H. Nealson, and J. T. Staley. The abundance and biological activity of manganese- oxidizing bacteria and *Metallogenium*-like morphotypes in Lake Washington, USA. FEMS Microbiol. Ecol. 45:21–29.

Mandernack, K. W. 1992. Oxygen isotopic, mineralogical, and field studies. Ph.D. thesis. University of California, San Diego.

Mann, P. J. G., and J. H. Quastel. 1946. Manganese metabolism in soils. Nature (Lond.) 158:154–156.

Mann, S., N. H. C. Sparks, G. H. E. Scott, and E. W. De Vrind-De Jong. 1988. Oxidation of manganese and formation of Mn_3O_4 (hausmannite) by spore coats of a marine *Bacillus* sp. Appl. Environ. Microbiol. 54:2140–2143.

Margolis, J. V., and R. G. Burns. 1976. Pacific deep-sea manganese nodules: their distribution, composition, and origin. Annu. Rev. Earth Planet. Sci. 4:229–263.

McKenzie, R. M. 1967. The sorption of cobalt by manganese minerals in soils. Aust. J. Soil Res. 5:235–246.

McKenzie, R. M. 1970. The reaction of cobalt with manganese minerals. Aust. J. Soil Res. 8:97–106.

Mercz, T. I., and J. C. Madgwick. 1982. Enhancement of bacterial manganese leaching by microalgal growth products. Proc. Australas. Inst. Min. Metall., No. 283, September, pp. 43–46.

Mero, J. L. 1962. Ocean-floor manganese nodules. 57:747–767.

Mills, V. H., and C. I. Randles. 1979. Manganese oxidation in *Sphaerotilus discophorus* particles. J. Gen. Appl. Microbiol. 25:205–207.

Mineral Facts and Problems, 1965. Bull. 630. Bureau of Mines, U.S. Department of the Interior, Washington, DC.

Mirchink, T. G., K. M. Zaprometova, and D. G. Zvyagintsev. 1970. Satellite fungi of manganese-oxidizing bacteria. Mikrobiologiya 39: 379–383.

Moese, J. R., and H. Brantner. 1966. Mikrobiologische Studien an manganoxydierenden Bakterien. Ztlbltt. Bakt. Parasitenk. Infektionskr. Hyg. II Abat. 120:480–495.

Molisch, H. 1910. Die Eisenbakterien. Gustav Fischer Verlag, Jena.

Moore, W. S. 1981. Iron-manganese banding in Oneida Lake ferromanganese nodules. Nature (Lond.) 292:233–235.

Moore, W. S., W. E. Dean, S. Krishnaswami, and D. V. Borole. Growth rates of manganèse nodules in Oneida Lake, New York. Earth Planet Sci. Lett. 46:191–200.

Mulder, E. G. 1972. Le cycle biologique tellurique et aquatique du fer et du manganèse. Rev. Ecol. Biol. Sol 9:321–348.

Murray, J. W., L. S. Balistrieri, and B. Paul. 1984. The oxidation state of manganese in marine sediments and ferromanganese nodules. Geochim. Cosmochim. Acta 48 1237–1247.

Murray, J. W., J. G. Dillard, R. Giovanoli, H. Moers, and W. Stumm. 1985. Oxidation of Mn(II): intial mineralogy, oxidation state and age-ing. Geochim. Cosmochim. Acta 49:463–470.

Mustoe, G. E. 1981. Bacterial oxidation of manganese and iron in a modern cold spring. Geol. Soc. Am. Bull. Part 1 92:147–153.

Myers, C. R., and K. H. Nealson. 1988a. Bacterial manganese reduction and growth with manganese oxide as the sole electron acceptor. Science 240:1319–1321.

Myers, C. R., and K. H. Nealson. 1988b. Microbial reduction of manganese oxides: Interactions with iron and sulfur. Geochim. Cosmochim. Acta 52:2727–2732.

Nealson, K. H. 1978. The isolation and characterization of marine bacteria which catalyze manganese oxidation. In: W. E. Krumbein, ed. Environmental Biogeochemistry, Vol. 3. Methods, Metals and Assessment. Ann Arbor Science Publishers, Ann Arbor, MI, pp. 847–858.

Nealson, K. H., and J. Ford. 1980. Surface enhancement of bacterial manganese oxidation: implications for aquatic environments. Geomicrobiol. J. 2:21–37.

Nealson, K. H., and D. Saraffini. 1994. Iron and manganese in anaerobic respiration: environmental significance, physiology, regulation. Annu. Rev. Microbiol. 48:311–343.

Nealson, K. H., B. M. Tebo, and R. A. Rosson. 1988. Occurrence and mechanisms of microbial oxidation of manganese. Adv. Appl. Microbiol. 33:299–318.

Neufeld, C. A. 1904. Z. Unters. Nahrungs-Genussmitt. 7:478 (as cited by Moese and Brantner, 1966).

Novozhilova, M. I., and F. S. Berezina. 1984. Iron- and manganese-oxidizing microorganisms in grounds in the northwestern part of the Indian Ocean and the Red Sea. Mikrobiologiya 53:129–136 (Engl. transl. pp. 106–112).

Ostwald, J. 1981. Evidence for a biogeochemical origin of the Groote Eylandt manganese ores. Econ. Geol. 556–567.

Pacszczynski, A., V-B. Huynh, and R. Crawford. 1986. Comparison of ligninase-1 and peroxidase-M2 from the white rot fungus *Phanaerochete chrysosporium*. Arch. Biochem. Biophys. 244:750–765.

Park, P. K. 1968. Seawater hydrogen-ion concentration: Vertical distribution. Science 162:357–358.

Patrick, W. H. Jr., and R. E. Henderson. 1981. Reduction and reoxidation cycles of manganese and iron in flooded soil and in water solution. Soil Sci. Soc. Am. J. 45:855–859.

Pattan, J. N., and A. V. Mudhkolkar. 1990. The oxidation state of manganese in ferromanganese nodules and deep-sea sediments from the Central Indian Ocean. Chem. Geol. 85:171–181.

Perfil'ev, B. V., and D. R. Gabe. 1965. The use of the microbial- landscape method to investigate bacteria which concentrate manganese and iron in bottom deposits. *In*: B. V. Perfil'ev, D. R. Gabe, A. M. Gal'perina, V. A. Rabinovich, A. A. Saponitskii, E. E. Sherman, and E. P. Troshanaov. Applied Capillary Microscopy. The Role of Microorganisms in the Formation of Iron-Manganese Deposits. Consultants Bureau, New York, pp. 9–54.

Perfil'ev, B. V., and D. R. Gabe. 1969. Capillary methods of studying microorganisms. English translation by J. Shewan. Univesity of Toronto Press, Toronto, Ont.

Perkins, E. C., and F. Novielli. 1962. Bacterial leaching of manganese ores. U.S. Bur. Mines Rep. Invest. 6102, 11 pp.

Piper, D. Z., J. R. Basler, and J. L. Bischoff. 1984. Oxidation state of marine manganese nodules. Geochim. Cosmochim. Acta 48: 2347–2355.

Poldervaart, A. 1955. Chemistry of the earth's crust, pp. 119–144. *In*: A. Poldervaart, ed. Crust of the Earth. Special Paper 62 A Symposium. Geological Society of America, Boulder, CO.

Potter, R. M., and G. R. Rossman. 1977. Desert varnish: the importance of clay minerals. Science 196:1446–1448.

Reid, A. S. J., and M. H. Miller. 1963. The manganese cycle. II. Forms of soil manganese in equilibrium with solution manganese. Can. J. Soil Sci. 43:250–259.

Reyss, J. L., V. Marchig, and T. L. Ku. 1982. Rapid growth of a deep-sea manganese nodule. Nature (Lond.) 295:401–403.

Richards, F. A. 1965. Anoxic basins and fjords. *In*: J. P. Riley and G. Skirrow, eds. Chemical Oceanography, Vol. 1. Academic Press, New York, pp. 611–641.

Richardson, L. L., C. Aguilar, and K. H. Nealson. 1988. Manganese oxidation in pH and O_2 microenvironments produced by phytoplankton. Limnol. Oceanogr. 33:352–363.

Riemann, F. 1983. Biological aspects of deep-sea manganese nodule formation. Oceanol. Acta 6:303–311.

Robertson, L. A., E. W. J. van Niel, R. A. M. Torremans, and G. J. Kuenen. 1988. Simultaneous nitrification and denitrification in aerobic chemostat cultures of *Thiosphaera pantropha*. Appl. Enviorn. Microbiol. 54:2812–2818.

Robinson, W. O. 1929. Detection and significance of manganese dioxide in the soil. Soil Sci.27:335–349.

Rossman, R., and E. Callender. 1968. Manganese nodules in Lake Michigan. Science 162:1123–1124.

Rosson, R. A., and K. H. Nealson. 1982. Manganese binding and oxidation by spores of a marine bacillus. J. Bacteriol. 151:1027–1034.

Rosson, R. A., B. M. Tebo, and K. H. Nealson. 1984. Use of poisons in determination of microbial manganese binding rates in seawater. Appl. Environ. Microbiol. 47:740–745.

Rusin, P. A., L. Quintana, N. A. Sinclair, R. G. Arnold, and K. L. Oden. 1991a. Physiology and kinetics of manganese-reducing *Bacillus polymyxa* strain D1 isolated from manganiferous silver ore. Geomicrobiol. J. 9:13–25.

Rusin, P. A., J. E. Sharp, R. G. Arnold, and N. A. Sinclair. 1991b. Enhanced recovery of manganese and silver from refractory ores. *In*: R. W. Smith and M. Misra, eds. Mineral Bioprocessing. The Minerals, Metals and Materials Society, Warrendale, PA, pp. 207–218.

Sartory, A., and J. Meyer. 1947. Contribution à l'étude du métabolisme hydrocarboné des bactéries ferrigineuses. C.R. Acad. Sci. (Paris) 225:541–542.

Schmidt, W-D., and J. Overbeck. 1984. Studies of "iron bacteria" from Lake Pluss (West Germany). 1. Morphology, fine structure and distribution of *Metallogenium* sp. and *Siderocapsa germinata*. Z. Allg. Mikrobiol. 24:329–339.

Schmidt, J. M., W. P. Sharp, and M. P. Starr. 1981. Manganese and iron encrustations and other features of *Planctomyces crassus* Hortobagyi 1965, morphotype Ib of the *Blastocaulis-Planctomyces* group of budding and appended bacteria, examined by electron microscopy and X-ray micro-analysis. Curr. Microbiol. 5:241–246.

Schmidt, J. M., W. P. Sharp, and M. P. Starr. 1982. Metallic oxide encrustations of the nonprosthecate stalks of naturally occurring populations of *Planctomyces bekfii*. Curr. Microbiol. 7:389–394.

Schoettle, M., and G. M. Friedman. 1971. Fresh water iron-manganese nodules in Lake George, New York. Geol. Soc. Am. Bull. 82: 101–110.

Schulz-Baldes, A., and R. A. Lewin. 1975. Manganese encrustation of zygospores of a *Chlamydomonas* (Chlorophyta: Volvocales). Science 188:1119–1120.

Schweisfurth, R. 1968. Untersuchungen ueber manganoxidierende und -reduzierende Mikroorganismen. Mitt. Internat. Verein. Limnol. 14; 179–186.

Schweisfurth, R. 1969. Manganoxidierende Pilze. Ztlbltt. Bakteriol. Parasitenk. Infektionskr. Hyg. I Orig. 212:486–491.

Schweisfurth, R. 1970. Manganoxidierende Pilze. I. Vorkommen, Isolierungen und mikroskopische Untersuchungen. Z. Allg. Mikrobiol. 11: 415–430.

Schweisfurth, R., and G. v. Hehn. 1972. Licht- und elektronenmikroskopische Untersuchungen sowie Kulturversuche zum Metallogenium-Problem. Zbltt. Bakt. Parasitenk. Infektionskr. Hyg. I. Abt. Orig. A 220:357–361.

Schweisfurth, R., and R. Mertes. 1962. Mikrobiologische und chemische Untersuchungen ueber Bildung und Bekaempfung von Manganschlammablagerung einer Druckleitung fuer Talsperrenwasser. Arch. Hyg. Bakteriol. 146:401–417.

Seyfried, W. E., Jr., and D. R. Janecky. 1985. Heavy metal and sulfur transport during subcritical and supercritical hydrothermal alteration of basalt: Influence of fluid pressure and basalt composition and crystallinity. Geochim. Cosmochim. Acta 49:2545–2560.

Shanks, W. C. III, J. L. Bischoff, and R. J. Rosenbauer. Seawater sulfate reduction and sulfur isotope fractionation in basaltic systems: Interaction of seawater with fayalite and magnetite at 200–350°C. Geochim. Cosmochim. Acta 45:1977–1995.

Sherman, G. D., J. S. McHargue, and W. S. Hodgkiss. 1942. Determination of active manganese in soil. Soil Sci. 54:253–257.

Shternberg, L. E. 1967. Biogenic structures in manganese ores. Mikrobiologiya 36:710–712 (Engl. transl. pp. 595–597).

Sienko, M. J., and R. A. Plane. 1966. Chemistry: Principles and Properties. McGraw-Hill, New York.

Silverio, C. M. 1985. Microaerobic manganese dioxide reduction. NSTA Technol. J. Oct.-Dec., pp. 51–63.

Silverman, M. P., and D. G. Lundgren. 1959. Studies on the chemoautotrophic iron bacterium *Ferrobacillus ferrooxidans*. I. An improved medium and a harvesting procedure for securing high cell yields. J. Bacteriol. 77:642–647.

Sly, L. I., M. C. Hodkinson, and V. Arunpairojana. 1988. Effect of water velocity on the early development of manganese-depositing biofilm in a drinking-water distribution system. FEMS Micorbiol. Ecol. 53: 175–186.

Soehngen, N. L. 1914. Umwandlung von Manganverbindungen unter dem Einfluss mikrobiologischer Prozesse. Ztlbltt. Bakteriol. Parasitenk. Infektionskr. Hyg. Abt. II 40:545–554.

Sokolova-Dubinina G. A., and Z. P. Deryugina. 1967a. On the role of microorganisms in the formation of rhodochrosite in Punnus-Yarvi lake. Mikrobiolobiya 36:535–542 (Engl. transl. pp. 445–451).

Sokolova-Dubinina, G. A., and Z. P. Deryugina. 1967b. Process of iron-manganese concretion formation in Lake Punnus-Yarvi. Mikrobiologiya 36:1066–1076.

Sokolova-Dubinina, G. A., and Z. P. Deryugina. 1968. Influence of limnetic environmental conditions on microbial manganese ore formation. Mikrobiologiya 37:143–153 (Engl. transl. pp. 123–128).

Sokolova-Dubinina, G. A., and Z. P. Deryugina. 1971. Electron microscope study of iron-manganese concretions from Lake Punnus-Yarvi. Dokl. Akad. Nauk SSSR 201:714–716 (Engl. transl. pp. 738–740).

Sorem, R. K., and A. R. Foster. 1972. Internal structure of manganese nodules and implications in beneficiation. In: D. R. Horn, ed. Ferromanganese Deposits on the Ocean Floor. Office of the International Decade of Ocean Exploration. National Science Foundation, Washington, DC, pp. 167–181.

Sorokin, Yu.I. 1971. Microflora of iron-manganese concretions from the ocean floor. Mikrobiologiya 40:563–566 (Engl. transl. pp. 493–495).

Sorokin, Yu.I. 1972. Role of Biological factors in the sedimentation of iron, manganese, and cobalt and in the formation of nodules. Oceanology (Oekonologiya) 12:1–11.

Sparrow, L. A. and N. C. Uren. 1978. Oxidation and reduction of Mn in acidic soils: effect of temperature and soil pH. Soil Biol. Biochem. 19:143–148.

Stabel, H. H., and J. Kleiner. 1983. Endogenic flux of manganese to the bottom of Lake Constance. Arch. Hydrobiol. 98:307–316.

Stone, A. T. 1987. Microbial metabolites and the reductive dissolution of manganese oxides: Oxalate and pyruvate. Geochim. Cosmochim. Acta 51:919–925.

Sugio, T., T. Katagiri, M. Moriyama, Y. L. Zhen, K. Inagaki, and T. Tano. 1988a. Existence of a new type of sulfide oxidase which utilizes ferric iron as an electron acceptor in Thiobacillus ferrooxidans. Appl. Envrion. Microbiol. 54:153–157.

Sugio, T., Y. Tsujitu, K. Hirayama, K. Inagaki, and T. Tano. 1988b. Mechanism of tetravalent manganese reduction with elemental sulfur. Agric. Microbiol. Chem. 52:185–190.

Tambiev, S. B., and L. L. Demina. 1992. Biogeochemistry and fluxes of manganese and some other metals in regions of hydrothermal activities (Axial Mountain, Juan de Fuca Ridge and Guaymas Basin, Gulf of California). Deep-Sea Res. 39:687–703.

Taylor, R. M., and R. M. McKenzie. 1966. The association of trace elements with manganese minerals in Australian soils. Aust. J. Soil Res. 4:29–39.

Taylor, R. M., R. M. McKenzie, and K. Norrish. 1964. The mineralogy and chemistry of manganese in some Australian soils. Aust. J. Soil Res. 2:235–248.

Taylor-George, S., F. Palmer, J. T. Staley, D. J. Borns, B. Curtiss, and J. B. Adams. 1983. Fungi and bacteria involved in Desert Varnish formation. Microb. Ecol. 9:227–245.

Tebo, B. M., and S. Emerson. 1985. Effect of oxygen tension, Mn(II) concentration, and temperature on the microbially catalyzed Mn(II) oxidation rate in a marine fjord. Appl. Environ. Microbiol. 50: 1268–1273.

Tebo, B. M., K. H. Nealson, S. Emerson, and L. Jacobs. 1984. Microbial mediation of Mn(II) and Co(II) precipitation at the O_2/H_2S interfaces in two anoxic fjords. Limnol. Oceanogr. 29:1247–1258.

Tebo, B. M., K. Mandernack, and R. A. Rosson. 1988. Manganese oxidation by a spore coat or exosporium protein from spores of a manganese(II) oxidizing marine bacillus. Abstr. Ann. Meeting, Am. Soc. Microbiol. I-121, p. 201.

Ten Khak-mun. 1967. Iron- and manganese-oxidizing microorganisms in soils of South Sakhalin. Mikrobiologiya 36:337–344 (Engl. transl. 276–281).

Thamdrup, B., K. Finster, J. Wuergler Hansen, and F. Bak. 1993. Bacterial disproportionation of elemental sulfur coupled to chemical reduction of iron or manganese. Appl. Environ. Microbiol. 59: 101–108.

Thiel, G. A. 1925. Manganese precipitated by microorganisms. Econ. Geol. 20:301–310.

Timonin, M. I., 1950a. Soil microflora and manganese deficiency. Trans. 4th Int. Congr. Soil Sci. 3:97–99.

Timonin, M. I. 1950b. Soil microflora in relation to manganese deficiency. Sci. Agric. 30:324–325.

Timonin, M. I., W. I. Illman, and T. Hartgerink. 1972. Oxidation of manganous salts of manganese by soil fungi. Can. J. Microbiol. 18: 793–799.

Tipping, E., D. W. Thompson, and W. Davidson. 1984. Oxidation products of Mn(II) oxidation in lake waters. Chem. Geol. 44:359–383.

Tipping, E., J. G. Jones, and C. Woof. 1985. Lacustrine manganese oxides: Mn oxidation states and relationships to "Mn depositing bacteria." Arch. Hydrobiol. 105:161–175.

Tortoriello, R. C. 1971. Manganic oxide reduction by microorganisms in fresh water environments. Ph.D. thesis. Rensselaer Polytechnic Institute, Troy, NY.

Trimble, R. B. 1968. MnO$_2$-reduction by two strains of marine ferromanganese nodule bacteria. M.S. thesis. Rensselaer Polytechnic Institute, Troy, NY.

Trimble, R. B., and H. L. Ehrlich. 1968. Bacteriology of manganese nodules. III. Reduction of MnO$_2$ by two strains of nodule bacteria. Appl. Microbiol. 16:695–702.

Trimble, R. B., and H. L. Ehrlich. 1970. Bacteriology of manganese nodules. IV. Induction of an MnO$_2$-reductase system in a marine Bacillus. Appl. Microbiol. 19:966–972.

Troshanov, E. P. 1968. Iron- and manganese-reducing microorganisms in ore-containing lakes of the Karelian isthmus. Mikrobiologiya 37: 934–940 (Engl. transl. pp. 786–791).

Troshanov, E. P. 1969. Conditions affecting the reduction of iron and manganese by bacteria in the ore-bearing lakes of the Karelian Isthmus. Mikrobiologiya 38:634–643 (Engl. transl. pp. 528–535).

Trost, W. R. 1958. The chemistry of manganese deposits. Mines Branch Research Report, R8. Dept. of Mines and Technical Surveys, Ottawa, Canada, 125 pp.

Tuttle, J. H., and H. L. Ehrlich. 1986. Coexistence of inorganic sulfur metabolism and manganese oxidation in marine bacteria. Abst. Ann. Meeting, American Soc. Microbiol., I21, p.168.

Tyler, P. A. 1970. Hyphomicrobia and the oxidation of manganese in aquatic ecosystems. Antonie v. Leeuwenhoek 36:567–578.

Tyler, P. A., and K. C. Marshall. 1967a. Microbial oxidation of manganese in hydro-electric pipelines. Antonie v. Leeuwenhoek 33:171–183.

Tyler, P. A., and K. C. Marshall. 1967b. Hyphomicrobia—A significant factor in manganese problems. J. Am. Water Works Assoc. 59: 1043–1048.

Uren, N. C., and G. W. Leeper. 1978. Microbial oxidation of divalent manganese. Soil Biol. Biochem. 10:85–87.

Van Veen, W. L. 1972. Factors affecting the oxidation of manganese by *Sphaerotilus discophorus*. Antonie v. Leeuwenhoek 38:623–626.

Van Veen, W. L. 1973. Biological oxidation of manganese in soils. Antonie v. Leeuwenhoek 39:657–662.

Van Waasbergen, L. G., J. A. Hoch, and B. M. Tebo. 1993. Genetic analysis of the marine manganese-oxidizing *Bacillus* sp. SG-1: Protoplast transformation, Tn917 mutagenesis, and identification of chromosomal loci involved in manganese oxidation. J. Bacteriol. 175: 7594–7603.

Varentsov, I. M. 1972. Geochemical studies on the formation of iron-manganese nodules and crusts in recent basins. I. Eningi-Lampi Lake, Central Karelia. Usta. Mineral. Petrogr. Szeged. 20:363–381.

Varentsov, I. M., and N. V. Pronina. 1973. On the study of mechanisms of iron-manganese ore formation in Recent basins: the experimental data on nickel and cobalt. Mineral. Depos. (Berl.) 8:161–178.

Vavra, P. and L. Frederick. 1952. The effect of sulfur oxidation on the availability of manganese. Soil Sci. Soc. Am. Proc. 16:141–144.

Vojak, P. W. L., C. Edwards, and M. V. Jones. 1984. Manganese oxidation and sporulation by an estuarine *Bacillus* species. Microbios 41: 39–47.

Vojak, P. W. L., C. Edwards, and M. V. Jones. 1985a. A note on the enumeration of manganese-oxidizing bacteria in estuarine water and sediment samples. J. Appl. Bacteriol. 59:375–379.

Vojak, P. W. L., C. Edwards, and M. V. Jones. 1985b. Evidence for microbiological manganese oxidation in the River Tamar estuary, South West England. Estuarine Coastal Shelf Sci. 20:661–671.

Von Damm, K. L., J. M. Edmond, B. Grant, C. I. Measures, B. Walden, and R. F. Weiss. 1985. Chemistry of submarine hydrothermal solutions at 21°N, East Pacific Rise. Geochim. Cosmochim. Acta 49: 2197–2220.

Von Wolzogen Kuehr, C. A. H. 1927. Manganese in waterworks. J. Am. Water Works. Assoc. 18:1–31.

Yurchenko, V. A., G. I. Karavaiko, V. I. Remizov, T. M. Klyushnikova, and A. D. Tarusin. 1987. Role of the organic acids produced by *Acinetobacter calcoaceticus* in manganese leaching. Prikl. Biokhim. Mikrobiol. 23:404–412.

Zapffe, C. 1931. Deposition of manganese. *Econ. Geol.* 26:799–832.

Zavarzin, G. A. 1961. Symbiotic culture of a new microorganism oxidizing manganese. Mikrobiologiya 30:393–395 (Engl. transl. pp. 343–345).

Zavarzin, G. A. 1962. Symbiotic oxidation of manganese by two species of *Pseudomonas*. Mikrobiologiya 31:586–588. (Engl. transl. pp. 481–482).

Zavarzin, G. A. 1964. *Metallogenium symbioticum*. Z. Allg. Mikrobiol. 4:390–395.

Zindulis, J., and H. L. Ehrlich. 1983. A novel Mn^{2+}-oxidizing enzyme system in a freshwater bacterium. Z. Allg. Mikrobiol. 23:457–465.

ZoBell, C. E. 1946. Marine Microbiology. Chronica Botanica, Waltham, MA.

16

Geomicrobial Interactions with Chromium, Molybdenum, Vanadium, and Uranium

16.1 MICROBIAL INTERACTIONS WITH CHROMIUM

Occurrence of Chromium   Chromium is not a very plentiful element in the Earth's crust, but it is fairly widespread. Its average crustal abundance of 122 ppm (Fortescue, 1980) is less than that of manganese. Average concentrations in rock range from 4 to 90 mg/kg, in soil around 70 mg/kg, in fresh water around 1 µg/kg and in seawater around 0.3 µg/kg (Bowen, 1979). Its chief mineral occurrence is as chromite, in which the chromium has an oxidation state of $+3$. It is a spinel whose end members are $MgCr_2O_4$ and $FeCr_2O_4$. The chromium in the mineral can be partially replaced by Al and Fe. Chromite is of igneous origin. Other chromium minerals of minor occurrence include eskolite (Cr_2O_3), daubreélite ($FeS \cdot Cr_2S_3$), crocoite ($PbCrO_4$), uvarovite, which is also known as garnet [$Ca_3Cr_2(SiO_4)_3$], and others (Smith, 1972).

Chemically and Biologically Important Properties   The element chromium is a member of the first transition series of elements in the periodic table together with scandium, titanium, vanadium, manganese, iron, cobalt, nickel, copper, and zinc. The chief valence states of chromium are 0, $+2$, $+3$, and $+6$. A $+5$ oxidation state is known, which appears to be of significance in at least some biochemical reductions of Cr(VI) (Shi and Dalal, 1990; Suzuki et al., 1992). The geomicrobially important valence states are $+3$ and $+6$.

Chromium in the hexavalent state is very toxic, in part because of its high solubility as chromate (CrO_4^{2-}) and dichromate ($Cr_2O_7^{2-}$) at physiological pH. At sufficient concentration, Cr(VI) can be mutagenic and carcinogenic. Chromium in the trivalent state is less toxic, in part probably because it is less soluble in this oxidation state at physiological pH. At neutral pH, Cr^{3+} tends to precipitate as a hydroxide [$Cr(OH)_3$] or a hydrated oxide. Chromate and dichromate are strong oxidizing agents.

Chromium has been reported to be nutritionally essential in trace amounts (Miller and Neathery, 1977; Mertz, 1981). Its metabolic function in biochemical terms is still unclear.

Cr as $Cr_2(SO_4)_3$, K_2CrO_4, or $K_2Cr_2O_7$ is inhibitory to the growth of bacteria at appropriate concentrations (e.g., Forsberg, 1978; Wong et al., 1982; Bopp et al., 1983). When taken into a cell, hexavalent chromium can act as a mutagen in prokaryotes and eukaryotes (Nishioka, 1975; Petrelli and DeFlora, 1977; Venitt and Levy, 1974) and as a carcinogen in animals (Gruber and Jennette, 1978; Sittig, 1985). In bacteria, chromate can be taken into the cell via the sulfate uptake system, which involves active transport (Ohtake et al., 1987; Silver and Walderhaug, 1992). Whether dichromate is transported by the same system is not clear. Bacterial resistance to Cr(VI) has been observed. In some instances, at least, the genetic trait is plasmid-borne (Summers and Jacoby, 1978; Bopp et al., 1983; Cervantes, 1991; Silver and Walderhaug, 1992). In *Pseudomonas fluorescens* strain LB300, resistance to chromate (CrO_4^{2-}) was found to be due to decreased chromate uptake (Bopp, 1980; Ohtake et al., 1987); In *Pseudomonas ambigua*, chromate resistance was attributed to formation of a thickened cell envelope which reduced permeability of Cr(VI) and to ability to reduce Cr(VI) to Cr(III) (Horitsu et al., 1987). The basis for resistance to dichromate ($Cr_2O_4^{2-}$) and chromite (Cr^{3+}) has not been clearly established. The resistance mechanism for dichromate need not be the same as for chromate since *Pseudomonas fluorescens* LB 300, which is resistant to chromate, is much more sensitive to dichromate (Bopp, 1980; Bopp et al., 1983). Some bacteria have an ability to accumulate chromium. In at least some cases, the accumulation may be due to adsorption (Coleman and Paran, 1983; Johnson et al., 1981; Marques et al., 1982).

Mobilization of Chromium with Microbially Generated Lixiviants *Thiobacillus thiooxidans* and *T. ferrooxidans* have been found to solubilize only a limited amount of chromium contained in chromite, with sulfuric acid generated by the oxidation of sulfur (Ehrlich, 1983). Similarly, acid produced during iron oxidation by *T. ferrooxidans* was able to solubilize only limited amounts of chromium from chromite (Wong et al., 1982). On the other hand, chromium can be successfully leached from some solid industrial wastes with biogenically formed sulfuric acid (Bosecker, 1986).

Biooxidation of Chromium (III)   No observations of enzymatic oxidation of Cr(III) to Cr(VI) have been reported. However, nonenzymatic oxidation of Cr(III) to Cr(VI) may occur in soil environments, where biogenic (or abiogenic) Mn(III) or Mn(IV) oxides may oxidize Cr(III) to Cr(VI) (Bartlett and James, 1979), for instance:

$$2Cr^{3+} + 3MnO_2 + 2H_2O \Rightarrow 2CrO_4^{2-} + 3Mn^{2+} + 4H^+ \qquad (16.1)$$

Such oxidation can be detrimental if the Cr(VI) produced reaches toxic levels.

Bioreduction of Chromium (VI)   Several different bacterial species have been shown to reduce Cr(VI) to Cr(III) (Romanenko and Koren'kov, 1977; Lebedeva and Lyalikova, 1979; Shimada, 1979; Kvasnikov et al., 1983; Gvozdyak et al., 1986; Horitsu et al., 1987; Bopp and Ehrlich, 1988; Wang et al., 1989; Ishibashi et al., 1990; Shen and Wang, 1993; Llovera et al., 1993; Lovley and Phillips, 1994; Gopalan and Veeramani, 1994). They include *Achromobacter eurydice*, *Aeromonas dechromatica*, *Agrobacterium radiobacter* strain EPS-916, *Arthrobacter* spp., *Bacillus subtilis*, *B. cereus*, *Desulfovibrio vulgaris* (Hildenborough) ATCC 29579, *Escherichia coli* strains K-12 and ATCC 33456, *Enterobacter cloacae HO1*, *Flavobacterium devorans*, *Sarcina flava*, *Micrococcus roseus*, and *Pseudomonas* spp. Whether all these strains reduce Cr(VI) enzymatically is not clear. Sulfate-reducing bacteria can also reduce chromate with the H_2S they produce from sulfate (Bopp, 1980), but *D. vulgaris* can do it enzymatically as well. Of those bacteria which reduce Cr(VI) enzymatically, some of the facultative strains reduce it only anaerobically whereas others will do it aerobically or anaerobically. Many bacterial strains reduce Cr(VI) as a form of respiration, but at least one (*Pseudomonas ambigua* G-1) reduces it as a means of detoxification.

   Pseudomonas fluorescens strain LB300, isolated from the upper Hudson River (New York State, U.S.A.), can reduce chromate aerobically with glucose or citrate as electron donor (Bopp and Ehrlich, 1988; DeLeo and Ehrlich, 1994). Conditions permitting aerobic reduction include batch culturing with shaking at 200 rpm and continuous culturing with stirring and forced aeration (DeLeo and Ehrlich, 1994). It converts chromate to Cr^{3+} in batch culture when growing in a liquid glucose-mineral salt medium (Vogel-Bonner medium) and in continuous culture (chemostat) when growing in a citrate-yeast extract-tryptone medium buffered with phosphate (Fig. 16.1A–C). Anaerobically, *Pseudomonas fluorescens* strain LB300 was found to reduce chromate only when growing with acetate as energy source (electron donor). Furthermore, whereas aerobically *P. fluorescens* LB300 will reduce chromate at an initial concentration as high as 314 μg ml^{-1}, anaerobically it reduces chromate only at concentrations below 50 μg ml^{-1} (Bopp and Ehrlich, 1988; DeLeo and Ehrlich, 1994). Other bacteria that can reduce chromate aerobically and anaerobically include *Escherichia coli* ATCC 33456 (Shen and Wang, 1993) and *Agrobacterium radiobacter* EPS-916 (Llovera et al., 1993). Reduction of chromate by *E. coli* ATCC 33456 is, however, partially repressed by oxygen through uncompetitive inhibition (Shen and Wang, 1993). Reduction of chromate by resting cells of *A. radiobacter* EPS-916 proceeded initially at similar rates aerobically and anaerobically but subsequently slowed significantly in air (Llovera et al., 1994). *Pseudomonas putida* PRS2000 reduces chromate aerobically more rapidly than anaerobically (Ishibashi et al., 1990). *Pseudomonas* sp. (C7) has so far been tested only aerobically (Gopalan and Veeramani, 1994).

By contrast, *Pseudomonas dechromaticans*, *Pseudomonas chromatophila*, *Enterobacter cloacae* OH1, and *Desulfovibrio vulgaris* reduce Cr(VI) only anaerobically with organic electron donors, or H_2 in the case of *D. vulgaris* (Romanenko and Koren'kov, 1977; Lebedeva and Lyalikova, 1979; Komori et al., 1989; Lovley and Phillips, 1994). Except for *E. cloacae* OH1, they cannot use glucose as reductant. *P. dechromaticans* and *P. chromatophila* appear to be able to reduce both chromate and dichromate.

Cell extracts of *P. fluorescens* LB300 reduce chromate with added glucose or NADH (Fig. 16.2). One or more plasma membrane components appear to be required (Bopp and Ehrlich, 1988). *E. cloacae* HO1 also uses a membrane-bound respiratory system to reduce chromate, but it functions only under anaerobic conditions (Wang et al., 1991). By contrast, most of the chromate-reducing activity in *E. coli* ATCC 33456 appears to be soluble; i.e., it does not involve plasma membrane components, but it is mediated by NADH (Shen and Wang, 1993). The chromate-reducing activity of *P. putida* PRS2000 also does not depend on plasma membrane components and mediates the reduction via NADH or NADPH (Ishibashi et al., 1990). *D. vulgaris* ATCC 29579 uses its cytochrome c_3 as its Cr(VI) reductase coupled to hydrogenase when using H_2 as reductant (Lovley and Phillips, 1994).

Although in situ microbial Cr(VI) reducing activity has not been reported, it would seem to be an important measurement to make since

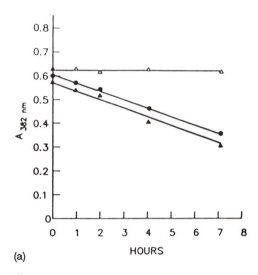

(a)

Figure 16.1   Chromate reduction by resting and growing cells of Cr^R *P. fluorescens* LB300. (a) Resting cells grown with and without chromate; open triangles, chromate-grown cells in absence of electron donor (results were the same for cells grown without chromate and assayed in the absence of an electron donor—no chromate reduction was observed); closed circles, chromate grown cells with 0.5% (wt/vol) glucose; closed triangles, cell grown without chromate and assayed with 0.5% (wt/vol) glucose. Chromate was not reduced by spent medium from either chromate-grown cells or cells grown without chromate or by assay buffer containing either 0.25 or 0.5% glucose. Chromate concentration measured as absorbance at 382 nm after cell removal. (b) Growing cells in VP broth at an initial K_2CrO_4 concentration of 40 μg/ml; growth of the culture was measured photometrically as turbidity at 600 nm; chromate concentration was measured as absorbance at 382 nm after first removing cells from replicate samples by centrifugation followed by filtration. (c) Chromate reduction by cells growing in citrate-chromate medium in a chemostat at a dilution rate of 1.17 ml h^{-1}; open triangles, chromate concentration in uninoculated reactor; open circles, chromate concentration in inoculated reactor; open squares, cell concentration in inoculated reactor. (a and b from Bopp and Ehrlich, Chromate resistance and reduction in *Pseudomonas fluorescens* strain LB300, Arch. Microbiol. 150:426 copyright © 1988 Springer-Verlag, by permission; c from DeLeo and Ehrlich, Reduction of hexavalent chromium by *Pseudomonas fluorescens* LB300 in batch and continuous cultures, Appl. Microbiol. Biotechnol. 40:756 copyright © 1994 Springer-Verlag, by permission.)

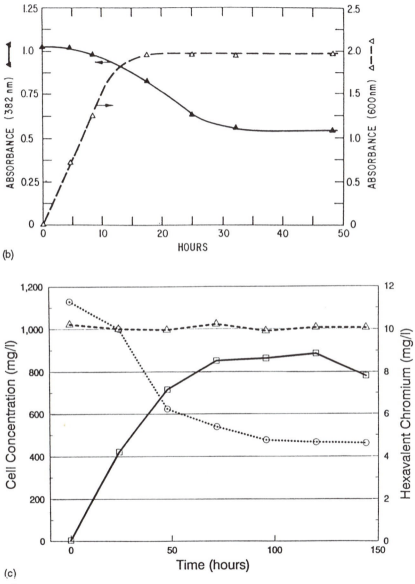

(b)

(c)

Figure 16.1 Continued.

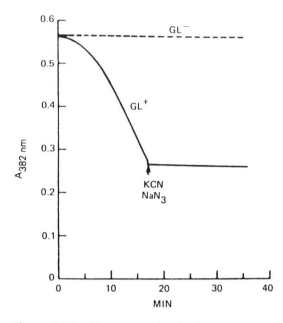

Figure 16.2 Chromate reduction by cell extract from *Pseudomonas fluorescens* strain LB300. GL⁻: without added glucose; GL⁺: with glucose from time 0 and KCN and NaN₃ at times indicated. Chromate concentration measured as absorbance at 382 nm after cell removal. (From Bopp and Ehrlich, Chromate resistance and reduction in *Pseudomonas fluorescens* strain LB300, Arch. Microbiol. 150: 426 Copyright © 1988 Springer-Verlag, by permission.)

a number of different bacteria possess the ability. Natural levels of Cr(VI) in most environments can be expected to be low. However, anthropogenic pollution may cause significant elevation of environmental chromium concentrations.

Lebedeva and Lyalikova isolated a strain of *Pseudomonas chromatophila* from the effluent of a chromite mine in Yugoslavia which contained the mineral crocoite (PbCrO₄) in its oxidation zone. At this site, the organism clearly found a plentiful source of Cr(VI). The isolate was shown to use a range of carbon/energy sources anaerobically for chromate reduction. These included ribose, fructose, benzoate, lactate, acetate, succinate, butyrate, glycerol, ethylene glycol, but not glucose or hydrogen. In the laboratory, crocoite reduction was demonstrated anaerobically with lactate as electron donor.

Applied Aspects of Chromium (VI) Reduction   The first practical application of microbial Cr(VI) reduction as a bioremediation process was explored by Russian investigators. They presented evidence that indicates that bacterial chromate reduction can be harnessed in waste water and sewage treatment to remove chromate (Romanenko et al., 1976; Pleshakov et al., 1981; Serpokrylov et al., 1985; Simonova, et al., 1985). The process also has potential for application in the treatment of tanners and, especially, electroplating wastes, and in situ bioremediation.

Extensive research in the use of *Enterobacter cloacae* HO1 in bioremediation of Cr(VI)-containing wastewaters was performed in Japan (Ohtake et al., 1990; Komori et al., 1990a,b; Fuji et al., 1990; Yamamoto et al., 1993).

16.2 MICROBIAL INTERACTION WITH MOLYBDENUM

Occurrence and Properties of Molybdenum   Molybdenum is an element of the second transition series in the periodic table. In mineral form, it occurs extensively as molybdenite (MoS_2). The minerals wulfenite ($PbMoO_4$) and powellite ($CaMoO_4$) are often associated with the oxidation zone of molybdenite deposits (Holliday, 1965). Molybdite (MoO_3) is another molybdenum-containing mineral that may be encountered in nature. The abundance of Mo has been reported as 2–4 g/ton in basaltic rock, 2.3 g/ton in granitic rocks, and 0.001–0.005 g/ton in ocean waters (Enzmann, 1972).

The oxidation states in which molybdenum can exist include 0, $+2$, $+3$, $+4$, $+5$, and $+6$. Of these, the $+4$ and the $+6$ state are the most common, but the $+5$ state is of biological significance. Molybdenum oxyanions of the $+6$ oxidation state tend to polymerize, the complexity of the polymers depending on the pH of the solution (Latimer and Hildebrand, 1942).

Molybdenum is a biologically important trace element. A number of enzymes feature it in their structure, e.g., nitrogenase, nitrate reductase (Brock and Madigan, 1991), sulfite reductase, and arsenite oxidase (Anderson et al., 1992). Molybdate is also an effective inhibitor of bacterial sulfate reduction (Oremland and Capone, 1988).

Microbial Oxidation and Reduction   Molybdenite (MoS_2) is aerobically oxidizable as an energy source by *Acidianus brierleyi* (Brierley and Murr, 1973) (see also Chapter 18) with the formation of molybdate and sulfate. *T. ferrooxidans* can also oxidize molybdenite but is poisoned by the resulting molybdate (Tuovinen et al., 1971) unless it is rendered insoluble by Fe^{3+}, for instance. Sugio et al. (1992) reported that *T. ferrooxidans* AP-19–3 contains an enzyme that oxidizes molybdenum blue (Mo^{5+}) to molybdate (Mo^{6+}). They purified the molybdenum oxidase, an enzyme complex, and showed that cytochrome oxidase was an important component. The function of this enzyme in the organism remains unclear in view of its sensitivity to molybdate.

Molybdate was first shown to be reduced by *Sulfolobus* sp. by Brieley and Brierley in 1982. In a more detailed study, molybdate was shown to be reduced aerobically to molybdenum blue (containing Mo^{5+}) by *Thiobacillus ferrooxidans* using sulfur as electron donor (Sugio et al., 1988). The enzyme that reduced the Mo^{6+} was identified to be sulfur:ferric ion oxidoreductase. Molybdate has also been shown to be reduced anaerobically by *Enterobacter cloacae* strain 48 to molybdenum blue with glucose as electron donor (Ghani et al., 1993). The reduction appears to be mediated via NAD and b-type cytochrome.

16.3 MICROBIAL INTERACTION WITH VANADIUM

Occurrence and Properties of Vanadium   Vanadium belongs to the first transition series of elements in the periodic table. In mineral form, it often occurs in complex forms of sulfide, silicate, vanadinite, and uranium vanadate (DeHuff, 1965). Its average abundance in granites is 72, in basalts 270, and in soil 90 mg kg^{-1}. Its average concentration in freshwater is 0.0005 and in seawater 0.0025 ng m^3 (Bowen, 1979).

Vanadium occurs in the oxidation states of 0, +2, +3, +4, and +5. Pentavalent vanadium in solution occurs as VO_3^- (vandate) and is colorless. Tetravalent vanadium in solution occurs as VO^{2+} and is deep blue. Trivalent vanadium (V^{3+}) forms a green solution, and divalent vanadium (V^{2+}) a violet solution (Dickerson et al., 1979).

As a trace element in prokaryotes, vanadium has been found to occur in place of molybdenum in certain nitrogenases (Brock and Madigan, 1991). It also occurs in the oxygen-carrying blood pigment of ascidian worms.

Bacterial Oxidation of Vanadium   Enzymatic oxidation of vanadium compounds has so far not been reported.

Bacterial Reduction of Vanadium   Five different bacteria have been reported to be able to reduce vanadate. The first three were *Veillonella* (*Micrococcus*) *lactilyticus*, *Desulfovibrio desulfuricans*, and *Clostridium pasteurianum*, which were shown by Woolfolk and Whiteley (1962) to be able to reduce vanadate to vanadyl with hydrogen,

$$VO_3^- + H_2 \Rightarrow VO(OH) + OH^- \qquad (16.2)$$

The fourth and fifth organisms were a new isolate assigned to the genus *Pseudomonas* (Yurkova and Lyalikova, 1990). One, isolated from a waste stream from a ferrovanadium factory, was named *P. vanadiumreductans*, and the other, isolated from seawater in Kraternaya Bay, Kuril Islands, *P. issachenkovii*. Both are gram-negative, motile, nonsporeforming rods that can grow as facultative chemolithotrophs and facultative anaerobes. Anaerobically, chemolithotrophic growth was observed with H_2 and CO as alternative energy sources, CO_2 as carbon source, and vanadate as terminal electron acceptor. However, the organisms can also grow organotrophically under anaerobic conditions with glucose, maltose, ribose, galactose, lactose, arabinose, lactate, proline, histidine, threonine, and serine as carbon/energy source and vanadate as terminal electron acceptor. *P. issachenkovii* can also use asparagine as a carbon/energy source. Vanadate reduction involved transformation of pentavalent vanadium to tretravalent and trivalent vanadium. The tetravalent oxidation state was identified by development of a blue color in the medium and the trivalent state by formation of a black precipitate and its reaction with tairon reagent. An equation to describe the overall reduction of vanadium by lactate in these experiments presented by the authors is

$$2NaVO_3 + NaC_3H_5O_3 \Rightarrow V_2O_3 + NaC_2H_3O_2 + NaHCO_3 \quad (16.3)$$
$$+ NaOH$$

It accounts for the alkaline pH developed by the medium during growth that started at pH 7.2.

16.4 MICROBIAL INTERACTION WITH URANIUM

Occurrence and Properties of Uranium Uranium is one of the naturally occurring radioactive elements. Its abundance in the Earth's crust is only 0.0002%. It is found in more than 150 minerals, but the most important are the igneous minerals pitchblende and coffinite and the secondary mineral carnotite. (Baroch, 1965). It is found in small amounts in granitic rocks (4.4 mg kg^{-1}) and even lesser amounts in basalt (0.43 mg kg^{-1}). In freshwater it has been reported in concentrations of 0.0004 and in seawater 0.0032 mg kg^{-1} (Bowen, 1979).

Uranium can exist in the oxidation states 0, $+3$, $+4$, $+5$, and $+6$ (Weast and Astle, 1982). The $+4$ and $+6$ oxidation states are of greatest significance microbiologically. In nature, the $+4$ oxidation state usually manifests itself in insoluble forms of uranium, e.g., UO_2. The $+6$ oxidation state predominates in nature in soluble and hence mobile form, e.g., UO_2^{2+} (Haglund, 1972). In radioactive decay of an isotopic uranium mixture, alpha, beta, and gamma radiation are emitted, but the overall rate of decay is very slow because the dominant isotopes have very long half-lives (Stecher, 1960). This slow rate of decay probably accounts for the ability of bacteria to interact with uranium species without experiencing lethal radiation damage.

Thiobacillus ferrooxidans has been shown to be able to oxidize tetravalent U^{4+} to hexavalent UO^{2+} in a reaction that yields enough energy to enable the organism to fix CO_2. *T. acidophilus* was also found to oxidize U^{4+} but without energy conservation. (DiSpirito and Tuovinen, 1981, 1982a,b; see also Chapter 18). Experimental demonstration of growth of *T. ferrooxidans* with U^{4+} as sole energy source has not succeeded to date.

A number of organisms have been shown to be able to reduce hexavalent uranium (UO_2^{2+}) to tetravalent uranium (UO_2). The first demonstration was with *Veillonella* (*Micrococcus*) *lactilyticus* using H_2 as electron donor under anaerobic conditions (Woolfolk and Whiteley, 1962).

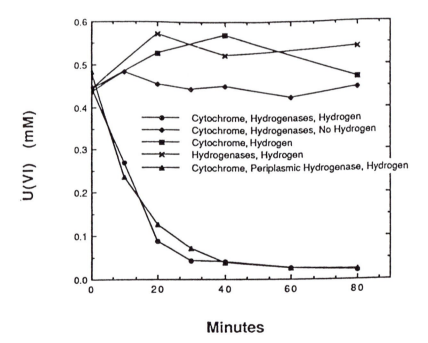

Figure 16.3 Reduction of U(VI) by electron transfer from H_2 to U(VI) via hydrogenase and cytochrome c_3. As noted, pure periplasmic hydrogenase or a protein fraction containing two hydrogenases was used. (From Lovley et al., 1993b, by permission.)

Recently, four other anaerobic bacteria have been shown to be able to reduce U(VI) to U(IV). They are *Geobacter metallireducens* strain GS15 (Lovley et al., 1991; 1993a), bacterial strain BrY (Caccavo et al., 1992), *Desulfovibrio desulfuricans* (Lovley and Phillips, 1992), and *D. vulgaris* (Lovley et al., 1993b). In *Desulfovibrio*, cytochrome c_3 appears to be the U(VI) reductase (Fig. 16.3) (Lovley et al., 1993b). The electron donors used by these organisms may be organic or, in some instances, H_2. The isolation of some of these organisms from freshwater and marine sediments suggest that this microbial activity may play or may have played a significant role in the immobilization of uranium and its accumulation in sedimentary rock.

16.5 SUMMARY

Enzymatic oxidation of Cr(III) by bacteria has not so far been demonstrated. Nonenzymatic oxidation of Cr(III), which is dependent on biogenic (bacterial, fungal) formation of Mn(IV), which then oxidizes Cr(III) to Cr(VI) chemically, may occur in soil.

Aerobic and anaerobic reduction of Cr(VI) by bacteria has been demonstrated. The process is in many instances a form of respiration. Various organic electron donors may serve, but not all act equally well aerobically and anaerobically. Chromate and dichromate are not necessarily reduced equally well. The ability to reduce chromate does not always correlate with chromate tolerance.

Although chromite is not very susceptible to leaching by acid formed by acidophiles like *T. ferrooxidans* and *T. thiooxidans*, chromium in some solid, inorganic, industrial wastes can be leached by sulfuric acid formed by *T. thiooxidans* in sulfur oxidation.

Cr(VI) reduction to Cr(III) is beneficial ecologically because it lowers chromium toxicity. At equivalent concentrations, Cr(VI) is more toxic than Cr(III). Cr(III) also tends to precipitate as a hydroxo compound around neutrality, the pH range at which all known Cr(VI) reducers operate.

Bacteria have been discovered that can enzymatically oxidize Mo(IV) and Mo(V) to Mo(VI) in air. Other bacteria have been found that can enzymatically reduce Mo(VI) to Mo(V), some aerobically, others anaerobically.

Vanadate (V^{6+}) has been found to be reduced anaerobically by a number of bacteria to vanadyl (V^{3+}). At least two of these organisms can use vanadate as terminal electron acceptor during chemolithotrophic growth with H_2 as electron donor.

Tetravalent uranium can be oxidized to hexavalent uranium, an oxidation that serves as a source of energy to *T. ferrooxidans*. Anaerobically, hexavalent uranium can be reduced to tetravalent uranium by an number of bacteria using either H_2 or one of a variety of organic electron donors.

These bacterial oxidations and reductions can play an important role in the mobilization or immobilization of the respective metals in soils and sediments.

REFERENCES

Anderson, G. L., J. Williams, and R. Hille. 1992. The purification and characterization of arsenite oxidase from *Alcaligenes faecalis*, a molybdenum-containing hydroxylase. J. Biol. Chem. 267:23674–23682.

Baroch, C. T. 1965. Uranium. *In*: Mineral Facts and Problems. Bulletin 630. Bureau of Mines. U.S. Department of the Interior. Washington, DC, pp. 1007–1037.

Bartlett, R. J., and B. R. James. 1979. Behavior of chromium in soils: III. Oxidation. J. Environ. Qual. 8:31–35.

Bopp, L. H. 1980. Chromate resistance and chromate reduction in bacteria. Ph.D. thesis. Rensselaer Polytechnic Institute, Troy, NY.

Bopp, H. L., and H. L. Ehrlich. 1988. Chromate resistance and reduction in *Pseudomonas fluorescens* strain LB300. Arch. Microbiol. 150: 426–431.

Bopp, L. H., A. M. Chakrabarty, and H. L. Ehrlich. 1983. Chromate resistance plasmid in *Pseudomonas fluorescens*. J. Bacteriol. 155: 1105–1109.

Bosecker, K. 1986. Bacterial metal recovery and detoxification of industrial waste. *In*: H. L. Ehrlich and D. S. Holmes, eds. Workshop on Biotechnology for the Mining Metal-Refining and Fossil Fuel Processing Industries. Biotech. Bioeng. Symp. 16. Wiley, New York, pp. 105–120.

Bowen, H. J. M. 1979. Environmental Chemistry of the Elements. Academic Press, London.

Brierley, C. L., and J. A. Brierley. 1982. Anaerobic reduction of molybdenum by *Sulfolobus* species. Zbl. Bakteriol. Hyg. I Abt. Orig. C3: 289–294.

Brierley, C. L., and L. E. Murr. 1973. A chemoautotrophic and thermophilic chemoautotrophic microbe. Science 179:488–490.

Brock, T. D., and M. T. Madigan. 1991. Biology of Microorganisms. 6th ed. Prentice Hall, Englewood Cliffs, NJ.

Caccavo, Jr., F., R. P. Blakemore, and D. R. Lovley. 1992. A hydrogen-oxidizing, Fe(III)-reducing microorganism from the Great Bay Estuary, New Hampshire. Appl. Environ. Microbiol. 58:3211–3216.

Cervantes, C. 1991. Bacterial interactions with chromium. Antonie v. Leeuwenhoek 59:229–233.

Coleman, R. N., and J. H. Paran. 1983. Accumulation of hexavalent chromium by selected bacteria. Environ. Technol. Lett. 4:149–156.

DeHuff, G. L. 1965. Vanadium. *In*: Mineral Facts and Problems. Bulletin 630. Bureau of Mines. U.S. Department of the Interior. Washington, DC, pp. 1039–1053.

DeLeo, P. C., and H. L. Ehrlich. 1994. Reduction of hexavalent chromium by *Pseudomonas fluorescens* LB300 in batch and continuous cultures. Appl. Microbiol. Biotechnol. 40:756–759.

Dickerson, R. E., H. B. Gray, and G. P. Haight, Jr. 1979. Chemical Principles. 3rd ed. Benjamin/Cummings Publishing Company, Menlo Park, CA.

DiSpirito, A. A., and O. H. Tuovinen. 1981. Oxygen uptake coupled with uranous sulfate oxidation by *Thiobacillus ferrooxidans* and *T. acidophilus*. Geomicrobiol. J. 2:275–291.

DiSpirito, A. A., and O. H. Tuovinen. 1982a. Uranous ion oxidation and carbon dioxide fixation by *Thiobacillus ferrooxidans*. Arch. Microbiol. 133:28–32.

DiSpirito, A. A., and O. H. Tuovinen. 1982b. Kinetics of uranous and ferrous iron oxidation by *Thiobacillus ferrooxidans*. Arch. Microbiol. 133:33–37.

Ehrlich, H. L. 1983. Leaching of chromite ore and sulfide matte with dilute sulfuric acid generated by *Thiobacillus thiooxidans* from sulfur. *In*: G. Rossi and A. E. Torma, eds. Recent Progress in Biohydrometallurgy. Associazione Mineraria Sarda. Iglesias, Italy, pp. 19–42.

Enzmann, R. D. 1972. Molybdenum and element geochemistry. *In*: R. W. Fairbridge, ed. The Encyclopedia of Geochemistry and Environmental Sciences. Encyclopedia of Earth Sciences Series, Vol. IVA. Van Nostrand Reinhold, New York, pp. 753–759.

Forsberg, C. W. 1978. Effects of heavy metals and other trace elements on the fermentative activity of the rumen microflora and growth of functionally important rumen bacteria. Can. J. Microbiol. 24:298–306.

Fortescue, J. A. C. 1980. Environmental Geochemistry. Springer-Verlag, New York.

Gopalan, R., and H. Veeramani. 1994. Studies on microbial chromate reduction by *Pseudomonas* sp. in aerobic continuous suspended culture. Biotechnol. Bioeng. 43:471–476.

Ghani, B., M. Takai, N. Z. Hisham, N. Kishimoto, A. K. M. Ismail, T.

Tano, and T. Sugio. 1993. Isolation and characterization of a Mo^{6+}-reducing bacterium. Appl. Environ. Microbiol. 59:1176–1180.

Gruber, J. E., and K. W. Jennette. 1978. Metabolism of the carcinogen chromate by rat liver microsomes. Biochem. Biophys. Res. Commun. 82:700–706

Gvozdyak, P. I., N. F. Mogilevich, A. F. Ryl'skii, and N. I. Grishchenko. 1986. Reduction of hexavalent chromium by collection strains of bacteria. Mikrobiologiya 55:962–965 (Engl. transl. pp. 770–773).

Haglund, D. S. 1972. Uranium: Element and geochemistry. *In*: R.W. Fairbridge, ed. The Encyclopedia of Geochemistry and Environmental Sciences. Encyclopedia of Earth Sciences Series, Vol. IVA. Van Nostrand Reinhold, New York, pp. 1215–1222.

Holliday, R. W. 1965. Molybdenum. *In*:Mineral Facts and Problems. Bulletin 630. Bureau of Mines, U.S. Department of the Interior, Washington, DC, pp. 595–606.

Horitsu, H., H. Nishida, H. Kato, and M. Tomoyeda. 1978. Isolation of potassium-chromate tolerant bacterium and chromate uptake by the bacterium. Agric. Biol. Chem. 42:2037–2043.

Horitsu, H., S. Futo, Y. Myazawa, S. Ogai, and K. Kawai. 1987. Enzymatic reduction of hexavalent chromium by hexavalent chromium tolerant *Pseudomonas ambigua* G-1. Agric. Biol. Chem. 51: 2417–2420.

Ishibashi, Y., C. Cervantes, and S. Silver. 1990. Chromium reduction by *Pseudomonas putida*. Appl. Environ. Microbiol. 56:2268–2270.

Johnson, I., N. Flower, and M. W. Loutit. 1981. Contribution of periphytic bacteria to the concentration of chromium in the crab *Helice crassa*. Microb. Ecol. 7:245–252.

Komori, K., P-C. Wang, K. Toda, and H. Ohtake. 1989. Factors affecting chromate reduction in *Enterobacter cloacae* strain HO1. Appl. Microbiol. Biotechnol. 31:567–570.

Komori, K., A. Rivas, K. Toda, and H. Ohtake. 1990a. Biological removal of toxic chromium using an *Enterobacter cloacae* strain that reduces chromate under anaerobic conditions. Biotechnol. Bioeng. 35: 951–954.

Komori, K., A. Rivas, K. Toda, and H. Ohtake. 1990b. A method for removal of toxic chromium using dialysis-sac cultures of a chromate reducing strain of *Enterobacter cloacae*. Appl. Microbiol. Biotechnol. 33:117–119.

Kvasnikov, E. I., V. V. Stepnyuk, T. M. Klyushnikova, N. S. Serpokrylov, G. A. Simonova, T. P. Kasatkina, and L. P. Panchenko. 1985. A new chromium-reducing, gram-variable bacterium with mixed type of flagellation. Mikrobiologiya 54:83–88 (Engl. transl. pp. 69–75).

Latimer, W. M., and J. H. Hildebrand. 1942. Reference Book of Inorganic Chemistry. Rev. ed. Macmillan, New York, pp. 357–360.

Lebedeva, E. V., and N. N. Lyalikova. 1979. Reduction of crocoite by *Pseudomonas chromatophila* nov. sp. Mikrobiologiya 48:517–522 (Engl. tansl. pp. 405–410).

Llovera, S., R. Bonet, M. D. Simon-Pujol, and F. Congregado. 1993. Chromate reduction by resting cells of *Agrobacterium radiobacter* EPS-916. Appl. Environ. Microbiol. 59:3516–3518.

Lovley, D. R., and E. J. P. Phillips. 1992. Reduction of uranium by *Desulfovibrio desulfuricans*. Appl. Environ. Microbiol. 58:850–856.

Lovley, D. R., and E. J. P. Phillips. 1994. Reduction of chromate by *Desulfovibrio vulgaris* and its c_3 cytochrome. Appl. Environ. Microbiol. 60:726–728.

Lovley, D. R., E. J. P. Phillips, Y. A. Gorby, and E. R. Landa. 1991. Microbial reduction of uranium. Nature (Lond.) 350:413–416.

Lovley, D. R., S. J. Giovanni, D. C. White, J. E. Champine, E. J. P. Phillips, Y. A. Groby, and S. Goodwin. 1993a. *Geobacter metallireducens* gen. nov. sp. nov., a microorganism capable of coupling the complete oxidation of organic compounds to the reduction of iron and other metals. Arch. Microbiol. 159:336–344.

Lovley, D. R., P. K. Widman, J. C. Woodward, and E. J. P. Phillips. 1993b. Reduction of uranium by cytochrome c_3 of *Desulfovibrio vulgaris*. Appl. Environ. Microbiol. 59:3572–3576.

Marques, A. M. M. J. Espuny Tomas, F. Congregado, and M. D. Simon-Pujol. 1982. Accumulation of chromium by *Pseudomonas aeruginosa*. Microbios Lett. 21:143–147.

Mertz, W. 1981. The essential trace elements. Science 213:1332–1338.

Miller, W. J., and M. W. Neathery. 1977. Newly recognized trace mineral elements and their role in animal nutrition. BioScience 27:674–679.

Nishioka, H. 1975. Mutagenic activites of metal compounds in bacteria. Mutat. Res. 31:185–190.

Ohtake, H., C. Cervantes, and S. Silver. 1987. Decreased chromate uptake in *Pseudomonas fluorescens* carrying a chromate resistance plasmid. J. Bacteriol. 169:3853–3856.

Ohtake, H., E. Fuji, and K. Toda. 1990. Reduction of toxic chromate in an industrial effluent by use of a chromate-reducing strain of *Enterobacter cloacae*. Environ. Technol. 11:663–668.

Oremland, R. S., and D. G. Capone. Use of "specific" inhibitiors in biogeochemistry and microbial ecology. Adv. Microb. Ecol. 10:285–383.

Petrelli, F. L., and S. DeFlora. 1977. Toxicity and mutagenicity of hexavalent chromium on *Salmonella typhimurium*. Appl. Environ. Microbiol. 33:805–809.

Pleshakov, V. D., V. N. Koren'kov, V. N. Zhukov, N. S. Serpokrylov, I. A. Lempert, and I. I. Pankrova. 1981. Biochemical removal of chromium(VI) compounds from wastewaters. USSR patent SU 835,978, June 7, 1981.

Romanenko, V. I., and V. N. Koren'kov. 1977. A pure culture of bacteria utilizing chromates and bichromates as hydrogen acceptors in growth under anaerobic conditions. Mikrobiologiya 46:414–417 (Engl. transl. pp. 329–332).

Romanenko, V. I., S. I. Kuznetsov, and V. N. Koren'kov. 1976. Koren'kov method for biological purification of wastewater. USSR patent SU 521,234.

Serpokrylov, N. S., I. M. Zhukov, G. A. Simonova, E. I. Kvasnikov, T. M. Klyushnikova, T. P. Kasatkina, and V. P. Kostyukov. 1985. Biological removal of chromium(VI) compounds from wastewater by anaerobic microorganisms in the presence of an organic substrate. USSR patent SU 1,198,020, December 1985.

Shen, H., and Y.-T. Wang. 1993. Characterization of enzymatic reduction of hexavalent chromium b *Escherichia coli* ATCC 33456. Appl. Environ. Microbiol. 59:3771–3777.

Shi, X., and N. S. Dalal. 1990. One-electron reduction of chromate by NADPH-dependent glutathione reductase. J. Inorg. Biochem. 40: 1–12.

Shimada, A. 1979. Effects of sixvalent chromium on growth and enzyme production of chromium-resistant bacteria. Abstr. Annu. Meet. Am. Soc. Microbiol., Q[H]25, p. 238.

Silver, S., and Mark Walderhaug. 1992. Gene regulation of plasmid- and chromosome-determined inorganic ion transport in bacteria. Microbiol. Rev. 56:195–228.

Simonova, G. A., N. S. Serpokrylov, and L. L. Tokareva. 1985. Adaptation characteristics of a culture of *Aeromonas dechromatica* KS-11 in microbial removal of hexavalent chromium from water. Izv. Sev. Kavk. Nauchn. Tsentra Vyssh. Shk. Estestv. Nauki No. 4, pp. 89–91

Sittig, M. 1985. Handbook of Toxic and Hazardous Chemicals and Carcinogens. Noyes Publications, Park Ridge, NJ.

Smith, C. H. 1972. Chromium: element and geochemistry. *In*: R. W. Fairchild, ed. The Encyclopedia of Geochemistry and Environmental Science. Encylopedia of Earth Sciences Series, Vol. IVA. Van Nostrand Reinhold, New York, pp. 167–170.

Stecher, P. G., ed. 1960. The Merck Index of Chemical and Drugs. 7th ed. Merck & Co., Rahway, NJ, p. 1493.

Sugio, T., Y. Tsujita, T. Katagiri, K. Inagaki, and T. Tano. 1988. Reduction of Mo^{6+} with elemental sulfur by *Thiobacillus ferrooxidans*. J. Bacteriol. 170:5956–5959.

Sugio, R., K. Hirayama, K. Inagaki, H. Tanaka, and T. Tano. 1992. Molybdenum oxidation by *Thiobacillus ferrooxidans*. Appl. Environ. Microbiol. 58:1768–1771.

Summers, A. O., and G. A. Jacoby. 1978. Plasmid-determined resistance to boron and chromium compounds in *Pseudomonas aeruginosa*. Antimicrob. Agents Chemother. 13:637–640.

Suzuki, T., N. Miyata, H. Horitsu, K. Kawai, K. Takamizawa, Y. Tai, and M. Okazaki. 1992. NAD(P)H-dependent chromium (VI) reductase of *Pseudomonas ambigua* G-1: a Cr(V) intermediate is formed during the reduction of Cr(VI) to Cr(III). J. Bacteriol. 174:5340–5345.

Tuovinen, O. H., S. I. Niemelae, and H. G. Gyllenberg. 1971. Tolerance of *Thiobacillus ferrooxidans* to some metals. Antonie v. Leeuwenhoek 37:489–496.

Venitt, S., and L. S. Levy. 1974. Mutagenicity of chromates in bacteria and its relevance to carcinogenesis. Nature (Lond.) 250:493–495.

Wang, P., T. Mori, K. Komori, M Sasatsu, K. Toda, and H. Ohtake. 1989. Isolation and characterization of an *Enterobacter cloacae* strain that reduces hexavalent chromium under anaerobic conditions. Appl. Environ. Microbiol. 55:1665–1669.

Wang, P-C., K. Toda, H. Ohtake, I. Kusaka, and I. Yabe. 1991. Membrane-bound respiratory system of *Enterobacter cloacae* strain HO1 grown anaerobically with chromate. FEMS Microbiol. Lett. 78:11–16.

Weast, R. C., and M. J. Astle. 1982. CRC Handbook of Chemistry and Physics. 63rd ed. CRC Press, Boca Raton, FL, pp. B44-B45.

Wong, C., M. Silver, and D. J. Kushner. 1982. Effects of chromium and manganese on *Thiobacillus ferrooxidans*. Can. J. Microbiol. 28:536–544.

Woolfolk, C. A., and H. R. Whiteley. 1962. Reduction of inorganic compounds with molecular hydrogen by *Micrococcus lactilyticus*. J. Bacteriol. 84:647–658.

Yamamoto, K., J. Kato, I. Yano, and H. Ohtake. 1993. Kinetics and modeling of hexavalent chromium reduction in *Enterobacter cloacae*. Biotechnol. Bioeng. 41:129–133.

Yurkova, N. A., and N. N. Lyalikova. 1990. New vanadate-reducing facultative chemolithotrophic bacteria. Mikrobiologiya 59:968–975 (Engl. transl. pp. 672–677).

17

Geomicrobiology of Sulfur

17.1 OCCURRENCE OF SULFUR IN THE EARTH'S CRUST

The abundance of sulfur in the Earth's crust has been estimated to be around 520 ppm (Goldschmidt, 1954). It is thus one of the more common elements in the biosphere. Its concentration in rocks, including igneous and sedimentary rocks, can range from 270 to 2,400 ppm (Bowen, 1979). In freshwater, average sulfur concentrations are around 3.7 ppm and in seawater 905 ppm (Bowen, 1979). In field soils in humid, temperate regions, the total sulfur concentration may range from 100 to 1,500 ppm, of which 50–500 ppm is soluble in weak acid or water (Lawton, 1955). Most of the sulfur in soil of pastureland in humid to semiarid climates is organic, whereas that in drier soils is contained in gypsum ($CaSO_4 \cdot 2H_2O$), epsomite ($MgSO_4 \cdot 7H_2O$), and in lesser amounts in sphalerite (ZnS), chalcopyrite ($CuFeS_2$), and pyrite or marcasite (FeS_2) (Freney, 1967).

17.2 GEOCHEMICALLY IMPORTANT PROPERTIES OF SULFUR

Inorganic sulfur occurs most commonly in the -2, 0, $+2$, $+4$, and $+6$ oxidation states (Roy and Trudinger, 1970). Table 17.1 lists geomicrobially

Table 17.1 Geomicrobially Important Forms of Sulfur and Their Oxidation States

Compound	Formula	Oxidation state(s) of sulfur
Sulfide	S^{2-}	-2
Polysulfide	S_n^{2-}	$-2, 0$
Sulfur[a]	S_8	0
Hyposulfite (dithionite)	$S_2O_4^{2-}$	$+3$
Sulfite	SO_3^{2-}	$+4$
Thiosulfate[b]	$S_2O_3^{2-}$	$-1, +5$
Dithionate	$S_2O_6^{2-}$	$+6$
Trithionate	$S_3O_6^{2-}$	$-2, +6$
Tetrathionate	$S_4O_6^{2-}$	$-2, +6$
Pentathionate	$S_5O_6^{2-}$	$-2, +6$
Sulfate	SO_4^{2-}	$+6$

[a] Occurs in an octagonal ring in crystalline form.
[b] Outer sulfur has an oxidation state of -1; the inner sulfur has an oxidation state of $+5$.

important forms and their various oxidation states. In nature, the -2, 0, and +6 oxidation states are most common, represented by sulfide, elemental sulfur, and sulfate, respectively. However, in some environments (e.g., chemoclines in aquatic environments, in some soils and sediments, etc.) some other mixed oxidation states (e.g., thiosulfate, tetrathionate) may also occur, though in lesser amounts.

Some chemical reactions involving elemental sulfur may have a geo-microbiological impact. For instance, elemental sulfur (usually written $S°$, but is really S_8 because it consists of eight sulfur atoms in a ring) can react reversibly with sulfite to form thiosulfate, which is readily oxidized or reduced by various microbes (Roy and Trudinger, 1970):

$$S° + SO_3^{2-} \Leftrightarrow S_2O_3^{2-} \tag{17.1}$$

The forward reaction is favored by neutral to alkaline pH, whereas the back reaction is favored by acid pH. Thiosulfate is very unstable in aqueous solution below pH 5.

Elemental sulfur also reacts with sulfide, forming polysulfides (Roy and Trudinger, 1970), a reaction that may play an important role in elemental sulfur metabolism:

$$S_8° + HS^- \Rightarrow HSS_n^- + S_{8-n}° \tag{17.2}$$

The value of n may equal 2, 3,and higher. It is related to the sulfide concentration.

Polythionates, starting with trithionate ($S_3O_6^{2-}$), may be viewed as disulfonic acid derivatives of sulfanes (Trudinger and Roy, 1970). They may be formed as by-products of the oxidation of sulfide and sulfur to sulfate and in disproportionation reactions. Polythionates are also metabolizable by some microorganisms.

17.3   BIOLOGICAL IMPORTANCE OF SULFUR

Sulfur is an important element for life. In the cell it is especially important in stabilizing protein structure and in the transfer of hydrogen by enzymes in redox metabolism. For some prokaryotes, reduced forms of sulfur can serve as energy sources and/or sources of reducing power. Oxidized forms, especially sulfate but also sulfur, can serve other prokakryotes as electron acceptors. It is the oxidation and reduction reactions involving sulfur and sulfur compounds which are especially important geomicrobiologically.

17.4   MINERALIZATION OF ORGANIC SULFUR COMPOUNDS

As part of the carbon cycle, microbes degrade organic sulfur compounds such as the amino acids cysteine, cystine, and methionine, and agar-agar (a sulfuric acid ester of a linear galactan), tyrosine-O-sulfate, and so on. Desulfurization is usually the first step in this degradation. Desulfurization of cysteine by bacteria may occur anaerobically by the reaction (Freney, 1967; Roy and Trudinger, 1970),

$$\text{HSCH}_2\underset{\substack{|\\ \text{NH}_2}}{\text{CHCOOH}} + \text{H}_2\text{O} \xrightarrow[\text{desulfhydrase}]{\text{cysteine}} \text{H}_2\text{S} + \text{NH}_3 \tag{17.3}$$

$$+ \text{CH}_3\text{COCOOH}$$

or by the reaction (Freney, 1967; Roy and Trudinger, 1970)

$$\text{HSCH}_2\underset{\substack{|\\ \text{NH}_2}}{\text{CHCOOH}} \underset{\substack{-\text{H}_2\text{O}\\ \text{serine}\\ \text{sulfhydrase}}}{\overset{+\text{H}_2\text{O}}{\rightleftharpoons}} \text{HOCH}_2\underset{\substack{|\\ \text{NH}_2}}{\text{CHCOOH}} + \text{H}_2\text{S} \tag{17.4}$$

The sulfur of cysteine may also be released aerobically. The reaction sequence in that instance is not completely certain and may differ with different types of organisms (Freney, 1967; Roy and Trudinger, 1970). Although alanine-3-sulfinic acid [HO_2S-$CH_2CH(NH_2)COOH$] has been postulated as a key intermediate by some workers, others have questioned it, at least for rat liver mitochondria (Wainer, 1964, 1967).

Methionine is decomposed by extracts of *Clostridium sporogenes* or *Pseudomonas* sp. as follows (Freney, 1967):

$$\text{Methionine} \Rightarrow \alpha\text{-ketobutyrate} + CH_3SH + NH_3 \tag{17.5}$$

17.5 SULFUR ASSIMILATION

Inorganic sulfur is usually assimilated into organic compounds as sulfate by plants and most microorganisms. One possible pathway of assimilation in bacteria may be the reduction of sulfate to sulfide and its subsequent reaction with serine to form cysteine, as in *Salmonella typhimurium* (see Freney, 1967, p. 239).*

$$ATP + SO_4^{2-} \xrightarrow{\text{ATP sulfurylase}} APS + PP_i \tag{17.6}$$

$$APS + ATP \xrightarrow{\text{APS kinase}} PAPS + ADP \tag{17.7}$$

$$PAPS + 2e^- \xrightarrow[\text{NADH}]{\text{PAPS reductase}} SO_3^{2-} + PAP \tag{17.8}$$

$$SO_3^{2-} + 7H^+ + 6e^- \xrightarrow[\text{NADH}]{SO_3^{2-} \text{ reductase}} HS^- + 3H_2O \tag{17.9}$$

$$HS^- + \text{serine} \xrightarrow{\text{cysteine synthase}} \text{cysteine} + H_2O \tag{17.10}$$

This sequence has also been found in *Bacillus subtilis*, *Aspergillus niger*, *Micrococcus aureus*, and *Enterobacter aerogenes* (Roy and Trudinger, 1970). Reaction 17.10 may be replaced by the following sequence:

$$\text{Serine} + \text{acetyl-SCoA} \Longrightarrow CH_2OOCCH_3 + \text{CoASH} \tag{17.11}$$

$$\begin{array}{l} | \\ CHNH_2 \\ | \\ COOH \\ {\scriptstyle o\text{-acetylserine}} \end{array}$$

---

* APS, adenosine 5'-sulfatophosphate; PAPS, adenosine 3'-phosphate-5'-sulfatophosphate; PP_i, inorganic pyrophosphate; PAP, adenosine 3',5'-diphosphate.

O-acetylserine + $H_2S \Rightarrow$ cysteine + acetate + H^+ (17.12)

This latter sequence has been observed in *Escherichia coli* and *Salmonella typhimurium* (Roy and Trudinger, 1970). The reduction of sulfate to "active thiosulfate" and its incorporation into serine to form cysteine is also possible for some organisms, such as *E. coli* (Freney, 1967).

Sulfate reduction occurs not only as part of an **assimilatory** process but also as a **dissimilatory** or **respiratory** process. The latter occurs in the great majority of known instances only in special anaerobic bacteria whereas the former occurs in aerobic as well as anaerobic bacteria and other organisms. The majority of dissimilatory sulfate reducers are eubacteria, but at least two archaebacteria, *Archeoglobus fulgidus* (Stetter et al., 1987; Stetter, 1988), and *A. profundus* (Burggraf et al., 1990) are now known in this physiological group. Almost without exception, assimilatory sulfate reduction does not consume more sulfate than is needed for assimilation, so that excess sulfide is not produced. The only known exceptions are a strain of *Bacillus megaterium* (Bromfield, 1953) and *Pseudomonas zelinskii* (Shturm, 1948). Unlike dissimilatory sulfate reduction, assimilatory sulfate reduction is thus a form of transitory sulfur immobilization involving small amounts of sulfur per cell; the fixed sulfur is returned to the sulfur cycle upon the death of the organism that assimilated it.

17.6 GEOMICROBIALLY IMPORTANT TYPES OF BACTERIA THAT REACT WITH SULFUR AND SULFUR COMPOUNDS

Oxidizers of Reduced Sulfur Most geomicrobially important microorganisms which oxidize reduced forms of sulfur in relatively large quantities are bacteria. They include representatives of eubacteria and archaebacteria. They comprise aerobes, facultative organisms, and anaerobes and are mostly obligate or facultative autotrophs, or mixotrophs. Among the eubacterial aerobes, one of the most important group terrestrially is that of the Thiobacillaceae (Table 17.2). This group includes obligate and facultative autotrophs as well as mixotrophs. Among the archaebacterial aerobes, one of the most widely studied group consists of the genera *Sulfolobus* and *Acidianus* (Table 17.2). Another eubacterial group of hydrogensulfide oxidizers that is of some importance in freshwater and marine environments are the Beggiatoacea (Fig. 17.1). Most known members of this group use hydrogen sulfide mixotrophically or heterotrophically. In the latter instance, they employ H_2S oxidation as protection against metabolically produced H_2O_2 in the absence of catalase (Kuenen and Beudeker, 1982; Nelson and Castenholz, 1981), but at least one marine strain,

Table 17.2 Some Aerobic Sulfur-Oxidizing Bacteria[a]

Autotrophic	Mixotrophic	Heterotrophic
Thiobacillus thioparus	*Thiobacillus*	*Thiobacillus*
Thiobacillus neapolitanus	*intermedius*	*perometabolis*
Thiobacillus tepidarius	*Thiobacillus versutus*[e]	*Beggiatoa* spp.
Thiobacillus denitrificans[b]	*Thiobacillus*	
Thiobacillus novellus[c]	*organoparus*	
Thiobacillus thermophilica[c]	*Pseudomonas* spp.	
Sulfobacillus thermosulfidooxidans[c]		
Thiobacillus thiooxidans		
Thiobacillus albertis		
Thiobacillus ferrooxidans		
Beggiatoa alba MS-81-6		
Sulfolobus acidocaldarius[d]		
Acidianus brierleyi[d]		
Thermothrix thiopara[c]		

[a] A more complete survey of aerobic sulfur-oxidizing bacteria can be found in Starr et al. (1981) and Holt (1989).
[b] Facultative anaerobe.
[c] Facultative autotroph.
[d] Archaebacterium.
[e] Can also grow autotrophically and heterotrophically.

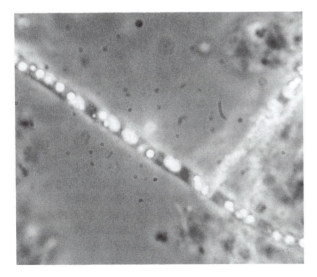

Figure 17.1 Trichome of *Beggiatoa* with sulfur granules in a pond water enrichment (×5,240). (Courtesy E. J. Arcuri.)

Beggiatoa alba MS-81–6, can grow autotrophically (Nelson and Jannasch, 1983). Other hydrogen-sulfide oxidizers found in aquatic environments include *Thiovulum* (autotrophic) (e.g., Wirsen and Jannasch, 1978), *Achromatium, Thiothrix, Thiobacterium*, (La Riviere and Schmidt, 1981), and *Thiomicrospira* (Kuenen and Tuovinen, 1981). Of all these groups, only the thiobacilli produce sulfate directly without accumulating elemental sulfur when oxidizing H_2S at normal oxygen tension. The other groups accumulate sulfur ($S°$) which they may oxidize further to sulfate when the supply of H_2S is limited or depleted.

Among eubacteria, *Thiobacillus thioparus* oxidizes $S°$ slowly to H_2SO_4. It is inhibited as the pH drops below 4.5. *T. halophilus* is another neutrophilic, but extremely halophilic, obligate chemolithotroph that oxidizes elemental sulfur to sulfate (Wood and Kelly, 1991). *T. thiooxidans, T. albertis* (Bryant et al., 1983), and *T. ferrooxidans* oxidize elemental sulfur to sulfuric acid more readily, and being acidophilic, may drop the pH to below 1.0. All these organisms are strict autotrophs.

The archaebacteria *Sulfolobus* spp. and *Acidianus* spp. are also able to oxidize sulfur to sulfuric acid. Both genera are thermophilic. *Sulfolobus acidocaldarius* will oxidize sulfur between 55 and 85°C (70–75° optimum) in a pH range of 0.9–5.8 (pH 2–3 optimum)(Brock et al., 1972). The organisms are facultative autotrophs. *Acidianus brierleyi* (formerly *Sulfolobus brierleyi*) has traits similar to *S. acidocaldarius* but can also reduce $S°$ with H_2 and has a different GC (guanine + cytosine) content (31 mol% versus 37 mol%)(Brierley and Brierley, 1973; Segerer et al., 1986).

Other thermophilic bacteria capable of oxidizing sulfur have also been observed. Several are incompletely characterized. Some were isolated from sulfurous hot springs, others from ore deposits. One of these has been described as a motile rod capable of forming endospores in plectridia. It is a facultative autotroph capable of oxidizing various sulfides and organic compounds besides sulfur. It was named *Thiobacillus thermophilica* Imshenetskii (Egorova and Deryugina, 1963). Another is an aerobic, facultative thermophile, capable of sporulation, which is capable not only of oxidizing elemental sulfur but also Fe^{2+} and metal sulfides. It was named *Sulfobacillus thermosulfidooxidans* (Golovacheva and Karavaiko, 1978). Still another is a gram-negative, facultatively autotrophic *Thiobacillus* sp. capable of growth at 50 and 55°C with a pH optimum of 5.6 (range 4.8–8) (William and Hoare, 1972). Other thermophilic thiobacillus-like bacteria have been isolated which can grow on thiosulfate at 60 and 75°C at pH 7.5, and still others which can grow at 60 and 75°C and a pH of 4.8 (LeRoux et al., 1977).

A number of heterotrophs have also been reported to be able to oxidize reduced sulfur in the form of elemental sulfur, thiosulfate, and tetrathionate. They include bacteria and fungi. Many of those that oxidize elemental sulfur oxidize it to thiosulfate while others oxidize thiosulfate or tetrathionate to sulfuric acid (Guittoneau, 1927; Guitonneau and Keiling, 1927; Grayston and Wainwright, 1988; see also Roy and Trudinger, 1970, pp. 248–249). Some marine pseudomonads and others can gain useful energy from the thiosulfate oxidation by using it as supplemental energy source (Tuttle et al, 1974; Tuttle and Ehrlich, 1986).

Two examples of eubacterial, facultatively anaerobic sulfur oxidizers are *Thiobacillus denitrificans* (e.g., Justin and Kelly, 1978) and *Thermothrix thiopara* (Caldwell et al., 1976; Brannan and Caldwell, 1980), the former a mesophile and the latter a thermophile. Anaerobically both organisms use nitrate as terminal electron acceptor and reduce it to oxides of nitrogen and dinitrogen with nitrite being an intermediate product. They can use sulfur in various oxidation states as energy source. *T. thiopara* is a facultative autotroph.

The strictly anaerobic sulfur oxidizers are represented by photosynthetic purple and green bacteria (Pfennig, 1977) and certain cyanobacteria (Table 17.3). Some of these bacteria may grow aerobically, but without oxidizing reduced sulfur compounds. The purple sulfur bacteria (Chromatiaceae) (Fig. 17.2) are obligate anaerobes which oxidize reduced sulfur, especially H_2S, by using it as a source of reducing power for CO_2 assimilation. Despite the terminology, several purple nonsulfur bacteria (Rho-

Table 17.3. Some Anaerobic Sulfur-Oxidizing Bacteria[a]

Photolithotrophs	Chemolithotrophs
Chromatium spp.	*Thermothrix thiopara*[b,c]
Chlorobium spp.	*Thiobacillus denitrificans*[c]
Ectothiorhodospira spp.	
Rhodopseudomonas spp.[b]	
Chloroflexus aurantiacus[b]	
Oscillatoria sp.[c]	
Lyngbya spp.[c]	
Aphanothece spp.[c]	
Microcoleus spp.[c]	
Phormidium spp.[c]	

[a] For a more complete description of anaerobic sulfur-oxidizing bacteria, see Starr et al. (1981) and Holt (1984).
[b] Facultatively autotrophic.
[c] Facultatively anaerobic.

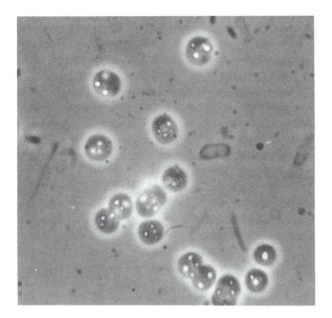

Figure 17.2   Unidentified purple sulfur bacteria (probably *Chromatium* sp.) in an enrichment culture ($\times 5,385$). Note the conspicuous sulfur granules in the spherical cells.

dospirillaceae) can also grow autotrophically on H_2S as a source of reducing power for CO_2 assimilation, but for the most part, they tolerate only low concentrations of sulfide in contrast to purple sulfur bacteria. In the laboratory, purple nonsulfur can also grow photoheterotrophically, using reduced carbon as a major carbon source. Most purple sulfur bacteria when growing on H_2S, oxidize it to S° which they deposit **intracellularly** (Fig. 17.2), but *Ectothiorhodospira* spp. are an exception in depositing it extracellularly. Under conditions of H_2S limitation, these strains oxidize the sulfur they accumulate further to sulfate. Among the purple nonsulfur bacteria *Rhodopseudomonas palustris* and *R. sulfidophila* do not form sulfur as an intermediate from H_2S but oxidize sulfide directly to sulfate (Hansen and van Gemerden, 1972; Hansen and Veldkamp, 1973). In contrast, *Rhodospirillum rubrum*, *Rhodpseudomonas capsulata*, and *R. spheroides* do form elemental sulfur from sulfide, which they deposit extracellularly (Hansen and van Gemerden, 1972). *R. sulfidophila* differs from most purple nonsulfur bacteria in being more tolerant of a high concentration of sulfide. Green sulfur bacteria (Chlorobiaceae) are strict anaerobic photoautotrophs that oxidize H_2S by using it as a source of reducing power in CO_2 fixation. They deposit the sulfur ($S°$) they produce **extracellularly**. Under H_2S limitation, they oxidize the sulfur further to sulfate. At least a few strains of *Chlorobium limicola* forma *thiosulfatophilum* do not accumulate $S°$ but oxidize H_2S directly to sulfate (Ivanov, 1968, p. 137; Paschinger, 1974). Many photosynthetic anaerobic bacteria can also use thiosulfate as electron donor in place of hydrogen sulfide.

Filamentous, gliding green bacteria (Chloroflexaceae) grow photoheterotrophically under anaerobic conditions, but at least some can also grow photoautotrophically with H_2S as electron donor under anaerobic conditions. (Brock and Madigan, 1988).

A few filamentous cyanobacteria, including some members of the genera *Oscillatoria*, *Lyngbya*, *Aphanothece*, *Microcoleus*, and *Phormidium*, which are normally classified as oxygenic photoautotrophs, can grow photosynthetically under anaerobic conditions with H_2S as a source of reducing power (Cohen et al., 1975; Garlick et al., 1977). They oxidize the H_2S to elemental $S°$ and deposit it extracellularly. In the dark, they can rereduce the sulfur they produce using internal reserves of polyglucose as reductant (Oren and Shilo, 1979). At this time there is no evidence that these organisms can oxidize the sulfur they produce anaerobically further to sulfate under H_2S limitation.

Reducers of Oxidized Sulfur   A geomicrobially and geochemically very important group of bacteria are the sulfate reducing bacteria. The known examples include mostly eubacteria, but two archaebacterial examples were recently discovered *Archaeoglobus fulgidus* (Stetter et al., 1987; Speich and Trueper, 1988) and *A. profundus* (Burggraf et al., 1990). More than twenty-eight years ago the sulfate reducers were thought to be represented by only three eubacterial genera, *Desulfovibrio* and *Desulfotomaculum* (originally classified as *Clostridium* because of its ability to form endospores), and *Desulfomonas*, which were very specialized nutritionally in that among organic energy sources, they could only use lactate, pyruvate, fumarate, malate, and ethanol. Furthermore, none of the organisms then known were able to degrade their organic energy sources beyond acetate (Postgate, 1984), with the result that at the time the importance of this group in anaerobic mineralization in sulfate-rich environments was unappreciated. This restricted view of sulfate reducers changed rapidly after 1976, with the discovery of a sulfate-reducer, *Desulfotomaculum acetoxidans* (Widdel and Pfennig, 1977; 1981), which is able to oxidize acetate anaerobically to CO_2 and H_2O with sulfate. Subsequently the discovery of a wide variety of other sulfate reducers was reported, which differed in the nature of the energy sources they were capable of utilizing. Various aliphatic, aromatic and heterocyclic compounds were found to be attacked and in many cases completely mineralized, each by a specific group (e.g., Pfennig et al., 1981; Imhoff-Stuckle and Pfennig, 1983; Braun and Stolp, 1985; Bak and Widdel, 1986a,b; Szewzyk and Pfennig, 1987; Platen et al., 1990; Zellner et al., 1990; Qatabi et al., 1991; Schnell and Schink, 1991; Boopathy and Daniels, 1991; Aeckersberg et al., 1991; Tasaki et al., 1991, 1992; Kuever et al., 1993; Rueter et al., 1994). Many sulfate-reducers are now also known to be able to use H_2 as an energy source. Most require an organic carbon source, but a few can grow autotrophically on hydrogen. Table 17.4 presents a list of some of the different eubacterial types and their nutritional ranges. Whereas most sulfate reducers discovered to date are mesophilic, thermophilic types are now also known (Pfennig et al., 1981; Zeikus et al., 1983). Morphologically, sulfate reducers are a very varied group, including cocci, sarcinae, rods, vibrios (Fig. 17.3), spirilla, and filaments. The eubacterial representatives are of a gram-negative cell type.

Table 17.4 Some Eubacterial Sulfate-Reducing Bacteria[a]

Heterotrophs	Autotrophs[b]
Desulfovibrio desulfuricans[c,d]	*Desulfovibrio baarsii*
Desulfovibrio vulgaris	*Desulfobacter hydrogenophilus*
Desulfovibrio gigas	*Desulfosarcina variabilis*
Desulfovibrio fructosovorans	*Desulfonema limicola*
Desulfomonas pigra	
Desulfotomaculum nigrificans	
Desulfotomaculum acetoxidans	
Desulfotomaculum orientis[d]	
Desulfobacter postgatei	
Desulfobulbus propionicus	
Desulfobacterium phenolicum[e]	
Desulfobacterium indolicum[f]	
Desulfobacterium catecholicum[g]	
Desulfovibrio sulfodismutans	

[a] For more detailed listings and descriptions of sulfate reducers, see Pfennig et al. (1981) and Postgate (1984).
[b] Autotrophic growth on H_2 and CO_2.
[c] Some strains can grow mixotrophically on H_2 and CO_2 + acetate.
[d] At least one strain can grow autotrophically on H_2 and CO_2.
[e] Bak and Widdel (1986b).
[f] Bak and Widdel (1986a).
[g] Szewzyk and Pfennig (1987).

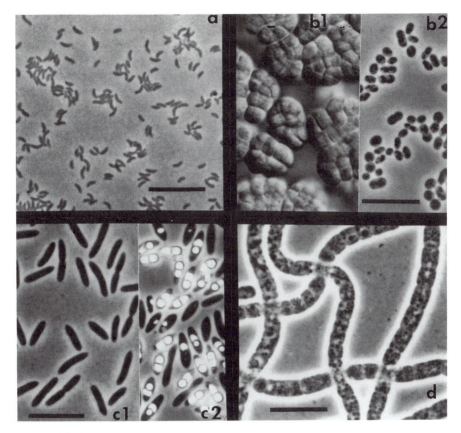

Figure 17.3 Sulfate reducing bacteria. (a) *Desulfovibrio desulfuricans* (phase contrast); (b1,b2) *Desulfosarcina variabilis*—(b1) sarcina packets (interference contrast), (b2) free-living cells (phase contrast); (c1,c2) *Desulfotomaculum acetoxidans*—(c1) vegetative cells (phase contrast), (c2) cells with spherical spores and gas vacuoles (phase contrast); (d) *Desulfonema limicola* (phase contrast). (From Chapter 74, The Dissimilatory Sulfate-Reducing Bacteria by N. Pfennig, F. Widdell, and H. G. Trüper, in The Prokaryotes. Handbook on Habitats, Isolation, and Identification of Bacteria. Vol. 1, M. P. Starr, H. Stolp, H. G. Trüper, A. Balows, and H. G. Schlegel, eds. Springer-Verlag, Berlin, Copyright © 1981, by permission.)

The two archaebacterial sulfate reducers that were discovered recently (Stetter et al., 1987; Burggraf et al., 1990) are all extremely thermophilic, anaerobic, gram-negative, irregularly shaped cocci. *Archeoglobus fulgidus* was found to grow naturally in hydrothermal systems at temperatures between 70 and 100°C in the vicinities of Vulcano and Stufe di Nerone in Italy. Under laboratory conditions, the cultures grow anaerobically in marine mineral salts medium supplemented with yeast extract. In this medium they produce a large amount of hydrogen sulfide and some methane. Thiosulfate, but not elemental sulfur, can act as alternative electron acceptor. Energy sources include hydrogen, some simple organic molecules, as well as glucose, yeast extract, and other complex substrates. Cells contain a number of compounds such as 8-OH-5-deazaflavin and methanopterin previously found only in methanogens, which are also archaebacteria, but 2-mercaptoethanesulfonic acid and factor F_{430} which are found in methanogens, were absent (Stetter et al., 1987).

Archeoglobus profundus was isolated from the Guaymas hot vent area (Gulf of California, also known as Sea of Cortez). It grows anaerobically at temperatures between 65 and 90°C (opt. 82°C) in a pH range of 4.5–7.5 at a NaCl concentration in the range of 0.9–3.6%. Unlike *A. fulgidus*, it is an obligate mixotroph that requires H_2 as energy source. Its organic carbon requirement can be satisfied by acetate, lactate, pyruvate, yeast extract, meat extract, peptone, or acetate-containing crude oil. As for *A. fulgidus*, sulfate, thiosulfate and sulfite can serve as electron acceptors for growth. Although S° is reduced by resting cells, it inhibits growth (Burggraaf et al., 1990).

The presence of as-yet-unidentified, extremely thermophilic sulfate reducers was detected in hot deep-sea sediments at the hydrothermal vents of the tectonic spreading center of the Guaymas Basin (Sea of Cortez or Gulf of California). Sulfate-reducing activity was measurable between 100 and 110°C (opt. 103–106°C). The responsible organisms are probably examples of archaebacteria (Jørgensen et al., 1992).

Some clostridia have been found to be strong sulfite reducers. They cannot reduce sulfate, however. An example is *Clostridium pasteurianum* (McCready et al., 1975; Laishley and Krouse, 1978). The geomicrobial significance of this trait is not clear because sulfite normally does not occur free in the environment in significant amounts in the absence of manmade pollution. An exception may occur around volcanic vents that may emit SO_2. This gas, which may also be formed when sulfur-containing fossil fuels are combusted, forms H_2SO_3 when dissolving in water. *Shewanella putrefaciens* from the Black Sea is another organism that cannot reduce sulfate but can reduce thiosulfate, sulfite, and elemental sulfur to sulfide (Perry et al., 1993). The organism forms 20–50% of the bacterial population in the suboxic zone of the Black Sea where its ability to reduce partially oxidized sulfur compounds appears to play a significant role in the sulfur cycle.

A number of different bacteria have been found to be able to use elemental sulfur anaerobically as terminal electron acceptor, reducing the sulfur to H_2S (Pfennig and Biebl, 1981). Some of these bacteria can grow autotrophically on sulfur using H_2 or methane as energy sources. They are generally thermophilic archaebacteria (e.g., Stetter et al., 1983; Segerer et al., 1986). Others grow heterotrophically on the sulfur, using organic carbon as complex as sugars or as simple as acetate (Belkin et al, 1986; Pfennig and Biebl, 1976) to reduce the sulfur in their energy generating process. This group includes both eubacteria and archaebacteria. A few, but by no means all, strains of sulfate-reducing bacteria of the genus *Desulfovibrio* have the ability to use sulfur in place of sulfate as terminal electron acceptor for growth, but strains of *Desulfotomaculum* and *Desulfomonas* are unable to do so (Biebl and Pfennig, 1977).

Some fungi can also reduce sulfur to sulfide with an electron donor such as glucose (e.g., Ehrlich and Fox, 1967). As expected, the activity was greater anaerobically than aerobically.

17.7 PHYSIOLOGY AND BIOCHEMISTRY OF MICROBIAL OXIDATION OF REDUCED FORMS OF SULFUR

Sulfide

Aerobic Attack   Many aerobic bacteria which oxidize sulfide are obligate or facultative chemosynthetic autotrophs (chemolithotrophs) which, when growing in the autotrophic mode, use sulfide as energy source to assimilate CO_2. Most oxidize the sulfide to sulfate regardless of the level of oxygen tension [e.g., *Thiobacillus thiooxidans* (London and Rittenberg, 1964)], but some, like *Thiobacillus thioparus* form elemental sulfur (S°) if the pH of their milieu is initially alkaline and the rH_2* is 12, i.e., if the milieu is partially reduced due to an oxygen tension below saturation. Thus, *T. thioparus* T5, isolated from a microbial mat, produces elemental sulfur in continuous culture in a chemostat under conditions of oxygen limitation. In this case, small amounts of thiosulfate together with even smaller amounts of tetrathionate and polysulfide are also formed (van den Ende and van Gemerden, 1993). In batch culture under oxygen limitation, *T. thioparus* has been observed to produce initially a slight increase in pH

---

* $rH_2 = -\log(H_2) = (E_h/0.029) + 2pH$, because $E_h = -0.029\log(H_2) + 0.058\log(H^+)$.

followed by a drop to about 7.5 in 4 days and a arise in rH_2 to about 20 (Sokolova and Karavaiko, 1968). The reaction leading to the formation of elemental sulfur can be summarized as follows:

$$HS^- + \tfrac{1}{2}O_2 + H^+ \Rightarrow S° + H_2O \tag{17.13}$$

Under conditions of high oxygen tension (at or near saturation), *T. thioparus* will oxidize soluble sulfide all the way to sulfate (London and Rittenberg, 1964; Sokolova and Karavaiko, 1968; van den Ende and van Gemerden, 1993):

$$HS^- + 2O_2 \Rightarrow SO_4^{2-} + H^+ \tag{17.14}$$

Thiovulum sp. is another example of a eubacterium that oxidizes sulfide to sulfur under reduced oxygen tension (Wirsen and Jannasch, 1987).

London and Rittenberg (1964) (see also Vishniac and Santer, 1957) suggested that the intermediate steps in the oxidation of sulfide to sulfate were

$$4S^{2-} \Rightarrow 2S_2O_3^{2-} \Rightarrow S_4O_6^{2-} \Rightarrow SO_3^{2-} + S_3O_6^{2-} \Rightarrow$$
$$4SO_3^{2-} \Rightarrow SO_4^{2-} \tag{17.15}$$

However, this reaction sequence does not explain the formation of elemental sulfur at reduced oxygen tension. Unless this occurs by way of a specialized pathway, which seems doubtful, a more attractive model of the pathway that explains both processes, S° and SO_4^{2-} formation, in a unified way is one proposed by Roy and Trudinger (1970):

$$S^{2-} \Rightarrow X \Rightarrow SO_3^{2-} \Rightarrow SO_4^{2-}$$
$$\updownarrow \tag{17.16}$$
$$S°$$

Here X represents a common intermediate in the oxidation of sulfide to sulfite and sulfur. Roy and Trudinger visualized X as a derivative of glutathione or a membrane-bound thiol. It may also be representative of the "intermediary sulfur" described by Pronk et al. (1990). The scheme of Roy and Trudinger permits integration of a mechanism for elemental sulfur oxidation into a unified pathway for oxidizing reduced forms of sulfur.

Sorokin (1970) has questioned the sulfide-oxidizing ability of thiobacilli, believing that they oxidize only thiosulfate resulting from chemical oxidation of sulfide by oxygen, and that any elemental sulfur formed by thiobacilli from sulfide is due to the chemical interaction of bacterial oxidation products with S^{2-} and $S_2O_3^{2-}$, as previously proposed by Nathanson (1902) and Vishniac (1952). This view is not generally accepted today. Indeed Vainshtain (1977) and others have presented clear evidence to the contrary.

Anaerobic Attack   Most bacteria that oxidize sulfide anaerobically are photosynthetic autotrophs (Chlorobiaceae, Chromatiaceae, some Rhodospirillaceae, and a few Cyanobacteria), but a few, like the facultative anaerobes *Thiobacillus denitrificans* and *Thermothrix thiopara*, are chemosynthetic autotrophs. In the presence of nonlimiting concentrations of sulfide, most photosynthetic autotrophs oxidize sulfide to sulfur, using the reducing power from this reaction in the assimilation of CO_2. However some exceptions exist that never form elemental sulfur (see section 17.6). When elemental sulfur is formed, it is usually accumulated intracellularly by purple sulfur bacteria and extracellularly by green sulfur bacteria and cyanobacteria. Elemental sulfur accumulated extracellularly by *Chlorobium* appears to be readily available to the cell which formed it but not to other individuals in a population of the same organism or to other photosynthetic bacteria which can oxidize elemental sulfur. The sulfur is apparently partially attached to the cell surface (van Gemerden, 1986). The details of the biochemistry of sulfide oxidation by the photosynthetic autotrophs remain to be explored.

The chemosynthetic autotroph, *Thiobacillus denitrificans*, can oxidize sulfide to sulfate anaerobically with nitrate as terminal electron acceptor. As the sulfide is oxidized, nitrate is reduced via nitrite to nitric oxide (NO), nitrous oxide (N_2O) and (N_2) (Baalsrud and Baalsrud, 1954; Milhaud et al., 1958; Peeters and Aleem, 1970; Adams et al., 1971; Aminuddin and Nicholas, 1973). Acetylene has been found to cause the accumulation of sulfur rather than sulfate in a gradient culture of a strain *T. denitrificans* using nitrous oxide as terminal electron acceptor. In the absence of acetylene, the gradient culture, unlike a batch culture, did not even accumulate sulfur transiently. It was suggested that acetylene prevents the transformation of S^0 to SO_3^{2-} in this culture system (Daalsgaard and Bak (1992). Polysufide ($S_{n-1}SH^-$) but not free sulfur appears to be an intermediate in sulfide oxidation to sulfate by this organism (Aminuddin and Nicholas, 1973). The polysulfide appears to be oxidized to sulfite and thence to sulfate (Aminuddin and Nicholas, 1973, 1974). This reaction sequence is like that proposed in Eq. (17.16).

Oxidation of Sulfide by Heterotrophs and Mixotrophs   Hydrogen sulfide oxidation is not limited to autotrophs. Most strains of *Beggiatoa* (Fig. 17.1) grow mixotrophically or heterotrophically on sulfide. In the former instance, the organisms derive energy from oxidation of the H_2S. In the latter instance, they apparently use sulfide oxidation to eliminate metabolically formed hydrogen peroxide in the absence of catalase (see also section 17.6). *Beggiatoa* deposits sulfur granules resulting from sulfide oxidation in its cells external to the cytoplasmic membrane in invaginated, double-layered membrane pockets (Strohl and Larkin, 1978). The sulfur can be further oxidized to sulfate under sulfide limitation (Pringsheim, 1967). One strain of *Beggiatoa* proven to be autotrophic, has been isolated from the marine environment (Nelson and Jannasch, 1983; see also Jannasch et al., 1989; see also earlier section). The heterotrophs *Sphaerotilus natans* (prokaryote), *Alternaria*, and yeast (eukaryotes; fungi) have also been reported to oxidize H_2S to elemental sulfur (Skerman et al., 1957a, 1957b). It has not been established whether these organisms derive useful energy from this oxidation.

Elemental Sulfur

Aerobic Attack   Elemental sulfur may be enzymatically oxidized to sulfuric acid by certain eubacteria and archaebacteria. The overall reaction may be written

$$S° + \tfrac{3}{2}O_2 + H_2O \Rightarrow H_2SO_4 \tag{17.17}$$

Cell extract of *Thiobacillus thiooxidans* to which catalytic amounts of glutathione were added oxidized sulfur to sulfite (Suzuki and Silver, 1966). In the absence of formaldehyde as a trapping agent for sulfite, thiosulfate was recovered in the reaction mixture (Suzuki, 1965), but this was an artifact due to chemical reaction of sulfite with residual sulfur (Suzuki and Silver, 1966) [see Reaction (17.1)]. Sulfite was also shown to accumulate when sulfur was oxidized by *T. thiooxidans* in the presence of the inhibitor 2-*n*-heptyl-4-hydroxyquinoline *N*-oxide, which has been shown to inhibit sulfite oxidation. The stoichiometry when the availability of sulfur was limited was 1 mol sulfite accumulated per mol each of sulfur and oxygen consumed (Suzuki et al., 1992). A sulfur-oxidizing enzyme in *Thiobacillus thioparus* used glutathione as a cofactor to produce sulfite (Suzuki and Silver, 1966). The enzyme in both organisms contained non-heme iron and was classed as an oxygenase. The mechanism of sulfur oxidation is consistent with the model described by Reaction (17.16). The glutathione in this instance forms a polysulfide [compound X in Eq. (17.16)] with the substrate sulfur, which is then converted to sulfite by introduction of molecular oxygen. This reaction appears not to yield useful energy to the cell. Sulfur oxidation to sulfite that does not involve oxygenase but an oxidase with a potential for energy conservation has also been considered. Some experimental evidence supports such a mechanism (see Pronk et al., 1990).

Anaerobic Oxidation of Elemental Sulfur   Few details are known as yet as to how elemental sulfur is oxidized in anaerobes, especially photosynthetic autotrophs. *Thiobacillus denitrificans* appears to follow the reaction sequence in Eq. (17.16) except that oxidized forms of nitrogen substitute for oxygen as terminal electron acceptors.

T. ferrooxidans has the capacity to oxidize elemental sulfur anaerobically using ferric iron as terminal electron acceptor (Brock and Gustafson, 1976; Corbett and Ingledew, 1987). This anaerobic oxidation yields enough energy to support growth at a doubling time of about 24 h (Pronk et al., 1991, 1992).

Disproportionation of Sulfur Anaerobic marine enrichment cultures consisting predominantly of slightly curved bacterial rods were recently shown to contain chemolithotrophic bacteria that were able to grow on sulfur by disproportionating it into H_2S and SO_4^{2-}, but only in the presence of sulfide scavengers such as FeOOH, $FeCO_3$, or MnO_2 (Thamdrup et al., 1993). The disproportionation reaction can be summarized as follows,

$$4S° + 4H_2O \Rightarrow SO_4^{2-} + 3H_2S + 2H^+ \tag{17.18}$$

The added Fe(II) scavenges the sulfide by forming FeS, whereas the MnO_2 scavenges sulfide in a redox reaction in which MnO_2 is reduced to Mn^{2+} by the sulfide producing SO_4^{2-}, with S° a probable intermediate. The scavenging action is needed to propel the reaction in the direction of sulfur disproportionation. In this reaction, three pairs of electrons from one atom of sulfur are transferred via an as-yet-undefined electron transport pathway to three other atoms of sulfur, generating the H_2S in Reaction (17.18). The transfer of these three pairs of electrons is also the source of the energy conserved by the organism for growth and reproduction.

Sulfite Oxidation

Oxidation by Aerobes Sulfite may be oxidized by two different mechanisms, one of which involves substrate-level phosphorylation while the other does not, although both can yield useful energy through oxidative phosphorylation by the intact cell (see, e.g., review by Wood, 1988). In substrate-level phosphorylation, sulfite reacts oxidatively with adenylic acid (AMP) to give adenosine 5′-sulfatophosphate (APS):

$$SO_3^{2-} + AMP \xrightarrow{\text{APS reductase}} APS + 2e^- \tag{17.19}$$

The sulfate of APS is then exchanged for phosphate:

$$APS + P_i \xrightarrow{\text{ADP sulfurylase}} ADP + SO_4^{2-} \tag{17.20}$$

ADP can then be converted to ATP as follows:

$$2\text{ADP} \xrightarrow{\text{adenylate kinase}} \text{ATP} + \text{AMP} \tag{17.21}$$

Hence the oxidation of 1 mol of sulfite yields 0.5 mol of ATP formed by substrate-level phosphorylation. However, most energy as ATP is gained from shuttling the electrons in Eq. (17.19) through the membrane-bound electron transport system to oxygen (Davis and Johnson, 1967).

A number of thiobacilli appear to use an AMP-independent sulfite oxidase system (Roy and Trudinger, 1970, p. 214). These systems do not all seem to be alike. The AMP-independent sulfite oxidase of autotrophically grown *Thiobacillus novellus* may use the following electron transport pathway (Charles and Suzuki, 1966):

$$\text{SO}_3^{2-} \Rightarrow \text{cytochrome c} \Rightarrow \text{cytochrome oxidase} \Rightarrow \text{O}_2 \tag{17.22}$$

The sulfite oxidase of *T. neapolitanus* can be pictured as a single enzyme complex that may react with sulfite and either with AMP to give APS followed by oxidation to sulfate or with water followed by oxidation to sulfate (Roy and Trudinger, 1970). The enzyme complex then transfers the reducing power that is generated to oxygen. Sulfite-oxidizing enzymes that do not require the presence of AMP have also been detected in *T. thiooxidans*, *T. denitrificans*, and *T. thioparus*. *T. concretivorus* (now considered to be a strain of *T. thiooxidans*) has been reported to shuttle electrons from SO_3^{2-} oxidation via the following pathway to oxygen (Moriarty and Nicholas, 1970b):

$$\begin{aligned} \text{SO}_3^{2-} &\Rightarrow \text{(flavin?)} \Rightarrow \text{coenzyme Q}_8 \Rightarrow \text{cytochrome b} \\ &\Rightarrow \text{cytochrome c} \Rightarrow \text{cytochrome a}_1 \Rightarrow \text{O}_2 \end{aligned} \tag{17.23}$$

Oxidation by Anaerobes   *Thiobacillus denitrificans* is able to form APS reductase (Bowen et al., 1966) that is not membrane-bound, as well as a membrane-bound, AMP-independent sulfite oxidase (Aminuddin and Nicholas, 1973, 1974a,b). Both enzyme systems appear to be active in anaerobically grown cells (Aminuddin, 1980). The electron transport pathway under anaerobic conditions terminates in cytochrome d, whereas under aerobic conditions it terminates in cytochromes aa_3 and cytochrome d. Nitrate but not nitrite acts as electron acceptor anaerobically when sulfite is the electron donor (Aminuddin and Nicholas, 1974b).

Thiosulfate Oxidation   Most chemosynthetic autotrophic bacteria that can oxidize sulfur can also oxidize thiosulfate to sulfate. The photosynthetic, autotrophic, purple and green sulfur bacteria and some purple nonsulfur bacteria oxidize thiosulfate to sulfate as a source of reducing power for CO_2 assimilation (e.g., Trueper, 1978; Neutzling et al., 1985). However, the mechanism of thiosulfate oxidation is probably not the same in all these organisms. The chemosynthetic, aerobic autotroph *Thiobacillus thioparus* will transiently accumulate elemental sulfur outside its cells when growing in an excess of thiosulfate in batch culture, but only sulfate when growing in limited amounts of thiosulfate. *T. denitrificans* will do the same anaerobically with nitrate as terminal electron acceptor (Schedel and Trueper, 1980). The photosynthetic purple bacteria may also accumulate sulfur transiently, but some green sulfur bacteria (Chlorobiaceae) do not (see discussion by Trueper, 1978). Several of the purple nonsulfur bacteria (Rhodospirillaceae) when growing photoautotrophically with thiosulfate do not accumulate sulfur in their cells (Neutzling et al, 1985). Some mixotrophic bacteria oxidize thiosulfate only to tetrathionate.

Thiosulfate is a reduced sulfur compound with sulfur in a mixed valence state. If the outer or sulfane sulfur of $S:SO_3^{2-}$ has a valence of -2, the inner or sulfone sulfur has a valence of $+6$ (but see present view on oxidation states in Table 17.1), and the two sulfurs are covalently linked, thiosulfate oxidation may proceed a described by Charles and Suzuki (1966a), in which thiosulfate is cleaved as follows,

$$S_2O_3^{2-} \Rightarrow SO_3^{2-} + S° \tag{17.24a}$$

The sulfite is then oxidized to sulfate:

$$SO_3^{2-} + H_2O \Rightarrow SO_4^{2-} + 2H^+ + 2e \tag{17.24b}$$

and the sulfur is oxidized to sulfate via sulfite as previously described:

$$S° \Rightarrow SO_3^{2-} \Rightarrow SO_4^{2-} \tag{17.24c}$$

Alternatively, thiosulfate oxidation may be preceded by a reduction reaction resulting in the formation of sulfite from the sulfone and sulfide from the sulfane sulfur:

$$S_2O_3^{2-} + 2e \Rightarrow S^{2-} + SO_3^{2-} \tag{17.25}$$

These products are then each oxidized to sulfate (Peck 1962a). In the latter case it is conceivable that sulfur could accumulate transiently by the mechanism suggested by Eq. 17.16, but sulfur could also result from asymmetric hydrolysis of tetrathionate resulting from direct oxidation of thiosulfate [see Roy and Trudinger (1970) for detailed discussion],

$$2S_2O_3^{2-} + 2H^+ + \tfrac{1}{2}O_2 \Rightarrow S_4O_6^{2-} + H_2O \tag{17.26}$$

$$S_4O_6^{2-} + OH^- \Rightarrow S_2O_3^{2-} + S° + HSO_4^- \tag{17.27}$$

The direct oxidation reaction may involve the enzymes thiosulfate oxidase and thiosulfate–cytochrome c reductase, a thiosulfate activating enzyme

(Aleem, 1965; Trudinger, 1961). The thiosulfate oxidase may use glutathione as coenzyme (see summary by Roy and Trudinger, 1970, and Wood, 1988).

Thiosulfate may also be cleaved by the enzyme rhodanese, which is found in most sulfur-oxidizing bacteria. It can transfer sulfane sulfur to acceptor molecules such as cyanide to form thiocyanate, for instance. This enzyme may also play a role in thiosulfate oxidation. In anaerobically growing *Thiobacillus denitrificans* strain RT, for instance, rhodanese initiates thiosulfate oxidation by causing formation of sulfite from the sulfone sulfur which is then oxidized to sulfate. The sulfane sulfur accumulates transiently as elemental sulfur outside the cells, and when the sulfone sulfur is depleted, the sulfane sulfur is rapidly oxidized to sulfate (Schedel and Trueper, 1980). In another strain of *T. denitrificans*, however, thiosulfate reductase rather than rhodanese catalyzes the initial step of thiosulfate oxidation, and both sulfane and sulfone sulfur are attacked concurrently (Aleem, 1970). *Thiobacillus versutus* (formerly *Thiobacillus* A_2) seems to oxidize thiosulfate to sulfate by a unique pathway (Lu and Kelly, 1983) that involves a thiosulfate-oxidizing multienzyme system that has a periplasmic location (Lu, 1986). No free intermediates appear to be formed from either the sulfane or the sulfone sulfur of thiosulfate.

Pronk et al. (1990) summarize evidence that supports a model in which *T. ferrooxidans*, *T. thiooxidans*, and *T. acidophilus* oxidize thiosulfate by forming tetrathionate in an initial step:

$$2S_2O_3^{2-} \Rightarrow S_4O_6^{2-} + 2e \qquad (17.28)$$

followed by a series of hydrolytic and oxidative steps whereby tetrathionate is transformed into sulfate with transient accumulations of intermediary sulfur from sulfane-monosulfonic acids (polythionates). Thiosulfate dehydrogenase from *T. acidophilus*, which catalyzes the oxidation of thiosulfate to tetrathionate, has been purified and partially characterized (Meulenberg et al., 1993).

Disproportionation of Thiosulfate It has been demonstrated experimentally that some bacteria, like *Desulfovibrio sulfodismutans*, can obtain energy anaerobically by disproportionating thiosulfate into sulfate and sulfide (Bak and Cypionka, 1987; Bak and Pfennig, 1987; Jørgensen, 1990a,b).

$$S_2O_3^{2-} + H_2O \Rightarrow SO_4^{2-} + HS^- + H^+$$
$$(-5.22 \text{ kcal mol}^{-1} \text{ or } 21.9 \text{ kJ mol}^{-1}) \qquad (17.29)$$

The energy from this reaction enables them to assimilate carbon from a combination of CO_2 and acetate. Energy conservation by thiosulfate disproportionation seems, however, paradoxical if the oxidation state of the sulfane-sulfur is -2 and that of the sulfone-sulfur is $+6$, because no redox reaction would be required to generate a mole of sulfate and sulfide each per mole of thiosulfate. A solution to this paradox is provided by the report of Vairavamurthy et al. (1993), who demonstrated spectroscopically that the charge density of the sulfane-sulfur in thiosulfate is really -1 and that of the sulfone-sulfur is really $+5$. Based on this finding, the formation of sulfide and sulfate by disproportionation of thiosulfate would indeed require a redox reaction.

   D. sulfodismutans can also generate useful energy from the disproportionation of sulfite and dithionite to sulfide and sulfate (Bak and Pfennig, 1987). The overall reaction for sulfite disproportionation is

$$4SO_3^{2-} + H^+ \Rightarrow 3SO_4^{2-} + HS^-$$
$$(-14.1 \text{ kcal mol}^{-1} \text{ or } 58.9 \text{ kJ mol}^{-1})$$

(17.30)

For dithionite disproportionation, the reaction is

$$8S_2O_4^{2-} + 8H_2O \Rightarrow 6HS^- + 10SO_4^{2-} + 10H^+$$
$$(-24.7 \text{ kcal mol}^{-1} \text{ or } 103.5 \text{ kJ mol}^{-1})$$

(17.31)

   D. sulfodismutans can also grow on lactate, ethanol, propanol, and butanol as energy sources and sulfate as terminal electron acceptor, like typical sulfate reducers, but growth is slower than by disproportionation of partially reduced sulfur compounds. Bak and Pfennig (1987) suggest that from an evolutionary standpoint, *D. sulfodismutans*–type sulfate reducers could be representative of the progenitors of typical sulfate reducers.

   Perry et al. (1993) suggest that *Shewanella putrefaciens* MR-4, which they had isolated from the Black Sea, disproportionates thiosulfate to either sulfide plus sulfite or elemental sulfur plus sulfite. They never detected any sulfate among the products from these reactions. These disproportionations are, however, endogonic ($+7.39$ and 3.84 kcal mol^{-1} or $+30.98$ and $+16.10$ kJ mol^{-1} at pH 7, 1 atm, and 25°C, respectively). Perry and co-workers suggest that in *S. putrefaciens* MR-4, these reactions must be coupled to exogonic reactions such as carbon oxidation.

   Thiosulfate disproportionation seems to play a significant role in the sulfur cycle of marine sediments (Jørgensen, 1990a). In Kysing Fjord (Denmark) sediment, thiosulfate was identified as a major intermediate product of anaerobic sulfide oxidation that was simultaneously reduced to sulfide, oxidized to sulfate, and disproportionated to sulfide and sulfate. This occurred at a rapid rate as reflected by a small thiosulfate pool. The metabolic fate of thiosulfate oxidation in these experiments was determined by adding differentially labeled ^{35}S-thiosulfate and following the consumption of the thiosulfate and the isotopic distribution in sulfide and sulfate formed from the sulfane- and sulfone-sulfur atoms of the labeled thiosulfate over time in separate experiments. According to Jørgensen (1990a), the disproportionation reaction can explain the observed large difference in $^{34}S/^{32}S$ in sulfate and sulfides in the sediments. These findings were extended to anoxic sulfur transformations in further experiments with Kysing Fjord sediments and in new experiments with sediments from Braband Lake, Aarhus Bay, and Aggersund by Elsgaard and Jørgensen (1992). They showed a significant contribution made by thiosulfate disproportionation in anaerobic production of sulfate from sulfide. Addition of nitrate stimulated anoxic oxidation of sulfide to sulfate. Additon of iron as lepidochrocite (FeOOH) caused partial oxidation of sulfide to pyrite and sulfur and precipitation of iron sulfides.

Tetrathionate Oxidation   Although bacterial oxidation of tetrathionate has been reported, the mechanism of oxidation is still not certain (see Roy and Trudinger, 1970; Kelly, 1982). It may involve dismutation and hydrolysis reactions. A more detailed scheme has been described by Pronk et al. (1990), already mentioned above in connection with thiosulfate oxidation.

17.8   AUTOTROPHIC AND MIXOTROPHIC GROWTH ON REDUCED FORMS OF SULFUR

Energy Coupling in Bacterial Sulfur Oxidation   All evidence to date indicates that to conserve biochemically useful energy, chemosynthetic autotrophic and mixotrophic bacteria which oxidize reduced forms of sulfur feed the reducing power (electrons) into a membrane-bound electron transport system whether oxygen, nitrate or nitrite is the terminal electron acceptor (Peeters and Aleem, 1970; Moriarty and Nicholas, 1970b; Sadler and Johnson, 1972; Aminuddin and Nicholas, 1974b; Loya et al., 1982; Lu and Kelly, 1983; Smith and Strohl, 1991; Kelly et al., 1993; also see review by Kelly, 1982). The components of the electron transport system, i.e., cytochromes, quinones, and nonheme iron proteins, are not identical in all organisms, however. Whatever the transport chain makeup, it is the oxidation state of a particular sulfur compound being oxidized, or more exactly the midpoint potential of its redox couple at physiological pH, that determines the entry point of the electrons removed during the oxidation into the electron transport chain. Thus, the electrons from elemental sulfur are generally thought to enter the transport chain at the level of a cytochrome bc_1 complex or equivalent. Actually, as pointed out earlier, the first step in the oxidation of sulfur to sulfate is the formation of sulfite by an oxygenation involving direct interaction with oxygen without involvement of the cytochrome system. Only in the subsequent oxidation of sulfite to sulfate is the electron transport system directly involved starting at the level of a cytochrome bc_1 complex or equivalent. Also, as discussed earlier, sulfite may be oxidized by an AMP-dependent or -independent pathway. In either case, electrons are passed into the electron transport system at the level of a cytochrome bc_1 complex or equivalent. In the AMP-dependent pathway, most of the energy coupling can be assumed to be chemiosmotic; i.e., on average 1 or 2 mol of ATP can be formed per electron pair passed to oxygen by the electron transport system, but in addition another 0.5 mol of ATP can be formed via substrate-level phosphorylation [Eqs. (17.18–17.20)]. By contrast only 1 or 2 mol of ATP can be formed on average per electron pair passed to oxygen by the AMP-independent pathway.

Chemiosmosis is best explained if it is assumed that the sulfite oxidation half-reaction occurs at the exterior of the plasma membrane (in the periplasm):

$$SO_3^{2-} + H_2O \Rightarrow SO_4^{2-} + 2H^+ + 2e^- \tag{17.32}$$

and the oxygen reduction half-reaction on the inner surface of the membrane (cytoplasmic side):

$$\tfrac{1}{2}O_2 + 2H^+ + 2e^- \Rightarrow H_2O \tag{17.33}$$

In *Thiobacillus versutus*, a thiosulfate-oxidizing, multienzyme system has been located in the periplasm (Lu, 1986).

The pH gradient resulting from sulfite oxidation and any proton pumping associated with electron transport together with any electrochemical gradient provide the proton motive force for ATP generation by F_0F_1 ATPase. Proton translocation during thiosulfate oxidation has been observed in *Thiobacillus versutus* (Lu and Kelly, 1988). Involvement of energy coupling via chemiosmosis is also indicated for *Thiobacillus neapolitanus* using thiosulfate as energy source. The evidence for this is (a) inhibition of CO_2 uptake by the uncouplers carbonyl cyanide *m*-chlorophenylhydrazone (CCCP) and carbonylcyanide *p*-trifluoromethoxyphenylhydrazone (FCCP), and (b) an increase in transmembrane electrochemical potential and CO_2 uptake in response to nigericin (Holthuijzen et al., 1987).

Reduced Forms of Sulfur as Sources of Reducing Power for CO_2 Fixation by Autotrophs

Chemosynthetic Autotrophs Reduced sulfur is not only an energy source but also a source of reducing power for chemosynthetic autotrophs that oxidize it. Since the midpoint potential for pyridine nucleotides is very reduced compared to midpoint potential for reduced sulfur couples that could serve as potential electron donors, **reverse electron transport** from the electron-donating sulfur substrate to pyridine nucleotide is required (see Chapter 6). Electrons thus travel up the electron transport chain, i.e., against the redox gradient, to NADP with consumption of ATP providing the needed energy. This applies to both aerobes and those anaerobes that use nitrate as terminal electron acceptor (denitrifiers).

Photosynthetic Autotrophs   In purple sulfur and nonsulfur bacteria, **reverse electron transport**, a light-independent sequence, is used to generate reduced pyridine nucleotide (NADPH) using ATP from photophosphorylation to provide the needed energy. In green sulfur bacteria as well as cyanobacteria, **light-energized electron transport** is used to generate reduced pyridine nucleotide (NADPH) (Stanier et al., 1986). See also discussion in Chapter 6.

CO_2 Fixation by Autotrophs

Chemosynthetic Autotrophs   Insofar as studied, thiobacilli (eubacteria) generally fix CO_2 by the Calvin-Benson cycle (see Chapter 6), i.e., by means of the ribulose 1,5-bisphosphate carboxylase pathway. In at least some thiobacilli, the enzyme is detected both in the cytosol and in cytoplasmic polyhedral bodies called carboxysomes (Shively et al., 1973). The carboxysomes appear to contain no other enzyme and may represent a means of regulating the level of the carboxylase in the cytosol (Beudeker et al., 1980, 1981; Holthuijzen et al. 1986a,b). *Sulfolobus*, an archaebacterium, assimilates carbon via a reverse, i.e., reductive, tricarboxylic acid cycle (Brock and Madigan, 1984), like green sulfur bacteria (see Chapter 6).

Photosynthetic Autotrophs   Cyanobacteria and purple sulfur bacteria, as well as purple non-sulfur bacteria growing photoautrophically on reduced sulfur, fix CO_2 by the Calvin-Benson cycle, i.e., via the ribulose 1,5-bisphosphate carboxylase pathway (see Chapter 6). Green sulfur bacteria, on the other hand, use a reverse, i.e., reductive, tricarboxylic-acid-cycle mechanism (Stanier et al., 1986), but *Chloroflexus aurantiacus* uses a 3-hydroxypropionate cycle. See discussion in Chapter 6.

Mixotrophy

Free-Living Bacteria   Some sulfur-oxidizing chemosynthetic autotrophs can also grow mixotrophically (e.g., Smith et al., 1980). Among oxidizers of reduced sulfur, *Thiobacillus versutus* (formerly *Thiobacillus* A_2) is a good model for studying autotrophy, mixotrophy, and heterotrophy. It can even grow anaerobically on nitrate (e.g., Wood and Kelly, 1983; Claassen et al., 1987). The organism can use each of these forms of metabolism, depending on medium composition (see review by Kelly, 1982). Another more recently studied example is *Thiobacillus acidophilus* growing on tetrathionate (Mason and Kelly, 1988).

Thiobacillus intermedius, which grows poorly as an autotroph in a thiosulfate-mineral salts medium, grows well in this medium if it is supplemented with yeast extract, glucose, glutamate, or other organic additives (London, 1963; London and Rittenberg, 1966). The organic matter seems to repress the CO_2 assimilating mechanism in this organism but not its energy-generating one (London and Rittenberg, 1966). *T. intermedius* also grows well heterotrophically in a medium containing glucose plus yeast extract or glutathione but not in a glucose-mineral salts medium minus thiosulfate (London and Rittenberg, 1966). It needs thiosulfate or organic sulfur compounds because it cannot assimilate sulfate (Smith and Rittenberg, 1974). A nutritionally similar organism is *Thiobacillus organoparus*, an acidophilic, facultative, heterotrophic bacterium. It was first isolated from acid mine water in copper deposits in Alaverdi (former Armenian S.S.R.). It was found to grow autotrophically and mixotrophically with reduced sulfur compounds (Markosyan, 1973).

Thiobacillus perometabolis cannot grow at all autotrophically in thiosulfate mineral salts medium but requires the addition of yeast extract, casein hydrolyzate, or an appropriate single organic compound in order to utilize thiosulfate as an energy source (London and Rittenberg, 1967). Growth on yeast extract or casein hydrolyzate is much less luxuriant in the absence of added thiosulfate.

Some marine pseudomonads, which are ordinarily considered to grow heterotrophically, have been shown to grow mixotrophically on reduced sulfur compounds (Tuttle et al., 1974). Growth of the cultures on yeast extract was stimulated by the addition of thiosulfate. The bacteria oxidized it to tetrathionate. The growth stimulation by thiosulfate oxidation manifested itself in increased organic carbon assimilation. A number of other heterotrophic bacteria, actinomycetes, and filamentous fungi are also able to oxidize thiosulfate to tetrathionate (Trautwein, 1921; Starkey, 1934; Guitonneau, 1927; Guitonneau and Keilling, 1927), but whether the growth of any is enhanced by this oxidation is unknown at this time. Even if not, these organisms may play a role in promoting the sulfur cycle in soil (Vishniac and Santer, 1957).

Unusual Consortia   Very unusual consortia in the form of associations of invertebrates and autotrophic sulfide-oxidizing bacteria have been discovered in submarine hydrothermal vent communities (Jannasch, 1984; Jannasch and Mottl, 1985; Jannasch and Taylor, 1984). Vestimentiferan tube worms (*Riftia pachyptila*) that grow around the submarine vents, especially white smokers, and that lack a mouth and digestive tract, harbor special organelles in their body cavity called collectively a trophosome. These structures when viewed in section under a transmission electron microscope are seen to be tightly packed bacteria (Cavanaugh et al., 1981). Metabolic evidence indicates that these are chemosynthetic, autotrophic bacteria (Felbeck, 1981; Felbeck et al., 1981; Rau, 1981; Williams et al., 1988). Some bacteria have been cultured from trophosome, but whether any of them are the important symbionts of the worm remains to be established (Jannasch and Taylor, 1984). They are considered to be autotrophic sulfur-oxidizing bacteria which share the carbon they fix with the worm. The worm absorbs hydrogen sulfide and oxygen from the water through a special organ at its head called an obtracular plume (Jones, 1981) and transmits these via its circulatory system to the trophosomes. Its blood contains hemoglobin for binding oxygen and another special protein (Arp and Childress, 1983; Powell and Somero, 1983) that binds hydrogen sulfide reversibly to prevent reaction of the sulfide with the hemoglobin and its consequent destruction. The bound hydrogen sulfide and oxygen are released at the site of the trophosome.

Somewhat less intimate consortia are formed between giant clams and mussels (Mollusca) and autotrophic sulfide oxidizing bacteria around the hydrothermal vents. The bacteria in these instances reside not in the gut of the animals but on their gills (see Jannasch and Taylor, 1984, for discussion; also Rau, 1981; Rau and Hedges, 1979). These looser consortia between autotrophic sulfide-oxidizing bacteria appear not to be restricted to hydrothermal vent communities but also occur in shallow water environments rich in hydrogen sulfide (Cavanaugh, 1983).

17.9 ANAEROBIC RESPIRATION USING OXIDIZED FORMS OF SULFUR AS ELECTRON ACCEPTORS

Sulfate Reduction   Various forms of oxidized sulfur can serve as terminal electron acceptors in the respiration of some bacteria under anaerobic conditions. These sulfur compounds include sulfate, thiosulfate, and sulfur, among others.

Biochemistry of Dissimilatory Sulfate Reduction   A variety of strictly anaerobic bacteria respire using sulfate as terminal electron acceptor. Many are taxonomically quite unrelated and include eubacteria as well as archaebacteria (see earlier section). Insofar as is now known, the mechanism by which they reduce sulfate follows a very similar, but not necessarily identical pattern, in all. As presently understood, the enzymatic reduction of sulfate requires an initial activation by ATP to form adenine phosphato-sulfate and pyrophosphate:

$$\text{ATP sulfurylase } SO_4^{2-} + \text{ATP} \Rightarrow \text{APS} + PP_i \qquad (17.34)$$

In members of the genus *Desulfovibrio* the pyrophosphate is hydrolyzed to inorganic phosphate (P_i), which helps to pull the reaction in the direction of APS:

$$PP_i + H_2O \xrightarrow{\text{Pyrophosphatase}} 2P_i \qquad (17.35)$$

The energy in the anhydride bond of pyrophosphate is thus not available to *Desulfovibrio*. In contrast, this energy is conserved by members of the genus *Desulfotomaculum*. They do not hydrolyze the pyrophosphate but use it as a substitute for ATP (Liu et al., 1982). This also has the effect of pulling reaction 17.34 in the direction of APS.

Unlike in assimilatory sulfate reduction, APS, once formed, is reduced directly to sulfite and adenylic acid (AMP):

$$\text{APS} + 2e^- \xrightarrow{\text{APS reductase}} SO_3^{2-} + \text{AMP} \qquad (17.36)$$

The APS reductase, unlike PAPS reductase, does not require NADP as a cofactor, but like PAPS reductase, contains bound FAD and iron (for further discussion see, for instance, Peck, 1993).

The subsequent details in the reduction of sulfite to sulfide are not fully agreed upon. One line of experimental evidence suggests a multistep process involving trithionate and thiosulfate as intermediates (Kobiyashi et al., 1969; modified by Akagi et al., 1974; Drake and Akagi, 1978):

$$3HSO_3^- + 2H^+ + 2e^- \xrightarrow{\text{bisulfite reductase}} S_3O_6^{2-} + 2H_2O + OH^- \qquad (17.37)$$

$$S_3O_6^{2-} + H^+ + 2e^- \xrightarrow{\text{trithionate reductase}} S_2O_3^{2-} + HSO_3^- \qquad (17.38)$$

$$S_2O_3^{2-} + 2H^+ + 2e^- \xrightarrow{\text{thiosulfate reductase}} HS^- + HSO_3^- \qquad (17.39)$$

In most *Desulfovibrio* cultures, the bisulfite reductase seems to be identical to desulfoviridin (Kobayashi et al., 1972; Lee and Peck, 1971). However, in *D. desulfuricans* strain Norway 4, which lacks desulfoviridin, desulforubidin appears to be the bisulfite reductase (Lee et al., 1973), and in *Desulfovibrio thermophilus*, it has been identified as desulfofuscidin (Fauque et al., 1990). In *Desulfotomaculum nigrificans*, a carbon monoxide–binding pigment, called P582 by Trudinger (1970), accounts for bisulfite reductase activity, which, according to Akagi et al. (1974), leads to the formation of trithionate, with thiosulfate and sulfide accumulating as endogenous side products. Inducible sulfite reduction has also been observed with *Clostridium pasteurianum*, a bacterium that is not a dissimilatory sulfate reducer. It can reduce sulfite to sulfide. In the absence of added selenite, whole cells do not release detectable amounts of trithionate or thiosulfate when reducing sulfite, but in the presence of selenite they do. Selenite was found to inhibit thiosulfate reductase but not trithionate reductase in whole cells, but inhibited both in cell extracts (Harrison et al., 1980). A purified sulfite reductase from *C. pasteurianum* produced sulfide from sulfite. It was also able to reduce NH_2OH, SeO_3^{2-}, and NO_2^-, but did not reduce trithionate or thiosulfate (Harrison et al., 1984). Several physical and chemical properties of this enzyme differed from those of bisulfite reductases in sulfate reducers. Its role in *C. pasteurianum* may be in detoxification when excess sulfite is present (Harrison et al., 1984). Peck (1993) refers to the enzymes involved in the transformation of bisulfite to sulfide collectively as bisulfite reductase. Distinct sulfite reductase, trithionate reductase, and thiosulfate reductase were also identified by Peck and LeGall (1982). However, at the time they did not visualize a major role for these enzymes in sulfite reduction to sulfate.

Chambers and Trudinger (1975) have questioned whether the trithionate pathway of sulfate reduction is the major pathway of *Desulfovibrio* spp. They found that results of experiments with isotopically labeled $^{35}SO_3^{2-}$, $^{35}SSO_3^{2-}$, and $S^{35}SO_3^{2-}$ could not be reconciled with the trithionate pathway, but were more consistent with a pathway involving the assimilatory kind of sulfite reductase. Their view was supported by Peck and LeGall (1982). However, Vainshtein et al. (1981), after reinvestigating this problem, concluded that the findings of Chambers and Trudinger (1975) were the result of using a heavy cell concentration (limiting bisulfite concentration) which did not permit transient accumulation of thiosulfate. Le Gall and Fauque (as cited by Fauque et al., 1991) concluded in 1988 that a direct pathway from sulfite to sulfate is used by *Desulfovibrio* and a trithionate pathway by *Desulfotomaculum*. Our understanding of the details of sulfite reduction to sulfide by sulfate reducing bacteria remains incomplete at this time.

Sulfur Isotope Fractionation   Sulfate-reducing bacteria can distinguish between ^{32}S and ^{34}S isotopes of sulfur; i.e., they can bring about isotope fractionation (Harrison and Thode, 1957; Jones and Starkey, 1957). Both isotopes are stable. The ^{32}S isotope of sulfur is the most abundant (average 95.1%) and the ^{34}S isotope is the next most abundant (average 4.2%). The

$^{32}S/^{34}S$ ratio of natural sulfur compounds ranges between 21.3 and 23.2. Meteoritic sulfur has a $^{32}S/^{34}S$ ratio of 22.22. Since this ratio appears to be relatively constant from sample to sample, it is often used as a reference standard against which to compare sulfur isotope ratios of other materials, which may be either enriched or depleted in ^{34}S. Under conditions of slow growth, sulfate reducers attack $^{32}SO_4^{2-}$ more readily than $^{34}SO_4^{2-}$ (Jones and Starkey, 1957, 1962) (Table 17.5A,B). The nature of the electron donor may also affect the degree of isotope fractionation (Kemp and Thode, 1968). The degree of isotope fractionation is calculated in terms of $\delta^{34}S$ values expressed in parts per thousand (‰):

$$\delta^{34}S = \frac{^{34}S/^{32}S_{sample} - ^{34}S/^{32}S_{\text{meteoritic or standard}}}{^{34}S/^{32}S_{\text{meteoritic or standard}}} \times 1,000 \qquad (17.40)$$

Table 17.5A Sulfide Production and Fractionation of Stable Isotopes of Sulfur by *Desulfovibrio desulfuricans* Cultivated at 28°C (Rapid Growth)

Sample number	Incubation period (h)[a]	Sulfide S in PbS (mg)	Sulfate reduced[b] (%)	Number of isotope determinations	$\delta^{34}S$
1	44	996	6.3	4	−5.4
2	8	2168	20.0	4	−4.9
3	4	1931	32.2	2	−3.1
4	5	1448	41.4	2	−3.1
5	4	1394	50.2	2	−5.4
6	3	1248	58.1	2	−5.4
7	9	317	60.1	2	−6.7
8	14	191	61.3	2	−8.9
9	41	103	62.0	2	−9.8
10	68	115	62.7	2	−12.9
11	59.5	387	65.2	2	−7.2
12	25.5	901	70.9	2	−3.1
13	7	615	74.8	2	−0.5
14	6	474	77.8	2	+0.9
15	24	856	83.2	2	+0.5
16	43	106	83.9	2	−4.9

[a] Periods were calculated from the time that sulfide first appeared in the culture substrate. This was 60 h after the medium was inoculated.
[b] The initial sulfate S in the substrate was 3,943 ppm.
Source: Adapted from Jones and Starkey (1957), by permission.

Table 17.5B   Cultivated at Low Temperatures (Slow Growth)

Sample number	Incubation period (hr)[a]	Sulfide S in PbS (mg)	Sulfate reduced (%)	$\delta^{34}S$
17	200	21.2	4.4	−22.1
18	142	65.8	4.5	−25.9
19	120	112.7	8.3	−25.9
20	120	167.8	10.0	−24.2
21	96	174.6	13.1	−24.2
22	120	180.8	16.0	−22.9
23	144	134.5	16.8	−21.6
24	120	102.0	18.4	−19.5

[a] Periods were calculated from the time that sulfide first appeared in the culture substrate; this was 18 h after the medium was inoculated.
Source: Adapted from Jones and Starkey (1957), by permission.

Harrison and Thode (1957) proposed that it was the rate-controlling S–O bond breakage in bacterial sulfate reduction (i.e., reduction of APS to sulfite and AMP) that is responsible for the isotope fractionation phenomenon.

Dissimilatory sulfate reduction is not the only process that may lead to sulfur isotope fractionation. Sulfite reduction by *Desulfovibrio* and *Saccharomyces cerevisiae* (Kaplan and Rittenberg, 1962, p. 81) and by *Clostridium pasteurianum* (Laishley and Krouse, 1978), sulfide release from cysteine by *Proteus vulgaris* (Kaplan and Rittenberg, 1962. 1964), assimilatory sulfate reduction by *Escherichia coli* and *S. cerevisiae* (Kaplan and Rittenberg, 1962) can also lead to sulfur isotope fractionation.

Isotopic analysis of sulfur minerals in nature has helped in deciding whether biogenesis was involved in their accumulation. Any given deposit must, however, be sampled at a number of locations since isotope enrichment values ($\delta S^{34}S$) generally fall in a narrow range or a wide range. Nonbiogenic $\delta^{34}S$ values generally fall in a narrow range and usually have a positive sign, whereas biogenic values tend to fall in a wide range and have a negative sign.

Reduction of Elemental Sulfur   Elemental sulfur can be used as terminal electron acceptor under anaerobic conditions in the respiration of certain eubacteria and archaebacteria (Schauder and Kroeger, 1993). The product of the reduction is sulfide. Polysulfide may be an intermediate in this reduction (Schauder and Mueller, 1993; Fauque et al., 1991). The eubacteria include *Thermotoga neapolitana* (Jannasch et al., 1988a), *Desulfuromonas acetoxidans*, *Desulfovibrio gigas*, and some other sulfate-reducing bacteria (Pfennig and Biebl, 1976; Biebl and Pfennig, 1977; Fauque et al., 1991). The archaebacteria include *Desulfurococcus*, *Thermoproteus* and *Thermofilum* (Jannasch et al., 1988b; Zillig et al., 1981, 1982a,b).

Although some anaerobic bacteria reduce sulfur to sulfide as a requisite part of their energy metabolism or to grow optimally, sulfur reduction in these instances appears to be a means of scavenging excess reducing power in what is otherwise a fermentation process. Examples are *Staphylothermus*, *Thermococcus*, and *Pyrococcus* (see Schauder and Kroeger, 1993), *Thermotoga* sp. FjSS3.B1 (Janssen and Morgan, 1992), and *Thermoproteus uzoniensis* (Bjonch-Osmolovskaya, 1990). The energy sources for these organsims are always organic, such as small peptides, starch, glucose, and so forth. Some fungi, e.g., *Rhodotorula* and *Trichosporon* (Ehrlich and Fox, 1967), can also reduce sulfur to H_2S with glucose as electron donor, and this is probably not a form of respiration.

The energy source for the sulfur-respiring archaebacteria is sometimes hydrogen and/or methane, but more often organic molecules such as glucose or small peptides, while that for the eubacteria may be simple organic compounds (e.g., ethanol, acetate, propanol) or more complex organics. In the case of the eubacterium *Desulfuromonas acetoxidans*, an electron transport pathway including cytochromes appears to be involved (Pfennig and Biebl, 1976). When acetate is used as energy source, oxidation proceeds anaerobically by way of the tricarboxylic acid cycle. The oxalacetate required for initiation of the cycle is formed by carboxylation of pyruvate, which arises from carboxylation of some acetate (Gebhardt et al., 1985). Energy is gained in the oxidation isocitrate and 2-ketoglutarate. Membrane preparations were shown to oxidize succinate using sulfur or NAD as electron acceptors by an ATP-dependent reaction. Similar membrane preparations reduced fumarate to succinate with H_2S as electron donor by an ATP-independent reaction. Menaquinone mediated hydrogen transfer. Protonophores and uncouplers of phosphorylation inhibited reduction of sulfur but not of fumarate. The compound 2-*n*-nonyl-4-hydroxy-quinoline-*N*-oxide inhibited electron transport to sulfur or fumarate. Together these observations support the notion that sulfur reduction in *D. acetoxidans* involves a membrane-bound electron transport system and that ATP is formed chemiosmotically, i.e., by oxidative phosphorylation when growing on acetate (Paulsen et al., 1986).

Reduction of Thiosulfate   Growth and growth yield of some members of the anaerobic thermophilic and hyperthermophilic Thermotogales was recently shown to be stimulated in the presence of thiosulfate (Ravot et al., 1995). The test organisms included *Fervidobacterium islandicum*,

Thermosipho africanus, *Thermotoga maritima*, *Thermotoga neapolitana*, and *Thermotoga* sp. SEBR 2665. The last named was isolated from an oil field. All reduced the thiosulfate to sulfide. The Thermotogales in this group are able to ferment glucose among various energy-yielding substrates. The thiosulfate, like sulfur (see, e.g., Janssen and Morgan, 1992), appears to serve as an electron sink by suppressing H_2 accumulation in the fermentation of glucose, for instance. H_2 accumulation has an inhibitory effect on the growth of these organisms. The biochemical mechanism by which they reduce thiosulfate remains to be elucidated.

Terminal Electron Acceptors Other than Sulfate, Sulfite, Thiosulfate, or Sulfur   A few sulfate reducers can grow with Fe(III) (Coleman et al., 1993; Lovley et al., 1993), nitrate, nitrite (McCready et al., 1983; Keith and Herbert, 1983; Seitz and Cykpionka, 1986), fumarate (Miller and Wakerley, 1966), or, in the case of *Desulfomonile tiedjei*, chloroaromatics (DeWeerd et al., 1990, 1991) as terminal electron acceptors. A few strains of *Desulfovibrio* can even grow on pyruvate or fumarate without an external electron acceptor by generating hydrogen as one of the metabolic end products (Postgate, 1952, 1963). *Desulfovibrio gigas* and a few strains of *D. desulfuricans* grow on fumarate by disproportionating it. They reduce a portion of the fumarate to succinate and oxidize the remainder to malate and acetate (Miller and Wakerley, 1966).

When Fe(III) serves as terminal electron acceptor, it may be reduced to $FeCO_3$ (siderite) or Fe^{2+}. When NO_3^- and NO_2^- are terminal electron acceptors, they are reduced to ammonia (nitrate ammonification). When fumarate is the terminal electron acceptor, it is reduced to succinate. When a chloroaromatic like 3-chlorobenzoate is the terminal electron acceptor, it is reductively dechlorinated to benzoate and chloride. How many different sulfate-reducing bacteria are able to substitute any of these terminal electron acceptors has yet to be systematically explored. The fact that some sulfate reducers can avail themselves of such "substitute" terminally electron acceptors may explain why the presence of such organisms can be detected in environments, like most soils, in which the natural sulfate, sulfite, thiosulfate, or sulfur concentration is very low.

Oxygen Tolerance of Sulfate Reducers   In general, sulfate reducers are strict anaerobes. Yet they have shown limited oxygen tolerance (Wall et al., 1990; Abdollahi and Wimpenny, 1990; Marshall et al., 1993). Indeed, *Desulfovibrio desulfuricans D. vulgaris*, *D. sulfodismutans*, *Desulfobacterium autotrophicum*, *Desulfobulbus propionicus*, and *Desulfococcus multivorans* show an ability to use oxygen as terminal electon acceptor, i.e., to respire microaerophilically (below 10 µM dissolved O_2), without, however, being able to grow under these conditions (Dilling and Cypionka, 1990).

Recently evidence has been presented in support of aerobic sulfate reduction by bacteria that have as yet not been obtained in pure culture (Canfield and Des Marais, 1991; Jørgensen and Bak, 1991; Fruend and Cohen, 1992). Results so far are not always explained by the existence of anaerobic microniches.

17.10 AUTOTROPHY, MIXOTROPHY, AND HETEROTROPHY AMONG SULFATE-REDUCING BACTERIA

Autotrophy   Although the ability of *Desulfovibrio desulfuricans* to grow autotrophically with hydrogen (H_2) as energy source had been previously suggested, experiments by Mechalas and Rittenberg (1960) failed to demonstrate it. Seitz and Cypionka (1986), on the other hand, obtained autotrophic growth with hydrogen as energy source of *D. desulfuricans* strain Essex 6, but the growth yield was small when sulfate was the terminal electron acceptor. Better yields were obtained with nitrate or nitrite as terminal acceptors, presumably because the latter two acceptors did not need to be activated by ATP, which is a requirement for sulfate reduction. Nitrate and nitrite are reduced to ammonia by *Desulfovibrio* (McCready et al., 1983; Mitchell et al., 1986; Keith and Herbert, 1983). *Desulfotomaculum orientis* also has an ability to grow autotrophically with hydrogen as energy source and sulfate, thiosulfate, or sulfite as terminal electron acceptor (Cypionka and Pfennig, 1986). Under optimal conditions, better growth yields were observed with this organism than had been reported for *Desulfovibrio desulfuricans* (12.4 versus 9.7 g of dry cell mass per mol of sulfate reduced). This may be explainable on the basis that *Desulfotomaculum* can utilize the inorganic pyrophosphate generated in sulfate activation whereas *Desulfovibrio* cannot. *Desulfotomaculum orientis* gave better growth yields when thiosulfate or sulfite was the terminal electron acceptor than when sulfate was. The organism excreted acetate that was formed as part of its CO_2 fixation process (Cypionka and Pfennig, 1986). Acetate may be formed via the activated acetate pathway in which acetate is formed directly from two molecules of CO_2, as is the case in methanogens and acetogens (see Chapters 6 and 20) and as has now been shown in *Desulfovibrio baarsii*, which can also grow autotrophically on hydrogen and sulfate (Jansen et al., 1984), and in *Desulfobacterium autotrophicum* (Schauder et al. 1989). *Desulfobacter hydrogenophilus*, by contrast, assimilates CO_2 by a reductive citric acid cycle when growing autotrophically with H_2 as energy source and sulfate as terminal electron acceptor (Schauder et al., 1987). Other sulfate reducers that are able to grow autotrophically on hydrogen as energy source and sulfate as terminal electron acceptor include *Desulfonema limicola* and *Desulfosarcina variabilis* (Pfennig et al., 1981) and *Desulfobacterium autotrophicum* (Schauder et al., 1989).

Mixotrophy *Desulfovibrio desulfuricans* has been shown to grow mixotrophically with any one of several different substances as sole energy source, including hydrogen, formate, and isobutanol. The carbon in the organic energy sources was not assimilated. It was derived instead from substances as complex as yeast extract or as simple as acetate or acetate plus CO_2. Sulfate was the terminal electron acceptor in all instances (Mechalas and Rittenberg, 1960; Sorokin, 1966 a–d; Badziong and Thauer, 1978; Badziong et al., 1978; Brandis and Thauer, 1981). A strain used by Sorokin (1966a) when growing on hydrogen as energy source and acetate plus CO_2 as carbon source was able to derive as much as 50% of its carbon from CO_2, whereas on lactate plus CO_2 it derived only 30% of its carbon from CO_2. Badziong et al. (1978) using a different strain of *Desulfovibrio*, also found 30% of its carbon to be derived from CO_2 when growing on hydrogen and acetate plus CO_2.

Members of some other genera of sulfate reducing bacteria can also grow mixotrophically on hydrogen and acetate plus CO_2 (Pfennig et al., 1981). In all these instances, ATP is generated chemiosmotically from hydrogen oxidation in the periplasm.

Heterotrophy All sulfate reducers can grow heterotrophically with sulfate as terminal electron acceptor. In general, sulfate reducers specialize with respect to the carbon-energy sources they can utilize (see section 17.6; also Pfennig et al, 1981). When acetate serves as energy source or is derived by degradation from some other organic energy source to be completely oxidized, as for instance with *Desulfobacter postgatei*, it may be oxidized anaerobically via the tricarboxylic acid cycle (Brandis-Heep et al., 1983; Gebhardt et al., 1983; Moeller et al., 1987). More commonly, however, sulfate reducers oxidize acetate by reversal of the active acetate synthesis pathway (Schauder et al., 1986). Assimilation of acetate most likely involves initial carboxylation to pyruvate. ATP synthesis in the heterotrophic mode of sulfate reduction insofar as it is understood, is mainly by oxidative phosphorylation (chemiosmotically) involving transfer of hydrogen abstracted from organic substrates into the periplasm followed by its oxidation (Odom and Peck, 1981, but see Odom and Wall, 1987; Kramer et al., 1987). In the case of lactate, this hydrogen transfer from the cytoplasm to the periplasm across the plasma membrane appears to be energy driven (Pankhania et al., 1988). Some ATP may be formed by substrate-level phosphorylation.

17.11 BIODEPOSITION OF NATIVE SULFUR

Types of Deposits Deposits of elemental sulfur of biogenic origin, and in most cases of abiogenic origin, have resulted from the oxidation of H_2S:

$$H_2S + \tfrac{1}{2}O_2 \Rightarrow S° + H_2O \qquad (17.41)$$

In some fumaroles, sulfur may also form abiogenically through the interaction of H_2S and SO_2:

$$2H_2S + SO_2 \Rightarrow 3S° + 2H_2O \qquad (17.42)$$

Most known native sulfur deposits are not of volcanogenic origin. Indeed, only 5% of known reserves are the result of volcanism (Ivanov, 1968, p. 139). Biogenic sulfur accumulations in sedimentary deposits may originate syngenetically or epigenetically. In **syngenetic** formation, sulfur is deposited contemporaneously with the enclosing host rock or sediment during its sedimentation. In **epigenetic** formation, sulfur is laid down in cracks and fissures of preformed host rock. This sulfur may originate from a diagenetic process in which a sulfate component of the host rock is converted to sulfur, or it may involve the conversion of dissolved sulfate or sulfide in a solution percolating through cracks and fissures of host rock. Syngenetic sulfur deposits are generally formed in limnetic environments, whereas epigenetic sulfur deposits tend to form in terrestrial environments.

If the source of elemental sulfur is sulfate, the microbial transformation is a two stage process. The first stage involves dissimilatory sulfate reduction to sulfide (elemental sulfur is not an intermediate in this process), and the second stage involves oxidation of the sulfide to elemental sulfur.

Examples of Syngenetic Sulfur Deposition

Cyrenaican Lakes, Libya, North Africa A typical example of contemporary syngenetic sulfur deposits is found in the sediments of the Cyrenaican lakes Ain ez Zauia, Ain el Rabaiba, and Ain el Braghi. The origin of the sulfur in these lakes was first studied by Butlin and Postgate (1952) and Butlin (1953). The extensive native sulfur in the sediments of these lakes makes up as much as one half of the silt. The waters of these lakes have a strong odor of hydrogen sulfide and are opalescent, owing to a fine suspension of sulfur crystals. A fourth lake in the same general area, called

Ain amm el Gelud, also contains sulfuretted water but shows no evidence of sulfur in its sediment. Ain es Zauia was the most thoroughly studied by Butlin and Postgate. It is made up of two adjacent basins, 55 × 30 m and 90 × 70 m in expanse, respectively, and no deeper than 1.5 m. Other characteristics of the lake are summarized in Tables 17.6 and 17.7. The water in the lake is introduced by warm springs (Butlin, 1953). The border of Ain ez Zauia, as well as those of the other two lakes with sulfur deposits, featured a characteristic red-colored, carpetlike, gelatinous material which extended several yards into shallow water in some places. The underside of this red, gelatinous material showed a green and black material. Some of the red material was found floating in the water in the form of red, bulbous formations. The red-colored material was massive growth of the photosynthetic purple sulfur bacterium *Chromatium* and the green material consisted of growth of the green photosynthetic sulfur bacterium *Chlorobium*. Many sulfate-reducing bacteria were also detected in the lakes. From these observations Butlin and Postgate (1952) inferred that the sulfate reducers were responsible for reducing the sulfate in the water to hydrogen sulfide, utilizing some of the carbon fixed by the photoautotrophic bacteria as carbon and energy source. Their model can be visualized as a cycle in which the photosynthetic bacteria oxidize the hydrogen sulfide, produced by the sulfate reducing bacteria, to elemental sulfur, assimilating CO_2 photosynthetically in the process. The sulfate-reducing bacteria, in turn, use fixed carbon produced by the photosynthetic bacteria to reduce sulfate in the lake to sulfide.

Although it was recognized by Butlin and Postgate that some of the hydrogen sulfide could undergo autoxidation, this process was held unimportant because Ain amm el Gelud, which contains sulfuretted water, contains no significant amount of sulfur in its sediment and also lacks noticeable growth of photosynthetic bacteria. Butlin and Postgate were

Table 17.6 Physical Characteristics of Lake Ain ez Zauni

Surface area	7,950 m²
Maximum depth	1.5 m
Surface temperature	30°C
Bottom temperature	32°C
Air temperature	16°C
Sulfur production per year	100 tons

Source: Ivanov, 1968.

Table 17.7 Chemical Composition of the Waters of
Lake Ain ez Zauni

H_2S in surface water: 15–20 mg liter^{-1}			
H_2S in bottom water: 108 mg liter^{-1}			
Total solids: 25.25 g liter^{-1}			
Ca	1,179 mg liter^{-1}	Cl	13,520 mg liter^{-1}
Mg	336 mg liter^{-1}	HCO_3	145 mg liter^{-1}
Na	7,636 mg liter^{-1}	SO_4	1,848 mg liter^{-1}
K	320 mg liter^{-1}	NO_3	3 mg liter^{-1}
NH_3	8 mg liter^{-1}	SiO_3	70 mg liter^{-1}

Source: Ivanov, 1968.

able to reconstruct an artificial system in the laboratory with pure and mixed cultures of sulfate-reducing and photosynthetic sulfur bacteria that reproduced the process they postulated for sulfur deposition in the Cyrenaican lakes. Significantly, however, they found it best to supplement their artificial lake water with 0.1% sodium malate to achieve good sulfur production. This led to questions about the correctness of their model for biogenesis of sulfur in these lakes.

Ivanov (1968) pointed out that Butlin and Postgate's model did not account for all the carbon needed for sulfur production from sulfate in the Cyrenaikan lakes. He argued that a cyclical mechanism in which the photosynthetic sulfur bacteria produce the organic carbon with which the sulfate-reducing bacteria reduce sulfate to sulfide and which the photosynthetic sulfur bacteria then turn into S°, suffers from carbon limitation. He showed that each turn of a cycle produces only one-fourth or less of the hydrogen sulfide that was produced in the just preceding cycle. Thus, the photosynthetic bacteria produce only one-fourth or less of the carbon in each succeeding cycle that they produced in the just preceding cycle. This is best illustrated by the following two reactions:

$$2CO_2 + 4H_2S \Rightarrow 2(CH_2O) + 4S° + 2H_2O \tag{17.43}$$

$$2(CH_2O) + SO_4^{2-} + 2H^+ \Rightarrow H_2S + 2H_2O + 2CO_2 \tag{17.44}$$

The first of these equations illustrates the photosynthetic reaction and the second illustrates sulfate reduction. It is seen that to produce the organic carbon (CH_2O) needed to reduce sulfate, four times as much H_2S is consumed as is produced in sulfate reduction. Ivanov (1968) therefore argued that most of the sulfide turned into sulfur by the photosynthetic bacteria is introduced into the lake by the warm springs and does not result from sulfate reduction. He noted that Butlin and Postgate (Butlin, 1953) had actually demonstrated that many artesian wells in the area contained sulfuretted water with sulfate reducing bacteria. Ivanov, however, did not consider the possibility that these wells might also inject H_2 into the lakes, which sulfate reducers could employ either autotrophically or mixotrophically as an alternate energy source and reductant of sulfate in a carbon-sparing action.

Ivanov (1968) also suggested that a portion of the sulfur in the lake may be produced by non-photosynthetic sulfur bacteria and by autoxidation. No matter what the source of the H_2S, biogenesis of the sulfur in the Cyrenaican lakes has been confirmed on the basis of isotope analysis (Macnamara and Thode, 1951; Harrison and Thode, 1958; Kaplan et al., 1960). Figure 17.4 summarizes the different biological reactions by which sulfur may be generated in these lakes.

Lake Sernoye   This lake is located in the Kuibyshev Oblast in the central Volga region, of Russia. It is an artificial, relatively shallow reservoir fed by the Sergievsk sulfuretted springs (Ivanov, 1968). The water output of these springs is around 6,000 m^3 day^{-1}. Waters contain 83–86 mg of H_2S per liter and have a pH of 6.7. The water temperature in summer ranges around 8°C. The lake drains into Molochni Creek. The waters which enter Molochni Creek are reported to be opalescent, owing to suspended native sulfur in them. The sulfur originates from oxidation of H_2S in the lake. Much of the lake sediment contains about 0.5% native sulfur, but some sediment contains as much as 2–5%. The lake freezes over in winter, at which time no significant oxidation of H_2S occurs. This fact is reflected by the stratified occurrence of sulfur in the lake sediment. Pure sulfur crystals which are paragenetic with calcite crystals have been found in some sediment cores (Sokolova, 1962). Most sulfur in the lake is deposited around the sulfuretted springs. At these locations, masses of purple and green sulfur bacteria are seen. Impression smears have shown the presence of *Chromatium* and large numbers of rod-shaped bacteria, which on culturing reveal themselves to be mostly thiobacilli. A study of H_2S oxidation in Lake Sernoye waters around the springs using $Na_2{}^{35}S$ revealed that the microflora of the lake made a significant contribution (more than 50%). The study measured differentially chemical and biological sulfide oxidation in the dark as well as biological, light-dependent sulfide oxidation. About the same amount of native sulfur was precipitated in the dark as in the light in these experiments, but more sulfate was formed in the light. These results suggested that most of the H_2S in the lake that is biologically oxidized to native sulfur is attacked by thiobacilli, in particular

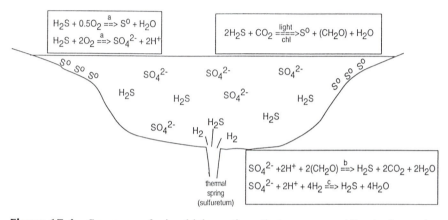

Figure 17.4 Summary of microbial reactions that can account for the formation of sulfur in the Cyrenaikan Lakes, Libya. (a) Caused by thiobacilli in oxidizing zones of the lakes [suggested by Ivanov (1968)], or by autoxidation; (b) caused by sulfate reducers like *Desulfovibrio desulfuricans*; (c) caused by autotrophic sulfate reducers like those listed in Table 17.4 [not reported by either Butlin and Postgate (1952) or Ivanov (1968)]; (chl) bacterial chlorophyll.

Thiobacillus thioparus (Sokolova, 1962). The photosynthetic bacteria appeared to oxidize H_2S for the most part directly to sulfate. They were found to be of the type that is physiologically like *Chromatium thiosulfatophilum*. An average dark production of sulfur during the summer months has been estimated to be 150 kg (Ivanov, 1968).

Lake Eyre This Australian lake represents another locality in which evidence of syngenetic sulfur deposition has been noted. In shallow water on the southern bank of this lake, sulfur nodules have been found by Bonython (see Ivanov, 1968, pp. 146–150). The nodules are oval to spherical and usually covered with crusts of crystalline gypsum on the outside while being cavernous on the inside (Baas Becking and Kaplan, 1956). Their composition includes (in percent by dry weight): $CaSO_4$, 34.8; $S°$,

62–63; NaCl, 0.8; Fe_2O_3, 0.45; $CaCO_3$, 0.32; organic carbon, 1.8; and moisture, 7.54 (Baas Becking and Kaplan, 1956). Most nodules as well as the water and muds in the lake were found to contain active sulfate-reducing bacteria and thiobacilli (Baas Becking and Kaplan, 1956). Flagellates and cellulytic, methane-forming, and other bacteria were also found to abound. Baas Becking and Kaplan (1956) at first proposed that the nodules were forming in the present, with the photosynthetic flagellates providing organic carbon which cellulytic bacteria convert into a form utilizable by sulfate-reducing bacteria for the reduction of sulfate in gypsum in the surrounding sedimentary rock. The H_2S was then thought to be subjected to chemical and biological (thiobacilli) oxidation to native sulfur. The nodule structure was seen to result from the original dispersion of the gypsum in septaria in which gypsum is replaced by sulfur. A difficulty with this model, as Ivanov (1968) argues, is that the present oxidation-reduction potential of the ecosystem is $+280$ to $+340$ mV, which is too high for intense sulfate reduction. Sulfate reduction commonly requires a redox potential no higher than around -110 mV.

Radiodating of some nodules has shown them to be 19,600 old (Baas Becking and Kaplan, 1956). The $^{32}S/^{34}S$ ratio of the sulfur and of the gypsum of the outer crust of the nodules was found to be very similar (22.40–22.56 and 22.31–22.53, respectively), whereas that of the gypsum of the surrounding rock was found to be 22.11. This clearly suggests that the gypsum of the nodule crust is a secondary formation, biogenically produced through the oxidation of the sulfur in the nodules which was itself biogenically produced, but in Quaternary time, the age of the surrounding sedimentary deposit. It is possible, therefore, that the sulfur in the nodules derived from H_2S microbiologically generated from the primary gypsum deposit and released into the water and converted to sulfur by organisms like photosynthetic bacteria or *Beggiatoa* residing on the surface of some kind of concretion.

Solar Lake   An example of a lake in which sulfur is produced biogenically but not permanently deposited in the sediment is Solar Lake in the Sinai on the western shore of the Gulf of Aqaba. It is a small, hypersaline pond (7000 m^2 surface area, 4–6 m depth), which has undergone very extensive limnological investigations (Cohen et al., 1977a,b,c). It is tropical and has a chemocline (O_2/H_2S interface) and a thermocline, which is inverted in winter (i.e., the hypolimnion is warmer than the epilimnion). The chemocline, which is 0–10 cm thick and located at a depth of 2–4 m below the surface, undergoes diurnal migration over a distance of 20–30 cm. The chief cause of this migration is the activity of the cyanobacteria *Oscillatoria limnetica* and *Microcoleus* sp., whose growth extends from the epilimnion into the hypolimnion. Sulfate-reducing bacteria, including a *Desulfotomaculum acetoxidans* type, near the bottom in the anoxic hypolimnion generate H_2S from the SO_4^{2-} in the lake water. Some of this H_2S migrates upward to the chemocline. During the early daylight hours, H_2S in the chemocline and below it is oxidized to elemental sulfur by anaerobic photosynthesis of the cyanobacteria. After they have depleted the H_2S available to them, the cyanobacteria switch to aerobic photosynthesis, generating O_2. Thus during the daylight hours the chemocline gradually drops. After dark, when all photosynthesis by the cyanobacteria has ceased, H_2S generated by the sulfate reducers builds up in the chemocline, together with H_2S generated by the cyanobacteria when they reduce the S° they formed earlier with polyglucose they stored from oxygenic photosynthesis. Some of the S° is also reduced by bacteria such as *Desulfuromonas acetoxidans*. Thus during dark hours, the chemocline rises. The cycle is repeated with break of day. Some thiosulfate is found in the chemocline during daylight hours, primarily as a result of chemical oxidation of sulfide. This thiosulfate is reduced in the night hours by biological and chemical means. Sulfur thus undergoes a cyclical transformation in the lake such that elemental sulfur does not accumulate to a significant extent. The major driving force of the sulfur cycle is sunlight. [See Jørgensen et al., 1979a,b for further details of the sulfur cycle in this lake.]

Thermal Lakes and Springs   An example of syngenetic sulfur deposition in a thermal lake is Lake Ixpaca, Guatemala. This body of water is a crater lake that is supplied with H_2S from **solfataras** (fumarolic hot springs that yield sulfuretted waters) which discharge water at a temperature of 87–95°C (Ljunggren, 1960). The water in the lake has a temperature in the range 29–32°C. The H_2S concentration of the lake water was reported to be 0.10–0.18 g L^{-1}. Some of it is oxidized to native sulfur, rendering the water of the lake opalescent. A portion of this sulfur settles out and is incorporated into the sediment. Another, smaller portion of the sulfur is oxidized to sulfuric acid, acidifying the lake to a pH of 2.27. The sulfate content of the lake water was reported to range from 0.46 to 1.17 g L^{-1}. The sulfuric acid in the lake water is very corrosive. It decomposes igneous minerals such as pyroxenes and feldspars into clay minerals (e.g., pickingerite). Ljunggren (1960) found an extensive presence of *Beggiatoa* in the waters associated with the sediments of this lake. He implicated this microorganism in the conversion of H_2S into native sulfur. He apparently did not determine if the sulfuric acid might not also be biogenically formed, at least in part, as it is in some hot springs in Yellowstone National Park in the U.S.A. (Brock, 1978; Ehrlich and Schoen, 1967; Schoen and Ehrlich, 1968; Schoen and Rye, 1970). Nor did Ljunggren consider the possibility that other organisms in addition to *Beggiatoa* might be involved in the genesis of S° from H_2S. More recent studies of solfataras elsewhere have led to the discovery of a number of thermophiles, mostly archaebacteria such as *Acidianus* (Brierley and Brierley, 1973; Segerer et al., 1986), *Sulfolobus* (Brock et al., 1972), *Thermoproteus* (Zillig et al., 1981), *Thermofilum pendens* (Zillig et al., 1982a), *Desulfurococcus* (Zillig et al., 1982b), but also occasional eubacteria, such as *Thermothrix thiopara* (Caldwell et al., 1976), that have a capacity either to oxidize H_2S, S°, or thiosulfate, or to reduce S°.

Studies of some hot springs in Yellowstone National Park (Brock, 1978) showed that in most the H_2S emitted in their discharge appears to be chemically oxidized to native sulfur. A major exception is Mammoth Hot Springs in which H_2S is biochemically oxidized to sulfur as deduced from sulfur isotope fractionation studies. Physiological evidence for bacterial H_2S oxidation was also obtained at Boulder Springs in the Park. Unlike most of the springs, its water has a pH in the range of 8–9. Oxidation here can occur at a temperature as high as 93°C (80–90°C, optimum). The bacteria are mixotrophic in this case, being able to use H_2S or other reduced sulfur compounds as energy source and organic matter as carbon source (Brock et al., 1971). Further study of the oxidation of elemental sulfur to sulfuric acid in the acid hot springs in the Park revealed that *Thiobacillus thiooxidans* was responsible at temperatures below 55°C and *Sulfolobus acidocaldarius* at temperatures between 55 and 85°C (Fliermans and Brock, 1972; Mosser et al., 1973). Almost all sulfur oxidation in the hot, acid springs and hot, acid soils was biochemical since sulfur appeared to be stable in the absence of bacterial activity (Mosser et al., 1973).

S. acidocaldarius consists of spherical cells that form frequent lobes and lack peptidoglycan in their cell wall (Brierley, 1966; Brock et al.; 1972) (Fig. 14.5). The organism is acidophilic (opt. pH 2–3; range pH 0.9–5.8) and thermophilic (temperature optimum 70–75°; range 55–80°). It has a guanine-cytosine ratio 60–68 mol%. In growing cultures in the laboratory, the growth rate parallels the oxidation rate on elemental sulfur (Shivvers and Brock, 1973). For oxidation, the organisms attach themselves to sulfur crystals. The presence of yeast extract in the medium was found to partially inhibit sulfur oxidation but not growth. The growth rates of *S. acidocaldarius* in several hot springs in Yellowstone National Park exhibit steady-state doubling times in the order of 10–20 h in the water of small springs having volumes of 20–2,000 liters, and on the order of 30 days in large springs with 1×10^6-liter volumes. Exponential doubling times measured in the water of artificially drained springs are on the order of a few hours (Mosser et al., 1974).

In effluent channels from alkaline hot springs in Yellowstone National Park where the temperature does not exceed about 70°C, a bacterium called *Chloroflexus aurantiacus* was discovered (Brock, 1978; Pierson and Castenholz, 1974). It is characterized as a gliding, filamentous (0.5–0.7 μm in width, variable in length), phototrophic bacterium with a tendency to form orange mats below, and to a lesser extent above thin layers of cyanobacteria such as *Synechococcus* (Doemel and Brock, 1974, 1977). Its photosynthetic pigments include bacteriochlorophylls a and c and beta- and gamma-carotene. The pigments occur in *Chlorobium* vesicles. Anaerobically, the organism can grow photoautotrophically in the presence of sulfide and bicarbonate (Madigan and Brock, 1975), but it can also grow photoheterotrophically with yeast extract and certain other organic supplements. Aerobically, the organism is capable of heterotrophic growth in the dark. Although showing some physiological resemblance to Rhodospirillaceae (purple nonsulfur bacteria) and even greater resemblance to the *Chlorobiaceae* (green sulfur bacteria), phylogenetically it is related to neither (Brock and Madigan, 1988).

The mats formed by *Chloroflexus* in some of the hot springs in Yellowstone National Park may be models for the formation of ancient stromatolites. They often incorporate detrital silica in the form of siliceous sinter from the geyser basins and in time are transformed into structures recognizable as stromatolites.

Sulfate in the mats can be reduced to sulfide by sulfate reducers below the upper 3 mm, and the sulfide can be converted to elemental sulfur by *Chloroflexus* in the mats (Doemel and Brock, 1976). This is another example where below 70°C at least some elemental sulfur in hot springs or their effluent may be of biogenic origin.

Examples of Epigenetic Sulfur Deposits

Sicilian Sulfur Deposits   An example of epigenetic sulfur deposition in which microbes must have played a role is found on the volcanic Mediterranean island of Sicily. Isotopic study (Jensen, 1962) showed that the sulfur in these sulfur deposits is significantly enriched in ^{32}S relative to the associated sulfate. This finding signifies that the sulfur could not have originated from volcanic activity but must have been the result of microbial sulfate reduction in evaporite deposits that originated from the Mediterranean Sea. The biological agents must have been a consortium of dissimilatory sulfate reducing bacteria, which reduced evaporite sulfate to sulfide and chemosynthetic sulfide oxidizing bacteria that formed sulfur from the sulfide. Abiotic sulfide oxidation could also have contributed to sulfur formation. Organic carbon if used in the microbial reduction of the sulfate came presumably from organic detritus in the sediment (algal or other remains).

Salt Domes   Another example of biogenic, native sulfur of epigenetic origin in a sedimentary environment is that associated with salt domes (Fig. 17.5) as found on the Gulf Coast of the United States and Mexico (northern and western shores of the Gulf Coast, including those of Texas, Louisiana, and Mexico) (see, e.g., Ivanov, 1968, pp. 92ff; Matinez, 1991). Such salt domes reside directly over a central plug consisting of 90–95% rock salt [NaCl], 5–10% anhydrite [$CaSO_4$] and traces of dolomite [$CaMg(CO_3)_2$], barite [$BaSO_4$], and celestite [$SrSO_4$]. Petroleum may be entrapped in peripheral deformations. The domes consist mainly of anhydrite topped by calcite which may have exploitable petroleum associated with it. Between the calcite and anhydrite exists a zone containing gypsum [$CaSO_4 \cdot 2H_2O$], calcite, and anhydrite relicts. Sulfur is associated with the calcite in this intermediate zone. The salt domes originated from evaporites which formed in a period between the late Paleozoic (230–280 million years ago) to the Jurassic (Middle Mesozoic, 135–181 million years ago). A current theory (see Strahler, 1977) proposes that the salt domes on the Gulf Coast began as beds of evaporite along the continental margins of the newly emergent Atlantic Ocean about 180 million years ago. The evaporite derived from hypersaline waters with the aid of heat emanating from un-

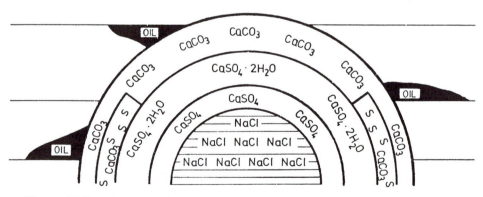

Figure 17.5   Diagrammatic representation of a salt dome. (After Ivanov, 1968.)

derlying magma reservoirs in these tectonically active areas. As the ocean basin broadened due to continental drift, and as the continental margins became more defined, turbidity currents began to bury the evaporite beds under ever-thicker layers of sediment. Ultimately, these sediment layers became so heavy that they forced portions of the evaporite, which has plastic properties, upward as fingerlike salt plugs through ever younger sediment strata. As these plugs intruded into the groundwater zone, they lost their more water-soluble constituents, particularly the rock salt, leaving behind relatively insoluble constituents, especially anhydrite, which became the cap rock. In time, some of the anhydrite was converted to more soluble gypsum and some was dissolved away. At that point, bacterial sulfate reduction is thought to have begun, lasting perhaps for a period of 1 million years. The active bacteria were most likely introduced into the dome structures from the native flora of the groundwater. The organic carbon needed for bacterial sulfate reduction is thought to have derived from adjacent petroleum deposits.

When the biological contribution to sulfur formation in the salt domes was first recognized (Jones et al., 1956), the only known sulfate reducer was *Desulfovibrio desulfuricans*, which uses lactate, pyruvate, and in most instances malate as energy sources, which it oxidized to acetate with sulfate as electron acceptor. To use carbon compounds associated with petroleum, it would have had to depend on other bacteria that could transform appropriate petroleum hydrocarbons, preferable anaerobically, into the carbon substrates usable by *Desulfovibrio*.

A more recent observation has indicated the existence of a sulfate reducer that can use methane (CH_4) as a source of energy (Panganiban and Hanson, 1976; Panganiban et al., 1979). Methane is a major gaseous constituent associated with petroleum deposits.

The most recent studies have revealed the existence of sulfate reducers with an ability to use some short-chain, saturated aliphatic (including chain lengths of $C_8–C_{16}$) or aromatic hydrocarbons, or heterocyclic compounds, many of which they can mineralize (see section 17.6; also Aeckersberg et al., 1991; Rueter et al., 1994). Many sulfate reducers have also been shown to be able to use H_2 as energy source. H_2 occurs in detectable amounts in oil wells. Thus, it is not necessary to postulate the past existence of a complex assmeblage of anaerobic, fermentative bacteria that converted petroleum hydrocarbons into energy sources for sulfate-reducing bacteria in salt domes.

Whichever sulfate reducing bacteria were active in the salt domes, they produced not only H_2S but also CO_2. The CO_2 arose from the mineralization of the organic carbon consumed for energy conservation in sulfate reduction. The H_2S was subsequently oxidized biologically and/or chemically to native sulfur, whereas the CO_2 was extensively precipitated as carbonate (secondary calcite) with the calcium from the anhydrite or gypsum attacked by the sulfate-reducing bacteria.

$$CaSO_4 + 2(CH_2O) \Rightarrow CaS + 2CO_2 + 2H_2O \tag{17.45}$$

$$CaS + CO_2 + H_2O \Rightarrow CaCO_3 + H_2S \tag{17.46}$$

$$H_2S + \tfrac{1}{2}O_2 \Rightarrow S° + H_2O \tag{17.47}$$

Mineralogical and isotopic study has shown that the sulfur and secondary calcite are physically associated in the cap rock (paragenetic). Their isotopic enrichment values indicate a biological origin (Jones et al., 1956; Thode et al., 1954). The sulfur exhibits enrichment in respect to ^{32}S and the secondary calcite in respect to ^{12}C (see discussion by Ivanov, 1968).

Gaurdak Sulfur Deposit   Epigenetic sulfur deposition in a mode somewhat similar to that associated with the salt domes in the United States took place in the Gaurdak Deposit of Turkmenistan (Ivanov, 1968). This deposit resides in rock of the Upper Jurassic in age and was probably emplaced in the Quaternary as plutonic waters picked up organic carbon from the Kugitang Suite containing bituminous limestone and sulfate from the anhydrite-carbonate rocks of the Gaurdak Suite. Sulfate-reducing bacteria which entered the plutonic waters attacked the the sulfate and reduced it to H_2S with the help of reduced carbon derived from the bituminous material. *Thiobacillus thioparus* oxidized the H_2S to S° at the interface where the plutonic water encountered infiltrating oxygenated surface water. Where the sulfur presently encounters oxygenated groundwater, intense biooxidation of sulfur to sulfuric acid has been noted, causing transformation of secondary calcite, formed during sulfate reduction in the initial phase of sulfur genesis, into secondary gypsum. The bacteria *T. thioparus* and *T. thiooxidans* have been found in significant numbers in sulfuretted waters in the sulfur deposits with paragenetic (secondary) calcite and in acidic sulfur deposits with secondary gypsum, respectively. Sulfur appears, therefore, to be deposited and degraded in the Gaurdak formation at the present time.

Shor-Su Sulfur Deposit   Another example of epigenetic, microbial sulfur deposition is the Shor-Su Deposit in the northern foothills of the Altai Range in the southeast corner of the West Siberian Plain. Here an extensive, folded sedimentary formation of lagoonal origin and mainly of Paleocene and Cretaceous age contains major sulfuretted regions in lower

Paleocene strata (Bukhara and Sazuk) of the second anticline and to a lesser extent in Quaternary conglomerates (Fig. 17.6) (see Ivanov, 1968, pp. 33–34). The sulfur of the main deposits occurs in heavily broken rock surrounded by gypsified rock. It contains some relict gypsum lenses. It is enclosed in a variety of cavernous rock and associated with calcite and celestite in the Bukhara stratum and in cavities and slitlike caves in the Suzak stratum. Petroleum and natural gas deposits are associated with the fourth anticline. This structure is hydraulically connected with the second anticline, which contains most of the sulfur. One basis for the claim of hydraulic connection between the two anticlines is that their pore waters are chemically very similar in composition. Sulfate-reducing bacteria occur in the plutonic waters that flow through the permeable strata from the fourth to the second anticline. It is believed that these bacteria have been reducing the sulfate that the plutonic waters picked up from the dissolution of some of the gypsum and anhydrite in the sur-

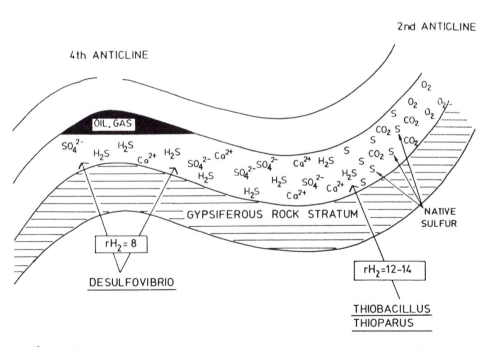

Figure 17.6 Diagrammatic representation of essential features of the Shor-Su deposit. (Based on a description by Ivanov, 1968.)

rounding rock. The bacteria are presumed to have been using petroleum hydrocarbons or derivatives from them as a source of energy (reducing power) and carbon for this process. The presence of sulfate-reducing bacteria has been reported in the waters of the second anticline and in any rock in which sulfur occurs (Ivanov, 1968). These bacteria were demonstrated to be able to reduce sulfate under in situ conditions at a measurable rate (0.009–0.179 mg of H_2S liter^{-1} day^{-1}). Native sulfur has been forming where rising plutonic water has been mixing with downward-seeping oxygenated surface water. In this zone of mixing of the two waters, *T. thioparus* was detected and shown to oxidize H_2S from the plutonic water to native sulfur.

Measurements have shown that sulfate reduction predominates where plutonic waters carry sulfate derived from surrounding gypsiferous rock and organic matter derived from associated petroleum. The waters at these sites in the deposit have an rH_2 that often is below 8, indicating strong reducing conditions. The H_2S is transported by the moving plutonic waters to a region in the second anticline where it encounters aerated surface waters. The waters here have an rH_2 of around 12–14 (16.5 maximum). In this environment, *T. thioparus* is favored. It causes conversion of H_2S into S°. Where the rH_2 exceeds 16.5, owing to extensive exposure to surface water as in the outcroppings of the western conglomerate of the Shor-Su, the sulfur is undergoing extensive oxidation by *T. thiooxidans*. The pH is found to drop from neutrality to less than 1 where the bacteria are most active. The sulfur in the main strata began to be laid down in the Quaternary, according to Ivanov (1968). Deposition continues to the present day. For this reason, events in the geological past can be reconstructed from current observations of bacterial distribution and activity in the Quaternary strata of the Shor-Su (Ivanov, 1968).

Kara Kum Sulfur Deposit   Spatially, a somewhat different mechanism of epigenetic sulfur deposition has been recognized in the Kara Kum Deposit north of Ashkhabad in Turkmenistan (Ivanov, 1968). Sulfate-reducing bacteria and H_2S-oxidizing bacteria have also been playingd a role in sulfur formation at this site. However, sulfate reduction has been taking place in a different stratum from that involving sulfur deposition, implying that these two activities are spatially separated. The H_2S has been transported to another site before conversion to sulfur. Hence, paragenetic (secondary) calcite is not found associated with the sulfur in this deposit, and, consequently, sulfuric acid formed at sites of outcropping of the sulfur deposit cannot form gypsum but reacts with sandstone, liberating aluminum and iron, which are reprecipitated as oxides in a more neutral environment.

17.12 MICROBIAL ROLE IN THE SULFUR CYCLE

As the foregoing discussion shows, microbes play an important role in inorganic as well as organic sulfur transformations. Figure 17.7 shows how these various biological interactions fit into the sulfur cycle in soil, sediment and aquatic environments. Although some of these transformations such as the aerobic oxidation of H_2S or $S°$ may proceed partly by an abiotic route, albeit often significantly more slowly than by a biotic route, at least two other transformations, the anaerobic oxidation of H_2S or $S°$ to sulfuric acid and the reduction of sulfate to H_2S do not proceed readily abiotically at atmospheric pressure in the temperature range that prevails at the Earth's surface. Sulfate reduction is now recognized to be an important mechanism of anaerobic mineralization of organic carbon in

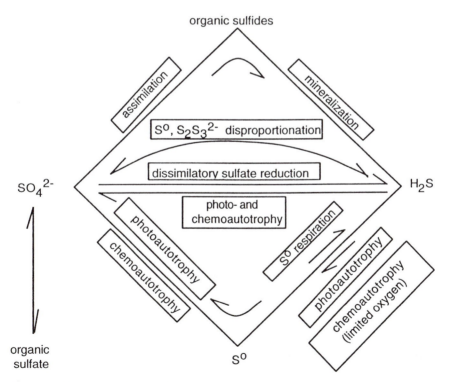

Figure 17.7 The sulfur cycle.

anaerobic estuarine and other coastal environments where plentiful sulfate is available from seawater (Skyring, 1987). Geochemically, sulfur-oxidizing and -reducing bacteria are important catalysts in the sulfur cycle in the biosphere.

17.13 SUMMARY

Sulfur, which occurs in organic and inorganic form in nature, is essential to life. Different organisms may assimilate it in organic or inorganic form. Plants and many microbes normally take it up as sulfate. Microbes are important in mineralizing organic sulfur compounds in soil and aqueous environments. The biochemistry of organic sulfur mineralization, as well as the synthesis of organic sulfur compounds, has been studied in some detail.

Inorganic sulfur may exist in various oxidation states in nature, most commonly as sulfide (-2), elemental sulfur (0), and sulfate ($+6$). Thiosulfate and tetrathionate, each with sulfur in mixed oxidation states, may also occur in significant amounts in some environments. Some microbes in soil and water play an important role in the interconversion of these oxidation states. These include several different eubacteria (even certain cyanobacteria under special conditions) and archaebacteria. Among bacteria that oxidize reduce forms of sulfur are chemolithotrophs, anoxygenic and oxygenic (cyanobacterial) photolithotrophs, mixotrophs, and heterotrophs. Most chemolithotrophs and mixotrophs use oxygen as oxidant, but a few chemolithotrophs can substitute nitrate or ferric iron when oxygen is absent. Some chemolithotrophs like $T.$ *thioparus* can oxidize H_2S to $S°$ under partially reduced conditions, but they form H_2SO_4 under fully oxidizing conditions. The anoxygenic photolithotrophic bacteria (purple and green bacteria) oxidize H_2S to $S°$ or H_2SO_4 to generate reducing power for CO_2 fixation and/or ATP. Certain cyanobacteria oxidize H_2S to $S°$ in the absence of oxygen for generating energy and reducing power for CO_2 fixation. Various chemolithotrophs and mixotrophs can oxidize $S°$ to H_2SO_4 aerobically in neutral or acid environments. Sulfur oxidation has been noted in mesophilic and thermophilic environments at temperatures exceeding 100°C. Thiosulfate is readily oxidized by some chemolithotrophs, mixotrophs, and heterotrophs. Some marine pseudomonads have been shown to use it as a supplemental energy source, oxidizing it to tetrathionate. Some bacteria can conserve energy by disproportionating elemental sulfur, dithionite, sulfite, or thiosulfate under anaerobic conditions to sulfide and sulfate.

Oxidized forms of elemental sulfur may be reduced by various microorganisms. Elemental sulfur is reduced to H_2S with or without energy conservation by some anaerobic eubacteria and archaebacteria. Among the eubacteria that conserve energy are *Desulfuromonas acetoxidans*, *Desulfovibrio gigas*, and some other sulfate-reducing bacteria. Two fungi, *Rhodotorula* and *Trichosporon*, have also been found to be able to reduce S° to H_2S, but most probably without energy conservation.

Sulfate may be reduced in sulfate respiration (dissimilatory sulfate reduction) by a number of specialized bacteria. Most known species are eubacteria, but at least two archaebacterial species have also been identified. This microbial activity is of major importance geologically because under natural conditions on the Earth's surface, sulfate cannot be reduced by purely chemical means because of the high activation-energy requirement of the process. Sulfate is reduced aerobically by various microbes and plants, but only in small amounts without any extracellular accumulation of H_2S (assimilatory sulfate reduction). The mechanisms of dissimilatory and assimilatory sulfate reduction differ biochemically.

Some reducers and oxidizers of sulfur and its compounds can distinguish between ^{32}S and ^{34}S and can bring about isotope fractionation. Geologically, this is useful in determining whether ancient sulfur deposits were biogenically or abiogenically formed.

Contemporary biogenic sulfur deposition involving sulfate-reducing and aerobic and anaerobic sulfide-oxidizing bacteria have been identified in several lacustrine environments. These represent syngenetic deposits. Bacterial oxidation of elemental sulfur to sulfuric acid in certain hot springs has also been reported.

Ancient epigenetic sulfur deposits of microbial origin have been identified in salt domes and other geologic formations associated with hydrocarbon (petroleum) deposits in various parts of the world. The sulfur in these instances arose from bacterial reduction of sulfate derived from anhydrite or gypsum followed by bacterial oxidation under partially reduced conditions to elemental sulfur. On full exposure to air, some of the sulfur is presently being oxidized by bacteria to sulfuric acid.

Less spectacular oxidative and reductive transformations of sulfur occur in soil, where they play an important role in maintenance of soil fertility.

REFERENCES

Adams, C. A., G. M. Warnes, and D. J. D. Nicholas. 1971. A sulfite-dependent nitrate reductase frm *Thiobacillus denitrificans* Biochim. Biophys. Acta 235:398–406.
Aeckersberg, F., F. Bak, and F. Widdel. 1991. Anaerobic oxidation of

saturated hydrocarbons to CO_2 by a new type of sulfate-reducing bacterium. Arch. Microbiol. 156:5–14.

Akagi, J. M., M. Chan, and V. Adams. 1974. Observations on the bisulfite reductase (P582) isolated from *Desulfotomaculum nigrificans*. J. Bacteriol. 120:240–244.

Aleem, M. I. H. 1965. Thiosulfate oxidation and electron transport in *Thiobacillus novellus*. J. Bacteriol. 90:95–101.

Aminuddin, M. 1980. Substrate level versus oxidative phosphorylation in the generation of ATP in *Thiobacillus denitrificans*. Arch. Microbiol. 128:19–25.

Aminuddin, M., and D. J. D. Nicholas. 1973. Sulfide oxidation linked to the reduction of nitrate and nitrite in *Thiobacillus denitrificans*. Biochim. Biophys. Acta 325:81–93.

Aminuddin, M., and D. J. D. Nicholas. 1974a. An AMP-independent sulfite oxidase from *Thiobacillus denitrificans*: Purification and properties. J. Gen. Microbiol. 82:103–113.

Aminuddin, M., and D. J. D. Nicholas. 1974b. Electron transfer during sulfide and sulfite oxidation in *Thiobacillus denitrificans*. J. Gen. Microbiol. 82:115–123.

Arp, A. J., and J. J Childress. 1983. Sulfide binding by the blood of the hydrothermal vent tube worm *Riftia pachyptila*. Science 219: 295–297.

Baas Becking, L. G. M., and I. R. Kaplan. 1956. The microbiological origin of the sulfur nodules of Lake Eyre. Trans. Roy. Soc. S. Aust. 79–62–65.

Badziong, W., and R. K. Thauer. 1978. Growth yields and growth rates of *Desulfovibrio vulgaris* (Marburg) growing on hydrogen plus sulfate and hydrogen plus thiosulfate as sole energy sources. Arch. Microbiol. 117:209–214.

Badziong, W., R. K. Thauer, and J. G. Zeikus. 1978. Isolation and characterization of *Desulfovibrio* growing on hydrogen plus sulfate as the sole energy source. Arch. Microbiol. 116:41–49.

Bak, F., and H. Cypionka. 1987. A novel type of energy metabolism involving fermentation of inorganic sulfur compounds. Nature (Lond.) 326:891–892.

Bak, F., and N. Pfennig. 1987. Chemolithotrophic growth of *Desulfovibrio sulfodismutans* sp. nov. by disproportionation of inorganic sulfur compounds. Arch. Microbiol. 147:184–189.

Bak, F., and F. Widdel. 1986a. Anaerobic degradation of indolic compounds by sulfate-reducing enrichment cultures, and description of *Desulfobacterium indolicum* gen. nov., sp. nov. Arch. Microbiol. 46:170–176.

Bak, F., and F. Widdel. 1986b. Anaerobic degradation of phenol and phenol derivatives by *Desulfobacterium phenolicum* sp. nov. Arch. Microbiol. 46:177–180.

Belkin, S., C. O. Wirsen, and H. W. Jannasch. 1986. A new sulfur-reducing, extremely thermophilic eubacterium from a submarine thermal vent. Appl. Environ. Microbiol. 51:1180–1185.

Beudeker, R. F., G. C. Cannon, J. G. Kuenen, and J. M. Shively. 1980. Relations between D-ribulose-1,5-bisphosphate carboxylase, carboxysomes and CO_2 fixing capacity in the obligate chemolithotroph *Thiobacillus neapolitanus* grown under different limitations in the chemostat. Arch. Microbiol. 124:185–189.

Beudeker, R. F., G. A. Codd, and J. G. Kuenen. 1981. Quantification and intracellular distribution of ribulose-1,5-bisphosphate carboxylase in *Thiobacillus neapolitanus*, as related to possible functions of carboxysomes. Arch. Microbiol. 129:361–367.

Biebl, H., and N. Pfennig. 1977. Growth of sulfate-reducing bacteria with sulfur as electron acceptor. Arch. Microbiol. 112:115–117.

Boopathy, R., and L. Daniels. 1991. Isolation and characterization of a furfural degrading sulfate-reducing bacterium from an anaerobic digester. Curr. Microbiol. 23:327–332.

Bowen, H. J. M. 1979. Environmental Chemistry of the Elements. Academic Press, London.

Bowen, T. J., F. C. Happold, and B. F. Taylor. 1966. Studies on adenosine 5'-phosphosulfate reductase from *Thiobacillus denitrificans*. Biochim. Biophys. Acta 118:566–576.

Brandis, A., and R. K. Thauer. 1981. Growth of *Desulfovibrio* species on hydrogen and sulfate as sole energy source. J. Gen. Microbiol. 126:249–252.

Brandis-Heep, A., N. A. Gebhardt, R. K. Thauer, F. Widdel, and N. Pfennig. 1983. Anaerobic acetate oxidation to CO_2 by *Desulfobacter postgatei*. 1. Demonstration of all enzymes reqired for the operation of the citric acid cycle. Arch. Microbiol. 136:222–229.

Brannan, D. K., and D. E. Caldwell. 1980. *Thermothrix thiopara*: growth and metabolism of a newly isolated thermophile capable of oxidizing sulfur and sulfur compounds. Appl. Environ. Microbiol. 40:211–216.

Braun, M., and H. Stolp. 1985. Degradation of methanol by a sulfate reducing bacterium. Arch. Microbiol. 142:77–60.

Brierley, C. L., and J. A. Brierley. 1973. A chemoautotrophic and thermophilic microorganism isolated from an acid hot spring. Can. J. Microbiol. 19:183–188.

Brock, T. D. 1978. Thermophilic Microorganisms and Life at High Temperatures. Springer-Verlag, New York.

Brock, T. D., and M. T. Madigan. 1988. Biology of Microorganisms. 5th ed. Prentice-Hall, Englewood Cliffs, NJ.

Brock, T. D., M. L. Brock, T. L. Bott, and M. R. Edwards. 1971. Microbial life at 90°C: the sulfur bacteria of Boulder Spring. J. Bacteriol. 107:303–314.

Brock, T. D., K. M. Brock, R. T. Belly, and R. L. Weiss. 1972. *Sulfolobus*: a new genus of sulfur-oxidizing bacteria living at low pH and high temperature. Arch. Microbiol. 84:54–68.

Bromfield, S. M. 1953. Sulfate reduction in partially sterilized soil exposed to air. J. Gen. Microbiol. 8:378–390.

Bryant, R. D., K. M. McGroarty, J. W. Costerton, and E. J. Laishley. 1983. Isolation and characterization of a new acidophilic *Thiobacillus* species (*T. albertis*). Can. J. Microbiol. 29:1159–1170.

Buchanan, R. E., and N. E. Gibbons, eds. 1974. Bergey's Manual of Determinative Bacteriology. 8th ed. Williams & Wilkins, Baltimore.

Butlin, K. 1953. The bacterial sulfur cycle. Research 6:184–191.

Burggraf, S., H. W. Jannasch, B. Nicolaus, and K. O. Stetter. 1990. *Archeoglobus profundus* sp. nov., represents a new species within the sulfate-reducing archaebacteria. Syst. Appl. Microbiol. 13:24–28.

Butlin, K. R., and J. R. Postgate. 1952. The microbiological formation of sulfur in the Cyrenaikan Lakes. *In*: Biology of Deserts. Institute of Biology. London.

Canfield, D. E., and D. J. Des Marais. 1991. Aerobic sulfate reduction in microbial mats. Science 251:1471–1473.

Caldwell, D. E., S. J. Caldwell, and J. P. Laycock. 1976. *Thermothrix thiopara* gen et sp. nov., a facultatively anaerobic, facultative chemolithotrophic living at neutral pH and high temperatures. Can. J. Microbiol. 22:1509–1517.

Cavanaugh, C. M. 1983. Symbiotic chemoautotrophic bacteria in marine invertebrates from sulfide-rich habitats. Nature (Lond.) 302:58–61.

Cavanaugh, C. M., S. L. Gardiner, M. L. Jones, H. W. Jannasch, J. B. Waterbury. 1981. Prokarkyotic cells in the hydrothermal vent tube worm *Riftia pachyptila* Jones: Possible chemoautotrophic symbionts. Science 213:340–342.

Chambers, L. A., and P. A. Trudinger. 1975. Are thiosulfate and trithionate intermediates in dissimilatory sulfate reduction? J. Bacteriol. 123:36–40.

Charles, A. M., and I. Suzuki. 1966a Mechanism of thiosulfate oxidation by *Thiobacillus novellus*. Biochim. Biophys. Acta 510–521.

Claassen, P. A. M., M. H. M.J. van den Heuvel, and A. J. B. Zehnder. 1987. Enzyme profiles of *Thiobacillus versutus* after aerobic and denitrifying growth: Regulation of isocitrate lyase. Arch. Microbiol. 147:30–36.

Cohen, Y., E. Padan, and M. Shilo. 1975. Facultative anoxygenic photosynthesis in the cyanobacterium *Oscillatoria limnetica*. J. Bacteriol. 123:855–861.

Cohen, Y., W. E. Krumbein, M. Goldberg, and M. Shilo. 1977a. Solar Lake (Sinai). 1. Physical and chemical limnology. Limnol. Oceanogr. 22:597–608.

Cohen, Y., W. E. Krumbein, and M. Shilo. 1977b. Solar Lake (Sinai). 2. Distribution of photosynthetic microorganisms and primary production. Limnol. Oceanogr. 22:609–620.

Cohen, Y., W. E. Krumbein, and M. Shilo. Limnol. Oceanogr. 1977c. Solar Lake (Sinai). 4. Stromatolitic cyanobacterial mats. Limnol. Oceanogr. 22:635–656.

Corbett, C. M., and W. J. Ingledew. 1987. Is Fe^{3+}/Fe^{2+} cycling an intermediate in sulfur oxidation by *Thiobacillus ferrooxidans*? FEMS Microbiol. Lett. 41:1–6.

Cypionka, H., and N. Pfennig. 1986. Growth yields of *Desulfotomaculum orientis* with hydrogen in chemostat culture. Arch. Microbiol. 143: 396–399.

Daalsgaard, T., and F. Bak. 1992. Effect of acetylene on nitrous oxide reduction and sulfide oxidation in batch and gradient cultures of *Thiobacillus denitrificans*. Appl. Environ. Microbiol. 58:1601–1608.

Davis, E. A., and E. J. Johnson. 1967. Phosphorylation coupled to the oxidation of sulfide and 2-mercaptoethanol in extracts of *Thiobacillus thioparus*. Can. J. Microbiol. 13:873–884.

DeWeerd, K. A., L. Mandelco, R. S. Tanner, C. R. Woese, and J. M. Suflita. 1990. *Desulfomonile tiedjei* gen. nov. spec. nov., a novel anarobic dehalogenating, sulfate-reducing bacterium. Arch. Microbiol. 154:23–30.

DeWeerd, K. A., F. Concannon, and J. M. Suflita. 1991. Relationship between hydrogen consumption, dehalogeneation, and the reduction of sulfur oxyanions by *Desulfomonile tiedjei*. Appl. Environ. Microbiol. 57:1929–1934.

Doemel, W. N., and T. D. Brock. 1974. Bacterial stromatolites: origin of laminations. Science 184:1083–1085.

Doemel, W. N., and T. D. Brock. 1976. Vertical distribution of sulfur species in benthic algal mats. Limnol. Oceanogr. 21:237–244.

Doemel, W. N., and T. D. Brock. 1977. Structrure, growth and decomposition of laminated algal-bacterial mats in alkaline hotsprings. Appl. Environ. Microbiol. 34:433–452.

Drake, H. L., and J. M. Akagi. 1978. Dissimilatory reduction of bisulfite by *Desulfovibrio vulgaris*. J. Bacteriol. 136:916–923.

Egorova, A. A., and Z. P. Deryugina. 1963. The spore forming thermophilic thiobacterium: *Thiobacillus thermophilica* Imschenetskii nov. spec. Mikrobiologiya 32:439–446.

Ehrlich, G. G., and R. Schoen. 1967. Possible role of sulfur-oxidizing bacteria in surficial acid alteration near hot springs. U.S. Geol. Survey Prof. Paper 575C, pp. C110–C112.

Ehrlich, H. L., and S. I. Fox. 1967. Copper sulfide precipitation by yeasts from acid mine-waters. Appl. Microbiol. 15:135–139.

Elsgaard, L., and B. B. Jorgensen. 1992. Anoxic transformations of radiolabeled hydrogen sulfide in marine and freswater sediments. Geochim. Cosmochim. Acta 56:2425–2435.

Fauque, G., A. R. Lino, M. Czechowski, L. Kang, D. V. DerVartanian, J. J. G. Moura, J. LeGall, and I. Moura. 1990. Purification and characterization of bisulfite reductase (desulfofuscidin) from *Desulfovibrio thermophilus* and its complexes with exogenous lignads. Biochim. Biophys. Acta 1040:112–118.

Fauque, G., J. LeGall, and L. L. Barton. 1991. Sulfate-reducing and sulfur-reducing bacteria. *In*: J.M. Shively and L.L. Barton, eds. Variations in Autotrophic Life. Academic Press, London, pp. 271–337.

Felbeck, H. 1981. Chemoautotrophic potential of the hydrothermal vent tube worm, *Riftia pachyptila* Jones (Vestmentifera). Science 213:336–338.

Felbeck, H., J.J. Childress, and G. N. Somero. 1981. Calvin-Benson cycle and sulfide oxidation enzymes in animals from sulfide-rich habitats. Nature (Lond.) 293:291–293.

Fliermans, C. B., and T. D. Brock. 1972. Ecology of sulfur-oxidizing bacteria in hot acid soils. J. Bacteriol. 111:343–350.

Freney, J. R. 1967. Sulfur-containing organics. *In*: A. D. McLaren and G. H. Petersen, eds. Soil Biochemistry. Marcel Dekker, New York, pp. 229–259.

Fruend, C., and Y. Cohen. 1992. Diurnal cycles of sulfate reduction under oxic conditions in cyanobacterial mats. Appl. Environ. Microbiol. 58:70–77.

Gebhardt, N. A., D. Linder, and R. K. Thauer. 1983. Anaerobic oxidation of CO_2 by *Desulfobacter postgatei*. 2. Evidence from ^{14}C-labelling studies for the operation of the citric acid cycle. Arch. Microbiol. 136:230–233.

Gebhardt, N. A., R. K. Thauer, D. Linder, P-M. Kaulfers, and N. Pfennig. 1985. Mechanism of acetate oxidation to CO_2 with elemental sulfur in *Desulfuromonas acetoxidans*. Arch. Microbiol. 141:392–398.

Goldschmidt, V. M. 1954. Geochemistry. Clarendon Press. Oxford, U.K., pp. 621–642.

Golovacheva, R.S., and G. I. Karavaiko. 1978. *Sulfobacillus*, a new genus of thermophilic sporeforming bacteria. Mikrobiologiya 47:815–822 (Engl. transl. pp. 658–665).

Grayston, S.J., and M. Wainwright. 1988. Sulfur oxidation by soil fungi including species of mycorrhizae and wood-rotting basidiomycetes. FEMS Microbiol. Ecol. 53:1–8.

Guittonneau, G. 1927. Sur l'oxydation microbienne du soufre au cours de l'ammonisation. C.R. Acad. Sci. (Paris) 184:45–46.

Guittonneau, G., and J. Keilling. 1927. Sur la solubilisation du soufre élémentaire et la formation des hyposulfides dans une terre riche en azote organique. C.R. Acad. Sci. (Paris) 184:898–901.

Hansen, T. A., and H. van Gemerden. 1972. Sulfide utilization by purple sulfur bacteria. Arch. Microbiol. 86:49–56.

Hansen, T. A., and H. Veldkamp. 1973. *Rhodopseudomonas sulfidophila* nov. spec a new species of the purple nonsulfur bacteria. Arch. Mikrobiol. 92:45–58.

Harrison, A. G., and H. Thode. 1957. The kinetic isotope effect in the chemical reduction of sulfate. Trans. Faraday Soc. 53:1–4.

Harrison, A. G., and H. Thode. 1958. Mechanism of the bacterial reduction of sulfate from isotope fractionation studies. Trans. Faraday Soc. 54:84–92.

Harrison, G. I., E. J. Laishley, and H. R. Krouse. 1980. Stable isotope fractionation by *Clostridium pasteurianum*. 3. Effect of SeO_3^{2-} on the physiology and associated sulfur isotope fractionation during SO_3^{2-} and SO_4^{2-} reductions. Can. J. Microbiol. 26:952–958.

Harrison, G., C. Curle, and E. J. Laishley. 1984. Purification and characterization of an inducible dissimilatory type of sulfite reductase from *Clostridium pasteurianum*. Arch. Microbiol. 138:172–178.

Holt, J. G., ed. 1984. Bergey's Manual of Systematic Bacteriology. Vol. 1. Williams and Wilkins, Baltimore.

Holt, J. G., ed. 1989. Bergey's Manual of Systematic Bacteriology. Vol. 3. Williams and Wilkins, Baltimore.

Holthuijzen, Y. A., J. F. L. van Breemen, W. N. Konings, and E.F.J. van Bruggen. 1986a. Electron microscopic studies of carboxysomes of *Thiobacillus neapolitanus*. Arch. Microbiol. 1986a 144:258–262.

Holthuijzen, Y. A., J. F. L. van Breemen, J. G. Kuenen, and W. N. Konings. 1986b. Protein composition of the carboxysomes of *Thiobacillus neapolitanus*. Arch. Microbiol. 144:398–404.

Holthuijzen, Y. A., F. F. M. van Dissel-Emiliani, J. G. Kuenen, and W.N. Konings. 1987. Energetic aspects of CO_2 uptake in *Thiobacillus neapolitanus*. Arch. Microbiol. 147:285–290.

Imhoff-Stuckle, D., and N. Pfennig. 1983. Isolation and characterization of a nicotinic acid-degrading sulfate-reducing bacterium, *Desulfococcus niacini* sp. nov. Arch. Microbiol. 136:194–198.

Ivanov, M. V. 1968. Microbiological Processes in the Formation of Sulfur Deposits. Israel Program for Scientific Translations. U.S. Department of Agriculture and National Science Foundation, Washington, DC.

Jannasch, H. W. 1984. Microbial processes at deep sea hydrothermal vents. *In*: P. A. Rona, K. Bostrom, L. Laubier, and K. L. Smith, Jr. Hydrothermal Processes at Seafloor Spreading Centers. Plenum Press, New York, pp. 677–709.

Jannasch, H. W., and M. J. Mottl. 1985. Geomicrobiology of deep-sea hydrothermal vents. Science 229:717–725.

Jannasch, H. W., and C. D. Taylor. 1984. Deep-sea microbiology. Annu. Rev. Microbiol. 38;487–514.

Jannasch, H. W., C. O. Wirsen, S. J. Molyneaux, and T. A. Langworthy. 1988a. Extremely thermophilic fermentative archaebacteria of the genus *Desulfurococcus* from deep-sea hydrothermal vents. Appl. Environ. Microbiol. 54:1203–1209.

Jannasch, H. W., R. Huber, S. Belkin, and K. O. Stetter. 1988b. *Thermotoga neapolitana* sp. nov. of the extremely thermophilic, eubacterial genus *Thermotoga*. Arch. Microbiol. 150:103–104.

Jannasch, H. W., D. C. Nelson, and C. O. Wirsen. 1989. Massive natural occurrence of unusually large bacteria (*Beggiatoa* sp.) at a hydrothermal deep-sea vent site. Nature (Lond.) 342:834–836.

Jansen, K., R. K. Thauer, F. Widdel, and G. Fuchs. 1984. Carbon assimilation pathways in sulfate reducing bacteria. Formate, carbon dioxide, carbon monoxide, and acetate assimilation by *Desulfovibrio baarsii*. Arch. Microbiol. 138:257–262.

Janssen, P. H., and H. W. Morgan. 1992. Heterotrophic sulfur reduction by *Thermotoga* sp. strain FjSS3B1. FEMS Microbiol. Lett. 96: 213–218.

Jones, G. E., and R. L. Starkey. 1957. Fractionation of stable isotopes of sulfur by microorganisms and their role in deposition of native sulfur. Appl. Microbiol. 5:111–118.

Jones, G. E., and R. L. Starkey. 1962. Some necessary conditions for fractionation of stable isotopes of sulfur by *Desulfovibrio desulfuricans*. *In*: M. L. Jensen, ed. Biogeochemistry of Sulfur Isotopes. N.S.F. Symposium. Yale University Press, New Haven, CT, pp. 61–79.

Jones, G. E., R. L. Starkey, H. W. Feeley, and J. L. Kulp. 1956. Biological origin of native sulfur in salt domes of Texas and Louisiana. Science 123:1124–1125.

Jørgensen, B. B. 1990a. A thiosulfate shunt in the sulfur cycle of marine sediments. Science 249:152–154.

Jørgensen, B. B. 1990b. The sulfur cycle of freshwater sediments: Role of thiosulfate. Limnol. Oceanogr. 35:1329–1342.

Jørgensen, B. B., and F. Bak. 1991. Pathways and microbiology of thiosulfate transformations and sulfate reduction in a marine sediment (Kattegat, Denmark). Appl. Environ. Microbiol. 57:847–856.

Jørgensen, B. B., J. G. Kuenen, and Y. Cohen. 1979a. Microbial transformations of sulfur compounds in a stratified lake (Solar Lake, Sinai). Limnol. Oceanogr. 24:799–822.

Jørgensen, B. B., N. P. Revsbech, H. Blackburn, and Y. Cohen. 1979b. Diurnal cycle of oxygen and sulfide microgradient and microbial photosynthesis in a cyanobacterial mat sediment. Appl. Environ. Microbiol. 38:46–58.

Jørgensen, B. B., M. F. Isaksen, and H. W. Jannasch. 1992. Bacterial sulfate reduction above 100°C in deep-sea hydrothermal vent sediments. Science 258:1756–1757.

Justin, P., and D. P. Kelly. 1978. Growth kinetics of *Thiobacillus denitrificans* in anaerobic and aerobic chemostat culture. J. Gen. Microbiol. 107:123–130.

Kaplan, I. R., and S. C. Rittenberg. 1962. The microbiological fractionation of sulfur isotopes. *In*: M. L. Jensen, ed. Biogeochemistry of Sulfur Isotopes. N.S.F. Symposium. Yale University Press, New Haven, CT, pp. 80–93.

Kaplan, I. R., and S. C. Rittenberg. 1964. Microbiological fractionation of sulfur isotopes. J. Gen. Microbiol. 34:195–212.

Kaplan, I. R., T. A. Rafter, and J. R. Hulston. 1960. Sulfur isotopic variations in nature. Part 8, Application to some biochemical problems. NZ J. Sci. 3:338–361.

Keith, S. M. and R. A. Herbert. 1983. Dissimilatory nitrate reduction by a strain of *Desulfovibrio desulfuricans*. FEMS Microbiol. Lett. 18: 55–59.

Kelly, D. P. 1982. Biochemistry of the chemolithotrophic oxidation of inorganic sulfur. Phil. Trans. R. Soc. Lond. B 298:499–528.

Kelly, D. P., W-P. Lu, and R. K. Poole. 1993. Cytochromes in *Thiobacillus tepidarius* and the respiratory chain involved in the oxidation of thiosulfate and tetrathionate. Arch. Microbiol. 87–95.

Kemp, A. L. W., and H. G. Thode. 1957. The mechanism of the bacterial reduction of sulfate and of sulfite from isotope fractionation studies. Geochim. Cosmochim. Acta 32:71–91.

Kramer, J. F., D. H. Pope, and J. C. Salerno. 1987. Pathways of electron transfer in *Desulfovibrio. In*: C. H. Kim, H. Tedeschi, J. J. Diwan and S. C. Salerno, eds. Advances in Membrane Biochemistry and Bioenergetics. Plenum Press, New York, pp. 249–258.

Kobiyashi, K. S., S. Tashibana, and M. Ishimoto. 1969. Intermediary formation of trithionate in sulfite reduction by a sulfate-reducing bacterium. J. Biochem. (Tokyo) 65:155–157.

Kobiyashi, K. S., E. Tukahashi, and M. Ishimoto. 1972. Biochemical studies on sulfate-reducing bacteria. XI. Purification and some properties of sulfite reductase, desulfoviridin. J. Biochem. (Tokyo) 72: 879–887.

Kuenen, J. G., and R. F. Beudeker. 1982. Microbiology of thiobacilli and other sulfur-oxidizing autotrophs, mixotrophs and heterotrophs. Phil. Trans. R. Soc. Lond. B298:473–497.

Kuenen, J. G., and O. H. Tuovinen. 1981. The genera *Thiobacillus* and *Thiomicrospira. In*: M. P. Starr, H. Stolp, H. G. Trueper, A. Balows, and H. G. Schlegel, eds. The Prokyrotes, a Handbook of Habitats, Isolation and Identification of Bacteria. Springer-Verlag, Berlin, pp. 1023–1036.

Kuever, J., J. Kulmer, S. Jannsen, U. Fischer, and K.-H. Blotevogel. 1993. Isolation and characterization of a new spore-forming sulfate-reducing bacterium growing by complete oxidation of catechol. Arch. Microbiol. 159:282–288l.

Laishley, E. J., and H. R. Krouse. 1978. Stable isotope fractionation by *Clostridium pasteurianum*. 2. Regulation of sulfite reductases by sulfur amino acids and their influence on sulfur isotope fractionation during SO_3^{2-} and SO_4^{2-} reduction. Can. J. Microbiol. 24:716–724.

La Riviere, J. W. M., and K. Schmidt. 1981. Morphologically conspicuous sulfur-oxidizing bacteria. *In*: M. P. Starr, H. Stolp, H. G. Trueper, A. Balows, and H. G. Schlegel, eds. The Prokaryotes, a Handbook of Habitats, Isolation and Identification of Bacteria. Springer-Verlag, Berlin, pp. 1037–1048.

Lee, J. P., and H. D. Peck Jr. 1971. Purification of the enzyme reducing bisulfite to trithionate from *Desulfovibrio gigas* and its identification as desulfoviridin. Biochem. Biophys. Res. Commun. 45:583–589.

Lee, J. P., C. S. Yi, J. LeGall, and H. D. Peck. 1973. Isolation of a new pigment, desulfoviridin, from *Desulfovibrio desulfuricans* (Norway strain) and its role in sulfite reduction. J. Bacteriol. 115:453–455.

LeRoux, N., D. S. Wakerley, and S. D. Hunt. 1977. Thermophilic thiobacillus-type bacteria from Icelandic thermal areas. J. Gen. Microbiol. 100:197–201.

Liu, C-L., N. Hart, and H. D. Peck, Jr. 1982. Inorganic pyrophosphate:

Energy source for sulfate-reducing bacteria of the genus *Desulfotomaculum*. Science 217;363–364.

London, J. 1963. *Thiobacillus intermedius* nov. sp. a novel type of facultative autotroph. Arch. Mikrobiol. 46:329–337.

London, J., and S. C. Rittenberg. 1964. Path of sulfur in sulfide and thiosulfate oxidation by thiobacilli. Proc. Natl. Acad. Sci. USA 52: 1183–1190.

London, J., and S. C. Rittenberg. 1966. Effects of organic matter on the growth of *Thiobacillus intermedius*. J. Bacteriol. 91:1062–1069.

London, J., and S. C. Rittenberg. 1967. *Thiobacillus perometabolis* nov. sp., a non-autotrophic thiobacillus. Arch. Mikrobiol. 59:218–225.

Lovley, D. R., E. E. Roden, E. J. P. Phillips, and J. C. Woodward. 1993. Enzymatic iron and uranium reduction by sulfate-reducing bacteria. Marine Geol. 113:41–53.

Loya, S., S. A. Yanofsky, and B. L. Epel. 1982. Characterization of cytochromes in lithotrophically and organotrophically grown cells of *Thiobacillus* A$_2$. J. Gen. Microbiol. 128:2371–2378.

Lu, W-P. 1986. A periplasmic location for the thiosulfate-oxidizing multienzyme system from *Thiobacillus versutus*. FEMS Microbiol. Lett. 34:313–317.

Lu, W-P., and D.P. Kelly. 1983. Purification and some properties of two principal enzymes of the thiosulfate-oxidizing multienzyme system from *Thiobacillus* A$_2$. J. Gen. Microbiol.129:3549–3564.

Lu, W-P., and D. P. Kelly. 1988. Respiration-driven proton translocation in *Thiobacillus versutus* and the role of the periplasmic thiosulfate-oxidizing enzyme system. Arch. Microbiol. 149:297–302.

Macnamara, J., and H. Thode. 1951. The distribution of S^{34} in nature and the origin of native sulfur deposits. Research 4:582–583.

Madigan, M. T., and T. D. Brock. 1975. Photosynthetic sulfide oxidation by *Chloroflexus aurantiacus*, a filamentous, photosynthetic, gliding bacterium. J. Bacteriol. 122:782–784.

Markosyan, G. E. 1973. A new mixotrophic sulfur bactgerium developing in acidic media *Thiobacillus organoparus* sp. n. Dokl. Akad. Nauk SSSR Ser. Biol. 211:1205–1208.

Marshall, C., P. Frenzel, and H. Cypionka, 1993. Influence of oxygen on sulfate reduction and growth of sulfate-reducing bacteria. Arch. Microbiol. 159:168–173.

Martinez, J. D. 1991. Salt domes. Amer. Scientist 79:420–431.

Mason, J., and D. P. Kelly. 1988. Mixotrophic and autotrophic growth of *Thiobacillus acidophilus* on tetrathionate. Arch. Microbiol. 149: 317–323.

McCready, R. G. L., E. J. Laishley, and H. R. Crouse. 1975. Stable isotope fractionation by *Clostridium pasteurianum*. 1. $^{34}S/^{32}S$: inverse isotope effects during SO_4^{2-} and SO_3^{2-} reduction. Can. J. Microbiol. 21:235–244.

McCready, R. G. L., W. D. Gould, and F. D. Cook. 1983. Respiratory nitrate reduction by *Desulfovibrio* sp. Arch. Microbiol. 135:182–185.

Mechalas, B. J., and S. C. Rittenberg. 1960. Energy coupling in *Desulfovibrio desulfuricans*. J. Bacteriol. 80:501–507.

Meulenberg, R., J. T. Pronk, W. Hazeu, J. P. van Dijken, J. Frank, P. Bos, and J. G. Kuenen. 1993. Purification and partial characterization of thiosulfate dehydrogenase from *Thiobacillus acidophilus*. J. Gen. Microbiol. 139:2033–2039.

Miller, J. D. A., and D. S. Wakerley. 1966. Growth of sulfate-reducing bacteria by fumarate dismutation. J. Gen. Microbiol. 43:101–107.

Mitchell, G. J., J. G. Jones and J. A. Cole. 1986. Distribution and regulation of nitrate and nitrite reduction by *Desulfovibrio* and *Desulfotomaculum* species. Arch. Microbiol. 144:35–40.

Moeller, D., R. Schauder, G. Fuchs, and R. K. Thauer. 1987. Acetate oxidation to CO_2 via a citric acid cycle involving an ATP-citrate lyase: a mechanism for the synthesis of ATP via substrate level phosphorylation in *Desulfobacter postgatei* growing on acetate and sulfate. Arch. Microbiol. 148:202–207.

Moriarty, D. J. W., and D. J. D. Nicholas. 1970b. Electron transfer during sulfide and sulfite oxidation by *Thiobacillus concretivorus*. Biochim. Biophys. Acta 216:130–138.

Mosser, J. L., A. G. Mosser, and T. D. Brock. 1973. Bacterial origin of sulfuric acid in geothermal habitats. Science 179:1323–1324.

Mosser, J. L., B. B. Bohlool, and T. D. Brock. 1974. Growth rates of *Sulfolobus acidocaldarius* in nature. J. Bacteriol. 118:1075–1081.

Nathansohn, A. 1902. Ueber eine neue Gruppe von Schwefelbakterien und ihren Stoffwechsel. Mitt. Zool. Sta. Neapel 15:655–680.

Nelson, D. C., and R. W. Castenholz. 1981. Use of reduced sulfur compounds by *Beggiatoa* sp. J. Bacteriol. 147:140–154.

Nelson, D. C., and H. W. Jannasch. 1983. Chemoautotrophic growth of a marine *Beggiatoa* in sulfide-gradient cultures. Arch. Microbiol. 136:262–269.

Neutzling, O., C. Pfleiderer, and H. G. Trueper. 1985. Dissimilatory sulfur metabolism in phototrophic "non-sulfur" bacteria. J. Gen. Microbiol. 131:791–798.

Odom, J. M., and H. D. Peck Jr. 1981. Hydrogen cycling as a general mechanism for energy coupling in the sulfate-reducing bacteria, *Desulfovibrio* sp. FEMS Microbiol. Lett. 12:47–50.

Odom, J. M. and J. D. Wall. 1987. Properties of a hydrogen-inhibited mutant of *Desulfovibrio desulfuricans* ATCC 27774. J. Bacteriol. 169:1335–1337.

Oren, A., and M. Shilo. 1979. Anaerobic heterotrophic dark metabolism in the cyanobacterium *Oscillatoria limnetica*: Sulfur respiration and lactate fermentation. Arch. Microbiol. 122:77–84.

Panganiban, A. T., and R. S. Hanson. 1977. Isolation of a bacterium that oxidizes methane in the absence of oxygen. Abstr., Ann. Meet. Amer. Soc. Microbiol., I59, p. 121.

Panganiban, A. T., Jr., T. E. Patt, W. Hart, and R. S. Hanson. 1979. Oxidation of methane in the absence of oxygen in lake water samples. Appl. Environ. Microbiol. 37:303–309.

Pankhania, I. P., A. M. Spormann, W. A. Hamilton, and R. K. Thauer. 1988. Lactate conversion to acetate, CO_2 and H_2 in cells suspensions of *Desulfovibrio vulgaris* (Marburg): indications for the involvement of an energy driven reaction. Arch. Microbiol. 150:26–31.

Paulsen, J., A. Kroeger, and R. K. Thauer. 1986. ATP-driven succinate oxidation in the catabolism of *Desulfuromonas acetoxidans*. Arch. Microbiol. 144:78–83.

Peck, H. D. Jr. 1962. Symposium on metabolism of inorganic compounds. V. Comparative metabolism of inorganic sulfur compounds in microorganisms. Bacteriol. Rev. 26:67–94.

Peck, H. D. Jr. 1993. Bioenergetic strategies of the sulfate-reducing bacteria. *In*: J. M. Odom and R. Singleton, Jr., eds. The Sulfate-reducing Bacteria: Contemporary Perspectives. Springer-Verlag, New York, pp. 41–76.

Peck, H. D. Jr., and J. LeGall. 1982. Biochemistry of dissimilatory sulfate reduction. Phil. Trans. R. Soc. Lond. B298:443–466.

Peeters, T., and M. I. H. Aleem. 1970. Oxidation of sulfur compounds and electron transport in *Thiobacillus denitrificans*. Arch. Mikrobiol. 71: 319–330.

Perry, K. A., J. E. Kostka, G. W. Luther III, and K. H. Nealson. 1993. Mediation of sulfur speciation by a Black Sea facultative anaerobe. Science 259:801–803.

Pfennig, N. 1977. Phototrophic green and purple bacteria: a comparative, systematic survey. Annu. Rev. Microbiol. 31:275–290.

Pfennig, N., and H. Biebl. 1976. *Desulfuromonas acetoxidans* gen. nov. and sp. nov., a new anaerobic, sulfur-reducing, acetate-oxidizing bacterium. Arch. Microbiol. 110:3–12.

Pfennig, N., and H. Biebl. 1981. The dissimilatory sulfur-reducing bacteria. *In*: M. P. Starr, H. G. Trueper, H. Stolp, H. G. Trueper, A. Balows, and H. G. Schlegel, eds. The Prokaryotes, a Handbook on

Habitats, Isolation, and Identification of Bacteria. Vol. 1. Springer-Verlag, Berlin, pp. 941–947.

Pfennig, N., F. Widdel, and H. G. Trueper. 1981. The dissimilatory sulfate-reducing bacteria. *In*: M. P. Starr, H. Stolp, H. G. Trueper, A. Balows, and H. G. Schlegel, eds. The Prokaryotes, a Handbook on Habitats, Isolation, and Identification of Bacteria. Vol. 1. Springer-Verlag, Berlin, pp. 926–940.

Pierson, B. K., and R. W. Castenholz. 1974. A photosynthetic gliding filamentous bacterium, of hot springs, *Chloroflexus aurantiacus* gen. and spec. nov. Arch. Microbiol. 100:5–24.

Platen, H., A. Temmes, and B. Schink. 1990. Anaerobic degradation of acetone by *Desulfococcus biacutus* spec. nov. Arch. Microbiol. 154: 355–361.

Postgate, J. R. 1952. Growth of sulfate reducing bacteria in sulfate-free media. Research 5:189–190.

Postgate, J. R. 1963. Sulfate-free growth of *Cl. nigrificans*. J. Bacteriol. 85:1450–1451

Postgate, J. R. 1984. The Sulfate-Reducing Bacteria. 2nd ed. Cambridge University Press, Cambridge.

Powell, M. A., and G. N. Somero. 1983. Blood components prevent sulfide poisoning of respiration of the hydrothermal vent tube worm *Riftia pachyptila*. Science 219:297–299.

Pringsheim, E. G. 1967. Die mixotrophie von *Beggiatoa*. Arch. Mikrobiol. 59:247–254.

Pronk, J. T., R. Meulenberg, W. Hazeu, P. Bos, and J. G. Kuenen. 1990. Oxidation of reduced inorganic sulfur compounds by acidophilic thiobacilli. FEMS Microbiol. Rev. 75:293–306.

Pronk, J. T., K. Liem, P. Bos, and J. G. Kuenen. 1991. Energy transduction by anaerobic ferric iron respiration in *Thiobacillus ferrooxidans*. Appl. Environ. Microbiol. 57:2063–2068.

Pronk, J. T., J. C. De Bruyn, P. Box, and J. G. Kuenen. 1992. Anaerobic growth of *Thiobacillus ferrooxidans*. Appl. Environ. Microbiol. 58: 2227–2230.

Qatabi, A. I., V. Niviere, and J. L. Garcia. 1991. *Desulfovibrio alcoholovorans* sp. nov., a sulfate-reducing bacterium able to grow on glycerol, 1,2- and 1,3-propanol. Arch. Microbiol. 155:143–148.

Rau, G. H. 1981. Hydrothermal vent clam and tube worm $^{13}C/^{12}C$: further evidence of nonphotosynthetic food sources. Science 213:338–340.

Rau, G. H., and J. I. Hedges. 1979. Carbon-13 depletion in a hydrothermal vent mussel: suggestion of a chemosynthetic food source. Science 203:648–649.

Reuter, P., R. Rabus, H. Wilkes, F. Aeckersberg, F. A. Rainey, H. W.

Jannasch, and F. Widdel. 1994. Anaerobic oxidation of hydrocarbons in crude oil by new types of sulfate-reducing bacteria. Nature (Lond.) 372:455–458.

Roy, A. B., and P. A. Trudinger. 1970. The Biochemistry of Inorganic Compounds of Sulfur. Cambridge University Press, Cambridge.

Sadler, M. H., and E. J. Johnson. 1972. A comparison of the NADH oxidase electron transport system of two obligately chemolithotrophic bacteria. Biochim. Biophys Acta 283:167–179.

Schauder, R., and A. Kroeger. 1993. Bacterial sulfur respiration. Arch. Microbiol. 159:491–497.

Schauder, R., and E. Mueller. 1993. Polysulfide as a possible substrate for sulfur-reducing bacteria. Arch. Microbiol. 160:377–382.

Schauder, R., B. Eikmanns, R. K. Thauer, F. Widdel, and G. Fuchs. 1986. Acetate oxidation to CO_2 in anaerobic bacteria via a novel pathway not involving reactions of the citric acid cycle. Arch. Microbiol. 145:162–172.

Schauder, R., F. Widdel, and G. Fuchs. 1987. Carbon assimilation pathways in sulfate-reducing bacteria. II. Enzymes of a reductive citric acid cycle in the autotrophic *Desulfobacter hydrogenophilus*. Arch. Microbiol. 148:218–225.

Schauder, R., A. Preuss, M. Jetten, and G. Fuchs. 1989. Oxidative and reductive acetyl CoA/carbon monoxide dehydrogenase pathway in *Desulfobacterium autotrophicum*. Arch. Microbiol. 151:84–89.

Schedel, M., and H.G. Trueper. 1980. Anaerobic oxidation of thiosulfate and elemental sulfur in *Thiobacillus denitrificans*. Arch. Microbiol. 124:205–210.

Schnell, S., and B. Schink. 1991. Anaerobic aniline degradation via reductive deamination of 4-aminobenzoyl-CoA in *Desulfobacterium anilini*. Arch. Microbiol. 155:183–190.

Schoen, R., and G. G. Ehrlich. 1968. Bacterial origin of sulfuric acid in sulfurous hot springs. 23rd Intern. Geol. Congr. 17:171–178.

Schoen, R., and R. O. Rye. 1970. Sulfur isotope distribution in solfataras, Yellowstone National Park. Science 170:1082–1084.

Segerer, A., A. Neuner, J. K. Kristiansson, and K. O. Stetter. 1986. *Acidianus infernus* gen. nov., sp. nov., and *Acidianus brierleyi* comb. nov.: Facultatively aerobic, extremely acidophilic thermophilic sulfur-metabolizing archaebacteria. Int. J. Syst. Bacteriol. 36:559–564.

Seitz, H-J., and H. Cypionka. 1986. Chemolithotrophic growth of *Desulfovibrio desulfuricans* with hydrogen coupled to ammonification of nitrate or nitrite. Arch. Microbiol. 146:63–67.

Shively, J. M., F. Ball, D. H. Brown, and R. E. Saunders. 1973. Functional organelles in prokaryotes: polyhedral inclusions (carboxysomes) of *Thiobacillus neapolitanus*. Science 182:584–586.

Shivvers, D. W., and T. D. Brock. 1973. Oxidation of elemental sulfur by *Sulfolobus acidocaldarius*. J. Bacteriol. 114:706–710.

Shturm, L. D. 1948. Sulfate reduction by facultative aerobic bacteria. Mikrobiologiya 17:415–418.

Skerman, V. B. D., G. Dementyeva, and B. Carey. 1957a. Intracellular deposition of sulfur by *Sphaerotilus natans*. J. Bacteriol. 73: 504–512.

Skerman, V. B. D., G. Dementyeva, and G. W. Skyring. 1957b. Deposition of sulfur from hydrogen sulfide by bacteria and yeasts. Nature (Lond.) 179:742.

Skyring, G. W. 1987. Sulfate reduction in coast ecosystems. Geomicrobiol. J. 5:295–374.

Smith, A. L., D. P. Kelly, A. P. Wood. 1980. Metabolism of *Thiobacillus* A_2 grown under autotrophic, mixotrophic, and heterotrophic conditions in chemostat culture. J. Gen. Microbiol. 121:127–138.

Smith, D. W., and S. C. Rittenberg. On the sulfur-source requirement for growth of *Thiobacillus intermedius*. Arch. Microbiol. 100:65–71.

Smith, D. W., and W. R. Strohl. 1991. Sulfur-oxidizing bacteria. *In*: J. M. Shiveley, and L. L. Barton. 1991. Variations in Autotrophic Life. Academic Press, London, pp. 121–146.

Sokolova, G. A. 1962. Microbiological sulfur formation in Sulfur Lake. Mikrobiologiya 31:324–327 (Engl. transl. pp. 264–266).

Sokolova, G. A., and G. I. Karavaiko. 1968. Physiology and geochemical activity of thiobacilli. English translation. U.S. Department of Commerce. Clearinghouse. Fed. Sci Tech. Info., Springfield, VA.

Sorokin, Yu.I. 1966a. Role of carbon dioxide and acetate in biosynthesis of sulfate-reducing bacteria. Nature (Lond.) 210:551–552.

Sorokin, Yu.I. 1966b. Sources of energy and carbon for biosynthesis by sulfate-reducing bacteria. Mikrobiologiya 35:761–766 (Engl. transl. pp. 643–647).

Sorokin, Yu. I. 1966c. Investigation of the structural metabolism of sulfate-reducing bacteria with ^{14}C. Mikrobiologiya 35:761–766 (Engl. transl. pp. 806–814).

Sorokin, Yu.I. 1966d. The role of carbon dioxide and acetate in biosynthesis in sulfate reducing bacteria. Dokl. Akad. Nauk SSSR 168:199.

Speich, N., and H. G. Trueper. 1988. Adenylsulfate reductase in a dissimilatory sulfate-reducing archaebacterium. J. Gen. Microbiol. 134: 1419–1425.

Stanier, R. Y., J. L. Ingraham, M. L. Wheelis, and P. R. Painter. 1986. The Microbial World. 5th ed. Prentice-Hall, Englewood Cliffs, NJ.

Starkey, R. L. 1934. The production of polythionates from thiosulfate by microorganisms. J. Bacteriol. 28:387–400.

Starr, M. P., H. Stolp, H. G. Trueper, A. Balows, and H. G. Schlegel, eds. 1981. The Prokaryotes, a Handbook of Habitats, Isolation and Identification of Bacteria. Springer-Verlag, Berlin.

Stetter, K. O. 1985. Thermophilic archaebacteria occurring in submarine hydrothermal areas. *In*: D. E. Caldwell, J. A. Brierley, and C. L. Brierley, eds. Planetary Ecology. Van Nostrand Reinhold, New York, pp. 320–332.

Stetter, K. O. 1988. *Archaeoglobus fulgidus* gen. nov., sp. nov.: a new taxon of extremely thermophilic archaebacteria. Syst. Appl. Microbiol. 10:172–173.

Stetter, K. O., G. Lauerer, M. Thomm, and A. Neuner. 1987. Isolation of extremely thermophilic sulfate reducers: evidence for a novel branch of archaebacteria. Science 236:822–824.

Strahler, A. N. 1977. Principles of Physical Geology. Harper & Row, New York.

Strohl, W. R., and J. M. Larkin. 1978. Enumeration, isolation, and characterization of *Beggiatoa* from freshwater sediments. Appl. Environ. Microbiol. 36:755–770.

Suzuki, I. 1965. Oxidation of elemental sulfur by an enzyme system of *Thiobacillus thiooxidans*. Biochim. Biophys. Acta 104:359–371.

Suzuki, I., and M. Silver. 1966. The initial pproduct and properties of the sulfur-oxidizing enzyme of thiobacilli. Biochim. Biophys. Acta 122: 22–33.

Suzuki, I., C. W. Chan, and T. L. Takeuchi. 1992. Oxidation of elemental sulfur to sulfite by *Thiobacillus thiooxidans* cells. Appl. Environ. Microbiol. 58:3767–3769.

Szewzyk R., and N. Pfennig. 1987. Complete oxidation of catechol by the strictly anaerobic sulfate-reducing *Desulfobacterium catecholicum* sp. nov. Arch. Microbiol. 147:163–168.

Tasaki, M., Y. Kamagata, K. Nakamura, and E. Mikami. 1991. Isolation and characterization of a thermophilic benzoate-degrading, sulfate-reducing bacterium, *Desulfotomaculum thermobenzoicum* sp. nov. Arch. Microbiol. 155:348–352.

Tasaki, M., Y. Kamagata, K. Nakamura, and E. Mikami. 1992. Utilization of methoxylated benzoates and formation of intermediates by *Desulfotomaculum thermobenzoicum* in the presence and absence of sulfate. Arch. Microbiol. 157:209–212.

Thamdrup, B., K. Finster, J. W. Hansen, and F. Bak. 1993. Bacterial disproportionation of elemental sulfur coupled to chemical reduction of iron and manganese. Appl. Environ. Microbiol. 59:101–108.

Thode, H. G., K. K. Wanless, R. Wallouch. 1954. The origin of native sulfur deposits from isotopic fractionation studies. Geochim. Cosmochim. Acta 5:286–298.

Trautwein, K. 1921. Beitrag zur Physiologie und Morphologie der Thionsoeurebakterien. Ztbltt. Bakteriol. Parasitenk. Infektionskr. Hyg. Abt. II 53:513–548.

Trudinger, P. A. 1961. Thiosulfate oxidation and cytochromes in *Thiobacillus* X. 2. Thiosulfate oxidizing enzyme. Biochem. J. 78:680–686.

Trudinger, P. A. 1970. Carbon monoxide-reacting pigment from *Desulfotomaculum nigrificans* and its possible relevance to sulfite reduction. J. Bacteriol. 104:158–170.

Trueper. H. G. 1978. Sulfur metabolism. *In*: R. K. Clayton and W. R. Sistrom, eds. The Photosynthetic Bacteria. Plenum Press, New York, pp. 677–690.

Tuttle, J. H., and H. L. Ehrlich. 1986. Coexistence of inorganic sulfur metabolism and manganese oxidation in marine bacteria. Abstr., Ann. Meet. Amer. Soc. Microbiol. I-21, p. 168.

Tuttle, J. H., P. E. Holmes, and H. W. Jannasch. 1974. Growth rate stimulation of marine pseudomonads by thiosulfate. Arch. Microbiol. 99:1–14.

Van den Ende, F. P., and H. van Gemerden. 1993. Sulfide oxidation under oxygen limitation by a *Thiobacillus thioparus* isolated from a marine microbial mat. FEMS Microbiol. Ecol. 13:69–78.

Van Gemerden, H. 1986. Production of elemental sulfur by green and purple sulfur bacteria. Arch. Microbiol. 146:52–56.

Vainshtein, M. B. 1977. Oxidation of hydrogen sulfide by thionic bacteria. Mikrobiologiya 46:1114–1116 (Engl. transl. pp. 898–899).

Vainshtein, M. B., A. G. Matrosov, V. B. Baskunov, A. M. Zyakun, and M. V. Ivanov. 1981. Thiosulfate as an intermediate product of bacterial sulfate reduction. Mikrobiologiya 49:855–858 (Engl. transl. pp. 672–675).

Vairavamurthy, A., B. Manowitz, G. W. Luther III, and Y. Jeon. 1993. Oxidation state of sulfur in thiosulfate and implications for anaerobic energy metabolism. Geochim. Cosmochim. Acta 57:1619–1623.

Vishniac, W. 1952. The metabolism of *Thiobacillus thioparus*. I. The oxidation of thiosulfate. J. Bacteriol. 64:363–373.

Vishniac, W., and M. Santer. 1957. The thiobacilli. Bacteriol. Rev. 21: 195–213.

Widdel, F., and N. Pfennig. 1977. A new anaerobic, sporing, acetate-oxidizing, sulfate-reducing bacterium, *Desulfotomaculum* (emend.) *acetoxidans*. Arch. Microbiol. 112:119–122.

Widdel, F., and N. Pfennig. 1981. Sporulation and further nutritional characteristics of *Desulfotomaculum acetoxidans*. Arch. Microbiol. 129: 401–402.

Williams, C. D., D. C. Nelson, B. A. Farah, H. W. Jannasch, and J. M. Shively. 1988. Ribulose bisphosphate carboxylase of the procaryotic symbiont of a hydrothermal vent tube worm: kinetics, activity, and gene hybridization. FEMS Microbiol. Lett. 50:107–112.

Williams, R. A. D. and D. S. Hoare. 1972. Physiology of a new facultative autotrophic thermophilic *Thiobacillus*. J. Gen. Microbiol. 70; 555–566.

Wirsen, C. O., and H. W. Jannasch. 1978. Physiological and morphological observations on *Thiovululm* sp. J. Bacteriol. 136:765–774.

Wood, A. P., and D. P. Kelly. 1983. Autotrophic, mixotrophic and heterotrophic growth with denitrification by *Thiobacillus* A_2 under anaerobic conditions. FEMS Microbiol. Lett. 16:363–370.

Wood, A. P., and D. Kelly. 1991. Isolation and characterization of *Thiobacillus halophilus* sp. nov., a sulfur-oxidizing autotrophic eubacterium from a Western Australian hypersaline lake. Arch. Microbiol. 156:277–280.

Wood, P. 1988. Chemolithotrophy. *In*: C. Anthony, ed. Bacterial Energy Transduction. Academic Press, London, pp. 183–230.

Zeikus, J. G., M. A. Dawson, T. E. Thompson, K. Ingvorsen, and E. C. Hatchikian. 1983. Microbial ecology of volcanic sulfidogenesis: isolation and characterization of *Thermodesulfobacterium commune* gen. nov. and sp. nov. J. Gen. Microbiol. 129:1159–1169.

Zellner, G., H. Kneifel, and J. Winter. 1990. Oxidation of benzaldehydes to benzoic acid derivatives by three *Desulfovibrio* strains. Appl. Environ. Microbiol. 56:2228–2233.

Zillig, W., J. Tu, and I. Holz. 1981. Thermoproteales—a third order of thermoacidophilic archaebacteria. Nature (Lond.) 293:85–86.

Zillig, W., K. O. Stetter, D. Prangishvilli, W. Schaefer, S. Wunderl, D. Janekovic, I. Holz, and P. Palm. 1982. Desulfurococcaceae: the second family of the extremely thermophilic, anaerobic, sulfur respiring Thermoproteales. Zbl. Bakt. Hyg. I. Abt. Orig. C 3:304–317.

Zillig, W., A. Gierl, G. Schreiber, S. Wunderl, D. Janekovic, K. O. Stetter, and H. P. Klenk. 1983. The archaebacterium *Thermofilum pendens*, a novel genus of the thermophilic, anaerobic sulfur respiring Thermoproteales. System. Appl. Microbiol. 4:79–87.

18

Biogenesis and Biodegradation of Sulfide Minerals on the Earth's Surface

18.1 INTRODUCTION

Sulfate-reducing bacteria play an important role in some sedimentary environments in the formation of certain sulfide minerals, especially iron pyrite. Other microbes play an even more pervasive role in nature in the oxidation of a wide range of metal sulfides, regardless of the mode of origin of these minerals. This oxidative microbial activity is being industrially exploited in the extraction of metals from metal sulfide ores and some others. The bioextractable ores currently include those of copper, precious metals (in particular, gold) and uranium. A great potential exists for bioextraction of a variety of other metal sulfide ores. A more widely used term for bioextraction is **bioleaching**. In this chapter, we examine biogenesis and bioleaching in some detail. Table 18.1 lists metal sulfide minerals of geomicrobial interest.

18.2 NATURAL ORIGINS OF METAL SULFIDES

Hydrothermal Origin (Abiotic)   Most metal sulfides of commercial interest are of igneous origin. Current theory explaining their formation invokes plate tectonics. It has played and is playing a central role in their

Table 18.1 Metal Sulfides of Geomicrobial Interest

Mineral or synthetic compound	Formula	References
Antimony trisulfide	Sb_2S_3	Silver and Torma (1974); Torma and Gabra (1977)
Argentite	Ag_2S	Baas Becking and Moore (1961)
Arsenopyrite	$FeAsS$	Ehrlich (1964a)
Bornite	Cu_5FeS_4	Cuthbert (1962); Bryner et al. (1954)
Chalcocite	Cu_2S	Bryner et al. (1954); Ivanov (1962); Razzell and Trussell (1963); Sutton and Corrick (1963, 1964); Fox (1967); Nielsen and Beck (1972)
Chalcopyrite	$CuFeS_2$	Bryner and Anderson (1957)
Cobalt Sulfide	CoS	Torma (1971)
Covellite	CuS	Bryner et al. (1954); Razzell and Trussell (1963)
Digenite	Cu_9S_5	Baas Becking and Moore (1961); Nielsen and Beck (1972)
Enargite	$3Cu_2S \cdot As_2S_5$	Ehrlich (1964a)
Galena	PbS	Silver and Troma (1974)
Gallium sulfide	Ga_2S_3	Torma (1978)
Marcasite, pyrite	FeS_2	Leathen et al. (1953); Silverman et al. (1961)
Millerite	NiS	Razzell and Trussell (1963)
Molybdenite	MoS_2	Bryner and Anderson (1957); Bryner and Jameson (1958); Brierley and Murr (1973)
Orpiment	As_2S_3	Ehrlich (1963)
Nickel sulfide	NiS	Torma (1971)
Pyrrhotite	Fe_4S_5	Freke and Tate (1961)
Sphalerite	ZnS	Ivanov et al. (1961); Ivanov (1962); Malouf and Prater (1961)
Tetrahedrite	$Cu_8Sb_2S_7$	Bryner et al. (1954)

formation. Terrestrial deposits of **porphyry copper ore** (small crystals of copper sulfides richly dispersed in host rock) are thought to have originated as a result of subduction of oceanic crust that had become somewhat enriched in copper by hydrothermal activity at midocean spreading centers. Subsequent formation of terrestrial deposits of porphyry sulfide ores is thought to have involved the following successive steps: (1) remelting of the subducted oceanic crust, (2) the rising of the resultant magma, (3) release of water with fracturing of incipient rock and the formation of hydrothermal solution containing hydrogen sulfide during progressive partial cooling of the magma, and finally (4) reformation of copper and other metal sulfides by crystallization of the cooling magma and/or from reaction of H_2S in the hydrothermal solution with metal constituents in the cooled magma in the fractured rock (see, e.g., Strahler, 1977; Bonatti, 1978; Tittley, 1981).

As happens even at present, the enrichment of the surficial deposits of metal sulfide in and on oceanic crust occurred at hydrothermally active regions at seafloor-spreading centers (midocean ridges) at depths of 2,500–2,600 m, as, for instance, in the eastern Pacific Ocean in the Galapagos Rift and on the East Pacific Rise (Ballard and Grassel, 1979; Corliss et al., 1979), and in the Atlantic Ocean on the the Mid-Atlantic Ridge (e.g., Klinkhammer et al., 1985). Metal sulfide deposits on the seafloor are evident where some hydrothermal vents ("black smokers"; see Chapters 2 and 15) in these areas have been found to discharge a brine solution having a temperature near 350°C which is metal-laden and charged with H_2S. Metal sulfides such as chalcopyrite ($CuFeS_2$) and sphalerite (ZnS) precipitate around the vents as the brine meets cold seawater and are often deposited in the form of hollow tubes (chimneys) at the mouth of the vents. The hydrothermal solution discharged by these vents originated from seawater which penetrated into porous volcanic rock (basalt) at the midocean spreading centers to depths as great as 10 km below the seafloor (Bonatti, 1978). As this water penetrated ever deeper into the rock, it absorbed heat diffusing away from underlying magma chambers and was subjected to increasing hydrostatic pressure. This caused the seawater to react with the basalt and pick up various metal species and hydrogen sulfide. The reactions responsible for these seawater modifications include among others the interaction of magnesium in the seawater with the rock to form new basic minerals and the release of acid (H^+) (Seyfried and Mottl, 1982), the leaching of metals from the basalt by the acid (Edmond et al., 1982; Marchig and Grundlach, 1982), and the formation of H_2S by reduction of sulfate in seawater and sulfur in the basalt by ferrous iron from the basalt (e.g., Shanks et al., 1981; Mottl et al., 1979; Styrt et al., 1981).

A quantitatively more significant deposition of metal sulfides occurs in the surficial crust where white smokers are noted. Here rising hydrothermal brine meets and mixes with cold seawater that penetrated superficially into the crustal basalt before the hydrothermal brine emerges from the vents. This mixing results in partial cooling of the hydrothermal solution and consequent precipitation of metal sulfides in the surficial crust before the brine emerges from vents, rather than external to the crust as in the case of black smokers. The emergent, metal-depleted brine still carries much hydrogen sulfide as well as iron and manganese in it but is much cooler than unmixed hydrothermal solution issuing from the black smokers. Figure 15.17 shows diagrammatically the origin of hydrothermal solution and metal sulfides at midocean spreading centers.

Sedimentary Metal Sulfides of Biogenic Origin   Among sedimentary metal sulfides of biogenic origin, iron sulfides are the most common. They are usually associated with reducing zones in sedimentary deposits in estuarine environments, which have a plentiful supply of sulfate. The presence of sulfate is important because the formation of these metal sulfides is usually the result of an interaction of iron compounds with H_2S that originated from bacterial reduction of the sulfate at these sites. The interaction of the H_2S with the iron compounds leads to formation of iron pyrite (FeS_2). Whether amorphous sulfide (FeS), mackinawite (FeS), and greigite (Fe_3S_4) are intermediates in the formation of the pyrite depends on prevailing environmental conditions (Schoonen and Barnes, 1991a,b; Luther, 1991). In at least one salt marsh (Great Sippewissett Marsh, Massachusetts, U.S.A.) where pyrite was formed, the pore waters were found to be undersaturated with respect to these compounds (Jørgensen, 1977; Fenchel and Blackburn, 1979; Howarth, 1979; Berner, 1984; Giblin and Howarth, 1984; Howarth and Merkel, 1984). Rapid and extensive microbial pyrite formation has been observed in salt marsh peat on Cape Cod (U.S.A.) (Howarth, 1979). Pyrite formation from biogenic H_2S has also been noted in organic-rich sediments at the Peru Margin of the Pacific Ocean (Mossmann et al., 1991), in Long Island Sound off the Atlantic coast of Connecticut and New York, U.S.A. (e.g., Westrich and Berner, 1984), along the Danish coast (Thode-Andersen and Jørgensen, 1989), in two seepage lakes (Gerrisfles and Kliplo), and in two moorland ponds in the Netherlands (Marnette et al., 1993).

Pyrite in many sedimentary environments does not represent a permanent sink for iron because it may be subject to seasonal reoxidation as conditions in the environment change from reducing to oxidizing (Luther et al., 1982; Giblin and Howarth, 1984; King et al., 1985; Giblin, 1988). Active growth of marsh grass may draw oxygen into the sediment by evapotranspiration (Giblin, 1988). Of all the biogenic sulfide formed in these environments, only a portion is consumed in the formation of pyrite and other metal sulfides, the rest being reoxidized as it enters oxidizing environments (Jørgensen, 1977). This oxidation may be biological or abiological (Fenchel and Blackburn, 1979).

Nonferrous sulfide deposits of sedimentary origin, especially biogenic ones, appear to be relatively rare. They are generally thought to have formed syngenetically. The metals in question were precipitated by hydrogen sulfide either of hydrothermal origin (abiotic formation) or of microbiological origin and then buried in contemporaneously formed sediment. The limiting conditions for sedimentary sulfide formation by bacteria as calculated by Rickard (1973) require a minimum of 0.1% carbon (dry weight) and an enriched source of metals such as a hydrothermal solution if more than 1% metal is to be deposited. Newer studies of microbial sulfate reduction have revealed that a significant amount or reducing power for sulfate reduction can be furnished by hydrogen, which would lower the requirement for organic carbon correspondingly (Nedwell and Banat, 1981; see also Chapter 17, section 17.9).

Examples of nonferrous sedimentary sulfide deposits, all but one of which of may have been biogenically formed, include the Permian Kupferschiefer of Mansfeld in Germany (Love, 1962; Stanton, 1972, p. 1139), Black Sea sediments (Bonatti, 1972, p. 51), the Roan Antelope Deposit in Zambia and Katanga (Cuthbert, 1962; Stanton, 1972, p. 1139), the Zechstein Deposit (Serkies et al., 1968), deposits in Peranatty Lagoon (Lambert et al., 1971), and the Pine Point Pb-Zn property in Northwest Territories, Canada (Powell and MacQueen, 1984). $\delta^{34}S$ analyses of the metal sulfides in the last example suggest an abiotic origin, the sulfide having been formed by a reaction between bitumen and sulfate at elevated temperature and pressure.

Although most instances of metal sulfide biogenesis in nature are associated with bacterial sulfate reduction, at least one case of biogenesis of galena has been attributed to the aerobic mineralizing of organic sulfur compounds by *Sarcina flava* Bary (Dévigne, 1968 a,b; 1973). The *Sarcina* was isolated from earthy concretions between crystals of galena in an accumulation in a karstic pocket located in the lead-zinc deposit of Djebel Azered, Tunisia. In laboratory experiments, the organism was shown to produce PbS from Pb^{2+} bound to sulfhydryl groups of peptone.

18.3 PRINCIPLES OF METAL SULFIDE FORMATION

Whether formed biogenically or abiotically, the metal sulfides result from an interaction between an appropriate metal ion and sulfide ion:

$$M^{2+} + S^{2-} \Rightarrow MS \tag{18.1}$$

The source of the sulfide in this reaction is what determines whether a metal sulfide is considered to be of biogenic origin, regardless of whether the sulfide resulted from bacterial sulfate reduction (see Chapter 17) or from bacterial mineralization of organic compounds (Dévigne, 1968a,b, 1973). Because of their relative insolubility, the metal sulfides form readily at ordinary temperatures and pressures. Table 18.2 lists solubility products for some common simple sulfides.

The following calculation will show that relatively low concentrations of metal ions are needed to form metal sulfides by reacting with H_2S at typical concentrations in some lakes. The ion activities in these calculations are taken as approximately equal to concentration because of the low concentrations involved. Let us examine, for instance, the case for iron in the formation of iron sulfide (FeS). The dissociation constant for FeS is

$$[Fe^{2+}][S^{2-}] = 1 \times 10^{-19} \tag{18.2}$$

The dissociation constant for H_2S is

$$[S^{2-}] = 1.1 \times 10^{-22} \frac{[H_2S]}{[H^+]^2} \tag{18.3}$$

since

$$\frac{[HS^-][H^+]}{[H_2S]} = 1.1 \times 10^{-7} \tag{18.4}$$

and

$$\frac{[S^{2-}][H^+]}{[HS^-]} = 1 \times 10^{-15} \tag{18.5}$$

Therefore, substituting Eq. (18.3) into Eq. (18.2), we obtain

$$[Fe^{2+}] = \frac{[H^+]^2}{[H_2S]} \frac{1 \times 10^{-19}}{1.1 \times 10^{-22}} = \frac{[H^+]^2}{[H_2S]} (9.1 \times 10^2) \tag{18.6}$$

Table 18.2 Solubility Products for Some Metal Sulfides

CdS	1.4×10^{-28}	FeS	1×10^{-19}	NiS	3×10^{-21}
Bi_2S_3	1.6×10^{-72}	PbS	1×10^{-29}	Ag_2S	1×10^{-51}
CoS	7×10^{-23}	MnS	5.6×10^{-16}	SnS	8×10^{-29}
Cu_2S	2.5×10^{-50}	Hg_2S	1×10^{-45}	ZnS	4.5×10^{-24}
CuS	4×10^{-38}	HgS	3×10^{-53}	H_2S	1.1×10^{-7}
				Hs^-	1×10^{-15}

Source: Latimer and Hildebrand, 1942.

Assuming that the bottom water of a lake contains about 34 mg of H_2S liter^{-1} (0.001 M) at pH 7, about 5.08×10^{-3} mg of Fe^{2+} liter^{-1} (9.1×10^{-8} M) will be precipitated by 3.4 mg of hydrogen sulfide per liter (10^{-4} M). The unused H_2S will ensure reducing conditions, which will keep the iron in the ferrous state. Since ferrous sulfide is one of the more soluble sulfides, metals whose sulfides have even smaller solubility products will form even more readily. Indeed the iron FeS will probably be transformed into FeS_2, which is more stable than FeS.

18.4 LABORATORY EVIDENCE IN SUPPORT OF BIOGENESIS OF METAL SULFIDES

Batch Cultures   Metal sulfides have been generated in laboratory experiments utilizing H_2S from bacterial sulfate reduction. Miller (1949; 1950) reported that sulfides of Sb, Bi, Co, Cd, Fe, Pb, Ni, and Zn were formed in a lactate-containing broth culture of *Desulfovibrio desulfuricans* to which insoluble salts of selected metals had been added. For instance, he found bismuth sulfide to be formed on addition of $(BiO_2)_2CO_3 \cdot H_2O$, cobalt sulfide on addition of $2CoCO_3 \cdot 3Co(OH)_2$, lead sulfide on addition of $2PbCO_3 \cdot Pb(OH)_2$ or $PbSO_4$, nickel sulfide on addition of $NiCO_3$ or $Ni(OH)_2$, and zinc sulfide on the addition of $2ZnCO_3 \cdot 3Zn(OH)_2$ as starting compounds. The reason for adding insoluble metal salts as starting compounds was to minimize metal toxicity for *D. desulfuricans*. Metal toxicity depends in part on the solubility of a metal compound. Obviously, for a metal sulfide to be formed, the metal sulfide must be even more insoluble than the source compound of the metal. Miller was not able to demonstrate copper sulfide formation from malachite $[CuCO_3 \cdot Cu(OH)_2]$, probably because malachite is too insoluble relative to copper sulfides in his medium. Miller (1949) also showed that Cd and Zn caused an increase in total sulfide generated from sulfate in batch culture when compared to the sulfide yield in the absence of the added metals. This was because uncombined sulfide at high enough concentration becomes itself toxic to sulfate reducers.

Baas Becking and Moore (1961) also undertook a study of biogenesis of sulfide minerals. Like Miller, they worked with batch cultures of sulfate-reducing bacteria. They worked with *Desulfovibrio desulfuricans* and *Desulfotomaculum* sp. (called by them *Clostridium desulfuricans*) growing in lactate or acetate/steel wool-containing medium. The steel wool in the acetate/steel wool-containing medium was meant to serve as a source of hydrogen for the bacterial reduction of sulfate.

$$Fe^° + 2H_2O \Rightarrow H_2 + Fe(OH)_2 \tag{18,7}$$

$$4H_2 + SO_4^{2-} + H^+ \xrightarrow{\text{bacteria}} HS^- + 4H_2O \qquad (18.8)$$

The media were saline to simulate marine (near-shore and estuarine) conditions under which the investigators thought the reactions could occur in nature. They formed ferrous sulfide from strengite ($FePO_4$) and from hematite (Fe_2O_3). Furthermore, they formed covellite (CuS) from malachite [$CuCO_3 \cdot Cu(OH)_2$]; argentite (Ag_2S) from silver chloride (Ag_2Cl_2) or silver carbonate ($AgCO_3$); galena (PbS) from lead carbonate ($PbCO_3$) or lead hydroxy carbonate [$PbCO_3 \cdot Pb(OH)_2$]; and sphalerite (ZnS) from smithsonite ($ZnCO_3$). All mineral products were identified by X-ray powder diffraction studies. Baas Becking and Moore were unable to form cinnabar from mercuric carbonate ($HgCO_3$), probably owing to the toxicity of Hg^{2+} ion, nor were they able to form alabandite (MnS) from rhodochrosite ($MnCO_3$), or bornite (Cu_5FeS_4) or chalcopyrite ($CuFeS_2$) from a mixture of cuprous oxide (Cu_2O) or malachite and hematite and lepidochrosite. They probably succeeded forming covellite from malachite where Miller (1950) failed because they performed their experiment in a saline medium (3% NaCl) in which Cl^- could complex Cu^{2+}, thereby increasing the solubility of Cu^{2+}. The starting materials that were the source of metal were all insoluble, as in Miller's experiments. Baas Becking and Moore found that in the formation of covellite and argentite, native copper and silver were respective intermediates that disappeared with continued bacterial H_2S production.

Leleu et al. (1975) synthesized ZnS by passing H_2S produced by unnamed strains of sulfate-reducing bacteria through a solution of $ZnSO_4$ contained in a separate vessel in one instance, and by having the biogenic H_2S react with $ZnSO_4$ added directly to the culture medium to a final concentration of 10^{-2} M. The ZnS formed under either experimental condition was identified as a sphalerite-wurtzite mixture by powder X-ray diffraction diagrams. The presence of Zn directly in the culture medium caused a lag in H_2S production.

Column Experiment: A Model for Biogenesis of Sedimentary Metal Sulfides The relatively high toxicity of many of the heavy metals for sulfate-reducing bacteria has been used as an argument that these organisms could not have been responsible for metal sulfide precipitation in nature (e.g., Davidson, 1962a,b). However, in a sedimentary environment, metal ions will be mostly adsorbed to sediment particles such as clays, or complexed by organic matter (Hallberg, 1978), which lessens their toxicity. Such adsorbed or complexed ions are still capable of reacting with sulfide and precipitating as metal sulfides, as has been shown experimentally by Temple and LeRoux (1964). They constructed a column in which a clay or ferric hydroxide slurry carrying adsorbed Cu^{2+}, Pb^{2+}, and Zn^{2+} ions was separated by an agar plug from a liquid culture of sulfate reducers actively generating hydrogen sulfide in saline medium. They also tested clay that was carrying Fe^{3+} in this setup. They found that in time, bands of precipitate formed in the agar plug separating metal-carrying slurry from the culture of sulfate reducing bacteria (Fig. 18.1). The bands formed as the upward-diffusing sulfide ion species and the downward-diffusing, desorbed metal ion species encountered each other in the agar. Differential desorption of metal ions from the adsorbent and their differential diffusion in the agar accounted for the discrete banding by the various sulfides. These results demonstrate that biogenesis of relatively large amounts of

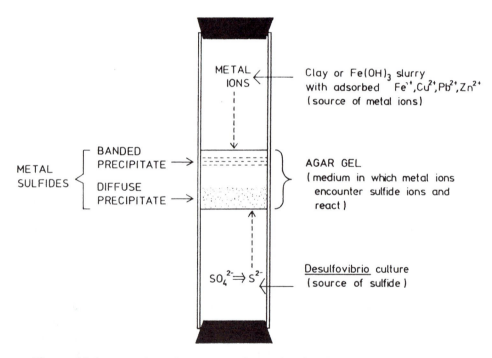

Figure 18.1 Temple and LeRoux column showing how sulfate reducers can precipitate metal sulfides by reaction of biogenic sulfide with metal ions. The adsorbents, clay or $Fe(OH)_3$ slurry, control the concentration of metal ions in solution, and the agar plug prevents physical contact of the sulfate reducers with metal ions. In nature, sediments can act as adsorbents of metal ions. They hold the metal-ion concentration in the interstitial water at such level that sulfate reducers are not poisoned.

sulfides in a sedimentary environment is possible, even in the presence of relatively large amounts of metal ions, provided that the metal ions are in a nontoxic form such as being adsorbed, complexed, or in the form of insoluble mineral oxide, carbonates or sulfate. As Temple (1964) has pointed out, syngenetic microbial metal sulfide production in nature is possible. Restrictions on the process, according to him, are not metal toxicity, but free movement of the bacterially generated sulfide and a need for metal-enriched zones in the sedimentary environment. Temple believes that on biochemical grounds, microbial sulfate reduction had evolved in the Precambrian. Sulfur isotope analyses of samples representing the early Precambrian in South Africa have indicated that extensive biogenic sulfate reduction first occurred 2,350 million years ago (Cameron, 1982).

18.5 BIOOXIDATION OF METAL SULFIDES

Regardless of whether they are of abiogenic or biogenic origin, metal sulfides in nature may be subject to microbial oxidation. This may take the form of **direct** or **indirect** interaction. In direct interaction, the microbes attack a metal sulfide directly by oxidizing it in its insoluble form, in most instances rendering it into a soluble sulfate salt, lead sulfate from galena (PbS) being an exception. In indirect interaction, the microbes are responsible for generating a reagent that causes solubilization of the metal sulfide, most commonly through oxidation, but, under some conditions, through complexation.

Direct Oxidation   In direct microbial action on sulfides, the crystal lattice of susceptible metal sulfides is attacked through enzymatic oxidation. To accomplish this, the microbes have to be in intimate contact with the mineral they attack. The mere dissociation of a mineral to yield oxidizable ion species in solution for microbial oxidation is insufficient. This can be shown by considering covellite (CuS), in which the only oxidizable constituent is the sulfide, as an example. *Thiobacillus ferrooxidans* is able to oxidize this mineral at pH 2 (see later in this section). Simple calculations show that at equilibrium at pH 2 in water, the dissociation of CuS will generate a concentration of HS^- equal to $10^{-15.53}$ M and of H_2S equal to $10^{-13.06}$ M. Since the most recent K_s value for sulfide oxidase in intact cells of *T. ferrooxidans* has been reported to be $10^{-5.30}$ M (Pronk et al., 1990), the mere dissociation of CuS into Cu^{2+}, HS^-, and H_2S cannot furnish nearly enough sulfide substrate to sustain its oxidation at a reasonable velocity, regardless of whether HS^- or H_2S is the actual substrate for sulfide oxidase. K_s values are a measure of half-maximal velocity of enzyme-catalyzed reactions. Evidence for attachment of *T. ferrooxidans* to mineral surfaces of chalcopyrite has been presented by McGoran et al. (1969) and Shrihari et al. (1991), to covellite particles by Pogliani et al. (1990), to galena crystals by Tributsch (1976), to pyrite crystals by Bennett and Tributsch (1978), Rodriguez-Leiva and Tributsch (1988), Mustin et al. (1992), and Murthy and Natarajan (1992), and to pyrite/arsenopyrite-containing auriferous ore by Norman and Snyman (1988).

Mustin et al. (1992) recognized four phases in the leaching of pyrite by *T. ferrooxidans* in a stirred reactor. The first phase, which lasted about 5 days, featured a measurable decrease in unattached bacteria. The small amount of dissolved ferric iron added with the inoculum was reduced by reacting with some of the pyrite. The second phase, which also lasted about 5 days, featured the start of pyrite dissolution with oxidation of its iron and sulfur, but with sulfur being preferentially oxidized. Unattached bacteria multiplied exponentially and the pH began to drop. The third phase, which lasted about 10 days, featured a significant increase in dissolved ferric iron, the ferrous iron concentration remaining low. Both iron and sulfur in the pyrite were being oxidized at high rates. However, the rate of sulfur oxidation decreased with time relative to iron oxidation, the ratio of sulfate/ferric iron becoming stoichiometric by day 18. Unattached bacteria continued to increase exponentially, and the pH continued to drop. The surface of the pyrite crystals began to show evidence of corrosion cracks. In the fourth and last phase, which lasted about 25 days, the dissolved Fe(III)/Fe(II) ratio decreased slightly, iron and sulfur in the pyrite continued to be strongly oxidized, the unattached bacteria reached a stationary phase, while the pH continued to drop to 1.3 by the 45th day. The surface of the pyrite particles now showed easily recognizable square or hexagonal corrosion pits. During the entire experiment, Mustin and co-workers also followed pH and electrochemical (redox) changes during the oxidation of the pyrite.

Bacterial attachment appears not to be random, but to occur at specific sites and even faces of crystals. Some evidence suggests that direct microbial attack is initiated at sites of crystal imperfections. Selective attachment of *Thiobacillus ferrooxidans* or *Sulfolobus acidocaldarius* to newly exposed pyrite crystals in coal is very rapid, i.e., about 90% complete in 2 and 5 min, respectively (Bagdigian and Myerson, 1986; Chen and Skidmore, 1987; 1988). Although the details of how microbes attack crystal lattices of metal sulfide crystals are not yet understood, a model has been proposed in which the bacterial cells act as catalytic conductors in transferring electrons from cathodic areas on the crystal surfaces of a metal sulfide to oxygen (Fig. 18.2). The electron transport system of the

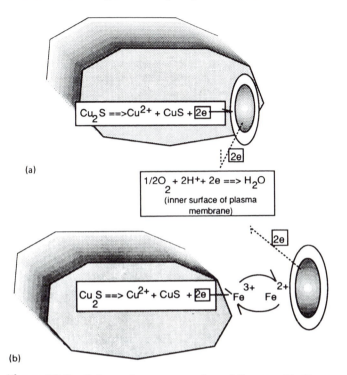

(a)

(b)

Figure 18.2 Schematic representation of direct and indirect oxidation of a particle of Cu_2S by *Thiobacillus ferrooxidans*. (a) Direct oxidation; in this model the bacterial cell acts essentially as a conductor of electrons from the crystal lattice of Cu_2S to oxygen while attached to the particle surface. (b) Indirect oxidation; in this model the bacterial cells generates and regenerates the oxidant (Fe^{3+}) which acts as a shuttle that carries electrons from the crystal lattice of Cu_2S to the bacterium which transfers the electrons to oxygen.

bacteria is the catalyst. Indeed, the bacteria can be viewed as cathodic extensions.

Direct evidence for enzymatic attack of synthetic covellite (CuS) through measurement of oxygen consumption and Cu^{2+} and SO_4^{2-} ion production in the presence and absence of the enzyme inhibitor trichloracetate (8 mM) has been obtained (Rickard and Vanselow, 1978). In the case of CuS, only the sulfide moiety of the mineral is attacked because the metal moiety is already as oxidized as possible. The oxidation of the mineral probably proceeds in two steps (Fox, 1967):

$$CuS + 0.5O_2 + 2H^+ \xrightarrow{\text{bacteria}} Cu^{2+} + S^\circ + H_2O \tag{18.9}$$

$$S^\circ + 1.5O_2 + H_2O \xrightarrow{\text{bacteria}} H_2SO_4 \tag{18.10}$$

By contrast, *Thiobacillus thioparus* promotes covellite oxidation only after **autoxidation** of the mineral to $CuSO_4$ and S° [similar to Reaction (18.9), but without bacterial catalysis] (Rickard and Vanselow, 1978). It is the bacterial catalysis of oxidation of S° to sulfate that helps the reaction by removing a product of the autoxidation of covellite.

In some instances, both an oxidizable metal moiety and the sulfide moiety may be attacked by separate enzymes, as, for example, in the case of chalcopyrite ($CuFeS_2$) (assuming the Fe of chalcopyrite to have an oxidation state of $+2$) (Duncan et al., 1967; Shrihari et al., 1991). Although Duncan and co-workers reported Fe and S to be simultaneously attacked, Shrihari and co-workers found that iron-grown *T. ferrooxidans* oxidized the sulfide sulfur of chalcopyrite by direct attack before oxidizing ferrous iron in solution to ferric iron. Once dissolved ferric iron attained a significant concentration, it promoted chemical oxidation of residual chalcopyrite. The overall reaction may be written as follows:

$$4CuFeS_2 + 17O_2 + 4H^+ \xrightarrow{\text{bacteria}} 4Cu^{2+} + 4Fe^{3+} + 8SO_4^{2-} \atop + 2H_2O \tag{18.11}$$

$$4Fe^{3+} + 12H_2O \Longrightarrow 4Fe(OH)_3 + 12H^+ \tag{18.12}$$

$$4CuFeS_2 + 17O_2 + 10H_2O \xrightarrow{\text{bacteria}} 4Cu^{2+} + 4Fe(OH)_3 \atop + 8SO_4^{2-} + 8H^+ \tag{18.13}$$

In other cases the oxidizable metal moiety may be attacked before the sulfide, as in the case of chalcocite (Cu_2S) oxidation (Fox, 1967; Nielsen and Beck, 1972) (Fig. 18.2):

$$Cu_2S + 0.5O_2 + 2H^+ \xrightarrow{\text{bacteria}} Cu^{2+} + CuS + H_2O \tag{18.14}$$

$$CuS + 0.5O_2 + 2H^+ \xrightarrow{\text{bacteria}} Cu^{2+} + S^\circ + H_2O \tag{18.15}$$

$$S^\circ + 1.5O_2 + H_2O \xrightarrow{\text{bacteria}} H_2SO_4 \tag{18.16}$$

Digenite (Cu_9S_5) may be an intermediate in the formation of CuS from Cu_2S (Nielsen and Beck, 1972).

The direct microbial oxidation of iron pyrite (FeS_2) is best summarized by the reaction

$$FeS_2 + 3.5O_2 + H_2O \xrightarrow{\text{bacteria}} Fe^{2+} + 2H^+ + 2SO_4^{2-} \tag{18.17}$$

The ferrous iron generated in this reaction is further oxidized by bacteria not attached to pyrite:

$$2Fe^{2+} + 0.5O_2 + 2H^+ \xrightarrow{\text{bacteria}} 2Fe^{3+} + H_2O \tag{18.18}$$

The resultant ferric iron then causes **chemical** oxidation of residual pyrite according to the reaction

$$FeS_2 + 14Fe^{3+} + 8H_2O \Longrightarrow 15Fe^{2+} + 2SO_4^{2-} + 16H^+ \tag{18.19}$$

whereby Fe^{2+} is regenerated. The bacterial oxidation of Fe^{2+} to Fe^{3+} becomes the rate-controlling reaction in this system (Singer and Stumm, 1970). This is because in the absence of the iron-oxidizing bacteria at the acid pH at which pyrite oxidation occurs, the rate of Fe^{2+} oxidation is extremely slow.

A number of different acidophilic, iron-oxidizing bacteria have been associated with direct metal sulfide oxidation. The most extensively studied are classified as *Thiobacillus ferrooxidans*, *Sulfolobus* spp., and *Acidianus brierleyi* (formerly *Sulfolobus brierleyi*). All three are aerobes and strongly acidophilic, growing best in a pH range from about 1.5 to 2.5. *T. ferrooxidans* is a mesophlic eubacterium whereas *Sulfolobus* and *Acidianus brierleyi* are thermophilic archaebacteria.

The less well-characterized acidophilic, moderately thermophilic (50–55°C) iron oxidizers, when cultivated in the laboratory, require small amounts of yeast extract, cysteine, or glutathione for growth. They are able to use iron as energy source and can grow on pyrite (Brierley and Lockwood, 1977; Brierley, 1978b; Brierley et al., 1978; Norris, 1990).

Nutritionally, *T. ferrooxidans*, *Sulfolobus*, and *Acidianus brierleyi* can grow autotrophically on iron and sulfur. However, in the case of the latter two organisms, their autotrophic growth is stimulated in laboratory culture by a trace of yeast extract. Appropriate metal sulfides are used as energy sources by all three organisms. Depending on the oxidation state of the metal moiety in the metal sulfide, both it and the sulfide may serve as energy sources. Thus in the case of chalcocite (Cu_2S) oxidation by *T. ferrooxidans*, as already described, the initial oxidation step involves the cuprous copper [Cu(I)] of the compound. The organism is able to derive energy from this oxidation, which it can use in CO_2 fixation (Nielsen and Beck, 1972). Cell extracts of *T. ferrooxidans* have been prepared which catalyze the oxidation of cuprous copper in Cu_2S but not of elemental sulfur (Imai et al., 1973). The oxidation is not inhibited by quinacrin (atebrine). It needs the addition of a trace of iron for proper activity. The effect of traces or iron on metal sulfide oxidation had been previously noted in experiments in which the addition of 9 mg of ferrous iron per liter of medium stimulated metal sulfide oxidation by whole cells of *T. ferrooxidans* (Ehrlich and Fox, 1967b).

T. ferrooxidans can use NH_4^+ and some amino acids as nitrogen sources (see Sugio et al., 1987; also Chapter 14). At least some strains are capable of nitrogen fixation (Mackintosh, 1978; Stevens et al., 1986).

T. ferrooxidans is very versatile in attacking metal sulfides. It has been reported to oxidize arsenopyrite (FeS_2FeAs_2), bornite (Cu_5FeS_4), chalcocite (Cu_2S), chalcopyrite ($CuFeS_2$), covellite (CuS), enargite ($3Cu_2$-$S·As_2S_5$), galena (PbS), millerite (NiS), orpiment (As_2S_3), pyrite (FeS_2), marcasite (FeS_2), sphalerite (ZnS), stibnite (Sb_2S_3), and tetrahedrite ($Cu_8Sb_2S_7$) (Silverman and Ehrlich, 1964). In addition, the oxidation of gallium sulfide, synthetic preparations of CoS, NiS, and ZnS, and pyrrhotite has been reported (Torma, 1971, 1978; Pinka, 1991; Bhatti et al., 1993).

The bacterial oxidation of PbS by *T. ferrooxidans* presents a special problem because the oxidation product, $PbSO_4$, is relatively insoluble. As oxidation of PbS proceeds, $PbSO_4$ is likely to accumulate on the crystal surface and block further access to PbS by bacteria and oxygen. In the laboratory, oxidation in a batch culture with agitation (Silver and Torma, 1974) or oxidation in a large volume of lixiviant in a continuous-flow reactor with stirring (Ehrlich, 1988) largely eliminates this problem.

Acidianus brierleyi is probably as versatile as *T. ferrooxidans* in its ability to oxidize various metal sulfides. In addition, it can oxidize MoS_2 (Brierley and Murr, 1973), which *T. ferrooxidans* can not because it is inhibited by molybdate ion (Tuovinen et al., 1971).

Metal-sulfide-oxidizing bacteria are naturally associated with metal sulfide-containing deposits, including ore bodies, bituminous coal seams, and the like. Under natural conditions, growth and activity of the microorganisms may be quite limited owing to one or more restricting environmental factors, such as limited access of air and/or moisture (Brock, 1975), limited access to the energy source (the metal sulfide crystals), limited nitrogen source, or unfavorable temperature (Ehrlich and Fox, 1967b; Ahonen and Tuovinen, 1989, 1992).

Indirect Oxidation   In indirect biooxidation of metal sulfides, a major role of the bacteria is the generation of a lixiviant which chemically oxidizes the sulfide ore. This lixiviant is acid ferric iron (Fe^{3+}). It may be generated from dissolved ferrous iron (Fe^{2+}) at pH values from 3.5 to 5.0 by *Metallogenium* in a mesophilic temperature range (Walsh and Mitchell, 1972a). At pH values below 3.5, ferric iron may be generated from ferrous iron by *Thiobacillus ferrooxidans* and *Leptospirillum ferrooxidans* Markosyan (see Balachova et al., 1974) in a mesophilic temperature range, and by *Sulfolobus* spp., *Acidianus brierleyi*, *Sulfobacillus thermosulfidooxidans*, and other as-yet-unidentified bacteria in a thermophilic temperature range (Brierley, 1978b; Brierley and Brierley, 1973; Brierley and Murr, 1973; Brierley and Lockwood, 1977; Balashova et al., 1974; Golovacheva and Karavaiko, 1979; Harrison and Norris 1985; Pivovarova et al., 1981; Segerer et al., 1986). Ferric iron may also be generated from iron pyrites (e.g., FeS_2) by *T. ferrooxidans* and other iron-oxidizing acidophiles as described previously. In whatever way it is formed, ferric iron in acid solution acts as oxidant on the metal sulfides (Ehrlich and Fox, 1967b):

$$MS + 2Fe^{3+} \Longrightarrow M^{2+} + S° + 2Fe^{2+} \tag{18.20}$$

where M may be any metal in an appropriate oxidation state, which does not always have to be divalent. It should be noted that in this type of chemical reaction the sulfide is only oxidized to sulfur ($S°$). Further chemical oxidation to sulfuric acid by oxygen is very slow, but is likely to be greatly accelerated by microorganisms like *T. thiooxidans*, *T. ferrooxidans*, *Sulfolobus spp.*, *and Acidianus brierleyi*. Elemental sulfur may form on the surface of metal sulfide crystals and interfere with their further chemical oxidation. The chemical oxidation of metal sulfides must occur in acid solution below pH 5.0 to keep enough ferric iron in solution. In nature, the needed acid may be formed chemically through autoxidation of sulfur, or more likely biologically through bacterial oxidation of sulfur, or iron either as ferrous iron in solution or as pyrite. In ferrous iron oxidation, the acid forms as follows:

$$2Fe^{2+} + 0.5O_2 + 2H^+ \xrightarrow{\text{bacteria}} 2Fe^{3+} + H_2O \tag{18.21}$$

$$Fe^{3+} + 3H_2O \Longrightarrow Fe(OH)_3 + 3H^+ \tag{18.22}$$

As this reaction normally takes place in the presence of sulfate, the ferric hydroxide may convert to the more insoluble jarosite, especially in the presence of *T. ferrooxidans* and probably other acidophilic iron oxidizers (Lazaroff, 1983; Lazaroff et al., 1982; Lazaroff et al., 1985; Carlson et al., 1992):

$$A^+ + 3Fe(OH)_3 + 2SO_4^{2-} \Longrightarrow AFe_3(SO_4)_2(OH)_6 + 3OH^- \tag{18.23}$$

where A^+ may represent Na^+, K^+, NH_4^+, or H_3O^+ (Duncan and Walden, 1972). The formation of jarosite decreases the ratio of protons produced per iron oxidized from 2:1 to 1:1. In pyrite oxidation, the acid forms as a result of Reactions (18.17), (18.18), and (18.19). Reaction (18.17) proceeds by autoxidation if the process is indirect. Like sulfur, jarosite may also form on the surface of metal sulfide crystals and block further oxidation.

It should be noted that in the case of pyrite oxidation by ferric iron, the sulfur is oxidized to sulfate while the iron remains in the ferrous state. The ferrous iron will subsequently be further oxidized to ferric iron by biological or abiological means with the production of additional acid. Since iron pyrites usually accompany other metal sulfides in nature, iron pyrite oxidation is an important source of acid for the oxidizing reactions of nonferrous metal sulfides, especially those which consume acid. In some cases the host rock in which metal sulfides, including pyrites, are contained, may itself react with acid and thus raise the pH of the environment enough to cause complete precipitation of ferric iron and thereby prevent any oxidation of metal sulfide by it.

18.6 BIOEXTRACTION OF METAL SULFIDE ORES BY COMPLEXATION

Some metal sulfide ores cannot be oxidized by acidophilic iron-oxidizing bacteria because they contain too much acid-consuming constituents in the host rock (gangue). The metals in such ores may be amenable to extraction by ligands formed by some microoganisms such as fungi (Burgstaller and Schinner, 1993). Wenberg et al. (1971) reported, for instance, the isolation of the fungus *Penicillium* sp. from a mine tailings pond of the White Pine Copper Co. in Michigan (U.S.A.), which produced unidentified metabolites in Czapek's broth containing sucrose, $NaNO_3$, and cysteine, methionine, or glutamic acid, which could leach copper from the sedimentary ores of the White Pine deposit. *T. ferrooxidans* could not be employed for leaching of this ore because of the presence of significant quantities of calcium carbonate that would neutralize the required acid. Similar findings were reported by Hartmannova and Kuhr (1974), who found not only *Penicillium* sp. but also *Aspergillus* sp. (e.g., *A. niger*) to be active in producing complexing compounds. Wenberg et al (1971) reported that if their fungus was grown in the presence of copper ore (sulfide or native copper minerals with basic gangue constituents), the addition of some citrate lowered the toxicity of the extracted copper for the fungus. Better results were obtained if the fungus was grown in the absence of ore, and the ore was then treated with spent medium. The principle of action of the fungi in these experiments is similar to that in the observations of Kee and Bloomfield (1961) in regard to the dissolution of the oxides of several trace elements (e.g., ZnO, PbO_2, MnO_2, $CoO \cdot Co_2O_3$) with anaerobically fermented plant material (Lucerne and Cocksfoot). It is also similar to the principle of action in the experiments of Parès (1964a,b,c), in which *S. marcescens, B. subtilis, B. sphaericus*, and *B. firmus* solubilized copper and some other metals that were associated with laterites and clays by generating appropriate chelates in special culture media. The chelates of the metal ions are stabler than the original insoluble form of the metals, which forces their solubliziation, as illustrated by the following equation:

$$MA + HCh \Longrightarrow MCh + H^+ + A^- \qquad (18.24)$$

where MA is a metal salt, HCh a chelating agent (ligand), MCh the resultant metal chelate, and A^- the counter ion of the original metal salt, S^{2-} in the case of the metal sulfides. The S^{2-} may undergo chemical or bacterial oxidation. The use of carboxylic acids in industrial leaching of ores has been proposed as a general process (Chemical Processing, 1965).

18.7 BIOLEACHING OF METAL SULFIDE AND URANINITE ORES

Metal Sulfide Ores   When metal sulfide ore bodies are exposed to moisture and air during mining activities, the ore minerals may slowly begin to undergo oxidation which may be accelerated by native microorganisms, especially acidophilic iron oxidizers. This leads to the formation of acid mine drainage which contains the metal solubilized by the oxidation. The microbial oxidation is harnessed on an industrial scale by mining companies to extract the metal, currently mostly from low-grade portions of the ore. Some successful field trials have now also been performed with high-grade copper ores. Low-grade sulfide ores generally contain metal values at concentrations below 0.5% (wt/wt). Their extraction is uneconomic by conventional means of milling and and subsequent ore enrichment (ore beneficiation) by flotation because of an unfavorable gangue/metal ratio. Rubblized, low-grade ore may be leached in place (McCready, 1988) or by constructing heaps with it to heights of tens of feet and watering it with lixiviant. Waste rock and mine tailings, by-products of mining and conventional metal recovery that still contain traces of metal values may be placed in heaps or dumps and similarly leached by watering them with lixiviant (Fig. 18.3A). The lixiviant is applied in a fine spray to avoid waterlogging, which would exclude needed oxygen. Simultaneous diffusion of oxygen into the ore in a heap or dump being leached is important because the microbial leaching process is aerobic. In some heap leaching operations, pipes are placed in strategic positions within the heap during its construction to optimize access of air to deeper portions of the heap. Depending on ore composition, the lixiviant may be water, acidified water, or acid ferric sulfate from other leach operations on the same mining property. The lixiviant promotes pyrite oxidation by appropriate acido-

(A)

Figure 18.3   Some steps in bioleaching of copper from sulfide ores. (A) Top of a leach dump showing transite pipes and hose for watering the dump with barren solution. The dark patches represent moistened areas in which oxidation has occurred. Note the thin streams of solution issuing from hoses in the distance. (B) Launder used in recovering copper by cementation from pregnant solutions from leach dumps. The copper recovered in this way has to be purified by smelting. Currently the preferred method of copper recovery from pregnant solution is electrolysis because it yields a pure product. (Courtesy of Duval Corporation.)

philic iron oxidizers resulting in their growth and multiplication. In time, their activity spreads to the metal sulfides of value resulting in direct oxidative attack by the microbes, and in indirect attack by the acid ferric sulfate which the bacteria generate from pyrite and any chalcopyrite present. Initial solutions from an operation may be recirculated without any treatment, but ultimately, as microbial and chemical activities continue, the lixiviant becomes charged with dissolved metal values and is called **pregnant solution**. It is collected in a special sump. It frequently harbors a variety of microorgansisms, including autotrophic and heterotrophic bacteria, fungi, and protozoa despite its very acid pH and high metal content (e.g., Ehrlich, 1964b). When the concentration of desired metal

(B)

values in the pregnant solution in the sump has become sufficient, they are stripped from the solution. This may be accomplished in one of several ways. A formerly widely used method for copper separation involved treatment of the pregnant solution with sponge iron in a basin called a launder (Fig. 18.3B). Copper was precipitated by the iron in a reaction called **cementation** as a result of the following spontaneous reaction:

$$Cu^{2+} + Fe^{\circ} \Longrightarrow Cu^{\circ} + Fe^{2+} \tag{18.25}$$

The copper metal formed in this way required further refinement by smelting.

The pregnant solution after stripping of the copper is called **barren solution**. It can be recirculated as lixiviant in the leaching operation. However, when cementation was used to strip the copper from solution, this barren solution became significantly enriched with Fe^{2+} as a result of the launder process. In many instances it was enriched to such an extent that some of the ferrous iron had to be removed from it before it was recirculated in the leach process. Without removal of the excess ferrous iron, the danger was excessive jarosite formation in the leaching operation and its precipitation on ore mineral surfaces and in the drainage channels of the ore heap or dump, with consequent plugging and thus obvious impeding of the leach process. Iron removal from acid barren solution was best accomplished by biooxidation in shallow lagoons called **oxidation ponds**. Here acidophilic iron oxidizers promoted the oxidation of the ferrous iron with concomitant acidification. A significant portion of the oxidized iron precipitated as basic ferric sulfates, including jarosite. The residual iron, mostly ferric, when reintroduced into a leaching operation causes indirect leaching. The acid in the lixiviant is likely to cause rapid weathering of the gangue with the liberation of aluminum from aluminosilicates. This weathering is important in exposing occluded metal sulfide crystals to lixiviant and the bacteria active in the leaching process. The liberated aluminum may be ultimately separated as $Al(OH)_3$ from pregnant solution by neutralizing it (Zimmerley et al., 1958; Moshuyakova et al., 1971). The $Al(OH)_3$ may be subsequently used in the manufacture of aluminum metal. High acidity in the lixiviant in leaching can play an important role in preventing metal ions formed in the leaching from being adsorbed by the host rock (gangue)(Ehrlich, 1977; Ehrlich and Fox, 1967b).

A currently preferred method of recovering metal values from pregnant solution involves **electrowinning** if the pregnant solution contains only one major metal value, or **solvent extraction**, if several different metal values are present, followed by electrowinning. Electrowinning involves deposition of a metal value on a cathode of the metal to be recovered in an electrolytic reaction in which the anode may be made of carbon. The product of electrowinning is usually of high purity and may not need further refining. These processes have the advantage of not raising the ferrous iron concentration in barren solution, but recovery of metal values by solvent extraction could introduce reagents into the barren solution that are inhibitory to bacteria-involved leaching.

The acidophilic iron-oxidizing bacteria developing in an ore leaching process play a dual role in solubilizing metal values, as already explained. They generate acid ferric sulfate lixiviant by attacking pyrite and by reoxidizing Fe^{2+}, and they also attack the mineral sulfides directly. Although it is usually not possible to assess to what extent they are involved in direct and in indirect oxidation, it is readily demonstrated in laboratory experiments that bacteria such as iron-oxidizing thiobacilli accelerate the metal sulfide oxidation process significantly. Owing to the fact that metal sulfide oxidation is an exothermic process, the interior temperature of some ore dumps or heaps, especially if not well ventilated, can rise as high as of 70–80°C, which is unfavorable for the growth of mesophilic bioleaching bacteria, such as *T. ferrooxidans* and *Leptospirillum ferrooxidans*. The concept has been developed that mesophiles such as *T. ferrooxidans* promote self-heating in ore leaching and thereby accelerate the chemical process (e.g., Lyalikova, 1960) and that its other important role is to regenerate ferric iron in the oxidation ponds. From this concept it was inferred that the leach process in dumps and heaps is mostly an abiotic (indirect) process. While it cannot be denied that *T. ferrooxidans* and *L. ferroxidans* are unable to live at temperatures in excess of about 37°C, acidophilic, iron-oxidizing thermophiles are now known, as previously discussed, and have been detected in the interior of leach dumps and heaps where the temperature has risen to the thermophilic range. These microorganisms, which catalyze reactions similar to *T. ferrooxidans*, operate optimally at the higher temperatures. Thus leaching in all parts of a dump or heap is most likely biological, involving direct and indirect action. It has also been suggested that the thermophiles may be responsible for regulating the internal temperature of active leach dumps and heaps (Murr and Brierley, 1978).

The primary copper mineral in ore bodies of magmatic-hydrothermal origin is chalcopyrite ($CuFeS_2$). This mineral tends to be somewhat refractory to chemical leaching by acidic ferric sulfate when compared to the secondary copper sulfide minerals chalcocite (Cu_2S) and covellite (CuS). The oxidation of chalcopyrite in nature can thus be significantly enhanced by *T. ferrooxidans* and *Acidianus brierleyi* (Razzell and Trussell, 1963; Brierley, 1974). Ferric iron, when present in excess of 1,000 ppm, has been found to inhibit chalcopyrite oxidation (Ehrlich, 1977; Duncan and Walden, 1972), probably because it precipitates as jarosite or adsorbs to the surface of residual chalcopyrite crystals and prevents further oxidation. Bacteria themselves may interfere by generating excess ferric iron from ferrous iron which precipitates or is adsorbed by the residual chalcopyrite.

A typical leach cycle in which copper is recovered either by cementation or by electrowinnig is diagrammed in Fig. 18.4 [see Brierley (1978a, 1982), and Lundgren and Malouf (1983), for further discussions of bacterial leaching].

Uraninite Leaching   The principles of bioleaching have also been applied on a practical scale to the leaching of uraninite ores, especially low-grade ores. This process may involve dump, heap, or in situ leaching (Wadden and Gallant, 1985; McCready and Gould, 1990). *Thiobacillus ferrooxidans* is the organism that has been harnessed for this process. Its action in this instance is chiefly indirect by generating an oxidizing lixiviant, acid ferric

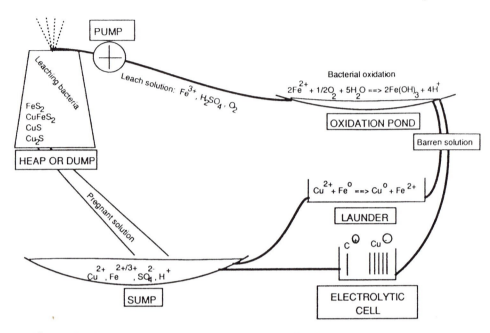

Figure 18.4 Schematic representation of a bioleach circuit for heap or dump leaching of copper-sulfide ore. In addition to copper recovery from pregnant solution by cementation in a launder or by electrowinning, copper can also be recovered by solvent extraction in combination with electrowinning. This is especially useful if the pregnant solution contains two or more base metals.

sulfate which oxidizes U(IV) in the uraninite to soluble U(VI). The overall reactions can be summarized as follows:

$$2Fe^{2+} + 0.5O_2 + 2H^+ \xrightarrow{\text{bacteria}} 2Fe^{3+} + H_2O \qquad (18.26)$$

Some of the resultant ferric iron hydrolyzes:

$$Fe^{3+} + 3H_2O \Longrightarrow Fe(OH)_3 + 3H^+ \qquad (18.27)$$

The ultimate product of this hydrolysis will more likely be a basic ferric sulfate such as jarosite rather than ferric hydroxide. The consequence of this is that the net yield of acid (protons) will be less than in reaction 18.27 (see reaction 18.23). The remaining dissolved ferric iron then reacts with the uraninite to form uranyl ions:

$$UO_2 + 2Fe^{3+} \Longrightarrow 2Fe^{2+} + UO_2^{2+} \qquad (18.28)$$

T. ferrooxidans will reoxidize the ferrous iron from this reaction, thus keeping the process going without a need for a continual external, fresh supply of ferric iron to the system. The dissolved uranium may be recovered from solution through concentration by ion exchange.

The acidophilic iron-oxidizing bacteria are often naturally associated with the ore body if it contains pyrite or other form of iron sulfide. Their growth may be stimulated in in situ leaching of depleted mines by intermittent spraying of nutrient-enriched solution on the floors, walls, and mud of mine stopes (Zajic, 1969). If the leachate should become anoxic and its pH rises, sulfate-reducing bacteria may develop in it. These bacteria can reprecipitate UO_2 as a result of reaction of UO_2^{2+} with H_2S:

$$UO_2^{2+} + H_2S \Longrightarrow UO_2 + 2H^+ + S° \tag{18.29}$$

Some recent observations have given indications that *T. ferrooxidans* can enzymatically catalyze oxidation of U(IV) to U(VI) and use some of the energy from this reaction to assimilate CO_2 (DiSpirito and Tuovinen, 1981, 1982a,b; see also Chapter 16). Living cells accumulate significantly less uranium than dead cells and bind it chiefly in their cell envelope (DiSpirito et al., 1983). Despite this capacity of *T. ferrooxidans* to oxidize U(IV) directly, bioleaching of uraninite is believed to involve chiefly the indirect mechanism.

Uranium bioleaching is another example of a natural process that is artificially stimulated. Under natural conditions, the reactions described here must occur on a very limited scale and thus cause only slow mobilization of uranium.

Mobilization of Uranium in Granitic Rocks by Heterotrophs *T. ferrooxidans* is not the only organism capable of uranium mobilization. Heterotrophic microorganisms such as some members of the soil microflora and bacteria from granites or mine waters (*Pseudomonas fluorescens, P. putida, Achromobacter*) can mobilize uranium in granite rocks, ore, and sand by weathering through mineral interaction with organic acids and chelators produced by the microorganisms (Magne et al., 1973, 1974; Zajic, 1969). Magne et al. found experimentally that the addition of thymol to percolation columns of uraniferous material fed with glucose solution selected a more efficient extractive flora in terms of greater production of oxalic acid. The authors suggested that in nature, phenolic and quinoid compounds of plant origin may serve the role of thymol. The authors also reported that microbes can precipitate uranium by digestion of soluble uranium complexes (Magne et al., 1974), that is, by microbial destruction of the organic moiety that complexes the uranium. The observations may explain how in nature uranium in granitic rock may be mobilized by bacteria and reprecipitated and concentrated elsewhere under the influence of microbial activity.

Industrial Versus Natural Bioleaching Industrial bacterial leaching of metal sulfide ores is the harnessing of naturally occurring, microbiological processes by creating selective and optimized conditions which allow leaching to occur at fast rates. In the absence of human intervention, these same processes occur only at very slow rates in highly localized situations, contributing in the case of sulfide ore to a very slow, gradual change from reduced to oxidized ore. This accounts for the relative stability of undisturbed ore bodies.

18.8   FORMATION OF ACID COAL-MINE DRAINAGE

When bituminous coal seams which contain pyrite inclusions are exposed to air and moisture during mining, the pyrites undergo oxidation leading to the formation of acid mine drainage. With the onset of pyrite oxidation, iron-oxidizing thiobacilli become readily detectable (Leathen, 1953). *Thiobacillus thiooxidans* also makes an appearance. Pyrite biooxidation proceeds by the reactions previously described (18.17–18.19) catalyzed principally by *T. ferrooxidans*. *T. thiooxidans*, which cannot oxidize ferrous iron, probably oxidizes elemental sulfur and other partially reduced forms of sulfur, which may form as intermediates in pyrite oxidation to sulfuric acid (e.g., Mustin et al., 1992; 1993) [Reaction (18.16)]. The chief products of pyrite oxidation thus are sulfuric acid and basic ferric sulfate (jarosite and amorphous basic iron sulfates). Streams which receive this mine drainage may exhibit pH values ranging from 2 to 4.5, and sulfate ion concentrations ranging from 1,000 to 20,000 mg L^{-1}, but a non-detectable ferrous iron concentration (Lundgren et al., 1972). It has been proposed that a *Metallogenium*-like organism which has been isolated from acid mine-drainage, may be the dominant iron-oxidizing organisms attacking the pyrite in exposed pyrite-containing coal seams until the pH drops below 3.5. Then the more acid-tolerant *T. ferrooxidans* and possibly *Leptospirillum ferrooxidans* take over (Walsh and Mitchell, 1972b). *L. ferrooxidans* will only be able to oxidize Fe^{2+} directly but not FeS_2. The *Metallogenium*-like organism does not, however, appear to be essential to lower the pH in a pyritic environment to make it favorable for the acidophilic iron oxidizers. *T. ferrooxidans* itself appears to be capable of it, at least in pyrite-containing coal or overburden (Kleinmann and Crerar, 1979). It may accomplish this by intial direct attack on pyrite, creating an acid microenvironment from which the organism and the acid it generates spread.

Microbial succession in coal spoil has been studied under laboratory conditions (Harrison, 1978). An artificial coal spoil consisting of a homogeneous mixture of 1 part of crushed, sifted coal plus 2 parts of shale and 8 parts of subsoil from the overburden of a coal deposit was heaped in a mound 50 cm in diameter and 25 cm high on a plastic tray. The mound was inoculated with 20 L of an emulsion of acid soil, drainage water and mud from a spoil from an old coal strip mine, which was poured on the bottom of the plastic tray. The inoculum was absorbed by the mound, migrating upward, presumably by capillary action. Evaporation losses during the experiment were made up by periodic additions of distilled water to the free liquid in the tray.

Initial samples taken at the base of the mound yielded evidence of the presence of heterotrophic bacteria. They were dominant and reached a population density of about 10^7 cells g^{-1} within 2 weeks. After 8 weeks, heterotrophs were still dominant, although the pH had dropped from 7 to 5. Between 12 and 20 weeks, this population decreased by about an order of magnitude, coinciding with a slight decrease in pH to just below pH 5 caused by a burst of growth by sulfur-oxidizing bacteria, which then died off progressively. Thereafter the heterotrophic population increased again to just below 10^7 g^{-1}.

In samples from near the summit of the mound, heterotrophs also predominated for the first 15 weeks but then decreased dramatically from 10^6 to 10^2 cells g^{-1}, concomitant with a drop in pH to 2.6. The pH drop was correlated with a marked rise in the population density of sulfur- and iron-oxidizing autotrophic bacteria, the former dominating briefly over the latter in the initial weeks. Protozoans, algae, an arthropod, and a moss were also noted, mostly at the higher pH values. *Metallogenium* of the type of Walsh and Mitchell (1972a) was not seen.

The sulfur-oxidizing bacteria were assumed to be utilizing elemental sulfur resulting from the oxidation of pyrite by ferric sulfate:

$$FeS_2 + Fe_2(SO_4)_3 \Longrightarrow 3FeSO_4 + 2S° \tag{18.30}$$

More specifically, as Mustin et al. (1992, 1993) have indicated, the sulfur may arise as a result of anodic reactions at the surface of pyrite crystals. It is also possible that at least some of the sulfur arises indirectly from microbial sulfate reduction in anaerobic zones of the coal spoil, yielding H_2S, which then becomes the energy source for thiobacilli like *T. thioparus* that oxidize it to sulfur. Anaerobic bacteria were not sought in this study.

After 7 weeks of incubation, a mineral efflorescence developed on the surface of the mound. It consisted mainly of sulfates of Mg, Ca, Na, Al, and Fe. The magnesium sulfate was in the form of a hexahydrate rather than epsomite. The metals were leached from the coal, but magnesium was also leached from the overburden material.

A thermophilic, acidophilic *Thermoplasma acidophilum* has been isolated from a coal refuse pile that had become self-heated (Darland et al., 1970). The organism lacks a true eubacterial cell wall and resembles mycoplasmas. It is an archaebacterium. Its growth temperature optimum was 59°C (range 45–62°C), and its optimum pH for growth was between 1 and 2 (range 0.96–3.5). The organism is a heterotroph, growing readily in a medium of 0.02% $(NH_4)_2SO_4$, 0.05% $MgSO_4$, 0.025% $CaCl_2 \cdot 2H_2O$, 0.3% KH_2PO_4, 0.1% yeast extract, and 1.0% glucose at pH 3.0. Its relation to the coal environment and its contribution, if any, to the acid mine drainage problem needs to be clarified.

18.9 SUMMARY

Metal sulfides may occur locally in high concentrations, in which case they constitute ores. Although most nonferrous sulfides are formed abiogenically through magmatic and hydrothermal processes, a few sedimentary deposits are of biogenic origin. More important, some sedimentary ferrous sulfide accumulations are biogenically formed. The microbial role in biogenesis of any of these sulfide deposits is the genesis of H_2S, usually from the bacterial reduction of sulfate, but in a few special cases possibly from the mineralization of organic sulfur compounds. Since metal sulfides are highly insoluble, spontaneous reaction of metal ions with the biogenic sulfide proceeds readily. Biogenesis of specific metal sulfide minerals has been demonstrated in the laboratory. These experiments require relatively insoluble metal compounds as starting materials to limit the toxicity of the metal ions to the sulfate-reducing bacteria. In nature, adsorption of the metal ions by sediment components serves a similar function in lowering the concentration of metal ions below their toxic level for sulfate reducers.

Metal sulfides are also subject to oxidation by bacteria, such as *Thiobacillus ferrooxidans*, *Sulfolobus* spp., and *Acidianus brierleyi*. The action may involve direct oxidative attack of the crystal lattice of a metal sulfide, or indirect oxidative attack by generating a lixiviant (acid ferric sulfate) which oxidizes the metal sulfide chemically. The indirect mechanism is of primary importance in the solubilization of uraninite (UO_2). Microbial oxidation of metal sulfides is industrially exploited in extracting metal values from low-grade metal sulfide ore and uraninite. Pyrite oxidation by these bacteria in bituminous coal seams exposed as a result of mining activity, on the other hand, is an industrially deleterious process; it is the source of acid mine drainage.

REFERENCES

Ahonen, L., and O. H. Tuovinen. 1989. Microbiological oxidation of ferrous iron at low temperatures. Appl. Environ. Microbiol. 55: 312–316.

Ahonen, L., and O. H. Tuovinen. 1992. Bacterial oxidation of sulfide minerals in column leaching experiments at suboptimal temperatures. Appl. Environ. Microbiol. 58:600–606.

Baas Becking, L. G. M., and D. Moore. 1961. Biogenic sulfides. Econ. Geol. 56:259–272.

Bagdigian, R. M. and A. S. Myerson. 1986. The adsorption of *Thiobacillus ferrooxidans* on coal surfaces. Biotech. Bioeng. 28:467–479.

Balashova, V. V., I. Ya. Vedenina, G. E. Markosyan, and G. A. Zavarzin. 1974. The autotrophic growth of *Leptospirillum ferrooxidans*. Mikrobiologiya 43:581–585 (Engl. transl. pp. 491–494).

Ballard, R. D., and J. F. Grassle. 1979. Strange world without sun. Return to oases of the deep. Natl. Geogr. Mag. 156:680–703.

Bennett, J. C., and H. Tributsch. 1978. Bacterial leaching patterns of pyrite crystal surfaces. J. Bacteriol. 134:310–317.

Berner, R. A. 1984. Sedimentary pyrite formation: An update. Geochim. Cosmochim. Acta 48:605–615.

Bhatti, T. M., J. M. Gigham, L. Carlson and O. H. Tuovinen. 1993. Mineral products of pyrrhotite oxidation by *Thiobacillus ferrooxidans*. Appl. Environ. Microbiol. 59:1984–1990.

Bonatti, E. 1972. Authigenisis of minerals—marine. *In*: R. W. Fairbridge, ed. The Encyclopedia of Geochemistry and Environmental Sciences. Encyclopedia of Earth Sciences Series. Vol. IVA. Van Nostrand Reinhold, New York, pp. 48–56.

Bonatti, E. 1978. The origin of metal deposits in the oceanic lithosphere. Sci. Am. 238:54–61.

Brierley, C. L. 1974. Leaching. Use of a high-temperature microbe. *In*: Solution Mining Symposium. Proc. 103rd AIME Ann. Meet., Dallas, Tx., Feb. 25–27, 1974, pp. 461–469.

Brierley, C. L. 1978a. Bacterial leaching. CRC Crit. Rev. 6:207–262.

Brierley, J. A. 1978b. Thermophilic iron-oxidizing bacteria found in copper leaching dumps. Appl. Environ. Microbiol. 36:523–525.

Brierley, C. L. 1982. Microbiological mining. Sci. Am. 247:44–53; 150.

Brierley, C. L., and J. A. Brierley. 1973. A chemoautotrophic and thermophlic microorganism isolated from an acid hot spring. Can. J. Microbiol. 19:183–188.

Brierley, J. A., and S. J. Lockwood. 1977. The occurrence of thermophilic iron-oxidizing bacteria in a copper leaching system. FEMS Microbiol. Lett. 2:163–165.

Brierley, C. L., and L. E. Murr. 1973. Leaching: use of a thermophilic chemoautotrophic microbe. Science 179:488–490.

Brierley, J. A., P. R. Norris, D. P. Kelly, and N. W. LeRoux. 1978. Characteristics of a moderately thermophilic and acidophilic iron-oxidizing *Thiobacillus*. Eur. J. Appl. Microbiol. Biotechnol. 5: 291–299.

Brock, T. D. 1975. Effect of water potential on growth and iron oxidation by *Thiobacillus ferrooxidans*. Appl. Microbiol. 29:495–501.

Bryner, L. C., and R. Anderson. 1957. Microorganisms in leaching sulfide minerals. Ind. Eng. Chem. 49:1721–1724.

Bryner, L. C., and A. K. Jameson. 1958. Microorganisms in leaching sulfide minerals. Appl. Microbiol. 6:281–287.

Bryner, L. C., J. V. Beck, B. B. Davis, and D. G. Wilson. 1954. Microorganisms in leaching sulfide minerals. Ind. Eng. Chem. 46:2587–2592.

Burgstaller, W., and F. Schinner. 1993. Leaching of metals with fungi. J. Biotechnol. 27:91–116.

Cameron, E. M. 1982. Sulfate and sulfate reduction in early Precambrian oceans. Nature (Lond.) 296:145–148.

Carlson, L., E. B. Lindstroem, K. B. Hallberg, and O. H. Tuovinen. 1992. Solid-phase products of bacterial oxidation of arsenical pyrite. Appl. Environ. Microbiol. 58:1046–1049.

Chemical Processing, 1965. Carboxylic acids in metal extraction. Chem. Proc. 11:24–25.

Chen, C-Y., and D. R. Skidmore. 1987. Langmuir adsorption isotherm for *Sulfolobus acidocaldarius* on coal particles. Biotech. Lett. 9: 191–194.

Chen, C-Y., and D. R. Skidmore. 1988. Attachment of *Sulfolobus acido-caldarius* cells in coal particles. Biotechnol. Progr. 4:25–30.

Corliss, J. B., J. Dymond, L. I. Gordon, J. M. Edmond, R. P. von Herzen, R. D. Ballard, K. Green, D. Williams, A. Bainbridge, K. Crane, and T. H. van Andel. 1979. Submarine thermal springs on the Galapagos Rift. Science 203:1073–1083.

Cuthbert, M. E. 1962. Formation of bornite at atmospheric temperature and pressure. Econ. Geol. 57:38–41.

Darland, G., T. D. Brock, W. Samsonoff, and S. F. Conti. 1970. A thermophilic, acidophilic mycoplasma isolated from a coal refuse pile. Science 170:1416–1418.

Davidson, C. F. 1962a. The origin of some strata-bound sulfide ore deposits. Econ. Geol. 57:265–274.

Davidson, C. F. 1962b. Further remarks on biogenic sulfides. Econ. Geol. 57:1134–1137.

Dévigne, J-P. 1968a. Précipitation du sulfure de plomb par un micrococcus tellurique. C.R. Acad. Sci. (Paris) 267:935–937.

Dévigne, J-P. 1968b. Une bactérie saturnophile, *Sarcina flava* Bary 1887. Arch. Inst. Past. (Tunis) 45:341–358.

Dévigne, J-P. 1973. Une métallogenèse microbienne probable en milieux sédimentaires: celle de la galène. Cah. Geol. no. 89, pp. 35–37.

DiSpirito, A. A., and O. H. Tuovinen. 1981. Oxygen uptake coupled with uranous sulfate oxidation by *Thiobacillus ferrooxidans* and *T. acidophilus*. Geomicrobiol. J. 2:275–291.

DiSpirito, A. A., and O. H. Tuovinen. 1982a. Uranous ion oxidation and carbon dioxide fixation by *Thiobacillus ferrooxidans*. Arch. Microbiol. 133:28–32.

DiSpirito, A. A., and O. H. Tuovinen. 1982b. Kinetics of uranous and ferrous iron oxidation by *Thiobacillus ferooxidans*. Arch. Microbiol. 133:33–37.

DiSpirito, A. A., J. W. Talnagi Jr., and O. H. Tuovinen. 1983. Accumulation and cellular distribution of uranium in *Thiobacillus ferroxidans*. Arch. Microbiol. 135:250–253.

Duncan, D. W., and C. C. Walden. 1972. Microbial leaching in the presence of ferric iron. Dev. Ind. Microbiol. 13:66–75.

Duncan, D. W., J. Landesman, and C. C. Walden. 1967. Role of *Thiobacillus ferrooxidans* in the oxidation of sulfide minerals. Can. J. Microbiol. 13:397–403.

Edmond, J. M., K. L. Von Damm, R. E. McDuff, and C.I. Measures. 1982. Chemistry of hot springs on the East Pacific Rise and their effluent dispersal. Nature (Lond.) 297:187–191.

Ehrlich, H. L. 1963. Bacterial action on orpiment. Econ. Geol. 58: 991–994.

Ehrlich, H. L. 1964a. Bacterial oxidation of arsenopyrite and enargite. Econ. Geol. 59:1306–1312.

Ehrlich, H. L. 1964b. Microorganisms in acid drainage from a copper mine. J. Bacteriol. 86:350–352.

Ehrlich, H. L. 1977. Bacterial leaching of low-grade copper sulfide ore with different lixiviants. *In*: W. Schwartz, ed. Conference—Bacterial Leaching. Gesellschaft fuer Biotechnologisch Forschung mbH, Braunschweig-Stoeckheim. Verlag Chemie, Weinstein, West Germany, pp. 145–155.

Ehrlich, H. L. 1988. Bioleaching of silver from a mixed sulfide ore in a stirred reactor. *In:* P. R. Norris and D. P. Kelly, eds. Biohydrometallurgy. Science and Technology Letters, Kew Surrey, U.K., pp. 223–231.

Ehrlich, H. L., and S. I. Fox. 1967b. Environmental effects on bacterial copper extraction from low-grade copper sulfide ores. Biotech. Bioeng. 9:471–485.

Fenchel, T. and T. H. Blackburn. 1979. Bacterial and Mineral Cycling. Academic Press, London.

Fox, S. I. 1967. Bacterial oxidation of simple copper sulfides. Ph.D. thesis. Rensselaer Polytechnic Institute, Troy, NY.

Freke, A. M., and D. Tate. 1961. The formation of magnetic iron sulfide by bacterial reduction of iron solutions. J. Biochem. Microbiol. Technol. Eng. 3:29–39.

Giblin, A. E. 1988. Pyrite formation in marshes during early diagenesis. Geomicrobiol. J. 6:77–97.

Giblin, A. E., and R. W. Howarth. 1984. Porewater evidence for a dynamic sedimentary iron cycle in salt marshes. Limnol. Oceanogr. 29:47–63.

Golovacheva, R. S., and G. I. Karavaiko. 1978. A new genus of thermophilic spore-forming bacteria, *Sulfobacillus*. Mikrobiologiya 47: 815–822 (Engl. transl. 658–665).

Hallberg, R. O. 1978. Metal-organic interactions at the redoxcline. *In*: W. E. Krumbein, ed. Environmental Biogeochemistry and Geomicrobiology. Vol 3. Methods, Metals, and Assessment. Ann Arbor Science Publishers, Ann Arbor, MI, pp. 947–953.

Harrison, A. P., Jr. 1978. Microbial succession and mineral leaching in an artificial coal spoil. Appl. Environ. Microbiol. 36:861–869.

Harrison, A. P. Jr., and P. R. Norris. 1985. *Leptospirillum ferrooxidans* and similar bacteria: some characteristics and genomic diversity. FEMS Microbiol. Lett. 30:99–102.

Hartmannova, V., and I. Kuhr. 1974. Copper leaching by lower fungi. Rudy 22:234–238.

Howarth, R. W. 1979. Pyrite: its rapid formation in a salt marsh and its importance in ecosystem metabolism. Science 203:49–51.

Howarth, R. W., and S. Merkel. 1984. Pyrite formation and the measurement of sulfate reduction in salt marsh sediments. Limnol. Oceanogr. 29:598–608.

Imai, K. H., H. Sakaguchi, T. Sugio, and T. Tano. 1973. On the mechanism of chalcocite oxidation by *Thiobacillus ferrooxidans*. J. Ferment. Technol. 51:865–870.

Ivanov, V. I. 1962. Effect of some factors on iron oxidation by cultures of *Thiobacillus ferrooxidans*. Mikrobiologiya 31:795–799 (Engl. transl. pp. 645–648).

Ivanov, V. I., F. I. Nagirynyak, and B. A. Stepanov. 1961. Bacterial oxidation of sulfide ores. I. Role of *Thiobacillus ferrooxidans* in the oxidation of chalcopyrite and sphalerite. Mikrobiologiya 30: 688–692.

Jørgensen, B. B. 1977. The sulfur cycle of a coastal marine sediment (Limfjorden, Denmark). Limnol. Oceanogr. 22:814–832.

Kee, N. S., and C. Bloomfield. 1961. The solution of some minor element oxides by decomposing plant materials. Geochim. Cosmochim. Acta 24:206–225.

King, G. M., B. L. Howes, and J. W. H. Dacey. 1985. Short-term end-products of sulfate reduction in a salt marsh; Formation of acid volatile sulfides, elemental sulfur, and pyrite. Geochim. Cosmochim. Acta 49:1561–1566.

Kleinmann, R. L. P., and D. A. Crerar. 1979. *Thiobacillus ferrooxidans* and the formation of acidity in simulated coal mine environments. Geomicrobiol. J. 1:373–388.

Klinkhammer, G., P. Rona, M. Greaves, and H. Elderfeld. 1985. Hydrothermal manganese plumes in the Mid-Atlantic Ridge Rift Valley. Nature (Lond.) 314:727–731.

Lambert, I. B., J. McAndrew, and H. E. Jones. 1971. Geochemical and bacteriological studies of the cupriferous enviroment at Pernatty Lagoon, South Australia. Australia. Inst. Min. Met., Proc., No.240, pp. 15–23.

Latimer, W. M., and J. H. Hildebrand. 1942. Reference Book of Inorganic Chemistry. Rev. ed. Macmillan, New York.

Lazaroff, N. 1983. The exclusion of D_2O from the hydration sphere of $FeSO_4 \cdot 7H_2O$ oxidized by *Thiobacillus ferrooxidans*. Science 222: 1331–1334.

Lazaroff, N., L. Melanson, E. Lewis, N. Santoro, and C. Pueschel. 1985. Scanning electron microscopy and infrared spectroscopy of iron sediments formed by *Thiobacillus ferrooxidans*. Geomicrobiol. J. 4: 231–268.

Lazaroff, N., W. Sigal, and A. Wasserman. 1982. Iron oxidation and precipitation of ferric hydroxysulfates by resting *Thiobacillus ferrooxidans* cells. Appl. Environ. Microbiol. 43:924–938.

Leathen, W. W., S. A. Braley, and L. D. McIntyre. 1953. The role of bacteria in the formation of acid from certain sulfuritic constituents associated with bituminous coal. II. Ferrous iron oxidizing bacteria. Appl. Microbiol. 1:65–68.

Leleu, M., T. Gulgalski, and J. Goni. 1975. Synthèse de wurtzite par voie bacterienne. Mineral. Deposita (Berl.) 10:323–329.

Love, L. G. 1962. Biogenic primary sulfide of the Permian Kupferschiefer and marl slate. Econ. Geol. 57:350–366.

Lundgren, D. G., and E. E. Malouf. 1983. Microbial extraction and concentration of metals. Adv. Biotechnol. Proc. 1:223–249.

Lundgren, D. G., J. R. Vestal, and F. R. Tabita. 1972. The microbiology of mine drainage pollution. *In*: R. Mitchell, ed. Water Pollution Microbiology. Wiley-Interscience, New York, pp. 69–88.

Luther, G. W., III. 1991. Pyrite synthesis via polysulfide compounds. Geochim. Cosmochim. Acta 55:2839–2849.

Luther, G. W., III, A. Giblin, R. W. Howarth, and R. A. Ryans. 1982. Pyrite and oxidized iron mineral phases formed from pyrite oxidation in salt marsh and estuarine sediments. Geochim. Cosmochim. Acta 46:2665–2669.

Lyalikova, N. N. 1960. Participation of *Thiobacillus ferrooxidans* in the oxidation of sulfide ores in pyrite beds of the Middle Ural. Mikrobiologiya 29:382–387.

Mackintosh, M. E. 1978. Nitrogen fixation by *Thiobacillus ferrooxidans*. J. Gen. Microbiol. 105:215–218.

Magne, R., J. Berthelin, and Y. Dommergues. 1973. Solubilisation de l'uranium dans les roches par des bactéries n'appartenant pas au genre *Thiobacillus*. C.R. Acad. Sci. (Paris) 276:2625–2628.

Magne, R.,, J. R. Berthelin, and Y. Dommergues. 1974. Solubilisation et insolubilisation de l'uranium des granites par des bactéries hétérotroph. *In*: Formation of Uranium Ore Deposits. Intern. Atomic Energy Commission, Vienna, Austria, pp. 73–88.

Malouf, E. E., and J. D. Prater. 1961. Role of bacteria in the alteration of sulfide minerals. J. Metals 13:353–356.

Marchig, V., and H. Gundlach. 1982. Iron-rich metalliferous sediments on the East Pacific Rise: prototype of undifferentiated metalliferous sediments on divergent plate boundaries. Earth Planet Sci. Lett. 58; 361–382.

Marnette, E. C. L., N. Van Breemen, K. A. Hordijk, and T. E. Cappenberg. 1993. Pyrite formation in two freshwater systems in the Netherlands. Geochim. Cosmochim. Acta 57:4165–4177.

McCready, R. G. L. 1988. Progress in the bacterial leaching of metals in Canada. *In*: P. R. Norris and D. P. Kelly, eds. BioHydroMetallurgy. Proc. Int. Symp. Warwick, 1987. Science and Technology Letters, Kew Surrey, Great Britain, pp. 177–195.

McCready, R. G. L., and W. D. Gould. 1990. Bioleaching of uranium. *In*: H.L. Ehrlich and C.L. Brierley, eds. Microbial Mineral Recovery. McGraw-Hill, New York, pp. 107–125.

McGoran, C. J. M., D. W. Duncan, and C. C. Walden. 1969. Growth of *Thiobacillus ferrooxidans* on various substrates. Can. J. Microbiol. 15:135–138.

Miller, L. P. 1949. Stimulation of hydrogen sulfide production by sulfate-reducing bacteria. Boyce Thompson Inst. Contr. 15:467–474.

Miller, L. P. 1950. Formation of metal sulfides through the activities of sulfate reducing bacteria. Boyce Thompson Inst. Contr. 16:85–89.

Moshuyakova, S. A., G. I. Karavaiko, and E. V. Shchetinina. 1971. Role of *Thiobacillus ferrooxidans* in leaching of nickel, copper, cobalt, iron, aluminum, magnesium, and calcium from ores of copper-nickel deposits. Mikrobiologyia 40:1100–1107 (Engl. transl. pp. 959–969).

Mossmann, J-R., A. C. Aplin, C. D. Curtis, and M. L. Coleman. 1991. Geochemistry of inorganic and organic sulfur in organic-rich sediments from the Peru Margin. Geochim. Cosmochim. Acta 55: 3581–3595.

Mottl, M. J., H. D. Holland, and R. F. Corr. 1979. Chemical exchange during hydrothermal alteration of basalt by seawater. II. Experimental results for Fe, Mn, and sulfur species. Geochim. Cosmochim. Acta 43:869–884.

Murr, L. E., and J. A. Brierley. 1978. The use of large-scale test facilities in studies of the role of microorganisms in commercial leaching operations. *In*: L. E. Murr, A. E. Torma, and J. A. Brierley, eds. Metallurgical Applications of Bacteria. Academic Press, New York, pp. 491–520.

Murthy, K. S. N., and K. A. Natarajan. 1992. The role of surface attachment of *Thiobacillus ferrooxidans* on the biooxidation of pyrite. Minerals and Metallurgical Processing, Feb., pp. 20–24.

Mustin, C., J. Berthelin, P. Marion, and P. de Donato. 1992. Corrosion and electrochemical oxidation of a pyrite by *Thiobacillus ferrooxidans*. Appl. Environ. Microbiol. 58:1175–1182.

Mustin, C., P. de Donato, and J. Berthelin. 1993. Surface oxidized species, a key factor in the study of bioleaching processes. *In*: A.E. Torma, J.E. Wey, and V.I. Lakshmanan, eds. Biohydrometallurgical Technologies, Vol. 1, Bioleaching Process. Minerals, Metals and Materials Society, Warrendale, PA, pp. 175–184.

Nedwell, D. B., and I. M. Banat. 1981. Hydrogen as an electron donor for sulfate-reducing bacteria in slurries of salt marsh sediment. Microb. Ecol. 7:305–313.

Nielsen, A. M., and J. V. Beck. 1972. Chalcocite oxidation and coupled carbon dioxide fixation by *Thiobacillus ferrooxidans*. Science 175: 1124–1126.

Norman, P. F., and C. P. Snyman. 1988. The biological and chemical leaching of an auriferous pyrite/arsenopyrite flotation concentrate: A microscopic examination. Geomicrobiol. J. 6:1–10.

Norris, P. R. 1990. Acidophilic bacteria and their activity in mineral sulfide oxidation. *In*: H. L. Ehrlich and C. L. Brierley, eds. Microbial Mineral Recovery. McGraw-Hill, New York, pp. 3–27.

Parès, Y. 1964a. Intervention des bactéries dans la solubilisation du cuivre. Ann. Inst. Past. (Paris) 107:132–135.

Parès, Y. 1964b. Action de *Serratia marcescens* dans le cycle biologique des métaux. Ann. Inst. Past. (Paris) 107:136–141.

Parès, Y. 1964c. Action d'*Agrobacterium tumefaciens* dans la mise on solution de l'or. Ann. Inst. Past. (Paris) 107:141–143.

Pinka, J. 1991. Bacterial oxidation of pyrite and pyrrhotite. Erzmetall 44: 571–573.

Pivovarova, T. A., G. E. Karkosyan, and G. I. Karavaiko. 1980. Morphogenesis and fine structure of *Leptospirillum ferrooxidans*. Mikrobiologiya 50:482–486 (Engl. transl. pp. 339–344).

Pogliani, C., C. Curutchet, E. Donati, and P. H. Tedesco. 1990. A need for direct contact with particle surfaces in the bacterial oxidation of covellite in the absence of a chemical lixiviant. Biotechnol. Lett. 12: 515–518.

Powell, T. G., and R. W. MacQueen. 1984. Precipitation of sulfide ores and organic matter: Sulfate reduction at Pine Point, Canada. Science 224:63–66.

Pronk, J. T., R. Meulenberg, W. Hazen, P. Bos, and J. G. Kuenen. 1990. Oxidation of reduced inorganic sulfur compounds by acidophilic thiobacilli. FEMS Microbiol. Rev. 75:293–306.

Razzell, W. E., and P. C. Trussell. 1963. Microbiological leaching of metallic sulfides. Appl. Microbiol. 11:105–110.

Rickard, D. T. 1973. Limiting conditions for synsedimentary sulfide ore formation. Econ. Geol. 68:605–617.

Rickard, P. A. D., and D. G. Vanselow. 1978. Investigations into the kinetics and stoichiometry of bacterial oxidation of covellite (CuS) using a polarographic oxygen probe. Can. J. Microbiol. 24:998–1003.

Rodriguez-Leiva, M., and H. Tributsch. 1988. Morphology of bacterial leaching patterns by *Thiobacillus ferrooxidans* on synthetic pyrite. Arch. Microbiol. 149:401–405.

Schoonen, M. A. A., and H. L. Barnes. 1991a. Reactions forming pyrite and marcasite from solution: I. Nucleation of FeS_2 below 100°C. Geochim. Cosmochim. Acta 55:1495–1504.

Schoonen, M. A. A., and H. L. Barnes. 1991b. Reactions forming pyrite and marcasite from solution: II. Via FeS precursors below 100°C. Geochim. Cosmochim. Acta 55:1505–1514.

Segerer, A., A. Neuner, J. K. Kristjansson, and K. O. Stetter. 1986. *Acidianus infernus* gen. nov., sp. nov., and *Acidianus brierleyi* comb. nov.: Facultatively aerobic, extremely acidophilic thermophilic sulfur-metabolizing archaebacteria. Arch. Microbiol. 36: 559–564.

Serkies, J., J. Oberc, and A. Idzikowski. 1967. The geochemical bearings of the genesis of Zechstein copper deposits in southwest Poland as exemplified by the studies on the Zechstein of the Leszczyna syncline. Chem. Geol. 2:217–232.

Seyfried, W. E. Jr., and M. J. Mottl. 1982. Hydrothermal alteration of basalt by seawater under seawater-dominated conditions. Geochim. Cosmochim. Acta 46:985–1002.

Shanks, W. C. III, J. L. Bischoff, and R. J. Rosenauer. 1977. Seawater sulfate reduction and sulfur isotope fractionation in basaltic systems: Interaction of seawater with fayalite and magnetite at 200–350°C. Geochim. Cosmochim. Acta 45:1995–1981.

Shrihari, R. K., K. S. Gandhi, and K. A. Natarajan. 1991. Role of cell attachment in leaching chalcopyrite mineral by *Thiobacillus ferrooxidans*. Appl. Microbiol. Biotechnol. 36:278–282.

Silver, M., and A. E. Torma. 1974. Oxidation of metal sulfides by *Thiobacillus ferrooxidans* grown on different substrates. Can. J. Microbiol. 20:141–147.

Silverman, M. P. and H. L. Ehrlich. 1964. Microbial formation and degradation of minerals. Adv. Appl. Microbiol. 6:153–206.

Silverman, M. P., M. H. Rogoff, and I. Wender. 1961. Bacterial oxidation of pyritic materials in coral. Appl. Microbiol. 9:491–496.

Singer, P. C., and W. Stumm. 1970. Acidic mine drainage: The rate-determining step. Science 167:1121–1123.

Stanton, R. L. 1972. Sulfides in sediments. *In*: R.W. Fairbridge, ed. The Encyclopedia of Geochemistry and Environmental Sciences. Encyclopedia of Earth Sciences Series, Vol. IVA. Van Nostrand Reinhold, New York, pp. 1134–1141.

Stevens, C. J., P. R. Dugan, and O. H. Tuovinen. 1986. Acetylene reduction (nitrogen fixation) by *Thiobacillus ferrooxidans*. Biotechnol. Appl. Biochem. 8:351–359.

Strahler, A. N. 1977. Principles of Physical Geology. Harper & Row, New York.

Styrt, M. M., A. J. Brackmann, H. D. Holland, B. C. Clark, V. Pisutha-Arnold, C. S. Eldridge, and H. Ohmoto. 1981. The mineralogy and the isotopic composition of sulfur in hydrothermal sulfide/sulfate deposits on the East Pacific Rise, 21°N latitude. Earth Planet. Sci. Lett. 53:382–390.

Sugio, T., S. Tanijiri, K. Fukuda, K. Yamargo, K. Inagaki, and T. Tano. 1987. Utilization of amino acids as a sole source of nitrogen by obligate chemoautotroph *Thiobacillus ferrooxidans*. Agric. Biol. Chem. 51:2229–2236.

Sutton, J. A., and J. D. Corrick. 1963. Microbial leaching of copper minerals. Mining Eng. 15:37–40.

Sutton, J. A., and J. D. Corrick. 1964. Bacteria in mining and metallurgy:

Leaching selected ores and minerals; experiments with *Thiobacillus thiooxidans*. Rept. Invest. RI 5839. Bureau of Mines, U.S. Department of the Interior, Washington, DC, 16 pp.

Temple, K. L. 1964. Syngenesis of sulfide ores. An explanation of biochemical aspects. Econ. Geol. 59:1473–1491.

Temple, K. L., and N. LeRoux. 1964. Syngenesis of sulfide ores: Desorption of adsorbed metal ions and their precipitation as sulfides. Econ Geol. 59:647–655.

Thode-Anderson, S., and B. B. Jørgensen. 1989. Sulfate reduction and the formation of ^{35}S-labeled FeS, FeS$_2$, and S° in coastal marine sediments. Limnol. Oceanogr. 34:793–806.

Tittley, S. R., 1981. Porphyry copper. Am. Scientist 69:632–638.

Torma, A. E. 1971. Microbiological oxidation of synthetic cobalt, nickel and zinc sulfides by *Thiobacillus ferrooxidans*. Rev. Can. Biol. 30: 209–216.

Torma, A. E. 1978. Oxidation of gallium sulfides by *Thiobacillus ferrooxidans*. Can. J. Microbiol. 24:888–891.

Torma, A. E., and G. G. Gabra. 1977. Oxidation of stibnite by *Thiobacillus ferrooxidans*. Antonie v. Leeuwenhoek 43:1–6.

Tributsch, H. 1976. The oxidative disintegration of sulfide crystals by *Thiobacillus ferrooxidans*. Naturwissenschaften 63:88.

Tuovinen, O. H., S. I. Niemelae, and H. G. Gyllenberg. 1971. Tolerance of *Thiobacillus ferrooxidans* to some metals. Antonie v. Leeuwenhoek 37:489–496.

Wadden, D., and A. Gallant. 1985. The in-place leaching of uranium at Denison Mines. Can. Metall. Q. 24:127–134.

Walsh, F., and R. Mitchell. 1972a. An acid-tolerant iron-oxidizing *Metallogenium*. J. Gen. Microbiol. 72:369–376.

Walsh, F., and R. Mitchell. 1972b. The pH-dependent succession of iron bacteria. Environ. Sci. Technol. 6:809–812.

Wenberg, G. M., F. H. Erbisch, and M. Volin. 1971. Leaching of copper by fungi. Trans. Soc. Min. Eng. AIME 250:207–212.

Westrich, J. T., and R. A. Berner. 1984. The role of sedimentary organic matter in bacterial sulfate reduction: the G model tested. Limnol. Oceanogr. 29:236–249.

Zajic, J. E. 1969. Microbial Biogeochemistry. Academic Press, New York.

Zimmerely, S. R., D. G. Wilson, and J. D. Prater. 1958. Cyclic leaching process employing iron oxidizing bacteria. U.S. patent 2,829,964.

19

Geomicrobiology of Selenium and Tellurium

19.1 OCCURRENCE IN THE EARTH'S CRUST

The elements selenium and tellurium, like sulfur, belong to group VI of the periodic table. All three have some properties in common, but selenium and tellurium, especially the latter, have some metallic attributes, unlike sulfur. Selenium and tellurium are much less abundant in the Earth's crust than is sulfur. Selenium amounts to only 0.05–0.14 ppm (Rapp, 1972, p. 1080) and tellurium to 10^{-2} to 10^{-5} ppm (Lansche, 1965). Both are associated with metal sulfides in nature and occur in distinct minerals [e.g., ferroselite ($FeSe_2$), challomenite ($CuSeO_3 \cdot 2H_2O$), hirsite (Ag_2Te), and tetradymite (Bi_2Te_2S)]. Selenium occurs in small amounts in various soils in concentrations in the range of 0.01–100 ppm. High concentrations are associated with arid, alkaline soils that contain some free $CaCO_3$ (Rosenfeld and Beath, 1964).

19.2 BIOLOGICAL IMPORTANCE

Some plants, such as *Astragalus* spp. and *Stanleya*, can accumulate large amounts of selenium in the form of organic selenium compounds. However, not all forms of selenium in the soil are available for assimilation by these plants.

Selenium is required nutritionally as a trace element by at least some microorganisms, plants and animals, including human beings (Stadtman, 1974; Miller and Neathery, 1977; Combs and Scott, 1977; Patrick, 1978; Mertz, 1981). It has been found to be an essential component of the enzyme glutathione peroxidase in mammalian red blood corpuscles (Rotruck et al., 1973). The enzyme catalyzes the reaction

$$2GSH + H_2O_2 \Rightarrow GSSG + 2H_2O \tag{19.1}$$

Selenium has also been found essential together with molybdenum in the structure of formate dehydrogenase in the bacteria *Escherichia coli*, *Clostridium thermoaceticum*, *C. sticklandii*, and *Methanococcus vannielii* among others (Pinsent, 1954; Lester and DeMoss, 1971; Shum and Murphy, 1972; Andreesen and Ljundahl, 1973; Enoch and Lester, 1972; Stadtman, 1974) and with tungsten in the formate dehydrogenase of *C. thermoaceticum* when grown in the presence of tungsten instead of molybdenum (Yamamoto et al., 1983). The enzyme catalyzes the reaction

$$HCOOH + NAD^+ \Rightarrow CO_2 + NADH + H^+ \tag{19.2}$$

Selenium has also been found to be essential to protein A of glycine reductase in clostridia (Stadtman, 1974), an enzyme that catalyzes the reaction

$$\underset{\underset{NH_2}{|}}{CH_2COOH} + R(SH)_2 + P_1 + ADP \Longrightarrow CH_3COOH + NH_3 + R\overset{\diagup S}{\underset{\diagdown S}{|}} + ATP$$

$$\tag{19.3}$$

For this reason, *Clostridium purinolyticum* exhibits an absolute requirement for selenium in its growth medium for fermentation of glycine (Duerre and Andreesen, 1982).

No biological requirement for tellurium has been observed up to now.

19.3 TOXICITY OF SELENIUM AND TELLURIUM COMPOUNDS

Both selenium and tellurium are toxic when present in excess, but the minimum toxic doses vary depending on the organism. As mentioned in section 19.2, some plants accumulate selenium to the extent of 1.5–2 g kg^{-1} of dry weight tissue (Stadtman, 1974). They usually grow in arid environments with unusually high concentrations of selenium in the soil. In the Kesterson National Wildlife Refuge in California (U.S.A), where extensive selenium intoxication of wild animals has been observed, selenate concentrations of 1.8 to 18 μM (0.14–1.4 ppm) have been reported, contrasted with normal concentrations of ~1.3–21.2 nM in the San Joaquin River, ~0.4–1.3 nM in the Sacramento River, and <2.0 nM in San Francisco Bay, all located in California (Zehr and Oremland, 1987). These concentrations are below minimum inhibitory concentrations (MIC) of selenate [Se(VI)] for three different selenium sensitive strains of bacteria from the same general area in California. Their MICs were found to range from 0.78 to 1.56 mM and for selenite [Se(IV)] from 1.56 to 25 mM. Selenium-resistant bacteria from the Kesterson Wildlife Refuge exhibited MICs of 50 to >200 mM selenate and 6.25 to >200 mM selenite (Burton et al., 1987). By contrast, both selenium-resistant and -sensitive organism from these same sites in California exhibited MIC's for tellurate in the range of 0.03–1 mM and for tellurite in the range of 0.03–4 mM (Burton et al., 1987). Selenium resistance and tellurium resistance appear to be regulated by different genes. In *Escherichia coli*, tellurium resistance appears to be mediated by the arsenical ATPase efflux pump. The genetic determinants for this pump reside on resistance plasmid R773 (Turner et al., 1992). Higher forms of life appear to be relatively more sensitive to Se than bacteria, even though Se is a nutritionally required trace element.

19.4 BIOOXIDATION OF REDUCED FORMS OF SELENIUM

Some inorganic forms of selenium have been reported to be oxidizable by microorganisms. *Micrococcus selenicus* isolated from mud (Breed et al., 1957), a rod-shaped bacterium isolated from soil (Lipman and Waksman, 1923), and a purple bacterium (Sapozhnikov, 1937) have been reported to oxidize $Se°$ to SeO_4^{2-}. A strain of *Bacillus megaterium* from top soil in a river alluvium was found to oxidize $Se°$ to SeO_3^{2-} and traces of SeO_4^{2-}. Red selenium was more readily attacked than gray selenium (Sarathchandra and Watkinson, 1981). *Thiobacillus ferrooxidans* has been reported to oxidize copper selenide (CuSe) to cupric copper (Cu^{2+}) and elemental selenium ($Se°$) (Torma and Habashi, 1972). The reaction may be written

$$CuSe + 2H^+ + 0.5O_2 \Rightarrow Cu^{2+} + Se° + H_2O \qquad (19.4)$$

19.5 BIOREDUCTION OF OXIDIZED SELENIUM COMPOUNDS

Various inorganic selenium compounds have been found to be reduced anaerobically by microorganisms. Crude cell extract of *Micrococcus lactilyticus* (also known as *Veillonella lactilyticus*) has been shown to reduce selenite but not selenate to Se°, and Se° to HSe⁻. The reductant was hydrogen. Cell extracts from strains of *D. desulfuricans* and *C. pasteurianum* were also found to reduce selenite with hydrogen. The enzyme hydrogenase mediated electron transfer from hydrogen in these reactions (Woolfolk and Whiteley, 1962). A variety of other known bacteria, actinomycetes, and fungi have been shown to reduce selenate and selenite to Se° (Bautista and Alexander, 1972; Lortie et al., 1992; Tomei et al., 1992; Zalokar, 1953). The bacteria include *Pseudomonas stutzeri*, *Wolinella succinogenes*, and *Micrococcus* sp. *Thiobacillus ferrooxidans* is able to reduce Se° (red form) to H_2Se anaerobically, albeit in small quantities (Bacon and Ingledew, 1989).

A newly discovered bacterium, *Thauera selenatis* can grow anaerobically with selenate or nitrate as terminal electron acceptor (Macy et al., 1993; Rech and Macy, 1992). In the absence of nitrate, it reduces selenate to selenite (DeMoll-Decker and Macy, 1993). The reductases for selenate and nitrate in this organism are distinct enzymes with different pH optima. Thus in contrast to the response of a selenate-reducing enrichment culture (Steinberg et al., 1992), nitrate does not inhibit selenate reduction by *T. selenatis*. Indeed, when present together, both selenate and nitrate are reduced simultaneously, with selenate being reduced to elemental selenium (DeMoll-Decker and Macy, 1993). The selenate reductase in this organism, which catalyzes the reduction of selenate to selenite, is found in its periplasm, whereas its nitrate reductase, which catalyzes the reduction of nitrate to nitrite, is found in its cytoplasmic membrane (Rech and Macy, 1992). Nitrite reductase is found in the periplasm of *T. selenatis* and plays a role in selenite reduction, besides catalyzing nitrite reduction (DeMoll-Decker and Macy, 1993). This helps to explain why *T. selenatis* produces elemental selenium in the presence of nitrate but selenite in its absence. Selenite does not support growth of *T. selenatis* (DeMoll-Decker, 1993).

A selenite reductase enzyme has been obtained from the fungus *Candida albicans* (Falcone and Nickerson, 1963; Nickerson and Falcone, 1963). It reduces selenite to Se°. A characterization of the enzyme has shown that it requires a quinone, a thiol compound (e.g., glutathione), a pyridine nucleotide (NADP), and an electron donor (e.g., glucose 6-phosphate) for activity. Electron transfer between NADP and quinone is probably mediated by flavin mononucleotide in this system. It is possible that this enzyme is part of an assimilatory SeO_4^{2-} and/or SeO_3^{2-} reductase system. How this enzyme compares to that in *T. selenatis* remains to be established.

In most studies of bacterial reduction of selenate and selenite, elemental selenium (red form) is usually found to be a major, if not the only, product. This is noteworthy because sulfate and sulfite cannot be directly reduced to $S°$ but are reduced to H_2S without intermediate formation of $S°$. Yet, selenium and sulfur are members of the same chemical family. The implication is that the enzymatic mechanisms of reduction for oxidized forms of these two elements are different. This cannot be universal, however, since *Desulfovibrio desulfuricans* subsp. *aestuarii* has been shown to reduce nanomolar but not millimolar quantities of selenate to selenide (Zehr and Oremland, 1987). Sulfate inhibited the reduction of selenate, suggesting but not proving that the mechanism of sulfate and selenate reduction in this case may be a common one. As Zehr and Oremland (1987) have pointed out, when sulfate is being reduced to H_2S in the presence of selenate, some of the H_2S formed may chemically reduce biogenically formed selenite to $Se°$. They found that in nature, the sulfate reducer can reduce selenate only if the ambient sulfate concentration was below 4 mM.

Other Products of Selenate and Selenite Reduction In *Escherichia coli*, a significant portion of selenite when reduced during glucose metabolism is deposited as $Se°$ on its cell membrane but not in its cytoplasm (Gerrard et al., 1974), while another portion is incorporated as selenide into organic compounds such as selenomethionine (Ahluwalia et al., 1968). Some soil microbes reduce selenate or selenite to dimethylselenide [$(CH_3)_2Se$] at elevated selenium concentrations (Kovalskii et al., 1968; Fleming and Alexander, 1972; Alexander, 1977; Doran and Alexander, 1977). Other volatile selenium compounds may also be formed, their relative quantities depending on reaction conditions (e.g., Reamer and Zoller, 1980). These compounds include dimethyl diselenide [$(CH_3)_2Se_2$) and dimethyl selenone [$(CH_3)_2SeO_2$].

Fungi are among microorganisms that have been found to be effective in forming methylated selenium compounds (Barkes and Fleming, 1974). *Alternaria alternata*, isolated from seleniferous water from a sample series collected from evopration ponds at Kesterson Reservoir, Lost Hills, and Peck Ranch in California, formed dimethylselenide more rapidly from selenate and selenite than from selenium sulfide (SeS_2) or various organic Se compounds. Methionine, a known biochemical methyl donor, and methyl cobalamin, a known methyl carrier in biochemical transmethylation, stimulated dimethylselenide formation by the fungus (Thompson-Eagle et al., 1989). Crude cell extracts and a supernatant fraction from the fungus *Pichia guillermondi* after centrifugation at 144,000 × g reduced selenite but not selenate (Bautista and Alexander, 1972). In a mechanism proposed by Reamer and Zoller (1980), all methylated forms of selenium arise by methylation of selenite and subsequent reductions and, where needed, further methylation. Dimethyl selenone is viewed as a precursor of dimethyl selenide, whereas methyl selenide (CH_3SeH) and/or CH_3SeOH is viewed as a precursor of dimethyl diselenide.

Selenium Reduction in the Environment Bacterial reduction of selenate and selenite has been detected in situ, especially in environments with significant soluble selenium. In Kesterson Reservoir, California, Maiers et al. (1988) reported that 4% of water samples, 92% of sediment samples, and 100% of the soil samples collected by them exhibited microbial selenate reduction. Of 100 mg selenate per liter, up to 75% was reduced to Se° (red), the rest to selenite. In the interstitial water of core samples from a wastewater evaporation pond in Fresno, California, selenate removal was stimulated by H_2 and by acetate addition and inhibited by O_2, NO_3^-, MnO_2, CrO_4^{2-}, and WO_4^{2-}, but not by SO_4^{2-}, MoO_4^{2-}, or FeOOH additions (Oremland et al., 1989). At other sites in California and also in Nevada, Steinberg and Oremland (1990) found measurable selenate-reducing activity in surficial sediment samples from bodies of freshwater to salinities of 250 g L^{-1} but not 320 g L^{-1}. Nitrate, nitrite, molybdate, and tungstate added separately to samples from two agricultural drains were inhibitory to different extents. Sulfate partially inhibited selenate reduction in a sample from a freshwater site but not in one from a site with water having a salinity of 60 g L^{-1}. These differences are likely reflections of differences in the types of selenate reducers in the different samples, and therefore, of differences in mechanisms of selenate reduction. Additional studies in the agricultural drainage region of western Nevada, revealed a selenate turnover rate of 0.04–1.8 h^{-1} at ambient Se oxyanion concentration (13–455 nM). Rates of removal of selenium oxyanions ranged from 14 to 155 μmol m^{-2} day^{-1} (Oremland et al., 1991). Formation of elemental selenium has a potential for selenium immobilization in soil and sediments under anaerobic conditions. Owing to the possibility of Se° reoxidation under aerobic conditions, remobilization may occur, however.

Methylation of selenium in aquatic environments has also been observed (e.g., Chau et al., 1976; Frankenberger and Karlson, 1992; 1995). This activity has a potential for Se removal from polluted soils and waters.

Ecologically, anaerobic selenate and selenite reduction to selenium represents a respiratory, energy-conserving process in some microorganisms and serves to detoxify the immediate environment for all organisms as long as anaerobic conditions are maintained. Selenium volatilization serves as a permanent detoxification process that can occur aerobically, although in at least one instance it was more effective anaerobically (Frankenberger and Karlson, 1995).

19.6 SELENIUM CYCLE

The existence of a selenium cycle in nature has been suggested by Shrift (1964). However, some of the details of this cycle are still obscure. The ultimate source of selenium must be igneous rocks, but whether microbes play a role in mobilizing the selenium from selenium-containing minerals is unknown. Similarly, little is known about the role that microbes may play in mobilizing selenium in soil and sediment. Such activity, when it occurs, is of great importance in understanding and controlling selenium pollution, as has occurred in the Kesterson Wildlife Refuge in California. The source of the selenium in that case appears to be the drainage of irrigation water applied to farm land in the San Joaquin Valley. It leached selenium from the soil. This drainage has been collecting in the wildlife refuge. Known biochemical steps of a selenium cycle are shown in Figure 19.1.

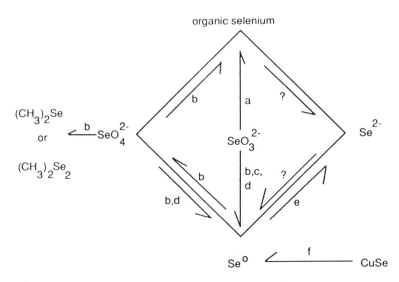

Figure 19.1 The selenium cycle. (a) *E. coli*; (b) bacteria; (c) actinomycetes; (d) fungi; (e) *M. lactilyticus*; (f) *T. ferrooxidans* (see also Doran and Alexander, 1977).

19.7   BIOOXIDATION OF   ːDUCED FORMS
OF TELLURIUM

Microbial oxidation of reduced forms of tellurium has so far not been reported. This may mean that this process does not occur in nature, but it is more likely that it has so far not been sought by investigators. Its geomicrobial importance is likely to be limited since the occurrence of tellurium is much rarer than that of selenium (see section 19.1).

19.8   BIOREDUCTION OF OXIDIZED FORMS
OF TELLURIUM

Microbial reduction of tellurates and tellurites to $Te°$ or $(CH_3)_2Te$ has been reported (Bautista and Alexander, 1972; Silverman and Ehrlich 1964; Nagai, 1965; Woolfolk and Whiteley, 1962). The fungus *Penicillium* sp. has been found to produce $(CH_3)_2Te$ from several inorganic tellurium compounds, provided only that reducible selenium compounds were also present (Fleming and Alexander, 1972). The amount of dialkyl telluride formed was related to the relative concentrations of Se and Te in the medium. Microbial reduction of oxidized forms of tellurium may represent detoxification reactions rather than a form of respiration, but this needs further investigation.

19.9   SUMMARY

Selenium, although a very toxic element, is nutritionally required by some bacteria, plants, and animals. Microorganisms have been described that can oxidize reduced selenium compounds. At least one, *Thiobacillus ferrooxidans*, can use selenide in the form of CuSe as a sole source of energy, oxidizing the compound to elemental selenium ($Se°$) and Cu^{2+}. Oxidized forms of inorganic selenium compounds can be reduced by microorganisms, including bacteria and fungi. Selenate and selenite may be reduced to one or more of the following: $Se°$, H_2Se, dimethylselenide [$(CH_3)_2Se$], dimethyl diselenide [$(CH_3)_2Se_2$], and dimethyl selenone [$(CH_3)_2SeO_2$]. The reductions are enzymatic. These microbial reactions contribute to a selenium cycle in nature. Microbial selenate and selenite reduction to elemental selenium in soil and sediment is a form of selenium immobilization, which is potentially reversible. Microbial selenate and selenite reduction to volatile forms of selenium in soil, sediment, and water columns of bodies of water is a form of selenium removal, which is permanent.

Tellurium occurs in such low concentration in nature that it does not seem geomicrobiologically important. Nevertheless, microbial tellurate and tellurite reduction to elemental tellurium (Te°) and dimethyltelluride [(CH$_3$)$_2$Te] has been observed. Microbial oxidation of tellurides has not been reported.

REFERENCES

Ahluwalia, G. S., Y. R. Saxena, and H. H. Williams. 1968. Quantitation studies on selenite metabolism in *Escherichia coli*. Arch. Biochem. Biophys. 124:79–84.

Alexander, M. 1977. Introduction to Soil Microbiology. 2nd ed. Wiley, New York.

Andreesen, J. R., and L. G. Ljungdahl. 1973. Formate dehydrogenase of *Clostridium thermoaceticum*: Incorporation of selenium-75, and the effects of selenite, molybdate, and tungstate on the enzyme. J. Bacteriol. 116:869–873.

Bacon, M., and W. J. Ingledew. 1989. The reductive reactions of *Thiobacillus ferrooxidans* on sulfur and selenium. FEMS Microbiol. Lett. 58:189–194.

Barkes, L., and R. W. Fleming. 1974. Production of dimethylselenide gas from inorganic selenium by eleven fungi. Bull. Environ. Contam. Toxicol. 12:308–311.

Bautista, E. M., and M. Alexander. 1972. Reduction of inorganic compounds by soil microorganisms. Soil Sci. Soc. Am. Proc. 36:918–920.

Breed, R. S., E. G. D. Murray, and N. R. Smith. 1957. Bergey's Manual of Determinative Bacteriology. 7th ed. Williams & Wilkins, Baltimore, MD.

Burton, G. A. Jr., T. H. Giddings, P. DeBrine, and R. Fall. 1987. High incidence of selenite-resistant bacteria from a site polluted with selenium. Appl. Environ. Microbiol. 53:185–188.

Chau, Y. K., P. T. S. Wong, B. A. Silverberg, P. L. Luxon, and G. A. Bengert. 1976. Methylation of selenium in the aquatic environment. Science 192:1130–1131.

Combs, G. F. Jr., and M. L. Scott. 1977. Nutritional interrelationships of vitamin E and selenium. BioScience 27:467–473.

DeMoll-Decker, H., and J. M. Macy. 1993. The periplasmic nitrite reductase of *Thauera selenatis* may catalyze the reduction of selenite to elemental selenium. Arch. Microbiol. 160:241–247.

Doran, J. W., and M. Alexander. 1977. Microbial transformations of selenium. Appl. Environ. Microbiol. 33:31–37.

Duerre, P., and J. R. Andreesen. 1982. Selenium-dependent growth and glycine fermentation by *Clostridium purinolyticum*. J. Gen. Microbiol. 128:1457–1466.

Enoch, H. G., and R. L. Lester. 1972. Effects of molybdate, tungstate, and selenium compounds on formate dehydrogenase and other enzymes in *Escherichia coli*. J. Bacteriol. 110:1032–1040.

Falcone, G., and W. J. Nickerson. 1963. Reduction of selenite by intact yeast cells and cell-free preparations. J. Bacteriol. 85:754–762.

Fleming, R. W., and M. Alexander. 1972. Dimethyl selenide and dimethyl telluride formation by a strain of *Penicillium*. Appl. Microbiol. 24: 424–429.

Frankenberger, W. T., and U. Karlson. 1992. Dissipation of soil selenium by microbial volatilization. *In* D.C. Adriano, ed. Biogeochemistry of Trace Metals. Lewis Publishers, Boca Raton, FL, pp. 365–381.

Frankenberger and Karlson, 1995. Soil management factors affecting volatilization of selenium from dewatered sediments. Geomicrobiol. J. 12:265–277.

Gerrard, T. L., J. N. Telford, and H. H. Williams. 1974. Detection of selenium deposits in *Escherichia coli* by electron microscopy. J. Bacteriol. 119:1057–1060.

Kovalskii, V. V., V. V. Ermakov, and S. V. Letunova. 1968. Geochemical ecology of microorganisms in soils with different selenium content. Mikrobiologiya 37:122–130 (Engl. transl. pp. 103–109).

Lansche, A. M. 1965. Tellurium. *In*: Mineral Facts and Problems. Bureau of Mines, U.S. Department of the Interior. Washington, DC, pp. 935–939.

Lester, R. L., and J. A. DeMoss. 1971. Effects of molybdate and selenite on formate and nitrate metabolism in *Escherichia coli*. J. Bacteriol. 105:1006–1014.

Lipman, J. G., and S. A. Waksman. 1923. The oxidation of selenium by a new group of autotrophic microorganisms. Science 57:60.

Lortie, L., W. D. Gould, S. Rajan, R. G. L. McCready, and J.-J. Cheng. 1992. Reduction of selenate and selenite to elemental selenium by a *Pseudomonas stutzeri* isolate. Appl. Environ. Microbiol. 58: 4042–4044.

Macy, J. M., S. Rech, G. Auling, M. Dorsch, E. Stackebrandt, and L. Sly. 1993. *Thauera selenatis* gen. nov. sp. nov., a member of the beta-subclass of *Proteobacteria* with a novel type of anaerobic respiration. Int. J. Syst. Bacteriol. 43:135–142.

Maiers, D. T., P. L. Wichlacz, D. L. Thompson, and D. F. Bruhn. 1988. Selenate reduction by bacteria from a selenium-rich environment. Appl. Environ. Microbiol. 54:2591–2593.

Mertz, W. 1981. The essential trace elements. Science 213:1332- 1338.

Miller, W. J. and M. W. Neathery. 1977. Newly recognized trace mineral elements and their role in animal nutrition. BioScience 27:674–679.

Nagai, S. 1965. Differential reduction of tellurite by growing colonies of normal yeasts and respiration deficient mutants. J. Bacteriol. 90: 220–222.

Nickerson, W. J., and G. Falcone. 1963. Enzymatic reduction of selenite. J. Bacteriol. 85:763–771.

Oremland, R. S., J. T. Hollibaugh, A. S. Maest, T. S. Presser, L. G. Miller, and C. W. Culbertson. 1989. Selenate reduction to elemental selenium by anaerobic bacteria in sediments and culture: Biogeochemical significance of a novel sulfate-independent respiration. Appl. Environ. Microbiol. 55:2333–2343.

Oremland, R. S., N. A. Steinberg, T. S. Presser, and L. G. Miller. 1991. In situ bacterial selenate reduction in the agricultural drainage systems of western Nevada. Appl. Environ. Microbiol. 57:615–617.

Patrick, R. 1978. Effects of trace metals in the aquatic ecosystem. Am. Sci. 66:185–191.

Pinsent, J. 1954. The need for selenite and molybdate in the coli-aerogenes group of bacteria. Biochem. J. 57:10–16.

Rapp, G., Jr. 1972. Selenium: element and geochemistry. *In*: R. W. Fairbridge, ed. The Encyclopedia of Geochemistry and Environmental Sciences. Encyclopedia of Earth Sciences Series. Vol. IVA. Van Nostrand Reinhold, New York, pp. 1079–1080.

Reamer, D. C., and W. H. Zoller. 1980. Selenium biomethylation products from soil and sewage. Science 208:500–502.

Rech, S., and J. M. Macy. 1992. The terminal reductases for selenate and nitrate respiration in *Thauera selenatis* are two distinct enzymes. J. Bacteriol. 174:7316–7320.

Rosenfeld, I., and O. A. Beath. 1964. Selenium, Geobotany, Biochemistry, Toxicity, and Nutrition. Academic Press, New York.

Rotruck, J. T., A. L. Pope, H. E. Ganther, A. B. Swanson, D. G. Hafeman, and W. G. Hoekstra. 1973. Selenium: Biochemical role as a component of glutathione peroxidase. Science 179:588–590.

Sapozhnikov, D. I. 1937. The substitution of selenium for sulfur in the photoreduction of carbonic acid by purple sulfur bacteria. Mikrobiologiya 6:643–644.

Sarathchandra, S. U., and J. H. Watkinson. 1981. Oxidation of elemental selenium to selenite by *Bacillus megaterium*. Science 211:600–601.

Shrift, A. 1964. Selenium in nature. Nature (Lond.) 201:1304–1305.

Shum, A. D., and J. C. Murphy. 1972. Effects of selenium compounds on formate metabolism and coincidence of selenium-75 incorporation and formic dehydrogenase activity in cell-free preparations of *Escherichia coli*. J. Bacteriol. 110:447–449.

Silverman, M. P., and H. L. Ehrlich. 1964. Microbial formation and degradation of minerals. Adv. Appl. Microbiol. 6:153–206.

Stadtman, T. C. 1974. Selenium biochemistry. Science 183:915- 922.

Steinberg, N. A., and R. S. Oremland. 1990. Dissimilatory selenate reduction potentials in a diversity of sediment types. Appl. Environ. Microbiol. 56:3550–3557.

Steinberg, N. A., J. S. Blum, L. Hochstein, and R. S. Oremland. 1992. Nitrate is a preferred electron acceptor for growth of freshwater selenate-respiring bacteria. Appl. Environ. Microbiol. 58:426–428.

Thompson-Eagle, E. T., W. T. Frankenberger, Jr., and U. Karlson. 1989. Volatilization of selenium by *Alternaria alternata*. Appl. Environ. Microbiol. 55:1406–1413.

Tomei, F. A., L. L. Barton, C. L. Lemansky, and T. G. Zocco. 1992. Reduction of selenate and selenite to elemental selenium by *Wollinella succinogenes*. Can. J. Microbiol. 38:1328–1333.

Torma, A. E., and F. Habashi. 1972. Oxidation of copper (II) selenide by *Thiobacillus ferrooxidans*. Can. J. Microbiol. 18:1780–1781.

Turner, R. J., Y. Hou, J. H. Weiner, and D. E. Taylor. 1992. The arsenical ATPase efflux pump mediates tellurite resistance. J. Bacteriol. 174: 3092–3094.

Woolfolk, C. A., and H. R. Whiteley. 1962. Reduction of inorganic compounds with molecular hydrogen by *Micrococcus lactilyticus*. I. Stoichiometry with compounds of arsenic, selenium, tellurium, transition and other elements. J. Bacteriol. 84:647–658.

Yamamoto, I., T. Saiki, S.-M. Liu, and L. G. Ljungdahl. 1983 Purification and properties of NADP-dependent formate dehydrogenase from *Clostridium thermoaceticum*, a tungsten-selenium-iron protein. J. Biol. Chem. 258:1826–1832.

Zalokar, M. 1953. Reduction of selenite by *Neurospora*. Arch. Biochem. Biophys. 44:330–337.

Zehr, J. P., and R. S. Oremland. 1987. Reduction of selenate to selenide by sulfate-respiring bacteria: experiments with cell suspensions and estuarine sediments. Appl. Environ. Microbiol. 53:1365–1369.

20

Geomicrobiology of Fossil Fuels

20.1 INTRODUCTION

Although most organic carbon in the biosphere is continually recycled, a not insignificant amount has been trapped in special sedimentary formations, where it is inaccessible to mineralization by microbes until it becomes reexposed to water and in many cases air through natural causes or human intervention. The trapped organic carbon exists in various forms. The degree of its chemically reduced state is related to the length of time it has been trapped and any secondary chemical changes that it has undergone during this time. Some of it has value as a fuel, a source of energy for industrial and other human activity, and is exploited for this purpose. Because of the great age of this material, it is known as **fossil fuel**. The remainder is chiefly kerogen and bitumen, some of which can be coverted to fuel by human intervention. Fossil fuels include methane gas, natural gas (which is largely methane), petroleum, oil shale, coal, and peat. They are generally considered to have had a microbial origin (Ourisson et al., 1984).

20.2 NATURAL ABUNDANCE OF FOSSIL FUELS

A major portion of the carbon at the Earth's surface is in the form carbonate (Fig. 20.1). It represents a major sink for carbon. The other sink is the trapped organic carbon which is not directly accessible for microbial mineralization. The carbonate carbon is not an absolute sink unless deeply buried because it is in a steady-state relationship with dissolved carbonate/bicarbonate and atmospheric CO_2, which in turn are in a steady-state relationship with organic carbon in living and dead biomass. The passage of carbon from one compartment into another is under biological control (Fig. 20.1) (Fenchel and Blackburn, 1979).

20.3 METHANE

Methane at atmospheric pressure and ambient temperature is a colorless, odorless, and flammable gas. Since its auto-ignition temperature is 650°C, it does not catch fire spontaneously. It is sparingly soluble in water (3.5 ml/100 ml of water) but readily soluble in organic solvents, including liquid hydrocarbons. Biogenic accumulations may occur in nature when methane is formed in consolidated sediment from which it cannot readily escape. Its formation comes about when organic matter in the sediment is undergoing anaerobic microbial mineralization in the absence of significant amounts of nitrate, Fe(III), Mn(IV), or sulfate. If gas pressure due to methane builds up sufficiently in anaerobic lake or coastal sediment, it may be seen to escape in the form of large gas bubbles that break at the water surface to release their methane into the air (Martens, 1976; Zeikus, 1977). In marshes, escaping methane may be ignited (by biogenic phosphene?) to burn continually as so-called will-o'-the-wisps.

Many of the methane accumulations on Earth are of biogenic origin. Methane may occur in association with peat, coal, and oil deposits or independent of them. That which occurs in association with coal and oil was probably microbially generated in the early stages of their formation, although some may have been formed abiotically in later diagenetic phases. Methane associated with coal deposits can be the cause of serious mine explosions when accidentally ignited. It is called **coal damp** by coal miners.

Biogenic methane formation is a unique biochemical process which appears to have arisen very early in the evolution of life. Indeed, the methanogenesis which results from the reduction of CO_2 by H_2 may represent the first or one of the first autotrophic processes on Earth (see Chapter 3).

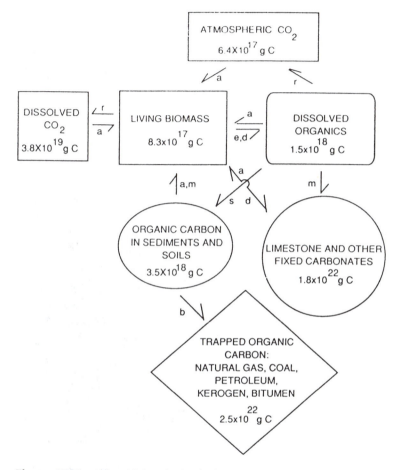

Figure 20.1 Microbial and physical processes contributing to carbon transfer among different compartments in the biosphere. (a) Microbial assimilation; (b) burial; (d) decomposition; (e) excretion; (m) microbial mineralization; (r) respiration; (s) sedimentation. Quantitative estimates from Fenchel and Blackburn (1979) and Bowen (1979).

Methanogens All methane-forming bacteria, i.e., **methanogens**, are archaebacteria. As a group, they are very diverse phylogenetically (Jones et al., 1987; Boone et al., 1993). They also show great diversity in morphology, including rods, spirilla, cocci, and sarcinae (Fig. 20.2). The feature that they share in common is that of being strict anaerobes that form methane in their respiration. The large majority of them are obligate or

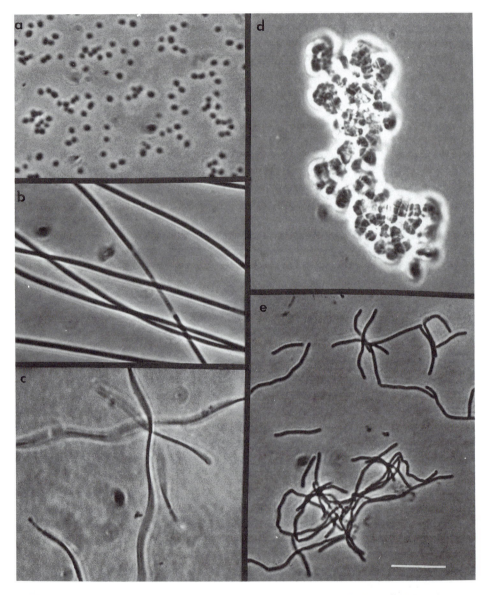

Figure 20.2   Morphologies of different methanogens. (a) *Methanoculleus bour-
gensis* (formerly *Methanogenium bourgense*); (b) *Methanosaeta* (formerly *Methanothrix) concilii;* (c) *Methanospirillum hungatei* OGC 16; (d) *Methanosarcina barkeri* OGC 35; (e) *Methanobacterium formicicum* OGC 55. Scale mark represents
10 μm and applies to all panels. (Courtesy of David R. Boone and Ron Willis.)

facultative autotrophs. They can get their energy from the reduction of carbon dioxide with hydrogen or its equivalent (formate or CO), while they get their carbon exclusively by assimilating carbon dioxide. But at least two, *Methanosaeta* (formerly *Methanothrix*) *soehngii* (Zehnder et al., 1980; Huser et al., 1982) and *Methanosaeta* (formerly *Methanothrix*) *concilii* (Patel, 1984), are heterotrophs requiring acetate as energy and carbon source and for this reason are known as **acetotrophic** or **aceticlastic** methanogens. At least one methanogen can use $S°$ in addition to CO_2 as terminal electron acceptor (Stetter and Gaag, 1983). As a group, most methanogens are nutritionally restricted to the following energy sources: H_2, CO, HCOOH, methanol, methylamines, and acetate (Atlas, 1988; Brock and Madigan, 1988). But Widdel (1986) found two exceptions. He reported that a freshwater strain of *Methanospirillum* and a strain of *Methanogenium* were each able to grow on 2-propanol and 2-butanol besides H_2 and formate. Zellner et al. found in 1989 that *Methanobacterium palustre* was able to grow on 2-propanol as energy source besides H_2 and formate. It was able to oxidize but not grow on 2-butanol. In 1990, Zellner et al. found that *Methanogenium liminatans* can use 2-propanol, 2-butanol, and cyclopentanol as energy sources in addition to H_2 and formate. Finster et al. found in 1992 that strain MTP4 can use methanethiol and dimethylsulfide besides methylamines, methanol, and acetate as energy sources. Yang et al. found in 1992 that *Methanococcus voltae*, *M. maripaludis*, and *M. vannielii* can each use pyruvate as an energy source in the absence of H_2.

To be able to draw on the wide range of oxidizable carbon compounds that may be available in their environment, methanogens in nature associate with heterotrophic fermenters and/or anaerobic respirers that do not completely mineralize their organic energy sources (see, e.g., Jain and Zeikus, 1989; Sharak Genthner et al., 1989; Grbić-Galić, 1990). Some methanogens, to optimize access to their needed, microbially generated energy sources, form intimate consortia (**syntrophic associations**) with other anaerobic bacteria that can furnish them with these nutrients (metabolites), which they form among their metabolic end products (e.g., Bochem et al., 1982; MacLeod et al., 1992; McInerney et al., 1979; Winter and Wolfe, 1979, 1980; Zinder and Koch, 1984; Wolin and Miller, 1987). Frequently the metabolites that are the basis for these syntrophic associations are not readily detectable when all the members of a consortium are growing together in mixed culture. This is because they are consumed as quickly as they are formed. When hydrogen is the metabolite, the process is called **interspecies hydrogen transfer** (Wolin and Miller, 1987).

Among the most widely recognized genera are *Methanobacterium*, *Methanothermobacter*, *Methanobrevibacterium*, *Methanococcus*, *Meth-*

anomicrobium, *Methanogenium*, *Methanospirillum*, *Methanosarcina*, *Methanoculleus*, and *Methanosaeta* (see Atlas, 1988; Brock and Madigan, 1988; Bhatnagar et al., 1991; Boone et al., 1993). Methanogens may be mesophilic or thermophilic. They are found in diverse anaerobic habitats (Zinder, 1993), including marine environments such as salt marsh sediments (Oremland et al., 1982; Jones et al., 1983), coastal sediments (Gorlatov et al., 1986; Sansone and Martens, 1981), anoxic basins (Romesser et al., 1979), geothermally heated seafloor (Huber et al., 1982), hydrothermal vent effluent on the East Pacific Rise (Jones et al., 1983), sediment effluent channel of the Crystal River Nuclear Power Plant (Florida, U.S.A.) (Rivard and Smith, 1982), lakes (Deuser et al., 1973; Jones et al., 1982; Giani et al., 1984), soils (Jakobsen et al., 1981), desert environments (Worakit et al., 1986), solfataric fields (Wildgruber et al., 1982; Zabel et al., 1984), oil deposits (Nazina and Rozanova, 1980; Rubinshtein and Oborin, 1986; Stetter et al., 1993), the digestive tracts of insects and higher animals, especially ruminants and other herbivores (Atlas, 1988; Breznak, 1982; Brock and Madigan, 1988; Wolin, 1981; Zimmerman et al., 1982), and as endosymbionts (van Bruggen et al., 1984; Fenchel and Finlay, 1992). Thus, despite their obligately anaerobic nature, methanogens are fairly ubiquitous.

Methanogens play an important but not exclusive role in anaerobic mineralization of organic carbon compounds in soil and aquatic sediments, especially freshwater sediments (Wolin and Miller, 1987). In marine sediments, where methanogens have to share hydrogen and/or acetate as sources of energy with sulfate-reducing bacteria, they tend to be outcompeted by the sulfate reducers because of the latter's higher affinity for hydrogen and acetate (Abrams and Nedwell, 1978; Kristjanssen et al., 1982; Schoenheit et al., 1982; Robinson and Tiedje, 1984). Thus in many estuarine or coastal anaerobic muds, sulfate-reducing activity and methanogenesis occur usually in spatially separated zones in the sediment profile, with the zone exhibiting sulfate-reducing activity overlying that exhibiting methanogenesis (e.g., Martens and Berner, 1974; Sansone and Martens, 1981). Some indirect evidence suggests that sulfate reducing bacteria may also use methane as electron donor (e.g., Oremland and Taylor, 1978). Whereas this does not explain the incompatibility, it would explain an absence of a very low concentration of methane in sulfate-reducing zones overlying methanogenic zones.

Under two special circumstances, methanogenesis and sulfate reduction can be compatible in an anaerobic marine environment. One circumstance is the existence of an excess supply of a shared energy source (H_2 or acetate) (Oremland and Taylor, 1978). The other circumstance is one where sulfate reducers and methanogens use different energy sources, namely decaying plant material and methanol or trimethyl amine, respectively (Oremland et al., 1982). In anaerobic freshwater sediments and soils where sulfate, nitrate, ferric oxide, and manganese oxide concentrations are very low, methanogenesis is usually the dominant mechanism of organic carbon mineralization. Yet even here, certain sulfate reducing bacteria may grow in the same niche as methanogens. Indeed, they may form a consortium with them. In the absence of sulfate, these sulfate reducers ferment suitable organic carbon with the production of H_2 as one of the products of their energy metabolism, which the methanogens then use in their energy metabolism forming methane (e.g., Bryant et al., 1977).

Methanogenesis and Carbon Assimilation by Methanogens

Methanogenesis   Methane formation represents a form of anaerobic respiration in which CO_2 is frequently but not always the terminal electron acceptor. Thus, in the simplest case where hydrogen is the energy source (electron donor), it reduces CO_2 to methane according to the overall reaction,

$$4H_2 + CO_2 \Longrightarrow CH_4 + 2H_2O \tag{20.1}$$

This reaction is exothermic ($\Delta G°$, -33 kcal) and yields energy which can be used by the organism to do metabolic work. Reactions utilizing formic or acetic acid, methanol or methylamines as electron acceptor instead of CO_2 follow essentially the same principle:

$$HCOOH + 3H_2 \Longrightarrow CH_4 + 2H_2O \ (\Delta G°, -42 \text{ kcal}) \tag{20.2}$$

$$CH_3COOH + 4H_2 \Longrightarrow 2CH_4 + 2H_2O \ (\Delta G°, -49 \text{ kcal}) \tag{20.3}$$

$$CH_3OH + H_2 \Longrightarrow CH_4 + H_2O \ (\Delta G°, -26.92 \text{ kcal}) \tag{20.4}$$

$$CH_3NH_2 + H_2 \Longrightarrow CH_4 + NH_3 \ (\Delta G°, -9 \text{ kcal}) \tag{20.5}$$

Here each of the carbons is reduced to methane by hydrogen.

In a few instances, secondary alcohols were found to serve as electron donors, with CO_2 as the terminal electron acceptor. The CO_2 was therefore the source of the methane formed (Widdell, 1986; Zellner et al., 1989). In these reactions, the alcohols were replacing H_2 as the reductant of CO_2.

Some methanogens can also form methane from carbon monoxide, formic acid, methanol, acetate, or methylamines without H_2 as electron donor. In these instances they perform disproportionations (fermentations) in which some of the substrate molecules act as electron donors (energy source) and others as electron acceptors. For instance (Brock and

Madigan, 1988; Mah et al., 1978; Smith and Mah, 1978; Zeikus, 1977),

$$4HCOOH \Longrightarrow CH_4 + 3CO_2 + 2H_2O \, (\Delta G°, \, -35 \, kcal) \qquad (20.6)$$

$$4CH_3OH \Longrightarrow 3CH_4 + CO_2 + 2H_2O \, (\Delta G°, \, -76 \, kcal) \qquad (20.7)$$

$$CH_3COOH \Longrightarrow CH_4 + CO_2 \, (\Delta G°, \, -9 \, kcal) \qquad (20.8)$$

$$4CH_3NH_2 + 2H_2O \Longrightarrow 3CH_4 + CO_2 + 4NH_3$$
$$(\Delta G°, \, -75 \, kcal) \qquad (20.9)$$

$$4CO + 2H_2O \Longrightarrow CH_4 + 3CO_2 \, (\Delta G°, \, -44.5 \, kcal) \qquad (20.10)$$

Some methanogens can even form methane from pyruvate by disproportionation (Yang et al., 1992). Resting cells of *Methanococcus* spp. grown in a pyruvate-containing medium in a N_2 atmosphere were shown to transform pyruvate to acetate, methane, and CO_2 according to the following stoichiometry (Yang et al., 1992):

$$4CH_3COCOOH + 2H_2O \Longrightarrow 4CH_3COOH + 3CO_2 + CH_4$$
$$(20.11)$$

This stoichiometry is attained if the organisms oxidatively decarboxylate pyruvate:

$$4CH_3COCOOH + 4H_2O \Longrightarrow 4CH_3COOH + 4CO_2 + 8(H)$$
$$(20.12)$$

and use the reducing power [8(H)] to reduce one-fourth of the CO_2 to CH_4:

$$CO_2 + 8(H) \Longrightarrow CH_4 + 2H_2O \qquad (20.13)$$

Bock et al. (1994) found that a spontaneous mutant of *Methanosarcina barkeri* could grow by fermenting pyruvate to methane and CO_2 with the following stoichiometry:

$$CH_3COCOOH + 0.5H_2O \Longrightarrow 1.25CH_4 + 1.75CO_2 \qquad (20.14)$$

To achieve this stoichiometry, the authors proposed the following mechanism based on known enzyme reactions in methanogens. Pyruvate is oxidatively decarboxylated to acetyl~SCoA and CO_2:

$$CH_3COCOOH + CoASH \Longrightarrow CH_3CO\sim SCoA + CO_2 + 2(H)$$
$$(20.15)$$

The available reducing power [2(H)] from this reaction is then used to reduce one-fourth of the CO_2 formed to methane:

$$2(H) + 0.25CO_2 \Longrightarrow 0.25CH_4 + 0.5H_2 \tag{20.16}$$

and the acetyl~SCoA is decarboxylated to methane and CO_2:

$$CH_3CO\text{\textasciitilde}SCoA + H_2O \Longrightarrow CH_4 + CO_2 + CoASH \tag{20.17}$$

The standard free energy yield at pH 7 ($\Delta G^{\circ\prime}$) was calculated to be 22.9 kcal mol^{-1} or 96 kJ mol^{-1} methane produced (Bock et al., 1994).

Although Reactions (20.1)–(20.11) look very disparate, they share a common metabolic pathway (Fig. 20.3). The reason why methanogens differ in which of these methane-forming reactions they can perform is that they do not all possess the same key enzymes that permit entry of particular methanogenic substrates into the common pathway (Brock and Madigan, 1988; Stanier et al.,1986; Vogels and Visser, 1983; Zeikus et al., 1985). The pathway involves stepwise reduction of carbon from the +4 to the −4 oxidation state via bound formyl, methylene, and methyl carbon. The operation of the methane-forming pathway requires some unique coenzyme and carrier molecules (Table 20.1). Coenzyme M (2-mercapto-ethylsulfonate) is unique to methanogens and may be used to identify them as methane formers. The large majority of methanogens synthesize this molecule de novo.

Bioenergetics of Methanogenesis As an anaerobic respiratory process, methane formation is performed to yield useful energy to the cell. Evidence to date indicates that ATP is generated by a chemiosmotic energy-coupling mechanism (e.g., Mountford, 1978; Doddema et al., 1978; 1979;

Table 20.1 Unusual Coenzymes in Methanogens

Coenzyme[a]	Function
Methanofuran	CO_2 reduction factor in first step of methanogenesis
Methanopterin (coenzyme F_{342})	Formyl and methene carrier in methanogenesis
Coenzyme M (2-mercaptoethane sulfonate)	Methyl carrier in methanogenesis
Coenzyme F_{430}	Hydrogen carrier for reduction of methyl coenzyme M
Coenzyme F_{420} (nickel-containing tetrapyrrole)	Mediates electron transfer between hydrogenase or formate and NADP; reductive carboxylation of acetyl ~CoA and succinyl ~CoA

[a] For structures of these coenzymes, see Brock and Madigan (1988) and Blaut et al. (1992).

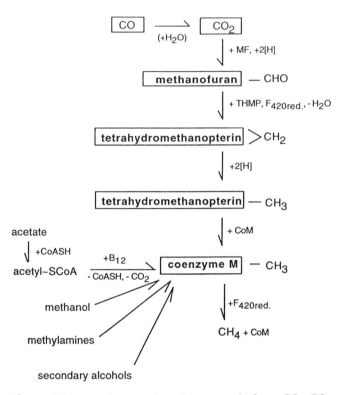

Figure 20.3 Pathways of methanogenesis from CO, CO_2, acetate, methanol, and methylamines.

Blaut and Gottschalk, 1984; Sprott et al., 1985; Brock and Madigan, 1988; Gottschalk and Blaut, 1990; Blaut et al. 1990, 1992; Mueller et al., 1993). The chemiosmotic coupling mechanism may involve protons or sodium ions. Membrane-associated electron transport constituents required in chemiosmotic energy conservation involving proton coupling in *Methanosarcina* strain Gö 1 include reduced factor F_{420} dehydrogenase, an unknown electron carrier, cytochrome b, and heterodisulfide reductase (see Blaut et al., 1992). The heterodisulfide consists of coenzyme M covalently linked to 7-mercaptoheptanoylthreonine by a disulfide bond (Blaut et al., 1992). A proton-translocating ATPase associated with the membrane catalyzes ATP synthesis in this organism.

    An example of a methanogen that employs sodium ion coupling is *Methanococcus voltae* (Dybas and Konisky, 1992; Chen and Konisky,

1993). It appears to employ a Na^+-translocating ATPase that is insensitive to proton translocation inhibitors. A scheme for pumping sodium ions from the cytoplasm to the periplasm that depends on a membrane-bound methyl tranferase has been proposed by Blaut et al. (1992).

Carbon Fixation by Methanogens   When methanogens grow autotrophically, their carbon source is CO_2. The mechanism by which they assimilate CO_2 is different from that of most autotrophs. Most eubacterial autotrophs use the pentose diphosphate pathway (Calvin-Benson cycle). Green sulfur bacteria use a reverse tricarboxylic acid cycle, and methane-oxidizing bacteria use either the hexulose-monophosphate or the serine pathway (see next section). Methanogens, like acetogens (see Chapter 6) and some sulfate-reducing bacteria (see Chapter 17), assimilate C by reducing one of two CO_2 molecules to methyl carbon and a second to a formyl carbon, which they then couple to the methyl carbon to form acetyl~SCoA (see Fig. 6.8 in Chapter 6). To form the important metabolic intermediate pyruvate, they next carboxylate the acetyl~SCoA reductively. All other cellular constituents are then synthesized from pyruvate.

Microbial Methane Oxidation

Methanotrophs   Methane can be used as a primary energy source by a number of bacteria. Some of these are obligate **methanotrophs**, others are facultative (Higgins et al., 1981). Methane is also oxidized by some yeasts (Higgins et al., 1981), and to a small extent by some methanogenic archaebacteria (Zehnder and Brock, 1979). Except for the methanogens, all known methanotrophs are aerobes. Some evidence, mostly indirect, points to the existence of anaerobic methanotrophs other than methanogens (e.g., Reeburgh, 1980; Martens and Klump, 1984; Iverson and Jorgenson, 1985; Gal'chenko et al., 1986; Henrichs and Reeburgh, 1987; Ward et al., 1987). So far only one of these anaerobic, methane-oxidizing strains has been cultured (Panganiban et al., 1979). Obligate methanotrophs include the genera *Methylomonas*, *Methylococcus*, *Methylobacter*, *Methylosinus*, and *Methylocystis* (Fig. 20.4). All are gram-negative and feature intracytoplamic membranes. On the basis of the organization of these membranes, the obligate methanotrophs can be assigned to two groups (Davies and Whittenbury, 1970). **Type I** contains stacked membranes whereas **Type II** contains paired membranes concentric with the plasma membrane and forming vesicle-like or tubular structures (Fig. 20.5). Facultative methanotrophs feature internal membranes with an appearance like those of the Type II obligate methanotrophs. All methano-

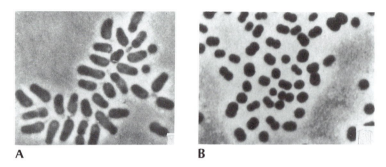

A   **B**

Figure 20.4 Methane-oxidizing bacteria (methanotrophs) (× 19,000). (A) *Methylosinus trichosporium* in rosette arrangement. Organisms are anchored by visible hold-fast material. (B) *Methylococcus capsulatus*. (From R. Whittenbury, K. C. Phillips, and J. F. Wilkinson, Enrichment and isolation and some properties of methane-utilizing bacteria, J. Gen. Microbiol. 61:205–218, 1970, by permission.)

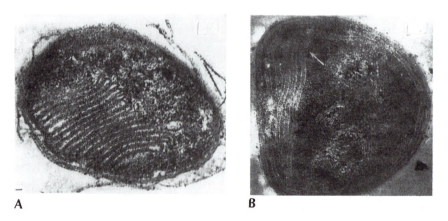

A   **B**

Figure 20.5 Fine structure of methane-oxidizing bacteria (× 80,000). (A) Section of *Methylococcus* (subgroup minimus) showing Type I membrane system. (B) Peripheral arrangement of membranes in *Methylosinus* (subgroup sporium) characteristic of Type II membrane systems. (From S. L. Davies and R. Whittenbury, Fine structure of methane and other hydrocarbon-utilizing bacteria, J. Gen. Microbiol. 61:227–232, 1970, by permission.)

trophs can also use methanol as a primary energy source, but not all methanol oxidizers can grow with methane as primary energy source. Methanol oxidizers which cannot oxidize methane are called **methylotrophs**.

Methanotrophs are important for the carbon cycle in returning methane, which is always generated anaerobically, to the reservoir of CO_2 (e.g., Vogels 1979). Obligate methanotrophs are found at aerobic/anaerobic interfaces in soil and aquatic environments that are crossed by methane (e.g., Alexander, 1977; Reeburgh, 1976; Ward et al., 1987; Sieburth et al., 1987), and also in coal and petroleum deposits (Ivanov et al., 1979; Kuznetsov et al., 1963). Some methanotrophs are also important intracellular symbionts in marine seep mussels and other benthic invertebrates. Cavanaugh et al. (1987) found evidence of the presence of such symbiotic methanotrophs in the epithelial cells of the gills of some mussels from reducing sediments at hypersaline seeps at abyssal depths in the Gulf of Mexico at the Florida Escarpment. MacDonald et al. (1990) made similar observations in mussels that occurred in a large bed surrounding a pool of hypersaline water rich in methane at 650 m on the continental slope south of Louisiana, U.S.A. Transmission electron microscopic examination by Cavanaugh et al. (1987) showed the symbionts to feature typical intracytoplasmic membranes of the Type I methanotrophs. They possessed the key enzymes associated with methane oxidation in that group (see below). The basis for the symbiosis between the invertebrate host and the methanogen is the sharing of fixed carbon derived from the methane taken up by the host and metabolized by the methanogen (Childress et al., 1986), just as symbiotic H_2S-oxidizing bacteria in some invertebrates of hydrothermal vent communities share with their host the carbon they fix from CO_2 taken up by the host (see Chapter 18). Whether the source of the methane issuing from Gulf of Mexico seeps is biogenic or abiotic is unclear at this time. The invertebrate fauna in the vicinity of hydrocarbon seeps on the Louisiana slope in the Gulf of Mexico features intracellular methanotrophic bacterial symbionts only in mussels. Vestimentiferan worms and three clam species in these locations feature autotrophic bacterial sulfur-metabolizing symbionts (Brooks et al., 1987). Some animals at the Oregon subduction zone have also formed associations with methanotrophs that enable them to feed on methane (Kulm et al., 1986).

Biochemistry of Methane Oxidation Obligate methanotrophs can use methane, methanol, and methylamines as energy sources by oxidizing them to CO_2, H_2O, and NH_3, respectively. When methane is the energy source, the following steps are involved in its oxidation:

$$CH_4 \xrightarrow{\frac{1}{2}O_2} CH_3OH \xrightarrow{\frac{1}{2}O_2} HCHO \xrightarrow{\frac{1}{2}O_2} HCOOH \xrightarrow{\frac{1}{2}O_2} CO_2 + H_2O$$

$$(20.18)$$

The first step in this reaction sequence is catalyzed by a mono-oxygenase, which causes the direct introduction of an atom of molecular oxygen into the methane molecule (Anthony, 1986). This step is generally considered not to yield useful energy to the cell. A report by Sokolov (1986), however, suggests the contrary, at least with *Methylomonas alba* BG8 and *Methylosinus trichosporium* OB3b (Sokolov, 1986). Since monooxygenase requires reduced pyridine nucleotide (NADH + H$^+$) in its catalytic process to provide electrons for reduction of one of the two oxygen atoms in O_2 to H_2O (the other atom is introduced into methane to form methanol), a proton motive force is generated in the electron transfer from NADH, which the cell may be able to couple to ATP synthesis. The enzyme that catalyzes methanol oxidation is methanol dehydrogenase, which in *Methylococcus thermophilus*, as in other methanotrophs, does not use pyridine nucleotide as cofactor (Anthony, 1986; Sokolov et al., 1981). Instead, the enzyme contains pyrroloquinone (+ 90 mV) as its prosthetic group, which feeds electrons from methanol into the electron transport chain. The formaldehyde resulting from methanol oxidation is oxidized to formate [Reaction (20.18)] (see Roitsch and Stolp, 1985). Formaldehyde oxidation to formate may involve an NAD$^+$ linked dehydrogenase or a pyrroloquinone-linked dehydrogenase (Stanier et al., 1988). Whatever the mechanism of formaldehyde oxidation, the reducing power is fed into the electron transport system for energy generation. The formate dehydrogenase is an NAD$^+$ coupled enzyme which oxidizes formate to CO_2 and H_2O and feeds electrons into the electron transport system for energy generation. ATP synthesis in methanotrophs appears to be mainly or entirely by chemiosmosis (Anthony, 1986).

It should be noted that the methane monooxygenase of methanotrophs is not a very specific enzyme. It can also catalyze NH_3 oxidation (O'Neill and Wilkinson, 1977). In this instance, the monooxygenase hydroxylates ammonia to NH_2OH. Ammonia-oxidizing autotrophic bacteria can similarly oxidize methane to methanol (Jones and Morita, 1983). However, just as ammonia oxidizers cannot grow on methane as energy source, methanotrophs cannot grow on ammonia as energy source. This is because they lack the enzyme sequences for methanol and hydroxylamine oxidation, respectively.

Carbon Assimilation by Methanotrophs   All autotrophically grown methanotrophs of Types I and II assimilate some carbon (up to 30%) in the form of CO_2 (Romanovskaya et al., 1980). The enzyme involved in the fixation

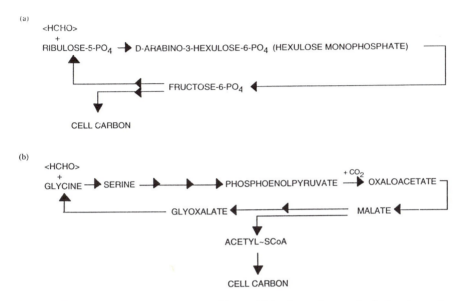

Figure 20.6 Alternative pathways of formaldehyde-carbon assimilation in methanotrophs. The hexulose monophosphate (HuMP) pathway is used by Type I methanotrophs and the serine pathway is used by Type II methanotrophs.

appears to be phosphoenolpyruvate carboxylase (PEP carboxylase). The mechanism of fixation of the remaining carbon depends on the methanotroph type. Both types assimilate it at the formaldehyde oxidation state. Type I fixes this carbon via an assimilatory, cyclical ribulose monophosphate pathway (Fig. 20.6a) whereas Type II fixes it by way of the cyclical serine pathway (Fig. 20.6b). In the ribulose monophosphate pathway, 3-phosphoglyceraldehyde is the key intermediate in carbon assimilation, whereas in the serine pathway it is acetyl~CoA (Brock and Madigan, 1988; Stanier et al., 1986; Gottschalk, 1986). Reducing power for assimilation derives from methane dissimilation and may require reverse electron transport to generate the needed $NADPH + H^+$. Methylotrophs generally use the serine pathway for carbon assimilation derived from C_1 compounds.

Obligate methanotrophs stand somewhere between typical autotrophs and heterotrophs in their carbon assimilation mechanism (Quayle and Ferenci, 1978).

The Position of Methane in the Carbon Cycle In order to maintain a sufficient pool of biologically available carbon in the biosphere, carbon has to be continually recycled from organic to inorganic carbon. This is accomplished both aerobically and anaerobically. Quantitatively, the aerobic process makes the greater contribution, but the anaerobic process is not negligible; indeed, anaerobic mineralization in sediment has been estimated to equal approximately the rate of burial of organic carbon (Henrichs and Reeburgh, 1987). Anaerobic mineralization involves fermentation processes coupled with anaerobic forms of respiration, including nitrate reduction (nitrate respiration, denitrification, nitrate ammonification) (e.g., Sørensen, 1987), iron and manganese respiration (Lovley and Phillips, 1988; Myers and Nealson, 1988; Ehrlich, 1993; Lovley 1993), iron reduction (Lovley, 1987; 1993), sulfate respiration (Skyring, 1987), and methanogensis (Wolin and Miller, 1987; Young and Frazer, 1987). Each of these forms of respiration is dominant where the respective electron acceptor is dominant and other environmental conditions are optimal. In some instances, more than one form of anaerobic respiration may occur concurrently in the same general environment, provided there is no competition for the same electron donor or other growth-limiting substance (Ehrlich, 1993).

In anoxic soils (paddy soils) or anoxic freshwater or marine sediments where extensive methanogenesis occurs, a small portion of the methane is oxidized to CO_2 without the benefit of oxygen, but a larger portion of the methane which is not trapped escapes into an oxidizing environment and is extensively oxidized to CO_2 by aerobic methanotrophs (Higgins et al., 1981). A small amount of methane may be used as energy and carbon source by special marine invertebrates. Any methane that is not biooxidized or otherwise combusted or trapped in natural sedimentary reservoirs escapes this biological attack and enters the atmosphere where it may be chemically oxidized in the troposphere (Vogels, 1979). The various paths for methanogenesis and methane oxidation are summarized in Figure 20.7.

20.4 PEAT

Nature of Peat Although peat and coal are two different substances, their modes of origin have included common initial steps. Indeed, peat formation may have been an intermediate step in coal formation. Peat is a form of organic soil or histosol. It is mostly derived from plant remains that have accumulated in marshes and bogs (Fig. 20.8). According to Francis (1954), these remains have come from (1) sphagnum, grasses, and heather (high moor peat); (2) reeds, grasses, sedges, shrubs, and bushes (low moor peat); (3) trees, branches, and debris of large forests in low-

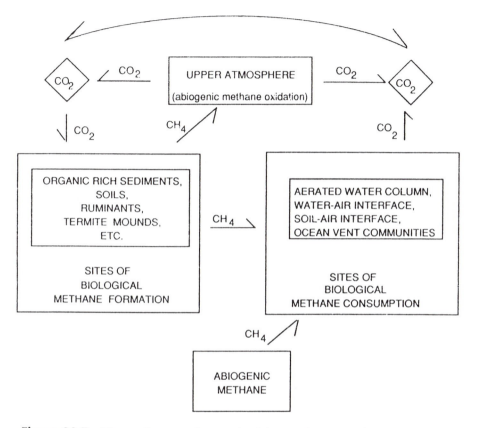

Figure 20.7 The methane cycle, emphasizing methanogenesis and methanotrophy.

lying wet ground (forest peat); or (4) plant debris accumulated in swamps (sedimentary or lake peat). In all these instances, plant growth outstripped the decay of the plant remains, assuring a continual supply of new raw material for peat formation.

Role of Microbes in Peat Formation Initially, the plant remains may have undergone attack by some of their own enzymes but soon were attacked by fungi, which degraded the relatively resistant polymers such as cellulose, hemicellulose, and lignin. Bacteria degraded the more easily oxidizable substances and the breakdown products of fungal activity that were not consumed by the fungi themselves. Fungal activity continued for as

Figure 20.8 Peat. (A) Section of ditch near Vestburg, showing light sphagnum peat over dark peat: (a) living sphagnum, (b) sphagnum peat, (c) shrub remains, (d) sedge rootstocks, (e) pond lily rootstocks; (f) laminated peat. (B) View of surface near ditch, showing corresponding vegetative zones: (a) shrub zone, (b) grass zone, (c) sedge zone, (d) pond lily zone. (From Davis, 1907, Plate XII from Annual Report, Geological Survey of Michigan, by permission.)

long as the organisms had access to air, but as they and their remaining substrate became buried and conditions became anaerobic, bacterial fermentation and anaerobic respiration set in and continued until arrested by accumulation of inhibitory (toxic) wastes, lack of sufficient moisture, depletion of suitable electron acceptors for anaerobic respiration (i.e., nitrate, sulfate, Fe(III) and/or manganese (IV), carbon dioxide, etc.), and other factors (Francis, 1954; Kuznetsov et al., 1963; Rogoff et al., 1962). Combined with these limiting factors, the overwhelming accumulation of organic debris must also be taken into account. It was more than the system could handle before conditions for continued mineralizing activity became unfavorable.

Uppermost aerobic layers of peat may harbor a viable microflora in places even today, indicating that peat formation may be occurring at the present time (Kuznetsov et al., 1963). Viable anaerobic bacteria and actinomycetes are present even in deep layers of some peats (Rogoff et al., 1962; Zvyagintsev et al., 1993). Differences in methane production in various peats are a reflection of differences in their origin with respect to source materials and environmental conditions (e.g., Yavitt and Lang, 1990; Brown and Overend, 1993).

During its formation, peat becomes enriched in lignin, ulmins, and humic acids. The first of these compounds is a relatively resistant polymer of woody tissue and the second and third are complex material from the incomplete breakdown of plant matter, including lignin. Peat also contains other compounds that are relatively resistant to microbial attack, such as resins and waxes from cuticles, stems, and spore exines of the peat-forming plants. When compared to the C, H, O, N, and S content of the original undecomposed plant material, peat is slightly enriched in carbon, nitrogen, and sulfur, but depleted in oxygen and sometimes hydrogen (Francis, 1954). This enrichment in C, N, and S over oxygen may be explained in part in terms of volatilization of the less-resistant components as a result of microbial attack and the buildup of residues (resins, waxes, lignin) that are relatively oxygen-deficient owing to hydrocarbon-like and aromatic properties. The plant origin of peat is still clearly recognizable by examination of its structure.

20.5 COAL

The Nature of Coal   Coal has been defined by Francis (1954) as "a compact, stratified mass of mummified plants which have been modified chemically in varying degrees, interspersed with smaller amounts of inorganic matter." Peat can be distinguished from coal in chemical terms by its much lower carbon content (51–59% dry wt) and higher hydrogen content (5.6–6.1% dry wt) than found in coal (carbon, 75–95% dry wt; hydrogen 2.0–5.8% dry wt) (Francis, 1954, p. 295). The average carbon content of typical wood has been given as 49.2% carbon (dry wt) and its average hydrogen content as 6.1% (dry wt) (Francis, 1954). Coalification can thus be seen to have resulted in an enrichment in carbon and a slight depletion in hydrogen of the substance that gave rise to coal. Coal is generally found buried below layers of sedimentary strata (**overburden**). Its geological age is generally advanced. Significant deposits formed in the Upper Paleozoic between 300 and 210 million years ago, in the Mesozoic between 180 and 100 million years ago, and in the Tertiary between 60 and 2.5 million years ago. Peats generally have developed from about 1 million years ago up to the present.

Coal is classified by rank. According to the ASTM classification system, four major classes are recognized (Mineral Facts and Problems, 1965). They are, starting with the least developed coal, lignitic coal, sub-bituminous coal, bituminous coal, and anthracite coal. Lignitic coal is the least-developed coal, structurally resembling peat and having the highest moisture and lowest carbon content (59.5–72.3% dry wt; Francis, 1954, p.335) as well as the lowest heat value of any of the coals. This coal formed in Tertiary times. Subbituminous coal has a somewhat lower moisture and higher carbon content (72.3–80.4% dry wt). It also has a higher heat content than lignitic coal. Bituminous coal (Fig. 20.9) has a carbon content ranging from 80.4 to 90.9% (dry wt) and a high heat value. Both types of bituminous coal are mostly of Paleozoic and Mesozoic age. Anthracite coals have very low moisture content, few volatiles, and a high carbon content (92.9–94.7% dry wt). They are of Paleozoic age. Cannel coal is a special variety of bituminous coal which was derived mainly from wind-blown spores and pollen rather than from woody plant tissue.

Role of Microbes in Coal Formation As mentioned earlier, coal deposits developed at special periods in geologic time. In these periods, the climate, landscape features, and biological activity were favorable. Large amounts of plant debris accumulated in swamps or shallow lakes because, owing to climatic conditions, plant growth was very profuse and provided a continual supply. At times, accumulated debris would become buried and covered by clay and sand under water before a new layer of plant debris accumulated. As more sediment accumulated, subsidence followed and is believed to have been an integral part in the formation of coal deposits (Francis, 1954).

Bacteria and fungi are generally believed to have played an important role in coalification only in the initial stages. Their role was similar to that

Figure 20.9 A section of the Pittsburgh coal seam in the Safety Research coal mine of the U.S. Bureau of Mines in Pittsburgh, PA (U.S.A.). Although not evident in this black-and-white photograph, extensive brown iron stains were present on portions of the face of the coal seam shown here. These stains are evidence of acid mine-drainage emanating from the fracture at the upper limit of the coal seam. The acid drainage resulted from microbial oxidation of exposed iron pyrite inclusions in the seam (see text). (Courtesy of the U.S. Bureau of Mines)

in peat formation. They destroyed the easily metabolizable substances, such as sugars, amino acids, and volatile acids, in a short time, and degraded more slowly the more stable polymers, such as cellulose, hemicellulose, lignins, waxes, and resins. Many of the latter were degraded only very incompletely before microbial activity ceased for reasons similar to those in peat formation (see previous section). Hyphal remains, sclerotia, and fungal spores have been identified in some coal remains. Initial microbial attack is believed to have been aerobic and mainly fungal in nature. Later attack occurred under progressively more anaerobic conditions and mostly by bacteria. Tauson believed, however, that in coal formation the anaerobic phase was abiological (Kuznetsov et al., 1963). Conversion of the residue of microbial activity (peat?) to coal is presently believed to have been due to physical and chemical agencies of unidentified nature, but probably involved heat and pressure which resulted in loss of volatile components.

Coal as a Microbial Substrate   Coal is not a very suitable nutrient to support microbial growth because according to early views, it contains inhibitory substances ("antibiotics") that may suppress it. These antibiotics have been thought to be associated with the waxy or resinous part of coal, extractable with methanol (Rogoff et al., 1962). In the first experiments with coal slurries, only marginal bacterial growth was obtainable, the limiting factors being the presence of inhibitory substances and lack of assimilable nutrients (Koburger, 1964). Growth of *Escherichia freundii* and *Pseudomonas rathonis* in such slurries improved if the coal had first been treated with H_2O_2. In much more recent experiments with run-of-the-mine bituminous coal from Pennsylvania, U.S.A., in which coal particles in a size range from 0.5 to 13 mm were wetted in glass columns with air-saturated distilled water at 10- or 14-day intervals, a bacterial community developed consisting mainly of autotrophic bacteria (iron and sulfur oxidizers) and to a lesser extent of heterotrophic bacteria. Progressive acid production by the autotrophs was thought to limit the development of heterotrophs. Observed changes in the rate of acetate metabolism may have been a reflection of microbial successions among the heterotrophs (Radway et al., 1987; 1989). The bacteria in these experiments lived at the expense of impurities in coal, such as iron pyrite.

Two basidiomycete fungi, including *Trametes versicolor* (also known as *Polysporus versicolor* and *Coriolus versicolor*) and *Poria monticola*, have been shown to grow directly on crushed lignite coal as well as in minimal lignite-noble agar medium (Cohen and Gabriele, 1982). With time, the cultures growing on lignite exuded a black liquid that was a product of lignite attack. Infrared spectra of the exudate gave indication that conjugated aromatic rings from the lignite had been structurally modified. It must be stressed that the coal in this case was lignite (a low rank coal) and not bituminous coal. Other fungi, including *Paecilomyces*, *Penicillium* spp., *Phanerochete chrysosporium*, *Candida* sp., and *Cunninghamella* sp., as well as *Streptomyces* sp. (actinomycete), have also been shown to grow on and degrade lignite and, in some cases, even bituminous coal (see Cohen et al., 1990; Stewart et al., 1990). In the case of *Trametes versicolor*, Cohen et al. (1987) and Pyne et al. (1987) at first attributed lignite solubilization to a protein secreted by the fungus that had polyphenol oxidase activity (syringaldazine oxidase) (Pyne et al., 1987). Subsequently Cohen et al. (1990) and Fredrickson et al. (1990) reported that a ligand produced by the fungus and identified as ammonium oxalate (Cohen et al., 1990) was the real solubilizing agent. The oxalate acted as a siderophore by removing iron from the test substrate leonardite (oxidized lignite). This later conclusion seems a little puzzling because it implies that iron(III) plays a central role in holding the leonardite structure together and that no modification of the organic skeleton of leonardite takes place.

White rot fungi, of which *Polyporus versicolor* and *Phanerochete chrysosporium* are examples, are known to produce two kinds of enzymes, lignin peroxidase and manganese-dependent peroxidase, both of which catalyze lignin attack (e.g., Paszczynski et al., 1986). Indeed, Stewart et al. (1990) reported that *P. chrysosporium* degraded lignite and a bituminous coal from Pennsylvania, albeit weakly. Some bacteria also can form such extracellular enzymes that are active on low-rank coals (Crawford and Gupta, 1993). Moreover, dibenzothiophene-degrading, aerobic bacteria have been found that are able to break down part of the carbon framework of liquefied bituminous coal (suspension of pulverized coal) and in the process remove some of the sulfur bound in the framework (Stoner et al., 1990). It would therefore seem reasonable that such types of enzymes play a role in lignite attack. Depending on the coal, it is probably a combination of enzymic and nonenzymic processes that leads to transformation of low-rank coals. The interested reader is referred to Crawford (1993) for further details on these processes.

Lignin and lignin derivatives can also be biologically attacked under anaerobic conditions (Young and Frazer, 1987). It would be of interest to study this action on lignite.

Microbial Desulfurization of Coal Bituminous coal may contain significant amounts of sulfur in inorganic (pyrite, marcasite, elemental sulfur, sulfate) and/or organic form. The total sulfur can range from 0.5% to 11% (Finnerty and Robinson, 1986). The proportion of pyritic and organic sulfur in the total sulfur depends on the source of the coal. Iron pyrite or marcasite (FeS_2) came to be included in some bituminous coal seams during their formation. Some or all of this iron disulfide may well have been formed biogenically, in which case it is a reflection of the anaerobic biogenic phase of coalification, representative of sulfate respiration (see Chapters 17 and 18) and also mineralization of organic sulfur compounds such as sulfur-containing amino acids. A model system illustrating how pyrite may have formed in coal is provided by the sulfur transformations presently occurring in the formation of Everglades peat (Casagrande and Siefert, 1977; Altschuler et al., 1983). The source of the sulfur in this case is organic.

The presence of pyritic sulfur in coal lowers its commercial value because on combustion of such coal, air pollutants such as SO_2 are generated. Pyritic sulfur can be removed in various ways (Bos and Kuenen,

1990; Blazquez et al., 1993). In all cases, the coal must first be pulverized to expose the pyrite. The pyrite particles can then be separated by differential flotation in which pyrite flotation is suppressed. Flotation suppression can be achieved chemically or biologically. In the latter case, *Thiobacillus ferrooxidans*, which attaches rapidly and selectively to pyrite particles, is the suppression agent (Pooley and Atkins, 1983; Bagdigian and Myerson, 1986; Townsley et al., 1987). But pyritic sulfur can also be removed by oxidizing it. This can be accomplished by the action of pyrite-oxidizing bacteria such as *Thiobacillus ferrooxidans* and *Sulfolobus* spp. or *Acidianus* spp., and *Metallosphaera*. (Dugan, 1986; Andrews et al., 1988; Merrettig et al., 1989; Larsson et al., 1990; Baldi et al., 1992; Clark et al., 1993)

Bituminous coal may also contain organically bound sulfur. Its presence is undesirable because on combustion it, too, contributes to air pollution. Microbiological methods are being sought to remove it, but progress has been slow (e.g., Dugan, 1986; Finnerty and Robinson, 1986; Crawford and Gupta, 1990; Mormile and Atlas, 1988; Van Afferden et al., 1990; Stoner et al., 1990; Omori et al., 1992; Olson et al., 1993; Izumi et al., 1994). Dibenzothiophene is used as a model structure for the organically bound sulfur in coal. Since the sulfur is likely to occur in different types of compounds bound in the structure of bituminous coal, whose degradation may require different enzyme systems, it may be that no one organism in nature can attack the whole range of these substances. Genetic engineering is being applied to find a solution to this problem. In any approach taken to organic sulfur removal from coal, it is important to find a way to remove the sulfur without significant loss of carbon, which would lower the caloric value of the coal.

20.6 PETROLEUM

Nature of Petroleum Petroleum is a mixture of aromatic and aliphatic hydrocarbons and various other heterocyclics including oxygen-, nitrogen-, and sulfur-containing compounds. The hydrocarbons include gaseous ones of the paraffinic series such as methane, ethane, propane, and butane besides longer-chained, nongaseous ones. Some of the heterocyclics, such as porphyrin derivatives, may contain metals such as vanadium or nickel bound in their structure. Petroleum accumulations are found in some folded, porous, sedimentary rock strata, such as limestone or sandstone, or in other fractured rock such as fissured shale or igneous rock. In petroleum geology these rock formations are collectively known as **reservoir rock**. The age of reservoir rocks may range from Late Cambrian (500-million years) to the Pliocene (1 to 13 million years). Very extensive petroleum reservoirs are found in rock of Tertiary age (70 million years) (North, 1985).

Most petroleum derived mainly from planktonic debris that was deposited on the floor of depressions of shallow seas and ultimately buried under heavy layers of sediment, deposited perhaps by turbidity currents. Over geological time the organic matter became converted to petroleum and natural gas (chiefly methane) (North, 1985).

Many theories have been advanced to explain the origin of petroleum and associated natural gas (see, e.g., Beerstecher, 1954; Robertson, 1966; North, 1985). None of these have been fully accepted. Some theories invoke heat or pressure or both as agents that promoted abiological conversion of planktonic residues to the hydrocarbons and other constituents of petroleum and natural gas. The source of the heat has been viewed as the natural radioactivity in the Earth's interior but was more likely heat diffusing from magma chambers underlying tectonically active areas. Other theories have invoked inorganic catalysis with or without the influence of heat and pressure and with or without prior acid or alkaline hydrolysis. Still other theories have proposed that petroleum constitutes a residue of naturally occurring hydrocarbons in the planktonic remains after all the other components have been biologically destroyed. It has even been proposed that biological agents produced the hydrocarbons by aerobic or anaerobic reduction of fatty acids, proteins, and amino acids, carbohydrates, carotenoids, sterols, glycerol, chlorophyll, and lignin-humus complexes, together with appropriate decarboxylations and deaminations. Finally, a theory has been put forward that biogenically formed methane from the planktonic debris became polymerized under high temperature and pressure and in the possible presence of inorganic catalysts (e.g., Mango, 1992), or, as an alternative, that bacteria modified the planktonic materials to substances closely resembling petroleum components, which were then converted to petroleum and natural gas constituents by heat and pressure. Abiotic methane could also have been a source and could have been polymerized in the same way. Chemical reaction of methane with liquid hydrocarbons has been noted in the laboratory at high temperature and pressure (1,000 atm, 150–259°C) (Gold et al., 1986). A hydrothermal mound area in the southern rift of the Guaymas Basin in the Gulf of California (Sea of Cortez) appears to be a site where active abiotic transformation of organic matter into petroliferous substances including gasoline-range aliphatic and aromatic hydrocarbons is currently occurring (Simoneit and Lonsdale, 1982; Didyk and Simoneit, 1989). This site may also be a source of methane used by methanotrophic consortia (see section 20.3).

Role of Microbes in Petroleum Formation At present it is generally thought that bacteria have played a role in the initial stage of petroleum formation, but what this role was remains obscure, except in the case of methane formation. ZoBell (1952, 1963) suggested that the planktonic debris was fermented, leading to compounds enriched in hydrogen and depleted in oxygen, sulfur, and phosphorus. Davis (1967, p. 23) has visualized microbial processes not unlike those in peat formation, involving initially aerobic attack of the sedimented planktonic debris followed by anaerobic activity after initial burial. This activity may have included hydrolytic, decarboxylating, deaminating, and sulfate-reducing reactions, resulting in the accumulation of marine humus, i.e., stabilized organic matter. Progressively deeper burial resulted in compaction and in cessation of microbial activity, accompanied by the evolution of small amounts of hydrocarbon substance plus petroleum precursors. The biotic reactions were followed by an abiotic phase of longer duration during which the microbially produced precursors were transformed under the influence of heat and pressure into the range of hydrocarbons associated with petroleum. This sequence of biotic and abiotic reactions is supported by observations on light hydrocarbon formation in marine sediments (Hunt, 1984; Hunt et al., 1980). Clays could have catalytically promoted further chemical reductions of petroleum precursors.

Role of Microbes in Petroleum Migration to Reservoir Rock As hydrocarbons accumulated during petroleum formation, the more volatile compounds generated an increasing gas pressure that helped to force the more liquid components through porous rock (sandstone, limestone, fractured rock) to anticlinal folds. The hydrocarbons were trapped in the apex of these folds below a stratum of impervious rock to form a petroleum reservoir. We tap such reservoirs today for our petroleum supply. The migration of petroleum from the source rock to the reservoir rock was probably helped by groundwater movement and by the action of natural detergents such as fatty acid soaps and other surface-active compounds of microbial origin. ZoBell (1952) has suggested that bacteria themselves may help to liberate oil from rock surfaces and thereby promote its migration by dissolving carbonate and sulfate minerals to which oil may adhere and by generating CO_2, whose gas pressure could help to force migration of petroleum. Bacterially produced methane may lower the viscosity of petroleum liquid by dissolving in it and thus help its migration (ZoBell, 1952). The important microbial contributions to petroleum formation thus come in the initial action on the source material (planktonic biomass) and in the final stages by promoting the migration of petroleum from the source rock to the reservoir rock. In addition, sulfate-reducing bacteria may play a role in sealing an oil deposit in the reservoir rock by deposition of secondary CaCO₃ at the interface between the oil-bearing stratum and the stratal waters (Ashirov and Sazanova, 1962; Davis 1967; Kuznetsov et al., 1963; see also Chapter 8).

Viable bacteria have been detected in brines associated with petroleum reservoirs to which access was gained by drilling. They were assigned to three specific groups. One included the sulfate reducers *Desulfovibrio desulfuricans* and *Desulfotomaculum nigrificans* (Kuznetsov et al., 1963; Nazina and Rozanova, 1978). The second group included the methanogens *Methanobacterium mazei*, *Sarcina methanica*, and *Methanobacterium omelianskii* (the second organism of the three would today be assigned to *Methanosarcina* and the third to *Methanobacterium* MOH). The third group included the photoheterotroph *Rhodopseudomonas palustris* (Rozanova, 1971), which is capable of anaerobic respiration in the dark (Madigan and Gest, 1978; Yen and Marrs, 1977). The petroleum-associated brines may be connate seawater whose mineral content was somewhat altered through contact with enclosing rock strata (see Chapter 5). Such brines may be low in sulfate but high in chlorides and not very conducive to microbial growth. But sulfate-containing groundwaters and alkaline carbonate waters, on mixing with the brines, can provide a milieu suitable for the activity of sulfate-reducing bacteria (Table 20.2). These waters furnish needed moisture, and in the case of sulfate-reducing bacteria the terminal electron acceptor sulfate for their respiration. By current understanding of methanogenesis (see section 20.3), the methane bacteria in these brines probably rely mainly on H_2 for energy and CO_2 as terminal electron acceptor and as source of carbon. The sulfate-reducing bacteria may obtain their energy source directly from petroleum, as has been claimed by some, but may also depend on other bacteria to produce the compounds they need as carbon and energy sources (Ivanov, 1967; Chapter 17). If the plutonic waters associated with an oil reservoir are hydraulically connected with infiltrating surface waters, it is also possible that carbon and energy sources for the sulfate reducers derive at least in part from aerobic bacteria in oxidizing strata (Jobson et al., 1979). The observation by Panganiban and Hanson (1976) that at least one sulfate reducer can use methane for energy and acetate for carbon makes a petroleum

Table 20.2 Composition of Petroleum-Associated Brines (Percent Equivalents)

Cl^-	7.4–49.90	Ca^{2+}	0.33–11.02
SO_4^{2-}	0.03–10.06	Mg^{2+}	0.04–4.70
CO_3^{2-}	0.03–42.2	K^+ and Na^+	34.28–49.34

Source: After Kuznetsov et al. (1963), p. 17.

reservoir a not impossible direct source of an energy substrate for sulfate reducers. This notion is also supported by the recent discovery of a sulfate reducer that uses saturated hydrocarbon as an energy source (Aeckersberg et al., 1991).

Microbes in Secondary and Tertiary Oil Recovery   When a petroleum reservoir is first tapped for commercial exploitation, the initial oil is recovered by being forced to the surface by gas pressure from the volatile components of the oil and by pumping. This action, however, recovers only a part of the total oil in a reservoir. To recover additional oil, a reservoir may be flooded with water by injection to force out additional oil (**secondary oil recovery**). Even secondary recovery will yield no more than 30–35% of the oil (North, 1985). To recover even a part of the remaining oil, which is more viscous, **tertiary** or **enhanced oil recovery** treatment is necessary. This may involve a thermal method (e.g., steam injection) to reduce viscosity (e.g., North, 1985) or other chemical or physical methods (see, e.g., Orr and Taber, 1984), or it may involve biological methods such as generation of surface-active agents of microbial origin to facilitate mobility of the oil, or generation of gas pressure by fermentation to force movement of the oil, or a combination of both processes (McInerney and Westlake, 1990; Tanner et al., 1991; Adkins et al., 1992). It may also involve the promotion of selective plugging by microbes of high permeability zones in oil reservoirs to increase volumetric sweep efficiency and the microscopic oil displacement efficiency (Raiders et al., 1989; McInerney and Westlake, 1990). Volumetric sweep efficiency refers to the ability of injected water to recover oil from the less permeable zones.

Water injection into oil reservoirs stimulates microbial activity by sulfate reducers as well as methanogens and fermenting bacteria (Rozanova, 1978; Nazina et al., 1985; Belyaev et al., 1990a,b). Since the injected water carries oxygen, hydrocarbon-oxidizing bacteria have also been detected (Belyaev et al., 1990a,b). Indeed a succession of organisms may occur, with the aerobic hydrocarbon oxidizers and others producing the substrates (e.g., acetate and higher fatty acids) for fermenting and sulfate reducing bacteria, and the fermenting bacteria producing hydrogen for use by methanogens and sulfate reducers (Nazina et al., 1985). In the Bondyuzh oil field (former Tatar S.S.R.), maximal numbers of aerobic hydrocarbon-oxidizing bacteria were found at the interface of injection- and stratal- waters. The destructive effect of these bacteria on petroleum, in one case where this was studied, appeared limited by the salt concentration of the stratal waters (Gorlatov and Belyaev, 1984).

For tertiary oil recovery that involves use of surface active agents of microbial origin, xanthan gums of the bacterium *Xanthomonas camp-*

estris or glucan polymers of fungi, such as *Sclerotium*, *Stromantinia*, and *Helotium* (Compere and Griffith, 1978), may be generated in separate processes and then injected into an oil well to facilitate oil recovery. As an alternative approach, appropriate organisms may be introduced into the oil well to produce a surface-active agent in situ (Finnerty et al., 1984). Promotion of tertiary oil recovery may also involve injection of dilute molasses solution into an oil well and subsequent injection of a gas-producing culture such as *Clostridium acetobutylicum*, which ferments the molasses to the solvents acetone, butanol, and ethanol and large amounts of CO_2 and H_2. The solvents help to lower the viscosity of the oil and the gases provide pressure to move the oil (Yarbrough and Coty, 1983).

Removal of Organic Sulfur from Petroleum   As in the case of coal, petroleum that contains significant amounts of organic sulfur may not be usable as a fuel because of air pollution by the volatile sulfur compounds like SO_2 that result from its combustion. The feasibility of removing this sulfur microbiologically is being actively explored (e.g., Foght et al., 1990). The experimental approach being taken is similar to that in the investigations to remove organic sulfur from coal. The model substance to evaluate microbial desulfurizing activity is dibenzothiophene. The best microbial agents for any industrially applicable process will be those that remove the sulfur without significant concomitant oxidation of the carbon.

Microbes in Petroleum Degradation   When natural petroleum reservoirs become industrially exploited, constituents in the oil, whether the oil is still in its reservoir or removed from it, become susceptible to microbial attack. This attack may be aerobic or anaerobic. An extensive literature has built up around this subject, largely because such microbial attack can be used in the management of oil pollution. Only the major principles of microbial petroleum degradation by oxidation will be discussed here.

Although hydrocarbon oxidation is often considered a strictly aerobic process because the initial attack usually involves an oxygenation, clear evidence now exists for anaerobic degradation of some oil constituents. Old claims, to which reference has already been made, that sulfate-reducing bacteria in oil-well brines are able to derive energy and/or carbon from petroleum constituents, especially methane, can be found in the literature (Davis, 1967, p. 243; Davis and Yarbrough, 1966; Panganiban et al., 1979). It was also suggested that even if sulfate reducers were not able to attack the hydrocarbons themselves, satellite organisms might be able to convert such compounds to products utilizable by the sulfate reducers (Dutova, 1962). Kuznetsova and Gorlenko (1965) reported anaerobic attack of hydrocarbons by a strain of *Pseudomonas* in a mineral salts medium with petroleum as the only carbon source. The bacterial population in these experiments increased a maximum of a millionfold; the redox potential dropped from $+40$ to -110 mV. Similarly, Kvasnikov et al. (1973) obtained growth of *Clostridium* (*Bacillus*) *polymyxa* anaerobically with *n*-alkanes as the sole source of carbon. Simakova et al. (1968) found that methane and high-paraffinaceous oil were more intensely attacked aerobically than anaerobically. They found very similar products, including fatty acids of high and low molecular weight, amino acids, alcohols, and aldehydes, under either condition. However, hydroxy acids were only formed aerobically. This work showed that petroleum degradation is much slower anaerobically than aerobically. Ward and Brock (1978) observed very slow anaerobic conversion of [1-^{14}C]hexadecane added to reducing sediments and bottom water from Lake Mendota, Wisconsin, U.S.A. They reported that 13.7% of the [1-^{14}C]hexadecane was converted in the sediment to [14]CO$_2$ and ^{14}C-cell carbon in 375 h of incubation. Aerobically the hexadecane was degraded much more rapidly in the same sediment samples. According to Zehnder and Brock (1979), methanogens are able to oxidize small amounts of the methane they form anaerobically. The methane oxidation mechanism in these organisms appears to differ from the methane-forming mechanism. The slow rate of anaerobic hydrocarbon degradation, when it occurs, helps to explain in part why petroleum has remained preserved over the eons of time. However, prolonged periods of an absence of degradative activity of any kind must have been a more important factor in its preservation. A case of in situ microbial conversion of petroleum into bitumen has recently been attributed to Alberta (Canada) oil sands, based on laboratory simulation (Rubinstein et al., 1977).

Current State of Knowledge of Aerobic and Anaerobic Petroleum Degradation by Microbes   A variety of bacteria and fungi are able to metabolize hydrocarbons (Atlas, 1981, 1984). Some examples are listed in Table 20.3. The mode of attack of hydrocarbons by microorganisms depends on the kind of microorganisms involved and the environmental conditions. Aerobically, alkanes may be attacked monoterminally to form an alcohol by an oxygenation (Doelle, 1975; Atlas, 1981; Gottschalk, 1986):

$$\mathrm{RCH_2CH_3} \xrightarrow{\frac{1}{2}O_2} \mathrm{RCH_2CH_2OH} \xrightarrow{-2H} \mathrm{RCH_2CHO}$$

$$\xrightarrow{+H_2O,\,-2H} \mathrm{RCH_2COOH} \tag{20.19}$$

This is followed sequentially by oxidating to an aldehyde and then a corresponding carboxylic acid [reaction (20.19)]. The carboxylic acid is oxidized to acetate, which is then oxidized to CO$_2$ and H$_2$O.

Table 20.3 Microorganisms Capable of Aerobic Hydrocarbon Metabolism

Organism	Substrates	Mode of attack	References[a]
Pseudomonas oleovorans	Octane	Desaturation	Abbott and Hou (1973)
P. fluorescens, P. aeruginosa	Aromatic hydrocarbons	Oxidation	Van der Linden and Thijsse (1965)
Nocardia salmonicolor	Hexadecane	Desaturation	Abbott and Casida (1968)
Yeasts			Ahearn et al. (1971)
Trichosporon	*n*-Paraffins		Barna et al. (1970)
Arthrobacter	*n*-Alkane aromatics		Klein et al. (1968); Stevenson (1967)
Mycobacterium	Butane		Nette et al. (1965); Phillips and Perry (1974)
Brevibacterium erythrogenes	Alkane	Oxidation	Pirnik et al. (1974)
Nocardia	Mono- and dicyclic hydrocarbons	Oxidation	Raymond et al. (1967)
Cladosporium	*n*-Alkane		Teh and Lee (1973); Walker and Cooney (1973)
Graphium	Ethane		Volesky and Zajic (1970)

[a] For recent reviews, see Assinder and Williams (1990), Cerneglia (1984), and Atlas (1984).

Alkanes may also be monoterminally attacked to form a ketone (Frederricks, 1967) or a hydroperoxide (Stewart et al., 1959). They may also be attacked diterminally (Doelle, 1975). For instance, *Pseudomonas aeruginosa* can attack 2-methylhexane at either end of the carbon chain, forming a mixture of 5-methylhexanoic and 2-methylhexanoic acid (Foster, 1962). Furthermore, alkanes may be desaturated terminally or subterminally, forming alkenes (Chouteau et al., 1962; Abbott and Casida, 1968). Subterminal desaturation may proceed as follows:

$$\text{Hexadecane} \xrightarrow{\text{n(2H)}} \begin{array}{l} \text{8-hexadecene} \\ \text{7-hexadecene} \\ \text{6-hexadecene} \end{array} \qquad (20.20)$$

Alkenes may be attacked by formation of epoxides, which may then be further metabolized (Abbott and Hon, 1973); diols may be formed in the process. In all the foregoing processes atmospheric oxygen acts as terminal electron acceptor and as the source of oxygen in oxygenation.

Until very recently, anaerobic attack of alkanes and alkenes was believed possible only if these compounds carried one or more substituents, in particular halogens. Dichloromethane (CH_2Cl_2) can be degraded anaerobically. In these reactions, the initial attack cannot involve oxygenation, but involves instead an initial dechlorination step. In the case of dichloromethane, the proposed reaction carried out by a consortium of two different bacterial strains is (Braus-Stromeyer et al., 1993)

$$2CH_2Cl_2 + 2H_2O \Longrightarrow 2(HCHO) + 4HCl \qquad (20.21)$$

This reaction is followed by an oxidation of the formaldehyde-like intermediate (HCHO) to formic acid. The formic acid is subsequently converted to acetate in an acetogenic reaction (Braus-Stromeyer et al., 1993).

In the case of tetra- and trichloroethylene, anaerobic dechlorination by a reductive process has been observed. In this process, the chlorinated hydrocarbon serves as terminal electron acceptor, but complete mineralization of the dechlorinated ethylene has so far not been observed in laboratory experiments (e.g., Ensley, 1991; Freedman and Gossett, 1989).

A clear demonstration that a saturated, unsubstituted alkane can be mineralized by bacteria was presented by Aeckersberg et al. (1991). These investigators isolated a sulfate-reducing bacterium, strain HxD3, from precipitate of an oil-water separator in an oil field near Hamburg, Germany. This organism mineralized hexadecane using sulfate as oxidant (terminal electron acceptor). The nature of the intial attack of hexadecane remains to be elucidated. The overall reaction of hexadecane mineralization followed the following stoichiometry (Aeckersberg et al., 1991):

$$C_{16}H_{34} + 12.25SO_4^{2-} + 8.5H^+ \Longrightarrow 16HCO_3^- + 12.25H_2S + H_2O$$

$$(20.22)$$

A wide range of aromatic compounds are aerobically and anaerobically degradable by microbes (Tables 20.3 and 20.4). **Aerobic scission** of the ring structure of the aromatic compounds involves oxygenation either between two adjacent oxygenated carbon atoms (ortho fission) or adjacent to one of them (meta fission) (Dagley, 1975). In the case of benzene, aerobic degradation of the ring structure involves an initial hydroxylation catalyzed by a mixed-function or monooxygenase to form catechol followed by action of dioxygenase to cleave the catechol ring to form *cis, cis*-muconate by ortho fission. This product can then be degraded enzymatically in several steps to acetate (Doelle, 1975), which is oxidized to CO_2 and H_2O. Naphthalene, anthracene, and phenanthrene and derivatives can be degraded by a similar mechanism by attacking each ring in succession (Doelle, 1975). In some instances, benzene derivatives are

Table 20.4 Bacteria Capable of Anaerobic Metabolism of Aromatic and Heterocyclic Hydrocarbons

Organism	Substrates	Mode of attack	References[a]
Rhodopseudomonas palustris	Benzoate, hydroxybenzoate	Reductive ring cleavage	Dutton and Evans (1969)
Desulfobacterium phenolicum	Phenol and derivatives	Anaerobic degradation	Bak and Widdel (1986a)
Desulfobacterium indolicum	Indolic compounds	Anaerobic degradation	Bak and Widdel (1986b)
Desulfobacterium catecholicum	Catechol	Anaerobic degradation	Szewzyk and Pfennig (1987)
Desulfococcus niacini	Nicotinic acid	Anaerobic degradation	Imhoff-Stuckle and Pfennig (1983)

[a] For recent reviews, see Evans and Fuchs (1988), Reineke and Knackmuss (1988), and Higson (1992).

attacked by meta fission instead of ortho fission as in the previous examples (Doelle, 1975). **Anaerobic scission** of the ring structure involves ring saturation, hydration, and dehydrogenation (Evans and Fuchs, 1988; Colberg, 1990; Grbić-Galić, 1990). Some aromatic hydrocarbons like benzoate can also be biodegraded anaerobically by photometabolism of certain Rhodospirillaceae (purple nonsulfur bacteria) (Table 20.4). Ring cleavage is by hydration to pimelate (Dutton and Evans, 1969).

Unlike chlorinated alkanes and alkenes, chlorinated aromatics have been shown to be completely degradable anaerobically, but by a consortium of bacteria (e.g., Sharak Genthner et al., 1989; Colberg, 1990; Grbić-Galić, 1990). In one kind of consortium, *Desulfomonile tiedjei* DCB-1 acts on a chlorinated hydrocarbon by using it as terminal electron acceptor in a respiratory process that involves dechlorination (Shelton and Tiedje, 1984; Dolfing, 1990; Mohn and Tiedje, 1990, 1991). Subsequent mineralization of the dechlorinated aromatic product depends on other anaerobic organisms in the consortium.

The ability of an organism to attack hydrocarbons aerobically does not necessarily mean that it can use such a compound as a sole source of carbon and energy. Many cases are known in which hydrocarbons are oxidized in a process known as **cooxidation**, wherein another compound, which may be quite unrelated, is the carbon and energy source but which somehow permits the simultaneous oxidation of the hydrocarbon. Examples are the oxidation of ethane to acetic acid, the oxidation of propane to propionic acid and acetone, and the oxidation of butane to butanoic acid and methyl ethyl ketone by *Pseudomonas methanica* growing on methane, as first shown by Leadbetter and Foster (1959). Methane is the only hydrocarbon on which this organism can grow. Another example is the oxidation of alkyl benzenes by a strain of *Micrococcus cerificans* growing on *n*-paraffins (Donos and Frankenfeld, 1968). Still other examples have been summarized by Horvath (1972).

Chain length and branching of aliphatic hydrocarbons can affect microbial attack. Recent in-situ observations revealed rapid microbial degradation of pristane and phytane (Atlas and Cerniglia, 1995). Some bacteria that attack alkanes of chain lengths C_8-C_{20} may not be able to attack alkanes of chain length C_1-C_6, whereas others cannot grow on alkanes of chain lengths greater than C_{10} (Johnson, 1964). Fungi are known that can grow on alkanes of chain lengths up to C_{34}. It has also been noted that certain placements of methyl and propyl groups in the alkane carbon chain lessen or prevent utilization of the compounds (McKenna and Kallio, 1964).

Use of Microbes in Prospecting for Petroleum   Prospecting for petroleum through detection of hydrocarbon-utilizing microorganisms has been proposed. The basis for this method is detection of microseepage of petroleum or some of its constituents, especially the more volatile components, in the ground overlying a deposit using the pressure of hydrocarbon-utilizing microorganisms as indicators. It involves enriching soil, sediment, and water samples from a suspected seepage area for microbes that can metabolize gaseous hydrocarbons and demonstrating hydrocarbon consumption (Davis, 1967). An enrichment medium consisting of a mineral salts solution with added volatile hydrocarbons (ethane, propane, butane, isobutane) is satisfactory. Methane-oxidizing bacteria are poor indicators in petroleum prospecting because methane can occur in the absence of petroleum deposits, and, moreover, some methane-oxidizing bacteria are unable to oxidize other aliphatic hydrocarbons. Dectection of bacteria that can oxidize ethane and longer-chain hydrocarbons, on the other hand, provides presumptive evidence for a hydrocarbon seep and an underlying petroleum reservoir (Davis, 1967). It is assumed that ethane and propane formed in anaerobic fermentation are produced in quantities too small to select for a hydrocarbon utilizing microflora. Likely organisms active in soil enrichments from hydrocarbon seeps may include *Mycobacterium paraffinicum* and *Streptomyces* spp.

Hydrocarbon enrichment cultures may be prepared using ^{14}C-labeled hydrocarbon. This allows easy quantification of the activity of hydrocarbon-oxidizing bacteria in water and sediments (Caparello and LaRock, 1975). With this method, the hydrocarbon-oxidizing potential of a sample can be correlated with the hydrocarbon burden of the environment from which the sample came.

Microbes and Shale Oil North (1985) describes oil shales as either bituminous, nonmarine limestones or marl-stones containing kerogen. **Tar sands** are consolidated or unconsolidated rock coated with bituminous material (North, 1985). **Bitumens** are solid hydrocarbons that are soluble in organic solvents and fusible below ~150°C. **Kerogens**, which are insoluble in organic solvents, are intermediate products in the diagenetic transformation of organic matter in sediments and are considered a precursor in petroleum formation. As in the formation of peat, coal, and petroleum, fungi and bacteria probably played a role in the early stages of transformation of the source material (mostly terrestrial microbial biomass). Later stages involved physicochemical processes. However, in the case of the oil-sand bitumens of Alberta, the origin appears to be partial biodegradation of petroleum leaving behind the high-viscosity components (Rubinstein et al., 1977).

Bitumen and kerogen can be converted to a petroleum-like substance by heat treatment (e.g., retorting) (North, 1985). Separation from host rock, especially if it is limestone, can be facilitated if the limestone is dissolved. This can be achieved, at least on a laboratory scale, by acid formed by microbes (e.g., sulfuric acid formed by *Thiobacillus thiooxidans*) (Meyer and Yen, 1976).

Although raw shale oil is considered relatively resistant to microbial attack, recent reports indicate otherwise. Both aerobic and anaerobic attack have been observed, but the anaerobic attack proceeded at a much slower rate (Roffey and Norqvist, 1991; Wolf and Bachofen, 1991; Ait-Langomazino et al., 1991). Hydrogenated shale oil was found readily metabolized by some gram-negative organisms (*Alcaligenes* and *Pseudomonas* or *Pseudomonas*-like organisms) (Westlake et al., 1976).

20.7 SUMMARY

Not all organic carbon in the biosphere is continually being recycled. Some is trapped in special sedimentary formations, where it is inaccessible to microbial attack. The forms in which this trapped carbon appears are methane, peat, coal, and petroleum.

Most methane in sedimentary formations is of biogenic origin. It may occur by itself or in association with coal or petroleum deposits. Its biogenic formation is a strictly anaerobic process involving methanogenic bacteria that may reduce CO_2, formate, methanol, methylamines, or acetate with H_2, or transform acetate to methane and CO_2 in the absence of H_2. The autotrophic methanogens use a unique mechanism for CO_2 assimilation that involves reduction of one CO_2 to methyl carbon and a second CO_2 to a formyl carbon and then coupling the two to form acetate. The acetate is subsequently carboxylated to form pyruvate, which becomes the key intermediate for forming the building blocks for all the cell constituents. Methanogenesis can occur mesophilically and thermophilically.

Methane may be oxidized and assimilated by a special group of microorganisms called methanotrophs. This process is generally aerobic, although some evidence of anaerobic methane oxidation has appeared. At least one of the organisms responsible for anaerobic methane consumption is a sulfate reducer. Limited anaerobic methane oxidation by a methanogen has also been observed. Carbon assimilation by aerobic methanotrophs may be via the hexulose monophosphate pathway or the serine pathway, in each case involving integration of the carbon at the oxidation level of formaldehyde that is produced as an intermediate in methane oxidation. Both types of methanotrophs derive some of their carbon also from CO_2.

Peat is the result of partial degradation of plant remains accumulating in marshes and bogs. Aerobic attack by enzymes in the plant debris and by fungi and some bacteria initiates the process. It is followed by anaerobic attack by bacteria during burial resulting from continual sedimentation until inhibited by accumulating wastes, lack of sufficient moisture, and so on. A viable microbial flora can usually be detected in peat at the present time, even though it formed over a geologically extended period. Coal is thought to have formed like peat, except that in the advanced stages, as a result of deeper burial, it was subject to physical and chemical influences that converted peat to coal. Different ranks of coal exist which differ from each other largely in carbon content and heat value. It is questionable whether coal itself harbors an indigenous flora. Bituminous coal has pyrite or marcasite associated with it. Upon exposure to air and moisture during mining, this iron disulfide becomes subject to attack by acidophilic, iron-oxidizing thiobacilli and is the source of acid mine drainage.

Whereas peat and coal derived from terrestrial plant matter, petroleum and associated natural gas (mostly methane) is derived from phytoplankton remains that accumulated in depressions of shallow seas. Microbial attack altered these remains biochemically until complete burial by accumulating sediment stopped the organisms. In tectonically active areas, the buried and biochemically altered organic matter became subject to further alteration by heat from magmatic activity and pressure from overlying sediment. These chemical alterations may have been catalyzed by clay minerals. The final products of these transformations were petroleum hydrocarbons. At least some of the natural gas associated with petroleum may represent biogenic methane formed in the initial stages of plankton debris fermentation. At its site of formation, petroleum is highly dispersed. As a result of gas (natural gas, CO_2) and hydrostatic pressure as well as lubrication by bacteria and some of their products of the surfaces of the sediment matrix, matured petroleum may be forced to migrate through pervious sedimentary strata until caught in a trap such as an anticlinal fold. It is such petroleum-filled traps that constitute commercially exploitable petroleum reservoirs. Sulfate-reducing bacteria may assist in the trapping of petroleum by laying down impervious calcite layers. This calcite may, however, also interfere with petroleum recovery.

At least some petroleum hydrocarbons are oxidizable in air by certain bacteria and fungi. Anaerobic bacterial attack of unsubstituted alkanes as well as chlorinated alkanes and alkenes and of a wide range of aromatic compounds has also been demonstrated. Hydrocarbon-utilizing microorganisms may be used as indicators in prospecting for petroleum.

REFERENCES

Abbott, B. J., and L. E. Casida, Jr. 1968. Oxidation of alkanes to internal monoalkenes by a nocardia. J. Bacteriol. 96:925–930.

Abbott, B. J., and C. T. Hou. 1973. Oxidation of 1-alkanes to 1,2-epoxyalkanes by *Pseudomonas oleovorans*. Appl. Microbiol. 26:86–91.

Abram, J. W., and D. B. Nedwell. 1978. Inhibition of methanogenesis by sulfate reducing bacteria competing for transferred hydrogen. Arch. Microbiol. 117:89–92.

Adkins, J. P. 1992. Adkins, J. P., R. S. Tanner, E. O. Udegbunam, M. J. McInerney, and R. M. Knapp. 1992. Microbially enhanced oil recovery from unconsolidated limestone cores. Geomicrobiol. J. 10: 77–86.

Aeckersberg, F., F. Bak, and F. Widdel. 1991. Anaerobic oxidation of saturated hydrocarbons to CO_2 by a new type of sulfate-reducing bacterium. Arch. Microbiol. 156:5–14.

Ahearn, D. G., S. P. Meyers, and P. G. Standard. 1971. The role of yeasts in the decomposition of oils in marine environments. Dev. Ind. Microbiol. 12:126–134.

Ait-Langomazino, N., R. Sellier, G. Jouquet, and M. Trescinski. 1991. Microbial degradation of bitumen. Experientia 47:533–539.

Alexander, M. 1977. Introduction to Soil Microbiology. 2nd ed. Wiley, New York.

Altschuler, Z. S., M. M. Schnepfe, C. C. Silber, and F. O. Simon. 1983. Sulfur diagenesis in Everglades peat and origin of pyrite in coal. Science 221:221–227.

Andrews, G., M. Darroch, and T. Hansson. 1988. Bacterial removal of pyrite from concentrated coal slurries. Biotech. Bioeng. 32:813–820.

Anthony, C. 1986. Bacterial oxidation of methane and methanol. Adv. Microb. Physiol. 27:113–210.

Ashirov, K. B., and I. V. Sazanova. 1962. Biogenic sealing of oil deposits in carbonate reservoirs. Mikrobiologiya 31:680–683 (Engl. transl. pp. 555–557).

Assinder, S. J., and P. A. Williams. 1990. The TOL plasmids: Determinants of the catabolism of toluene and the xylenes. Adv. Microb. Physiol. 31:1–69.

Atlas, R. M.. 1981. Microbial degradation of petroleum hydrocarbons: an environmental perspective. Microbiol. Rev. 45:180–209.

Atlas, R. M. 1984. Petroleum Microbiology. McGraw-Hill, New York.

Atlas, R. M. 1988. Microbiology. Fundamentals and Applications. 2nd ed. Macmillan, New York.

Atlas, R. M., and C. E. Cerniglia. 1995. Bioremediation of petroleum pollutants. Bioscience 45:332–338.

Bagdigian, R. M., and A. S. Myerson. 1986. The adsorption of *Thiobacillus ferrooxidans* on coal surfaces. Biotech. Bioeng. 28:467–479.

Bak, F., and F. Widdel. 1986a. Anaerobic degradation of phenol and phenol derivatives by *Desulfobacterium phenolicum* sp. nov. Arch. Microbiol. 177–180.

Bak, F., and F. Widdel. 1986b. Anaerobic degradation of indolic compounds by sulfate-reducing enrichment cultures, and description of *Desulfobacterium indolicum* gen. nov., sp. nov. Arch. Microbiol. 46:170–176.

Baldi, F., T. Clark, S. S. Pollack, and G. J. Olson. 1992. Leaching pyrites of various reactivities by *Thiobacillus ferrooxidans*. Appl. Environ. Microbiol. 58:1853–1856.

Barna, P. K., S. D. Bhagat, K. R. Pillai, H. D. Singh, J. N. Barnah, and M. S. Iyengar. 1970. Comparative utilization of paraffins by a *Trichosporon* species. Appl. Microbiol. 20:657–661.

Beerstecher, E. 1954. Petroleum Microbiology. Elsevier Press, Houston, TX.

Belyaev, S. S., E. P. Rozanova, I. A. Borzenkov, I. A. Charakhch'yan, Yu. M. Miller, M. Yu. Sokolov, and M. V. Ivanov. 1990a. Characteristics of microbiological processes in a water-flooded oilfield in the middle Ob' region. Mikrobiologiya 59:1075–1081 (Engl. transl. pp. 754–759).

Belyaev, S. S., I. A. Borzenkov, E. I. Milekhina, I. A. Charakhch'yan, and M. V. Ivanov. 1990b. Development of microbiological processes in reservoirs of the Romashkino oil field. Mikrobiologiya 59: 1118–1126 (Engl. transl. pp. 786–792).

Bhatnagar, L., M. K. Jain, and J. G. Zeikus. 1991. Methanogenic bacteria. *In*: J. M. Shively and L. L. Barton, eds. Variations in Autotrophic Life. Academic Press, London, pp. 251–270.

Blaut, M., and G. Gottschalk. 1984. Protonmotive force-driven synthesis of ATP during methane formation from molecular hydrogen and for-maldehyde or carbon dioxide in *Methanosarcina barkeri*. FEMS Microbiol. Lett. 24:103–107.

Blaut, M., S. Peinemann, U. Deppenmeier, and G. Gottschalk. 1990. Energy transduction in vesicles of the methanogenic strain Gö 1. FEMS Microbiol. Rev. 87:367–372.

Blaut, M., V. Mueller, and G. Gottschalk. 1992. Energetics of methanogenesis studied in vesicular systems. J. Bioenerg. Biomembr. 24: 529–546.

Blazquez, M. L., A. Ballesterl, F. Gonzalez, and J. L. Mier. 1993. Coal biodesulfurization. Biorecovery 2:155–157.

Bochem, H. P., S. M. Schoberth, B. Sprey, and P. Wengler. 1982. Thermophilic biomethanation of acetic acid: morphology and ultrastructure of a granular consortium. Can. J. Microbiol. 28:500–510.

Bock, A.-K., A. Prieger-Kraft, and P. Schoenheit. 1994. Pyruvate—a novel substrate for growth and methane formation in *Methanosarcina barkeri*. Arch. Microbiol. 161:33–46.

Boone, D. R., W. B. Whitman, and P. Rouvière. 1993. Diversity and taxonomy of methanogens. *In*: J. G. Ferry, ed. Methanogenesis. Chapman and Hall, New York, pp. 35–80.

Bos, P. and J. G. Kuenen. 1990. Microbial treatment of coal. *In*: H.L. Ehrlich and C.L. Brierley, eds. Microbial Mineral Recovery. McGraw-Hill, New York, pp. 343–377.

Bowen, H. J. M. 1979. Environmental Chemistry of the Elements. Academic Press, London.

Braus-Stromeier, S. A., R. Hermann, A. M. Cook, and T. Leisinger. 1993. Dichloromethane as the sole carbon source for an acetogenic mixed culture and isolation of a fermentative, dichloromethane-degrading bacterium. Appl. Environ. Microbiol. 59:3790–3797.

Breznak, J. A. 1982. Intestinal microbiota of termites and other xylophagus insects. Annu. Rev. Microbiol. 36:323–343.

Brock, T. D., and M. T. Madigan. 1988. Biology of Microorganisms. 5th ed. Prentice-Hall, Englewood Cliffs, NJ.

Brooks, J. M., M. C. Kennicutt II, C. R. Fisher, S. A. Macko, K. Cole,

J. J. Childress, R. R. Bridigare, and R. D. Vetter. 1987. Deep-sea hydrocarbon seep communities: Evidence for energy and nutritional carbon sources. Science 238:1138–1142.

Brown, D. A., and R. P. Overend. 1993. Methane metabolism in raised bogs of northern wetlands. Geomicrobiol. J. 11:35–48.

Bryant, M. P., L. L. Campbell, C. A. Reddy, and M. R. Crabill. 1977. Growth of *Desulfovibrio* in lactate or methanol media low in sulfate in association with H_2-utilizing methanogenic bacteria. Appl. Environ. Microbiol. 33:1162–1169.

Caparello, D. M., and P. A. LaRock. 1975. A radioisotope assay for quantification of hydrocarbon biodegradation potential in environmental samples. Microb. Ecol. 2:28–42.

Casagrande, D., and K. Siefert. 1977. Origins and sulfur in coal: Importance of the ester sulfate content of peat. Science 195:675–676.

Cavanaugh, C. M., P. R. Levering, J. S. Maki, R. Mitchell, and M.E. Lidstrom. 1987. Symbiosis of methylotrophic bacteria and deep-sea mussels. Nature (Lond.) 325:346–348.

Cerneglia, C. E. 1984. Microbial metabolism of polycyclic aromatic hydrocarbons. Adv. Appl. Microbiol. 30:31–71.

Chen, W., and J. Konisky. 1993. Characterization of a membrane-associated ATPase from *Methanococcus voltae*, a methanogenic member of the Archaea. J. Bacteriol. 175:5677–5682.

Childress, J. J., C. R. Fisher, J. M. Brooks, M. C. Kennicutt II, R. Bidigare, and A. E. Anderson. A methanotrophic marine molluscan (Bivalvia, Mytilidae) symbiosis: Mussels fueled by gas. Science 233:1306–1308.

Chouteau, J. E. Azoulay, and J. C. Senez. 1962. Dégradation bactérienne des hydrocarbones paraffiniques. IV. Identification par spectrophotométrie infrarouge du hept-1-ene produit à partir du n-heptane par des suspensions nonproliférantes de *Pseudomonas aeruginosa*. Bull. Soc. Chim. Biol. 44:1670–1672.

Clark, T. R., F. Baldi, and G. J. Olson. Coal depyritization by the thermophilic archeon *Metallosphaera sedula*. Appl. Environ. Microbiol. 59:2375–2379.

Cohen, M. S., and P. D. Gabriele. 1982. Degradation of coal by the fungi *Polysporus versicolor* and *Poria monticola*. Appl. Environ. Microbiol. 44:23–27.

Cohen, M. S., W. C. Bowers, H. Aronson, and E. T. Gray, Jr. 1987. Cell-free solubilization of coal by *Polysporus versicolor*. Appl. Environ. Microbiol. 53:2840–2843.

Cohen, M. S., K. A. Feldman, Cynthia S. Brown, and E. T. Gray, Jr.

1990. Isolation and identification of the coal-solubilizing agent produced by *Trametes versicolor*. Appl. Environ. Microbiol. 56: 3285–3291.

Colberg, P. J. S. 1990. Role of sulfate in microbial transformations of environmental contaminants: chlorinated aromatic compounds. Geomicrobiol. J. 8:147–165.

Compere, A. L., and W. L. Griffith. 1978. Production of high viscosity glucans from hydrolyzed cellulosics. Dev. Ind. Microbiol. 19: 601–607.

Crawford, D. L., ed. 1993. Microbial Transformations of Low Rank Coals. CRC Press, Boca Raton, FL.

Crawford, D. L., and R. K. Gupta. 1990. Oxidation of dibenzothiophene by *Cunninghamella elegans*. Curr. Microbiol. 21:229–231.

Crawford, D. L., and R. K. Gupta. 1993. Microbial depolymerization of coal. *In*: D.L. Crawford, ed. Microbial transformations of Low Rank Coals. CRC Press, Boca Raton, FL, pp. 65–92.

Dagley, S. 1975. Microbial degradation of organic compounds in the biosphere. Sci. Am. 63:681–689.

Davies, S. L., and R. Whittenbury. 1970. Fine structure of methane and other hydrocarbon-utilizing bacteria. J. Gen. Microbiol. 61:227–232.

Davis, C. A. 1907. Peat: Essays on the origin and Distribution in Michigan. Wynkoop Hallenbeck Crawford, Co. State Printers, Lansing, MI.

Davis, J. B. 1967. Petroleum Microbiology. Elsevier, Amsterdam.

Davis, J. B., and H. F. Yarbrough. 1966. Anaerobic oxidation of hydrocarbons by *Desulfovibrio desulfuricans*. Chem. Geol. 1:137–144.

Deuser, W. G., E. T. Degens, and G. R. Harvey. 1973. Methane in Lake Kivu: new data bearing on origin. Science 181:51–54.

Didyk, B. M., and B. R. T. Simoneit. 1989. Hydrothermal oil of Guaymas Basin and implications for petroleum formation mechanisms. Nature (Lond.) 342:65–69.

Doddema, H. J., T. J. Hutten, C. van der Drift, and G. D. Vogels. 1978. ATP hydrolysis and synthesis by the membrane-bound ATP synthetase complexes of *Methanobacterium thermoautotrophicum*. J. Bacteriol. 136:19–23.

Doddema, H. J., C. van der Drift, G. D. Vogels, and M. Veenhuis. 1979. Chemiosmotic coupling in *Methanobacterium thermoautotrophicum*: hydrogen-dependent adenosine 5'-triphosphate synthesis by subcellular particles. J. Bacteriol. 140:1081–1089.

Doelle, H. W. 1975. Bacterial Metabolism. 2nd ed. Academic Press, New York.

Dolfing, J. 1990. Reductive dechlorination of 3-chlorobenzoate is coupled to ATP production and growth in an anaerobic bacterium, strain DCB-1. Arch. Microbiol. 153:264–266.

Donos, J. D., and J. W. Frankenfeld. 1968. Oxidation of alkyl benzenes by a strain of *Micrococcus cerificans* growing on n-paraffins. Appl. Microbiol. 16:532–533.

Dugan, P. R. 1986. Microbiological desulfurization of coal and its increased monetary value, pp. 185–293. *In*: H. L. Ehrlich and D. S. Holmes, eds. Workshop on Biotechnology for the Mining, Metal-Refining and Fossil Fuel Processing Industries. Biotech. Bioeng. Symp. No. 16. Wiley, New York.

Dutova, E. N. 1962. The significance of sulfate-reducing bacteria in prospecting for oil as exemplified in the study of ground water in Central Asia. *In*: S. I. Kuznetsov, ed. Geologic Activity of Microorganisms. Consultants Bureau, New York, pp. 76–78.

Dutton, P. L., and W. C. Evans. 1969. The metabolism of aromatic compounds by *Rhodospeudomonas palustris*. A new, reductive, method of aromatic ring metabolism. Biochem. J. 113:525–536.

Dybas, M., and J. Konisky. 1992. Energy transduction in the methanogen *Methanococcus voltae* is based on a sodium current. J. Bacteriol. 174:5575–5583.

Ehrlich, H. L. 1993. Bacterial mineralization of organic carbon under anaerobic conditions. *In*: J-M. Bollag and G. Stotzky, eds. Soil Biochemistry. Vol. 8. Marcel Dekker, New York, pp. 219–247.

Ensley, B. D. 1991. Biochemical diversity of trichloroethylene metabolism. Annu. Rev. Microbiol. 45:283–299.

Evans, W. C., and G. Fuchs. 1988. Anaerobic degradation of aromatic compounds. Annu. Rev. Microbiol. 42:289–317.

Fenchel, T., and T. H. Blackburn. 1979. Bacteria and Mineral Cycling. Academic Press, London.

Fenchel, T., and B. J. Finlay. 1992. Production of methane and hydrogen by anaerobic ciliates containing symbiotic methanogens. Arch. Microbiol. 157:475–480.

Finnerty, W. R., M. E. Singer, and A. D. King. 1984. Microbial processes and the recovery of heavy petroleum, pp. 424–429. *In*: R. F. Meyer, J. C. Wynne, and J. C. Olson, eds. Future Heavy Crude Tar Sands. 2nd Int. Conf. McGraw-Hill, New York.

Finnerty, W. R., and M. Robinson. 1986. Microbial desulfurization of fossil fuels: a review, pp. 205–221. *In*: H. L. Ehrlich and D. S. Holmes, eds. Workshop the Mining, Metal-Refining and Fossil Fuel Processing Industries. Biotech. Bioeng. Symp. No. 16. Wiley, New York.

Finster, K., Y. Tanimoto, and F. Bak. 1992. Fermentation of methanediol and dimethylsulfide by a newly isolated methanogenic bacterium. Arch. Microbiol. 157:425–430.

Foght, J. M., P. M. Fedorak, M. R. Gray, and D. W. S. Westlake. 1990. Microbial desulfurization of petroleum. *In*: H. L. Ehrlich and C. L. Brierley, eds. Microbial Mineral Recovery. McGraw-Hill, New York, pp. 379–407.

Foster, J. W. 1962. Hydrocarbons as substrates for microorganisms. Antonie v. Leeuwenhoek 28:241–274.

Francis, W. 1954. Coal. Its Formation and Composition. Edward Arnold, London.

Fredericks, K. M. 1967. Products of the oxidation of n-decane by *Pseudomonas aeruginosa* and *Mycobacterium rhodochrous*. Antonie v. Leeuwenhoek 33:41–48.

Fredrickson, J. K., D. L. Stewart, J. A. Campbell, M. A. Powell, M. McMulloch, J. W. Pyne, and R. M. Bean. 1990. Biosolubilization of low-rank coal by *Trametes versicolor* siderophore-like product and other complexing agents. J. Indust. Microbiol. 5:401–406.

Freedman, D. L., and J. M. Gossett. 1989. Biological reductive dechlorination of tetrachloroethylene and trichloroethylene to ethylene under methanogenic conditions. Appl. Environ. Microbiol. 55: 2144–2151.

Gal'chenko, V. F., S. N. Gorlatov, and V. G. Tokarev. 1986. Microbial oxidation of methane in Bering Sea sediments. Mikrobiologiya 55: 669–673 (Engl. transl. pp. 526–530).

Giani, D., L. Giani, Y. Cohen, and W. E. Krumbein. 1984. Methanogenesis in the hypersaline Solar Lake (Sinai). FEMS Microbiol. Lett. 25:219–224.

Gold, T., B. E. Gordon, W. Streett, E. Bilson, and P. Patnaik. 1986. Experimental study of the reaction of methane with petroleum hydrocarbons in geological conditions. Geochim. Cosmochim. Acta 50: 2411–2418.

Gorlatov, S. N., and S. S. Balyaev. 1884. The aerobic microflora of an oil field and its ability to destroy petroleum. Mikrobiologiya 53:843–849 (Engl. transl. pp. 701–706).

Gorlatov, S. N., V. F. Gal'chenko, and V. G. Tokarev. 1986. Microbiological methane formation in deposits of the Bering Sea. Mikrobiologiya 55:490–495 (Engl. transl. pp. 380–385).

Gottschalk, G. 1986. Bacterial Metabolism. 2nd ed. Springer-Verlag, New York.

Gottschalk, G., and M. Blaut. 1990. Generation of proton and sodium motive forces in methanogenic bacteria. Biochim. Biophys. Acta 1018:263–266.

Grbić-Galić, D. 1990. Methanogenic transformation of aromatic hydrocarbons and phenols in groundwater aquifers. Geomicrobiol. J. 8: 167–200.

Henrichs, S. M., and W. S. Reeburgh. 1987. Anaerobic mineralization of organic matter: rates and the role of anaerobic processes in the oceanic carbon economy. Geomicrobiol. J. 5:191–237.

Higgins, I. J., D. J. Best, R. C. Hammond, and D. Scott. 1981. Methane-oxidizing microorganisms. Microbiol. Rev. 45:556–590.

Higson, F. K. 1992. Microbial degradation of biphenyl and its derivatives. Adv. Appl. Microbiol. 37:135–164.

Horvath, R. S. 1972. Microbial cometabolism and the degradation of organic compounds in nature. Bacteriol. Rev. 36:146–155.

Huber, H., M. Thomm, H. Koenig, G. Thies, and K. O. Stetter. 1982. *Methanococcus thermolithotrophicus*, a novel thermophilic lithotrophic methanogen. Arch. Microbiol. 132:47–50.

Hunt, J. M. 1984. Generation and migration of light hydrocarbons. Science 226:1265–1270.

Hunt, J. M., J. K. Whelan, and A. Y. Huc. 1980. Genesis of petroleum hydrocarbons in marine sediments. Science 209:403–404.

Huser, B. A., K. Wuhrmann, and A. J. B. Zehnder. 1982. *Methanothrix soehngii* gen. nov. spec.nov, a new acetotrophic non-hydrogen-oxidizing methane bacterium. Arch. Microbiol. 132:1–9.

Imhoff-Stuckle, D., and N. Pfennig. 1983. Isolation and characterization of a nicotinic acid-degrading sulfate-reducing bacterium, *Desulfococcus niacini* sp. nov. Arch. Microbiol. 136:194–198.

Ivanov, M. V. 1967. The development of geological microbiology in the U.S.S.R. Mikrobiologiya 36:849–859 (Engl. transl. pp. 715–722).

Ivanov, M. V., A. I. Nesterov, G. B. Nasaraev, V. F. Gal'chenko, and A. V. Nazarenko. 1978. Distribution and geochemical activity of methanotrophic bacteria in coal mine waters. Mikrobiologiya 47: 489–494.

Iversen, N., and B. B. Jorgensen. 1985. Anaerobic methane oxidation rates at the sulfate-methane transition in marine sediments from Kattegat and Skagerrak (Denmark). Limno. Oceanogr. 30:944–955.

Izumi, Y., T. Ohshiro, H. Ogino, Y. Hine, and M. Shimao. 1994. Selective desulfurization of dibenzothiophene by *Rhodococcus erythropolis* D-1. Appl. Environ. Microbiol. 60:223–236.

Jain, M. K., and J. G. Zeikus. 1989. Bioconversion of gelatin to methane by a coculture of *Clostridium collagenovorans* and *Methanosarcina barkeri*. Appl. Environ. Microbiol. 55:366–371.

Jakobsen, P., W. H. Patrick Jr., and B. G. Wiliams. 1981. Sulfide and methane formation in soils and sediments. Soil Sci. 132:279–287.

Jobson, A. M., F. D. Cook, and D. W. S. Westlake. 1979. Interaction of aerobic and anaerobic bacteria in petroleum biodegradation. Chem. Geol. 24:355–365.

Johnson, M. J. 1964. Utilization of hydrocarbons by microorganisms. Chem. Ind. 36:1532–1537.

Jones, J. G., B. M. Simon, and S. Gardener. 1982. Factors affecting methanogenesis and associated anaerobic processes in the sediments of a stratified eutrophic lake. J. Gen. Microbiol. 128:1–11.

Jones, R. D., and R. Y. Morita. 1983. Methane oxidation by *Nitrosococcus oceanus* and *Nitrosomonas europaea*. Appl. Environ. Microbiol. 45:401–410.

Jones, W. J., J. A. Leigh, F. Mayer, C. R. Woese, and R. S. Wolfe. 1983a. *Methanococcus jannaschii* sp. nov., an extremely thermophilic methanogen from a submarine hydrothermal vent. Arch. Microbiol. 136:254–261.

Jones, W. J., M. J. B. Paynter, and R. Gupta. 1983b. Characterization of *Methanococcus maripaludis* sp. nov., a new methanogen from salt marsh sediment. Arch. Microbiol. 135:91–97.

Jones, W. J., D. P. Nagle, Jr., and W. B. Whitman. 1987. Methanogens and the diversity of Archaebacteria. Microbiol. Rev. 51:135–177.

Klein, D. A., J. A. Davis, and L. E. Casida. 1968. Oxidation of n-alkanes to ketones by an *Arthrobacter* species. Antonie v. Leeuwenhoek 34: 495–503.

Koburger, J. A. 1964. Microbiology of coal: Growth of bacteria in plain and oxidized coal slurries. 39th Annu. Session of W. Virginia Acad. Sci. Proc. W. Virginia Acad. Sci. 36:26–30.

Kristjansson, J. K., P. Schoenheit, and R. K. Thauer. 1982. Different K_s values for hydrogen of methanogenic and sulfate reducing bacteria: An explanation for the apparent inhibition of methanogenesis by sulfate. Arch. Microbiol. 131:278–282.

Kulm, L. D., E. Suess, J. C. Moore, B. Carson, B. T. Lewis, S. D. Ritger, D. C. Kadko, T. M. Thornburg, R. W. Embley, W. D. Rugh, G. J. Massoth, M. G. Langseth, G. R. Cochrane, and R. L. Scamman. 1986. Oregon subduction zone: Venting, fauna, and carbonates. Science 231:561–566.

Kuznetsov, S. I., M. V. Ivanov, and N. N. Lyalikova. 1963. Introduction to Geological Microbiology. McGraw-Hill, New York.

Kuznetsova, V. A., and V. M. Gorlenko. 1965. The growth of hydrocarbon-oxidizing bacteria under anaerobic conditions. Prikl. Biokhim. Mikrobiol. 1:623–626.

Kvasnikov, E. I., V. V. Lipshits, and N. V. Zubova. 1973. Facultative anaerobic bacteria of producing petroleum wells. Mikrobiologiya 42: 925–930 (Engl. transl. pp. 823–827).

Larsson, L., G. Olsson, O. Holst, and H. T. Karlsson. 1990. Pyrite oxidation by thermophilic archaebacteria. Appl. Environ. Microbiol. 56: 697–701.

Leadbetter, E. R. and J. W. Foster. 1959. Oxidation products formed from gaseous alkanes by the bacterium *Pseudomonas methanica*. Arch. Biochem. Biophys. 82:491–492.

Lovley, D. R. 1987. Organic matter mineralization with the reduction of ferric iron: a review. Geomicrobiol. J. 5:375–399.

Lovley, D. R. 1993. Dissimilatory metal reduction. Annu. Rev. Microbiol. 47:263–290.

Lovley, D. R., and E. J. P. Phillips, 1988. Novel mode of microbial energy metabolism: organic carbon oxidation coupled to dissimilatory reduction of iron or manganese. Appl. Environ. Microbiol. 54: 1472–1480.

MacLeod, F. A., S. R. Guiot, and J. W. Costerton. 1990. Layered structure of bacterial aggregates produced in an upflow anaerobic sludge bed and filter reactor. Appl. Environ. Microbiol. 56:1598–1607.

Madigan, M. T. and H. Gest. 1978. Growth of a photosynthetic bacterium anaerobically in darkness, supported by "oxidant-dependent" sugar fermentation. Arch. Microbiol. 117:119–122.

Mah, R. A., M. R. Smith, and L. Baresi. 1978. Studies on an acetate-fermenting strain of *Methanosarcina*. Appl. Environ. Microbiol. 35: 1174–1184.

Martens, C. S. 1976. Control of methane sediment-water bubble transport by macroinfaunal irrigation in Cape Lookout Bight, North Carolina. Science 192:998–1000.

Martens, C. S., and R. A. Berner. 1974. Methane production in the interstitial waters of sulfate-depleted marine sediments. Science 185: 1167–1169.

Martens, C. S., and J. V. Klump. 1984. Biogeochemical cycling in an organic-rich coastal marine basin. 4. An organic carbon budget for sediments dominated by sulfate reduction and methanogenesis. Geochim. Cosmochim. Acta 48:1987–2004.

McInerney, M. J., and D. W. S. Westlake. 1990. Microbially enhanced oil recovery. *In*: H. L. Ehrlich and C. L. Brierley, eds. Microbial Mineral Recovery. McGraw-Hill, New York, pp. 409–445.

McInerney, M. J., M. P. Bryant, and N. Pfennig. 1979. Anaerobic bacterium that degrades fatty acids in syntrophic association with methanogens. Arch. Microbiol. 122:129–135.

McKenna, E. J., and R. E. Kallio. 1964. Hydrocarbon structure: its effect on bacterial utilization of alkanes, pp. 1–14. *In*: H Heukelakian and N. Dondero, eds. Principles and Applications in Aquatic Microbiology. Wiley, New York.

Merrettig, U., P. Wlotzta, and U. Onken. 1989. The removal of pyritic sulfur from coal by *Leptspirillum*-like bacteria. Appl. Microbiol. Biotechnol. 31:626–628.

Meyer, W. C., and T. F. Yen. 1976. Enhanced dissolution of oil shale by bioleaching with thiobacilli. Appl. Environ. Microbiol. 32:610–616.

Mineral Facts and Problems. 1965. *Bulletin 630. Bureau of Mines*. U.S. Department of the Interior. Washington, DC.

Mohn, W. W., and J. M. Tiedje. 1990. Strain DCB-1 conserves energy for growth from reductive dechlorination coupled to formate oxidation. Arch. Microbiol. 153:267–271.

Mohn, W. W., and J. M. Tiedje. 1991. Evidence for chemiosmotic coupling of reductive dechlorination and ATP sysnthesis in *Desulfomonile tiedjei*. Arch. Microbiol. 157:1–6.

Mountford, D. O. 1978. Evidence for ATP synthesis driven by a proton gradient in *Methanosarcina barkeri*. Biochem. Biophys. Res. Commun. 85:1346–1351.

Mueller, V., M. Blaut, and G. Gottschalk. 1993. Bioenergetics of methanogenesis. *In*: J. G. Ferry, ed. Methanogenesis. Chapman and Hall, New York, pp. 360–406.

Myers, C. R., and K. H. Nealson. 1988. Bacterial manganese reduction and growth with manganese oxide as the sole electron acceptor. Science 240:1319–1321.

Nazina, T. N. and E. P. Rozanova. 1978. Thermophilic sulfate-reducing bacteria from oil strata. Mikrobiologiya 47:142–148 (Engl. transl. pp. 113–118).

Nazina, T. N. and E. P. Rozanova, 1980. Ecological conditions for the occurrence of methane-producing bacteria in oil-bearing strata in Apsheron. Mikrobiologiya 49:123–129 (Engl. transl. pp. 104–109).

Nazina, T. N., E. P. Rozanova, and S. I. Kuznetsov. 1985. Microbial oil transformation processes accompanied by methane and hydrogen-sulfide formation. Geomicrobiol. J. 4:103–130.

Nette, I. T., N. N. Grechushkina, and I. L. Rabotnova. 1965. Growth of certain mycobacteria in petroleum and petroleum products. Prikl. Biokhim. Mikrobiol. 1:167–174.

North, F. K. 1985. Petroleum Geology. Allen & Unwin, Boston.

Olson, E. S., D. C. Stanley, and J. R. Gallagher. 1993. Characterization of intermediates in the microbial desulfurization of dibenzothiophene. Energy Fuels 7:159–164.

Omori, T. L. Monna, Y. Saiki, and T. Kodama. 1992. Desulfurization of dibenzothiophene by *Corynebacterium* sp. strain SY1. Appl. Environ. Microbiol. 58:911–915.

O'Neill, J. G., and J. F. Wilkinson. 1977. Oxidation of ammonia by methane-oxidizing bacteria and the effects of ammonia on methane oxidation. J. Gen. Microbiol. 100:407–412.

Oremland, R. S., and B. F. Taylor. 1978. Sulfate reduction and methanogenesis in marine sediments. Geochim. Cosmochim. Acta 42: 209–214.

Oremland, R. S., L. M. Marsh, and S. Polcin. 1982. Methane production and simultaneous sulfate reduction in anoxic, salt marsh sediments. Nature (Lond.) 296:143–145.

Orr, F. M. Jr., and J. J. Taber. 1984. Use of carbon dioxide in enhanced oil recovery. Science 224:563–569.

Ourisson, G., P. Albrecht, and M. Rohmer. 1984. Microbial origin of fossil fuels. Sci. Am. 251:44–51.

Panganiban, A., and R. S. Hanson. 1976. Isolation of a bacterium that oxidizes methane in the absence of oxygen. Abstr., Annu. Meet. Am. Soc. Microbiol. I59, p. 121.

Panganiban, A. T. Jr., T. E. Patt, W. Hart, and R. S. Hanson. 1979. Oxidation of methane in the absence of oxygen in lake water samples. Appl. Environ. Microbiol. 37:303–309.

Paszczynski, A., V-B. Huynh, and R. Crawford. 1986. Comparison of lignase-1 and peroxidase-M2 from the white-rot fungus *Phanerochaete chrysosporium*. Arch. Biochem. Biophys. 244:750–765.

Patel, G. B. 1984. Characterization and nutritional properties of *Methanothrix concilii* sp. nov., an aceticlastic methanogen. Can. J. Microbiol. 30:1383–1396.

Phillips, W. E. Jr., and J. J. Perry. 1974. Metabolism of n-butane and 2-butanone by *Mycobacterium vaccae*. J. Bacteriol. 120:987–989.

Pirnik, M. P., R. M. Atlas, and R. Bartha. 1974. Hydrocarbon metabolism by *Brevibacterium erythrogenes*: normal and branched alkanes. J. Bacteriol. 119:868–878.

Pooley, F. D., and A. S. Atkins. 1983. Desulfurization of coal using bacteria by both dump and process plant techniques, pp. 511–526. *In*: G. Rossi and A. E. Torma, eds. Recent Progress in Biohydrometallurgy. Associazione Mineraria Sarda. 09016 Iglesias, Italy.

Pyne, J. W. Jr., D. L. Stewart, J. Fredrickson, and B. W. Wilson. 1987. Solubilization of leonardite by an extracellular fraction from *Coriolus versicolor*. Appl. Environ. Microbiol. 53:2844–2848.

Quayle, J. R., and T. Ferenci. 1978. Evolutionary aspects of autotrophy. Microbiol. Rev. 42:251–273.

Radway, J., J. H. Tuttle, N. J. Fendinger, and J. C. Means. 1987. Microbially mediated leaching of low-sulfur coal in experimental coal columns. Appl. Environ. Microbiol. 53:1056–1063.

Radway, J., J. H. Tuttle, and N. J. Fendinger. 1989. Influence of coal source and treatment upon indigenous microbial communities. J. Indust. Microbiol. 4:195–208.

Raiders, R. A., R. M. Knapp, and M. J. McInerney. 1989. Microbial selective plugging and enhanced oil recovery. J. Indust. Microbiol. 4:215–230.

Raymond, R. L., V. W. Jamison, and J. O. Hudson. 1967. Microbial hydorcarbon cooxidation. I. Oxidation of mono- and dicyclic hydrocarbons by soil isolates of the genus *Nocardia*. Appl. Microbiol. 15: 857–865.

Reeburgh, W. S. 1976. Methane consumption in Cariaco Trench waters and sediments. Earth Planet Sci. Lett. 28:337–344.

Reeburgh, W. S. 1980. Anaerobic methane oxidation: rate depth distribution in Skan Bay sediments. Earth Planet Sci. Lett. 47:345–352.

Reineke, W., and H-J. Knackmuss. 1988. Microbial degradation of haloaromatics. Annu. Rev. Microbiol. 42:263–287.

Rivard, C. J., and P. H. Smith. 1982. Isolation and characterization of a thermophilic marine methanogenic bacterium, *Methanogenium thermophilicum* sp. nov. Int. J. Syst. Bacteriol. 32:430–436.

Robertson, R. 1966. The origins of petroleum. Nature (London) 212: 1291–1295.

Robinson, J. A., and J. M. Tiedje. 1984. Competition between sulfate-reducing and methanogenic bacteria for H_2 under resting and growing conditions. Arch. Microbiol. 137:26–32.

Roffey, R., and A. Norqvist. 1991. Biodegradation of bitumen used for nuclear waste disposal. Experientia 47:539–542.

Rogoff, M. H., I Wander, and R. B. Anderson. 1962. Microbiology of coal. U.S. Department of the Interior. Bureau of Mines. Information Circular 8075.

Roitsch, T., and H. Stolp. 1985. Distribution of dissimilatory enzymes in methane and methanol oxidizing bacteria. Arch. Microbiol. 143: 233–236.

Romanovskaya, V. A., E. S. Lyudvichenko, T. P. Kryshtab, V. G. Zhukov, I. G. Sokolov, and Yu. R. Malashenko. 1980. Role of exogenous carbon dioxide in metabolism of methane-oxidizing bacteria. Mikrobiologiya 49:687–693 (Engl. transl. pp. 566–571).

Romesser, J. A., R. S. Wolfe, F. Mayer, E. Spiess, and A. Walther-Mauruschat. *Methanogenium*, a new genus of marine methanogenic bacteria, and characterization of *Methanogenium cariaci* sp. nov. and *Methanogenium marisnigri* sp. nov. Arch. Microbiol. 121: 147–153.

Rozanova, E. P. 1971. Morphology and certain physiological properties of purple bacteria from oil-bearing strata. Mikrobiologiya 40:152–157 (Engl. transl. pp. 134–138).

Rozanova, E. P. 1978. Sulfate reduction and water-soluble organic substances in a flooded oil reservoir. Mikrobiologiya 47:495–500 (Engl. transl. pp. 401–405).

Rubinshtein, L. M., and A. A. Oborin. 1986. Microbial methane production in stratal waters of oilfields in the Perm area of the Cis-Ural region. Mikrobiologiya 55:674–678 (Engl. transl. pp. 530–534).

Rubinstein, I., O. P. Strausz, C. Spyckerelle, R. J. Crawford, and D. W. S. Westlake. 1977. The origin of the oil sand bitumens of Alberta: a chemical and a microbiological study. Geochim. Cosmochim. Acta 41:1341–1353.

Sansone, F. J., and C. S. Martens. 1981. Methane production from acetate and associated methane fluxes from anoxic coastal sediments. Science 211:707–709.

Schoenheit, P., J. K. Kristjansson, and R. K. Thauer. 1982. Kinetic mechanism for the ability of sulfate reducers to out-compete methanogens for acetate. Arch. Microbiol. 132:285–288.

Sharak Genthner, B. R., W. A. Price II, and P. H. Pritchard. 1989. Anaerobic degradation of chloroaromatic compounds in aquatic sediments under a variety of enrichment conditions. Appl. Environ. Microbiol. 55:1466–1471.

Shelton, D. R., and J. M. Tiedje. 1984. Isolation and characterization of bacteria in an anaerobic consortium that mineralizes 3-chlorobenzoic acid. Appl. Environ. Microbiol. 48:840–848.

Sieburth, J. McN., P. W. Johnson, M. A. Eberhardt, M. E. Sieracki, M. Lidstrom, and D. Laux. 1987. The first methane-oxidizing bacterium from the upper mixing layer of the deep ocean: *Methylomonas pelagica* sp. nov. Curr. Microbiol. 14:285–293.

Simakova, T. L., Z. A. Kolesnik, N. V. Strigaleva, et al. 1968. Transformation of high-paraffinaceous oil by microorganisms under anaerobic and aerobic conditions. Mikrobiologiya 37:233–238 (Engl. transl. pp. 194–198).

Simoneit, B. R. T., and P. F. Lonsdale. 1982. Hydrothermal petroleum in mineralized mounds at the seabed of Guaymas Basin. Nature (Lond.) 295:198–202.

Simpson, P. G., and W. B. Whitman. 1993. Anabolic pathways in methanogens. *In*: Ferry, ed. Methanogenesis. Chapman and Hall, New York, pp. 445–472.

Skyring, G. W. 1987. Sulfate reduction in coastal ecosystems. Geomicrobiol. J. 5:295–374.

Smith, M. R., and R. A. Mah. 1978. Growth and methane genesis by *Methanosarcina* strain 227 on acetate and methanol. Appl. Environ. Microbiol. 36:870–879.

Sokolov, I. G. 1986. Coupling of the process of electron transport to methane mono-oxygenase with the translocation of protons in methane-oxidizing bacteria. Mikrobiologiya 55:715–722 (Engl. transl. pp. 559–565).

Sokolov, I. G., Yu. R. Malashenko, and V. A. Romanovskaya. 1981. Electron transport chain of thermophilic methane-oxidizing culture *Methylococcus thermophilus*. Mikrobiologiya 50:13–20 (Engl. transl. pp. 7–13).

Sørensen, J. 1987. Nitrate reduction in marine sediment: pathways and interactions with iron and sulfur cycling. Geomicrobiology 5: 401–421.

Sprott, G. D., S. E. Bird, and I. J. MacDonald. 1985. Proton motive force as a function of the pH at which *Methanobacterium bryantii* is grown. Can. J. Microbiol. 31:1031–1034.

Stanier, R. Y., J. L. Ingraham, M. L. Wheelis, and P. R. Painter. 1986. The Microbial World. 5th ed. Prentice-Hall, Englewood Cliffs, NJ.

Stetter, K. O., and G. Gaag. 1983. Reduction of molecular sulfur by methanogenic bacteria. Nature (Lond.) 305:309–311.

Stetter, K. O., R. Huber, E. Boechl, M. Kurr, R. D. Eden, M. Fielder, H. Cash, and I. Vance. 1993. Hyperthermophilic archaea are thriving in deep North Sea and Alaskan oil reservoirs. Nature (Lond.) 365: 743–745.

Stevenson, I. L. 1967. Utilization of aromatic hydrocarbons by *Arthrobacter* spp. Can. J. Microbiol. 13:205–211.

Stewart, D. L., B. L. Thomas, R. M. Bean, and J. K. Fredrickson. 1990. Colonization and degradation of bituminous and lignite coals by fungi. J. Indust. Microbiol. 6:53–59.

Steward, J. E., R. E. Kallio, D. P. Stevenson, A. C. Jones, and D. O. Schissler. 1959. Bacterial hydrocarbon oxidation. I. Oxidation of n-hexadecane by a gram-negative coccus. J. Bacteriol. 78:441–448.

Stoner, D. L., J. W. Wey, K. B. Barrett, J. G. Jolley, R. B. Wright, and P. R. Dugan. 1990. Modification of water-soluble coal-derived products by dibenzothiophene-degrading microorganisms. Appl. Environ. Microbiol. 56:2667–2676.

Szewzyk, R., and N. Pfennig. 1987. Complete oxidation of catechol by the strictly anaerobic sulfate reducing *Desulfobacterium catecholicum* sp. nov. Arch. Microbiol. 147:163–168.

Tanner, R. S., E. O. Udegbunam, M. J. McInerney, and R. M. Knapp. 1991. Microbially enhanced oil recovery from carbonate reservoirs. J. Indust. Microbiol. 9:169–195.

Teh, J. S., and K. H. Lee. 1973. Utilization of *n*-alkanes by *Cladosporium resinae*. Appl. Microbiol. 25:454–457.

Townsley, C. C., A. S. Atkins, and A. J. Davis. 1987. Suppression of pyritic sulfur during flotation tests using the bacterium *Thiobacillus ferrooxidans*. Biotech. Bioeng. 30:1–8.

Van Afferden, M., S. Schacht, J. Klein, and H. G. Trueper. 1990. Degradation of dibenzothiophene by *Brevibacterium* sp. DO. Arch. Microbiol. 153:324–328.

Van Bruggen, J. J. A., K. B. Zwart, R. M. van Assema, C. K. Stumm, and G. D. Vogels. 1984. *Methanobacterium formicium*, an endosymbiont of the anaerobic ciliate *Metopus striatus* McMurrich. Arch. Microbiol. 139:1–7.

Van der Linden, A. C., and G. J. E. Thijsse. 1965. The mechanisms of microbial oxidations of petroleum hydrocarbons. Adv. Enzymol. 27: 469–546.

Vogels, G. D. 1979. The global cycle of methane. Antonie v. Leeuwenhoek 45:347–352.

Vogels, G. D., and C. M. Visser. 1983. Interconnection of methanogenic and acetogenic pathways. FEMS Microbiol. Lett. 20:291–297.

Volesky, B., and J. E. Zajic. 1970. Ethane and natural gas oxidation by fungi. Dev. Ind. Microbiol. 11:184–185.

Walker, J. D., and J. J. Cooney. 1973. Pathway of *n*-alkane oxidation in *Cladosporium resinae*. J. Bacteriol. 115:635–639.

Ward, B. B., K. A. Kilpatrick, P. C. Novelli, and M. I. Scranton. 1987. Methane oxidation and methane fluxes in the ocean surface layer and deep anoxic layers. Nature (Lond.) 327:226–229.

Ward, D. M., and T. D. Brock. 1978. Anaerobic metabolism of hexadecane in sediments. Geomicrobiol. J. 1:1–9.

Westlake, D. W. S., W. Belicek, A. Jobson, and F. D. Cook. 1976. Microbial utilization of raw and hydrogenated shale oils. Can. J. Microbiol. 22:221–227.

Whittenbury, R., K. C. Phillips, and J. F. Wilkinson. 1970. Enrichment, isolation and some properties of methane-utilizing bacteria. J. Gen. Microbiol. 61:205–218.

Widdel, F. 1986. Growth of methanogenic bacteria in pure culture with 2-propanol and other alcohols as hydrogen donors. Appl. Environ. Microbiol. 51:1056–1062.

Wildgruber, G., M. Thomm, H. Koenig, K. Ober, T. Ricchiuto, and K. O. Stetter. *Methanoplanus limicola*, a plate-shaped methanogen representing a novel family, the Methanoplanaceae. Arch. Microbiol. 132:31–36.

Williamson, I. A. 1967. Coal Mining Geology. Oxford University Press, New York.

Winter, J., and R. S. Wolfe. 1979. Complete degradation of carbohydrate to carbon dioxide and methane by syntrophic cultures of *Acetobacterium woodii* and *Methanosarcina barkeri*. Arch. Microbiol. 121: 97–102.

Winter, J. U., and R. S. Wolfe. 1980. Methane formation from fructose by syntrophic associations of *Acetoberium woodii* and different strains of methanogens. Arch. Microbiol. 73–79.

Wolf, M., and R. Bachofen. 1991. Microbial degradation of bitumen. Experientia 47:542–548.

Wolin, M. J. 1981. Fermentation in the rumen and human large intestine. Science 213:1463–1468.

Wolin, M. J., and T. L. Miller. 1987. Bioconversion of organic carbon to CH_4 and CO_2. Geomicrobiol. J. 5:239–259.

Worakit, S., D. R. Boone, R. A. Mah, M.-E. Abdel-Samie, and M. M. El-Halwagi. 1986. *Methanobacterium alcaliphilum* sp. nov., an H_2-utilizing methanogen that grows at high pH values. Int. J. Syst. Bacteriol. 36:380–382.

Yang, Y-L., J. Lapado, and W. B. Whitman. 1992. Pyruvate oxidation by *Methanococcus* spp. Arch. Microbiol. 158:271–275.

Yarbrough, H. F., and V. F. Coty. 1983. Microbially enhanced oil recovery from the Upper Cretaceous Nacatoch formation, Union County, Arkansas, pp. 149–153. *In*: E. C. Donaldson and J. B. Clark, eds. Proc. Int. Conf. Microb. Enhancement Oil Recovery (Conf. −8205140), NTIS, Springfield, VA.

Yavitt, J. B., and G. E. Lang. 1990. Methane production in contrasting wetland sites: response to organic-chemical components of peat and to sulfate reduction. Geomicrobiol. J. 8:27–46.

Yen, H. C., and B. Marrs. 1977. Growth of *Rhodopseudomonas capsulata* under anaerobic dark conditions with dimethyl sulfoxide. Arch. Biochem. Biophys. 181:411–418.

Young, L. Y., and A. C. Frazer. 1987. The fate of lignin and lignin-derived compounds in anaerobic environments. Geomicrobiol. J. 5:261–293.

Zabel, H. P., H. Koenig, and J. Winter. 1984. Isolation and characterization of a new coccoid methanogen, *Methanogenium tatii* spec. nov. from a solfataric field on Mount Tatio. Arch. Microbiol. 137: 308–315.

Zehnder, A. J. B., and T. D. Brock. 1979. Methane formation and methane oxidation by methanogenic bacteria. J. Bacteriol. 137:420–432.

Zehnder, A. J. B., B. A. Huser, T. D. Brock, and K. Wuhrmann. 1980. Characterization of an acetate-decarboxylating, non-hydrogen-oxidizing methane bacterium. Arch. Microbiol. 124:1–11.

Zeikus, J. G. 1977. The biology of methanogenic bacteria. Bacteriol. Rev. 41:514–541.

Zeikus, J. G., R. Kerby, and J. A. Krzycki. 1985. Single-carbon chemistry of acetogenic and methanogenic bacteria. Science 227:1167–1173.

Zellner, G, K. Bleicher, E. Braun, H. Kneifel, B. J. Tindall, E. Conway de Macario, and J. Winter. 1989. Characterization of a new mesophilic, secondary alcohol-utilizing methanogen, *Methanobacterium palustre* spec. nov. from a peat bog. Arch. Microbiol. 151:1–9.

Zellner, G., U. B. Sleytr, P. Messner, H. Kneifel, and J. Winter. 1990. *Methanogenium liminatans* spec. nov., a new coccoid, mesophilic methanogen able to oxidize secondary alcohols. Arch. Microbiol. 153:287–293.

Zimmerman, P. R., J. P. Greenberg, S. O., Windiga, and P. J. Crutzen. 1982. Termites: a potentially large source of atmospheric methane, carbon dioxide and molecular hydrogen. Science 218:563–565.

Zinder, S. H. 1993. Physiological ecology of methanogens. *In*: J. G. Ferry, ed. Methanogenesis. Chapman and Hall, New York, pp. 128–206.

Zinder, S. H., and M. Koch. 1984. Non-aceticlastic methanogenesis from acetate: acetate oxidation by a thermophlic syntrophic coculture. Arch. Microbiol. 138:263–272.

ZoBell, C. E. 1952. Part played by bacteria in petroleum formation. J. Sed. Petrol. 22:42–49.

ZoBell, C. E. 1963. The origin of oil. Intern. Sci. Technol., August, pp. 42–48.

Zvyagintsev, D. G., G. M. Zenova, and I. G. Shirokykh. 1993. Distribution of actinomycetes with the vertical structure of peat bog ecosystems. Microbiolgiya 62:548–555 (Engl. transl. pp. 339–342).

Glossary

Acetogen A bacterial culture that forms acetic acid as the sole product in the reduction of CO_2 with hydrogen or in the fermentation of a sugar.

Acidophilic bacteria Bacteria that need an acid environment in which to grow.

Actinomycete Mycelium-forming bacterium; found in soil and aquatic environments.

Adenosine 5'-diphosphate (ADP) A phosphate anhydride that serves as acceptor of high-energy phosphate in energy conservation, or may be a product in enzymatic phosphorylation with ATP; it contains one high-energy phosphate bond.

Adenosine 5'-triphosphate A phosphate anhydride that conserves metabolic energy. It contains two high-energy phosphate bonds.

Adenosine triphosphatase (ATPase) An enzyme located on the inner surface of the plasma membrane of bacteria and the inner membrane of mitochondria that is involved in ATP synthesis or degradation.

Adipocere a waxy substance from dead organic matter; a soap.

ADP Adenosine 5'-diphosphate.

Adventitious organisms Organisms introduced naturally from an adjacent habitat; they may or may not be able to grow or survive in the new habitat.

Aerobe An organism that lives in air and uses oxygen as terminal electron acceptor in its respiratory process.

Aerobic heterotroph An organism that uses organic substances as carbon and energy sources in air (oxygen-respiring).

Agar (*also* **agar agar**) A polysaccharide heteropolymer derived from the walls of certain red algae and used for gelling bacteriological culture media.

Agar-shake culture A bacterial culture method in which the bacterial inoculum is completely mixed in an agar medium in a test tube.

Allochthonous Introduced from another place.

Aluminosilicate A mineral containing a combination of aluminum and silicate.

Amictic lake A lake that never turns over.

Ammonification A biochemical process that releases amino nitrogen from organic compounds such as proteins and amino acids as ammonia.

Amorphous Non-crystalline.

AMP Adenylic acid; adenosine monophosphate.

Amphibole A ferromagnesian mineral with two infinite chains of silica tetrahedra linked to each other; the double chains are cross-linked by Ca, Mg, and Fe.

Amphoteric Having both acidic and basic properties.

Anabolism That part of metabolism which deals with synthesis and poly-merization of biomolecules.

Anaerobe An organism that grows in the absence of oxygen; it may be oxygen tolerant or intolerant.

Anaerobic heterotrophy A form of nutrition using organic energy and carbon sources in the absence of air (oxygen).

Anaerobic respiration A respiratory process in which nitrate, sulfate, sulfur, carbon dioxide, ferric oxide, manganese oxide, or some other externally supplied, reducible inorganic or organic compounds, substi-tute for oxygen as terminal electron acceptor.

Anhydrite A calcium sulfate mineral, $CaSO_4$.

Anodic surface A surface exhibiting a net positive charge.

Anticyclones In oceanography, small closed-current systems of water de-rived from a surface current with a warm core surrounded by colder water, spinning in a clockwise rotation (*see also* rings; meddies).

APS Adenosine phosphosulfate; adenosine sulfatophosphate.

Aragonite A calcium carbonate mineral with orthorhombic structure.

Archaea See Archaebacteria.

Archaebacteria A group of bacteria that includes methanogens, *Sulfolo-bus, Acidianus, Halobacterium,* and *Thermoplasma,* among other gen-era, which have unique cell envelope and plasma membrane structure and a unique type of ribosomal RNA that distinguishes them from eubacteria.

Archaeobacteria See Archaebacteria.

Aridisol A mature desert soil.

Arroyo A dried-up river bed in a desert region through which water flows after a rain storm; a wash.

Arsenopyrite An iron-arsenic sulfide mineral, FeAsS.

Arsine AsH_3.

Ascomycetes Fungi that deposit their sexual spores in sacs (asci), e.g., *Neurospora*.

Asparagine The amide of aspartic acid, a dicarboxylic amino acid.

Assimilation Uptake and incorporation of nutrients by cells.

Asthenosphere Upper portion of the Earth's mantle, which is thought to have a plastic consistency and upon which the crustal plates float.

Atmosphere The gaseous envelope around the Earth.

ATP Adenosine 5'-triphosphate.

Augite A pyroxene type of mineral.

Authigenic Formed *de novo* from dissolved species in the case of minerals.

Autochthonous Generated in place; indigenous.

Autotroph An organism capable of growth exclusively at the expense of inorganic nutrients.

Bacterioneuston The bacterial population located in a thin film at the air–water interface in a natural body of water.

Bacterioplankton Unattached bacterial forms in an aqueous environment.

Bacterium A prokaryotic single- or multicelled organism. Single cells may appear as rods, spheres, spirals, or other shapes.

Baltica A former continent encompassing Russia west of the Urals, Scandinavia, Poland, and northern Germany.

Banded iron formation (BIF) A sedimentary deposit featuring alternating iron-oxide-rich (Fe_2O_3 and Fe_3O_4) layers and iron-oxide-poor cherty layers; thought to have originated 3.3–2 billion years ago.

Barium psilomelane A complex manganese (IV) oxide.

Barophile An organism capable of growth at elevated hydrostatic pressure.

Barren solution Pregnant solution after its valuable metals have been removed.

Basaltic rock Rock of volcanic origin showing very fine crystallization due to rapid cooling. Basalt is rich in pyroxene and feldspars.

Basidiomycetes Fungi that form sexual spores on basidia (club-shaped cells) and feature septate mycelium (e.g., mushrooms).

Benthic   Located at the bottom of a body of water.

Betaine $HOOCCH_2N^+(CH_3)_3$.

Binary fission   Cell division in which one cell divides into two cells of approximately equal size.

Bioherm   A large mineral aggregate of biological origin; a microbialite.

Bioleaching   A microbial process whereby metal values are extracted from ore by solubilizing them through oxidation, reduction, or complexation.

Biosphere   The portion of the Earth's surface inhabited by living organisms.

Birnessite   A manganese (IV) oxide mineral, δMnO_2.

Bisulfite HSO_3^-.

Calcareous ooze   A sediment having calcareous structures from foraminifera, coccolithophores, or other $CaCO_3$-depositing organisms as major constituents.

Calcite   A calcium carbonate mineral with rhombohedral structure.

Capillary culture method   The use of a glass capillary with optically flat sides inserted into soil or sediment for culturing microbes from these sources in situ; developing microbes in the capillaries may be observed directly under a microscope.

Catabolism   That part of metabolism which involves degradation of nutrients and energy conservation from their oxidation.

Catalase   An enzyme capable of catalyzing the reaction $H_2O_2 = H_2O + \frac{1}{2}O_2$; it can also catalyze the reduction of H_2O_2 with an organic hydrogen donor or inorganic electron donor.

Cathodic surface   A surface exhibiting a net negative charge.

Celestite   A strontium sulfate mineral, $SrSO_4$.

Cellulolytic   Capable of enzymatic hydrolysis of cellulose.

Centric geometry   Cyclindrical, in reference to diatoms.

Chalcopyrite   A copper-iron sulfide mineral, $CuFeS_2$.

Chasmolithic   Living inside preformed pores or cavities in rock.

Chemocline   A chemical gradient zone in a water column that separates a more dilute and less dense phase from a more concentrated phase.

Chemolithotroph   An autotroph that derives energy from the oxidation of inorganic matter.

Chemostat   A culture system permitting microbial growth under steady-state conditions.

Chlorophyll   A light-harvesting and energy-transducing type of pigment of photosynthetic organisms.

Chloroplasts   Photosynthetic organelles in eukaryotic cells.

Choline $HOCH_2CH_2N^+(CH_3)_3$.

Coccolithophore A chrysophyte alga whose surface is covered with $CaCO_3$ platelets (coccoliths).

Colony counting A method of microbial enumeration by counting colonies (visible aggregates of microbes) on and/or in agar medium in a petri dish or test tube.

Conjugation A method of transfer of genetic information between cells that requires cell-to-cell contact.

Connate waters Saline water trapped in rock strata in the geologic past, usually having undergone chemical alteration through reaction with the enclosing rock.

Consortium An association of two or more different microbes which exhibit a metabolic interdependence.

Constitutive enzyme An enzyme that is always present in an active form in a cell, whether needed or not.

Contaminant An organism introduced during experimental manipulation of a habitat.

Continental drift Migration of continents on the Earth's surface as a result of crustal plate motions.

Continental margin The edge of a continent.

Continental rise Gently sloping sea floor at the base of the continental slope.

Continental shelf Gently sloping sea floor between the shore and the continental slope.

Continental slope Steeply sloping sea floor at the outer edge of the continental shelf.

Convergence The confluence of two water masses.

Co-oxidation Simultaneous microbial oxidation of two compounds which may be quite unrelated and only one of which supports growth.

Copiotroph A bacterium that requires a nutrient-rich environment.

Coriolis force An apparent force that seems to deflect a moving object to the right in the northern hemisphere and to the left in the southern hemisphere of the Earth.

Crustal plates Portions of the Earth's crust which have irregular shapes and sizes and which contact and interact with each other while floating on the asthenosphere.

Cyanobacteria Oxygenic, photosynthetic bacteria, formerly known as blue-green algae.

Cysteine $HSCH_2CH(NH_2)COOH$.

Cytochrome system An electron transport system used in biological oxidations (respiration) which includes iron porphyrin proteins called cytochromes.

Dehydrogenase   An enzyme that catalyzes removal or addition of hydrogen.

Denitrification   A process in which nitrate is reduced to dinitrogen, nitrous oxide, and nitric oxide.

Deoxyribonucleic acid   A biopolymer consisting of purine and pyrimidine bases, deoxyribose, and phosphate and having genetic information encoded in it.

Desert varnish   A manganese- or iron-rich coating on a rock surface.

Desferrisiderophore   A siderophore that does not contain ferric iron complexed by it.

Deuteromycetes   Fungi that do not form sexual spores.

Diagenesis   A process of transformation or alteration of rocks or minerals.

Diatom   A chrysophyte alga grouped with the Bacillarophyceae, which is encased in a siliceous wall.

Diatomaceous ooze   A sediment having diatom frustules as a major constituent.

Dimethyl arsinate

$$CH_3AsO(OH)$$
$$|$$
$$CH_3$$

the acid form; also known as cacodylic acid.

Dimethylmercury $(CH_3)_2Hg$.

Dimictic lake   A lake that turns over twice a year.

Disproportionation reaction   A stoichiometric chemical reaction in which part of the reactant undergoes oxidation and the rest undergoes reduction, e.g., $2H_2O_2 \rightarrow 2H_2O + O_2$.

Dithiothreitol $HSCH_2CH(OH)CH_2SH$.

Divergence   The separation of two water masses.

DNA   Deoxyribonucleic acid.

Dolomite   A $CaMg(CO_3)_2$ mineral.

Dunite   An ultrabasic rock rich in olivine.

Dystrophic   Referring to waters with an oversupply of organic matter that is only incompletely decomposed because of an insufficiency of oxygen, phosphorus, and/or nitrogen.

Earth's core   The innermost portion of the Earth, consisting mostly of Fe and some Ni.

Earth's mantle   The portion of the Earth overlying the core, consisting mainly of O, Mg, and Si with lesser amounts of Fe, Al, Ca, and Na.

Enargite   A copper-arsenic sulfide mineral, Cu_3AsS_4.

Endolithic   Living inside rock (limestone) as a result of boring into it.

Endosymbionts   Cells that live inside other cells for mutual benefit.

Enrichment culture A culture method that selects for a desired organism(s) by providing special nutrients and/or physical conditions that favor its (their) development.

Entisol An immature desert soil.

Epigenetic Referring to emplacement of a mineral in cracks of fissures of pre-existing rock.

Epilimnion The portion of a lake above the thermocline.

Epiphytes Organisms attached to the surface of other living organisms or inanimate objects.

Eubacteria Bacterial prokaryotes, the majority of which feature a peptidoglycan cell wall and phosphoglycerol esters (phospholipids) in their plasma membrane; bacteria which are not archaebacteria.

Eukaryotic cell A cell with a true nucleus, mitochondria, and chloroplasts (if photosynthetic).

Euphotic zone That part of a water column which is penetrated by sunlight in sufficient quantity to permit photosynthesis.

Euryhaline Capable of growth over a wide range of salinities.

Eutrophic Referring to a nutrient-rich state of natural water.

Facultative chemolithotroph A bacterium that can grow heterotrophically or chemolithotrophically, depending on growth conditions.

Facultative microorganism A microorganism capable of living with or without oxygen.

Fauna A term used in ecology to denote an assemblage of organisms that may include members of one or more of the following groups: protozoans, invertebrates, and vertebrates, even though protozoans do not belong to the Animalia in modern systematics.

Fecal pellet Compacted fecal matter packaged in a membrane by the organism that excretes it.

Feldspar A type of mineral consisting of anhydrous aluminosilicates of Na, K, Ca, and Ba.

Fermentation A process of intramolecular oxidation/reduction operating without an externally supplied terminal electron acceptor; a biochemical dismutation.

Ferrisiderophore A siderophore that contains ferric iron complexed by it.

Flora A term used in ecology to denote an assemblage of organisms that may include members of one or more of the following groups: prokaryotes, algae, fungi, and plants, even though the first three groups are not considered plants in modern systematics.

Fluorescence microscopy A microscopy method making use of natural or artificial fluorescence of objects upon irradiation with UV light.

Foraminifer(a) Amoeboid protozoan(s) that mostly form a calcareous test (shell) about them; some form tests by cementing sand grains or other inorganic detrital structures to their cell surface (e.g., arenaceous foraminifera).

Fungus(i)   Mycelial or, occasionally, single-celled eukaryotic organism(s), possessing a cell wall but no chloroplasts; yeasts, molds, mildews, and mushrooms are examples.

Galena   A lead sulfide mineral, PbS.

Gangue   A term of technical slang that refers to the host rock of an ore which encloses the metal-containing minerals of an ore.

Garnet   A silicate mineral of Ca, Mg, Fe, or Mn; it is hard and vitreous.

Generation time   The average time required for cell doubling.

Geomicrobiology   The study of microbes and the role they have played and are playing in a number of fundamental geologic processes.

Gleying   A process in some soils involving microbial reduction of ferric iron manifested by a color change from brownish to grayish and a development of stickiness; often associated with water logging of soil.

Glutathione

$$CH_2SH$$
$$|$$
$$HOOC(H_2N)CHCH_2CH_2CONHCH$$
$$|$$
$$CONHCH_2COOH$$

Goethite   An iron oxide mineral, $Fe_2O_3 \cdot H_2O$.

Gondwana   A former continent encompassing Africa, South America, Australia, Antarctica, and India.

Gram-negative   Referring to a differential staining reaction of bacteria in which a counterstain, usually safranin, is retained by the cell.

Gram-positive   Referring to a differential staining reaction of bacteria in which crystal violet is retained by the cell.

Granite   Rock of volcanic origin showing coarse crystallization due to slow cooling of magma; granite is rich in quartz and feldspars.

Grandiorite   A volcanic rock intermediate between granite and diorite, showing coarse crystallization.

Gravitational water   A film of water surrounding pellicular water, which moves by gravity, responds to hydrostatic pressure, and may freeze.

GSH   Reduced glutathione.

GSSG   Oxidized glutathione.

Guyot   Flat-topped seamount.

Gypsum   A calcium sulfate mineral, $CaSO_4 \cdot 2H_2O$.

Halophile   A microbe that grows preferentially at a high salt concentration.

Hematite An iron oxide mineral, Fe_2O_3.

Heterotroph An organism requiring one or more organic nutrients for carbon and for energy for growth.

Heulandite A type of zeolite mineral.

Histosol Organic soil.

Holozoic Feeding on living cells; predatory.

Homeostasis Maintenance of a state of equilibrium.

Hornblende A type of amphibole mineral.

Humic acid A humus fraction that is acid- and alcohol-insoluble.

Humus In soil, a mixture of substances derived from partial decomposition of plant, animal, and microbial remains and from microbial syntheses; in marine sediment of the open ocean, a mixture of substances derived from phytoplankton remains.

Hydrogenase An enzyme catalyzing the reaction $H_2 \leftrightarrow 2H^+ + 2e$.

Hydrosphere That portion of the Earth's surface which is covered by water; it includes the oceans, seas, lakes, rivers, and groundwater.

Hydrothermal solution A hot, metal-laden solution generated by reaction of water (e.g., seawater) with rock in the lithosphere in regions receiving heat from adjacent magma chambers.

Hygroscopic water A thin film of water covering a soil particle, which never freezes or moves as a liquid.

Hypersthene A type of pyroxene mineral.

Hypha(e) A branch of a mycelium; it is filamentous.

Hypolimnion The portion of a lake located below the thermocline.

Hypophosphite HPO_2^{2-}.

Igneous rock Rock of volcanic or magmatic origin.

Illite A group of micalike clay minerals, having a three-layered structure like montmorillonite in which Al may substitute for Si, and that contain significant amounts of Fe and Mg.

Indigenous organisms Organisms native to a habitat.

Inducible enzyme An enzyme that is formed by a cell only when needed.

Juvenile water Water from within the Earth that had never before reached the Earth's surface.

Kaolinite A type of clay $[Al_4SiO_{10}(OH)_8]$ featuring alternating aluminum oxide and tetrahedral silica sheets.

Karstic Referring to a landscape with sinkholes or cavities due to local dissolution of limestone.

Kazakhstania A former continent encompassing present-day Kazakhstan.

Labradorite A type of feldspar mineral related to plagioclase.

Laterization A soil transformation in which iron and aluminum oxides, silicates, and carbonates are precipitated, cementing soil particles together and thus destroying the porosity of the soil.

Laurasia   A former continent encompassing North America, Europe, and most of Asia.

Laurentia   A former continent encompassing most of North America, Greenland, Scotland, and the Chukotski Peninsula of eastern Russia.

Lentic waters   Static waters.

Lichen   An organism consisting of an intimate association of a fungus and a green alga or a cyanobacterium.

Lignin   A heteropolymer of units of substituted phenylpropane derivatives; an abundant constituent of wood.

Limestone   A rock type rich in $CaCO_3$.

Limonite   An amorphous iron oxide mineral, $Fe_2O_3 \cdot nH_2O$; now frequently described as FeOOH.

Lithification   A process of rock formation by compaction and/or cementation.

Lithosphere   The crust of the Earth.

Lotic waters   Flowing waters.

Macrofauna   The fauna excluding protozoa and microscopic invertebrates.

Magma   Molten rock.

Mannitol   A polyhydric alcohol which may be formed by reduction of fructose or mannose.

Meddies   In oceanography, small closed-current systems of water whose core is more saline than the surrounding water and which exhibit clockwise rotation (*see also* Rings and Anticyclones).

Mercaptoethanol $HOCH_2CH_2SH$.

Mesophile   A microorganism capable of growth in a temperature range from 10 to 45°C (optimal range between 25 and 40°C).

Mesotrophic   Referring to a nutritional state of natural water between oligotrophic and eutrophic.

Metabolism   Cellular biochemical activities collectively.

Metabolite   A metabolic reactant or product.

Metamorphic rock   Rock produced by alteration of igneous or sedimentary rock through action of heat and pressure.

Methanogen   A methane-forming bacterium (archaebacterium).

Methanotroph   A methane-oxidizing bacterium.

Methylotroph   A methanol-oxidizing microbe which can oxidize methanol but not methane.

Microcosm   An experimental setup that approximates important features of a natural environment, but on a small scale that can be manipulated, e.g., a soil- or sediment-percolation column.

Mineralization   In *microbial physiology,* the complete decomposition of an organic compound into CO_2, H_2O, and—if the corresponding elements are present in the organic compound—PO_4^{2-}, NO_3^- or NH_3, and SO_4^{2-} or H_2S. In *mineralogy*, the formation of a mineral.

Mixotroph A bacterium that uses simultaneously inorganic and organic energy sources and/or inorganic and organic carbon sources.

Molybdenite A molybdenum disulfide mineral, MoS_2.

Monomethylarsinate

$$CH_3AsO(OH)$$
$$|$$
$$OH$$

in the acid form.

Monomictic lake A lake that turns over once a year.

Montmorillonite A type of clay $[Al_2Si_4O_{10}(OH)_2 \cdot nH_2O]$ consisting of successive aluminum oxide sheets, each sandwiched between two sheets of silica tetrahedra.

Mycelium A network of hyphae produced by most fungi and some bacteria.

Nepheline A sodium aluminum silicate.

Nitrate ammonification Reduction of nitrate to ammonia via nitrite.

Nitrification A bacterial process in which ammonia is converted to nitrate autotrophically or heterotrophically; some fungi are also capable of heterotrophic nitrification.

Nitrogen fixation A bacterial process in which dinitrogen (N_2) is enzymatically converted to ammonia.

Nucleic acid A biopolymer containing purines, pyrimidines, pentose, or deoxypentose and phosphoric acid found in chromosomes, plasmids, ribosomes, plastids, and cytoplasm of cells.

Nucleotides Polymer units of nucleic acid consisting of a purine or pyrimide plus pentose or deoxypentose and phosphoric acid.

Ocean eddies Collectively, the oceanic, small, closed-current systems: rings, anticyclones, and meddies.

Ocean trench Deep cleft in the ocean floor; a site of subduction of oceanic crustal plate below a continental plate.

Ochre An iron oxide ore, $FeOOH$.

Oligotrophic Referring to a nutrient-poor state of natural water.

Olivine A mineral consisting of orthosilicate of magnesium and iron.

Organic soil A soil formed from accumulation and slow and incomplete decomposition of organic matter in a sedimentary environment.

Orogeny Mountain building.

Orpiment An arsenic sulfide mineral, As_2S.

Orthoclase A feldspar mineral.

Orthophosphate Monomeric phosphate, H_3PO_4.

Orthosilicate   Monomeric silicate, H_4SiO_4.

Oxidative phosphorylation   A process of ATP synthesis coupled to electron transport.

Oxisol   A soil type in tropic and subtropic humid climates.

Pangaea   A supercontinent including all major continents of today, existing from about 250 to 200 million years ago.

Panspermia   Transfer of life in the form of spores from another world (universe, planet) to Earth.

PAPS   3′-Phosphoadenosine phosphosulfate; 3′-phosphoadenosine phosphatosulfate.

Pectinolytic   Capable of enzymatic hydrolysis of pectin.

Pedoscope   A system of glass capillaries with optically flat sides for insertion into soil and subsequent microscopic inspection for microbial development in the capillary lumen.

Pellicular water   A film of water surrounding hygroscopic water that moves by intermolecular attraction and that may freeze.

Peloscope   A system of capillaries with optically flat sides for insertion into sediment and subsequent microscopic inspection for microbial development in the capillary lumen.

Pennate geometry   Symmetrical about a long and a short axis, in reference to diatoms.

Peptone   A mixture of peptides from a digest of beef muscle by pepsin; used in bacterial culture media.

Peridotite   An igneous granitoid rock, rich in olivines but lacking in feldspars.

Peroxidase   An enzyme that catalyzes the reduction of H_2O_2 by an oxidizable organic molecule.

Phagotrophic   Consuming whole cells by engulfment (phagocytosis).

Phosphatase   An enzyme catalyzing hydrolysis of phosphate esters.

Phosphine PH_3.

Phosphite HPO_3^{2-}.

Phosphorite   A calcium phosphate mineral; apatite.

Photolithotroph   An autotroph that derives energy from sunlight.

Photophosphorylation   A light-dependent process of ATP synthesis associated with photosynthesis.

Photosynthesis   A metabolic process using sunlight energy for the assimilation of carbon in the form of CO_2, HCO_3^-, or CO_3^{2-}.

Phycomycete   Aquatic or terrestrial fungus whose vegetative mycelium shows no septation (e.g., *Rhizopus*).

Phytoplankton   Photosynthetic plankton.

Plankton   Free-floating biota in an aqueous habitat.

Plasmid   An extrachromosomal bit of genetic substance (DNA).

Plutonic water   Deep, anoxic underground water, likely to contain significant amounts of sulfate and/or chloride.

Podzolic soil   A type of spodosol associated with humid, temperate climates; a naturally acidic forest soil.

Pure culture   A microbial culture that consists of one and only one species or strain.

Pregnant solution   A metal-laden effluent from an ore-leaching operation.

Primary producers   Organisms that transform (fix) CO_2 into organic carbon; include photo- and chemolithotrophs.

Prokaryotic cell   A cell lacking a true nucleus, mitochondria, and chloroplasts.

Proteolytic   Capable of enzymatic hydrolysis of proteins.

Psychrophile   A microorganism capable of growth in a temperature range from slightly below 0 to 20°C (optimum at 15°C or below).

Psychrotolerant   Capable of surviving but not growing at a temperature in the psychrophilic range.

Psychrotroph   A microorganism capable of growth in a temperature range from 0 to 30°C (optimum about 25°C).

Purines   A group of organic bases having the purine ring structure in common.

Pyridine nucleotide   Nicotinamide adenine dinucleotide or nicotinamide adenine dinucleotide phosphate; a hydrogen-carrying coenzyme.

Pyrimidines   A group of organic bases having the pyrimidine ring structure in common.

Pyrite (iron)   An iron disulfide mineral, FeS_2.

Pyroxene   A ferromagnesian mineral with silica tetrahedra linked in single chains and cross-linked mainly by Ca, Mg, and Fe.

Quartzite   A metamorphic rock derived from sandstone.

Radiolarian ooze   A sediment having radiolarian tests as a major constituent.

Red-bed deposit   A sedimentary deposit rich in ferric oxide; first appeared when the atmosphere of the Earth became oxidizing.

Respiration   Biological oxidations utilizing an electron transport system that may operate with either oxygen or another external inorganic or organic compound as terminal electron acceptor.

Reverse electron transport   The transfer of electrons by an electron transport system against the redox gradient, requiring the input of metabolic energy.

Rhizosphere   The zone in soil that surrounds the root system of a plant where special environmental conditions may prevail as a result of root secretion or uptake of specific inorganic and/or organic substances. It represents a special habitat for some microbes.

Rhodanese   An enzyme capable of catalyzing the reaction $CN^- + S_2O_3^{2-} = SCN^- + SO_3^-$ and the reductive cleavage of $S_2O_3^{2-}$.

Rhodochrosite   A mineral form of $MnCO_3$.

Rhyolite   An igneous rock rich in plagioclase feldspar.

Ribonucleic acid   Heteropolymer consisting of purine and pyrimidine bases, ribose, and phosphoric acid, different forms of which may serve as templates in protein synthesis (messenger RNA), as amino-acid-transfer RNA (to locate position in a peptide chain determined by messenger RNA), and as part of the structure of ribosomes.

Ribosome   Submicroscopic, intracellular particle consisting of ribonucleic acids and proteins, which is part of the protein-synthesizing system of cells.

Ribulose bisphosphate carboxylase   An enzyme that catalyzes carboxylation or oxygenation of ribulose diphosphate in many autotrophic bacteria and in algae and plants.

Rings   In oceanography, small closed-current systems, having a diameter as great as 300 km and a depth as great as 2 km, with a core of cold water surrounded by warmer water and having a counterclockwise rotation.

RNA   Ribonucleic acid.

Rock   Massive, solid inorganic matter, usually consisting of two or more intergrown minerals.

Rusticyanin   A copper-containing enzyme involved in Fe^{2+} oxidation in *Thiobacillus ferrooxidans*.

Saccharolytic   Capable of enzymatic hydrolysis or fermentation of sugars.

Salinity   A measure of the salt content of seawater based on its chlorinity.

Salt dome   The cap rock composed of anhydrite, gypsum, and calcite at the top of a salt plug; a geologic formation.

Sandstone   A rock formed from compacted and cemented sand.

Saponite   A montmorillonite types of clay in which Mg replaces Al.

Saprozoic   Feeding on dead organic matter.

Satellite microorganism   In the case of mixed bacterial cultures, an organism not identical to the dominant organism in the culture, which will give rise to distinctive colonies on appropriate solid media.

Sclerotium(a)   A vegetative, resting, food-storage body in higher fungi, composed of a compact mass of hardened mycelium.

Sediment   Finely divided mineral and organic matter that has settled to the bottom in a body of water.

Sedimentary rock   Rock formed from compaction and/or cementation of sediment.

Seismic activity   Earth tremors.

Shale   A laminate sedimentary rock formed from mud or clay.

Sheath   In bacteriology, an organic tubular structure around some bacterial organisms.

Siderite   A mineral form of $FeCO_3$.

Siderophore   An organic iron-chelating substance produced by certain microbes.

Silica   Silicon dioxides; quartz and opal are examples.

Silicate   A salt of silicic acid; a mineral containing silicate.

Slime molds   A group of microorganisms that have a life cycle including a motile swarmer stage and an aggregational phase, which may be multinucleate, leading to formation of a sessile fruiting body.

Sodium azide NaN_3, an inhibitor or cytochrome oxidase.

Soil horizon   A soil stratum as seen in a soil profile.

Soil profile   A vertical section through soil.

Solfatara   Fumarolic hot spring that yields sulfuretted water.

Spent culture medium   Culture medium after microbial growth has taken place in it.

Spodosol   Forest soil type in temperate climates.

Stenohaline   Capable of growth in only a narrow range of salinities.

Stromatolite   A laminated structure formed by filamentous organisms that grew in mats which have either entrapped inorganic detrital material or formed $CaCO_3$ deposits in which the organism became embedded; the organisms are most commonly cyanobacteria; in ancient stromatolites, the organic remains have frequently disappeared during replacement by silica.

Subduction   A process in which the edge of an oceanic crustal plate slips under a continental crustal plate manifested in the form of deep ocean trenches.

Substrate-level phosphorylation   A process of ATP synthesis involving high-energy phosphate bond formation on the substrate being oxidized.

Sulfate-reducing bacteria   Bacteria that convert sulfate to sulfide as a respiratory process.

Sulfhydryl compound   An organic compound with an —SH functional group.

Superoxide dismutase   An enzyme that catalyzes the dismutation of superoxide (O_2^-) into H_2O_2 and O_2.

Syngenetic   Referring to deposition of an ore mineral contemporaneously with the enclosing sediment or rock.

Synergism   The interaction of two or more microorganisms, resulting in a reaction that none of the organisms could carry out alone.

Talc   A hydrous magnesium silicate mineral.

Tectonic activity   Interaction of crustal plates of the Earth.

Teichoic acid   A glycerol- or ribitol-based polymeric constituent of the cell walls of gram-positive bacteria.

Tetrathionate $S_4O_6^{2-}$.

Thermocline   A zone in a water column with a steep temperature gradient.

Thermophilic bacteria   Bacteria that grow at temperatures above 45°C; some have been shown capable of growing above the boiling point of water when under pressure.

Thiobacilli   Gram-negative rod-shaped bacteria, mostly chemolithotrophic, which can use H_2S, $S°$, and $S_2O_3^{2-}$ as energy sources.

Thiosulfate $S_2O_3^{2-}$.

Todorokite   A complex manganese (IV) oxide mineral.

Transduction   A method of transfer of genetic information between bacteria, involving a bacterial virus as the transmitting agent.

Transpiration   Loss of water by evaporation through the stomata (pores) of leaves.

Travertine   A porous limestone that may be formed by rapid $CaCO_3$ precipitation by cyanobacteria.

Tricarboxylic acid cycle   A cyclic sequence of biochemical reactions in which acetate is completely oxidized in one turn of the cycle.

Trithionate $S_3O_6^{2-}$.

Trophosome   A structure in the coelomic cavity of some vestimentiferan worms consisting of symbiotic hydrogen-sulfide-oxidizing bacteria which share with the worm the carbon they fix chemoautotrophically.

Tundra soil   A soil type occurring in high northern latitudes.

Turbidity current   A strong current of a sediment suspension moving rapidly downslope; may exert scouring action as it moves over rock surfaces.

Ultramafic rock   An igneous rock, usually rich in olivine and pyroxenes.

Upwelling   An upward movement of a mass of deep, cold ocean water, which may bring nutrients (nitrate, phosphate) into surface waters.

Vermiculite   A micaceous mineral.

Wad   A complex manganese (IV) oxide.

Wadi   See Arroyo.

Water potential.   A measure of water availability, for instance in soil.

Weathering   A breakdown process of rock.

Wollastonite   A calcium silicate mineral, $CaSiO_3$.

Zeolite   A hydrated silicate of aluminum containing alkali metals.

Zooplankton   Non-photosynthetic plankton.

Zygospore   A sexual spore formed by certain algae and fungi.

Index